Active and Passive Vibration Damping

Active and Passive Vibration Damping

Amr M. Baz
University of Maryland, USA

WILEY

This edition first published 2019

Registered Offices
John Wiley & Sons, Inc., 111 River Street, Hoboken, NJ 07030, USA
John Wiley & Sons Ltd, The Atrium, Southern Gate, Chichester, West Sussex, PO19 8SQ, UK

Editorial Office
The Atrium, Southern Gate, Chichester, West Sussex, PO19 8SQ, UK

For details of our global editorial offices, customer services, and more information about Wiley products visit us at www.wiley.com.

Wiley also publishes its books in a variety of electronic formats and by print-on-demand. Some content that appears in standard print versions of this book may not be available in other formats.

Library of Congress Cataloging-in-Publication Data

Names: Baz, Amr Mahmoud Sabry, 1945– author.
Title: Active and passive vibration damping / Amr M. Baz.
Description: Hoboken, NJ, USA : John Wiley & Sons, Inc., [2019] | Includes bibliographical references and index. |
Identifiers: LCCN 2018027360 (print) | LCCN 2018030715 (ebook) | ISBN 9781118537589 (Adobe PDF) | ISBN 9781118537602 (ePub) | ISBN 9781118481929 (hardcover)
Subjects: LCSH: Damping (Mechanics) | Vibration.
Classification: LCC TA355 (ebook) | LCC TA355 .B368 2018 (print) | DDC 620.3/7–dc23
LC record available at https://lccn.loc.gov/2018027360

Cover Design: Wiley
Cover Image: © oxygen / Moment/ Getty Images

Set in 10/12pt Warnock by SPi Global, Pondicherry, India

Printed and bound by CPI Group (UK) Ltd, Croydon, CR0 4YY

10 9 8 7 6 5 4 3 2 1

To the memory of my father and sister.

Contents

Preface

This book is intended to present the basic principles and potential applications of passive and active vibration damping technologies. The presentation encompasses a mix between the associated physical fundamentals, governing theories, and optimal design strategies of various configurations of vibration damping treatments. Utilization of smart materials to augment the vibration damping of passive treatments is the common thread that is pursued, in depth, throughout the book.

The focus has been on developing a deeper understanding of the science behind various phenomena that govern the control of structural vibration using appropriate damping techniques. It is my intention, in writing this book, to explain in a simple yet comprehensive manner such scientific basics with particular focus on viscoelastic damping materials and the means for controlling passively and actively their energy dissipation characteristics. The book was developed throughout the years during my teaching of various classes on passive and active vibration and noise control. My research in these areas has enriched my teaching and broadened my understanding of these topics. I have tried to blend simple theory with basic engineering practice to enable the students and practicing engineers to understand the science and apply it with confidence. My guide in this effort has been the saying of Albert Einstein:

> *"Why does this applied science, which saves work and makes life easier, bring us so little happiness? The simple answer runs: Because we have not yet learned to make sensible use of it".*

So, in this book, I have attempted applying the theories to various applications, introducing a wide variety of examples and presenting detailed computer simulations, to make the implementation real and practical.

The book includes 12 chapters divided into two parts. The first part is devoted to outlining the basics of vibration damping and this coverage is divided into six chapters. In the second part, various configurations of advanced vibration damping treatments are presented in four chapters that include applications to different structural systems. Part I starts with an introductory chapter on the field of passive and active vibration damping, followed by Chapter 2 that covers the classical models of viscoelastic damping materials. Chapter 3 presents the important characterization methods of viscoelastic materials both in the frequency and time domains. Advanced modeling techniques of viscoelastic materials are covered in Chapter 4. These methods, which include the Prony series, Gola–Hughes–MacTavish, augmented-temperature field, and fractional derivative methods, are presented to enable modeling the dynamics of structures treated with viscoelastic materials by using the finite element method in both the time and frequency domains. The use of modal strain energy as a metric for predicting the modal loss factor of structures treated with damping materials and for optimal design of damping treatments is discussed in Chapter 5. Estimation of the energy dissipation characteristics of various configurations of passive and active damping treatments is described in Chapter 6 for rods, beams, and plates. Part II presents in Chapter 7 the application of passive and active constrained layer damping treatments to beams, plates, and shells. Chapter 8 deals with modeling of various advanced damping treatments such as: stand-off, functionally graded, active piezoelectric damping composites, and magnetic damping treatments. In Chapters 9 and 10, shunted piezoelectric and periodic treatments are described, respectively, as undamped treatments that behave as conventional damping treatments with potentially tunable characteristics. Chapter 11 presents a wide variety of passive and active nanoparticle damping composites and Chapter 12 looks at the problem of power flow in damped structural systems.

The book has a large number of numerical examples to reinforce the understanding of the theories covered, provide means for exercising the knowledge gained, and emphasize the learning of strategies for the design and application of active and passive vibration damping systems. The examples are supported by a set of MATLAB software modules to enable the designers of vibration damping systems extend the theories presented to various applications.

Each chapter of the book will end with a number of problems that cover the different aspects of theoretical analysis, design, and applications of vibration damping technologies.

In this multi-prong coverage approach, the book is targeted to senior undergraduate students, graduate students, researcher, and practicing engineers who are interested in gaining an in-depth exposure to the field of vibration damping. The presentation and supporting tools associated with the book will enable the readers of having hands-on experience to the analysis, design, optimization, and application of this exciting technology to a wide range of situations.

Writing this book would have been virtually impossible without the tireless support of many students, colleagues, and friends who have enriched my life in many ways. These contributions are apparent throughout the book. In particular, I would like to mention the invaluable inputs and contributions from Professors Wael Akl and Adel Al Sabbagh of Ain Shams University in Cairo, Egypt. Also, thanks are due to Prof. Osama Aldraihem of King Saud University in Saudi Arabia for his collaborations over the years and contributions to Chapter 11 and Prof. Massimo Ruzzene of Georgia Tech for many years of very fruitful collaborations.

Thanks are also due to my colleagues and former students who have pioneered the field of active vibration damping and control including: Dr. Mohamed Raafat, Dr. Soon-Neo Poh of the NSWC Center, Prof. Jeng-Jong Ro at Da-Yeh University in Taiwan, Dr. Tung Huei Chen of the NSWC Center, Dr. Chul-Hue Park at Korea Institute for Robot Industry Advancement (KIRIA), Dr. Charles Kim at NASA-Goddard, Dr. Zheng Gu at Zhejiang Tiatai Liangxin Co. in China, the late Dr. Jaeho Oh, Dr. Adel Omer of the Military Technical College in Cairo, Dr. Ted Shields of Northrop-Grumman, Dr. Peter Herdic of the NRL, Dr. William Laplante the Under Secretary of the Air Force, Prof. Ray Manas of Indian Institute of Technology, Kharagpur, Prof. Mustafa Arafa of the American University in Cairo, Prof. Mohammed Al-Ajami of Kuwait University, Prof.

Mohamed Tawfik of Cairo University, Dr. Mary Leibolt of NSWC, Prof. Mostafa Nouh of SUNY Buffalo, and Prof. John Crassidis of SUNY Buffalo. Thanks are also due to my former students Mr. Atif Chaudry of the US Patent Office and Mr. Giovanni Rosannova of NASA Wallop for their work in passive and active damping.

It is important to note that my work in the area of active vibration damping has been funded primarily by the Army Research Office (ARO) with Dr. Gary Anderson as the technical monitor, the Office of Naval Research (ONR) with Dr. Kam Ng as the technical monitor and by Dr. Turki S. Al-Saud the President of King Abdulaziz City for Science & Technology (KACST), Riyadh. Without their support, trust, and friendship, this work would have not been possible.

Special thanks are also due to the administration of the University of Maryland for providing me with the excellent scholarly environment that enabled me developing my professional career and of course writing this book. Important among those administrators are: former President Dan Mote, now the President of NAE, former Provost William Destler, now the President of RIT, former Provost Nariman Farvardin, now the President of Stevens Institute of Technology, current President of UMD President Wallace Loh, Dean of Engineering Darryll Pines, Prof. Davinder Anand, former Chair of ME Department, Prof. William Fourney, Associate Dean of Engineering, Prof. Avi Bar-Cohen, former Chair of ME Department, and Prof. Balakumar Balachandran the current Chair of the ME Department. Apart from their vast professional impact on me, I sincerely and equally value their friendship and collegiality.

Finally, writing this book has been enjoyable and possible because of the tireless support and sacrifice of my wonderful wife and my two great sons who are my true friends and heroes.

Amr M. Baz
College Park, MD
December 2018

List of Symbols

Symbol	Meaning	Units
a	Dimension of a plate side	m
a	Dilatation or contraction scaling parameter of wavelets Area	— m^2
A	Magnetic potential	Ampere
A_{ATF}	Affinity of the ATF model $(=-\partial f_{ATF}/\partial z)$	$N\, m^{-2}\, {}^{\circ}K^{-1}$
$[A_r^*]$	The correspondence concentration factors of phase, r	—
b	Dimension of a plate side	m
b	Translation parameter of wavelets	s
B	Input state-space matrix	—
B	Magnetic induction	Tesla
$B^*{,}_0$	Characteristics complex length of passive treatments	m
B_0	The magnetic flux density	Tesla
B_F	The structural susceptance matrix	$m\, (Ns)^{-1}$
$[B_r^*]$	The correspondence concentration factors of phase, r	—
c	The sound speed	$m\, s^{-1}$
c_c	The critical damping coefficient $(=2\sqrt{km})$	$Ns\, m^{-1}$
c_d	Damping coefficient of dissipative element	$Ns\, m^{-2}$
$[c^*]$	Complex stiffness matrix	$N\, m^{-2}$
$[c_r^*]$	Complex stiffness matrix of phase, r	$N\, m^{-2}$
C	Measurement state-space matrix Capacitance	— Farad
C_G	Control parameter	—
C^S	Strain-free capacitance	Farad

(*Continued*)

Symbol	Meaning	Units
C^T	Stress-free capacitance	Farad
d_{ij}	Piezo-strain constants in the *i* and *j* directions due to applied electric field in the *k* direction	$\mathrm{m\,V^{-1}}$
D	Energy dissipated during a full vibration cycle of the viscoelastic material	Nm
D	Denominator of a transfer function	—
D	Nano-particle diameter	m
D_a	Distance between neutral axis of entire sandwiched beam and piezo-actuator	m
D_i	Electrical displacement along the *ith* direction	$\mathrm{Coulomb\,m^{-2}}$
D_t^*	Complex bending stiffness ($= D_t(1 + i\eta_B)$)	$\mathrm{Nm^2}$
$[D_i]$	Stiffness matrix relating the stress and strain vectors	$\mathrm{N\,m^{-2}}$
e	Electron charge ($=1.60217662 \times 10^{-19}$ Coulombs)	Coulomb
e	Power flow error	$\mathrm{Nm\,s^{-1}}$
e_{31}	Piezoelectric charge/strain constant ($=d_{31}/s_{11}^E$)	$\mathrm{m^3\,(N\,V)^{-1}}$
E	Young's modulus	$\mathrm{N\,m^{-2}}$
E_i	Electrical field along the *ith* direction	$\mathrm{V\,m^{-1}}$
E_n	Total energy ($E_n = PE + KE$)	Nm
$E(t)$	Relaxation modulus	$\mathrm{N\,m^{-2}}$
E'	Storage modulus	$\mathrm{N\,m^{-2}}$
E''	Loss modulus	$\mathrm{N\,m^{-2}}$
E^*	Complex relaxation modulus	$\mathrm{N\,m^{-2}}$
E_0	Equilibrium modulus	$\mathrm{N\,m^{-2}}$
E_i	Relaxation strength	$\mathrm{N\,m^{-2}}$
E_∞	Instantaneous modulus of GMM Un-relaxed or high frequency modulus of elasticity	$\mathrm{N\,m^{-2}}$
E_iA_i	Longitudinal rigidity	N
E_iI_i	Flexural rigidity	$\mathrm{Nm^2}$
EQ	Product of elastic modulus and first moment of area	$\mathrm{Nm^2}$
f	Frequency	$\mathrm{rad\,s^{-1}}$, Hz
f_{ATF}	Helmholtz free energy density of the ATF model	$\mathrm{N\,m^{-2}}$
F	Force	N
F_c	Control force	N
$\mathbf{F_m}$	Magnetic forces	N
$\{F\}$	Force and moment vector	N, NM
g	The shear factor of constrained damping treatments	—
g_{31}	The piezoelectric voltage constant ($=d_{31}/\varepsilon_{33}$)	$\mathrm{m\,V^{-1}}$
G'	Storage modulus in shear	$\mathrm{N\,m^{-2}}$
G''	Loss modulus in shear	$\mathrm{N\,m^{-2}}$

Symbol	Meaning	Units
G^*	Complex modulus in shear	$\mathrm{N\,m^{-2}}$
G_F	The structural conductance matrix	m (Ns) $^{-1}$
h	Layer thickness	m
h_P	Plank constant (= 6.626×10^{-34})	$\mathrm{m^2\,kg\,s^{-1}}$
$\mathbf{H}$	Magnetic field	Ampere $\mathrm{m^{-1}}$
i	The "unit" imaginary number = $\sqrt{-1}$	—
I	Area moment of inertia	$\mathrm{m^4}$
I	Performance Index	—
$\mathbf{I}$	Current density	Amperes $\mathrm{m^{-2}}$
$\boldsymbol{I}_{x,y}$	Structural intensity	Nm (sm) $^{-1}$
J^*	Complex creep compliance	$\mathrm{m^2\,N^{-1}}$
J_j	Retardation strength	$\mathrm{N\,m^{-2}}$
$\hat{J}(i\omega)$	Fourier transform of creep compliance	m
J	Performance index	—
$\mathbf{J}$	Jacobian matrix	—
K, k	Stiffness	$\mathrm{N\,m^{-1}}$
$K_{d,p}$	Derivative and proportional controller gains	—
$\mathbf{K}_{\mathrm{geo}}{}^{\mathrm{e}}$	Element geometric matrix	—
K_g	Gain of the controller	—
$K_{v,D}$	Gain of velocity (or derivative) feedback controller	$\mathrm{Ns\,m^{-1}}$
k_{31}^2	The electro-mechanical coupling factor	—
k_B^*	Complex bending wave number (= $\left(m\omega^2/D_t^*\right)^{1/4}$)	$1\,\mathrm{m^{-1}}$
k_r	Ratio between derivative and proportional control gains	—
$k_{x,y}$	Wave numbers in the x and y directions	$1\,\mathrm{m^{-1}}$
$k_{r,i}$	Real and imaginary wave numbers	$1\,\mathrm{m^{-1}}$
$\bar{k}$	Dimensionless wave number (=$B_0 k$)	—
$[K]$	Stiffness matrix	$\mathrm{N\,m^{-1}}$
$[K_{e,s}]$	Elastic and structural stiffness matrices	$\mathrm{N\,m^{-1}}$
$[K_{I,R,v}]$	Imaginary, real, and VEM stiffness	$\mathrm{N\,m^{-1}}$
l_s	Sample thickness	m
L	Length, Laplace transform, Lagrangian	m
L	Electrical inductance	Henry
L_{ATF}	Proportionality constant of the ATF model	$\mathrm{m^2\,K^2}$ (Ns) $^{-1}$
$\bar{L}$	Dimensionless electrical inductance (=L/R^2C^e)	—
M, m	Mass, electron mass	kg

(*Continued*)

Symbol	Meaning	Units
$M_{c,e}$	Control and external moments	Nm
$\mathbf{M}$	Magnetization	Amperes m^{-1}
$M_{x,y}$	Moments along the x and y directions	Nm
M_{ij}	Torsion moment in the i-j plane	Nm
$[M]$	Mass matrix	kg
$\{\mathbf{Mg}\}$	Global magnetization vector	Amperes m^{-1}
N	Number of mini-oscillators, Number of finite elements, Numerator	—
$N_{ix}, N_{i\theta}$	The longitudinal and tangential forces	N
N_{px}	Piezoelectric longitudinal control forces generated along the x-axis	N
$N_{x,y}$	Normal forces along the x and y directions	N
N_{ij}	Shear force in the i-j plane	N
$[N]$	Shape function of the finite element model	—
p_i	Internal normal forces per unit length	N m^{-1}
P	Axial load	m
P_F	Active power (=*real* $[S_P]$)	Nm s^{-1}
$P_{Fi,\ Fr}$	Instantaneous and reference active power flow	Nm s^{-1}
$\{q\}$	Modal displacement vector	mrad
q_i	Externally applied body forces per unit length	N m^{-1}
$\{q_{i,r}\}$	Imaginary and real modal displacement vectors	—
Q	Electrical charge	Coulomb
Q_F	Reactive power (=*imag* $[S_P]$)	Nm s^{-1}
$Q_{x,y}$	The shear forces along the x,y directions	N
r	Scaled attenuation factor	—
R	Shell radius	m
R	Electrical resistance	ohm
R_a	Electrical resistance across filler particle	ohm
R_c	Contact electrical resistance between two filler particles	ohm
R_n	Eigenvector matrix of the n non-zero eigenvalues Λ	—
ΔR	Change in piezo-resistance of a conducting polymer	ohm
$[R_i]$	Rotation matrix	—
s	Laplace complex number	rad s^{-1}
s	Separation distance between adjacent nano-particles	m
s_{11}^D	Compliance, in direction 1, at constant electric displacement, D	m^2 N^{-1}
s_{11}^E	Compliance, in direction 1, at constant electric field, E	m^2 N^{-1}

Symbol	Meaning	Units
s^{SH}	Compliance of the shunted network	$m^2\ N^{-1}$
$[s^*]$	Overall compliance	$m^2\ N^{-1}$
$[s_r^*]$	Compliance of the *rth* phase	$m^2\ N^{-1}$
$[S^*]$	Eshleby strain tensor	—
$\mathbf{S}_{an}^e$	Magnetic stiffness matrix of the element	$N\ m^{-1}$
S_P	Complex vibrational power	$Nm\ s^{-1}$
t	Time	s
T	Temperature	°C
T	Kinetic energy	Nm
$T(t)$	A temporal function in, t	—
T_c	Internal axial tension	N
T_i	Stress on piezoelectric element along the *ith* direction	$N\ m^{-2}$
T_g	Glass transition temperature	°C
$[T]$	Transformation matrix	—
$[T_k]$	Transfer matrix for *kth* cell	—
$[T_r^*]$	Dilute concentration matrix	—
u,v,w	Deflections in the x, y, and z directions	m
$\hat{u}(x,\omega)$	Fourier transform of $u(x,t)$	ms
U	Potential energy	Nm
v, v_f	Volume fraction	—
$V_{c,s}$	Control and sensor voltages	V
$V_{x,y}$	Shear forces in the x and y directions	N
W	Energy dissipated of the viscoelastic material	Nm
$W(x)$	A spatial function in x	—
$W_{D,e}$	Dissipated and elastic energy	Nm
W_n	Nominal energy	Nm
W_{piezo}	Work done by the piezo-layer	Nm
$[W_r^*]$	Dilute concentration matrix	—
$\Delta W_{a,\ P}$	Dissipated energy due to active and passive damping	Nm
$\Delta W_{unconstrained}$	Dissipated energy due to unconstrained damping	Nm
x,y,z	Position	m
x_e	The "shear length" of a constrained damping treatment	m
X	Electrical reactance	ohm (Ω)
$\{X\}$	State vector	—
Y		—

(*Continued*)

Symbol	Meaning	Units
	The geometrical factor of a constrained damping treatment	
Y^D	Electrical admittance at constant electrical displacement, D	mho
Y^{EL}	Electrical admittance	mho
Y_F	Mobility	m (sN) $^{-1}$
Y^{SH}	Electrical shunted admittance	mho
z_i	*ith* internal degree of freedom of the VEM	—
Z^{EL}	Electrical impedance	ohm (Ω)
$\bar{Z}^{EL}$	Dimensionless electrical impedance (= Y^D/Y^{EL})	—
$(Z^{ME})^D$	Mechanical impedance at constant electrical displacement	Ns m^{-1}
Z^{ME}	Mechanical impedance	Ns m^{-1}
$(Z^{ME})^{SH}$	Mechanical impedance with electrical shunting	Ns m^{-1}
$\bar{Z}^{ME}$	Dimensionless mechanical impedance [=$(Z^{ME})^{SH}/(Z^{ME})^D$]	—

Greek Symbols

Symbol	Meaning	Units
α	Order of the fractional derivative	—
α	Attenuation factor	dB m^{-1}
α_n	Gain of GHM *nth* mini-oscillator	—
α_T	Temperature shift factor	—
β_i	The *ith* relative modulus of the GMM	—
β	weighting parameter of the Weighted Stiffness Matrix method (WSM)	—
γ	Shear strain, Localization factor, Lamé parameter	—
$\gamma_{a,\ p}$	Shear strain with active and passive treatments	—
γ_{ATF}	Effective modulus of the ATF model	N m^{-2} K^{-2}
$\Gamma(n)$	Gama function	—
$\langle\Gamma\rangle$	Average orientation	rad
δ	Phase shift due to damping (η = $tan\ \delta$),	rad
δ_{ATF}	Coupling term between the mechanical displacement and the augmented temperature fields	N m^{-2} K^{-1}
Δ	The cubic dilation	—
Δ_{ATF}	Relaxation resistance of the ATF model	—

Symbol	Meaning	Units
$\{\Delta\}$	Deflection vector	mrad
ε	strain	—
$\bar{\varepsilon}$	Strain function of the fractional derivative method	—
ε^A	Applied strain	—
ε^C	Constrained strain	—
ε^T	Uniform transformation strain	—
$\boldsymbol{\varepsilon}_{33}^T$	Permittivity (or dielectric constant) in direction 3	Farad m^{-1}
$\hat{\varepsilon}(\omega)$	Fourier transform of the strain $\varepsilon(t)$	s
$[\bar{\varepsilon}_r]$	The average strain field in phase r	—
$[\varepsilon^0]$	Uniform elastic strain	—
ζ	Damping ratio	—
ζ_n	Damping ratio of GHM *nth* mini-oscillator,	—
	Damping ratio of the *nth* mode of vibration	—
η	Loss factor	—
η_n	Loss factor of the *nth* mode	—
η_v	Loss factor of the *VEM*	—
θ	Euler angle	rad
κ	Curvature	1/m
λ	Time constant ($\lambda = c_d/E_s$)	s
λ_B	Bending wavelength	m
Λ	Non-zero eigenvalues	rad s^{-1}
μ	Coefficient of friction, Propagation parameter and Lamé parameter (=G)	— N m^{-2}
μ_0	The permeability of space ($=4\pi \times 10^{-7}$)	T m A^{-1}
μ_r	The relative permeability of a magnetic material	Henry m^{-1}
ρ	Density	kg m^{-3}
ρ	Resistivity	ohm.m (Ω·m)
ρ_i	Relaxation time constant	s
ρ_{ATF_i}	Relaxation time constant of the ATF model	s
σ	Stress	N m^{-2}
$[\bar{\sigma}_r]$	The average stress field in phase, r	N m^{-2}
$[\sigma^0]$	Uniform elastic stress	N m^{-2}
τ	Time constant	s
τ_d	The dissipative shear stress in the VEM	N m^{-2}
τ_j	Retardation time	s
τ_{ij}	Shear stress in the *i*-*j* plane	N m^{-2}

(*Continued*)

Symbol	Meaning	Units
υ	Poisson's ratio	—
ϕ	Transformer turning ratio of piezo-element that transforms voltage into force $\left(=, -d_{31}A, /\left(s_{11}^{E}L\right)\right)$	$\mathrm{N\,V^{-1}}$
ϕ	Euler angle	rad
ϕ	Height of potential barrier between adjacent particles	eV
ϕ_n^*	The complex *nth* mode shape	—
$\phi_{n_{i,r}}$	imaginary and real components of the *nth* mode shape	—
Φ	Magnetic flux	Webers
$[\Phi]$	Mode shape matrix	—
Ψ	Shear strain of the stand-off layer	rad
$\Psi(t)$	Wavelet function	—
ω	Frequency	$\mathrm{rad\,s^{-1}}$
ω_n	Natural frequency	$\mathrm{rad\,s^{-1}}$
	Frequency of GHM *nth* mini-oscillator	$\mathrm{rad\,s^{-1}}$
ω_r	Reduced frequency ($=\alpha_T\omega$)	$\mathrm{rad\,s^{-1}}$
ω^*	Dimensionless length of damping treatment ($=L/B_0$)	—
Ω	Dimensionless frequency for VEM $\left(=\sqrt{(mh_1/G)}\,\omega\right)$,	—
	Dimensionless frequency for resistive shunting ($=RC^S\omega$)	—

Subscripts

Symbol	Meaning
0	Initial value
d	Dissipative
e	Electrical
f	Friction
H	Hysteretic
i	Incident wave
o	Overall
p	Parallel, piezoelectric
r	Reflected waves
s	Elastic solid, series
sf	Strain free

Symbol	Meaning
stf	Stress free
S	Structural
t	Transmitted
v	Viscous

Superscripts

Symbol	Meaning
$*$	Complex conjugate
D	Constant electrical displacement
E	Constant electrical field
s	Constant strain
T	Transpose

Operators

Symbol	Meaning
$\lvert . \rvert$	Absolute value
$\lVert . \rVert$	Norm
$[.]^{-1}$	Inverse of the matrix [.] between the brackets
$[.]^{T}$	Transpose of the matrix [.] between the brackets
$\frac{d}{dx}(.)$	Differential operator with respect to x
$\frac{\partial}{\partial x}(.) = (.)_{,x}$	Partial differential operator with respect to x
$\dot{(.)} = \frac{d}{dt}(.)$	First derivative with respect to time
$\ddot{(.)} = \frac{d^2}{dt^2}(.)$	Second derivative with respect to time
$\delta(.)$	Variation of the quantity (.) between parentheses
$Re\,(.)$	Real part
$Im\,(.)$	Imaginary part

Abbreviations

ACLD	Active constrained layer damping
APDC	Active piezoelectric damping composites
ATF	Augmented temperature field
BVP	Boundary value problem
CLD	Constrained layer damping
DMTA	Dynamic mechanical thermal analysis
DOF	Degrees of freedom
DPM	Distributed-parameter model
EAP	Electroactive polymers
EDT	Engineered damping treatments
EMDC	Electromagnetic damping composites
FD	Fractional derivatives
FEM	Finite element method
FFT	Fast Fourier transform
FGM	Functionally graded material
GHM	Golla–Hughes–MacTavish model
G-L	Grunwald–Letnikov approach
GMC	Generalized method of cells
HTM	Halpin–Tsai method
IDOF	Internal degree of freedom of the VEM
IRS	Improved reduction system method
KE	Kinetic energy
LFA	Low frequency approximation method
LMS	Least mean square
MCLD	Magnetic constrained layer damping
MDR	Modal damping ratios
MMA	Method of moving asymptote
MR	Magnetorheological fluid
MSE	Modal strain energy
MTM	Mori–Tanaka method
MWCNT	Multi-walled carbon nanotubes
NSC	Negative stiffness composite

OC	Open circuit
P.E.	Potential energy
PCLD	Passive constrained layer damping
PVDF	Polyvinylidene fluoride
PZT	Lead zirconate titanate
R–L	The Reimann–Liouville approach
RVE	Representative volume element
SAFE	Semi-analytical finite element method
SC	Short circuit
SCM	Self-consistent method
SHPB	Split Hopkinson pressure bar
SOL	Stand-off layer
TTS	Time–temperature superposition
VAMUCH	Variational asymptotic method for unit cell homogenization
VEM	Viscoelastic material
WLF	Williams–Landel–Ferry formula
WSM	Weighted stiffness matrix method
WSTM	Weighted storage modulus method

Part I

Fundamentals of Viscoelastic Damping

1

Vibration Damping

1.1 Overview

Vibration control is recognized as an essential means for attenuating excessive amplitudes of oscillations, suppressing undesirable resonances, and avoiding premature fatigue failure of critical structures and structural components. The use of one form of vibration control or another in most of the newly designed structures is becoming very common in order to meet the pressing needs for large and light-weight structures. With such vibration control systems, the strict constraints imposed on present structures can be met to ensure their effective operation as quiet and stable platforms for manufacturing, communication, observation, and transportation.

1.2 Passive, Active, and Hybrid Vibration Control

Various passive, active, and hybrid vibration control approaches have been considered over the years employing a variety of structural designs, damping materials, active control laws, actuators, and sensors. Distinct among these approaches are the passive, active, and hybrid vibration damping methods.

It is important to note here that passive damping can be very effective in damping out high frequency excitations, whereas active damping can be utilized to control low frequency vibrations as shown in Figure 1.1. For effective control over broad frequency band, hybrid damping methods are essential.

1.2.1 Passive Damping

Passive damping treatments have been successfully used, for many years, to damp out the vibration of a wide variety of structures ranging from simple beams to complex space structures. Examples of such passive damping treatments include:

1.2.1.1 Free and Constrained Damping Layers

Both types of damping treatments rely in their operation on the use of a viscoelastic material (VEM) to extract energy from the vibrating structure as shown in Figure 1.2. In the free (or unconstrained) damping treatment, the vibrational energy is dissipated

Active and Passive Vibration Damping, First Edition. Amr M. Baz.

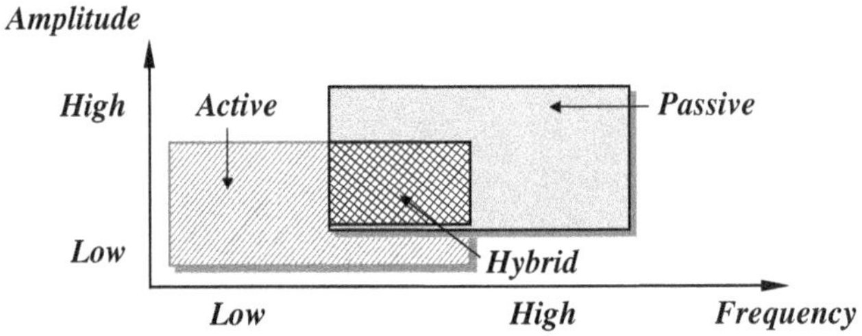

Figure 1.1 Operating range of various damping methods.

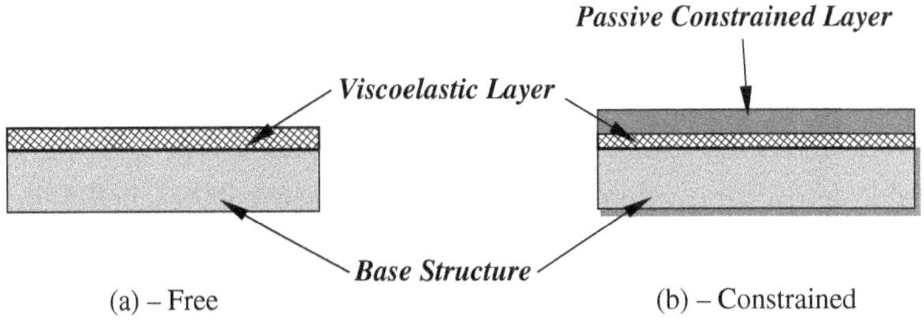

Figure 1.2 Viscoelastic damping treatments. (a) Free and (b) constrained.

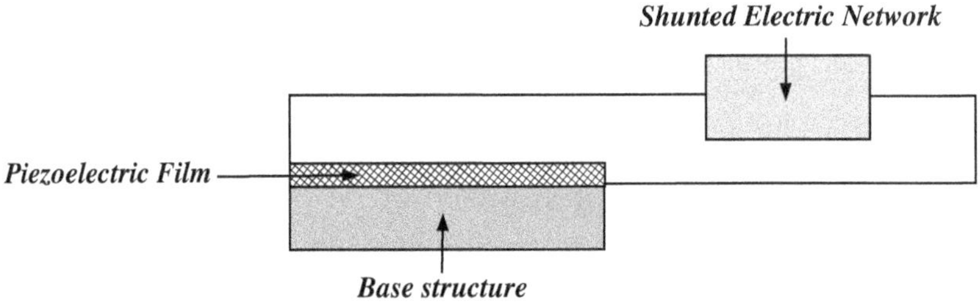

Figure 1.3 Shunted piezoelectric treatments.

by virtue of the extensional deformation of the VEM, whereas in the constrained damping treatment more energy is dissipated through shearing the VEM (Nashif et al. 1985).

1.2.1.2 Shunted Piezoelectric Treatments

These treatments utilize piezoelectric films, bonded to the vibrating structure, to convert the vibrational energy into electrical energy. The generated energy is then dissipated in a shunted electric network, as shown in Figure 1.3, which are tuned in order to maximize the energy dissipation characteristics of the treatments (Lesieutre 1998). The electric networks are usually resistive, inductive, and/or capacitive. Other configurations of the shunted piezoelectric treatments include the viscoelastic polymer composites loaded with shunted piezoelectric inclusions introduced by Aldraihem et al. (2007).

1.2.1.3 Damping Layers with Shunted Piezoelectric Treatments

In these treatments, as shown in Figure 1.4, a piezoelectric film is used to passively constrain the deformation of a viscoelastic layer, which is bonded to a vibrating structure. The film is used also as a part of a shunting circuit that is tuned to improve the damping characteristics of the treatment over a wide operating range (Ghoneim 1995).

1.2.1.4 Magnetic Constrained Layer Damping (MCLD)

These treatments rely in their operation on arrays of specially arranged permanent magnetic strips that are bonded to viscoelastic damping layers. The interaction between the magnetic strips can improve the damping characteristics of the treatments by virtue of enhancing either the compression or the shear of the viscoelastic damping layers as shown in Figure 1.5.

In the compression MCLD configuration of Figure 1.5a, the magnetic strips (1 and 2) are magnetized across their thickness. Hence, the interaction between the strips generates magnetic forces that are perpendicular to the beam longitudinal axis. These forces subject the viscoelastic layer to across the thickness loading, which makes the treatment act like a Den–Hartog dynamic damper. In the shear MCLD configuration of Figure 1.5b, the magnetic strips (3 and 4) are magnetized along their length. Accordingly, the developed magnetic forces, which are parallel to the beam longitudinal axis, tend to shear the viscoelastic layer. In this configuration, the MCLD acts as conventional constrained layer

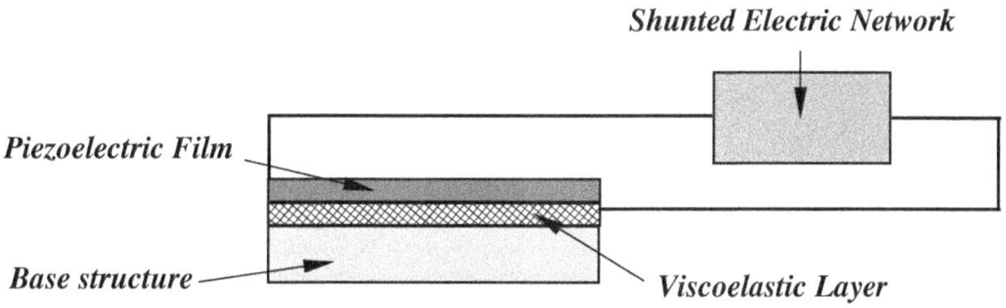

Figure 1.4 Damping layers with shunted piezoelectric treatments.

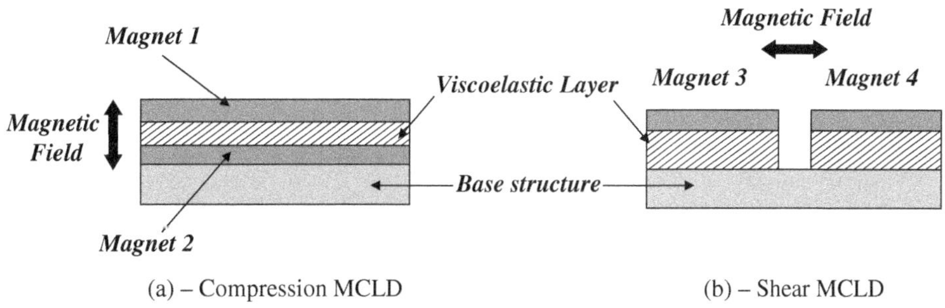

Figure 1.5 Configurations of the MCLD treatment. (a) Compression MCLD and (b) shear MCLD.

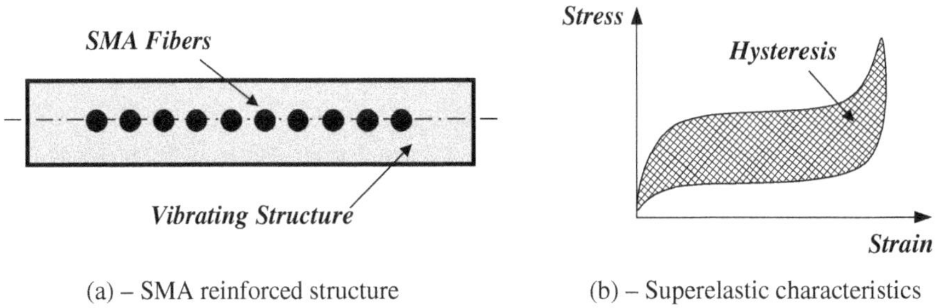

Figure 1.6 Damping with shape memory fibers. (a) SMA reinforced structure and (b) superelastic characteristics.

damping treatment whose shear deformation is enhanced by virtue of the interaction between the neighboring magnetic strips (Baz 1997; Oh et al. 1999).

1.2.1.5 Damping with Shape Memory Fibers

This damping mechanism relies on embedding superelastic shape memory fibers in the composite fabric of the vibrating structures as shown in Figure 1.6a. The inherent hysteretic characteristics of the Shape Memory Alloy (SMA), in its superelastic form, are utilized to dissipate the vibration energy. The amount of energy dissipated is equal to the area enclosed inside the stress–strain characteristics (Figure 1.6b). This passive mechanism has been successfully used in damping out the vibration of a wide variety of structures including large structures subject to seismic excitation (Greaser and Cozzarelli 1993).

1.2.2 Active Damping

Although the passive damping methods described here are simple and reliable, their effectiveness is limited to a narrow operating range because of the significant variation of the damping material properties with temperature and frequency. It is, therefore, difficult to achieve optimum performance with passive methods alone particularly over wide operating conditions.

Hence, various active damping methods have been considered. All of these methods utilize control actuators and sensors of one form or another. The most common types are made of piezoelectric films bonded to the vibrating structure as shown in Figure 1.7.

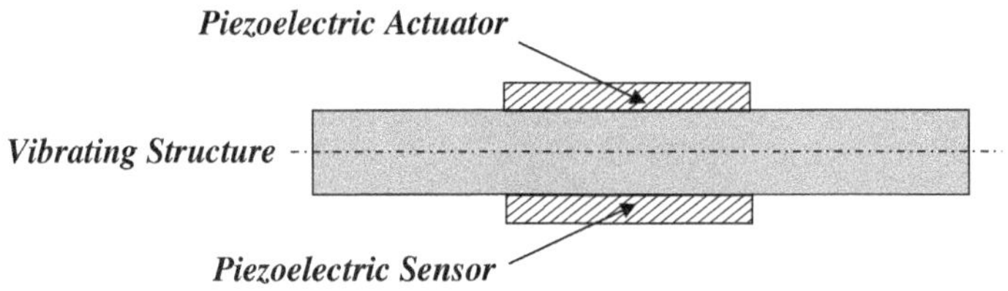

Figure 1.7 Active damping.

This active control approach has been successfully used in damping out the vibration of a wide variety of structures ranging from simple beams to more complex space structures (Preumont 1997; Forward 1979).

1.2.3 Hybrid Damping

Because of the limited control authority of the currently available active control actuators, and because of the limited effective operating range of passive control methods, treatments that are a hybrid combination of active damping and passive damping treatments have been considered. Such hybrid treatments aim to use various active control mechanisms to augment the passive damping in a way that compensates for its performance degradation with temperature and/or frequency. Also, these treatments combine the simplicity of passive damping with the effectiveness of active damping in order to ensure optimal blend of the favorable attributes of both damping mechanisms.

Among the most commonly used hybrid treatments are:

1.2.3.1 Active Constrained Layer Damping (ACLD)

This class of treatments is a blend between a passive constrained layer damping and active piezoelectric damping as shown in Figure 1.8. Here, the piezo-film is actively strained in such a manner to enhance the shear deformation of the viscoelastic damping layer in response to the vibration of the base structure (Baz 1996, 2000; Crassidis et al. 2000).

1.2.3.2 Active Piezoelectric Damping Composites (APDC)

In this class of treatments, an array of piezo-ceramic rods embedded across the thickness of a viscoelastic polymeric matrix are electrically activated to control the damping characteristics of the matrix that is directly bonded to the vibrating structure as shown in Figure 1.9. The figure displays two arrangements of the APDC. In the first arrangement, the piezo-rods are embedded perpendicular to the electrodes to control the compressional damping (Reader and Sauter 1993) and in the second arrangement, the rods

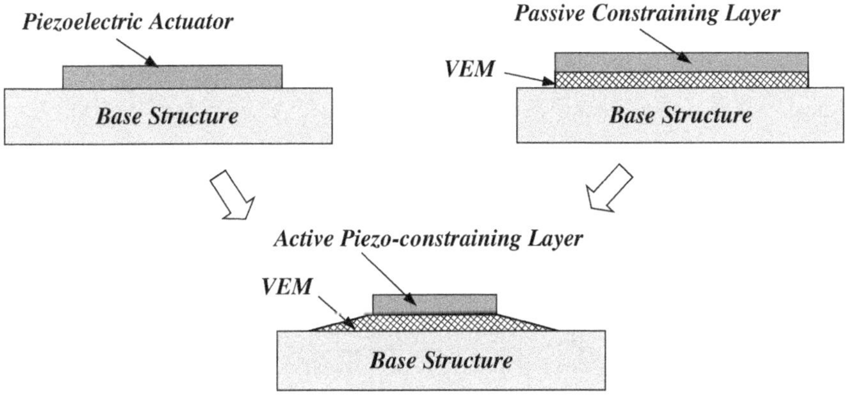

Figure 1.8 Active constrained layer damping treatment.

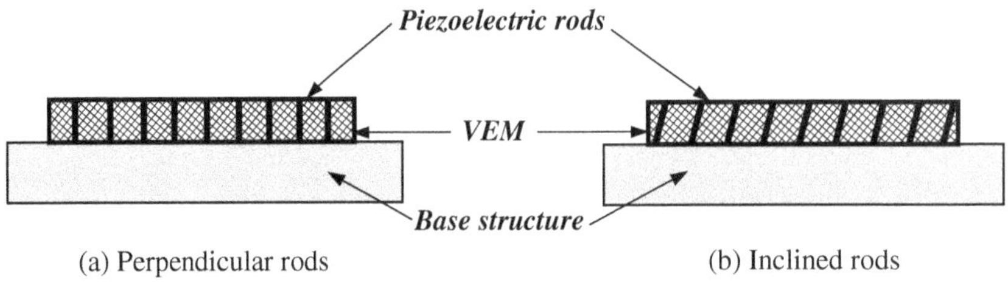

Figure 1.9 Active piezoelectric damping composites. (a) Perpendicular rods and (b) inclined rods.

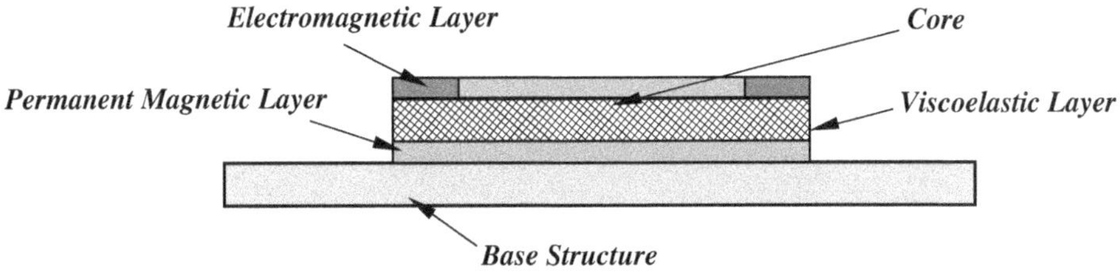

Figure 1.10 Electromagnetic damping composite (EMDC).

are obliquely embedded to control both the compressional and shear damping of the matrix (Baz and Tampia 2004; Arafa and Baz 2000).

1.2.3.3 Electromagnetic Damping Composites (EMDC)

In this class of composites, a layer of viscoelastic damping treatment is sandwiched between a permanent and electromagnetic layer as shown in Figure 1.10. The entire assembly is bonded to the vibrating surface to act as a smart damping treatment. The interaction between the magnetic layers, in response to the structural vibration, subjects the viscoelastic layer to compressional forces of proper magnitude and phase shift. These forces counterbalance the transverse vibration of the base structure and enhance the damping characteristics of the VEM. Accordingly, the electromagnetic damping composite (EMDC) acts in effect as a tunable Den–Hartog damper with the base structure serving as the primary system, the electromagnetic layer acting as the secondary mass, the magnetic forces generating the adjustable stiffness characteristics, and the viscoelastic layer providing the necessary damping effect (Baz 1997; Omer and Baz 2000; Ruzzene et al. 2000; Baz and Poh 2000; Oh et al. 2000).

1.2.3.4 Active Shunted Piezoelectric Networks

In this class of treatments, shown schematically in Figure 1.11, the passive shunted electric network is actively switched on and off in response to the response of the structure/network system in order to maximize the instantaneous energy dissipation characteristics and minimize the frequency-dependent performance degradation (Lesieutre 1998; Tawfik and Baz 2004; Park and Baz 2005; Thorp et al. 2005).

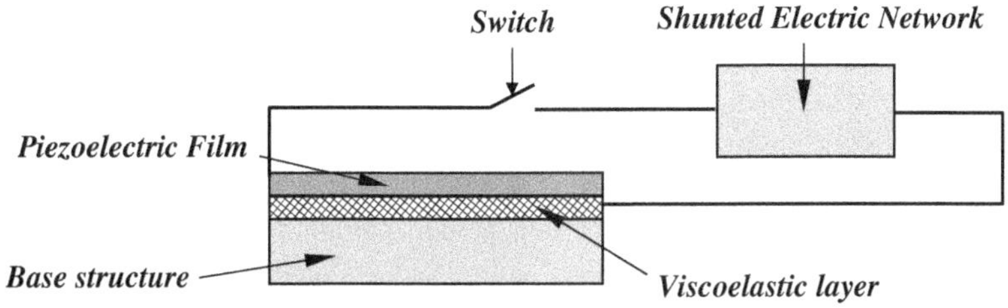

Figure 1.11 Damping layers with shunted piezoelectric treatments.

1.3 Summary

This chapter has presented a brief description of the main vibration control methods that have been successfully applied to damping out the vibration of a wide variety of structures. Analysis and performance characteristics of these vibration damping control methods will be presented in the remaining chapters.

References

Aldraihem, O., Baz, A., and Al-Saud, T.S. (2007). Hybrid composites with shunted piezoelectric particles for vibration damping. *Journal of Mechanics of Advanced Materials and Structures* 14: 413–426.

Arafa, M. and Baz, A. (2000). Dynamics of active piezoelectric damping composites. *Journal of Composites Engineering: Part B* 31: 255–264.

Baz A. Active Constrained Layer Damping, US Patent 5,485,053, filed October 15 1993 and issued January 16 1996.

Baz A. "Magnetic constrained layer damping", Proceedings of 11th Conference on Dynamics & Control of Large Structures, Blacksburg, VA (May 1997), pp. 333–344.

Baz, A. (2000). Spectral finite element modeling of wave propagation in rods using active constrained layer damping. *Journal of Smart Materials and Structures* 9: 372–377.

Baz, A. and Poh, S. (2000). Performance characteristics of magnetic constrained layer damping. *Journal of Shock & Vibration* 7 (2): 18–90.

Baz, A. and Tampia, A. (2004). Active piezoelectric damping composites. *Journal of Sensors and Actuators: A. Physical* 112 (2–3): 340–350.

Crassidis, J., Baz, A., and Wereley, N. (2000). H_∞ control of active constrained layer damping. *Journal of Vibration & Control* 6 (1): 113–136.

Forward, R.L. (1979). Electronic damping of vibrations in optical structures. *Applied Optics* 18 (5): 1.

Ghoneim H. "Bending and twisting vibration control of a cantilever plate via electromechanical surface damping". Proceedings of the Smart Structures and Materials Conference (ed. C. Johnson), Vol. SPIE-2445, pp. 28–39, 1995.

Greaser E. and Cozzarelli F., "Full cyclic hysteresis of a Ni-Ti shape memory alloy", Proceedings of DAMPING '93 Conference, San Francisco, CA, Wright Laboratory Document no. WL-TR-93-3105, Vol. 2, pp. ECB-1–28, 1993.

Lesieutre, G.A. (1998). Vibration damping and control using shunted piezoelectric materials. *The Shock and Vibration Digest* 30 (3): 187–195.

Nashif, A., Jones, D., and Henderson, J. (1985). *Vibration Damping.* New York: Wiley.

Oh, J., Ruzzene, M., and Baz, A. (1999). Control of the dynamic characteristics of passive magnetic composites. *Journal of Composites Engineering, Part B* 30: 739–751.

Oh, J., Poh, S., Ruzzene, M., and Baz, A. (2000). Vibration control of beams using electromagnetic compressional damping treatment. *ASME Journal of Vibration & Acoustics* 122 (3): 235–243.

Omer, A. and Baz, A. (2000). Vibration control of plates using electromagnetic compressional damping treatment. *Journal of Intelligent Material Systems & Structures* 11 (10): 791–797.

Park, C.H. and Baz, A. (2005). Vibration control of beams with negative capacitive shunting of interdigital electrode piezoceramics. *Journal of Vibration and Control* 11 (3): 331–346.

Preumont, A. (1997). *Vibration Control of Active Structures.* Dordrecht, The Netherlands: Kluwer Academic Publishers.

Reader W. and Sauter D., "Piezoelectric composites for use in adaptive damping concepts", Proceedings of DAMPING '93, San Francisco, CA (February 24–26, 1993), pp. GBB 1–18.

Ruzzene, M., Oh, J., and Baz, A. (2000). Finite element modeling of magnetic constrained layer damping. *Journal of Sound & Vibration* 236 (4): 657–682.

Tawfik, M. and Baz, A. (2004). Experimental and spectral finite element study of plates with shunted piezoelectric patches. *International Journal of Acoustics and Vibration* 9 (2): 87–97.

Thorp, O., Ruzzene, M., and Baz, A. (2005). Attenuation of wave propagation in fluid-loaded shells with periodic shunted piezoelectric rings. *Journal of Smart Materials & Structures* 14 (4): 594–604.

2
Viscoelastic Damping

2.1 Introduction

Viscoelastic damping treatments have been extensively used in various structural applications to control undesirable vibrations and associated noise radiation in a simple and reliable manner (Nashif et al. 1985; Sun and Lu 1995). In this chapter, particular emphasis is placed on studying the dynamic characteristics of such damping treatments and outlining the different mathematical models used to describe the behavior of these treatments over a wide range of operating frequencies and temperatures. Particular focus is given to ascertain the merits and drawbacks of the classical models by Maxwell, Kelvin–Voigt, and Zener (Zener 1948; Flugge 1967; Christensen 1982; Haddad 1995; Lakes 1999, 2009) both in the time and frequency domains.

2.2 Classical Models of Viscoelastic Materials

These models include the of Maxwell, Kelvin–Voigt, and Poynting–Thomson models (Haddad 1995; Lakes 1999, 2009). In these models, the dynamics of **V**isco**E**lastic **M**aterials (VEMs) are described in terms of series and/or parallel combinations of viscous dampers and elastic springs as shown in Figure 2.1. The dampers are included to capture the viscous behavior of the VEM, whereas the springs are used to simulate the elastic behavior of the VEM.

2.2.1 Characteristics in the Time Domain

The dynamic characteristics of Maxwell and Kelvin–Voigt models in the time domain are summarized in Table 2.1.

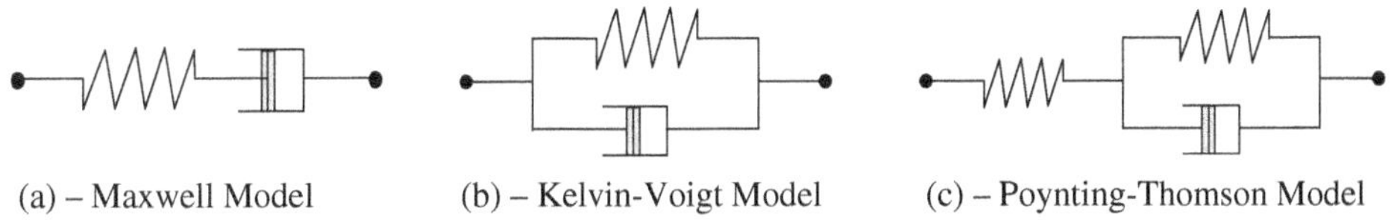

Figure 2.1 Classical models of VEMs. (a) Maxwell model, (b) Kelvin–Voigt model, and (c) Poynting–Thomson model.

Active and Passive Vibration Damping, First Edition. Amr M. Baz.

Table 2.1 The dynamic equations of Maxwell and Kelvin–Voigt models.

Model	Maxwell model	Kelvin–Voigt model
Stresses and strains of components	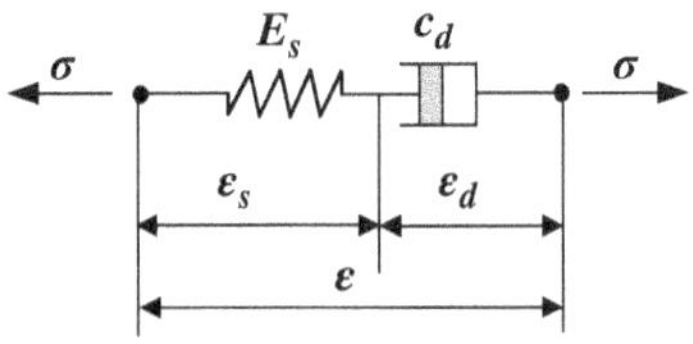	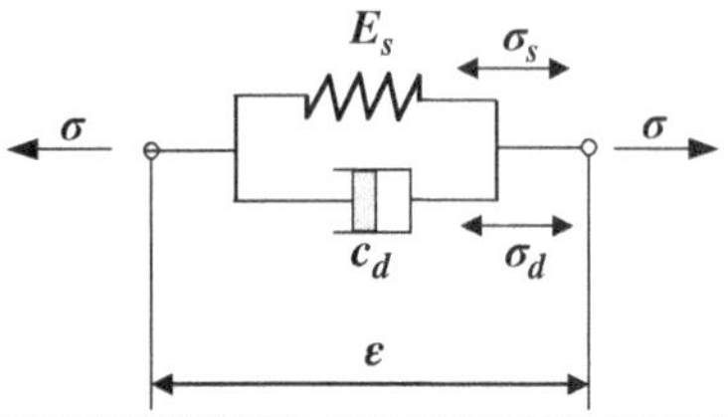
Equilibrium and kinematic equations	• Stress σ is same for spring and damper • Strain ε is sum of strains of spring and damper: $\sigma = \sigma_s = \sigma_d$ (2.1) and $\varepsilon = \varepsilon_s + \varepsilon_d$ (2.3)	• Strain ε is same for spring and damper • Stress σ is sum of stresses of spring and damper: $\sigma = \sigma_s + \sigma_d$ (2.2) and $\varepsilon = \varepsilon_s = \varepsilon_d$ (2.4)
Constitutive equations	**Spring**: $\sigma = E_s\varepsilon_s$ (2.5) **Damper**: $\sigma = c_d\dot{\varepsilon}_d$ (2.7)	**Spring**: $\sigma_s = E_s\varepsilon$ (2.6) **Damper**: $\sigma_d = c_d\dot{\varepsilon}$ (2.8)
Model equation	Substituting Eqs. (2.5) and (2.7) into Eq. (2.3) gives: $\lambda\dot{\sigma} + \sigma = c_d\dot{\varepsilon}$ (2.9) where $\lambda = c_d/E_s$	Substituting Eqs. (2.6) and (2.8) into Eq. (2.2) gives: $\sigma = E_s\varepsilon + c_d\dot{\varepsilon}$ (2.10)

(E_s = Young's modulus of elastic element, c_d = damping coefficient of dissipative element)

One can note that the stress–strain equations of the Maxwell and Kelvin–Voigt models can generally be written as follows:

$$P\sigma = Q\varepsilon \tag{2.11}$$

where P and Q are differential operators given by:

$$P = \sum_{i=0}^{p} \alpha_i \frac{d^i}{dt^i} \text{ and } Q = \sum_{j=0}^{q} \beta_j \frac{d^j}{dt^j} \tag{2.12}$$

Hence, for Maxwell model, $p = 1$, $q = 1$, $\alpha_0 = 1$, $\alpha_1 = \lambda$, $\beta_0 = 0$ and $\beta_1 = c_d$ while for the Kelvin–Voigt model, $p = 0$, $q = 1$, $\alpha_0 = 1$, $\beta_0 = E_s$ and $\beta_1 = c_d$.

The ability of both the Maxwell and the Kelvin–Voigt models to predict the characteristics of realistic VEM will be determined by considering the behavior under creep and relaxation loading conditions.

2.2.2 Basics for Time Domain Analysis

The initial and final value theorems of the Laplace transform are essential to the complete understanding of the behavior of viscoelastic models in the time domain. Appendix 2.A summarizes the two theorems and presents the necessary proofs.

Application of these two theorems to Maxwell and Kelvin–Voigt models is summarized in Tables 2.2 and 2.3 when these models are subjected to creep and relaxation

Table 2.2 Initial and final values of stresses and strains of Maxwell and Kelvin–Voigt models when subjected to creep loading.

Model	Maxwell model	Kelvin–Voigt model
Model	$\lambda\dot{\sigma} + \sigma = c_d\dot{\varepsilon}$	$\sigma = E_s\varepsilon + c_d\dot{\varepsilon}$
Loading conditions	The stress is constant $\sigma = \sigma_0$ and the initial and final values of the strain ε are predicted. 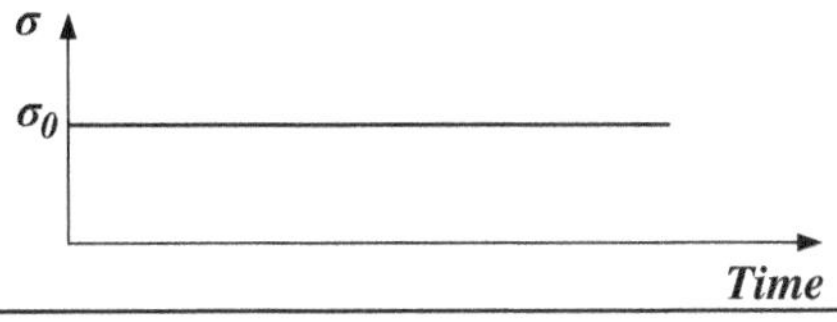	
The strain in the Laplace *s* domain	$\varepsilon = \frac{\lambda s + 1}{c_d s}\sigma = \frac{\lambda s + 1}{c_d s^2}\sigma_0$	$\varepsilon = \frac{1}{E_s(\lambda s + 1)}\sigma = \frac{1}{E_s s(\lambda s + 1)}\sigma_0$
Initial value	$\varepsilon_0 = \lim_{s\to\infty} s\varepsilon = \lim_{s\to\infty} \frac{\lambda + 1/s}{c_d}\sigma_0 = \frac{\sigma_0}{E_s}$	$\varepsilon_0 = \lim_{s\to\infty} s\varepsilon = \lim_{s\to\infty} \frac{1}{E_s(\lambda s + 1)}\sigma_0 = 0$
Final value	$\varepsilon_\infty = \lim_{s\to 0} s\varepsilon = \lim_{s\to 0} \frac{\lambda s + 1}{c_d s}\sigma_0 = \infty$	$\varepsilon_\infty = \lim_{s\to 0} s\varepsilon = \lim_{s\to 0} \frac{1}{E_s(\lambda s + 1)}\sigma_0 = \frac{\sigma_0}{E_s}$

Table 2.3 Initial and final values of stresses and strains of Maxwell and Kelvin–Voigt models when subjected to relaxation loading.

Model	Maxwell model	Kelvin–Voigt model
Model	$\lambda\dot{\sigma} + \sigma = c_d\dot{\varepsilon}$	$\sigma = E_s\varepsilon + c_d\dot{\varepsilon}$
Loading conditions	The strain is constant $\varepsilon = \varepsilon_0$ and the initial and final values of the stress σ are predicted. 	
The stress in the Laplace *s* domain	$\sigma = \frac{c_d s}{\lambda s + 1}\varepsilon = \frac{c_d}{\lambda s + 1}\varepsilon_0$	$\sigma = E_s(\lambda s + 1)\varepsilon = \frac{E_s(\lambda s + 1)}{s}\varepsilon_0$
Initial value	$\sigma_0 = \lim_{s\to\infty} s\sigma = \lim_{s\to\infty} \frac{c_d s}{\lambda s + 1}\varepsilon_0 = E_s\varepsilon_0$	$\sigma_0 = \lim_{s\to\infty} s\sigma = \lim_{s\to\infty} E_s\varepsilon_0 = E_s\varepsilon_0$
Final value	$\sigma_\infty = \lim_{s\to 0} s\sigma = \lim_{s\to 0} \frac{c_d s}{\lambda s + 1}\varepsilon_0 = 0$	$\sigma_\infty = \lim_{s\to 0} s\sigma = \lim_{s\to 0} E_s\varepsilon_0 = E_s\varepsilon_0$

loading, respectively. These two theorems provide the means for determining the initial and final limits of the VEM response under different loading conditions. This feature enables the correct calculation of the time response, between these two limits, when the differential equations describing these models are solved as will be demonstrated later.

Table 2.2 indicates that the Maxwell model experiences an initial strain when the creep load is applied, which is typical in VEMs. However, this strain tends to become unbounded as time grows. This feature is not observed or supported experimentally. As for the Kelvin–Voigt model, the initial value theorem indicates zero initial strain, which is rather unrealistic and a bounded final strain of σ_0/E_s that is observed in a realistic VEM.

Table 2.3 indicates that the Maxwell model experiences an initial stress when the relaxation strain is applied and that stress is completely relieved as time progresses. Both of these characteristics are typical in VEM. As for the Kelvin–Voigt model, the initial and the final values remain constant $E_s\varepsilon_0$, which is rather unrealistic behavior of a VEM.

2.2.3 Detailed Time Response of Maxwell and Kelvin–Voigt Models

Tables 2.4 and 2.5 summarize the detailed behavior characteristics of Maxwell and Kelvin–Voigt models in the time domain between the initial and final values predicted in Tables 2.2 and 2.3.

Tables 2.4 and 2.5 indicate that the Maxwell model predicts unrealistic creep characteristics as the strain tends to be unbounded even for finite stress levels or the strain tends to remain constant when the stress is removed. The Kelvin–Voigt model also yields unrealistic relaxation characteristics with the stress remaining constant with time, indicating that the VEM does not exhibit any stress relaxation. Therefore, neither the Maxwell nor the Kelvin–Voigt model replicates the behavior of realistic VEM.

Note that these predictions, particularly at $t = 0$, are in agreement with the predictions of the initial and final value theorems listed in Tables 2.2 and 2.3.

In order to avoid the drawbacks and limitations of both the Maxwell and Kelvin–Voigt models, several other spring-damper arrangements have been considered. For example, a damper with series and parallel springs is considered to combine the attractive attributes and compensate for the deficiencies of both the Maxwell and Kelvin–Voigt models. The resulting model is the *Poynting–Thomson* model, shown in Figures 2.1c and 2.2a. Other common models are also displayed in Figure 2.2 such the "three-parameter model" and the "standard solid model" (Zener 1948).

Figure 2.3a,b show the most widely used spring-mass configurations of VEM models that are employed extensively, particularly, in commercial finite element packages. These two configurations are, namely, the generalized Maxwell model and the generalized Kelvin–Voigt model.

These two generalized n classical models are assembled in parallel or series to model the complex behavior of realistic VEMs. These models are augmented with additional springs E_0, either in parallel or series, to eliminate the drawbacks associated with the classical models as outlined in Tables 2.2–2.5.

Table 2.4 The creep characteristics of Maxwell and Kelvin–Voigt models.

Model	Maxwell model	Kelvin–Voigt model
Loading conditions	The stress is constant $\sigma = \sigma_0$ and the time history of the strain is predicted 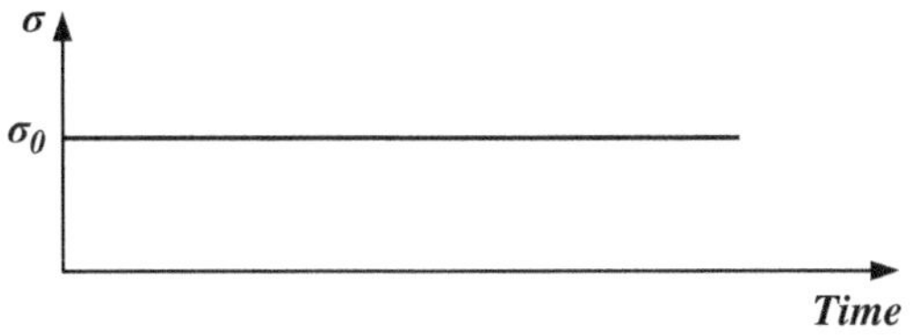	
Response	• As the initial strain $\varepsilon = \varepsilon_0$[a] at time $t = 0$, • Hence: $\varepsilon = \frac{\sigma_0}{c_d}t + \frac{\sigma_0}{E_s} = \frac{\sigma_0}{E_s}(1 + t/\lambda)$ (2.13) 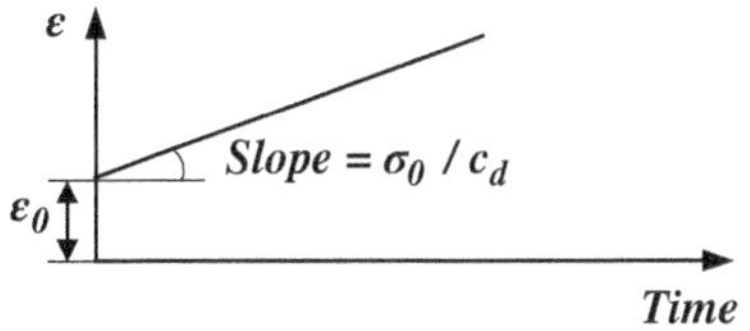 • Unbounded strain for bounded stress	• As the initial strain $\varepsilon = 0$[a] at time t = 0, • Hence: $\varepsilon = \frac{\sigma_0}{E_s}\left[1 - e^{-t/\lambda}\right]$ (2.14) 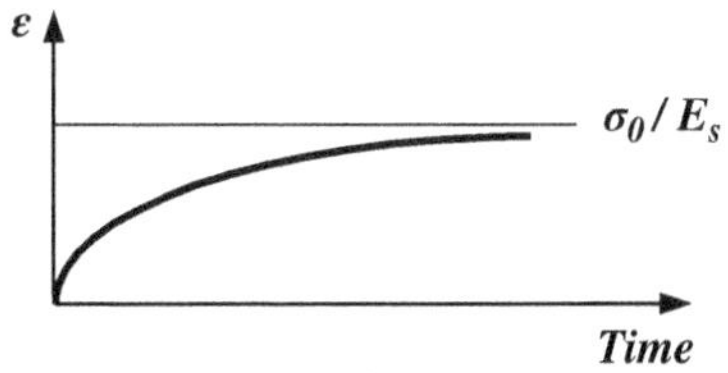 • Bounded strain for bounded stress
Unloading Conditions	The stress is reduced back to zero at time $t = t_1$ and the time history of the strain 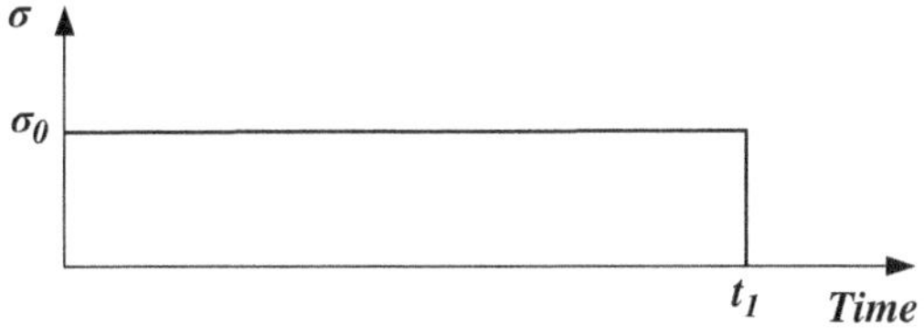	
Response	• At $t = t_1$, $\varepsilon_1 = \frac{\sigma_0}{c_d}t_1 + \frac{\sigma_0}{E_s}$, • Hence, when $\sigma = 0$: $\dot{\varepsilon} = 0$ with solution $\varepsilon = \varepsilon_1 =$ constant (2.15) 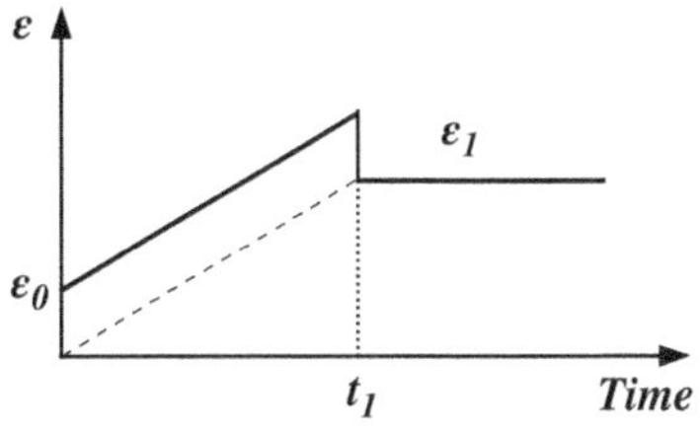 • No contraction after stress removal	• At $t = t_1$, $\varepsilon_1 = \frac{\sigma_0}{E_s}\left[1 - e^{-t_1/\lambda}\right]$, • Hence, when $\sigma = 0$: $E_s\varepsilon + c_d\dot{\varepsilon} = 0$ or $\lambda\dot{\varepsilon} + \varepsilon = 0$ with solution $\varepsilon = \varepsilon_1 e^{-(t-t_1)/\lambda}$ (2.16) 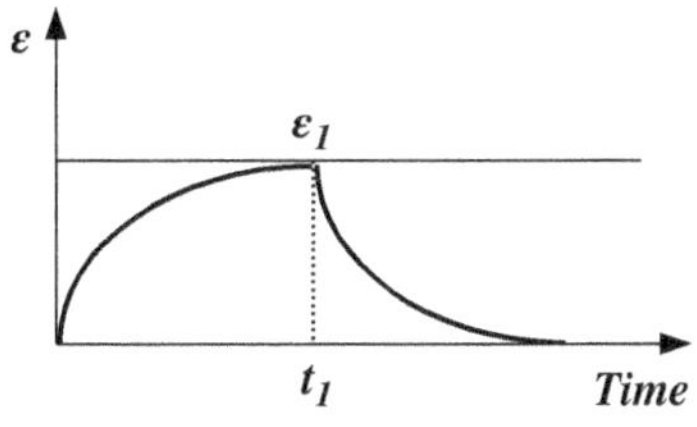 • Complete strain relief after stress removal

[a] Table 2.2 (using initial value theorem).

Table 2.5 Relaxation characteristics of Maxwell and Kelvin–Voigt models.

Model	Maxwell model	Kelvin–Voigt model
Loading conditions	The strain is constant $\varepsilon = \varepsilon_0$ and determines the time history of the stress 	
Response	• As the initial stress $\sigma = \sigma_0{}^a$ at time $t = 0$: • Hence: $\lambda\dot{\sigma} + \sigma = 0$ with solution $\sigma = E_s e^{-t/\lambda}\varepsilon_0$ (2.17) 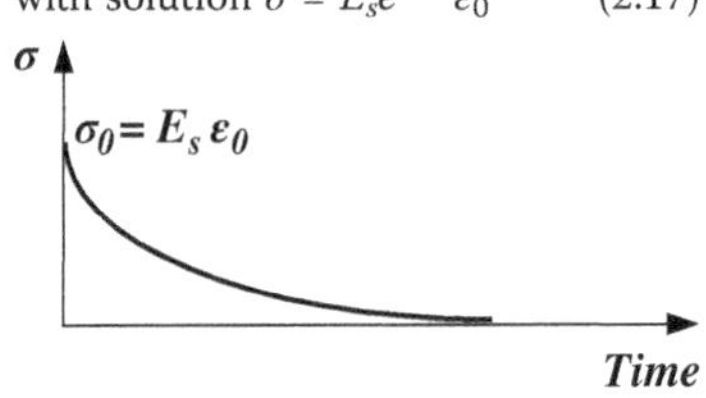 • stress decays to zero without any residual stress	• As the initial strain $\varepsilon = \varepsilon_0{}^a$ at time $t = 0$: • Hence: $\sigma = E_s\varepsilon_0 = \text{constant}$ (2.18) 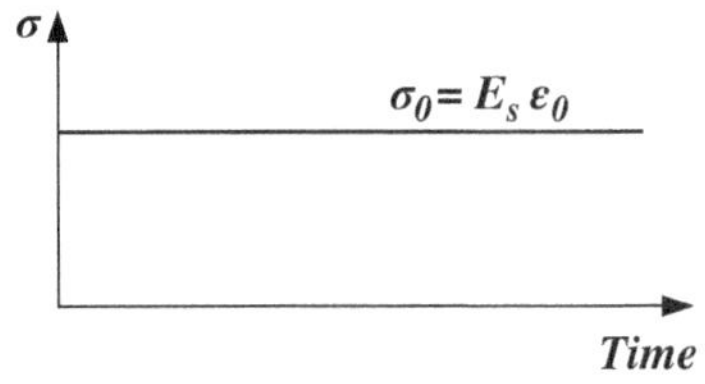 • stress remains constant, that is, VEM exhibits no relaxation

[a] Table 2.3 (using initial value theorem).

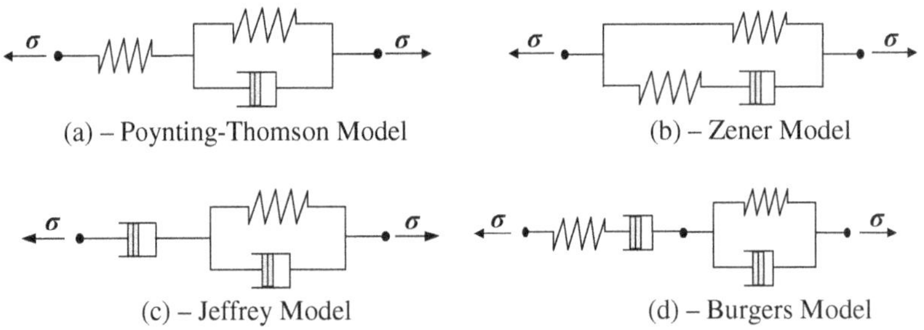

Figure 2.2 Other common viscoelastic models. (a) Poynting–Thomson model, (b) Zener model, (c) Jeffrey model, and (d) Burgers model.

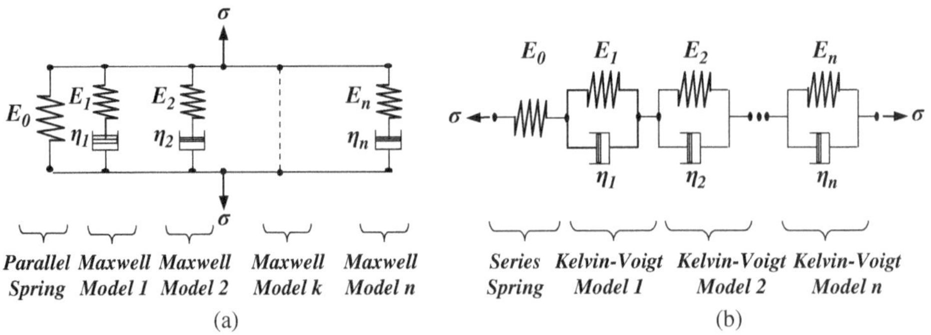

Figure 2.3 Generalized Maxwell (a) and Kelvin–Voigt (b) models.

2.2.4 Detailed Time Response of the Poynting–Thomson Model

The stress σ across the series spring of the Poynting–Thomson model, shown in Figure 2.4, is given by:

$$\sigma = E_s \varepsilon_s \tag{2.19}$$

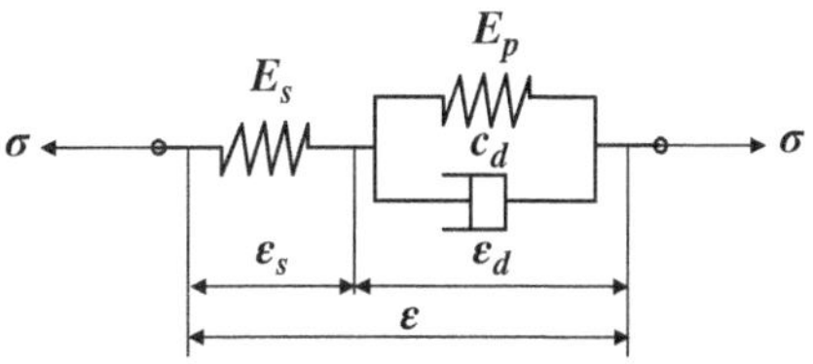

Figure 2.4 Poynting–Thomson viscoelastic model.

and the stress σ across the damper and the parallel spring is given by:

$$\sigma = E_p \varepsilon_d + c_d \dot{\varepsilon}_d \tag{2.20}$$

Using the Laplace transformation, yields

$$\varepsilon_s = \sigma / E_s \text{ and } \varepsilon_d = \sigma / \left(E_p + c_d s\right) \tag{2.21}$$

Hence, the total strain ε across the Poynting–Thomson model is

$$\varepsilon = \varepsilon_s + \varepsilon_d = \left[\frac{\left(E_s + E_p\right) + c_d s}{E_s\left(E_p + c_d s\right)}\right] \sigma \tag{2.22}$$

In the time domain, this equation becomes

$$\left(E_s + E_p\right)\sigma + c_d \dot{\sigma} = E_s E_p \varepsilon + E_s c_d \dot{\varepsilon} \tag{2.23}$$

From Eqs. (2.11), (2.12), and (2.23), $p = 1$, $q = 1$, $\alpha_0 = (E_s + E_p)$, $\alpha_1 = c_d$, $\beta_0 = E_s E_p$, and $\beta_1 = E_s c_d$.

a) *The creep characteristics of the Poynting–Thomson model* are obtained as follows:

i) Determine the initial and final values strain:

For stress $\sigma = \sigma_0$, then Eq. (2.22) reduces to:

$$\varepsilon = \left[\frac{\left(E_s + E_p\right) + c_d s}{E_s\left(E_p + c_d s\right)}\right] \frac{\sigma_0}{s}$$

Then,

$$\varepsilon_0 = \lim_{s \to \infty} s\varepsilon = \lim_{s \to \infty} \left[\frac{\left(E_s + E_p\right) + c_d s}{E_s\left(E_p + c_d s\right)}\right] \sigma_0 = \frac{\sigma_0}{E_s}$$

and

$$\varepsilon_\infty = \lim_{s \to 0} s\varepsilon = \lim_{s \to 0} \left[\frac{\left(E_s + E_p\right) + c_d s}{E_s\left(E_p + c_d s\right)}\right] \sigma_0 = \frac{\sigma_0}{E_\infty}$$

where $E_\infty = \dfrac{E_s E_p}{\left(E_s + E_p\right)}$.

ii) Determine the time history of the strain:

The time history of the strain is determined by solving Eq. (2.23) such that at $t = 0$, $\sigma = \sigma_0$, and the initial strain $\varepsilon_0 = \sigma_0/E_s$. Hence, Eq. (2.23) reduces to:

$$E_s c_d \dot{\varepsilon} + E_s E_p \varepsilon = (E_s + E_p)\sigma_0$$

This equation has a solution:

$$\varepsilon = \frac{\sigma_0}{E_\infty}\left[1 + \frac{E_\infty - E_s}{E_s} e^{-t/\lambda}\right] \tag{2.24}$$

where $\lambda = c_d/E_p$ and $E_\infty = E_s E_p/(E_s + E_p)$. Note that Eq. (2.24) has the initial and final values ε_0 and ε_∞ at $t = 0$ and $t = \infty$.

Figure 2.5 shows the strain–time characteristics as predicted by Eq. (2.24).

b) *The Relaxation characteristics of the Poynting–Thomson model* are obtained as follows:

i) Determining the initial and final values stress:

For strain $\varepsilon = \varepsilon_0$, then Eq. (2.22) reduces to:

$$\sigma = \left[\frac{E_s(E_p + c_d s)}{(E_s + E_p) + c_d s}\right]\frac{\varepsilon_0}{s}$$

Then,

$$\sigma_0 = \lim_{s\to\infty} s\sigma = \lim_{s\to\infty}\left[\frac{E_s(E_p + c_d s)}{(E_s + E_p) + c_d s}\right]\varepsilon_0 = E_s\varepsilon_0$$

and

$$\sigma_\infty = \lim_{s\to 0} s\varepsilon = \lim_{s\to 0}\left[\frac{E_s(E_p + c_d s)}{(E_s + E_p) + c_d s}\right]\varepsilon_0 = E_\infty\varepsilon_0$$

where $E_\infty = \dfrac{E_s E_p}{(E_s + E_p)}$.

ii) Determining the time history of the stress:

The time history of the stress can be determined by solving Eq. (2.23) such that at $t = 0$, $\varepsilon = \varepsilon_0$, and the initial stress $\sigma_0 = E_s\varepsilon_0$. Hence, Eq. (2.23) reduces to:

$$c_d\dot{\sigma} + (E_s + E_p)\sigma = E_s E_p \varepsilon_0$$

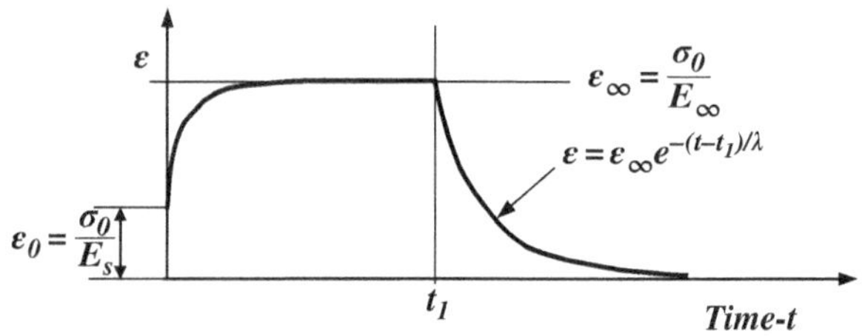

Figure 2.5 The creep characteristics of the Poynting–Thomson model.

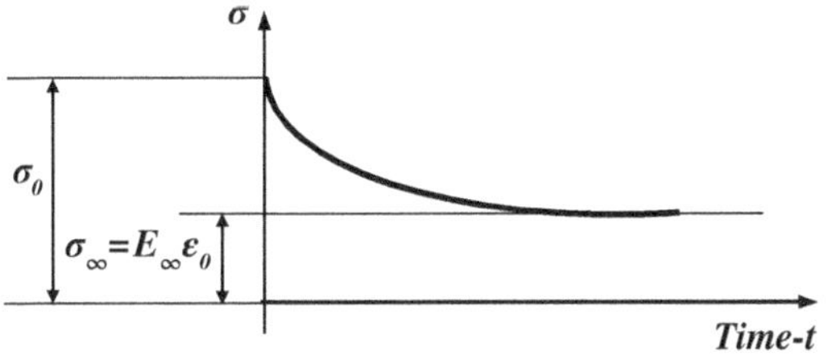

Figure 2.6 The relaxation characteristics of the Poynting–Thomson model.

This equation has the following solution:

$$\sigma = E_{\infty}\varepsilon_0\left[\left(1-e^{-t/\alpha}\right)\right] + E_s\varepsilon_0 e^{-t/\alpha} \tag{2.25}$$

where $\alpha = \dfrac{c_d}{\left(E_s + E_p\right)}$.

Figure 2.6 shows the stress–time characteristics as predicted by Eq. (2.25).

Table 2.6 summarizes the main characteristics of Maxwell, Kelvin–Voigt, and Poynting–Thomson models.

The characteristics summarized in Table 2.6 ascertain the ability of the Poynting–Thomson model to simulate a realistic behavior of VEMs. However, several

Table 2.6 Time domain characteristics of classical viscoelastic models.

Parameter	Maxwell	Kelvin–Voigt	Poynting–Thomson
Model		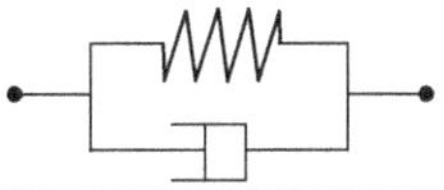	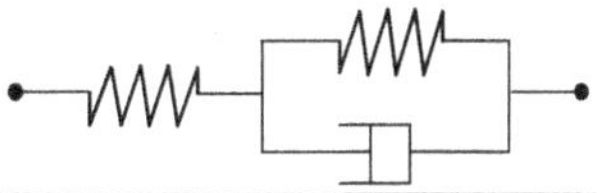
Dynamic equations	$\lambda\dot{\sigma} + \sigma = c_d\,\dot{\varepsilon}$	$\sigma = E_s\varepsilon + c_d\dot{\varepsilon}$	$\left(E_s + E_p\right)\sigma + c_d\dot{\sigma} = E_sE_p\varepsilon + E_sc_d\dot{\varepsilon}$
Creep characteristics	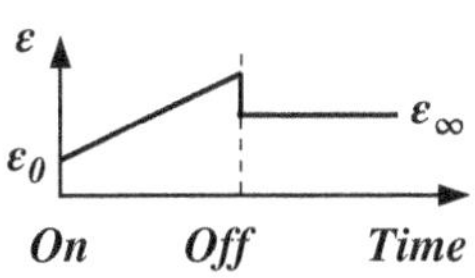	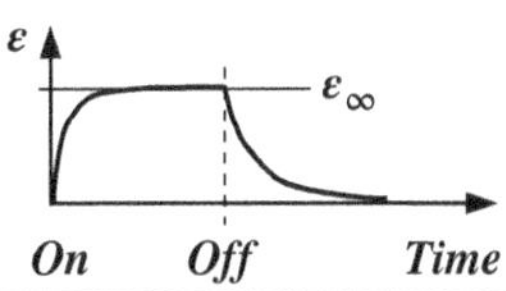	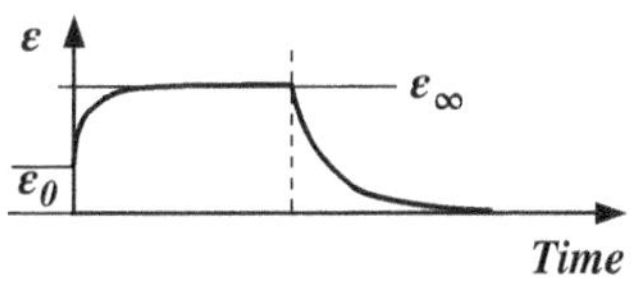
Comments	Unrealistic	Realistic	Realistic
Relaxation characteristics	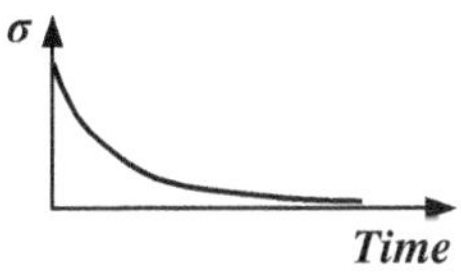	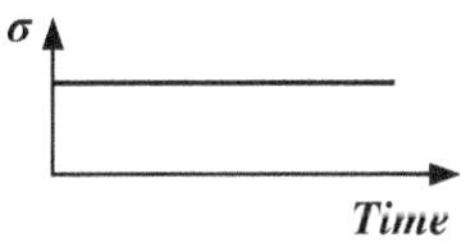	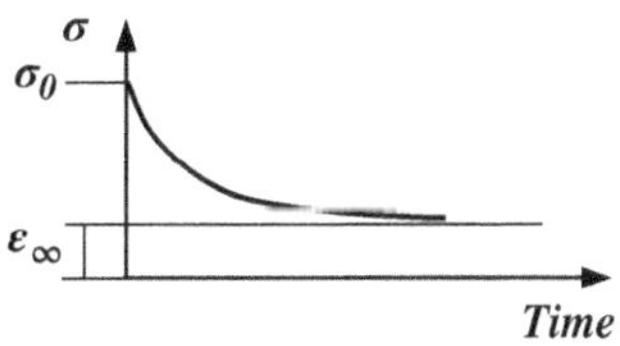
Comments	Realistic	Unrealistic	Realistic

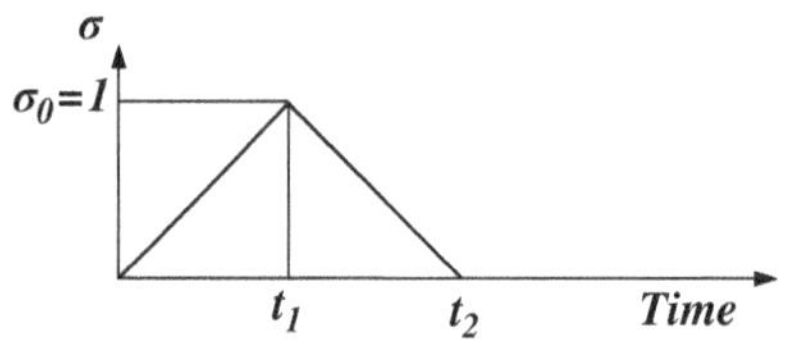

Figure 2.7 A ramp creep loading and loading cycle.

Table 2.7 Solutions of the constitutive equations.

Parameter	Maxwell	Kelvin–Voigt
Constitutive equation	$\dot{\varepsilon} = \dot{\sigma} + \sigma$	$\dot{\varepsilon} + \varepsilon = \sigma$
Equation during loading	$\dot{\varepsilon} = 1 + t$	$\dot{\varepsilon} + \varepsilon = t$
Initial condition (ε_0)	$\varepsilon_0 = 0$	$\varepsilon_0 = 0$
Response	$\varepsilon = \varepsilon_0 + t + {}^1/_2 t^2$	$\varepsilon = (\varepsilon_0 + 1)e^{-t} - 1 + t$
Equation during unloading	$\dot{\varepsilon} = 1 - t$	$\dot{\varepsilon} + \varepsilon = 2 - t$
Initial condition (ε_1)	$\varepsilon_1 = 1.5$	$\varepsilon_1 = e^{-1}$
Response	$\varepsilon = 1 + t - {}^1/_2 t^2$	$\varepsilon = 3 - t + e^{-t} - 2e^{1-t}$

combinations of Poynting–Thomson models are necessary to replicate the behavior of realistic VEMs.

Example 2.1 Plot the stress–strain characteristics for the Maxwell and Kelvin–Voigt models when the VEM is subjected to the loading and unloading cycle shown in Figure 2.7.

Assume that $E_s = 1$, $E_p = 1$, $c_d = 1$, $\varepsilon_0 = 0$, $t_1 = 1$, and $t_2 = 2$.

Solution

Table 2.7 lists the solutions of the constitutive equations of Maxwell and Kelvin–Voigt models for the given loading and unloading cycle.

Figure 2.8a,b displays the stress–strain characteristics of the Maxwell and Kelvin–Voigt models. The figures indicate that, according to the Maxwell model, the VEM is stiffer and dissipates less energy, as represented by the enclosed area, than that predicted by the Kelvin–Voigt model.

2.3 Creep Compliance and Relaxation Modulus

In Section 2.2, time domain relationships are derived for the different classical VEM models when these models are subjected to creep or relaxation loading. These relationships are obtained by solving the constitutive equations that describe the dynamics of the VEM models subject to initial conditions determined by applying the initial value theorem.

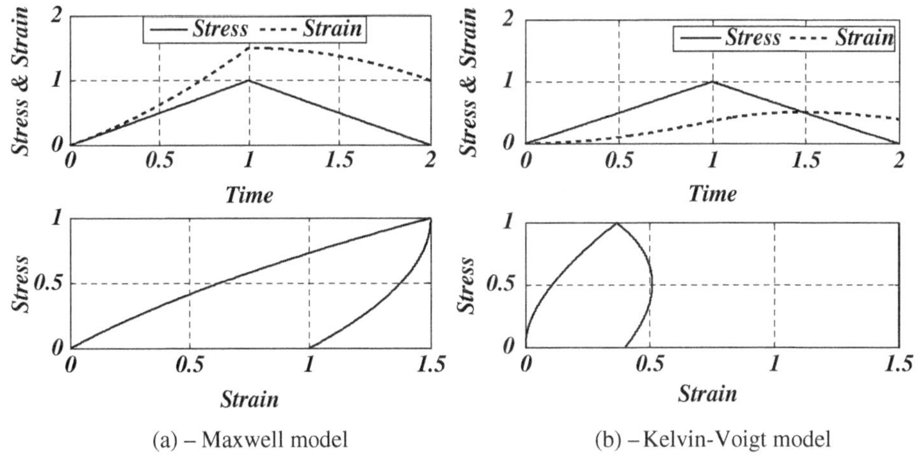

Figure 2.8 Stress–strain characteristics of Maxwell and Kelvin–Voigt models. (a) Maxwell model and (b) Kelvin–Voigt model.

Table 2.8 lists these relationships such that the ratio between the strain and amplitude of the creep stress ε/σ_0 is denoted by the symbol *J(t)* and called "the creep compliance" and the ratio between the stress and amplitude of the relaxation strain σ/ε_0 is designated by the symbol $E(t)$ and called "the relaxation modulus."

It is important to note that these two characteristic properties of VEM are time-dependent, unlike the corresponding properties for solids, which are constants.

Note also that the process involved in deriving these properties has been tedious and cumbersome as it consists of applying the initial and final value theorems followed by the exhaustive procedure for solving the constitutive equations subject to the initial values and then validating the solution against the obtained final values.

In this section, two other approaches are presented. In the first approach, the direct Laplace and inverse Laplace transformations are applied one after the other to the constitutive equations of the VEM. In the second approach, the topology of the VEM model is translated into a linear set of equations that can be reduced by using the Gauss elimination to determine both *J*(*t*) and *E*(*t*) simultaneously. The two approaches are implemented in a MATLAB environment to enhance their practicality and utility.

Table 2.8 The creep characteristics of Maxwell, Kelvin–Voigt, and Poynting–Thomson models.

Model	Maxwell model	Kelvin–Voigt model	Poynting–Thomson model
Creep compliance (*J*)	$J=\frac{\varepsilon}{\sigma_0}=\frac{1}{E_s}\left(1+\frac{t}{\lambda}\right)^a$	$J=\frac{\varepsilon}{\sigma_0}=\frac{1}{E_s}\left[1-e^{-t/\lambda}\right]^a$	$J=\frac{\varepsilon}{\sigma_0}=\frac{1}{E_\infty}\left[1+\frac{E_\infty-E_s}{E_s}e^{-t/\lambda}\right]$
Relaxation modulus (*E*)	$E=\frac{\sigma}{\varepsilon_0}=E_s e^{-t/\lambda}$	$E=\frac{\sigma}{\varepsilon_0}=E_s+c_d\,dirac(t)^b$ $=\frac{\sigma}{\varepsilon_{0+}}=E_s$	$E=\frac{\sigma}{\varepsilon_0}=E_\infty\left(1-e^{-t/\alpha}\right)+E_s e^{-t/\alpha}$ where $\alpha=\frac{c_d}{(E_s+E_p)}$, $\lambda=\frac{c_d}{E_p}$

[a] $\lambda = c_d/E_s$.

[b] $dirac(t) = \infty$ at $t = 0$ and 0 at $t = 0+$.

2.3.1 Direct Laplace Transformation Approach

The constitutive equations of the VEM are transformed into the Laplace domain to assume one of the following transfer function forms:

$$J^* = \frac{\varepsilon}{\sigma} \tag{2.26a}$$

and

$$E^* = \frac{\sigma}{\varepsilon} \tag{2.26b}$$

When the VEM is subjected to creep loading, the stress σ is replaced by its Laplace transform σ_0/s and Eq. (2.26a) reduces to:

$$J^* = s\frac{\varepsilon}{\sigma_0} = sJ(s) \tag{2.27a}$$

Similarly, when the VEM is subjected to relaxation loading, the strain ε is replaced by its Laplace transform ε_0/s and Eq. (2.26b) reduces to:

$$E^* = s\frac{\sigma}{\varepsilon_0} = sE(s) \tag{2.27b}$$

The inverse Laplace transform is then used to transform $J(s)$ and $E(s)$ into the time domain creep compliance $J(t)$ and relaxation modulus $E(t)$.

Table 2.9 lists the corresponding creep compliance $J(t)$ and relaxation modulus $E(t)$ for the Maxwell, Kelvin–Voigt, and Poynting–Thomson models.

Table 2.9 Time domain characteristics of classical viscoelastic models.

Operation	Maxwell	Kelvin–Voigt	Poynting–Thomson
Dynamic equations	$\lambda\dot{\sigma} + \sigma = c_d\,\dot{\varepsilon}$	$\sigma = E_s\varepsilon + c_d\dot{\varepsilon}$	$(E_s + E_p)\sigma + c_d\dot{\sigma}$ $= E_sE_p\varepsilon + E_sc_d\dot{\varepsilon}$
Laplace transform of the strain due to creep loading σ_0	$\frac{\varepsilon}{\sigma_0} = \frac{\lambda s + 1}{c_d s^2}$	$\frac{\varepsilon}{\sigma_0} = \frac{1}{E_s s(\lambda s + 1)}$	$\frac{\varepsilon}{\sigma_0} = \frac{(E_s + E_p) + c_d s}{(E_sE_p + E_sc_d s)s}$
Inverse Laplace transform of $\varepsilon/\sigma_0 = J$ using MATLAB	>> **syms** *L cd s t* >> **ilaplace** ((*L***s* + 1)/(*cd***s*^2),*s*,*t*) $J = \lambda/cd + t/cd$	>> **syms** *L* E *s t* >> **ilaplace**(1/(*E***s**(*L***s* + 1)),*s*,*t*) $J = 1/E - 1/(E^* \exp(t/\lambda))$	>> **syms** *Es Ep cd s t* >> **ilaplace** (((*Es* + *Ep*) + *cd***s*)/(*s**(*Es***Ep*+*Es***cd***s*)),*s*,*t*) *J* = (*Ep* + *Es*)/(*Ep***Es*) - 1/(*Ep***exp*((*Ep***t*)/*cd*))
Laplace transform of the stress due to relaxation loading ε_0	$\frac{\sigma}{\varepsilon_0} = \frac{c_d}{\lambda s + 1}$	$\frac{\sigma}{\varepsilon_0} = \frac{E_s(\lambda s + 1)}{s}$	$\frac{\sigma}{\varepsilon_0} = \frac{(E_sE_p + E_sc_d s)}{[(E_s + E_p) + c_d s]s}$
Inverse Laplace transform of $\sigma/\varepsilon_0 = R$ using MATLAB	>> **syms** *L cd s t* >> **ilaplace** ((*cd*)/(*L***s*+1),*s*,*t*) $R = cd/(\lambda^* \exp(t/\lambda))$	>> **syms** *L E s t* >> **ilaplace**(*E**(*L***s* + 1)/*s*,*s*,*t*) ***R*** = *E* + *E***L***dirac*(*t*)	>> **syms** *Es Ep cd s t* >> **ilaplace**((*Es***Ep* + *Es***cd***s*)/(*s**(*Es* + *Ep* + *cd***s*)),*s*,*t*) ***R*** = (*Ep***Es*)/(*Ep* + *Es*) + *Es*^2/(*exp*((*t**(*Ep* + *Es*))/*cd*)*(*Ep* + *Es*))

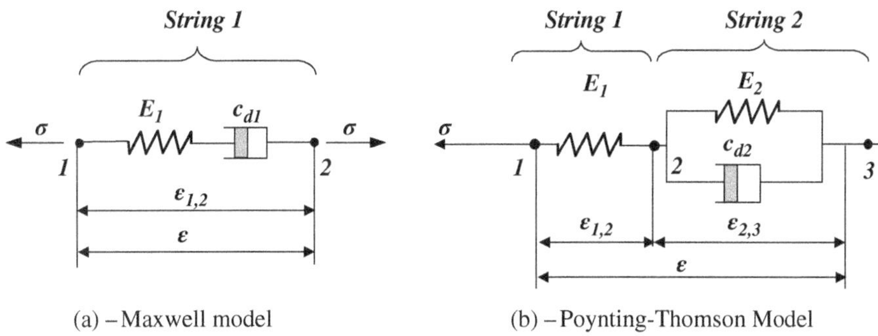

Figure 2.9 Topology of the Maxwell and Poynting–Thomson models. (a) Maxwell model and (b) Poynting–Thomson model.

2.3.2 Approach of Simultaneous Solution of a Linear Set of Equilibrium, Kinematic, and Constitutive Equations

This approach was developed by Vondřejc (2009) and translates the topology of the VEM model into a linear set of equilibrium, kinematic, and constitutive equations which can be reduced by using the Gauss elimination to determine both J and R simultaneously. The approach is implemented in a MATLAB environment to enhance its practicality and utility.

In this approach, the topology of the viscoelastic model is described by N serial strings extending between P points. Each string can consist of a spring and a damper in series. For example, Figure 2.9 shows the description of the topology of the Maxwell and Poynting–Thomson VEM models. In Figure 2.9a, the Maxwell model is described by one string and two points whereas in Figure 2.9b, the Poynting–Thomson model is defined by two strings and three points.

In a symbolic MATLAB environment, the topology description of the two models is given by the vectors B such that:

Maxwell Model: $B = [1, 2, E_1, cd_1]$,

Poynting–Thomson Model: $B = [1, 2, E_1, inf; 2, 3, E_2, inf; 2, 3, inf, cd_2]$;

Note that in this description of any string, the component value of a spring or a damper that is missing in the string is set equal to "inf."

The mathematical formulation of the *VEM* model, in the Laplace domain, is described as follows:

Constitutive equations $$-\sum_{j=b_i}^{e_i-1}\varepsilon_{j,j+1}+\sigma_i\left(\frac{1}{E_i}+\frac{1}{c_{d_i}s}\right)=0 \quad i=1,2,\ldots,N \tag{2.28}$$

Equilibrium equations $$-\sum_{j=1}^{P}\delta_{ib_j}\sigma_j+\sigma=0 \quad \text{For beginning points } b_j \tag{2.29}$$

$$\sum_{j=1}^{P}\left(\delta_{ib_j}-\delta_{ie_j}\right)\sigma_j=0 \quad i=1,2,\ldots,P-1 \tag{2.30}$$

$$-\sum_{k=1}^{P}\delta_{ie_k}\sigma_k+\sigma=0 \quad \text{For end points } e_k \tag{2.31}$$

Kinematic equations $$-\sum_{j=1}^{P-1}\varepsilon_{j,j+1}+\varepsilon=0 \tag{2.32}$$

Note that b_j and e_j denote the beginning and end points of the jth string.

In a matrix form, Eqs. (2.28) through (2.32) take the following form:

$$Ax = 0 \tag{2.33}$$

where x is a vector of the strains and stresses given by:

$$x = \{\varepsilon_{1,2} \;\; \varepsilon_{2,3} \ldots \varepsilon_{P-1,P} \;\; \sigma_1 \; \sigma_2 \ldots \sigma_N \;\; \sigma \;\; \varepsilon\} \tag{2.34}$$

where $\varepsilon_{j,\,j+1}$ = strain between points j and $j + 1$, σ_i = stress in the ith string, σ = stress applied to the entire VEM topology, and ε = total strain of the entire VEM topology.

Example 2.2 Derive expressions for the creep compliance and relaxation modulus for the Maxwell model using the approach of simultaneous solution of a linear set of equilibrium, kinematic, and constitutive equations described in Section 2.3.2.

Solution

From Eqs. (2.28) through (2.32), the system of equations describing the dynamics of the Maxwell model is given by:

$$\begin{bmatrix} -1 & C & 0 & 0 \\ 0 & -1 & 1 & 0 \\ 0 & -1 & 1 & 0 \\ -1 & 0 & 0 & 1 \end{bmatrix} \begin{Bmatrix} \varepsilon_{1,2} \\ \sigma_{1,2} \\ \sigma \\ \varepsilon \end{Bmatrix} = 0 \;\text{ or }\; Ax = 0 \tag{2.35}$$

where $C = \dfrac{1}{E_1} + \dfrac{1}{c_{d1}s}$.

Applying the Gauss elimination method to Eq. (2.35), it reduces to:

$$\begin{bmatrix} -1 & C & 0 & 0 \\ 0 & -1 & 1 & 0 \\ 0 & 0 & 0 & 0 \\ 0 & 0 & -C & 1 \end{bmatrix} \begin{Bmatrix} \varepsilon_{1,2} \\ \sigma_{1,2} \\ \sigma \\ \varepsilon \end{Bmatrix} = 0 \tag{2.36}$$

Expanding the last row of Eq. (2.36) gives:

$$C\sigma = \varepsilon \;\text{ or }\; \left(\frac{1}{E_1} + \frac{1}{c_{d1}s}\right)\sigma = \varepsilon \tag{2.37}$$

Hence, if the VEM is subjected to creep loading such that $\sigma = \sigma_0$, then Eq. (2.37) reduces to:

$$\varepsilon = \left[\frac{1}{E_1} + \frac{1}{c_{d1}s}\right]\frac{\sigma_0}{s} \tag{2.38}$$

Using MATLAB symbolic manipulation gives:

```
>> syms E1 cd1 sigma0 s t
>> ilaplace ((1/E1+1/(cd1*s))*sigma0/s, s, t)
J = sigma0/E1 + (sigma0*t)/cd1)
```

Note that the obtained creep compliance J matches that listed in Table 2.9.

Also, if the VEM is subjected to relaxation strain such that $\varepsilon = \varepsilon_0$, then Eq. (2.37) reduces to:

$$\sigma = \frac{\varepsilon_0}{s} / \left[\frac{1}{E_1} + \frac{1}{c_{d1}s}\right] \tag{2.39}$$

Using MATLAB symbolic manipulation gives:

```
>> syms E1 cd1 eps0 s t
>> ilaplace (1/(1/E1+1/(cd1*s))*eps0/s, s, t)
E = (E1*eps0)/exp((E1*t)/cd1)
```

The obtained relaxation modulus E matches that listed in Table 2.9.

2.4 Characteristics of the VEM in the Frequency Domain

Assume a VEM is subjected to sinusoidal stress σ and strain ε, at a frequency ω, such that:

$$\sigma = \sigma_0 e^{i\omega t} \text{ and } \varepsilon = \varepsilon_0 e^{i\omega t} \tag{2.40}$$

where σ_0 and ε_0 denote the amplitude of the stress and strain, respectively, with $i = \sqrt{-1}$.

Hence, for a VEM described by the Maxwell model, Eqs. (2.9) and (2.26b) give:

$$(1 + i\lambda\omega)\sigma_0 e^{i\omega t} = \lambda E_s \omega \varepsilon_0 i\, e^{i\omega t} \quad \text{or} \quad \sigma_0 = E_s \left[\frac{\omega^2\lambda^2}{1+\omega^2\lambda^2} + i\frac{\omega\lambda}{1+\omega^2\lambda^2}\right]\varepsilon_0.$$

In a compact form,

$$\sigma_0 = E'[1 + i\eta]\,\varepsilon_0 \tag{2.41}$$

where $E' = E_s\left[\frac{\omega^2\lambda^2}{1+\omega^2\lambda^2}\right]$ and $\eta = \frac{1}{\omega\lambda}$. The constitutive equation of the VEM, as given by Eq. (2.41), indicates that the material has a complex modulus $E^* = E'[1 + j\eta]$ that relates the stress and the strain. Note that:

a) the real part of the complex modulus = E' is called the *storage modulus*,
b) the imaginary part of the modulus = $E'\eta$ is called the *loss modulus* E'', and
c) the ratio between loss and the storage moduli is η is called the *loss factor*.

Figure 2.10 shows the effect of the excitation frequency on the storage modulus and the loss factor of the Maxwell model.

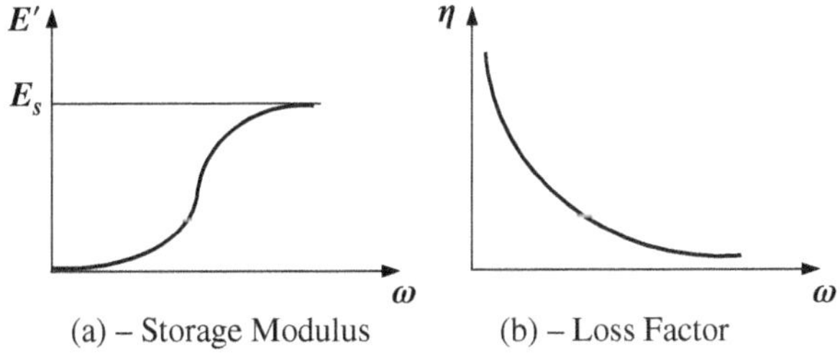

Figure 2.10 Effect of frequency on the storage modulus and loss factor of Maxwell model. (a) Storage modulus and (b) loss factor.

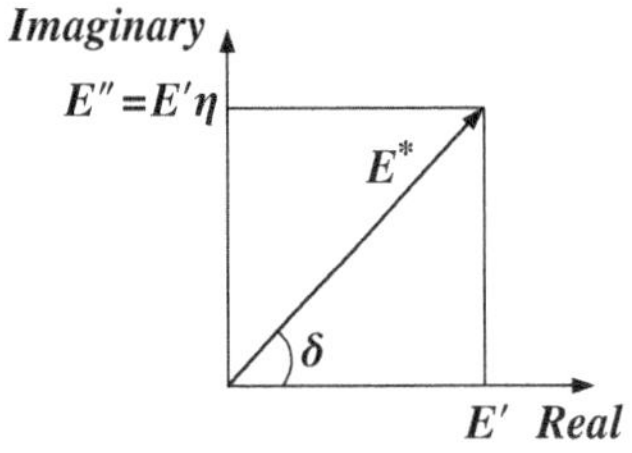

Figure 2.11 Graphical representation of complex modulus.

Note that the Maxwell model indicates that the VEM has zero storage modulus under static conditions ($\omega = 0$) and has a loss factor that is continuously decaying with frequency. These two characteristics contradict the behavior of realistic VEMs.

Figure 2.11 displays graphically the different components of the complex modulus $E^* = E'[1 + i\eta]$.

Note that the complex modulus makes an angle δ with the real axis such that:

$$\tan(\delta) = \eta \tag{2.42}$$

Because of this relationship, the loss factor is also called "tan delta" or the "loss tangent."

In a similar manner, the constitutive equations for the Kelvin–Voigt and Poynting–Thomson models can be determined in the frequency domain. Table 2.10 lists these equations and gives expressions for the corresponding storage modulus and loss factor for the different models.

The characteristics summarized in Table 2.10 ascertain the ability of the Poynting–Thomson model to simulate a realistic behavior of VEMs. However, several

Table 2.10 Frequency domain characteristics of classical viscoelastic models.

Parameter	Maxwell	Kelvin–Voigt	Poynting–Thomson
Model	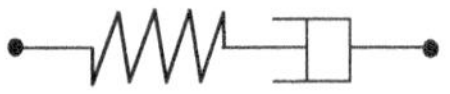	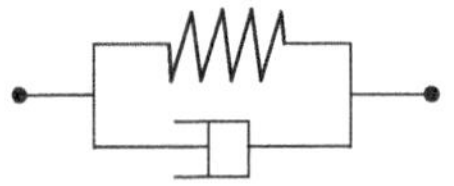	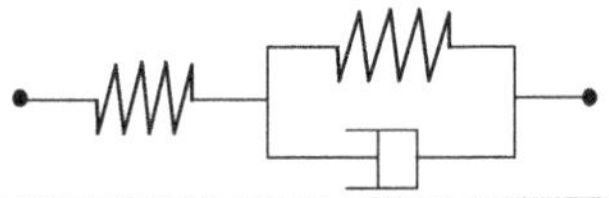
Storage modulus	$E' = E_s\left[\dfrac{\omega^2\lambda^2}{1+\omega^2\lambda^2}\right]$	$E' = E_s$	$E' = E_\infty\left[\dfrac{1+\alpha\beta\omega^2}{1+\alpha^2\omega^2}\right]^a$
	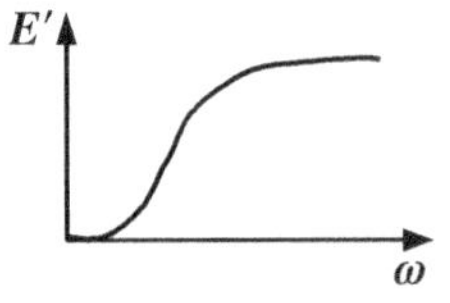	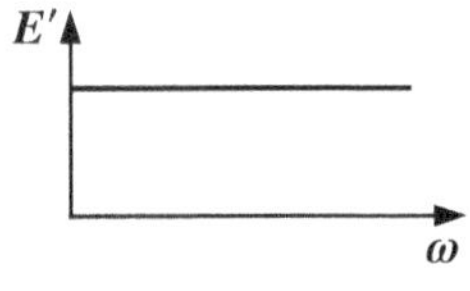	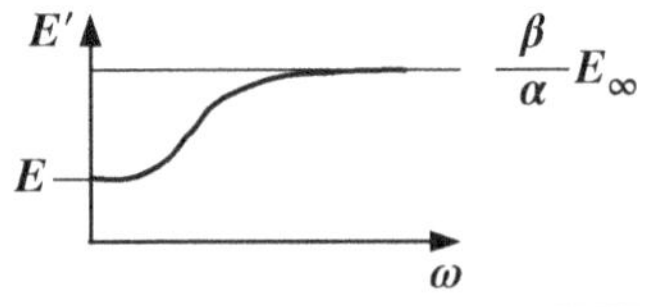
Comments	Unrealistic	Unrealistic	Realistic
Loss Factor	$\eta = 1/\omega\lambda$	$\eta = \omega\lambda$	$\eta = (\beta-\alpha)\omega/[1+\alpha\beta\omega^2]$
	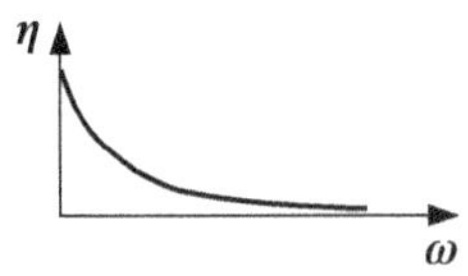	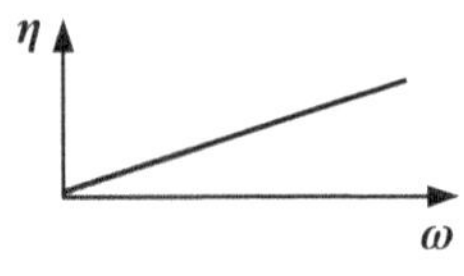	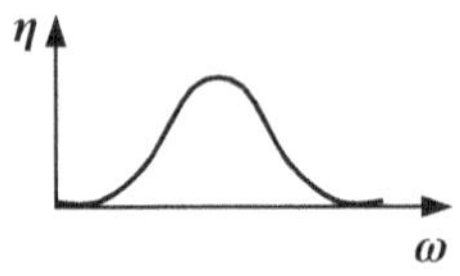
Comments	Unrealistic	Unrealistic	Realistic

[a] where $E_\infty = \dfrac{E_sE_p}{E_s+E_p}$, $\alpha = \dfrac{c_d}{E_s+E_p}$, and $\beta = \dfrac{c_d}{E_p}$.

combinations of Poynting–Thomson models are necessary to replicate the behavior of realistic VEMs.

2.5 Hysteresis and Energy Dissipation Characteristics of Viscoelastic Materials

2.5.1 Hysteresis Characteristics

Consider a VEM subjected to sinusoidal stress σ and strain ε given by

$$\sigma = \sigma_0 e^{i\omega t} \text{ and } \varepsilon = \varepsilon_0 e^{i\omega t} \tag{2.43}$$

with the stress and strain related by the following constitutive equation

$$\sigma = E'(1 + i\eta)\,\varepsilon \tag{2.44}$$

Combining Eqs. (2.43) and (2.44) gives

$$\begin{aligned}\sigma &= \sigma_0 \sin(\omega t)\\ &= E'\varepsilon_0 \sin(\omega t) + \eta E'\varepsilon_0 \cos(\omega t)\\ &= \sigma_e + \sigma_d\end{aligned} \tag{2.45}$$

where $\sigma_e = E'\varepsilon_0 \sin(\omega t)$ and $\sigma_d = \eta E'\varepsilon_0 \cos(\omega t)$ denote the elastic and dissipative components of the applied stress σ.

Then σ_d can be written as:

$$\begin{aligned}\sigma_d &= \eta E'\varepsilon_0 \cos(\omega t)\\ &= \pm\eta E'\sqrt{\varepsilon_0^2 - \varepsilon_0^2 \sin^2(\omega t)}\\ &= \pm\eta E'\sqrt{\varepsilon_0^2 - \varepsilon^2}\end{aligned} \tag{2.46}$$

Rearranging this equation reduces it to

$$(\sigma_d/\eta E')^2 + \varepsilon^2 = \varepsilon_0^2 \tag{2.47}$$

which is an equation of an ellipse as shown in Figure 2.12a. Figure 2.12b shows a plot of the elastic stress versus the strain and Figure 2.12c combines the elastic and dissipative stress components. Figure 2.12c displays accordingly the total stress σ acting on the VEM versus the strain ε.

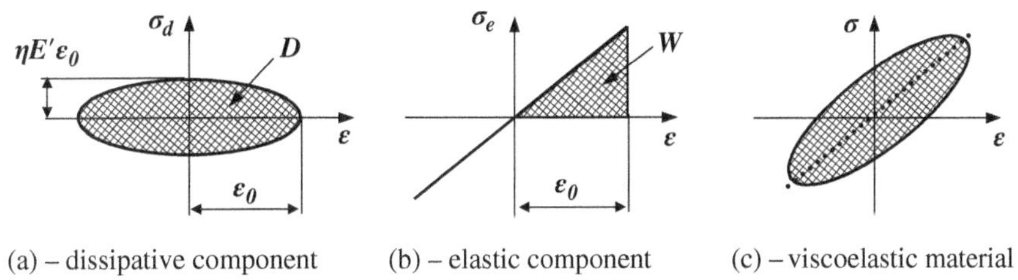

Figure 2.12 Stress–strain relationship for a viscoelastic material. (a) dissipative component, (b) elastic component, and (c) viscoelastic material.

Note that the dissipative component takes the form of a hysteresis loop. The area inside the loop quantifies the amount of energy D dissipated during the cyclic deformation of the VEM.

2.5.2 Energy Dissipation

The energy dissipated during a full vibration cycle of the VEM, at a frequency ω, per unit volume can be determined from

$$D = \int \sigma_d d\varepsilon = \int_0^{\frac{2\pi}{\omega}} \sigma_d \frac{d\varepsilon}{dt} dt \tag{2.48}$$

But as $\sigma_d = \eta E' \varepsilon_0 \cos(\omega t)$ and $\varepsilon = \varepsilon_0 \sin(\omega t)$, then Eq. (2.48) reduces to

$$\begin{aligned} D &= \int_0^{\frac{2\pi}{\omega}} \sigma_d \frac{d\varepsilon}{dt} dt \\ &= \int_0^{\frac{2\pi}{\omega}} [\eta E' \varepsilon_0 \cos(\omega t)]\,[\omega \varepsilon_0 \cos(\omega t)] dt \\ &= \pi \eta E' \varepsilon_0^2 \end{aligned} \tag{2.49}$$

2.5.3 Loss Factor

Two methods can be used to extract the loss factor from the hysteresis characteristics of the VEM. These methods are based on the following:

2.5.3.1 Relationship Between Dissipation and Stored Elastic Energies

Consider now the energy W stored in the elastic component, during one-quarter of a vibration cycle, which can be determined from:

$$W = \int \sigma_e d\varepsilon = \int_0^{\frac{\pi}{2\omega}} \sigma_e \frac{d\varepsilon}{dt} dt$$

With $\sigma_e = E' \varepsilon_0 \sin(\omega t)$, the equation reduces to

$$\begin{aligned} W &= \int_0^{\frac{\pi}{2\omega}} \sigma_e \frac{d\varepsilon}{dt} dt \\ &= \int_0^{\frac{\pi}{2\omega}} [E' \varepsilon_0 \sin(\omega t)]\,[\omega \varepsilon_0 \cos(\omega t)]\, dt \\ &= \frac{1}{2} E' \varepsilon_0^2 \end{aligned} \tag{2.50}$$

From Eqs. (2.49) and (2.50), the loss factor η can be determined from

$$\eta = \frac{D}{2\pi W} \tag{2.51}$$

Hence, Eq. (2.51) defines the physical meaning of the loss factor as the ratio between the dissipated energy and stored energy. Also, Figure 2.12 shows the graphical representation and physical meaning of both the dissipated and stored energies.

2.5.3.2 Relationship Between Different Strains

Equations (2.45) and (2.46) can be rewritten as:

$$\sigma = E'\varepsilon \pm \eta E' \sqrt{\varepsilon_0^2 - \varepsilon^2} \tag{2.52}$$

When the stress σ is set = 0, the corresponding strain ε_{sf} can be obtained from:

$$0 = E'\varepsilon_{sf} \pm \eta E' \sqrt{\varepsilon_0^2 - \varepsilon_{sf}^2}$$

or,

$$\varepsilon_{sf} = \frac{\eta}{\sqrt{1+\eta^2}}\varepsilon_0 \tag{2.53}$$

The $\sigma - \varepsilon$ relationship of the upper branch of hysteresis characteristics can be expressed form Eq. (2.45) as:

$$\sigma = E'\varepsilon + \eta E' \sqrt{\varepsilon_0^2 - \varepsilon^2} \tag{2.54}$$

The maximum stress is attained when:

$\dfrac{d\sigma}{d\varepsilon} = 0$ at $\varepsilon = \varepsilon_{\text{max}\sigma}$ given by:

$$\varepsilon_{\text{max}\sigma} = \varepsilon_0 / \sqrt{1+\eta^2} \tag{2.55}$$

Figure 2.13 displays the graphical interpretation of the strains ε_{sf} and $\varepsilon_{\text{max}\sigma}$. From Eqs. (2.53) and (2.55),

$$\frac{\varepsilon_{sf}}{\varepsilon_{\text{max}\sigma}} = \eta \tag{2.56}$$

Hence, the loss factor can be computed as the ratio between the two strains ε_{sf} and $\varepsilon_{\text{max}\sigma}$ as measured from the hysteresis characteristics.

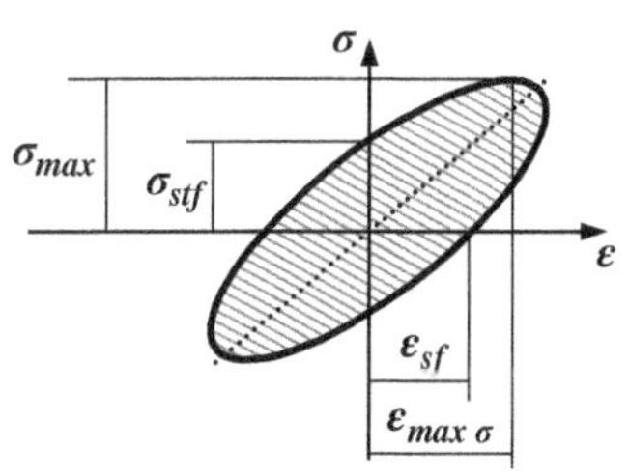

Figure 2.13 Graphical representation of the strains ε_{sf} and $\varepsilon_{\text{max}\sigma}$.

2.5.4 Storage Modulus

The storage modulus can be determined by considering the stress under a strain-free condition σ_{stf}. This value can be obtained by setting $\varepsilon = 0$ in Eq. (2.54), giving:

$$\sigma_{stf} = \pm \eta E' \varepsilon_0 \tag{2.57}$$

Once η and ε_0 are determined from Eqs. (2.56) and (2.53), then Eq. (2.57) can be used to compute the storage modulus E'.

Example 2.3 Plot the stress–strain characteristics for the Poynting–Thomson model when the VEM is subjected to sinusoidal stress such that $\sigma = \sin t$. Assume that $E_s = 1$, $E_p = 1$, and $c_d = 1$. Determine the loss factor and the storage modulus according to the methods described in Sections 2.5.3.2 and 2.5.4. Compare the results with the loss factor and the storage modulus expressions listed in Table 2.10.

Solution

The constitutive equation for the Poynting–Thomson model is:

$$\dot{\varepsilon} + \varepsilon = \dot{\sigma} + 2\sigma$$

For sinusoidal stress: $\sigma = \sin t$, this equation reduces to:

$$\dot{\varepsilon} + \varepsilon = 2\sin t + \cos t$$

This equation is integrated numerically, with respect to time, using MATLAB to extract the time history of the strain as function of the time history of the input stress. Then, the strain is plotted against the stress to yield the stress–strain characteristics shown in Figure 2.14.

a) From Figure 2.14,

$$\varepsilon_{\max\sigma} = \varepsilon_0 / \sqrt{1 + \eta^2} = 1.5$$

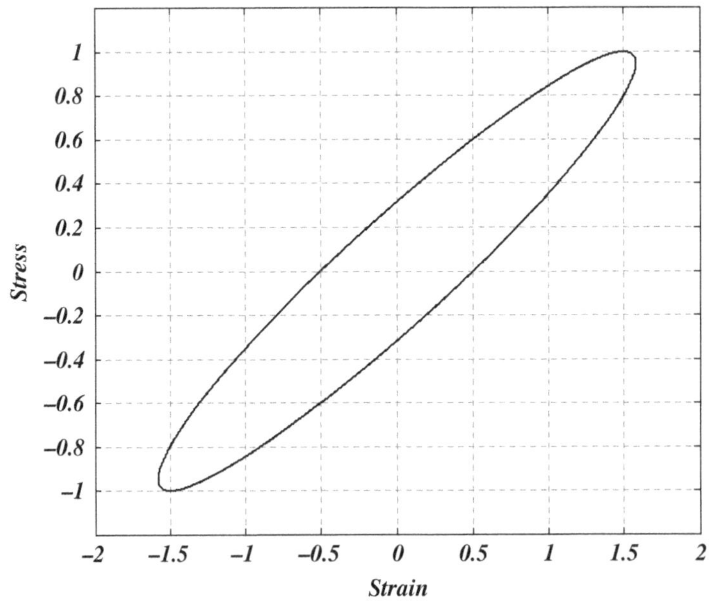

Figure 2.14 Stress–strain characteristics of a Poynting–Thomson model.

$$\varepsilon_{sf} = \frac{\eta}{\sqrt{1+\eta^2}}\varepsilon_0 = 0.5$$

Then,

$$\frac{\varepsilon_{sf}}{\varepsilon_{\max\sigma}} = \eta = \frac{0.5}{1.5} = 0.33$$

This yields $\varepsilon_0 = 1.581$.

Also, from Figure 2.14, $\sigma_{stf} = \pm\eta E'\varepsilon_0 = \pm 0.3148$, or

$$E' = 0.3148/(\eta\varepsilon_0)$$
$$= 0.3148/(0.333 \times 1.581) = 0.598$$

b) From Table 2.10,

$$\text{As}\, E = \frac{E_s E_p}{E_s + E_p} = \frac{1}{2},\ \alpha = \frac{c_d}{E_s + E_p} = \frac{1}{2},\ \beta = \frac{c_d}{E_p} = 1, \text{ and } \omega = 1.$$

Then,

$$E' = E\left[\frac{1+\alpha\beta\omega^2}{1+\alpha^2\omega^2}\right] = \frac{1}{2}\left[\frac{1+0.5\omega^2}{1+0.25\omega^2}\right] = \frac{1.5}{2\times 1.25} = 0.6$$

and

$$\eta = (\beta-\alpha)\omega/\left[1+\alpha\beta\omega^2\right] = 0.5/(1+0.5) = 0.333$$

Hence, the two methods yield exactly the same results.

Example 2.4 Plot the storage modulus and the loss factor as predicted by a Maxwell, Kelvin–Voigt, and Poynting–Thomson models that best fits the experimental behavior of the VEM Dyad 606 (Soundcoat, Deer Park, NY) at 37.8°C (100°F).

Solution

The storage modulus and the loss factor of the different VEM models, listed in Table 2.10, are plotted versus the frequency ω as shown in Figure 2.15. The plots are obtained for $\alpha = 1$, $\beta = 20$, $\lambda = 3$, $E_s = 500$, and $E = 500$.

The figure indicates clearly that all the three models are incapable of capturing the behavior of the Dyad606. However, the predictions of the Poynting–Thomson model qualitatively have the general trends but fail to quantitatively describe the behavior over a broad frequency range.

A combination of several Poynting–Thomson models is necessary to replicate the behavior of realistic VEMs.

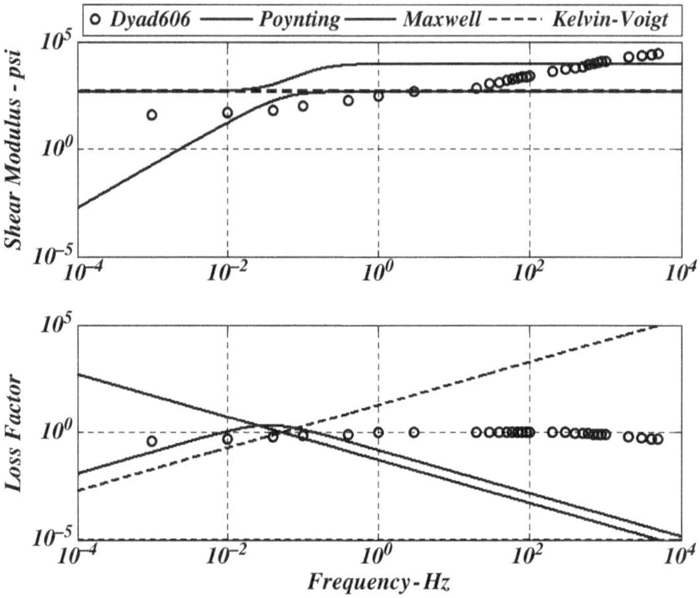

Figure 2.15 Storage modulus and loss factor of different VEM models.

2.6 Fractional Derivative Models of Viscoelastic Materials

As indicated in Section 2.5, the simple classical models of VEMs cannot replicate the dynamic behavior of real VEMs. Other alternative models have been considered to overcome such serious limitations. Among these models are fractional derivative (FD) model (Bagley and Torvik 1983), Golla–Hughes–MacTavish (GHM) model (Golla and Hughes 1985), and the Augmented Temperature Field model (Lesieutre and Mingori 1990; Lesieutre et al. 1996).

2.6.1 Basic Building Block of Fractional Derivative Models

The basic concepts of fractional calculus are summarized in Appendix 2.B.

In this section, the basic building block of FD models is the "spring-pot" that replaces the spring and dashpot elements used in the classical models. The spring-pot element is employed to simplify, improve the applicability, and reduce the number of parameters used to model the complex behavior of viscoelastic polymers.

The spring-pot element is a nonlinear FD element that has the following constitutive equation:

$$\sigma(t) = E\tau^{\alpha}\frac{d^{\alpha}\varepsilon(t)}{dt^{\alpha}} \tag{2.58}$$

Note that the stress $\sigma(t)$ applied to the element is dependent on the FD of order α of the strain $\varepsilon(t)$ where α ranging between 0 and 1. When $\alpha = 0$, the spring-pot element reduces to a linear spring and when $\alpha = 1$, the spring-pot element becomes a linear dashpot (damper) as shown in Figure 2.16.

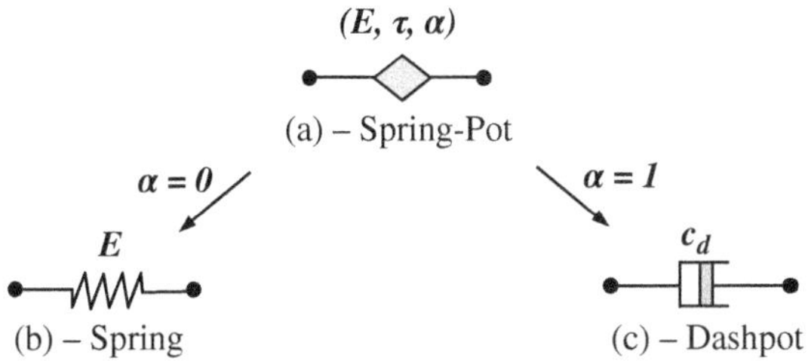

Figure 2.16 Representations of a spring-pot, spring, and dashpot. (a) Spring-pot, (b) spring, and (c) dashpot.

From Eq. (2.58), the storage and loss moduli of the spring-pot can be determined as follows:

$$\begin{aligned} E^*(\omega) &= E\,(i\omega\tau)^{\alpha} \\ &= E\,(\omega\tau)^{\alpha}\, e^{\frac{\pi}{2}\alpha i} \\ &= E\,(\omega\tau)^{\alpha}\left[\cos\left(\frac{\pi}{2}\alpha\right) + i\sin\left(\frac{\pi}{2}\alpha\right)\right] \\ &= E' + iE'' \end{aligned} \tag{2.59}$$

where $E' = E\,(\omega\tau)^{\alpha}\cos\left(\frac{\pi}{2}\alpha\right)$ and $E'' = E\,(\omega\tau)^{\alpha}\sin\left(\frac{\pi}{2}\alpha\right)$.

The relaxation modulus $E(t)$ of the spring-pot can be obtained by applying the inverse Fourier transform to Eq. (2.58) knowing that $E(s) = E^*/s$, as indicated in Eq. (2.27b). This yields the following expressions:

Relaxation modulus

$$\begin{aligned} E(t) &= \frac{2}{\pi}\int_0^{\infty}\left[\frac{1}{\omega}E'\sin(\omega t)\right]d\omega \\ &= \frac{2}{\pi}E\,\tau^{\alpha}\cos\left(\frac{\pi}{2}\alpha\right)\int_0^{\infty}\left[\omega^{\alpha-1}\sin(\omega t)\right]d\omega \\ &= \frac{E}{\Gamma(1-\alpha)}\left(\frac{t}{\tau}\right)^{-\alpha} \end{aligned} \tag{2.60}$$

and

Creep compliance

$$J(t) = \frac{E^{-1}}{\Gamma(1+\alpha)}\left(\frac{t}{\tau}\right)^{\alpha} \tag{2.61}$$

2.6.2 Basic Fractional Derivative Models

The basic FD models presented in this section are the FD Maxwell, FD Kelvin–Voigt, and FD Poynting–Thomson models.

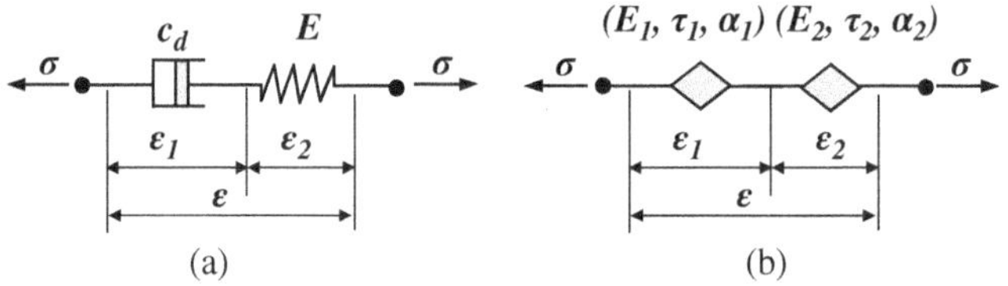

Figure 2.17 Classical (a) and fractional derivative (b) Maxwell models.

For example, consider the classical Maxwell model shown in Figure 2.17a is transformed to a FD Maxwell model shown in Figure 2.17b by replacing each component by a spring-pot with different parameters.

For the FD model and using the equivalent spring-pot Eq. (2.58), the strains ε_1 and ε_2 can be written as:

$$\varepsilon_1(t) = E_1^{-1}\tau_1^{-\alpha_1}\frac{d^{-\alpha_1}\sigma(t)}{dt^{-\alpha_1}} \text{ and } \varepsilon_2(t) = E_2^{-1}\tau^{-\alpha_2}\frac{d^{-\alpha_2}\sigma(t)}{dt^{-\alpha_2}} \tag{2.62}$$

But as, $\varepsilon = \varepsilon_1 + \varepsilon_2$, then:

$$\sigma(t) + \tau^{\alpha_1-\alpha_2}\frac{d^{\alpha_1-\alpha_2}\sigma(t)}{dt^{\alpha_1-\alpha_2}} = E\tau^{\alpha_1}\frac{d^{\alpha_1}\varepsilon(t)}{dt^{\alpha_1}} \tag{2.63}$$

where $\tau = (E_1\tau_1^{\alpha_1}/E_2\tau_2^{\alpha_2})^{1/(\alpha_1-\alpha_2)}$, $E = E_1(\tau_1/\tau_2)^{\alpha_1}$, and $\alpha_1 \geq \alpha_2$.

Fourier-transforming Eq. (2.63) gives:

$$E^* = \frac{E(i\omega\tau)^{\alpha_1}}{1 + (i\omega\tau)^{\alpha_1-\alpha_2}} \tag{2.64}$$

Following the same approach adopted in Section 2.6.1, it can be easily shown that the relaxation modulus and creep compliance of the FD Maxwell model are given by:

$$J(t) = \frac{E^{-1}}{\Gamma(1+\alpha_1)}\left(\frac{t}{\tau}\right)^{\alpha_1} + \frac{E^{-1}}{\Gamma(1+\alpha_2)}\left(\frac{t}{\tau}\right)^{\alpha_2} \tag{2.65}$$

Table 2.11 summarizes the complex moduli for the FD Maxwell, Kelvin–Voigt, and Zener models.

Example 2.5 Determine the storage modulus, loss modulus, and loss factor as predicted by the FD Zener model that is given by the following four-parameter FD model proposed by Bagley and Torvik (1983):

$$\sigma(t) + \tau^\alpha D^\alpha\sigma(t) = E_0\varepsilon(t) + E_\infty\tau^\alpha D^\alpha\varepsilon(t)$$

where $E_0 = E_sE_p/(E_s + E_p)$ = relaxed elastic modulus, $E_\infty = E_s$ = unrelaxed elastic modulus, $\tau = c_d/(E_s + E_p)$ = the relaxation time, and $0 < \alpha < 1$.

Solution

From Eq. (2.58), the complex modulus as predicted by the four-parameter FD model can be obtained by using Eq. (2.B.10), to give:

$$E^* = \frac{\sigma(s)}{\varepsilon(s)} = \frac{E_0 + E_\infty(\tau s)^\alpha}{1 + (\tau s)^\alpha}$$

Table 2.11 Frequency domain characteristics of fractional derivative viscoelastic models.

Parameter	Maxwell	Kelvin–Voigt	Zener
Model	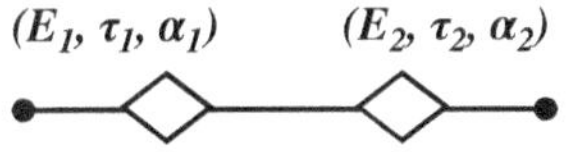	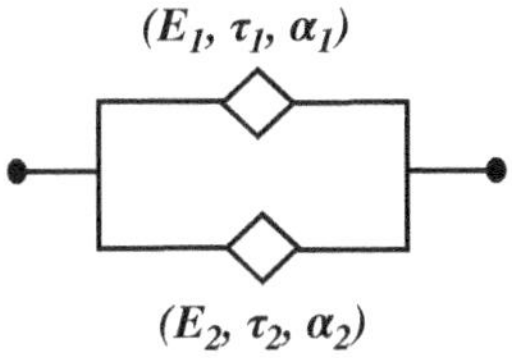	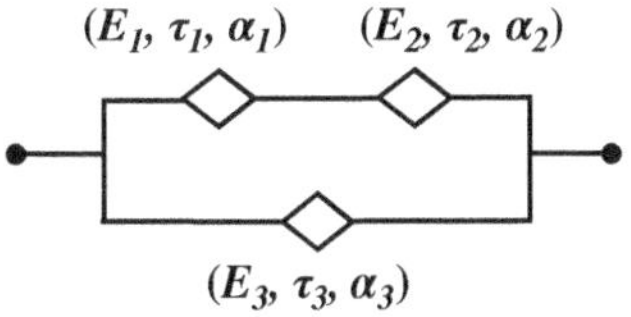
Storage modulus	$E^* = \dfrac{E(i\omega\tau)^{\alpha_1}}{1+(i\omega\tau)^{\alpha_1-\alpha_2}}$ [a]	$E^* = E(i\omega\tau)^{\alpha_1} + E(i\omega\tau)^{\alpha_2}$	$E^* = E_0\dfrac{(i\omega\tau)^{\alpha_2}}{1+(i\omega\tau)^{\alpha_2-\alpha_1}} + E(i\omega\tau)^{\alpha_3}$ [b]

[a] where $\tau = (E_1\tau_1^{\alpha_1}/E_2\tau_2^{\alpha_2})^{1/(\alpha_1-\alpha_2)}$, $E = E_1(\tau_1/\tau_2)^{\alpha_1}$, and $\alpha_1 \geq \alpha_2$.
[b] $E_0 = E_1(\tau_1/\tau)^{\alpha_1}$.

It can be easily shown that the storage and loss moduli of the four-parameter FD model are given by:

$$E'(\omega) = \frac{E_0 + (E_\infty + E_0)(\omega\tau)^\alpha \cos(\pi\alpha/2) + E_\infty(\omega\tau)^{2\alpha}}{1 + 2(\omega\tau)^\alpha \cos(\pi\alpha/2) + (\omega\tau)^{2\alpha}}$$

and

$$E''(\omega) = \frac{(E_\infty - E_0)(\omega\tau)^\alpha \sin(\pi\alpha/2)}{1 + 2(\omega\tau)^\alpha \cos(\pi\alpha/2) + (\omega\tau)^{2\alpha}}$$

Accordingly, the loss factor η is given by:

$$\eta = \frac{E''(\omega)}{E'(\omega)} = \frac{(E_\infty - E_0)(\omega\tau)^\alpha \sin(\pi\alpha/2)}{E_0 + (E_\infty + E_0)(\omega\tau)^\alpha \cos(\pi\alpha/2) + E_\infty(\omega\tau)^{2\alpha}}$$

This yields a value of α that can be estimated from (see Problem 2.9):

$$\alpha = \frac{2}{\pi}\sin^{-1}\left[\eta_{max}(E_\infty - E_0) \times \frac{2\sqrt{E_\infty E_0} + (E_\infty + E_0)\sqrt{1+\eta_{max}^2}}{\eta_{max}^2(E_\infty + E_0)^2 + (E_\infty - E_0)^2}\right]$$

Example 2.6 Plot the storage modulus and the loss factor as predicted by a FD model that best fits the experimental behavior of the *VEM* Dyad 606 (Soundcoat, Deer Park, NY) at 37.8°C (100°F). Compare the predictions with those of a Poynting–Thomson model.

Solution

The storage modulus and the loss factor of the FD and Poynting–Thomson VEM models are plotted versus the frequency ω as shown in Figure 2.18. The plots are obtained for

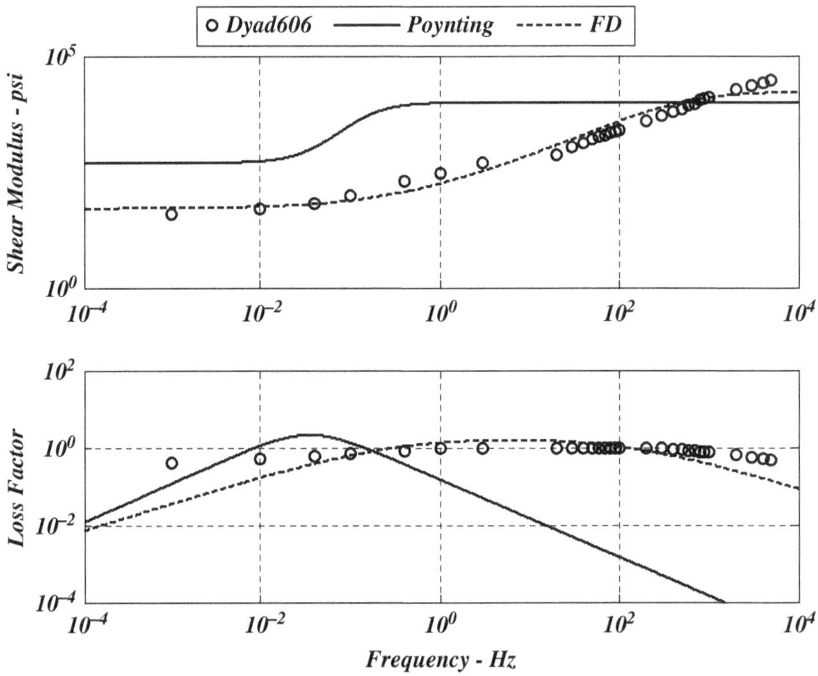

Figure 2.18 Storage modulus and loss factor of a fractional derivative and Poynting–Thomson models.

the Poynting–Thomson model with $\alpha = 1$, $\beta = 20$, $\lambda = 3$, and $E_s = 500\,psi$ while the FD model is given by:

$$\sigma = \frac{52.5 + 18000(0.004s)^{0.7}}{1 + (0.004s)^{0.7}}\varepsilon$$

The figure indicates clearly that the four-parameter FD model adequately replicates the physical behavior of Dyad 606 unlike the linear Poynting–Thomson model.

2.6.3 Other Common Fractional Derivative Models

In this section, some of the commonly used FD models are introduced. The main characteristics of these models are presented in the frequency ω domain as indicated in Table 2.12. These models vary in complexity by increasing the number of the included parameters from three (E_∞, Δ, τ) as in the Debye model to five parameters (E_∞, Δ, τ, α, and β) as in the Havriliak–Negami model (Pritz 2003, Ciambella et al. 2011).

Example 2.7 Determine the time response of the following spring-mass system, shown in Figure 2.19, which is damped by a FD damper of the order α:

$$m\ddot{x} + c\tau^{\alpha}D^{\alpha}x + kx = f$$

Assume that $m = 1$ kg, $k = 1\,\mathrm{N\,m^{-1}}$, $\tau = 1$ s, $f = 1$ N, and $\alpha = 0.75$. Assume also that $x(0) = 0$ and $\dot{x}(0) = 0$. Use the *Grunwald–Letnikov* (*G–L*) definition of FDs described in Appendix 2.B.5.

Table 2.12 Frequency domain characteristics of fractional derivative viscoelastic models.

Model	$E^*(\omega)$	Parameters
Debye	$E^* = E_\infty \left[1 + \dfrac{\Delta}{1+(i\omega\tau)}\right]^a$	3
Cole–Cole[b]	$E^* = E_\infty \left[1 + \dfrac{\Delta}{1+(i\omega\tau)^\alpha}\right]$	4
Cole–Davidson	$E^* = E_\infty \left[1 + \dfrac{\Delta}{[1+(i\omega\tau)]^\beta}\right]$	4
Havriliak–Negami	$E^* = E_\infty \left[1 + \dfrac{\Delta}{[1+(i\omega\tau)^\alpha]^\beta}\right]$	5

[a] where $\Delta = (E_0 - E_\infty)/E_\infty$ = relaxation strength, = relaxed elastic modulus, = unrelaxed elastic modulus.
[b] Friedrich and Braun (1992), Pritz (2003), and Ciambella et al. (2011).

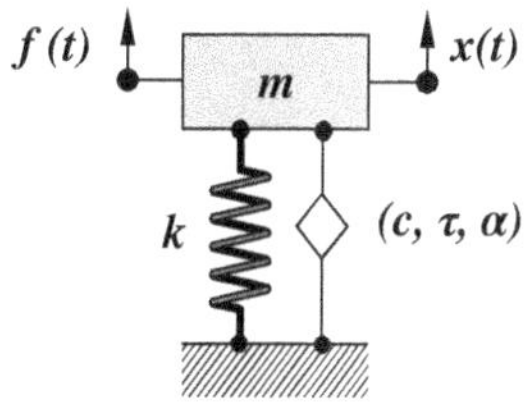

Figure 2.19 A spring-mass system with a fractional derivative damper.

Solution

According to the set parameters, the system equation reduces to:

$$\ddot{x} + D^{0.75}x + x = 1$$

or

$$\ddot{x} = -D^{0.75}x - x + 1$$

where,

$$D^{0.75}x = \lim_{N\to\infty}\left[\left(\frac{t}{N}\right)^{-0.75}\sum_{j=0}^{N-1} A_{j+1}\, f(t - jt/N)\right]$$

where A_{j+1} = Grunwald coefficients.

Figure 2.20a,b displays the Grunwald coefficients and the time response of the spring-mass system with a FD damper, respectively. It is evident, from Figure 2.20a, that the Grunwald coefficients vanish as the number of terms included in the summation of Eq. (2B.14) increases. This demonstrates clearly the "fading memory" characteristics of FDs. Figure 2.20b indicates that the system reaches, with an error of 3.14%, the desired reference command after 15 s.

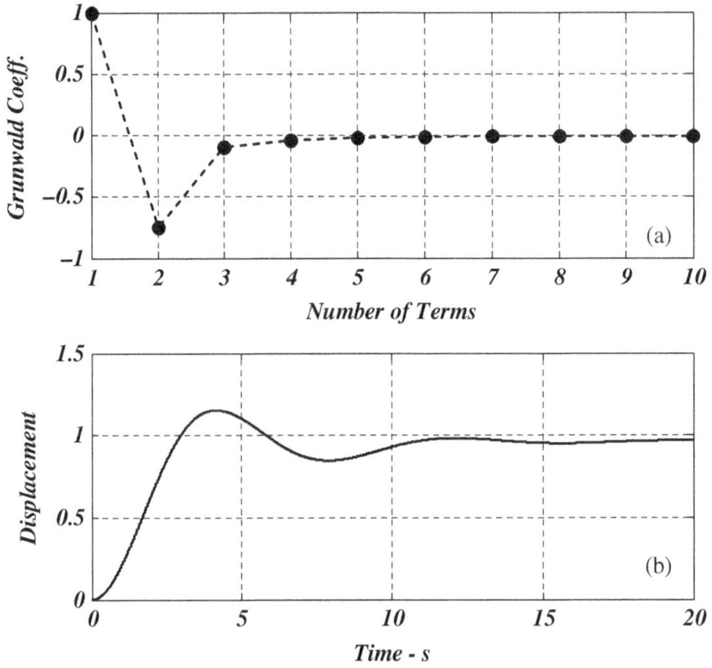

Figure 2.20 Time response of a spring-mass system with a fractional derivative damper.

Note that the approach adopted in Example 2.7 establishes the basis for time domain analysis of finite element models of structures treated with VEMs that are described by FD models.

2.7 Viscoelastic Versus Other Types of Damping Mechanisms

In this section, four important damping mechanisms are presented, including: viscous, hysteretic, structural, and friction damping in order to distinguish, compare, and relate their characteristics to those of viscoelastic damping materials.

Tables 2.13 and 2.14 summarize the main characteristics of these four damping mechanisms. Table 2.13 presents the physical representation of each mechanism, its mathematical model, force-displacement characteristics, and a typical time response behavior.

The energy dissipated per cycle by the different damping forces is calculated as follows:

$$D_i = 4 \int_0^{\pi/2\omega} F_d \dot{x} \, dt \tag{2.66}$$

where F_d denotes the damping force as listed in the third column of Table 2.13.

The equivalent viscous damping coefficient, for any damping mechanism, is obtained by equating the energy dissipated by the ith mechanism to that of the viscous damping

Table 2.13 Characteristics of viscous, hysteretic, structural, and friction damping.

Damping mechanism	Physical representation	Model	Force – displacement	Time response
Viscous[a,b]	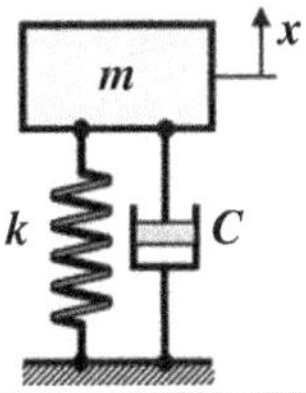	$F_d = c\dot{x}$	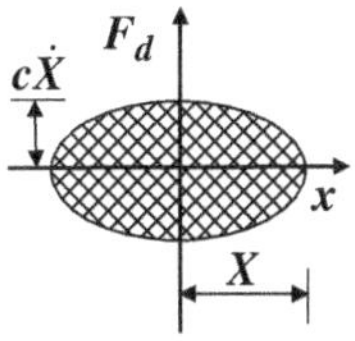	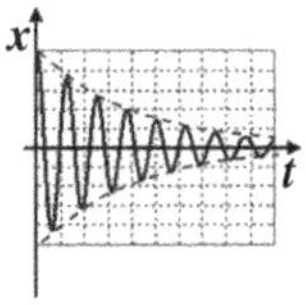
Hysteretic[a,b]	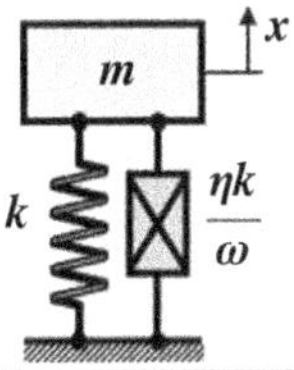	$F_d = \frac{\eta k}{\omega}\dot{x}$	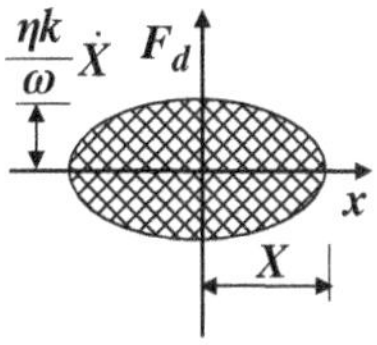	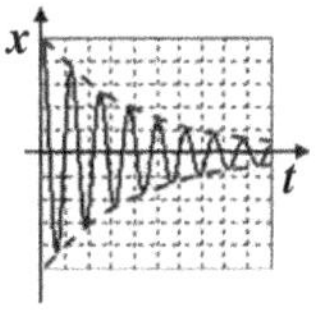
Structural[c,d]	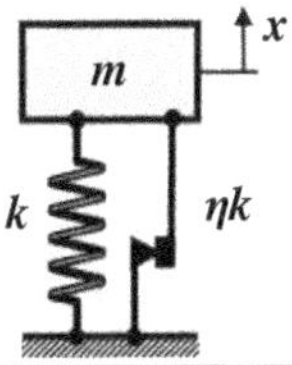	$F_d = \eta k\|x\|\,\mathrm{sgn}(\dot{x})$	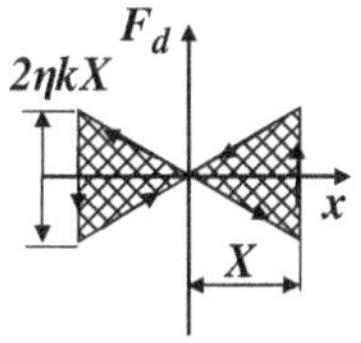	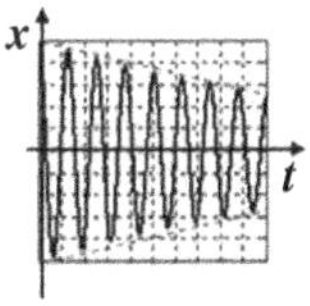
Friction[a,b]	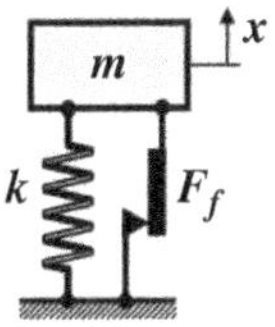	$F_d = F_f\,\mathrm{sgn}(\dot{x})$	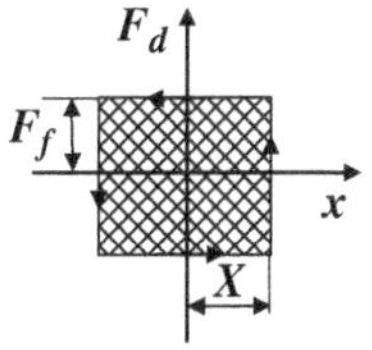	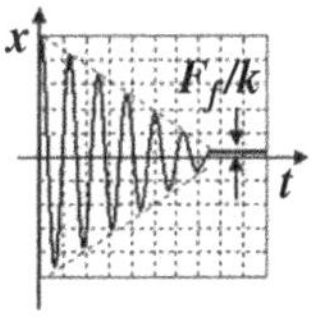

[a] Beards (1996).
[b] Rao (2010).
[c] Muravskii (2004).
[d] Gremaud (1987).

Table 2.14 Energy dissipation and damping ratios of viscous, hysteretic, structural, and friction damping.

Damping mechanism	Energy dissipation	Equivalent damping coefficient	Equivalent damping ratio
Viscous	$D_v = \pi c\omega X^2$	$c_v = c$	$\zeta_v = c/2\sqrt{km}$
Hysteretic	$D_H = \pi k\eta X^2$	$c_H = \frac{k\eta}{\omega}$	$\zeta_H = \frac{\eta}{2}$
Structural	$D_S = 2k\eta X^2$	$c_S = \frac{2k\eta}{\pi\omega}$	$\zeta_S = \frac{\eta}{\pi}$
Friction	$D_F = 4F_f X$	$c_F = \frac{4F_f}{\pi\omega X}$	$\zeta_F = \frac{2}{\pi}\left(\frac{F_f}{kX}\right)$
Viscoelastic	$D_{VEM} = \pi k\eta X^2$	$c_{VEM} = \frac{k\eta}{\omega}$	$\zeta_{VEM} = \frac{\eta}{2}$

mechanism D_v. Hence, the equivalent damping ratio of the ith mechanism is calculated as follows:

$$\zeta_i = c_i / 2\sqrt{km} \tag{2.67}$$

Equation (2.67) assumes that $\zeta_i = c_i/c_c$ where $c_c = 2\sqrt{km}$ is the critical damping coefficient for a single degree of freedom vibrating system.

Table 2.14 summarizes the energy dissipated per cycle, the equivalent viscous damping coefficients, and equivalent damping ratios of the different mechanisms in comparison with the corresponding values for viscoelastic materials.

2.8 Summary

This chapter has presented the classical models of VEMs. The merits and limitations of these models have been discussed both in the time and frequency domains. The energy dissipation characteristics of the VEMs have been presented with particular emphasis on the use of the unifying concept of the complex modulus. A brief description of the FD models has also been outlined to emphasize their utility and compactness. Measurement methods of the complex modulus of the VEMs will be presented in the next chapter, and extension of the classical models to more practical models that can be easily incorporated in the formulation of finite element method will be presented in Chapters 4–6.

References

Bagley, R.L. and Torvik, P.J. (1983). Fractional calculus-a different approach to the analysis of viscoelastically damped structures. *AIAA Journal* 21: 741–749.

Beards, C. (1996). *Structural Vibration: Analysis and Damping*. London: Arnold.

Christensen, R.M. (1982). *Theory of Viscoelasticity: An Introduction*, 2nde. New York: Academic Press Inc.

Ciambella J., Paolone A., Vidoli S., "Dynamic Behavior of Viscoelastic Solids at Low Frequency: Fractional vs Exponential Relaxation", 1–5. In Proceedings XX Congresso dell'Associazione Italiana di Meccanica Teorica e Applicata, Bologna 12–15 September 2011; F. Ubertini, *E. Viola*, S. de Miranda and G. Castellazzi (Eds.), ISBN 978-88-906340-1-7, 2011.

Flugge, W. (1967). *Viscoelasticity*. Waltham, MA: Blaisdell Publishing Company.

Friedrich, C. and Braun, H. (1992). Generalized Cole–Cole behavior and its rheological relevance. *Rheologica Acta* 31 (4): 309–322.

Galucio, A.C., Deü, J.-F., Mengué, S., and Dubois, F. (2006). An adaptation of the gear scheme for fractional derivatives. *Computer Methods in Applied Mechanics and Engineering* 195 (44–47): 6073–6085.

Golla, D.F. and Hughes, P.C. (1985). Dynamics of viscolelastic structures – a time domain finite element formulation. *ASME Journal of Applied Mechanics* 52: 897–906.

Gremaud G., "The Hysteretic Damping Mechanisms Related to Dislocation Motion", *Journal de Physique, Colloque C8*, Supplement au N012, Tome 48, December 1987.

Haddad, Y.M. (1995). *Viscoelasticity of Engineering Materials*. New York: Chapman & Hall.

Heymans, N. (1996). Hierarchical models for viscoelasticity: dynamic behaviour in the linear range. *Rheologica Acta* 35 (5): 508–519.
Iwan, W.D. (1964). An electric analog for systems containing Coulomb damping. *Experimental Mechanics* 4 (8): 232–236.
Lakes, R. (1999). *Viscoelastic Solids*. Boca Raton, FL: CRC Press.
Lakes, R. (2009). *Viscoelastic Materials*. Cambridge, UK: Cambridge University Press.
Lesieutre, G.A., Bianchini, E., and Maiani, A. (1996). Finite element modeling of one-dimensional viscoelastic structures using anelastic displacement fields. *Journal of Guidance, Control, and Dynamics* 19 (3): 520–527.
Lesieutre, G.A. and Mingori, D.L. (1990). Finite element modeling of frequency-dependent material damping using augmenting thermodynamic fields. *Journal of Guidance, Control, and Dynamics* 13 (6): 1040–1050.
Muravskii, G.B. (2004). On frequency independent damping. *Journal of Sound and Vibration* 274: 653–668.
Nashif, A., Jones, D., and Henderson, J. (1985). *Vibration Damping*. New York: Wiley.
Nise, N.S. (2015). *Control Systems Engineering, 7th Edn*. Hoboken, NJ: Wiley.
Oldham, K.B. and Spanier, J. (1974). *An Introduction to the Fractional Calculus and Fractional Differential Equations*. New York: Wiley.
Padovan, J. (1987). Computational algorithms for FE formulations involving fractional operators. *Computational Mechanics* 2: 271–287.
Podlubny, I. (1999). *Fractional Differential Equations*. San Diego, California: Academic Press.
Pritz, T. (2003). Five-parameter fractional derivative model for polymeric damping materials. *Journal of Sound and Vibration* 265: 935–952.
Sun, C. and Lu, Y.P. (1995). *Vibration Damping of Structural Elements*. Englewood Cliffs, NJ: Prentice Hall.
Rao, S.S. (2010). *Mechanical Vibrations*, 5the. New Jersey: Prentice Hall.
Vondřejc, J. (2009). *Constitutive models of linear viscoelasticity using Laplace transform*. Czech Republic, Prague: Department of Mechanics, Faculty of Civil Engineering, Czech Technical University.
Zener, C.M. (1948). *Elasticity and Anelasticity of Metals*. Chicago: University of Chicago Press.

2.A Initial and Final Value Theorems

The initial and final values of a function $x(t)$ are given by the following theorems (Nise 2015):

- *Initial value theorem:* $x(0) = \lim_{t\to 0} x(t) = \lim_{s\to\infty} sX(s)$
- *Final value theorem:* $x(\infty) = \lim_{t\to\infty} x(t) = \lim_{s\to 0} sX(s)$

Proof

From the definition of the Laplace transform **L**:

$$\mathbf{L}\left[\frac{d}{dt}x(t)\right] = \int_0^\infty \left[\frac{d}{dt}x(t)\right] e^{-st} dt = sX(s) - x(0)$$

Consider the following two extreme cases:

i) when $s \to 0$, this equation becomes

$$\lim_{s \to 0} \int_0^{\infty} \left[\frac{d}{dt}x(t)\right] e^{-st} dt = \int_0^{\infty} dx(t) = x(\infty) - x(0) = \lim_{s \to 0} [sX(s) - x(0)]$$

that is, $x(\infty) = \lim_{s \to 0} sX(s)$

ii) when $s \to \infty$, we have

$$\lim_{s \to \infty} \int_0^{\infty} \left[\frac{d}{dt}x(t)\right] e^{-st} dt = 0 = \lim_{s \to \infty} [sX(s) - x(0)]$$

that is, $x(0) = \lim_{s \to \infty} sX(s)$

2.B Fractional Calculus

2.B.1 Fractional Integration

The fractional integral of a function $f(x)$ with n-folds is defined as:

$$D^{-n}f(x) = \int_0^x dx_1 \int_0^{x_1} dx_2 \int_0^{x_2} dx_3 \ldots\ldots \int_0^{x_{n-1}} f(t)dt \tag{2.B.1}$$

Let

$$D^{-1}f(x) = \int_0^x f(t)dt \tag{2.B.2}$$

Then, integrating one more time gives:

$$\begin{aligned} D^{-2}f(x) &= \int_0^x dx_1 \int_0^{x_1} f(t)dt \\ &= \int_0^x f(t)dt \int_0^{x_1} dx_1 \\ &= \int_0^x f(t)(x-t)dt \end{aligned} \tag{2.B.3}$$

Integrating another time gives:

$$\begin{aligned} D^{-3}f(x) &= \int_0^x dx_1 \int_0^{x_1} dx_2 \int_0^{x_2} f(t)dt \\ &= \int_0^x dx_1 \left[\int_0^{x_1} dx_2 \int_0^{x_2} f(t)dt \right] \\ &= \int_0^x dx_1 \left[\int_0^{x_1} (x_1 - t)f(t)dt \right] = \int_0^x f(t)dt \left[\int_t^x (x_1 - t)dx_1 \right] \\ &= \frac{1}{2}\int_0^{x_1} f(t)(x-t)^2 dt \end{aligned} \tag{2.B.4}$$

After n integrations, yields the following expression, which is called "Reimann–Liouville Fractional Integral":

$$D^{-n}f(x) = \frac{1}{(n-1)!}\int_0^x (x-t)^{n-1} f(t)dt \tag{2.B.5}$$

But, as $\Gamma(n) = (n\text{-}1)!$ = gamma function, then this equation reduces to:

$$D^{-n}f(x) = \frac{1}{\Gamma(n)}\int_0^x (x-t)^{n-1} f(t)dt \tag{2.B.6}$$

2.B.2 Convolution Theorem

As the Laplace inverse of a product $G(s)$ $F(s)$ is given by the following convolution integral (Weber and Arfken 2003)[1]:

$$L^{-1}[G(s)F(s)] = \int_0^\infty g(x-t).f(t)dt \tag{2B.7}$$

If

$$G(s)F(s) = s^{-n}F(s)$$

Then,

$$L^{-1}(s^{-n}) = L^{-1}(G(s)) - \frac{t^{n-1}}{n-1!}$$

1 Weber H. and Arfken G., *Mathematical Methods for Physicists*, Academic Press, 2003.

or

$$L^{-1}[G(s)F(s)] = \frac{1}{(n-1)!}\int_0^{\infty} (x-t)^{n-1}.\, f(t)dt = D^{-n}f(x) \tag{2.B.8}$$

Equation (2.B.8) indicates that the fractional integration of $f(x)$ is the convolution integral resulting from the inverse Laplace transformation of $F(s)$ operated on by the n-fold integral operator s^{-n}.

Example 2.B.1 Find the half integral of $f(x) = x$.

Solution

With $n = 1/2$, Eq. (2.B.6) becomes:

$$D^{-1/2}f(x) = \frac{1}{\Gamma(^1/_2)}\int_0^x (x-t)^{-^1/_2}t\,dt$$

which can be solved symbolically using MATLAB as follows:

```
>> int((x-t)^-0.5*t,t,0,x)
ans = (4*x^(3/2))/3
```

that is, $D^{-1/2}f(x) = \frac{1}{\Gamma(^1/_2)}\frac{4}{3}x^{^3/_2} = 0.7523x^{^3/_2}$

2.B.3 Fractional Derivatives

Let $m = p - l$, then: $D^m f(x) = D^p D^{-l} f(x)$.

From Eq. (2.B.6),

$$D^m f(x) = D^p \frac{1}{\Gamma(l)}\int_0^x (x-t)^{l-1} f(t)dt$$

If $0 < m < 1$, then $p = 1$ or

$$D^{1-l}f(x) = \frac{d}{dx}\frac{1}{\Gamma(l)}\int_0^x (x-t)^{l-1} f(t)dt$$

Let $n = 1 - l$, then:

$$D^n f(x) = \frac{1}{\Gamma(1-n)}\frac{d}{dx}\int_0^x \frac{f(t)}{(x-t)^n}dt \tag{2.B.9}$$

which is called the "Reimann–Liouville Fractional Derivative."

Example 2.B.2 Find the half derivative of $f(x) = x$.

Solution

With $n = 1/2$, Eq. (2.B.9) becomes:

$$D^{1/2}f(x) = \frac{1}{\Gamma(^1/_2)}\int_0^x \frac{t}{(x-t)^{^1/_2}}dt$$

which can be solved symbolically using MATLAB as follows:

```
>> diff((int((x-t)^-0.5*t,t,0,x)),x)/gamma(0.5)
ans =1.1283*x^(1/2)
```

that is $D^{1/2}f(x) = \frac{d}{dx\Gamma(^1/_2)}\frac{1}{}\frac{4}{3}x^{^3/_2} = 1.1283x^{^1/_2}$

2.B.4 Laplace Transform of Fractional Derivatives

Let $n = 1 - l$, then $D^n f(t) = D^1 D^{-l} f(t)$.

Performing the Laplace transform, gives:

$$\mathbf{L}(D^n f(t)) = \mathbf{L}\left(\frac{d}{dt}D^{-l}f(t)\right)$$

$$= \int_0^\infty e^{-st}\left(\frac{d}{dt}D^{-l}f(t)\right)d\tau = \int_0^\infty \frac{d}{dt}(e^{-st}D^{-l}f(t))d\tau$$

Integration by parts gives:

$$\mathbf{L}(D^n f(t)) = \left(e^{-st}D^{-l}f(t)\right)_0^\infty + s\int_0^\infty \left(e^{-st}D^{-l}f(t)\right)d\tau$$

or

$$\mathbf{L}(D^n f(t)) = s^{1-l}F(s)$$

But as $n = 1 - l$, then:

$$\mathbf{L}(D^n f(t)) = s^n F(s) \tag{2.B.10}$$

2.B.5 Grunwald–Letnikov Definition of Fractional Derivatives

a) Integer Derivatives

Using the definition of integer derivatives in terms of a backward finite difference quotient, the following expressions can be extracted:

$$\frac{d^1 f(t)}{dt^1} = \lim_{\Delta t \to 0}\frac{1}{\Delta t}[f(t) - f(t-\Delta t)],$$

$$\frac{d^2f(t)}{dt^2} = \lim_{\Delta t \to 0} \frac{1}{(\Delta t)^2}[f(t) - 2f(t-\Delta t) + f(t-2\Delta t)],$$

and

$$\frac{d^3f(t)}{dt^3} = \lim_{\Delta t \to 0} \frac{1}{(\Delta t)^3}[f(t) - 3f(t-\Delta t) + 3f(t-2\Delta t) - f(t-3\Delta t)]$$

Hence, in general, the *nth* derivative can be written as:

$$\frac{d^nf(t)}{dt^n} = \lim_{\Delta t \to 0} \frac{1}{(\Delta t)^n} \sum_{j=0}^{n} (-1)^j \binom{n}{j} f(t - j\Delta t) \tag{2B.11}$$

Replacing Δt by t/N, then Eq. (2B.11) reduces to:

$$\frac{d^nf(t)}{dt^n} = \lim_{N \to \infty} \left[\left(\frac{t}{N}\right)^{-n} \sum_{j=0}^{N-1} (-1)^j \binom{n}{j} f(t - jt/N) \right] \tag{2B.12}$$

where the binomial coefficient $\binom{n}{j} = 0$ for $j > n$.

b) ***Fractional Derivatives***

To extend Eq. (2B.12) to any fractional order derivative, the following extended definition of the binomial coefficient is used:

$$\binom{a}{j} = \begin{cases} \dfrac{a(a-1)(a-2)...(a-j+1)}{j} & \textit{for } j > 0 \\ 1 & \textit{for } j = 0 \end{cases}$$

or

$$\begin{aligned} (-1)^j \binom{n}{j} &= (-1)^j \frac{n(n-1)(n-2)...(n-j+1)}{j!} \\ &= \frac{(j-n-1)(j-n-2)....(-n)}{j!} = \binom{j-n-1}{j} \\ &= \frac{\Gamma(j-n)}{\Gamma(-n)\Gamma(j+1)} \end{aligned} \tag{2B.13}$$

Substituting Eq. (2B.12) into Eq. (2B.13) gives:

$$\frac{d^nf(t)}{dt^n} = \lim_{N \to \infty} \left[\left(\frac{t}{N}\right)^{-n} \sum_{j=0}^{N-1} A_{j+1} \, f\left(t - j\frac{t}{N}\right) \right] \tag{2B.14}$$

where $A_{j+1} = \frac{\Gamma(j-n)}{\Gamma(-n)\Gamma(j+1)}$ = Grunwald coefficients. These coefficients can be given also by the following recursive relationships:

$$A_1 = 1$$

$$A_{j+1} = [(j-1-\alpha)/j]A_j \text{ for } 0 < j < \infty$$

and

$$A_\infty = 0 \tag{2B.15}$$

Note that the Grunwald coefficients in Eq. (2B.15) decay to zero as the number of terms N increases. This feature describes the "fading memory" phenomenon which indicates that the behavior of the VEM is governed primarily by the recent time history rather than by the most remote history. Due to this property, it is possible to truncate higher order terms of summation in Eq. (2B.15) in order to simplify and speed the computational effort.

Other numerical algorithms to approach the FDs are summarized by Oldham and Spanier (1974), Padovan (1987), Podlubny (1999), and Galucio et al. (2006).

Example 2.B.3 Find the half derivative of $f(x) = x$ using the G–L approach and compare the results with the predictions of the Reimann–Liouville (R–L) approach.

Solution

Figure 2.B.1 shows a comparison between the FDs predictions using the G–L and the R–L approaches. It is evident that close agreement is obtained between the

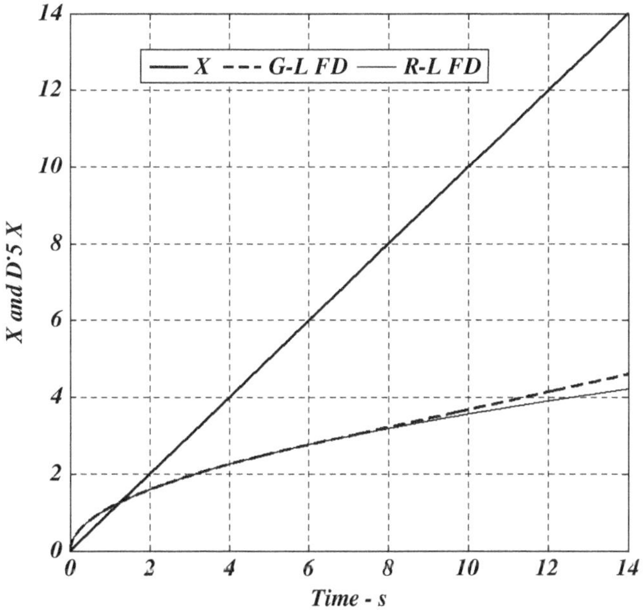

Figure 2.B.1 Comparison between the fractional derivative predictions using *Grunwald–Letnikov* (*G–L*) and the "Reimann-Liouville" (*R–L*) approaches.

two approaches. Note that the number of terms N used in the G–L summation is 12 000.

Problems

2.1 Consider the Poynting–Thomson model of the VEM shown in Figure P2.1a. Derive the constitutive equation of the model that relates the applied stress σ to the resulting strain ε. Determine the creep behavior, during loading and unloading of the model according the cycle shown in Figure P2.1b, by solving the constitutive equation. Assume that $\sigma = \sigma_0$ and $\varepsilon = \sigma_0/(E_p + E_s)$ at $t = 0$ while $\sigma = 0$ and $\varepsilon = \varepsilon_1$ at $t = t_1$.

Check the initial and final of the strain, during loading and unloading, using the initial and final value theorem (outlined in Appendix 2.A).

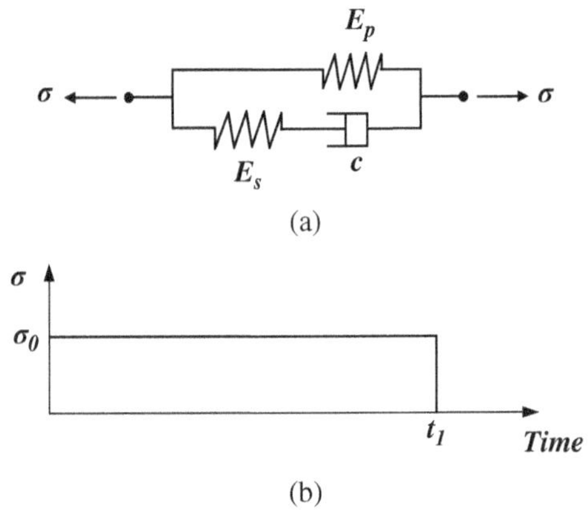

Figure P2.1 Poynting–Thomson model subjected to creep loading.

2.2 If a VEM is represented by the Poynting–Thomson model and is subjected to a step strain ε_0, show that the initial stress σ_0 experienced by the VEM is given by:

$$\sigma_0 = E_s\varepsilon_0$$

Use the initial value theorem.

2.3 Consider the three-parameter Jeffery model of the VEM shown in Figure P2.2a. Derive the constitutive equation of the model that relates the applied stress σ to the resulting strain ε. Determine the creep behavior, during loading and unloading of the model according the cycle shown in Figure P2.1b, by solving the constitutive equation. Assume that $\sigma = \sigma_0$ and $\varepsilon = 0$ at $t = 0$ while $\sigma = 0$ and $\varepsilon = \varepsilon_1$ at $t = t_1$. Discuss the results.

Determine also the relaxation behavior when subjected to the strain loading shown in Figure P2.2b, by solving the constitutive equation. Assume that $\varepsilon = \varepsilon_0$ at $t = 0$. Discuss the results.

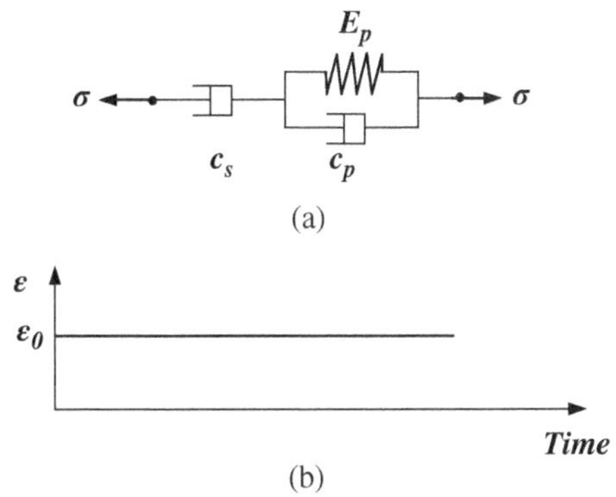

Figure P2.2 Jeffery model under relaxation loading.

2.4 Consider the three-parameter models of a VEM shown in Figure P2.3.

Derive the constitutive equations of the models that relate the applied stress σ to the resulting strain ε. For sinusoidal excitations, at a frequency ω, such that $\sigma = \sigma_0 e^{i\omega t}$ and $\varepsilon = \varepsilon_0 e^{i\omega t}$, use the derived constitutive equations to extract expressions for the complex modulus E, storage modulus E', and loss factor η for each model $[E = E'(1 + i\eta)]$.

If $E_s = 1$, $E_p = 1$, and $c = 1$, plot the effect of the frequency ω on the storage modulus $E'$and loss factor η for each model. Comment on the results.

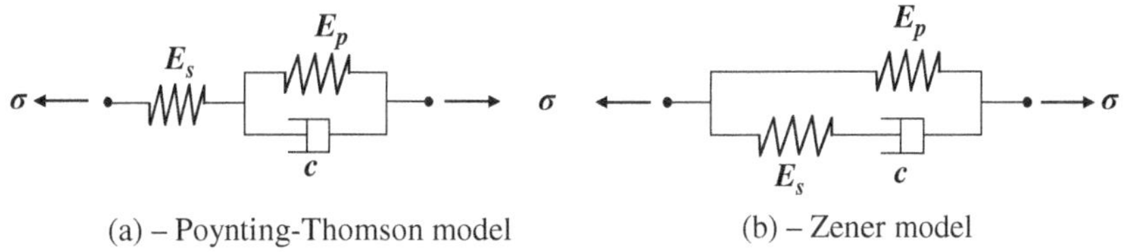

Figure P2.3 Three-parameter models (a) Poynting–Thomson model and (b) Zener model.

2.5 Derive the governing constitutive equations of Burgers model of a VEM, which is shown in Figure P2.4. Note that the model is combination between Maxwell and Kelvin–Voigt models.

For sinusoidal excitations at a frequency ω, such that $\sigma = \sigma_0 e^{i\omega t}$ and $\varepsilon = \varepsilon_0 e^{i\omega t}$, extract expressions for the complex modulus E, storage modulus E' and loss factor η for each model $[E = E'(1 + i\eta)]$.

If $E_s = 1$, $E_p = 1$, and $c = 1$, plot the effect of the frequency ω on the storage modulus E' and loss factor η for each model. Compare the results with those of the three-parameter models of Problem 2.4.

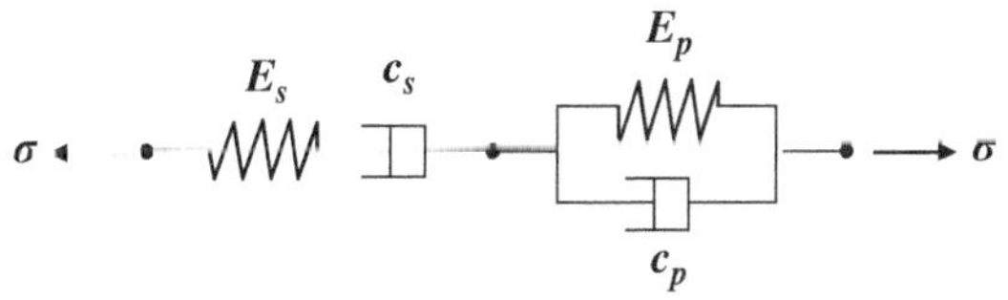

Figure P2.4 Burgers model.

2.6 For the viscoelastic hysteresis loop shown in Figure P2.5 show that:
a) the strain $\varepsilon_{\max} = \varepsilon_0$,
b) the stress $\sigma_{\max\varepsilon} = E'\varepsilon_0$, and
c) the loss factor $\eta = \sigma_{stf}/\sigma_{\max\varepsilon}$.

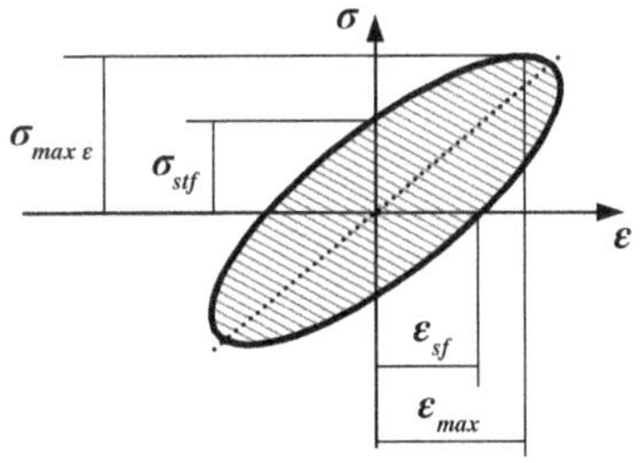

Figure P2.5 The characteristics of viscoelastic hysteresis.

2.7 Show that the complex modulus of the Jeffery model of VEMs with an added inertia element I, as shown in Figure P2.6, is given by:

$$\frac{\sigma}{\varepsilon} = E^* = \frac{I\eta_2(\eta_1 s + G)s^2}{I(\eta_1 + \eta_2)s^2 + (IG + \eta_1\eta_2)s + \eta_2 G}$$

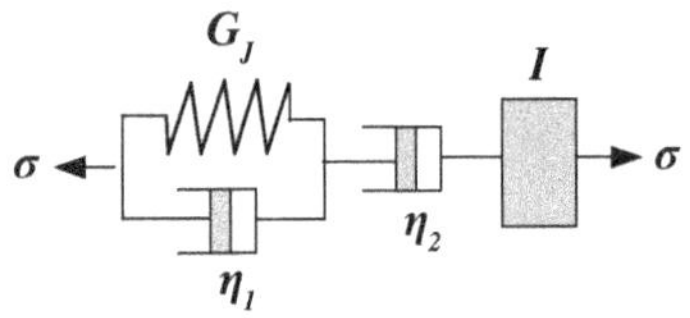

Figure P2.6 Jeffery model with added inertia.

2.8 Show that the complex modulus E^* of the hierarchical viscoelastic model of Figure P2.7, satisfies the following:

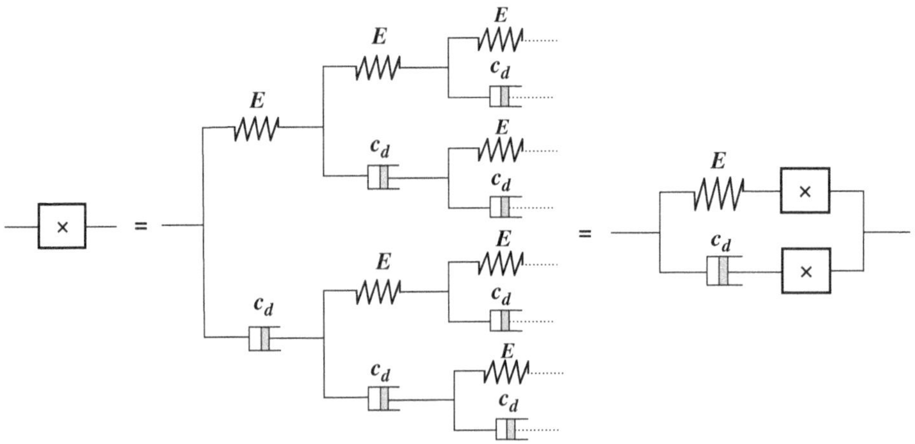

Figure P2.7 Hierarchical viscoelastic model.

$$E^* = \left(\frac{1}{E} + \frac{1}{E^*}\right)^{-1} + \left(\frac{1}{c_d s} + \frac{1}{E^*}\right)^{-1}$$

Show that this equation implies that: $E^* = \sqrt{c_d E s} = E(\lambda s)^{\frac{1}{2}}$ with $\lambda = c_d/E$, that is, the hierarchical viscoelastic model is actually a FD model (Heymans 1996).

2.9 The complex modulus as predicted by the four-parameter FD model is given by:

$$E^* = \frac{\sigma(s)}{\varepsilon(s)} = \frac{E_0 + E_\infty (\tau s)^\alpha}{1 + (\tau s)^\alpha}$$

Show that:

a) the storage and loss moduli of the model are given by:

$$E'(\omega) = \frac{E_0 + (E_\infty + E_0)(\omega\tau)^\alpha \cos(\pi\alpha/2) + E_\infty (\omega\tau)^{2\alpha}}{1 + 2(\omega\tau)^\alpha \cos(\pi\alpha/2) + (\omega\tau)^{2\alpha}}$$

and

$$E''(\omega) = \frac{(E_\infty - E_0)(\omega\tau)^\alpha \sin(\pi\alpha/2)}{1 + 2(\omega\tau)^\alpha \cos(\pi\alpha/2) + (\omega\tau)^{2\alpha}}$$

b) the loss factor η is given by:

$$\eta = \frac{E''(\omega)}{E'(\omega)} = \frac{(E_\infty - E_0)(\omega\tau)^\alpha \sin(\pi\alpha/2)}{E_0 + (E_\infty + E_0)(\omega\tau)^\alpha \cos(\pi\alpha/2) + E_\infty (\omega\tau)^{2\alpha}}$$

c) the value of α that can be estimated from:

$$\alpha = \frac{2}{\pi}\sin^{-1}\left[\eta_{max}(E_\infty - E_0) \times \frac{2\sqrt{E_\infty E_0} + (E_\infty + E_0)\sqrt{1 + \eta_{max}^2}}{\eta_{max}^2 (E_\infty + E_0)^2 + (E_\infty - E_0)^2}\right]$$

(Hint: $e^{\frac{\pi}{2}i} = \cos\frac{\pi}{2} + i\sin\frac{\pi}{2} = i$, and $e^{\frac{\pi}{2}\alpha i} = \cos\alpha\frac{\pi}{2} + i\sin\alpha\frac{\pi}{2}$)

2.10 Determine the parameters of the following two fractional models:

a) $G^* = \dfrac{\tau_\sigma(s)}{\gamma(s)} = \dfrac{G_0 + G_\infty (\tau s)^\alpha}{1 + (\tau s)^\alpha}$ (Parameters are: G_0, G_∞, τ, α)

b) $G^* = \dfrac{\tau_\sigma(s)}{\gamma(s)} = \dfrac{G_m (\tau s)^\alpha}{1 + (\tau s)^\alpha}$ (Parameters are: G_m, τ, α)

that best fit the actual experimental behavior of Dyad606 VEM of Soundcoat operating at 38°C/100°F. Note that G, τ_σ, and γ denote shear modulus, shear stress, and shear strain respectively. For these two models, derive the expressions for the:

a) The storage modulus G' and
b) The loss factor η.

The storage modulus and loss factor of Dyad606 are given in the table here. Compare and comment on the results.

Frequency (Hz)	0.001	0.01	0.040	0.10	0.20	0.40	0.70	1.0	3.0	5.0	10
G' (kpsi)	0.040	0.050	0.065	0.100	0.150	0.200	0.250	0.300	0.500	0.58	0.67
Loss Factor η	0.40	0.50	0.60	0.70	0.75	0.80	0.90	0.92	0.93	0.94	0.95

Frequency (Hz)	20	30	40	50	60	70	80	90	100	200	300
G' (kpsi)	0.75	1.1	1.3	1.6	1.8	2	2.2	2.3	2.6	4	5.2
Loss Factor η	0.95	0.96	0.98	0.99	1.	1.	1.	1.	1.	0.99	0.95

Frequency (Hz)	400	500	600	700	800	900	1000	2000	3000	4000	5000
G' (kpsi)	6.3	7.2	8.5	9.5	11	12	13	19	23	2.6	3
Loss Factor η	0.91	0.88	0.85	0.82	0.79	0.77	0.75	0.63	0.56	0.51	0.48

2.11 Show that the plots of loss modulus/E_s versus the storage modulus/E_s of the Maxwell, Kelvin–Voigt, and Poynting–Thomson models are as displayed in Figure P2.8. Note that these plots are called the "Nyquist diagrams" or "Cole–Cole" plots.

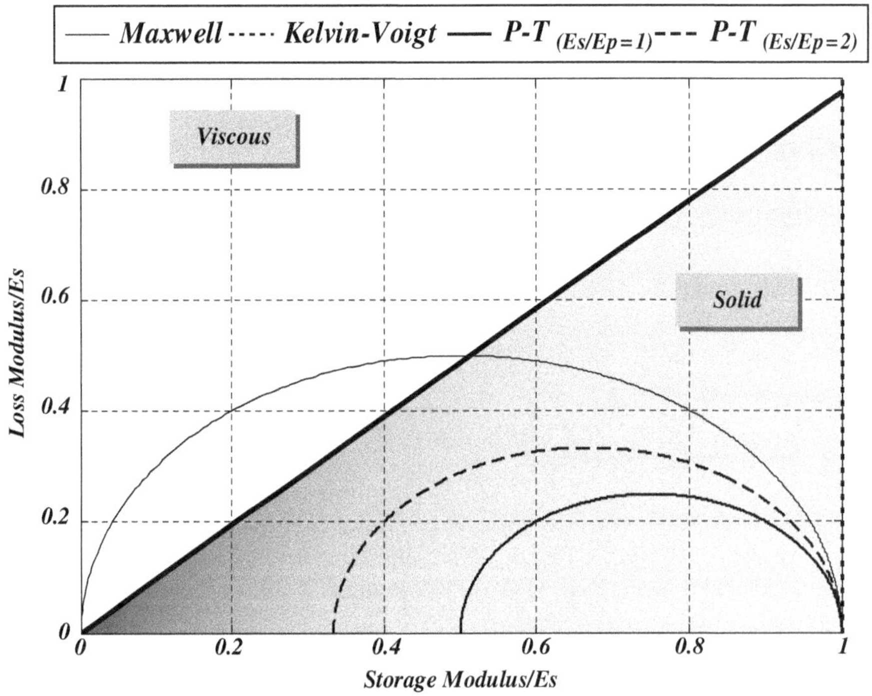

Figure P2.8 The Cole–Cole plots.

2.12 Show that the plots of loss factor versus the storage modulus/E_s of the Maxwell, Kelvin–Voigt, and Poynting–Thomson models are as displayed in Figure P2.9. Note that the plot is called the "Wicket" plot.

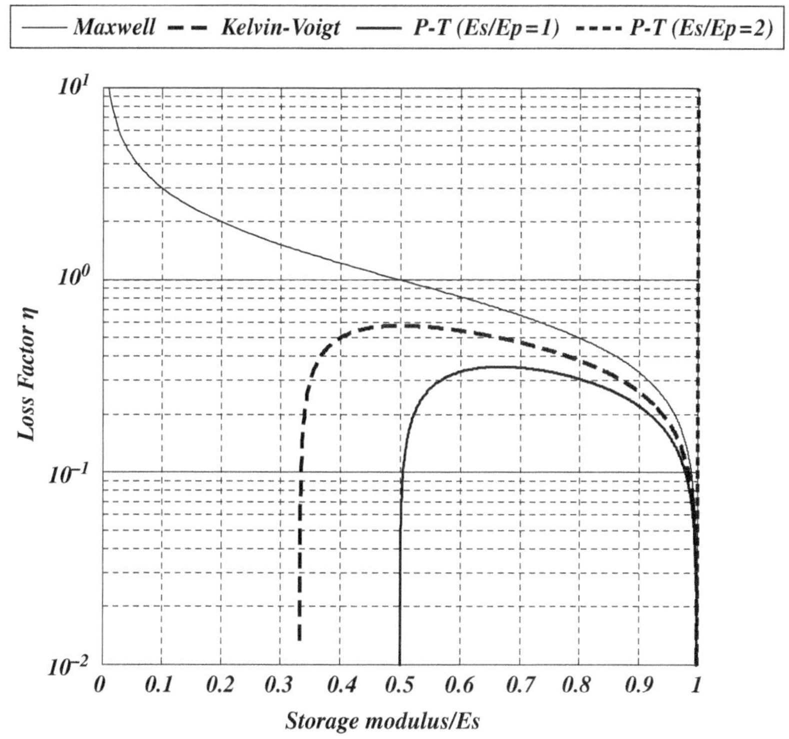

Figure P2.9 The "Wicket" plots.

2.13 If the relaxation modulus of a VEM is given by:
$E(t) = 1 + e^{-t}$ in GPa, where t = time in seconds.
Determine its creep compliance $J(t)$.

2.14 Show that the complex modulus E^* of the Ladder Viscoelastic Model in Figure P2.10, satisfies the following equation:

$$E^* \cong \left[(E_0)^{-1} + (\eta s + E^*)^{-1}\right]^{-1}$$

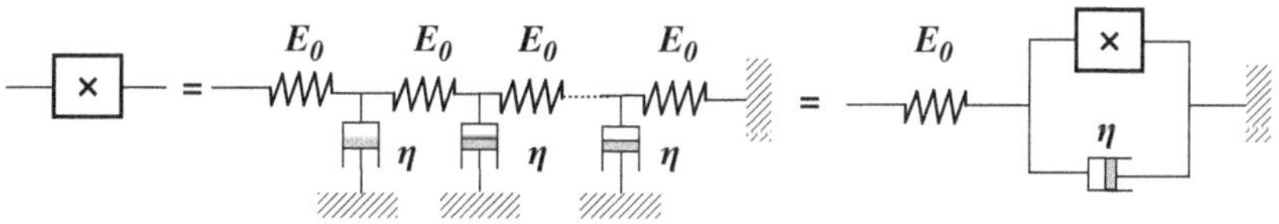

Figure P2.10 The ladder viscoelastic model.

Also show this equation implies that: $E^* \cong E_0\left[1 - \frac{1}{4(\lambda s)} + \frac{1}{8(\lambda s)^2} - \ldots\right]$,

with $\lambda = \eta/E_0$.

2.15 Consider the dynamic system shown in Figure P2.11 that simulates the dynamics of VEMs. Derive the following:

a) relationship between the force F and the net deflection (x_1) in the following Laplace domain form:

$F = K^* x_1$

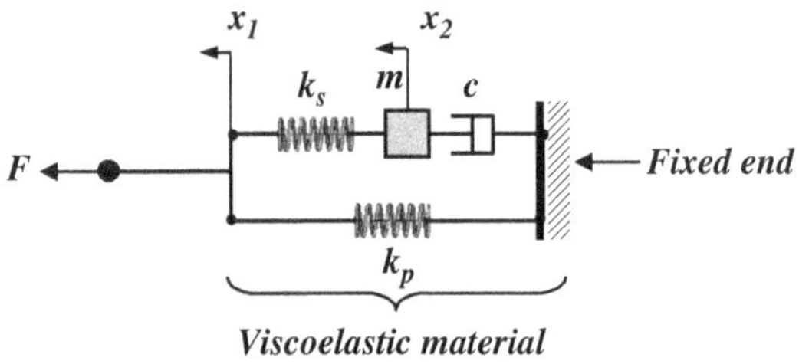

Figure P2.11 Dynamic model of the viscoelastic material.

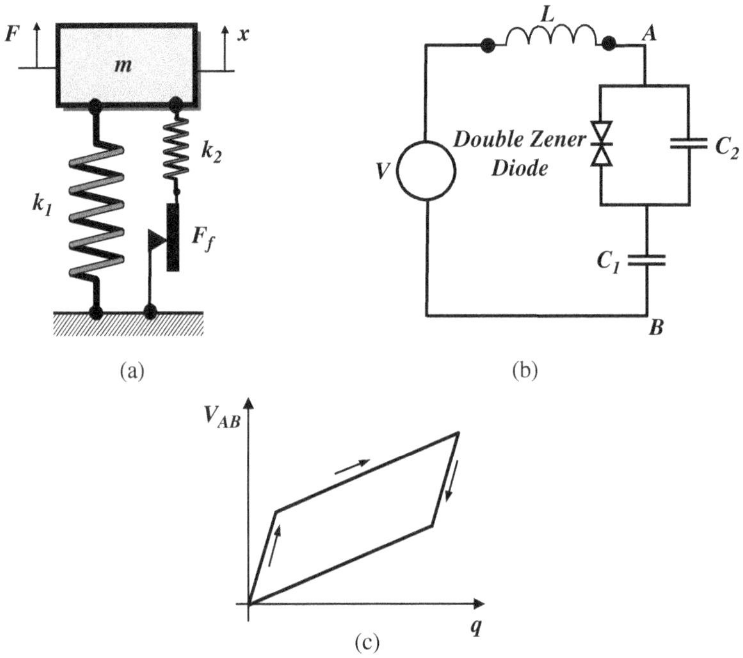

Figure P2.12 Mechanical system with coulomb damping.

where K^* = complex stiffness = $K'(1 + i\,\eta)$ with K' and η denoting the storage modulus and loss factor, respectively.

b) storage modulus of the VEM K' in terms of the frequency
c) loss factor η of the VEM in terms of the frequency

2.16 Show that the mechanical system with coulomb damping, which is displayed in Figure P2.12a, has the electrical analog circuit represented schematically in Figure P2.12b. Show also that the system has the hysteresis characteristics illustrated in Figure P2.12c.

Relate the parameters of the mechanical elements of Figure P2.12a to those of their electrical analog components of Figure P2.12b. Note that q denotes the charge (Iwan, 1964).

3

Characterization of the Properties of Viscoelastic Materials

3.1 Introduction

Characterization of the mechanical behavior of viscoelastic materials (VEMs) is essential to the proper design of effective damping treatments that are suitable for attenuating structural vibrations over a particular range of operating temperatures and frequencies. This chapter presents the behavior of typical VEMs as well as the important techniques for characterizing the mechanical properties of this class of materials both in the frequency and time domains. The principles of "time-frequency" and "time-temperature" superposition are introduced. These principles are applied to generate the widely used "master curves" that constitute the unified means and the industry norm for presenting the damping properties of VEMs.

3.2 Typical Behavior of Viscoelastic Materials

The behavior of typical VEMs depends primarily on the operating temperature, and frequency. Figure 3.1 displays the effect of temperature on the storage modulus and loss factor of most VEMs.

The figure indicates that the behavior of the VEM varies drastically over three distinct temperature regions. At the low temperature region, that is, the "glassy region," the VEM behaves like glass where the storage modulus assumes it highest value and drops slightly with increased temperatures. Within this region, the loss factor increases significantly as the temperature is increased. At the medium temperature region, the "transition region," the VEM softens and the storage modulus drops orders of magnitude over a very narrow temperature range. During such a transition zone, the loss factor attains its peak. Further increase of the temperature makes the VEM enter the "rubbery region" where the material behaves like a very soft rubber with much reduced storage modulus and loss factor (Christensen 1982; Nashif et al. 1985; Haddad 1995; Lakes 1999, 2009).

For optimal performance, the VEM should operate near its peak loss factor without compromising its structural integrity as measured by its storage modulus. Therefore, VEMs with broader damping peaks will behave more favorably over wider temperature range.

Figure 3.2 displays the effect of the operating frequency on both the storage modulus and loss factor of most VEMs.

Active and Passive Vibration Damping, First Edition. Amr M. Baz.

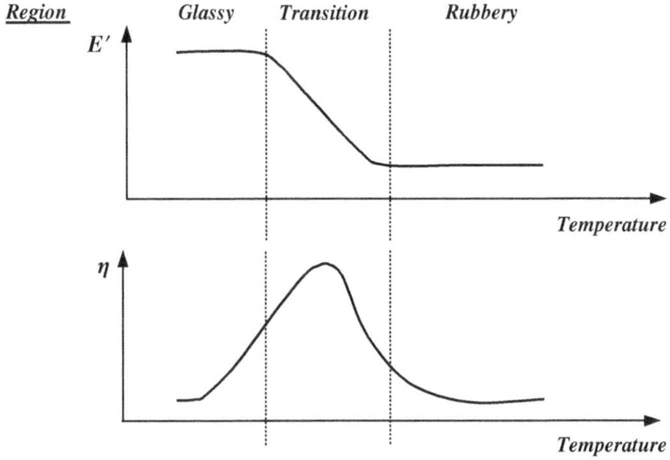

Figure 3.1 Effect of temperature on the storage modulus and loss factor of viscoelastic materials at constant frequency.

Note that increasing the frequency is accompanied with a stiffening effect of the VEM. However, there is a peak of the loss factor over the medium frequency zone (Christensen 1982; Nashif et al. 1985; Haddad 1995; Lakes 2009).

Close examination of Figures 3.1 and 3.2 indicates that the viscoelastic behavior with increased frequencies is the inverse of that with increased temperatures. Such unique characteristics provide the basis for the "temperature-frequency superposition principle," which is utilized for a simplified representation of the viscoelastic properties as will be presented in Section 3.4.

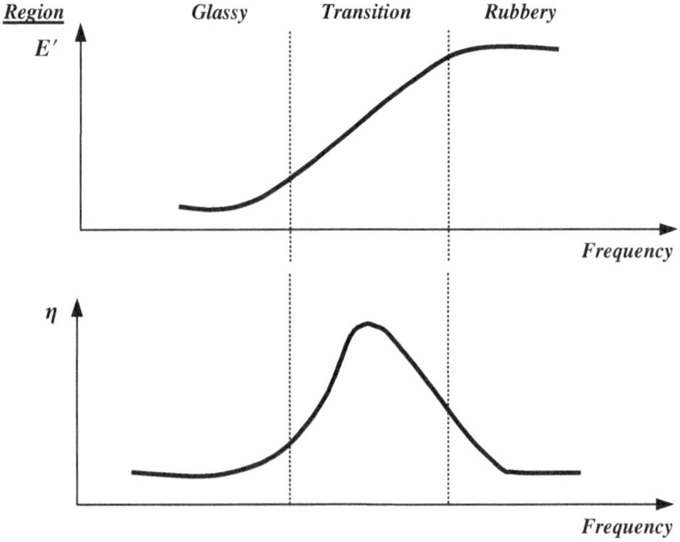

Figure 3.2 Effect of frequency on the storage modulus (top) and loss factor (bottom) of viscoelastic materials at constant temperature.

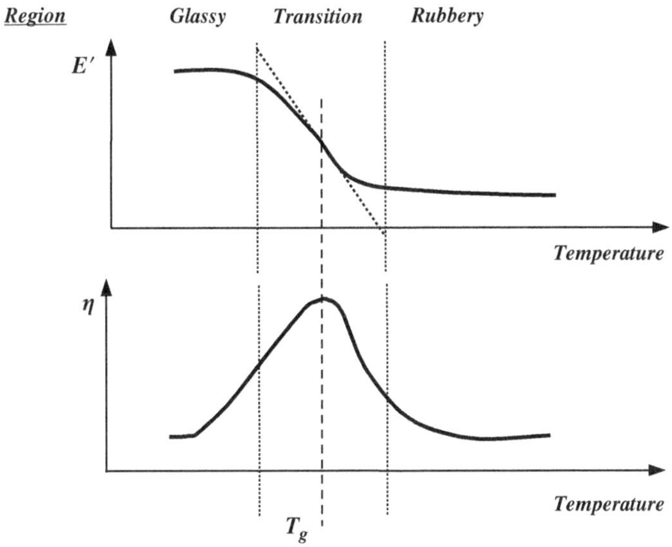

Figure 3.3 Glass transition temperature of viscoelastic materials.

Figure 3.3 defines an important parameter of VEMs, which is the glass transition temperature, T_g. This temperature approximately corresponds to the inflection point of the storage modulus E' and the peak of the loss factor characteristics.

The combined effect of the operating temperature and frequency on the properties of a typical VEM is shown in Figure 3.4.

Typical VEMs, whether made of polymers or elastomers, exhibit high damping characteristics at usually low stiffness as shown in the loss factor-storage modulus map of Figure 3.5. Such characteristics distinguish these damping materials from other materials, alloys, and ceramics that have high stiffness and relatively low loss factors (Lakes 1999; Cebon and Ashby 1994). As a general rule, there is an inverse correlation between damping and stiffness. For most polymers or elastomers, the product of their loss factor and modulus ($\eta|E^*|$) is usually less than 0.6 *GPa* (Ashby 1999 and Lakes 2009) as indicated by the dark line displayed in Figure 3.5.

3.3 Frequency Domain Measurement Techniques of the Dynamic Properties of Viscoelastic Material

There are several methods for measuring the dynamic properties of VEMs as reviewed by Ferry (1980), Dlubac et al. (1990), Garibaldi and Onah (1996), Lakes (2004, 2009). These methods rely in their operation on monitoring the harmonic response of specimens of the VEMs under consideration.

Table 3.1 and Figure 3.6 summarize the parameters and operating ranges of the most common harmonic response methods.

Emphasis is placed here on the describing the theory and principle of operation of the dynamic, mechanical, and thermal analyzer (DMTA) and Oberst beam methods. These

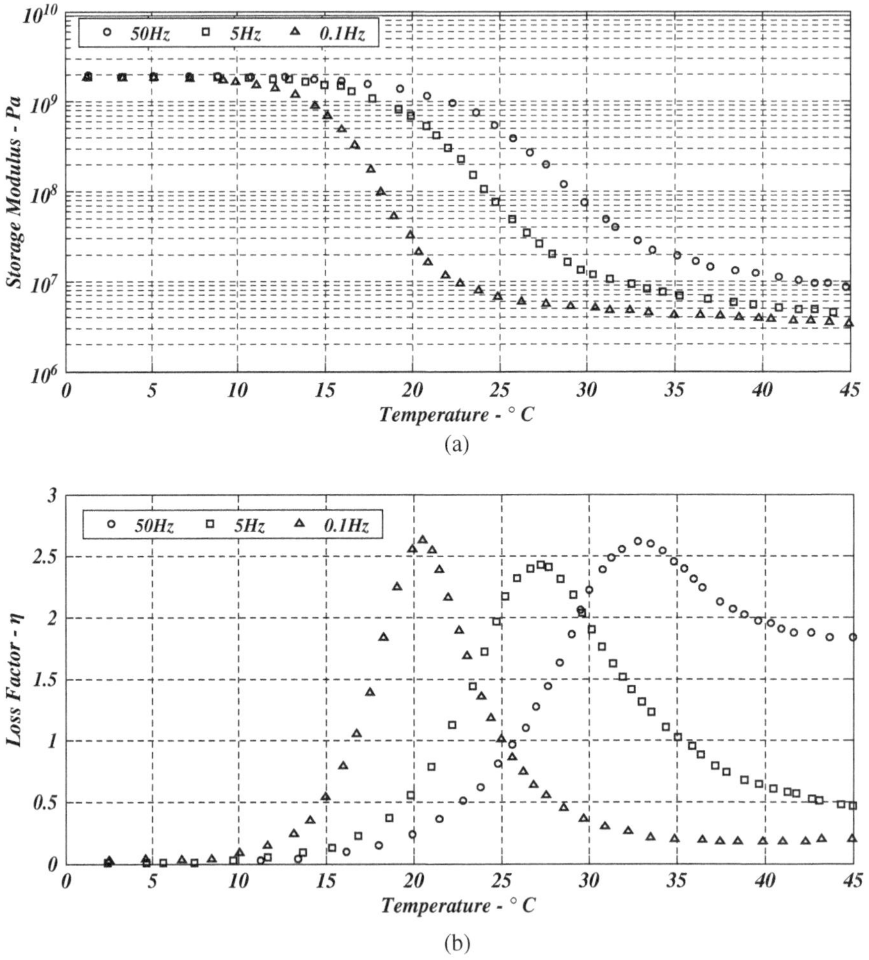

Figure 3.4 Effect of temperature and frequency on the storage modulus and loss factor of viscoelastic materials.

methods enable the measurements of the properties over a wide temperature and frequency ranges that are compatible with most of the vibration control applications. The DMTA method is classified as a non-resonant technique, whereas the Oberst beam method belongs to the class of resonant methods.

3.3.1 Dynamic, Mechanical, and Thermal Analyzer

Conventional DMTA of Polymer Laboratories provides an effective and accurate means for measuring the complex modulus of VEMs under a wide variety of temperatures ranging between −150°C and 300°C and frequencies ranging between 0.01 and 200 Hz. However, recently, McHugh (2008) developed an ultrasonic DMTA that is capable of measuring up to 6 MHz.

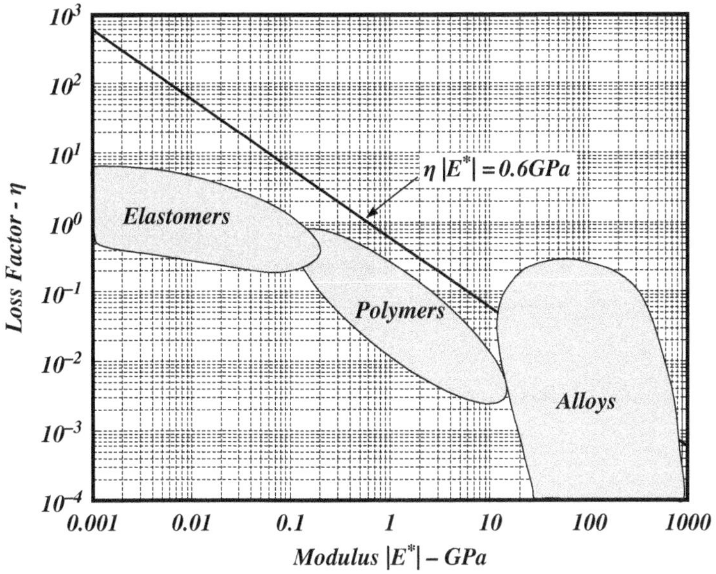

Figure 3.5 The modulus and loss factor map for some engineering materials (Ashby 1999 and Lakes 2009).

Table 3.1 Parameters and operating ranges of some harmonic response methods.

No	Method	Frequency (Hz)	Temp. (°C)	Storage modulus (Pa)	Loss factor	Sample size (mm)	Testing mode
1	DMTA[a] (Brown and Read 1984)	0.01–200	−150 to 300	10^5–10^{11}	10^{-4}–9.99	3-10-12	Shear, tension, and bending
2	Torqued cylinder (Magrab 1984)	50–1500	−40 to 70	10^6–10^{10}	0.05–1.2	50 diam.-50	Shear
3	Resonant bar (Madigosky and Lee 1983)	2.5–25,000	-60 to 70	10^4–10^{12}	0.01–5	6-6-150	Tension
4	Oberst beam (ASTM E756 1993)	10–10,000	−60 to 120	10^4–10^9	0.001–2	12.5-(175-250)-(1-3.125)	Bending of composite beams
5	Ultrasonic spectroscopy (Alig and Lellinger 2000)	10^4–10^{10}	−35 to 60	10^5–10^{11}	0.0001–0.2	25 diam.- 25	Shear

[a] DMTA = Dynamic, mechanical, and thermal analyzer.

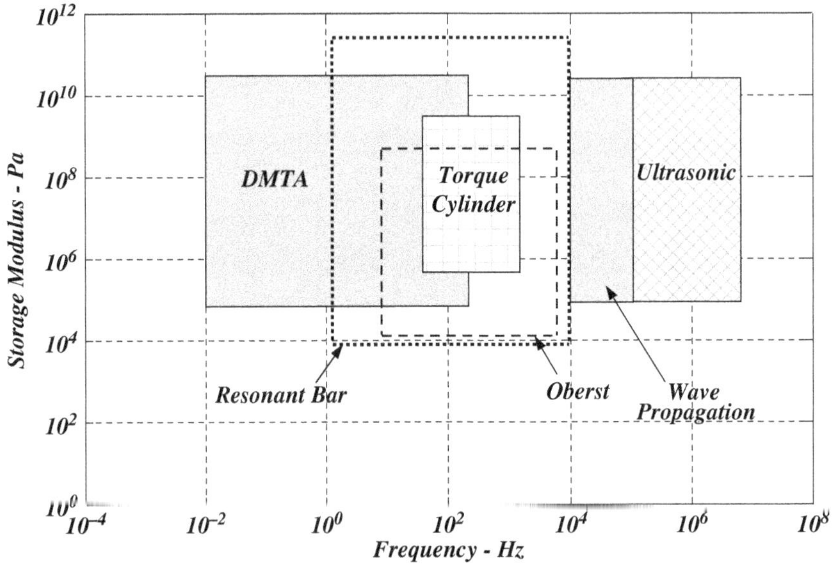

Figure 3.6 Parameters and operating ranges of some harmonic response methods.

In the DMTA, the VEM sample is clamped in tension, bending, or shear configuration and placed inside a temperature-controlled enclosure. The temperature of the enclosure is controlled between −150°C and 300°C. The specimen is then subjected to sinusoidal stresses using an electromechanical shaker. The resulting response (strain) of the sample is monitored using a displacement transducer. Figure 3.7 shows the DMTA system, main components, and different types of heads.

The theory governing the operation of the DMTA is based on the relationship between the stress and strain associated with VEMs. For a sinusoidal strain of magnitude ε_0 and frequency ω, the strain is given by:

$$\varepsilon = \varepsilon_0 \, e^{i\omega t} \tag{3.1}$$

with $i = \sqrt{-1}$ and accordingly the corresponding stress σ can be determined from

$$\sigma = (E' + iE'')\varepsilon = E'(1 + i\eta)\, \varepsilon_0 e^{i\omega t} \tag{3.2}$$

where E' and E'' are storage and loss modulus of the VEM. Also, η is the loss factor of the VEM given by:

$$\eta = E''/E'$$

From Figure 3.8, the loss factor can also be written as:

$$\eta = E''/E' = \tan(\delta) \tag{3.3}$$

Combining Eqs. (3.2) and (3.3) gives:

$$\sigma = E'\sqrt{(1+\eta^2)}\, e^{i\delta}\, \varepsilon_0 e^{i\omega t} = E'\varepsilon_0 \sqrt{(1+\eta^2)} e^{i(\omega t + \delta)}, \tag{3.4}$$

(a) – The *DMTA* system

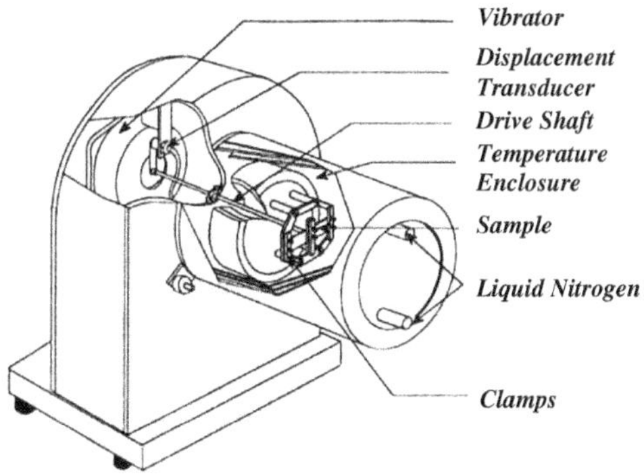

(b) – Main components of the *DMTA*

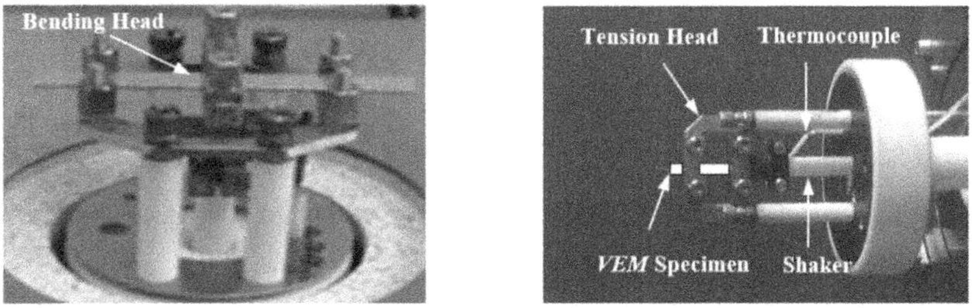

(c) - Different types of *DMTA* heads

Figure 3.7 The dynamic, mechanical, and thermal analyzer (DMTA). (a) The DMTA system, (b) Main components of the DMTA, and (c) Different types of DMTA heads.

or

$$\sigma = \sigma_0 \, e^{i(\omega t + \delta)}$$

where

$$\sigma_0 = E' \varepsilon_0 \sqrt{(1 + \eta^2)}. \tag{3.5}$$

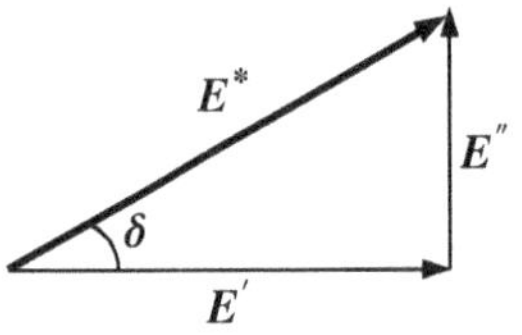

Figure 3.8 Graphical representation of the complex modulus.

Equations (3.1) and (3.4), indicate that the real strain and stress can be given by

$$\varepsilon = \varepsilon_0 \sin(\omega t) \text{ and } \sigma = \sigma_0 \sin(\omega t + \delta) \tag{3.6}$$

Equation (3.6) indicates that strain ε lags behind the stress σ by a phase angle δ, as is also shown in Figure 3.9.

Hence, by subjecting the VEM to a sinusoidal stress σ with magnitude σ_0 and measuring the time history of the resulting strain ε, one can measure the amplitude of the strain ε_0 and the phase angle δ. Equations (3.3) and (3.5) can then be used to compute the loss factor η and the storage modulus E' at the excitation frequency ω. This process can be repeated for different excitation frequencies and operating temperatures to determine the effect of the frequency and temperature on both the storage modulus and loss factor.

The obtained results are analyzed using the "temperature-frequency" superposition approach to generate the unified master curves for the VEM.

3.3.2 Oberst Test Beam Method

3.3.2.1 Set-Up and Beam Configurations

Oberst developed in 1952 a classical resonant method for measuring the damping properties of materials, including loss factor, η, Storage modulus, E', and shear modulus, G. The method is adopted by The American Society for Testing and Materials as a standard test procedure for quantifying the damping characteristics of VEMs (ASTM E 756-93).

Figure 3.10 shows a schematic drawing of the Oberst beam test set-up and Figure 3.11 displays the different configurations of the Oberst beam.

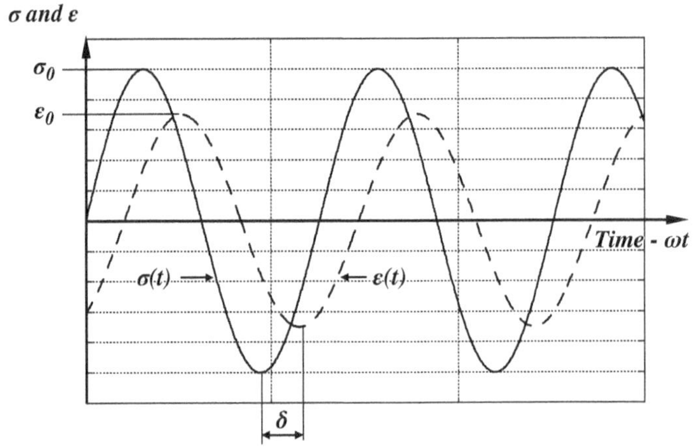

Figure 3.9 Time history of the stress and strain.

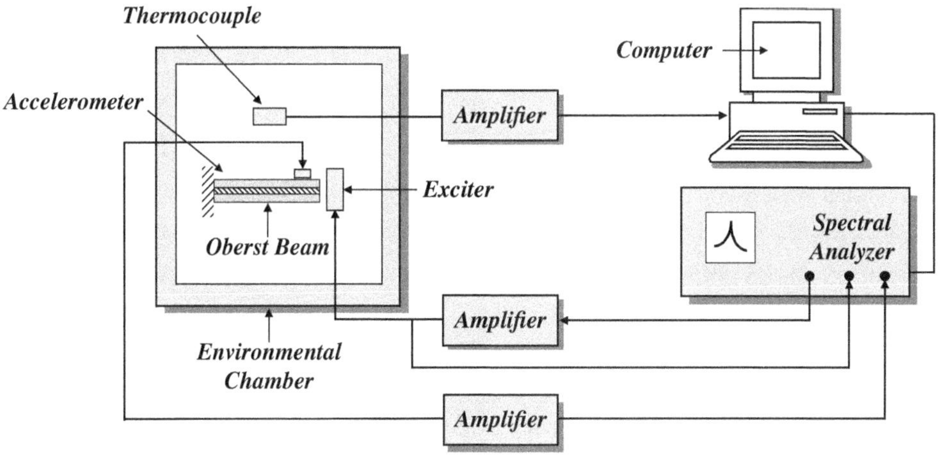

Figure 3.10 Oberst beam test set-up.

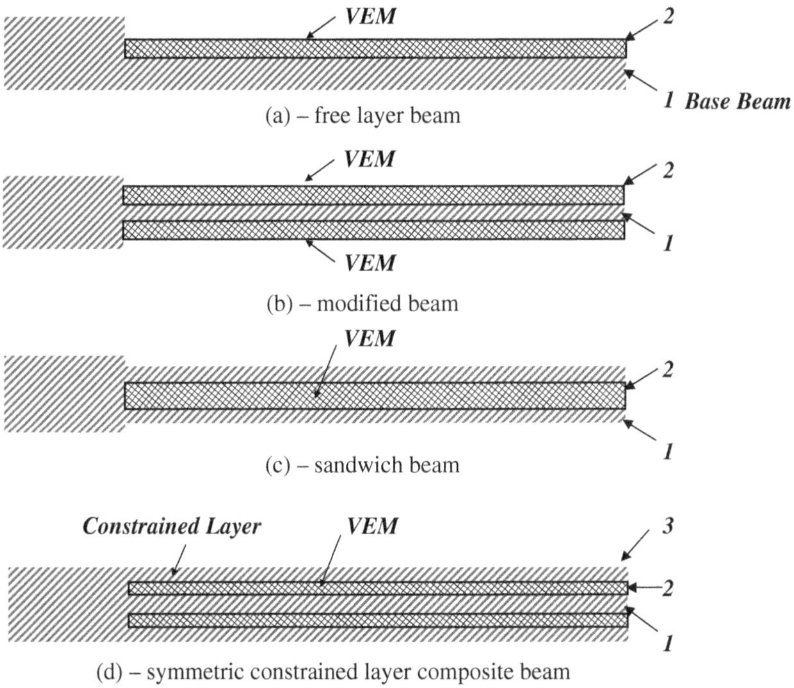

Figure 3.11 Different configurations of the Oberst beam. (a) Free layer, (b) Modified beam, (c) Sandwich, and (d) Symmetric constrained layer composite beam.

The proper beam configuration is selected according to the characteristics to be extracted as outlined in Table 3.2. The selected beam is installed inside an environmental chamber where the temperature is controlled over the desired operating range and the pressure can be reduced to vacuum to minimize the effect of air damping. The beam is then excited by a non-contact exciter, over a broad range of frequencies, and the

Table 3.2 Characteristics of Oberst test beams.

Configuration	Free layer beam	Modified beam	Sandwich beam	Symmetric constrained layer beam
Mode of operation	Tension	Tension	Shear	Shear
Parameters extracted	E' and η	E' and η	G' and η	$G'$and η
Complexity of test beam	Very simple	Simple	Complex	More complex
Complexity of parameter extraction	Complex	Simple	Simple	More complex

response is monitored by an accelerometer. The excitation can be a swept sinusoidal or random in nature. From the frequency response of the beam, the damping ratio η_{s_n} is determined, using the half-power method, at the different resonant frequencies f_{s_n} of the beam where n = frequency order.

3.3.2.2 Parameter Extraction

The basic equations needed to extract the damping parameters of the VEM for the different beam configurations will be developed in detail in Chapters 4 and 7. However, a brief summary is given in the following:

3.3.2.2.1 Free Layer Beam

Once the resonant frequency of the beam system f_{sn} and the base beam f_{bn} are measured experimentally, the dimensionless number Z is computed from:

$$Z^2 = (1 + \rho_2 h_2)\left(\frac{f_{s_n}}{f_{b_n}}\right)^2 \tag{3.7}$$

where f_{sn} = nth resonant frequency of beam system, f_{bn} = nth resonant frequency of metal base beam, $e_2 = E_{2_n}/E_{1_n}$ = modulus ratio of VEM to base beam, $h_2 = t_2/t_1$ = thickness ratio of VEM to base beam, and $\rho_2 = \rho_2/\rho_1$ = density ratio of VEM to base beam.

It will be shown in Chapter 6 that:

$$Z^2 = \frac{1 + 2e_2h_2(2 + 3h_2 + 2\,h_2^2) + e_2\,h_2^4}{(1 + e_2h_2)} \tag{3.8}$$

Equation (3.8) reduces to the following quadratic equation in e_2:

$$e_2^2h_2^4 + e_2h_2(4 + 6h_2 + 4h_2^2 - Z^2) + 1 - Z^2 = 0 \tag{3.9}$$

Equation (3.9) has a solution:

$$e_2 = \frac{-h_2(4 + 6h_2 + 4\,h_2^2 - Z^2) + \sqrt{h_2^2(4 + 6h_2 + 4\,h_2^2 - Z^2)^2 + 4\,h_2^4(Z^2 - 1)}}{2\,h_2^4} \tag{3.10}$$

From Eq. (3.10), $E_{2_n} = E'_n$ can be determined. Similarly, the loss factor η_{2_n} of the VEM can be determined from measurement of the loss factor of the beam/VEM system η_{s_n}:

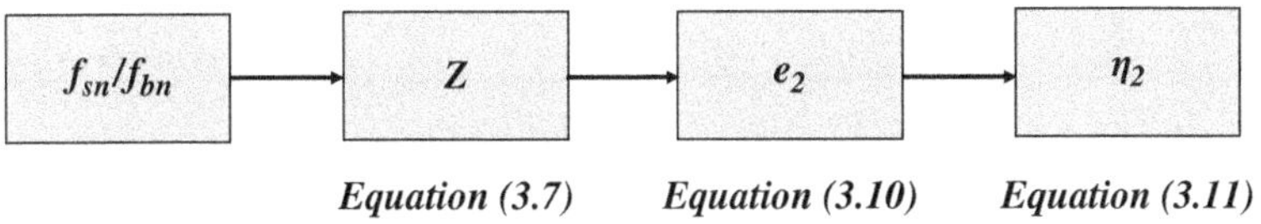

Figure 3.12 Extraction of damping parameters of free layer beam.

$$\frac{\eta_{2_n}}{\eta_{s_n}} = \frac{(1 + e_2 h_2)\left(1 + 4e_2 h_2 + 6e_2\, h_2^2 + 4e_2\, h_2^3 + e_2\, h_2^4\right)}{e_2 h_2\left(3 + 6h_2 + 4\, h_2^2 + 2e_2\, h_2^3 + e_2^2\, h_2^4\right)} \tag{3.11}$$

Accordingly, Figure 3.12 summarizes the steps to be followed to extract the parameters E'_n and η_{2_n}.

3.3.2.2.2 Modified Beam

In this case, the dimensionless parameter Z is given by:

$$Z^2 = (1 + \rho_2 h_2)\left(\frac{f_{s_n}}{f_{b_n}}\right)^2 = e_2\left(6h_2 + 12h_2^2 + 8h_2^3\right) + 1 \tag{3.12}$$

Note that Eq. (3.11) is much simpler than Eq. (3.8) for the free layer beam. This simplicity is due to the fact that the neutral axis of the modified beam remains the same as the base beam unlike the case of the free layer beam.

Equation (3.12) can be solved to yield e_2 as follows:

$$e_2 = \left(Z^2 - 1\right) / \left(6h_2 + 12h_2^2 + 8h_2^3\right) \tag{3.13}$$

Equation (3.13) gives $E_{2_n} = E'_n$ and the loss factor η_{2_n} of the VEM can be determined from measurement of the loss factor of the beam/VEM system η_{s_n}:

$$\frac{\eta_{2_n}}{\eta_{s_n}} = \left[\frac{1}{e_2\left(6h_2 + 12\, h_2^2 + 8\, h_2^3\right)} + 1\right] \tag{3.14}$$

3.3.2.2.3 Sandwich Beam

The shear modulus G'_{2_n} can be calculated from:

$$\frac{G'_{2_n} L^2}{E_1 H_1 H_2 \alpha_n} = \frac{(A - B) - 2(A - B)^2 - 2\left(A\eta_{s_n}\right)^2}{(1 - 2A + 2B)^2 + 4\left(A\eta_{s_n}\right)^2} \tag{3.15}$$

where L = beam length. Also, the parameters A and B are given by:

$$A = (2 + \rho_2 h_2)\left(\frac{f_{s_n}}{f_{b_n}}\right)^2 \text{ and } B = \frac{1}{6(1 + h_2)^2} \tag{3.16}$$

with α_n assuming the values listed in Table 3.3.

Similarly, the loss factor η_{2_n} is given by:

$$\frac{\eta_{2_n}}{\eta_{s_n}} = \frac{A}{(A - B) - 2(A - B)^2 - 2\left(A\eta_{s_n}\right)^2} \tag{3.17}$$

Table 3.3 Values of α_n.

n	1	2	3	4	5
α_n	3.516	22.035	61.697	120.90	199.86

3.3.2.2.4 Symmetric Constrained Layer Composite Beam

The shear modulus G'_{2_n} can be calculated from:

$$\frac{G_2 L^2}{E_3 H_3 H_2 \alpha_n} = \left(\frac{\alpha}{\alpha^2 + \beta^2}\right) - 1 = g \tag{3.18}$$

where the parameters α and β are given by:

$$\alpha = \frac{1}{D}\left(\frac{(EI)_s}{(EI)_b} - C\right),\ C = 1 + 2e_3 h_3^3 + D, \beta = \frac{\eta_s (EI)_s}{D(EI)_b},\ D = 6e_3 h_3 (1 + 2h_2 + h_3)^2 \tag{3.19}$$

Similarly, the loss factor η_{2_n} is given by:

$$\eta_{2_n} = \frac{\beta}{g(\alpha^2 + \beta^2)} \tag{3.20}$$

The obtained results E'_n (or G'_{2_n}) and η_{2_n} at different frequencies f_{sn} are analyzed using the temperature-frequency superposition approach to generate the unified master curves for the VEM.

3.4 Master Curves of Viscoelastic Materials

3.4.1 The Principle of Temperature-Frequency Superposition

The principle of temperature-frequency superposition is used to generate the unified master curves for the VEM. To illustrate this principle, consider the characteristics shown in Figure 3.13 where the measurements are carried out over a narrow test band extending between frequencies ω_1 and ω_2 as well as between temperatures T_{-1} and T_1.

In the figure, the storage modulus is plotted versus the frequency for different temperatures. Note that the characteristics measured at a low temperature T_{-1} can be made to coincide with that measured at a reference temperature T_0 when the former characteristics is shifted forward along the frequency axis by a shift factor α_{-1}. Similarly, the characteristics at a higher temperature T_1 can be shifted backward by a shift factor α_1 to coincide with that at T_0. Such a frequency shift process can be carried out over a narrow frequency band $(\omega_2 - \omega_1)$ extending it to a larger frequency band $(\alpha_{-1}\omega_2 - \alpha_1\omega_1)$ as shown in Figure 3.13b,c. Continuing this process over the entire frequency band of interest results in the superposition of all the characteristics measured at different temperatures on a single *master curve*, which represents the storage modulus E' as a function of the reduced frequency $\alpha_T\omega$. Figure 3.14 displays the variation pattern of the "temperature

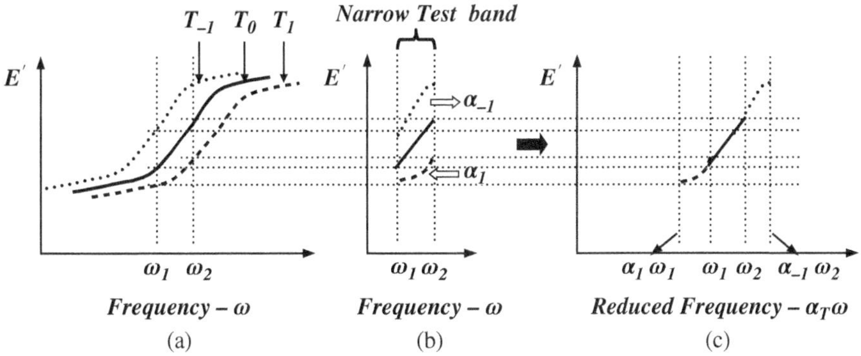

Figure 3.13 Temperature-frequency superposition of storage moduli.

Figure 3.14 Temperature shift factor.

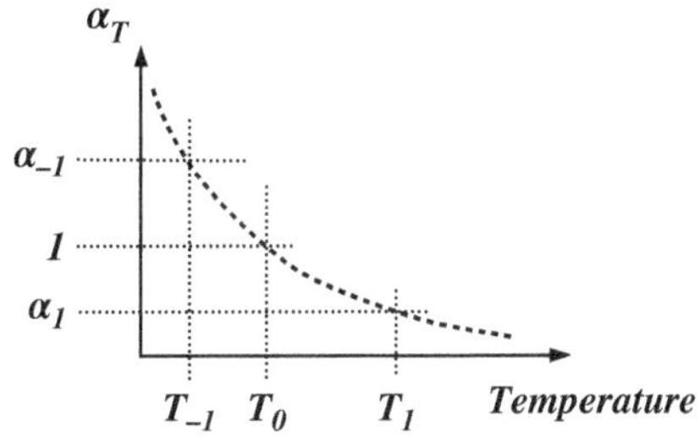

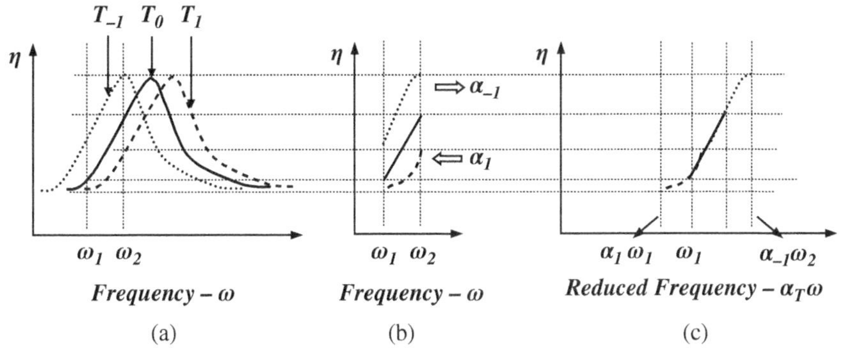

Figure 3.15 Temperature-frequency superposition of loss factors.

shift" factor α_T with temperature, which is necessary to achieve the desired superposition. Note that $\alpha_T = 1$ for the reference temperature T_0, $\alpha_T < 1$ for $T > T_0$, and $\alpha_T > 1$ for $T < T_0$.

The principle of temperature-frequency superposition applies equally to the generation of the "master curve" of the loss factor characteristics as shown in Figure 3.15.

The temperature shift factor, α_T, which is used to generate the master curves for both the storage modulus and loss factor, is based on the most common empirical formula from Williams–Landel–Ferry (WLF) given by:

$$\log_{10}(\alpha_T) = -\frac{C_1(T - T_g)}{C_2 + (T - T_g)} \tag{3.21}$$

Table 3.4 Values of the constants of the WLF formula (Ferry 1980). Reproduced with permission of John Wiley & Sons.

Polymer	C_1	C_2, °K	T_g, °K
Polyisobutylene	8.1	200.4	205
Butyl rubber	9.03	201.6	205
Polyurethane	16.7	68.0	221
Polystyrene	12.7	49.8	370
Polyethylmethacrylate	8.86	101.6	276
Universal constants	17.4	51.6	

where T and T_g are the temperature and the glass transition temperature, respectively. Also, C_1 and C_2 are constants that best fit the experimental data and minimize the deviations from the master curves. The WLF showed that for all amorphous polymers, the constants C_1 and C_2 have universal values of 17.44 and 51.6.

Origin of the WLF formula Although, the WLF formula seems to be empirical in nature, it has its roots embedded in the phenomenological behavior and the kinetic theory of VEMs (Lakes 2009). For example, the viscosity c_d of a viscoelatic material can be written according to the *Doolittle Equation* as follows:

$$c_d = a\, e^{b\left(V/V_f - 1\right)} \tag{3.22}$$

where V = total volume, V_f = free volume with a and b are constants.

Defining the free volume fraction f as:

$$f = V_f/V = f_g + \alpha_f\left(T - T_g\right) \tag{3.23}$$

where f_g and α_f denote the fractional free volume frozen at T_g and the thermal expansion coefficient of the free volume.

Let, the temperature shift factor be written as (see Problem 3.1)

$$\alpha_T = \left(c_d/c_{d_g}\right) \tag{3.24}$$

Equations (3.22) through (3.24), give

$$\log_{10}(\alpha_T) = \log_{10}\left(\frac{c_d}{c_{d_g}}\right) = 0.4343b\left(\frac{V}{V_f} - \frac{V}{V_g}\right) = -\frac{0.4343b}{f_g}\frac{\left(T - T_g\right)}{f_g/\alpha_f + \left(T - T_g\right)} \tag{3.25}$$

Note that Eq. (3.25) has the same form as the WLF formula (3.21) with $C_1 = 0.4343b/f_g$ and $C_2 = f_g/\alpha_f$. With b = 1, f_g = 0.025, and α_f = 0.00048 °C^{-1}, then C_1 = 17.44 and C_2 = 51.6.

It should be noted that the WLF equation has been found to fit, with reasonable accuracy, the characteristics of thermorheologically simple VEMs within the temperature range $(T - T_g) < 100$°C. Table 3.4 lists the values of the constants C_1 and C_2 for some of the commonly used polymers. For more comprehensive list of these constants for different types of polymers, the handbook of physical properties of polymers (Mark 2007) should be consulted.

Other commonly used temperature shift factor formulas include the Arrhenius Equation, given by:

$$\log_{10}(\alpha_T) = C\left(\frac{1}{T} - \frac{1}{T_g}\right) \tag{3.26}$$

where C is a constant depending on the activation energy of the material.

Example 3.1 Generate the master curves for the VEM that has the storage modulus and loss modulus shown in Figure 3.16 over the frequency range from 0 to 200 Hz (Cowans 2006). Plot also the corresponding temperature shift factor.

Solution

Adopting the approach presented in Section 3.5, the master curves are obtained with $C_1 = 12$ and $C_i = 200$. Figure 3.17 displays the master curves and Figure 3.18 shows the corresponding temperature shift factor with $T_0 = 313$ K.

In Figure 3.17, the shifted data of Cowans (2006) are displayed, in circles, along with the solid line (for the storage modulus) and dotted line (for the loss factor) as obtained by Drake and Terborg (1979). The data from Drake and Terborg (1979) are reported in the Appendix.

3.4.2 The Use of the Master Curves

Figure 3.19 displays how the master curves can be used to extract the storage modulus E' and the loss factor η for a given excitation frequency ω_1 and operating temperature T_1.

Intersection of the constant frequency line ω_1 and the constant temperature line T_1 yields point A, which can be used to generate the intersection points B and C with the storage modulus line and the loss factor line. Moving horizontally from points B and C gives the storage modulus E_1' and the loss factor η_1.

3.4.3 The Constant Temperature Lines

The constant temperature lines displayed on the master curve are generated from the relationship between the reduced frequency ω_r and the excitation frequency ω:

$$\omega_r = \alpha_T \omega \tag{3.27}$$

Taking the logarithm of both sides gives:

$$\log(\omega) = \log(\omega_r) - \log(\alpha_T) \tag{3.28}$$

For a given temperature, $\log(\alpha_T)$ = constant and Eq. (3.28) indicates that the relationship between the logarithm of the reduced frequency ω_r and the logarithm of the excitation frequency ω is linear with slope = 1. Therefore, the constant temperature line is also a constant temperature shift line and is a straight line in the $\log(\omega) - \log(\omega_r)$ plane.

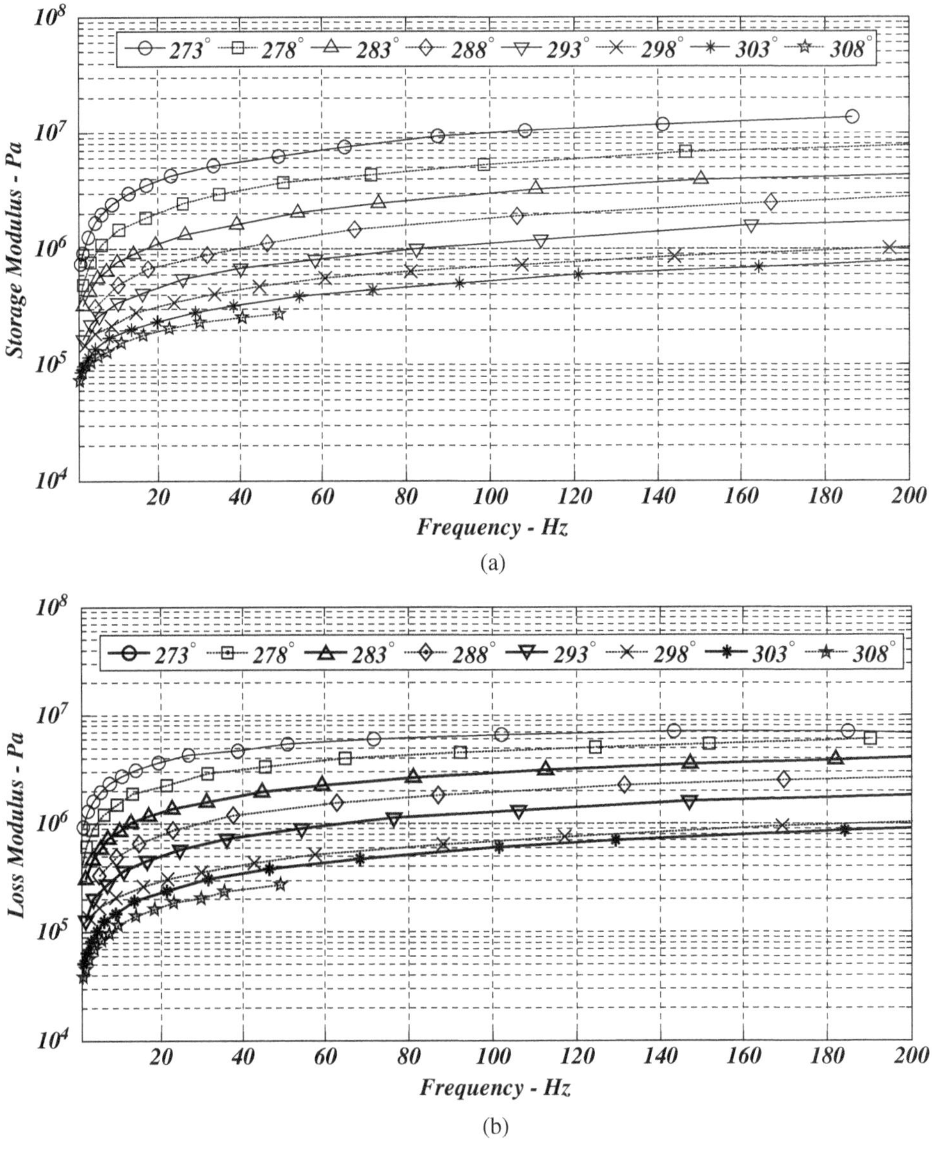

Figure 3.16 Storage and loss modulus of ISD-112.

3.5 Time-Domain Measurement Techniques of the Dynamic Properties of Viscoelastic Materials

Time-domain measurement techniques are equally as important as frequency domain measurements. Three time-domain techniques are presented in this section. These techniques include creep and relaxation measurements, the Hopkinson bar approach, and the wave propagation method. These techniques are either very slow, such as the creep

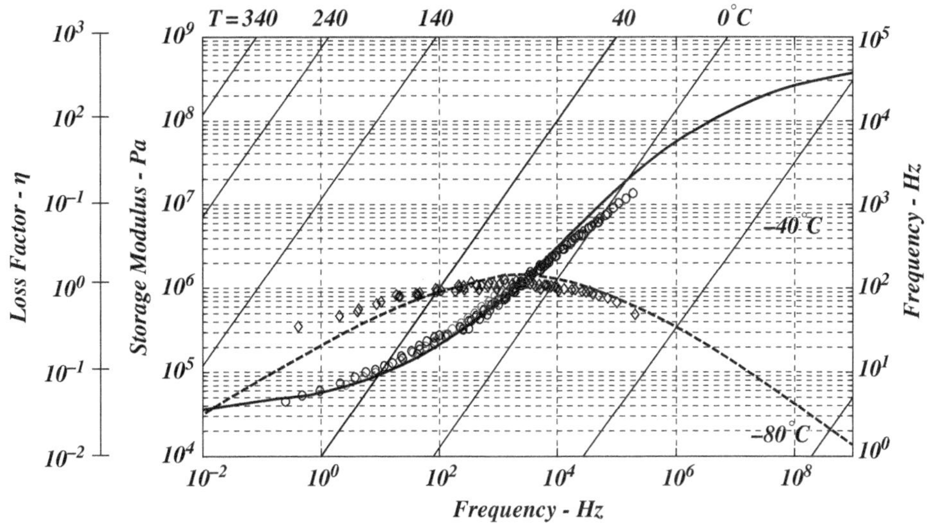

Figure 3.17 Master curves (—— storage modulus, ········· loss factor).

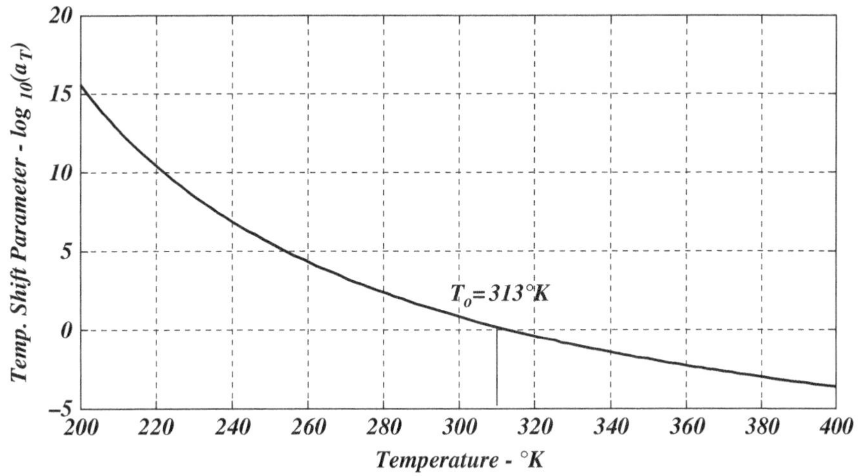

Figure 3.18 Temperature shift factor.

and relaxation methods, or fairly fast, such as the Hopkinson bar approach and the wave propagation method. Figure 3.20 indicates the spread of these techniques over the entire spectrum of the characterization methods.

3.5.1 Creep and Relaxation Measurement Methods

3.5.1.1 Testing Equipment

The creep and relaxation methods are by far the most commonly used and well-studied methods of characterization of the VEM properties. These methods rely on the use of

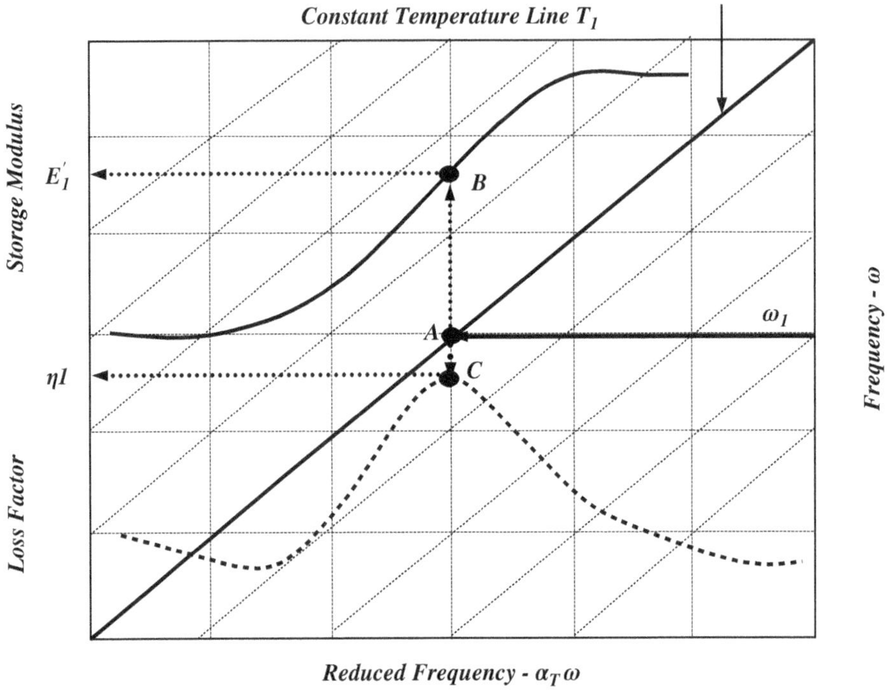

Figure 3.19 Extracting storage modulus and loss factor from the master curve.

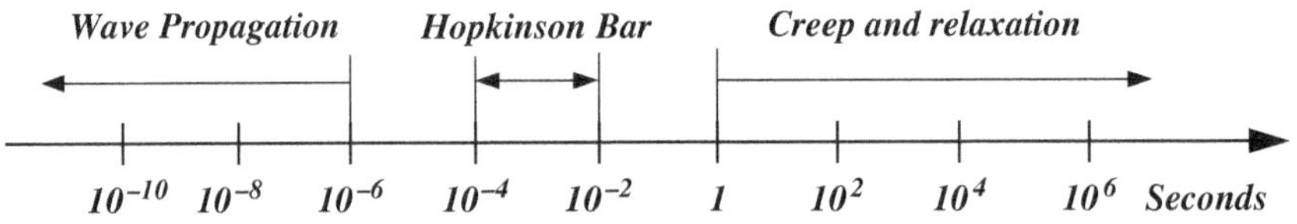

Figure 3.20 Spectrum of characterization methods (Ward 1983). Reproduced with permission of John Wiley & Sons.

conventional testing equipment such as the MTS (www.mts.com) and Instron (www.instron.us) universal machines. Other dedicated equipment such as TA Instruments (model DMA–QM800, New Castle, DE) or BOSE, Electro-Force Systems Group (model ELF 3200, Eden Prairie, MN) as shown in Figure 3.21. The equipment can be programmed to implement any desired creep or relaxation testing profiles while holding the VEM specimens in tension or shear configurations inside temperature-controlled environmental chambers.

3.5.1.2 Typical Creep and Relaxation Behavior

Figure 3.22 shows typical behavior of VEM undergoing creep and relaxation testing as measured at a constant ambient temperature. In Figure 3.22a, the material is subjected to a sudden stress σ_0 and the resulting creep strain $\varepsilon(t)$ is recorded as function of time. The

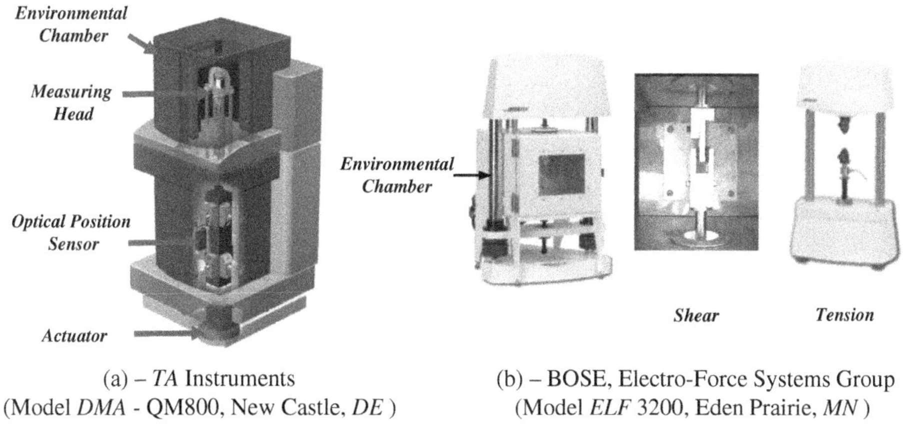

Figure 3.21 Typical creep and relaxation testing equipment. (a) *TA* Instruments (model DMA–QM800, New Castle, DE) and (b) BOSE, Electro-Force Systems Group (model ELF 3200, Eden Prairie, MN).

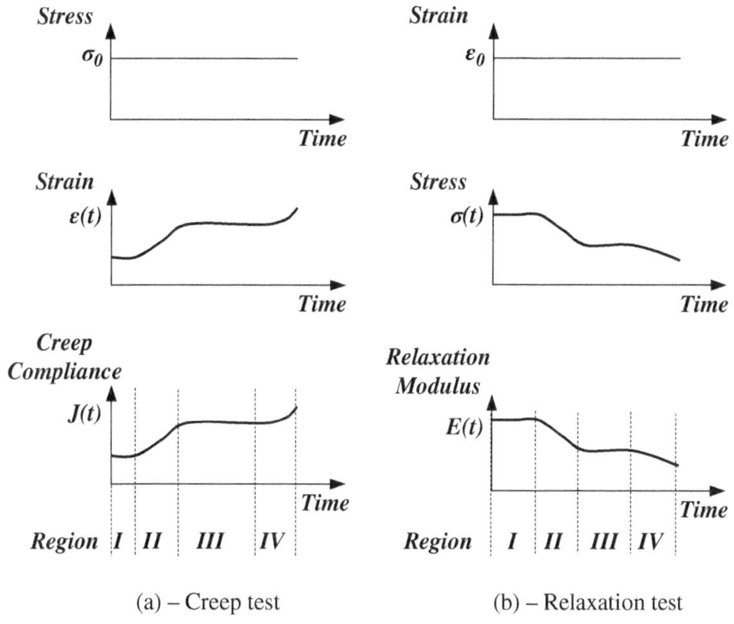

Figure 3.22 Typical creep and relaxation behavior (Region I = glassy, II = viscoelastic, III = rubbery, IV = flow). (a) Creep test and (b) relaxation test.

creep compliance $J(t) = \varepsilon(t)/\sigma_0$ is computed and displayed as function of time. Similarly, Figure 3.22b shows the VEM subjected to a sudden strain ε_0 and the resulting time history of the relaxation stress $\sigma(t)$ is recorded. The time history of the relaxation modulus $E(t) = \sigma(t)/\varepsilon_0$ is computed and displayed.

The creep compliance and relaxation modulus characteristics shown in Figure 3.22 define four clear regions that are: the glassy (I), the viscoelastic (II), the rubbery (III), and the flow (IV) regions.

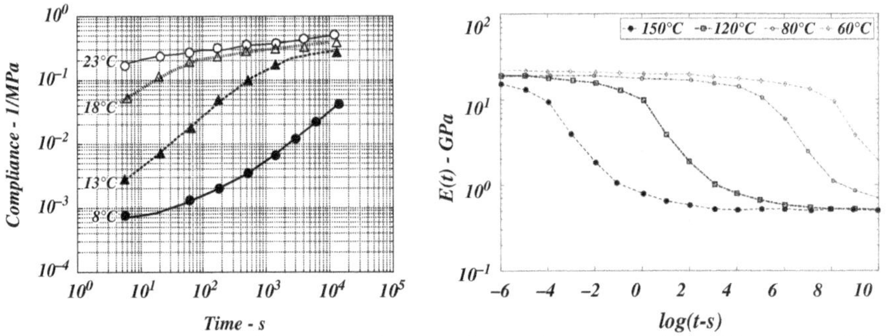

(a) – Creep compliance (Chen *et al.*, 1999) (b) – Relaxation modulus (Chae *et al.*, 2010)

Figure 3.23 Typical creep and relaxation behavior at different temperatures. (a) Creep compliance (Chen et al. 1999) and (b) relaxation modulus (Chae et al. 2010).

Figure 3.23a,b displays the creep compliance and relaxation modulus characteristics for two VEMs at different temperatures.

3.5.1.3 Time-Temperature Superposition

The time-temperature superposition (TTS) principle is very similar to the temperature-frequency superposition principle discussed in Section 3.4.1 to generate the master curves for the complex modulus as function of the reduced frequency and temperature.

In this section, a similar procedure is followed to exploit the TTS principle to generate the master curves for creep compliance or the relaxation modulus as functions of reduced time and temperature.

Usually, the creep and relaxation behavior of VEM are measured over a short period of time and a narrow temperature range. Typically, data are taken at higher temperatures over a fast time scale and the TTS is utilized to predict the behavior at lower temperatures and a much slower time scale.

Consider the relaxation test results shown in Figure 3.24a, which are measured over the time span t_1 and t_2 for three isotherms (T_{-1}, T_0, T_1) such that $T_{-1} < T_0 < T_1$ where T_0 is a reference temperature. A temperature shift factor α_T is applied to shift each isotherm along the time axis until it partially overlaps the reference isotherm. This process is continued with all the isotherms, higher and lower than the reference temperature, to generate the master curve shown in Figure 3.24b.

The temperature shift factor has its origin rooted in the kinetic theory of polymers. For example, according to the Rouse equation (Brinson and Brinson 2008), the relaxation time τ_T, at any temperature T, is given by:

$$\tau_T = Cc_{d_T}/\rho_T T \tag{3.29}$$

where C = constant, c_{d_T} = viscosity, ρ_T = density, and T = absolute temperature.

For a reference temperature T_0, Eq. (3.29) becomes:

$$\tau_0 = Cc_{d_0}/\rho_0 T_0 \tag{3.30}$$

Combining Eqs. (3.29) and (3.30) gives:

$$\tau_T/\tau_0 = (\rho_T T/\rho_0 T_0)c_{d_T}/c_{d_0} \tag{3.31}$$

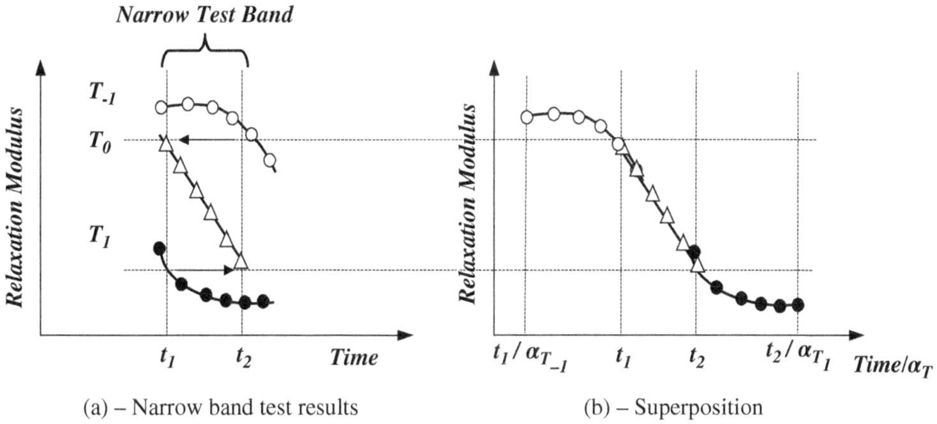

Figure 3.24 Time-temperature superposition applied to relaxation test results. (a) Narrow band test results and (b) superposition.

Let the temperature shift factor α_T be defined as:

$$\alpha_T = \tau_T / \tau_0 \tag{3.32}$$

Then,

$$\alpha_T = (\rho_T T / \rho_0 T_0) c_{d_T} / c_{d_0} \tag{3.33}$$

Note that the multiplier $(\rho_T T/\rho_0 T_0)$, which is commonly called the "vertical shift multiplier," is usually very close to 1 and Eq. (3.33) reduces to:

$$\alpha_T \cong c_{d_T} / c_{d_0} \tag{3.34}$$

Equation (3.34) is exactly the same as Eq. (3.24), so then following the same approach adopted in Section 3.4.1 and using Eqs. (3.22) through (3.25) yields the following WLF temperature shift factor:

$$\log_{10}(\alpha_T) = -\frac{C_1(T - T_g)}{C_2 + (T - T_g)} \tag{3.35}$$

To gain more insight in the physical meaning of the temperature shift factor, consider the relaxation response of the Maxwell model given by Eq. (2.17):

$$\sigma = E_s e^{-t/\lambda} \varepsilon_0 \tag{3.36}$$

Equation (3.36) suggests that the Maxwell model has a relaxation modulus $E(t, T)$, at any temperature T, given by:

$$E(t, T) = E_s e^{-t/\tau_T} \tag{3.37}$$

Combining Eqs. (3.32) and (3.37) gives:

$$E(t, T) = E_s e^{-t/\alpha_T \tau_0} \tag{3.38}$$

Define a "reduced time" t' such that:

$$t' = t / \alpha_T \tag{3.39}$$

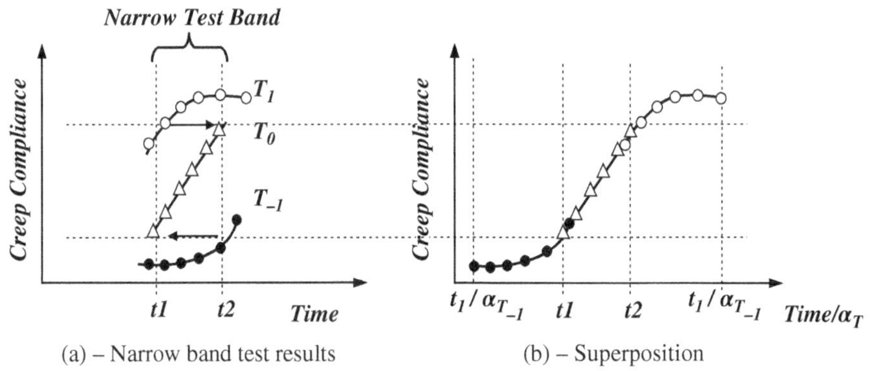

Figure 3.25 Time-temperature superposition applied to creep test results. (a) Narrow band test results and (b) superposition.

Then Eq. (3.38) reduces to:

$$E(t,T) = E_s e^{-t'/\tau_0} = E(t',T_0) \tag{3.40}$$

Equation (3.40) expresses mathematically the "principle of time-temperature equivalence," which indicates that the effect of changing the temperature is equivalent to shifting the time scale by applying a multiplicative factor $1/\alpha_T$. Note that the WLF formula, Eq. (3.35), implies that the shift factor $\alpha_T > 1$ for $T < T_g$, $\alpha_T = 1$ for $T = T_g$, and $\alpha_T < 1$ for $T > T_g$. If $T_g = T_0$, then the reduced time t' will be less than t for $T < T_0$, implying shifting backward on the time scale as can be seen in Figure 3.34b for the isotherm T_{-1}. Similarly, for $T > T_0$ the reduced time t' will be greater than t, which suggests a shift forward on the time scale as indicated in Figure 3.34b for the isotherm T_1.

In a similar manner, the temperature shift factor can be applied to the narrow band results, shown in Figure 3.25a, which are obtained from a creep test. This results in the master curve shown in Figure 3.25b.

Note that the reduced time is still given by t/α_T resulting in shift forward for temperatures $>T_0$ and shift forward for temperatures $<T_0$.

3.5.1.4 Boltzmann Superposition Principle

The Boltzmann superposition principle can be easily explained by considering the time histories of the creep stress applied to a VEM along with the resulting creep strains shown in Figure 3.26.

Mathematically, the strain ε_1 due to stress σ_1 is given by:

$$\varepsilon_1(t) = (\sigma_1 - 0)J(t - t_1) \tag{3.41}$$

where $J(t)$ = creep compliance. In here, it is assumed that the compliance is only function of time and not stress, which is true for linear viscoelasticity.

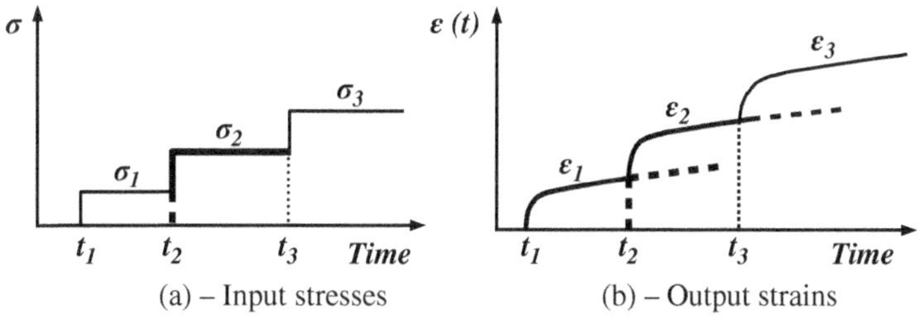

Figure 3.26 Superposition of the creep strains resulting from stepped creep stresses. (a) Input stresses and (b) output strains.

For example, for the Kelvin–Voigt model, Eq. (2.14) indicates that its creep response is given by:

$$\varepsilon = \frac{1}{E_s}\left[1 - e^{-t/\lambda}\right]\sigma_0 \tag{3.42}$$

that is, it has a creep compliance given by:

$$J(t) = \frac{1}{E_s}\left[1 - e^{-t/\lambda}\right] \tag{3.43}$$

When the stress increases to σ_2, the strain ε_2 becomes:

$$\varepsilon_2(t) = (\sigma_2 - \sigma_1)J(t - t_2) \tag{3.44}$$

Accordingly, for the general creep loading profile shown in Figure 3.26, the resulting creep response is:

$$\varepsilon(t) = (\sigma_1 - 0)J(t - t_1) + (\sigma_2 - \sigma_1)J(t - t_2) + (\sigma_3 - \sigma_2)J(t - t_2) + \ldots.. \tag{3.45}$$

Equation (3.45) can be written in the following compact form:

$$\varepsilon(t) = \sum_{i=-\infty}^{n} \Delta\sigma_i J(t - t_i) \tag{3.46}$$

Equation (3.46) is a mathematical representation of the discrete Boltzmann superposition principle whereby the time history of the strain is described by a superposition of the effects of the incremental stresses imposed on the VEM.

In the continuous time-domain, Eq. (3.46) becomes:

$$\varepsilon(t) = \int_{-\infty}^{t} \frac{\partial\sigma(\tau)}{\partial\tau} J(t - \tau)\, d\tau \tag{3.47}$$

In a similar manner, the effect of time varying relaxation strains on the time history of the resulting stress can be illustrated as shown in Figure 3.27.

Mathematically, the superposition of the stresses can be described, in terms of the incremental strains and the relaxation modulus $E(t)$ by:

$$\sigma(t) = \int_{-\infty}^{t} \frac{\partial\varepsilon(\tau)}{\partial\tau} E(t - \tau)\, d\tau \tag{3.48}$$

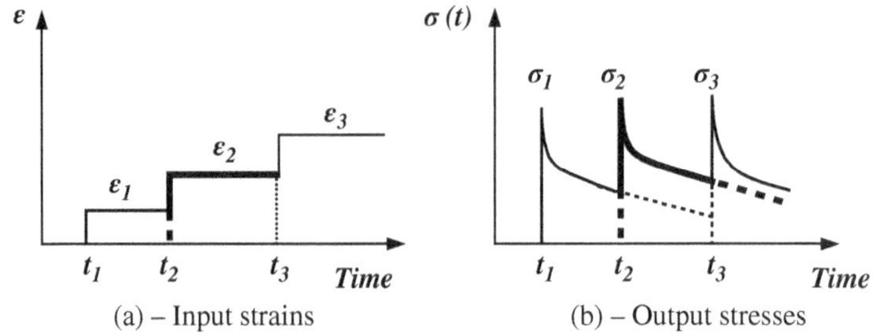

Figure 3.27 Superposition of the relaxation stresses resulting from stepped relaxation strains. (a) Input and (b) output stresses.

3.5.1.5 Relationship Between the Relaxation Modulus and Complex Modulus

This relationship can be established by substituting $a = t - \tau$ into Eq. (3.48), it becomes:

$$\sigma(t) = -\int_0^\infty \frac{\partial\varepsilon(t-a)}{\partial a} E(a)\, da \tag{3.49}$$

If the VEM is subjected to sinusoidal excitation such that:

$$\varepsilon(t) = \varepsilon_0 e^{i\omega t} \tag{3.50}$$

Combining Eqs. (3.49) and (3.50) gives:

$$\begin{aligned}\sigma(t) &= \int_0^\infty \varepsilon_0 \omega i e^{i\omega(t-a)} E(a)\, da \\ &= \varepsilon_0 e^{i\omega t}\int_0^\infty \omega i e^{-i\omega a} E(a)\, da \\ &= \varepsilon \int_0^\infty \omega[\sin(\omega a) + i\cos(\omega a)]\, E(a)\, da \\ &= [E' + iE'']\varepsilon\end{aligned} \tag{3.51}$$

where

$$E' = \int_0^\infty \omega \sin(\omega a) E(a)\, da \quad \text{and} \quad E'' = \int_0^\infty \omega\cos(\omega a) E(a)\, da \tag{3.52}$$

Equation (3.52) establishes how the storage and loss moduli are related to the relaxation modulus.

It is important here to note that as:

$$\sigma(t) = [E' + iE'']\varepsilon \tag{3.53}$$

Then, from Eqs. (3.50) and (3.53),

$$\begin{aligned}\sigma(t) &= E'(1+i\eta)\varepsilon_0 e^{i\omega t}\\ &= E'\sqrt{1+\eta^2}\varepsilon_0 e^{i(\omega t+\delta)}\\ &= |E^*|\varepsilon_0 e^{i(\omega t+\delta)} = \sigma_0 e^{i(\omega t+\delta)}\end{aligned} \tag{3.54}$$

where η = loss factor = $\tan\delta$, $|E^*| = E'\sqrt{1+\eta^2}$ = magnitude of the modulus, and $\sigma_0 = |E^*|\varepsilon_0$.

Equation (3.54) indicates that the strain lags the stress by a phase angle δ. This conforms to the *Causality Principle.*

Example 3.2 Using the relaxation modulus of Maxwell model $E(t) = E_s e^{-t/\lambda}$, determine the storage and loss moduli.

Solution

From Eq. (3.52),

$$E' = \int_0^\infty \omega \sin(\omega a) E(a)\, da$$

$$\text{that is}, E' = E_s\omega \int_0^\infty \sin(\omega a) e^{-a/\lambda}\, da = E_s \frac{(\omega\lambda)^2}{1+(\omega\lambda)^2}$$

$$\text{Also}, E'' = \int_0^\infty \omega \cos(\omega a) E(a)\, da$$

$$\text{that is}, E'' = E_s\omega \int_0^\infty \cos(\omega a)\, e^{-a/\lambda}\, da = E_s \frac{(\omega\lambda)}{1+(\omega\lambda)^2}$$

Accordingly, the loss factor η is:

$$\eta = \frac{1}{(\omega\lambda)}$$

These values match those listed in Table 2.6.

3.5.1.6 Relationship Between the Creep Compliance and Complex Compliance

This relationship can be established by integrating by parts Eq. (3.47) to yield:

$$\varepsilon(t) = \sigma(t)J(0) - \int_{-\infty}^{t} \sigma(\tau) \frac{\partial J(t-\tau)}{\partial \tau}\, d\tau \tag{3.55}$$

where $J(0)$ is the static compliance. Let $a = t - \tau$, then Eq. (3.55) reduces to:

$$\varepsilon(t) = \sigma(t)J(0) + \int_0^\infty \sigma(t-a)\frac{\partial J(a)}{\partial a}\,da \tag{3.56}$$

For sinusoidal stress such that:

$$\sigma(t) = \sigma_0 e^{i\omega t} \tag{3.57}$$

Equation (3.56) reduces to:

$$\begin{aligned}\varepsilon(t) &= \sigma(t)J(0) + \int_0^\infty \sigma_0 e^{i\omega(t-a)}\frac{\partial J(a)}{\partial a}\,da \\ &= \sigma(t)J(0) + \sigma\int_0^\infty [\cos(\omega a) - i\sin(\omega a)]\frac{\partial J(a)}{\partial a}\,da \\ &= (J' - iJ'')\sigma\end{aligned} \tag{3.58}$$

where

$$J' = J(0) + \int_0^\infty \cos(\omega a)\frac{\partial J(a)}{\partial a}\,da \quad \text{and} \quad J'' = \int_0^\infty \sin(\omega a)\frac{\partial J(a)}{\partial a}\,da \tag{3.59}$$

Therefore, the complex compliance is given by $J^* = (J' - iJ'')$. The importance of the negative sign in the complex compliance can be easily seen when $\varepsilon(t) = (J' - iJ'')\sigma$ is combined with Eq. (3.57) to yield,

$$\varepsilon(t) = \varepsilon_0 e^{i(\omega t - \delta)} \tag{3.60}$$

Hence, the strain lags the stress by a phase angle δ and this conforms to the *Causality Principle.*

Example 3.3 Using the creep compliance of Kelvin–Voigt model $J(t) = \frac{1}{E_s}\left[1 - e^{-t/\lambda}\right]$, determine the storage and loss moduli.

Solution

From Eq. (3.59) with $J(0) = 0$,

$$J' = \int_0^\infty \cos(\omega a)\frac{\partial J(a)}{\partial a}\,da$$

$$\text{i.e.}J' = \frac{1}{E_s\lambda}\int_0^\infty \cos(\omega a)e^{-a/\lambda}\,da = \frac{1}{E_s}\frac{1}{1+(\omega\lambda)^2}$$

$$\text{i.e.}J'' = \int_0^\infty \sin(\omega a)\frac{\partial J(a)}{\partial a}\,da \quad \text{or} \quad J'' = \frac{1}{E_s\lambda}\int_0^\infty \sin(\omega a)\,e^{-a/\lambda}\,da = \frac{1}{E_s}\frac{\omega\lambda}{1+(\omega\lambda)^2}$$

Accordingly, the loss factor η is:

$$\eta = J''/J' = \omega\lambda$$

The complex modulus can be extracted from the complex compliance as follows:

$$J^* = J' - iJ'' = \frac{1}{E_s}\frac{1 - i(\omega\lambda)}{1 + (\omega\lambda)^2} = \frac{1}{E_s}\frac{1}{\sqrt{1 + (\omega\lambda)^2}}e^{-i\delta}$$

$$\text{Then}, E^* = 1/J^* = E_s\sqrt{1 + (\omega\lambda)^2}e^{i\delta} = E_s(1 + i\omega\lambda),$$

which yields, $E' = E_s$, $E'' = \omega\lambda E_s$, *and* $\eta = \omega\lambda$
These values match those listed in Table 2.6.

3.5.1.7 Relationship Between the Creep Compliance and Relaxation Modulus

As Eq. (3.47) gives:

$$\varepsilon(t) = \int_0^t \frac{\partial\sigma(\tau)}{\partial\tau} J(t-\tau)\, d\tau$$

Then, applying the convolution theorem, listed in the Appendix, yields:

$$\varepsilon(s) = s\,J(s)\,\sigma(s) \tag{3.61}$$

Similarly as Eq. (3.48) gives:

$$\sigma(t) = \int_0^t \frac{\partial\varepsilon(\tau)}{\partial\tau} E(t-\tau)\, d\tau \tag{3.62}$$

Applying the convolution theorem gives:

$$\sigma(s) = s\,E(s)\,\varepsilon(s) \tag{3.63}$$

From Eqs. (3.61) and (3.63),

$$E(s)\,J(s) = 1/s^2 \tag{3.64}$$

Then, the convolution theorem gives:

$$\int_0^t E(\tau).J(t-\tau)\, d\tau = t$$

This implies that, for VEMs:

$$E(t) \neq 1/J(t) \tag{3.65}$$

which is unlike elastic materials where $E(t) = 1/J(t)$.

3.5.1.8 Alternative Relationship Between the Creep Compliance and Complex Compliance

As Eq. (3.61) gives:

$$\varepsilon(s) = s\,J(s)\,\sigma(s) \tag{3.66}$$

While as the complex compliance J^* is defined as:

$$\varepsilon(s) = J^*(s)\,\sigma(s) \tag{3.67}$$

Then, Eqs. (3.66) and (3.67) yield:

$$J^*(s) = sJ(s) \tag{3.68}$$

In other words,

$$J^*(s) = sJ(s) = s\int_0^\infty J(t)\,e^{-st}\,dt \tag{3.69}$$

For sinusoidal motions at frequency ω, Eq. (3.69) becomes:

$$J^*(i\omega) = i\omega\int_0^\infty J(t)\,e^{-i\omega t}\,dt \tag{3.70}$$

that is, the complex compliance is the Fourier transform of the creep compliance.

Example 3.4 For Kelvin–Voigt model show that the complex compliance is the Fourier transform of the creep compliance $J(t) = (1/E_s)(1 - e^{-t/\lambda})$.

Solution

Using MATLAB command "fourier" as follows:

```
>> syms t lam w Es
>> fourier (1/Es* (1-1/exp(t/lam)) *heaviside(t),w) *i*w
ans = - (i*w* (i/w + 1/(i*w + 1/lam))) /Es
```

giving:

$$J^* = \frac{1}{E_s}\left[\frac{1}{1+\omega^2\lambda^2} - i\frac{\omega\lambda}{1+\omega^2\lambda^2}\right]$$

This result matches the results obtained in Example 3.3.

3.5.1.9 Alternative Relationship Between the Relaxation Modulus and Complex Modulus

As Eq. (3.62) gives:

$$\sigma(s) = s\,E(s)\varepsilon(s) \tag{3.71}$$

But, the complex modulus E^* is defined as:

$$\sigma(s) = E^*(s)\,\varepsilon(s) \tag{3.72}$$

Combining Eqs. (3.71) and (3.72) results in:

$$E^*(s) = sE(s) \tag{3.73}$$

From the definition of the Laplace transform, Eq. (3.73) can be written as:

$$E^*(s) = sE(s) = s\int_0^\infty E(t)\,e^{-st}dt \tag{3.74}$$

For sinusoidal motions at frequency ω, Eq. (3.69) becomes:

$$E^*(i\omega) = i\omega\int_0^\infty E(t)\,e^{-i\omega t}dt \tag{3.75}$$

that is, the complex modulus is the Fourier transform of the relaxation modulus.

Example 3.5 For Maxwell model show that the complex modulus is the Fourier transform of the relaxation modulus $E(t) = E_s e^{-t/\lambda}$.

Solution

Using MATLAB command "fourier" as follows:

```
>> syms t lam w Es
>> fourier(Es/exp(t/lam)*heaviside(t),w)*i*w
 ans = (Es*i*w)/(i*w + 1/lam)
```

This gives: $E^* = E_S\left[\dfrac{\omega^2\lambda^2}{1+\omega^2\lambda^2} + i\dfrac{\omega\lambda}{1+\omega^2\lambda^2}\right]$.

Hence, the outcome matches the results obtained in Example 3.2.

3.5.1.10 Summary of the Basic Interconversion Relationship

In Section 3.5.1, the creep and relaxation methods of characterization of the damping properties of VEM have been described. The outputs of these characterization methods are the creep compliance and relaxation modulus. Both of these two parameters are time-domain parameters. The relationship between them is outlined in Section 3.5.1.7. Therefore, if creep tests are carried out, the relaxation modulus can be extracted using Eq. (3.64). The same equation can be used to extract the creep compliance if a relaxation tests are conducted (see Problem 3.5). Accordingly, either one of the two tests is sufficient to characterize the VEM. Hence, there is no need to carry out both tests.

In Section 3.5.1.5, the relationship between the relaxation modulus and the complex modulus are established by Eq. (3.52). Hence, characterization of the properties can be carried out either through relaxation tests or through sinusoidal tests, which are described in Section 3.4.

Similarly, in Section 3.5.1.6, the relationship between the creep compliance and the complex compliance or modulus is described by Eq. (3.59). Hence, characterization of the properties can be carried out either through creep tests or through sinusoidal tests.

Figure 3.28 summarizes these interconversion relationships. A more comprehensive study of the interconversion relationships are given, for example, by Gross (1953), Brinson and Brinson (2008), and Anderssen et al. (2008).

3.5.1.11 Practical Issues in Implementation of Interconversion Relationships

In this section, two approaches will be described to facilitate the implementation of the interconversion from time and frequency domain parameters. These approaches aim at avoiding the problems associated with the use of Fourier transform that results in inaccuracies, particularly if the time record is short. Other problems arise from the fact that the Fourier transform for the creep compliance, for example, is not a convergent integral since the compliance increases with increasing time.

Schwarzl (1970), Park and Schapery (1999), Schapery and Park (1999), Parot and Duperray (2008), and Evans et al. (2009) summarize several of the approximate and exact approaches for handling the interconversion problems in a practical manner. In here, two approaches will be outlined because of their utility and practicality. These approaches are presented in the following.

3.5.1.11.1 Indirect Conversion Method

This conversion approach is a well-established approach for obtaining frequency-dependent dynamic moduli from creep or relaxation measurements. In this approach, the experimental data are curve-fitted to a particular time-domain model, such as the generalized Maxwell model (GMM) or the generalized Kelvin–Voigt model, which are both shown in Figure 3.29. Subsequently, the resulting curve-fitted time-domain models are transformed to the frequency domain using Fourier transform in order to extract the complex viscoelastic modulus.

Following the procedures outlined in Chapter 2, it can be easily shown that these models are represented mathematically by the following Prony series:

$$E(t) = E_0 + \sum_{i=1}^{n} E_i e^{-t/\rho_i} \quad \textit{generalized Maxwell model} \tag{3.76}$$

and

$$J(t) = J_0 + \sum_{j=1}^{m} J_j \left(1 - e^{-t/\tau_j}\right) \quad \textit{generalized Kelvin – Voigt model} \tag{3.77}$$

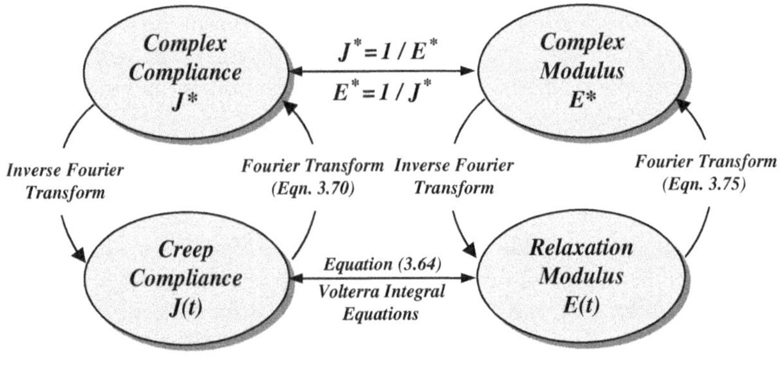

Figure 3.28 Interconversion relationships between different characterization methods.

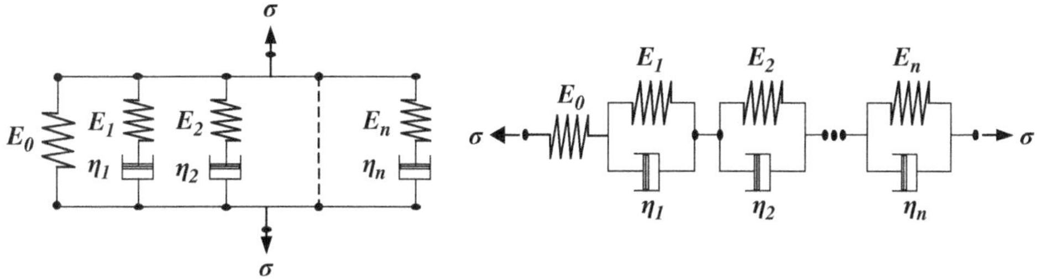

(a) Generalized Maxwell model (b) Generalized Kelvin-Voigt model

Figure 3.29 Generalized VEM models. (a) Generalized Maxwell model and (b) generalized Kelvin–Voigt model.

where E_0 = the equilibrium modulus, E_i = relaxation strength, and ρ_i = relaxation times = η_i/E_i. Also, J_0 = the glassy compliance, J_j = retardation strength, and τ_j = retardation times such that $J_0 = 1/E_0$, $J_j = 1/E_j$, $\tau_j = \eta_j/E_j$. The parameters E_0, E_i, ρ_i, J_0, J_j, and τ_j are all positive constants and are identified in order to optimally curve-fit the experimental data.

Note that the GMM will be used to fit relaxation modulus data and the generalized Kelvin–Voigt model will be employed to fit creep compliance data.

Once these parameters are obtained, Eqs. (3.76) and (3.77) are transformed to the frequency domain using Fourier transform using the approaches described in Examples 3.4 and 3.5 yielding the complex modulus E^* and the complex compliance J^*:

$$E^*(\omega) = E_0 + \sum_{k=1}^{n} E_k \frac{i\omega\rho_k}{i\omega\rho_k + 1} \tag{3.78}$$

and

$$J^*(\omega) = J_0 + \sum_{l=1}^{m} J_l\left(1 - \frac{i\omega\tau_l}{i\omega\tau_l + 1}\right) \tag{3.79}$$

Application of this approach can be best understood by considering the following example.

Example 3.6 If the experimental values of the relaxation modulus, storage modulus, loss modulus, and the loss factor of a VEM are as shown in Figure 3.30, then:

a) find the order and the coefficients of the GMM that best fit the relaxation modulus.
b) apply the Fourier transform on the obtained optimal GMM to extract the storage modulus, loss modulus, and the loss factor of the VEM. Compare the predictions of the GMM with the experimental values.

Solution

Consider the GMM given by Eq. (3.76), or:

$$E(t) = E_0 + \sum_{i=1}^{n} E_i e^{-t/\rho_i}$$

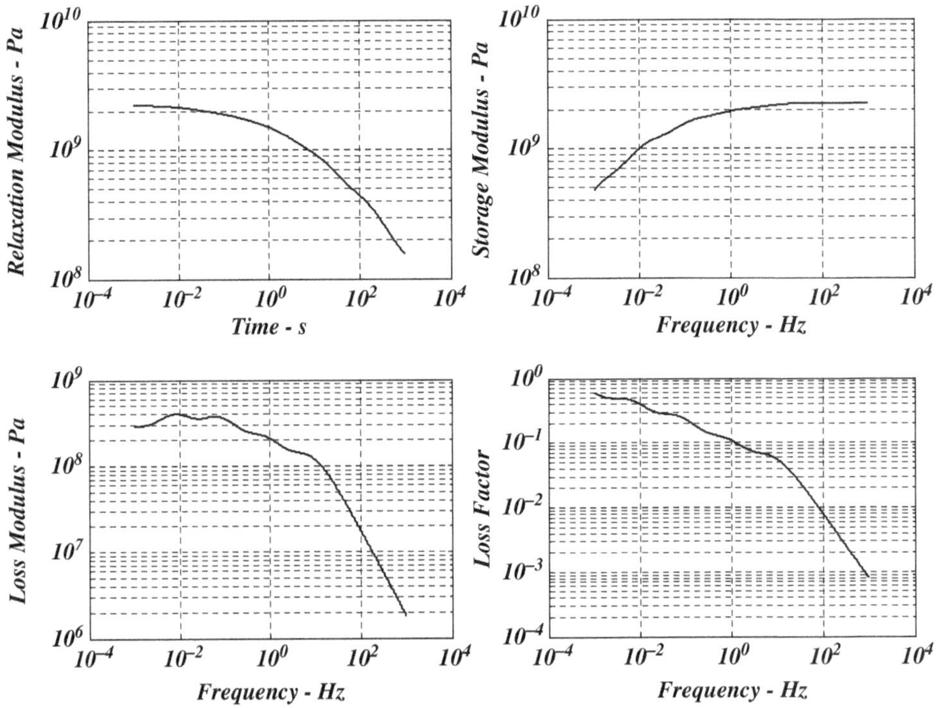

Figure 3.30 Experimental relaxation modulus, complex modulus, and loss factor of a VEM.

For a specific model order n, the parameters: E_0, E_i, and ρ_i are determined in order to replicate closely the experimental behavior of actual VEM. The optimal values of these parameters are selected to minimize the deviations between the predictions of the GMM and the experimental data. The optimization problem is formulated as follows:

$$\left\langle \begin{array}{l} \text{Determine } E_i \text{ and } \rho_i \\ \text{To minimize} F = \sum\limits_{Time} \left(E_{GMM}/E_{Exp} - 1\right)^2 \\ \text{such that } E_i \text{ and } \rho_i > 0 \;\; \text{For } i = 1,..,n \\ \text{and } E_0 + \sum\limits_{i=1}^{n} E_i = E(t=0) \end{array} \right\rangle \tag{3.80}$$

where $E(t = 0)$ and E_0 are the relaxation moduli of the VEM at $t = 0$ and ∞, respectively, as shown in Figure 3.31. Also, n denotes the number of the GMM terms.

In Eq. (3.80), the objective function F is formulated in order to minimize the sum of the prediction error of the GMM model of the relaxation modulus as computed per the entire experimental time range. The errors are cast in a normalized and quadratic form in order to make the optimization problem well-conditioned. Furthermore, constraints are imposed to guarantee that the parameters are all nonnegative and that the model can predict the relaxation modulus at $t = 0$ and ∞.

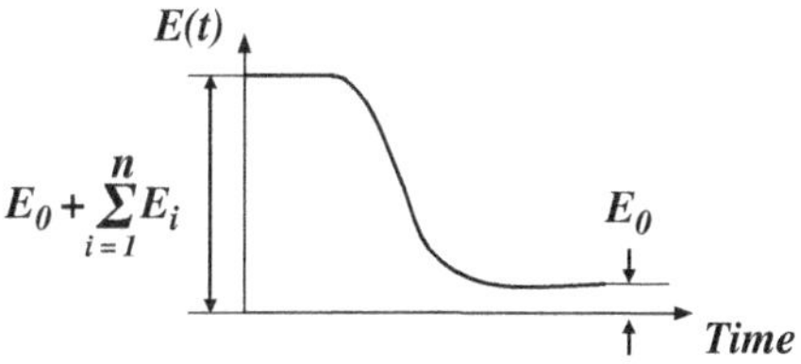

Figure 3.31 Limits of GMM relaxation modulus.

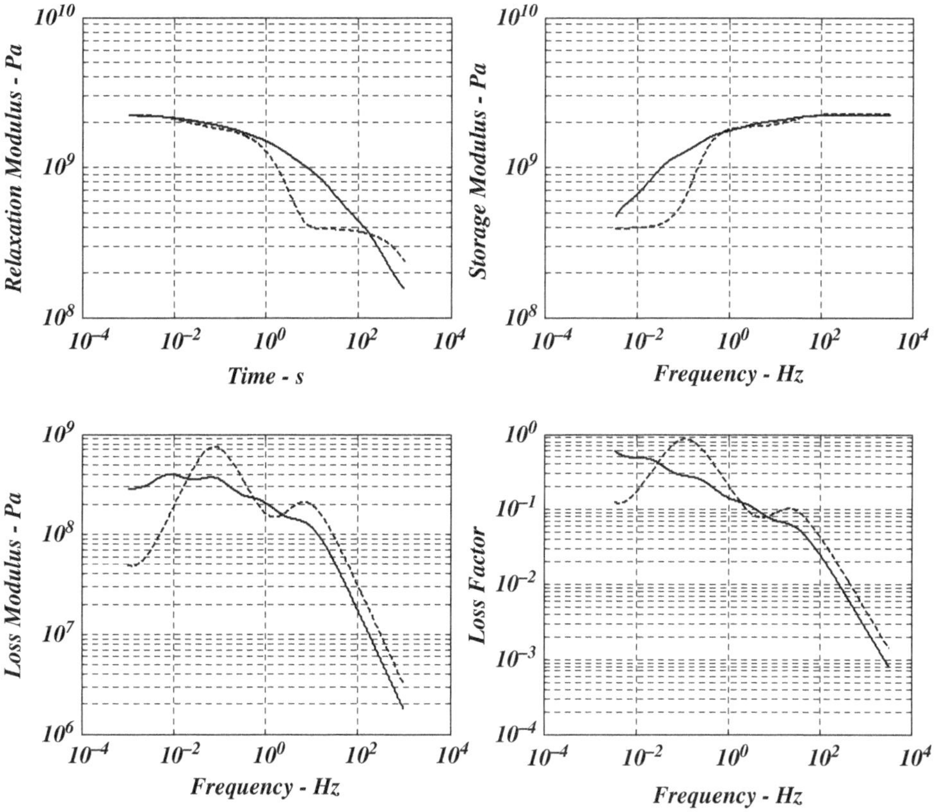

Figure 3.32 Comparisons between experiments and prediction of three-term GMM (—— experimental ············ GMM).

From the experimental data, E_0 = 2.24E6 Pa, and $E(t = 0)$ = 2.28E9 Pa. Then, for n = 3 terms, the relaxation time constants are selected such that: ρ_1, ρ_2, and ρ_3 = [2E-2 2E0 2E3] s to cover the entire experimental time range. Starting with an initial guess of E_1, E_2, and E_3 = [0.3E9 1.4E9 0.3E9] Pa, a MATLAB solution of the optimization problem using the "fmincon" subroutine of the Optimization Toolbox gives optimal values of E_1, E_2, and E_3 = [0.393E9 1.493E9 0.393E9] Pa. The corresponding value of the objective function F = 799.356. Figure 3.32 shows comparisons between the predictions of the GMM

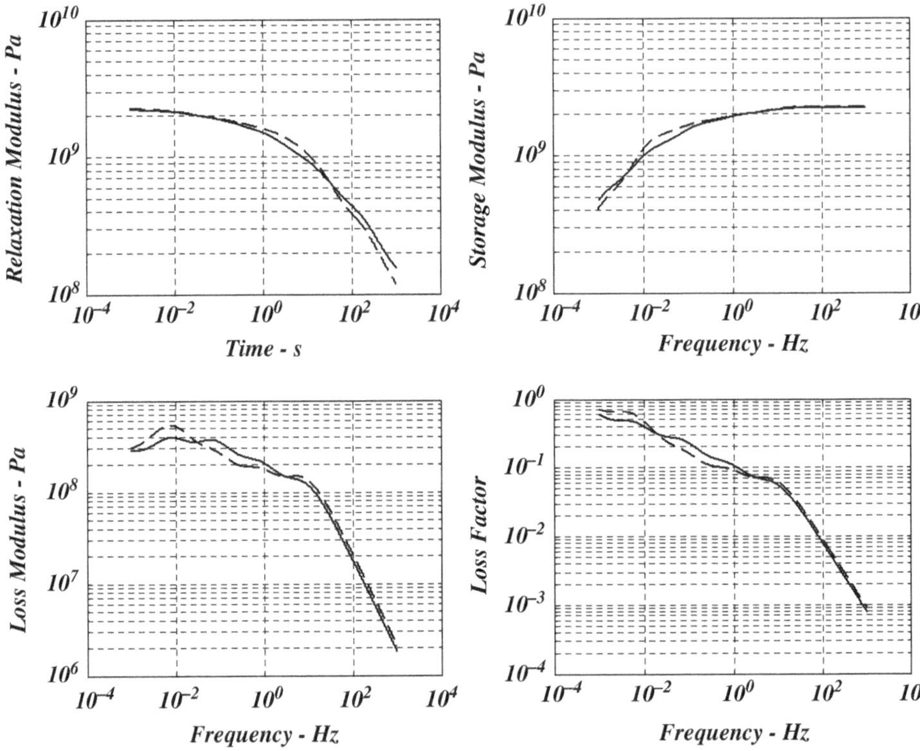

Figure 3.33 Comparisons between experiments and prediction of six-term GMM (—— Experimental .... GMM).

and the experimental values. It is evident that three terms are not enough to replicate the behavior of the VEM.

When n is increased to six terms, the relaxation time constants are selected such that: ρ_1 through ρ_6 = [2E-2 2E-1 2E0 2E1 2E2 2E3] s to cover the entire experimental time range. Starting with an initial guess of E_1 through E_6 = [0.2E9 0.2E9 0.3E9 0.9E9 0.3E9 1.5E9] Pa, the MATLAB Optimization Toolbox gives optimal values of E_1 through E_6 = [0.238E9 0.238E9 0.338E9 0.938E9 0.338E9 0.188E9] Pa. The corresponding value of the objective function F = 89.706. Figure 3.33 shows comparisons between the predictions of the GMM and the experimental values. It is evident that six terms adequately replicate the behavior of the VEM.

3.5.1.11.2 Direct Conversion Method

The indirect conversion method, described here, can be somewhat restrictive, as it requires the use of a prescribed model which may have a very large number of fitting parameters to adequately replicate the behavior of the VEM. It also may artificially hide the experimental noise, making the uncertainties difficult to quantify. Therefore, other conversion methods are considered. Distinct among these methods are those of Evans

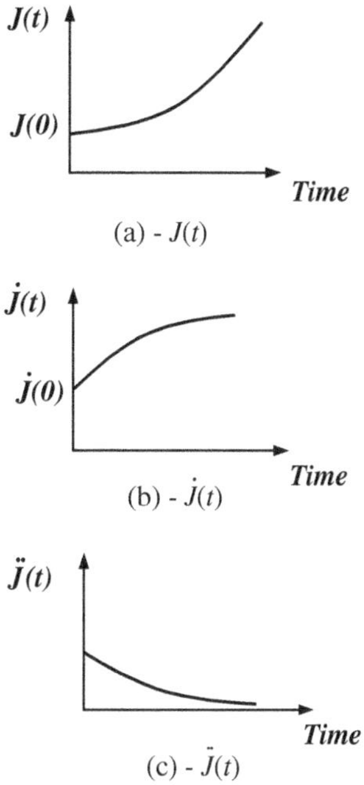

Figure 3.34 Typical creep compliance and its time derivatives.

et al. (2009) and Parot and Duperray (2008). In the first method, the conversion is carried out from the time to the frequency domain whereas in the second method, the creep compliance and relaxation modulus are calculated from complex modulus measurement data.

In here, the method of Evans et al. (2009) is presented whereby the experimental data of the creep compliance is used to extract directly the complex modulus of the VEM without the need for any curve fitting or any subsequent application of the Fourier transform. The method is based on the fact that as the creep compliance $J(t)$ grows with increasing time, makes its Fourier transform $\hat{J}(i\omega)$ a non-convergent integral. Note that as $\hat{J}(i\omega)$ is given by:

$$\hat{J}(i\omega) = \int_0^\infty J(t)\, e^{-i\omega t} dt \tag{3.81}$$

then, it would be inaccurate to calculate the complex modulus from Eq. (3.70) or:

$$E^* = \frac{1}{i\omega \hat{J}(i\omega)} \tag{3.82}$$

To avoid this problem, the Fourier transform is applied to the second derivative of $J(t)$ as it is a converging function as shown in Figure 3.34.

This can be achieved by integrating Eq. (3.81) by parts twice to give:

$$\begin{aligned}\hat{J}(i\omega) &= \int_0^\infty J(t)\,e^{-i\omega t}dt \\ &= -\frac{1}{i\omega}J(t)\,e^{-i\omega t}\Big|_0^\infty + \frac{1}{i\omega}\int_0^\infty \dot{J}(t)\,e^{-i\omega t}dt \\ &= \frac{1}{i\omega}J(0) + \frac{1}{i\omega}\int_0^\infty \dot{J}(t)\,e^{-i\omega t}dt \\ &= \frac{1}{i\omega}J(0) - \frac{1}{\omega^2}\dot{J}(0) - \frac{1}{\omega^2}\int_0^\infty \ddot{J}(t)\,e^{-i\omega t}dt \\ &= -\frac{1}{\omega^2}\left(i\omega J(0) + \dot{J}(0) + \int_0^\infty \ddot{J}(t)\,e^{-i\omega t}dt\right)\end{aligned} \tag{3.83}$$

As the Fourier transform requires $J(t)$ to be defined for values of $0 < t < \infty$ while the experimental data is finite, extrapolation to time $t = \infty$ requires the introduction of an additional parameter, which is the steady-state viscosity η that denotes the reciprocal of the slope of the creep compliance as shown in Figure 3.35.

The effect of the viscosity is accounted for in the description of the creep compliance such that:

$$J_C(t) = J(t) + \frac{t}{\eta} \tag{3.84}$$

Hence, the Fourier transform of $J_C(t)$ is $\hat{J}_C(i\omega)$, which is given by:

$$\hat{J}_C(i\omega) = \int_0^\infty \left(J(t) + \frac{t}{\eta}\right) e^{-i\omega t}dt = \int_0^\infty J(t)e^{-i\omega t}dt + \int_{t_N}^\infty \frac{t}{\eta}e^{-i\omega t}dt = \hat{J}(i\omega) - \frac{1}{\omega^2\eta}e^{-i\omega t_N} \tag{3.85}$$

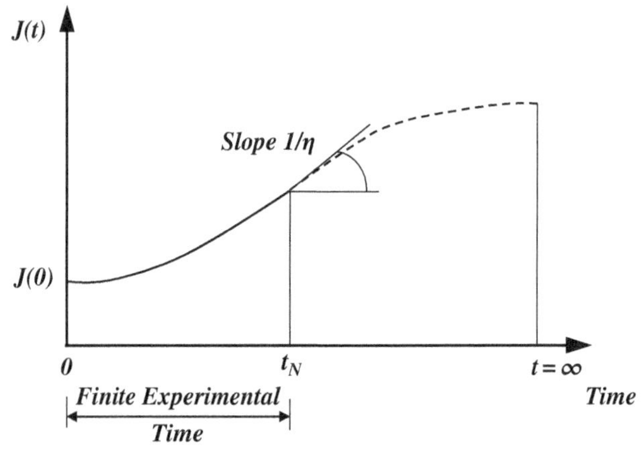

Figure 3.35 Creep compliance measurements over a finite experimental time.

Combining Eqs. (3.83) and (3.85) gives:

$$\hat{J}_C(i\omega) = -\frac{1}{\omega^2}\left(i\omega J(0) + \dot{J}(0) + \int_0^{\infty} \ddot{J}(t)\, e^{-i\omega t} dt + \frac{e^{-i\omega t_N}}{\eta}\right) \tag{3.86}$$

Hence, the complex modulus E^* can be calculated from:

$$E^* = \frac{1}{i\omega \hat{J}_C(i\omega)} \tag{3.87}$$

Substituting Eq. (3.86) into (3.87) gives:

$$E^*(i\omega) = i\omega \Bigg/ \left(i\omega J(0) + \dot{J}(0) + \int_0^{\infty} \ddot{J}(t)\, e^{-i\omega t} dt + \frac{e^{-i\omega t_N}}{\eta}\right) \tag{3.88}$$

Numerically, Eq. (3.88) can be evaluated as follows:

$$E^*(i\omega) = \frac{i\omega}{i\omega J(0) + \left(1 - e^{-i\omega t_1}\right)\frac{J(t_1) - J(0)}{t_1} + \sum_{k=2}^{N} \frac{J(t_k) - J(t_{k-1})}{t_k - t_{k-1}}\left(e^{-i\omega t_{k-1}} - e^{-i\omega t_k}\right) + \frac{e^{-i\omega t_N}}{\eta}} \tag{3.89}$$

Details of implementation of Eq. (3.89) are described by Evans et al. (2009).

Motivation behind augmenting *J(t)* by the viscosity term t/η can be appreciated by considering the creep compliance of the Burgers model of VEM which is in effect a Maxwell model connected in series to a Kelvin–Voigt model as shown in Figure 3.36.

It can be easily shown that the constitutive equation of the model is:

$$\varepsilon(s) = \left[\frac{1}{E_0} + \frac{1}{E_1 + \eta_1 s} + \frac{1}{\eta_0 s}\right]\sigma(s) \tag{3.90}$$

where s = Laplace complex number.

For a step creep stress σ_0, the resulting strain response $\varepsilon(t)$ is:

$$\varepsilon(t) = \left[\frac{1}{E_0} + \frac{1}{E_1}\left(1 - e^{-t/\lambda}\right) + \frac{t}{\eta_0}\right]\sigma_0 \,\text{with}\, \lambda = \eta_1/E_1 \tag{3.91}$$

Hence, the creep compliance is given by:

$$J(t) = \varepsilon(t)/\sigma_0 = \left[\frac{1}{E_0} + \frac{1}{E_1}\left(1 - e^{-t/\lambda}\right) + \frac{t}{\eta_0}\right] \tag{3.92}$$

The last term in the creep compliance matches the second term in Eq. (3.84).

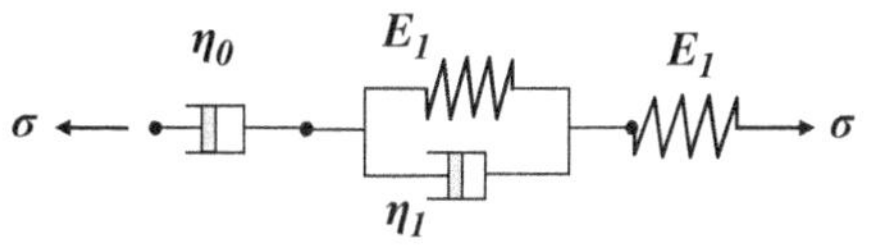

Figure 3.36 Burgers model of VEM.

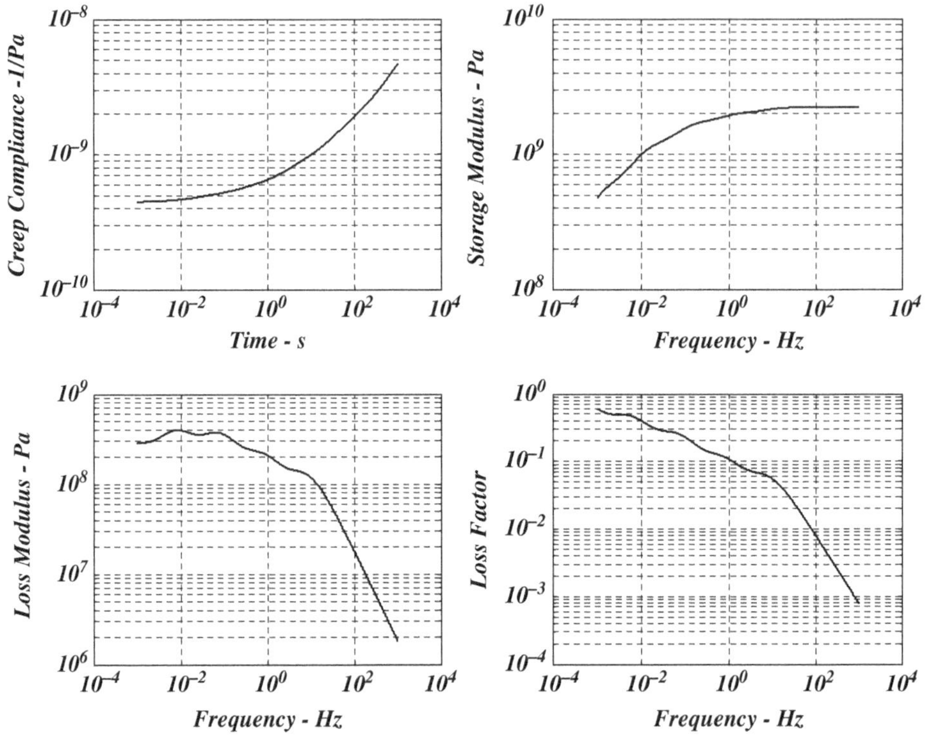

Figure 3.37 Experimental creep compliance, complex modulus, and loss factor of a VEM.

Example 3.7 The experimental values of the creep compliance, storage modulus, loss modulus, and the loss factor of a VEM are shown in Figure 3.37. Use the Evans et al. method to calculate the storage modulus, loss modulus, and the loss factor of the VEM. Compare the predictions of the Evans et al. method with the experimental values.

Solution

Applying Eq. (3.89) on the experimental creep compliance, shown in Figure 3.37, yields the storage modulus, loss modulus, and the loss factor characteristics shown in Figure 3.38.

It is evident that Evans et al. direct method replicates adequately the experimental results particularly in the frequency range between 0.01 and 100 Hz.

3.5.2 Split Hopkinson Pressure Bar Method

3.5.2.1 Overview

The split Hopkinson pressure bar (SHPB) is a widely accepted dynamic testing device for characterization of the properties of VEM at high strain rates reaching strain rates on the order of $10^4\,s^{-1}$. The device is developed by John and Bertram Hopkinson in the early

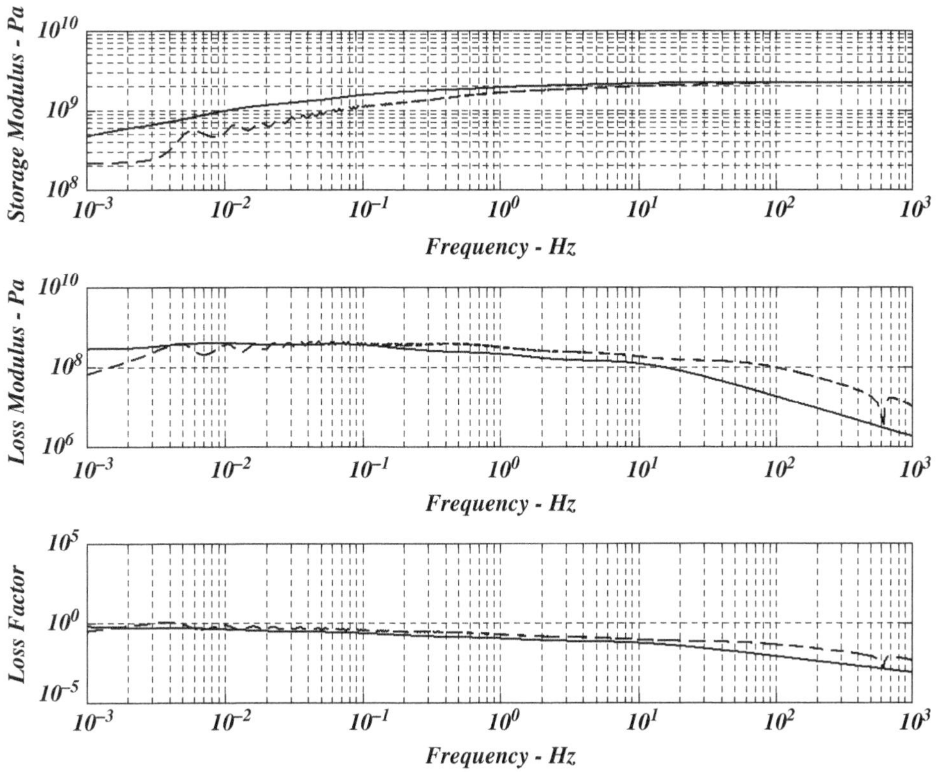

Figure 3.38 Comparisons between experiments and prediction of Evans *et al.* method (—— experimental .... Evans *et al.*).

1900 (Gama et al. 2004). Then, in 1949, Kolsky used the pressure bar technique in determining the dynamic compression stress–strain behavior of different materials.

During a SHPB test, the VEM test specimen is compressed between two bars which are called the incident bar and the transmitter bar as shown in Figure 3.39. The incident bar is then impacted by a striker bar which is launched at high speed using compressed gas. The stress–strain characteristics of the specimen are determined through measurements of resulting stress waves in the two bars as measured by strain gages that are bonded to the incident bar and the transmitter bar.

3.5.2.2 Theory of 1D SHPB

The stress–strain behavior of a VEM tested in a SHPB is based on the same principle of 1D wave propagation (Gray 2000). In this theory, it is assumed that the bars are thin, long, linear, and dispersion free.

Consider the SHPB shown in Figure 3.40. The wave propagation, in either bar, is governed by the following equation:

$$\frac{\partial^2 u}{\partial x^2} - \frac{1}{c^2}\frac{\partial^2 u}{\partial t^2} = 0 \tag{3.93}$$

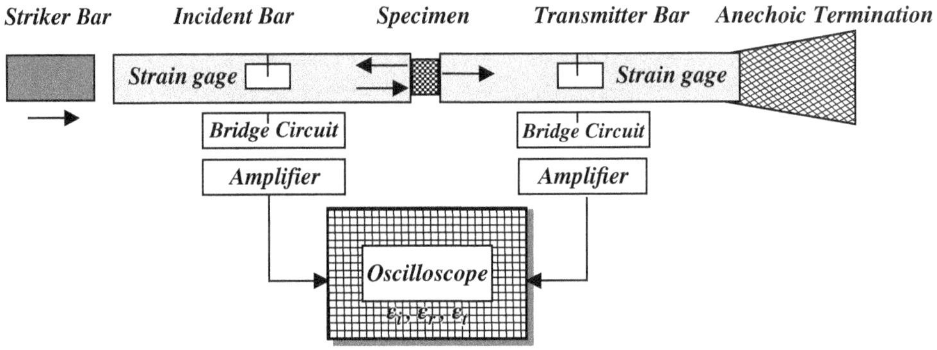

Figure 3.39 Schematic drawing of the split Hopkinson pressure bar (SHPB).

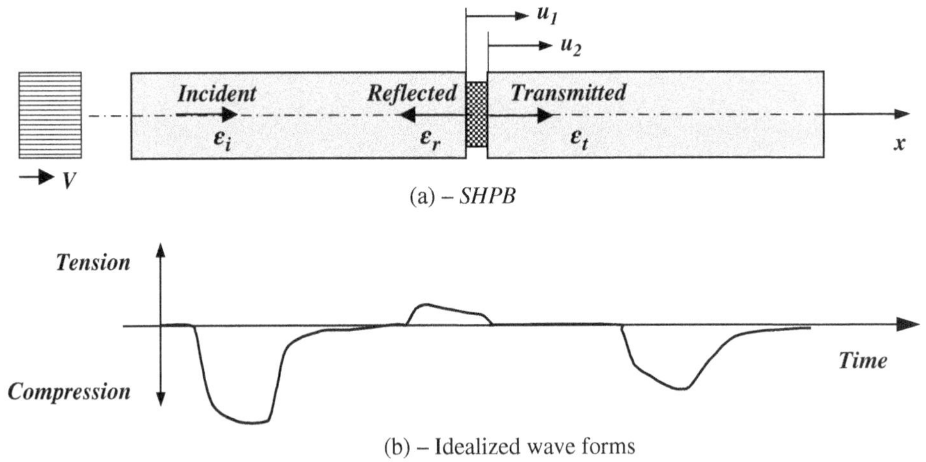

Figure 3.40 Wave propagation in the split Hopkinson pressure bar (SHPB). (a) SHPB and (b) idealized wave forms.

where u = displacement of bar at any location x, c = sound speed in bar = $\sqrt{E/\rho}$, E = Young's modulus of bar, ρ = density of bar, and t = time.

The general solution of Eq. (3.93) is given by:

$$u(x,t) = U(x)e^{i\omega t} \tag{3.94}$$

where $U(x)$ is a spatial function of x and $e^{i\omega t}$ is a temporal function of the time t with ω denoting the frequency of oscillation of the waves.

Substituting Eq. (3.94) into Eq. (3.93) gives:

$$U_{,xx}(x) + k^2 U = 0 \tag{3.95}$$

where $k = \dfrac{\omega}{c}$ = wave number. The solution of Eq. (3.95) is:

$$U(x) = \left(Ae^{-ikx} + Be^{ikx}\right)$$

$$\text{that is}, u = \left(Ae^{-ikx} + Be^{ikx}\right)e^{i\omega t} = u_i + u_r \tag{3.96}$$

i) the strain in incident rod is given by:

$$\varepsilon_1 = \frac{\partial u}{\partial x} \tag{3.97}$$

Differentiating Eq. (3.96) with respect to x, gives:

$$\frac{\partial u_1}{\partial x} = \left(-ikAe^{-ikx} + ikBe^{ikx}\right)e^{i\omega t} = \varepsilon_i + \varepsilon_r \tag{3.98}$$

Also, differentiating Eq. (3.96) with respect to time t, gives:

$$\dot{u}_1 = i\omega\left(Ae^{-ikx} + Be^{ikx}\right)e^{i\omega t} = -c\varepsilon_i + c\varepsilon_r \tag{3.99}$$

ii) the strain in transmitter rod is given by:

$$\dot{u}_2 = -c\varepsilon_t \tag{3.100}$$

Note that Eq. (3.100) is obtained in a similar fashion as Eq. (3.99) after setting the reflected component of the wave = 0 due to the presence of an anechoic termination at the end of the transmitter bar as shown in Figure 3.40.

iii) the strain rate in the specimen is given by:

$$\dot{\varepsilon}_s = \frac{(\dot{u}_1 - \dot{u}_3)}{l_s} \tag{3.101}$$

where l_s is the length of the specimen.

Combining Eqs. (3.99) through (3.101) gives:

$$\dot{\varepsilon}_s = \frac{c}{l_s}(-\varepsilon_i + \varepsilon_r + \varepsilon_t) \tag{3.102}$$

But, the compatibility of deflections implies that:

$$\varepsilon_t = \varepsilon_i + \varepsilon_r \tag{3.103}$$

Then, Eq. (3.102) reduces to:

$$\dot{\varepsilon}_s = \frac{2c}{l_s}\varepsilon_r \tag{3.104}$$

Also, the equilibrium of forces requires that:

$$F_1 = AE(\varepsilon_i + \varepsilon_r) = F_2 = AE\varepsilon_t \tag{3.105}$$

where A = area of bar.
Hence, the stress in the specimen is given by:

$$\sigma_s = \frac{F_1}{A_s} = \frac{F_2}{A_s} = \frac{AE}{A_s}\varepsilon_t \tag{3.106}$$

Accordingly, measuring the reflected strain ε_r and the transmitted strain ε_t can be used to calculate the strain ε_s and stress σ_s in the VEM as function of time. From such time histories, the stress–strain characteristics, the creep compliance, and complex modulus can be obtained.

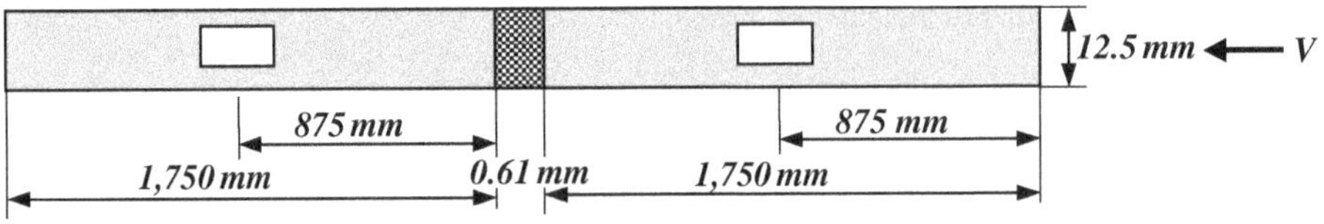

Figure 3.41 Experimental split Hopkinson pressure bar (SHPB).

Example 3.8 Consider the SHPB shown in Figure 3.41. The bar has a Young's modulus = 2.5 GPa and sound speed = 2870 m s^{-1}.

The SHPB is used to characterize the viscoelastic properties of a VEM specimen at different strain rates. The time history profiles of the strains measured at the incident and transmitted bars are shown in Figure 3.42 for four strain rates ranging from 2000 to 12,000 s^{-1}. Determine:

a) the stress–strain characteristics of the VEM at the different strain rates.
b) the energy dissipated by the VEM at the different strain rates.

Solution

The theory of the SHPB presented in Section 3.5.2.2 is employed to determine the stress–strain characteristics of the VEM according to the flow chart shown in Figure 3.43.

Figures 3.44 and 3.45 display the obtained time histories of the strain and stress, as well as stress–strain characteristics of the VEM for the considered four strain rates, respectively. Figure 3.45 indicates that increasing the strain rate applied to the VEM increases the area enclosed by the stress–strain curve, which is a direct measure of the energy dissipation by the VEM.

Table 3.5 lists the effect of the strain rate on the energy dissipation by the VEM.

3.5.2.3 Complex Modulus of a VEM from SHPB Measurements

The complex modulus of a VEM can be extracted from the time histories of the stress and the strain measurements as obtained from the SHPB measurements. Consider the typical time histories shown in Figure 3.46.

Using the discrete Boltzmann superposition principle, described in Section 3.5.1, the time history of the stress ($\sigma_1, \sigma_2, \ldots, \sigma_N$) is determined by a superposition of the effects of the incremental strains ($\varepsilon_1, \varepsilon_2, \ldots, \varepsilon_N$) imposed on the VEM as shown in Figure 3.46. Mathematically and for an infinite discretization of the strain time history, the resulting time history of the stress is given by Eq. (3.62) as follows:

$$\sigma(t) = \int_0^t \frac{\partial \varepsilon(t)}{\partial t} E(t-\tau)d\tau \tag{3.107}$$

As the time histories of the stress $\sigma(t)$ and the strain $\varepsilon(t)$ are known from the SHPB measurements, then the strain rate $\partial\varepsilon(t)/\partial t$ can be computed and Eq. (3.107) reduces to an equation in a single unknown $E(t)$. Hence, the relaxation modulus $E(t)$ can be extracted and the complex modulus can be determined using the approaches outlined in Sections 3.5.1.9 and 3.5.2.5.

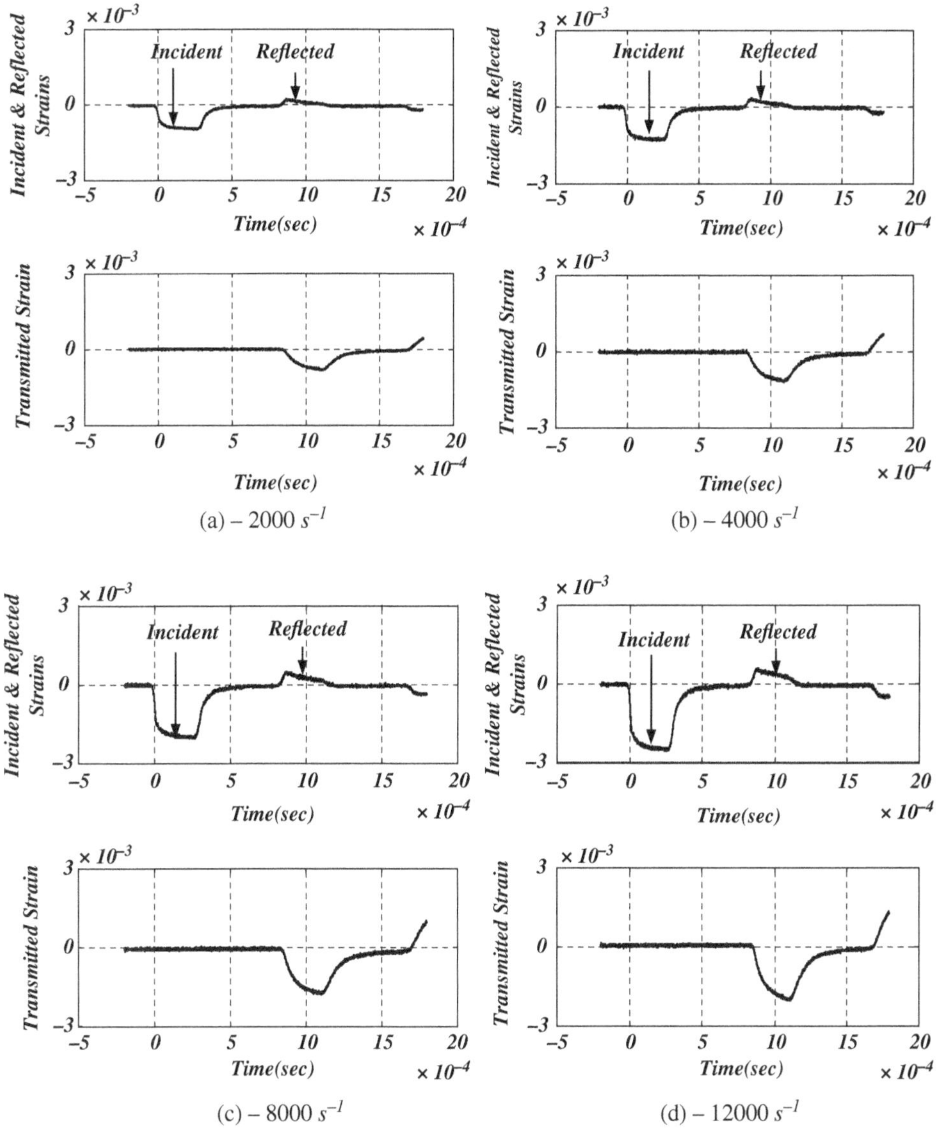

Figure 3.42 Strains in Hopkinson bar at different strain rates. (a) 2000 s^{-1}, (b) 4000 s^{-1}, (c) 8000 s^{-1}, and (d) 12,000 s^{-1}.

However, finding a solution for Eq. (3.107) is easier said than done. A viable approach for finding such a solution begins by assuming that $E(t)$ is represented by the following GMM such that:

$$E(t) = E_0 + \sum_{j=1}^{n} E_j e^{-t/\rho_j} \tag{3.108}$$

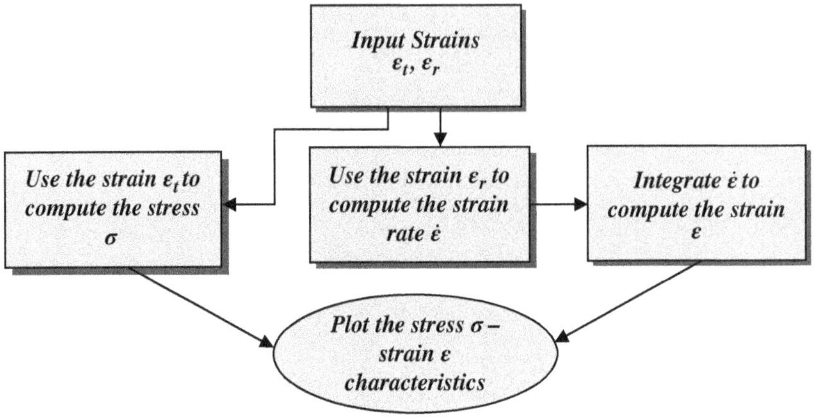

Figure 3.43 Computational algorithm of the stress–strain characteristics of VEM using the SHPB.

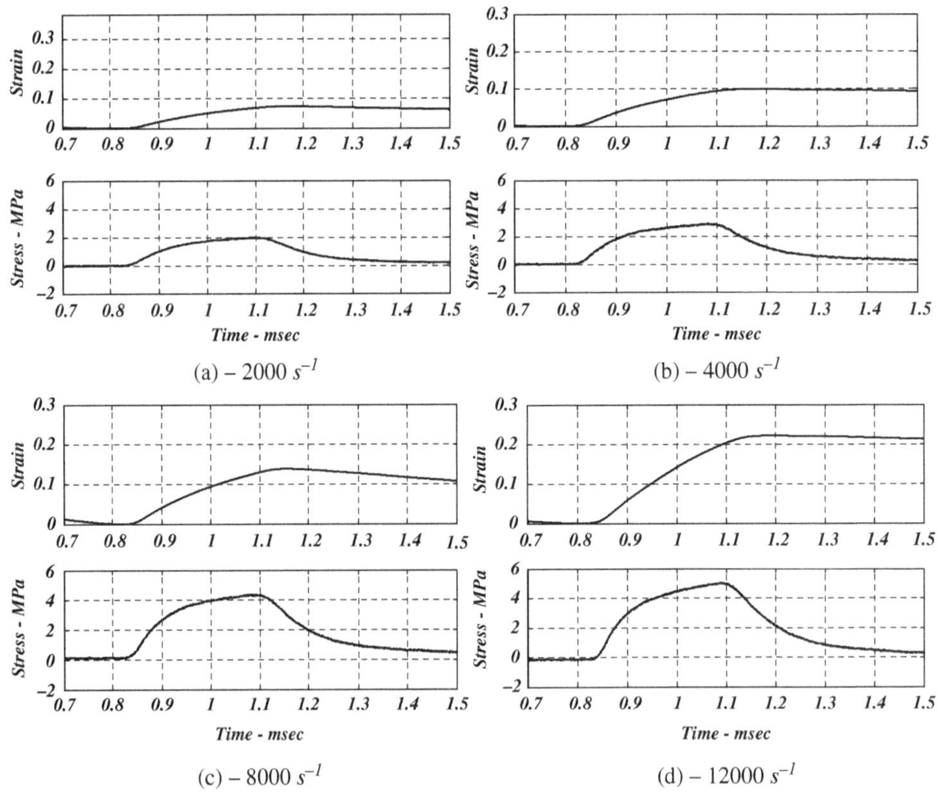

Figure 3.44 Time history of the stress and strain of the polymer at different strain rates.

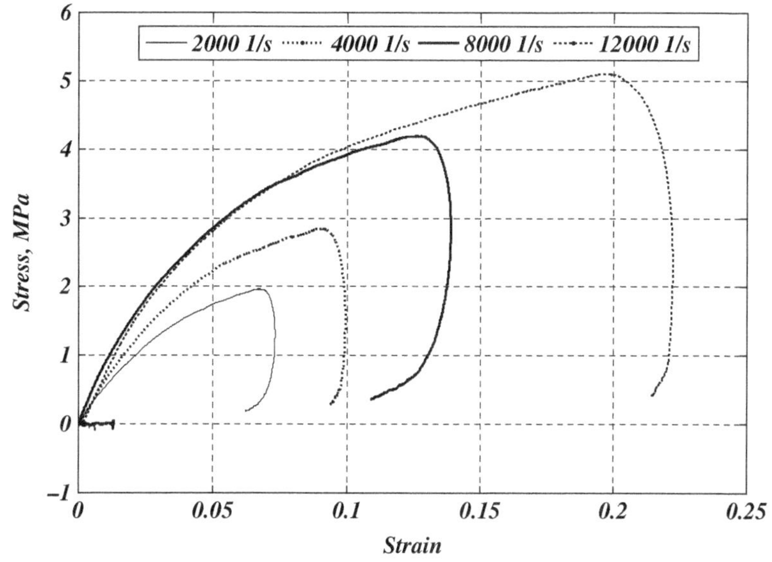

Figure 3.45 Stress–strain characteristics of the polymer at different strain rates.

Table 3.5 Comparison between the natural frequencies, modal damping ratios (MDR), and MDR per unit volume for the different viscoelastic treatments.

Strain rate (s^{-1})	2000	4000	8000	12,000
Energy dissipated ($Nm\,m^{-3}$)	36,000	112,000	273,000	450,000

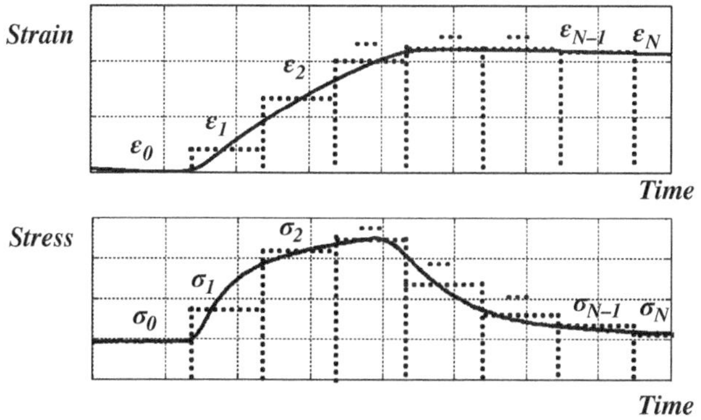

Figure 3.46 Superposition of the stresses resulting from stepped strains.

where E_0, E_j, and ρ_j are the equilibrium modulus, jth relaxation strength, and jth relaxation time, respectively. These unknown parameters of the *GMM* are determined by minimizing the square of the difference between the right and left hand sides of Eq. (3.107). A mathematical formulation of the optimization problem is given as follows:

$$\left\langle \begin{array}{l} \text{Find } E_0, E_j, \text{ and } \rho_j \text{ for } j = 1,..,n \\ \text{to minimize } F = \sum_{t=0}^{t=T}\left[\sigma(t) - \int_0^t \frac{\partial \varepsilon(\tau)}{\partial \tau}\left(E_0 + \sum_{j=1}^{n} E_j e^{-(t-\tau)/\rho_j}\right) d\tau\right]^2 \\ \text{such that } E_0 > 0, E_j > 0, \text{ and } \rho_j > 0 \text{ for } j = 1,..n \end{array} \right\rangle \tag{3.109}$$

Once the optimal parameters E_0, E_j, and ρ_j are determined, then the complex modulus $E^*(\omega)$ of the VEM can be determined by applying the Fourier transform to Eq. (3.108) to yield:

$$E^*(\omega) = E_0 + \sum_{j=1}^{n} E_j \frac{i\omega\rho_j}{i\omega\rho_j + 1}$$

Example 3.9 Use the time response characteristics of the stress and strain as obtained by the SHPB experiment and shown in Figure 3.44 to determine the complex modulus of the VEM.

Solution

The stress–strain characteristics shown in Figure 3.44 are used to determine the equivalent modulus $E(t)$ of the VEM using the following GMM such that:

$$E(t) = E_0 + E_1 e^{-t/\rho_1} + E_2 e^{-t/\rho_2} + E_3 e^{-t/\rho_3}$$

where E_1, E_2, and E_3 are determined by curve fitting the experimental results with $E_1 = 10MPa$, $\rho_1 = 10^{-2}s$, $\rho_2 = 10^{-3}s$, and $\rho_3 = 10^{-4}s$. Table 3.6 lists obtained values of the parameters E_0, E_1, E_2, and E_3 and that best fit the experimental results for the different strain rates. Figure 3.47 displays comparisons between the experimental and curve-fitted characteristics for the different strain rates.

Figure 3.48 shows the corresponding complex modulus of the VEM at the different strain rates. For all practical purposes, the neither the storage moduli nor the loss factors have changed significantly with the strain rate and the VEM maintains its linear behavior for the considered strain rates.

Example 3.10 Consider the SHPB shown in Figure 3.49. The bar has a Young's modulus = 70 MPa and density = 2700 kg m^{-3}. The SHPB is used to characterize a VEM that has a complex modulus described by:

Table 3.6 Values of the parameters E_1, E_2, and E_3 for the different strain rates.

Strain rate – (s^{-1})	2000	4000	8000	12,000
E_1 – (MPa)	1.2	1.1	3.0	1.1
E_2 – (MPa)	11.1	10.5	15.0	20.5
E_3 – (MPa)	57.0	63.0	60.0	45.0

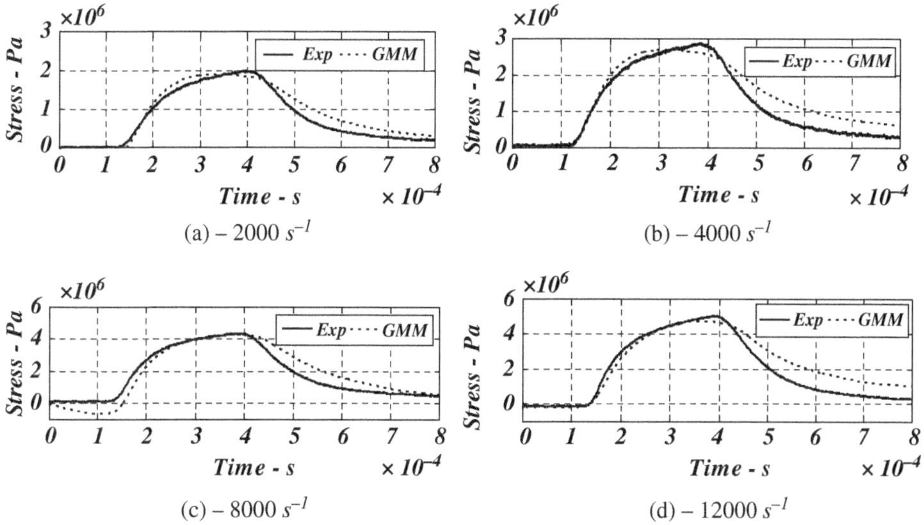

Figure 3.47 Comparisons between experimental and GMM moduli at different strain rates. (a) 2000 s^{-1}, (b) 4000 s^{-1}, (c) 8000 s^{-1}, and (d) 12,000 s^{-1}.

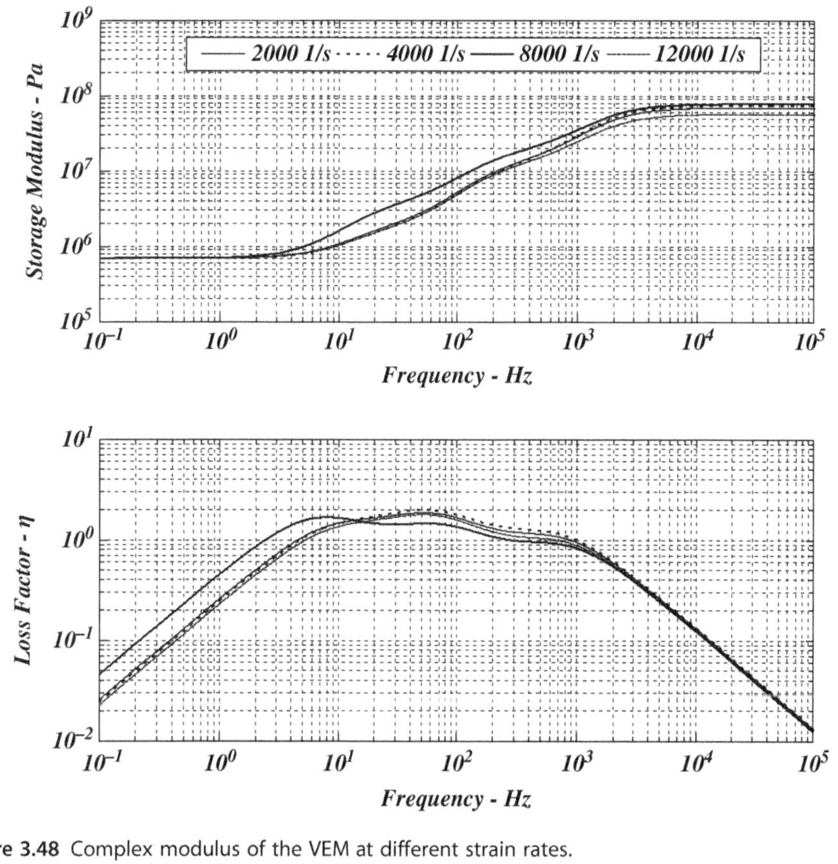

Figure 3.48 Complex modulus of the VEM at different strain rates.

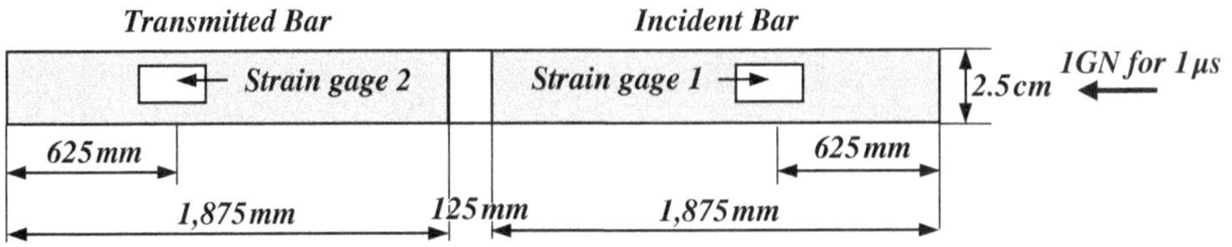

Figure 3.49 Experimental plastic split Hopkinson pressure bar (SHPB).

$$E = E_0\left[1 + \alpha\frac{s^2 + 2\omega_n s}{s^2 + 2\omega_n s + \omega_n^2}\right] \tag{3.110}$$

where $E_0 = 15.3MPa$, $\alpha = 39$, and $\omega_n = 19\,058$ rad/s.

Describe the dynamics of the SHPB using the theory of finite elements and account for the VEM using the complex modulus given by Eq. (3.110). Divide the SHPB into 30 elements. (*Hint*: use the approaches presented in Chapter 4).

Determine:

i) The incident and reflected as well as the transmitted strains as measured by strain gages 1 and 2, respectively, which are located at distances of 62.5 cm from the two ends of the bar. Note that the bar is impacted by a force of 1 MN.
ii) Stress–strain characteristics of the VEM.
iii) The equivalent relaxation modulus of the VEM.
iv) The parameters (E_0, E_1, E_2, ρ_1, and ρ_2) of a *GMM* that models the obtained equivalent relaxation modulus of the VEM such that:

$$E(t) = E_0 + E_1 e^{-t/\rho_1} + E_2 e^{-t/\rho_2}$$

v) The complex modulus of the VEM as obtained from the GMM.
vi) Compare the results with the given complex modulus described by Eq. (3.110).

Solution

Figure 3.50 summarizes the obtained results obtained by finite element modeling of the SHPB/VEM system. Figure 3.50a displays the incident and transmitted strains as measured by strain gages 1 and 2, respectively. Then, the theory of the SHPB presented in Section 3.5.2.2 is employed to determine the stress–strain characteristics of the VEM according to the flow chart shown in Figure 3.43. This yields the obtained stress–strain characteristics of the VEM shown in Figure 3.50b.

The stress–strain characteristics are then used to determine the modulus $E(t)$ of the VEM which is described by the two-mini oscillators *GMM*:

$$E(t) = E_0 + E_1 e^{-t/\rho_1} + E_2 e^{-t/\rho_2}$$

where the parameters (E_0, E_1, E_2, ρ_1, and ρ_2) that best fit the experimental results are found to be: $E_0 = 10MPa$, $E_1 = 1GPa$, $\rho_1 = 10^{-4}s$, and $E_2 = 1GPa$, $\rho_2 = 10^{-5}s$.

Figure 3.50d shows a comparison between the complex moduli of the VEM as obtained from the GMM and by the model given by Eq. (3.110). A close agreement is evident,

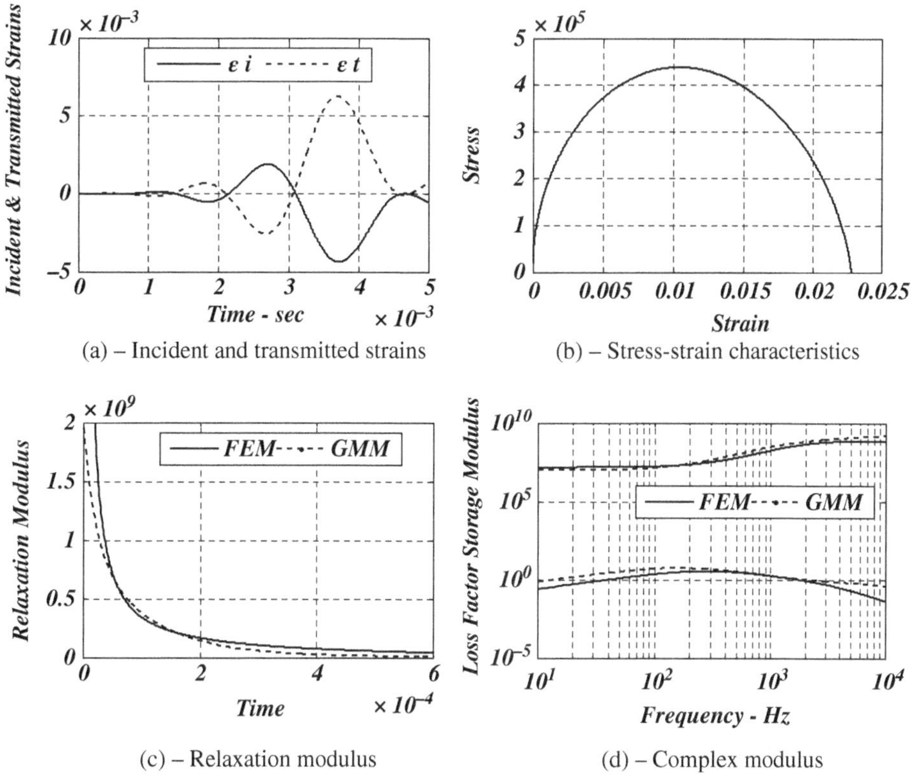

Figure 3.50 Characteristics of the VEM as obtained by finite element modeling of the split Hopkinson pressure bar (SHPB)/VEM system. (a) Incident and transmitted strains, (b) Stress–strain characteristics, (c) Relaxation modulus, and (d) Complex modulus.

indicating that the finite element model of the SHPB/VEM system is able to extract the correct parameters of the VEM.

3.5.3 Wave Propagation Method

Wave propagation methods utilize time-domain measurements of the strain along viscoelastic specimens to determine the complex modulus of these materials (Hillstrom et al. 2000) or the complex modulus and Poisson's ratio (Mousavi et al. 2004). These methods are well suited for the determination of complex modulus over a broadband range of frequencies ranging between 100 Hz and 10 kHz.

Axisymmetric viscoelastic waves in a uniform bar can be represented in the frequency domain by three complex-valued functions of frequency provided that the wavelengths are much larger than the diameter of the bar so that the conditions are approximately one-dimensional. One function is a wave propagation coefficient from which the complex modulus can be determined if the density of the material is known. The real part of this function is the damping coefficient, and the imaginary part is the wave number.

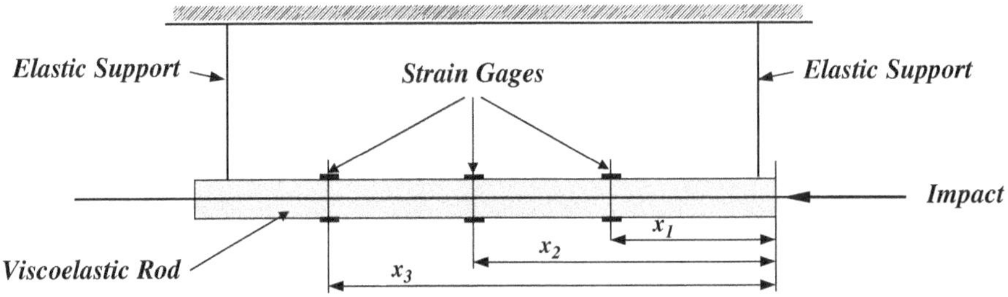

Figure 3.51 Viscoelastic bar subject to an axial impact.

The complex modulus of a material with linearly viscoelastic behavior is identified on the basis of strains that are known, from measurements and sometimes from a free end boundary condition, at three or more sections of an axially impacted bar specimen. The aim is to improve existing identification methods based on known strains at three uniformly distributed sections by increasing the number of sections considered and by distributing them non-uniformly. The increased number of sections results in an over-determined system of equations from which an approximate solution for the complex modulus is determined using the method of least squares. Through the non-uniform distribution of sections, critical conditions with accompanying large errors at certain frequencies are largely eliminated.

Consider the straight and uniform bar of Figure 3.51 which is made of a linearly VEM. The equation of longitudinal motion u (x,t) of the bar, at any location x and time t, is given by

$$\frac{\partial^2 u}{\partial x^2} = \frac{\rho}{E^*}\ddot{u} \tag{3.111}$$

where ρ = density and E^* = complex modulus such that $E^* = E' + jE''$, where E' and E'' are the storage and the loss moduli respectively.

Equation (3.111) can be transformed to the frequency domain by applying the Fourier transform to yield the following expression

$$\frac{\partial^2 \hat{u}}{\partial x^2} - \gamma^2 \hat{u} = 0 \tag{3.112}$$

where $\gamma^2 = -\dfrac{\rho\omega^2}{E^*}$ and $\hat{u}(x,\omega) = \hat{u} = \int_{-\infty}^{\infty} u e^{-j\omega t} dt$ is the Fourier transform of $u(x,t)$ with ω denoting the frequency.

Equation (3.112) has the following solution

$$\hat{u} = \bar{A}e^{-\gamma x} + \bar{B}e^{\gamma x} \tag{3.113}$$

which gives the strain $\hat{\varepsilon} = \dfrac{\partial \hat{u}}{\partial x}$ in the frequency domain as

$$\hat{\varepsilon} = -\gamma\bar{A}e^{-\gamma x} + \gamma\bar{B}e^{\gamma x} = Ae^{-\gamma x} + Be^{\gamma x} \tag{3.114}$$

Note that Eq. (3.114) includes three unknown parameters that are A, B, and γ. All these parameters are function of the frequency ω. To determine these parameters, three strain gages are needed to generate three equations in the three unknowns. Larger number of strain gages may be necessary to improve the estimation accuracy of these parameters by using the classical least squares approaches (Hillstrom et al. 2000).

In this chapter, the parameters are determined using three strain gages only, which are placed at distances x_1, x_2, and x_3 from the impacted end of the rod as shown in Figure 3.51. The time-domain strain signals $\varepsilon_i(t)$ of the three gages (i = 1–3) are measured and transformed to the frequency domain using the Fourier transform to yield: $\hat{\varepsilon}_i(\omega)$. Assembling the measured strains by using Eq. (3.114) gives:

$$\hat{\varepsilon}_1 = e^{-\gamma x_1}A + e^{\gamma x_1}B, \tag{3.115}$$

$$\hat{\varepsilon}_2 = e^{-\gamma x_2}A + e^{\gamma x_2}B, \tag{3.116}$$

and

$$\hat{\varepsilon}_3 = e^{-\gamma x_3}A + e^{\gamma x_3}B. \tag{3.117}$$

Let $x_2 = x_1 + h$ and $x_3 = x_2 + h$ with h denoting the axial spacing between the strain gages, then solving Eqs. (3.115) and (3.116) for A and B gives:

$$A = \frac{\zeta\hat{\varepsilon}_2 - \zeta^2\hat{\varepsilon}_1}{(1-\zeta^2)}e^{\gamma x_1} \text{ and } B = \frac{\hat{\varepsilon}_1 - \zeta\hat{\varepsilon}_2}{(1-\zeta^2)}e^{-\gamma x_1} \tag{3.118}$$

where

$$\zeta = e^{\gamma h} \tag{3.119}$$

Substituting Eq. (3.118) into Eq. (3.117), gives:

$$(\zeta^2 + 1)\hat{\varepsilon}_2 - \zeta\hat{\varepsilon}_1 = \zeta\hat{\varepsilon}_3 \tag{3.120}$$

which has the solution $\zeta = \psi \pm \sqrt{\psi^2 - 1}$ with $\psi = \frac{(\hat{\varepsilon}_1 + \hat{\varepsilon}_3)}{2\hat{\varepsilon}_2}$. Hence, the third parameter γ can be extracted from Eq. (3.119) as follows:

$$\gamma = \ln(\zeta)/h \tag{3.121}$$

and the complex modulus of the VEM can be computed from

$$E^* = -\rho\omega^2/\gamma^2 \tag{3.122}$$

Example 3.11 Consider a viscoelastic bar that is 0.21 m long and has a cross sectional area of 0.000 625 m^2. The rod has a density ρ of 1100 kg m^{-3}. The rod is provided with three strain gages, which are placed such that h = 0.06 m as shown in Figure 3.52. The time response of the three strain gages is shown in Figure 3.53. Determine the storage modulus and loss factor of the VEM.

Solution

Using Eqs. (3.115) through (3.117), Eqs. (3.120) through (3.122), the complex modulus of the VEM is calculated. The obtained results are displayed in Figure 3.54 along with a comparison with the exact damping characteristics of the VEM.

It can be seen that the wave propagation method adequately replicates the exact characteristics of the VEM.

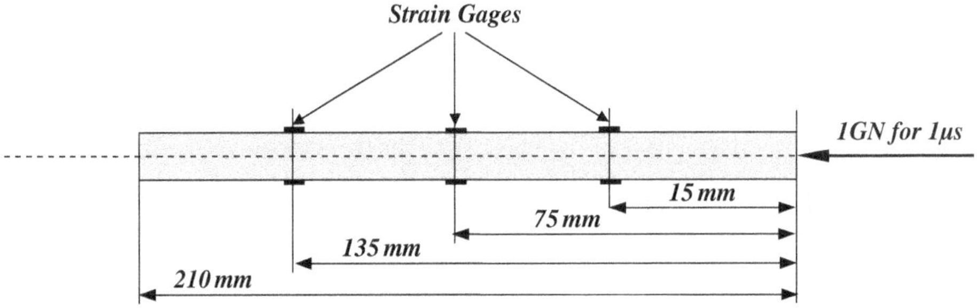

Figure 3.52 Arrangement of the strain gages.

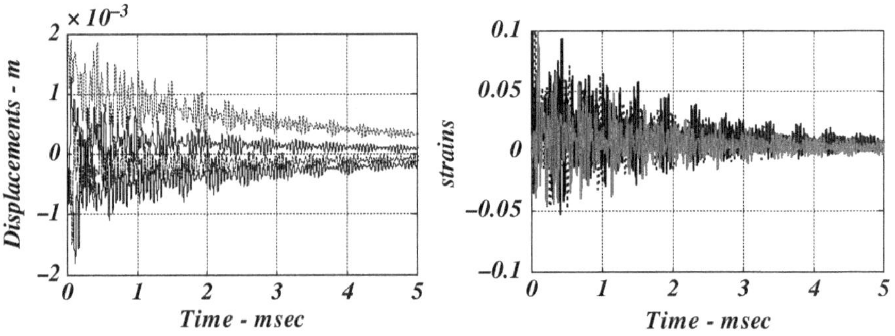

Figure 3.53 Time response of the three strain gages.

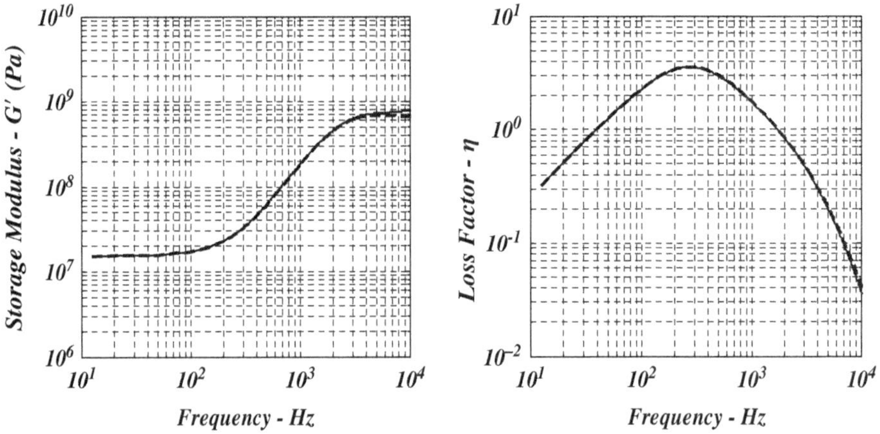

Figure 3.54 The identified storage modulus and loss factor of the VEM using the wave propagation method (············ exact ——— extracted).

3.5.4 Ultrasonic Wave Propagation Method

3.5.4.1 Overview

Ultrasonic waves are acoustic waves, propagating through a VEM, at frequencies in the range from 20 kHz to 2 GHz as indicated in Figure 3.55. However, the frequency range used in ultrasonic characterization of the viscoelastic properties of VEM is usually limited between 1 and 20 MHz. The ultrasonic characterization methods are adopted mainly because of their accuracy and relative simplicity. Furthermore, these methods enable the measurement of the phase velocity, attenuation, and the complex modulus of a wide variety of VEMs at high frequencies (Rose 1999).

Generally speaking, ultrasonic shear or longitudinal waves may be employed to characterize the behavior of polymers at higher frequencies. However, shear waves are generally difficult to use for measuring the characteristics of very soft gel type materials particularly at high temperatures because of the problems encountered to achieve good acoustic coupling with the specimens. Therefore, in this section, the emphasis is placed on the use of longitudinal ultrasonic waves to extract the mechanical properties of VEM.

3.5.4.2 Theory

The equation of longitudinal motion $u\ (x,t)$ of the bar, at any location x and time t, is given by Eq. (3.111) as follows (Qiao et al. 2011; Lakes 2009; Rose 1999):

$$E^* \frac{\partial^2 u}{\partial x^2} = \rho \frac{\partial^2 u}{\partial t^2} \tag{3.123}$$

where ρ = density and E^* = longitudinal complex modulus such that: $E^* = E' + jE''$, where E' and E'' are the storage and the loss moduli, respectively.

Consider a solution of "the wave" Eq. (3.123), which takes the following form:

$$u = u_0 e^{i(\omega t - kx)} \tag{3.124}$$

This solution separates the time and space variables and is cast in the form of a product of a temporal function $e^{i\omega t}$ and a spatial function e^{ikx} where ω denotes the frequency and k denotes the wave number.

Substituting Eq. (3.124) into Eq. (3.123) gives:

$$\omega^2 = \frac{E^*}{\rho} k^2$$

or

$$\omega = k\sqrt{\frac{E^*}{\rho}} = k\sqrt{\frac{E'(1+\eta i)}{\rho}} = k\sqrt{c^2(1+\eta i)} \tag{3.125}$$

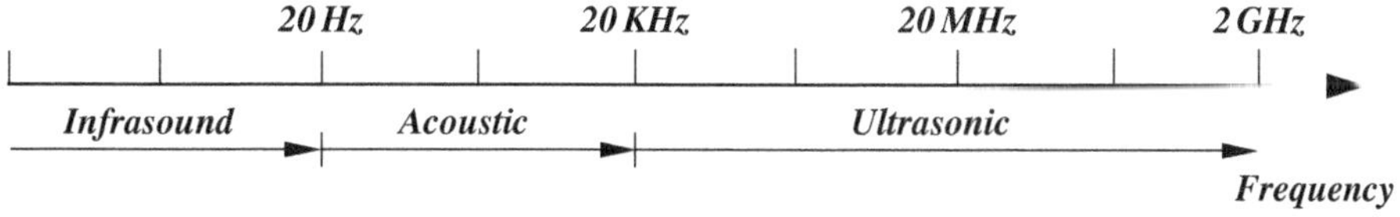

Figure 3.55 The range of ultrasonic waves.

where $c = \sqrt{\frac{E'}{\rho}}$ phase velocity and $\eta = E''/E'$ = loss factor. Equation (3.125) can be rewritten as:

$$k = \frac{\omega}{c} / \sqrt{(1+\eta i)} \cong \frac{\omega}{c}\left(1 - \frac{1}{2}\eta i\right) = \beta - \alpha i \tag{3.126}$$

with $\beta = \frac{\omega}{c}$ and $\alpha = \frac{\omega}{2c}\eta$.

Hence, Eq. (3.124) reduces to:

$$u = u_0 e^{-\alpha x} e^{i(\omega t - \beta x)} \tag{3.127}$$

Equation (3.123) can also be rewritten as:

$$c^* \frac{\partial^2 u}{\partial x^2} = \frac{\partial^2 u}{\partial t^2}$$

where c^* is the complex wave velocity, which can be written from Eq. (3.125) as follows:

$$\sqrt{E^*/\rho} = \frac{\omega}{k} = c^* = c' + ic'' = \sqrt{(E' + iE'')/\rho} \tag{3.128}$$

Equating the real and imaginary components on both sides of Eq. (3.128) yields:

$$E' = \rho\left(c'^2 - c''^2\right) \text{ and } E'' = 2\rho c' c'' \tag{3.129}$$

Accordingly, Eq. (3.124) can be rewritten as:

$$\begin{aligned} u &= u_0 e^{i\omega\left(t - \frac{k}{\omega}x\right)} = u_0 e^{i\omega\left(t - \frac{x}{c^*}\right)} \\ &= u_0 e^{i\omega\left(t - \frac{x}{c' + c''i}\right)} = u_0 e^{-\frac{\omega c'' x}{c'^2 + c''^2}} e^{i\omega t} e^{i\omega\left(t - \frac{c'x}{c'^2 + c''^2}\right)} \end{aligned} \tag{3.130}$$

Comparing Eqs. (3.127) and (3.130) gives:

$$\alpha = \frac{\omega c''}{c'^2 + c''^2} \quad \text{and} \quad \beta = \frac{c'}{c'^2 + c''^2} \tag{3.131}$$

Let $r = \frac{\alpha c}{\omega}$, then Eq. (3.131) yields:

$$r = \frac{cc''}{c'^2 + c''^2} \quad \text{and} \quad 1 = \frac{cc'}{c'^2 + c''^2} \tag{3.132}$$

Hence,

$$r = \frac{c''}{c'} = \frac{1}{2}\eta \tag{3.133}$$

Also, the second part of Eq. (3.132) gives:

$$c'^2 + c''^2 = cc' \tag{3.134}$$

Combining Eqs. (3.133) and (3.134) yields:

$$c'^2(1+r^2) = cc'$$

$$\text{that is, } c' = \frac{c}{(1+r^2)} \tag{3.135}$$

$$\text{and accordingly, } c'' = \frac{cr}{(1+r^2)} \tag{3.136}$$

Then, the real and imaginary components of the longitudinal complex modulus can be determined by substituting Eqs. (3.135) and (3.136) into Eq. (3.129) to yield the following expressions:

$$E' = \frac{\rho c^2(1-r^2)}{(1+r^2)^2} \quad \text{and} \quad E'' = \frac{2\rho c^2 r}{(1+r^2)^2} \tag{3.137}$$

Equation (3.137) represents the basic relationships necessary to extract the components of the complex modulus of the VEM in terms of the two parameters $c = \sqrt{\frac{E'}{\rho}}$ and $r = \frac{\alpha c}{\omega} = \frac{1}{2}\eta$, which denote the phase velocity and a scaled attenuation factor (or loss factor), respectively.

3.5.4.3 Measurement of the Phase Velocity and Attenuation Factor

The measurements are carried out using a test set-up similar to that shown schematically in Figure 3.56. In the set-up, the VEM test specimen is sandwiched between an ultrasonic transmitter and a receiver. The transmitter is excited by an input pulsed voltage generated by a pulser and the resulting signal received by the ultrasonic transducer is recorded using a high frequency storage oscilloscope. The recoding process is initiated by a trigger signal produced by the pulser indicating the start of the transmission of the input pulsed train. The high voltage trigger signal is attenuated before sending it to the oscilloscope using a high power attenuator.

This process is repeated using a thicker test specimen. The recorded signals of the two tests are superimposed on top of each other as shown in Figure 3.57 indicating a time delay t_d and amplitude attenuation from A_1 to A_2 due to the use of the thicker specimen.

Consequently, the phase velocity in the VEM specimen can be calculated from:

$$c = d/t_d$$

where d is the thickness difference between the second and first samples and t_d denotes the time shift observed between the two tests, as indicated in Figure 3.57. The attenuation factor can be calculated from the amplitudes of the transmitted signals using Eq. (3.127) as follows:

$$\textit{For sample 1}: A_1 = u_0 e^{-\alpha L} \tag{3.138}$$

$$\text{and } \textit{For sample 2}: A_2 = u_0 e^{-\alpha(L+d)} \tag{3.139}$$

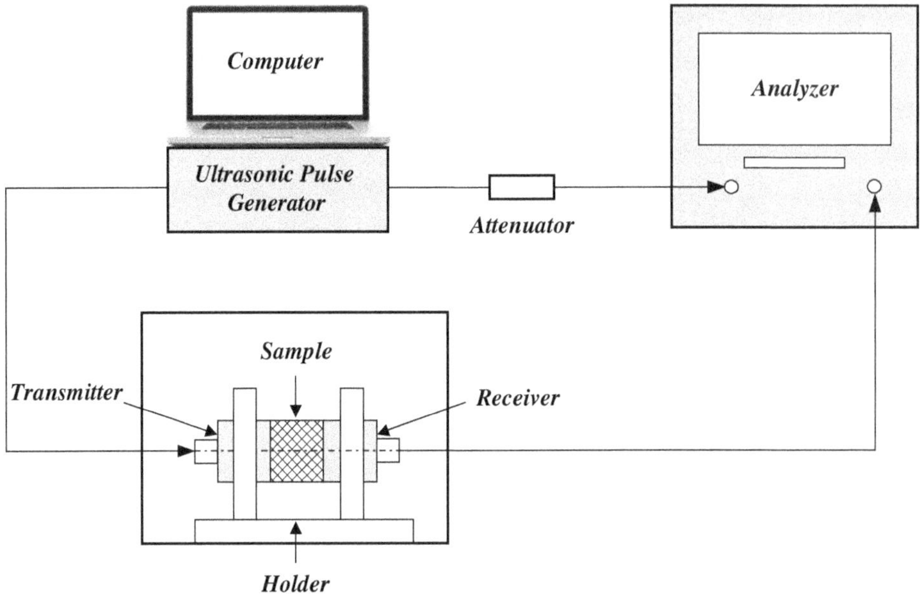

Figure 3.56 Ultrasonic characterization of the complex modulus of VEM: set up.

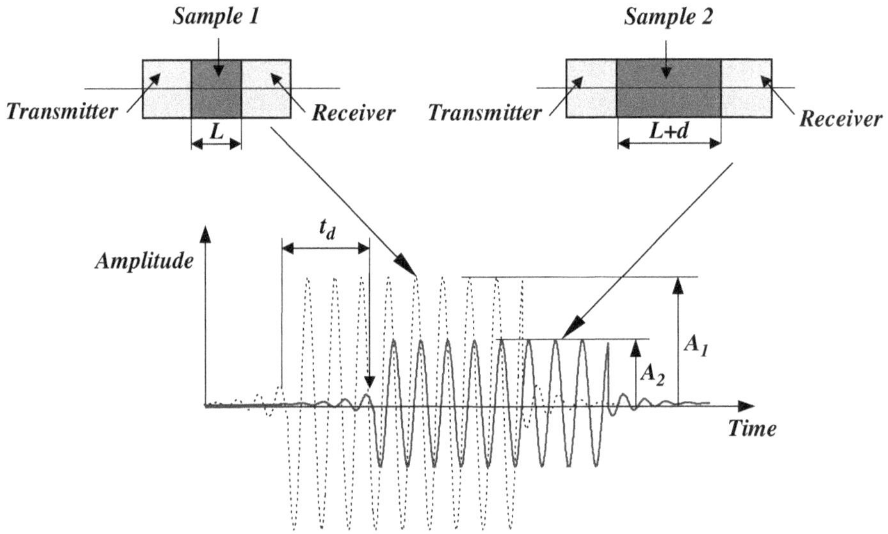

Figure 3.57 Ultrasonic characterization of the complex modulus of VEM: recorded signals of the two tests.

Dividing Eq. (3.138) by Eq. (3.139) and taking the natural log gives the following expression for the attenuation factor:

$$\alpha = \frac{1}{d}\ln\left(\frac{A_1}{A_2}\right) \text{ in Neper m}^{-1} \tag{3.140}$$

where A_1 and A_2 are the amplitudes of the received signals from the first and second test, respectively, as indicated in Figure 3.57. Note that "Neper" is a dimensionless metric that is used to express ratios. It is named after the microwave scientist John Napier. Another way of expressing the attenuation factor is given in decibels per meter as follows:

$$\text{or}\,\alpha = 20\frac{1}{d}\log_{10}\left(\frac{A_1}{A_2}\right) \quad \text{in dB m}^{-1} \tag{3.141}$$

Once α is determined from Eq. (3.140), then r can be calculated from $r = \frac{\alpha c}{\omega}$ and the components of the complex modulus E' and E'' can be computed using Eq. (3.137). This process is repeated for different excitation frequencies and ambient temperatures in order to generate the master curves of the VEM.

3.5.4.4 Typical Attenuation Factors

Figure 3.58 displays typical values of the attenuation factors for different polymers, epoxies, rubber, and other fluids as function of the excitation frequency. The displayed characteristics are bounded from below and above by the characteristics of water and air. The data presented in the figure are summarized from the work of Qiao et al. (2011), Challis et al. (2009), Zhang et al. (2005), He and Zheng (2001), and He (1999). Plotted also on the figure are the attenuation characteristics of the piezoelectric ceramic (Lead Zirconate Titanate (PZT-4)) indicating its high attenuation capability at very high frequencies.

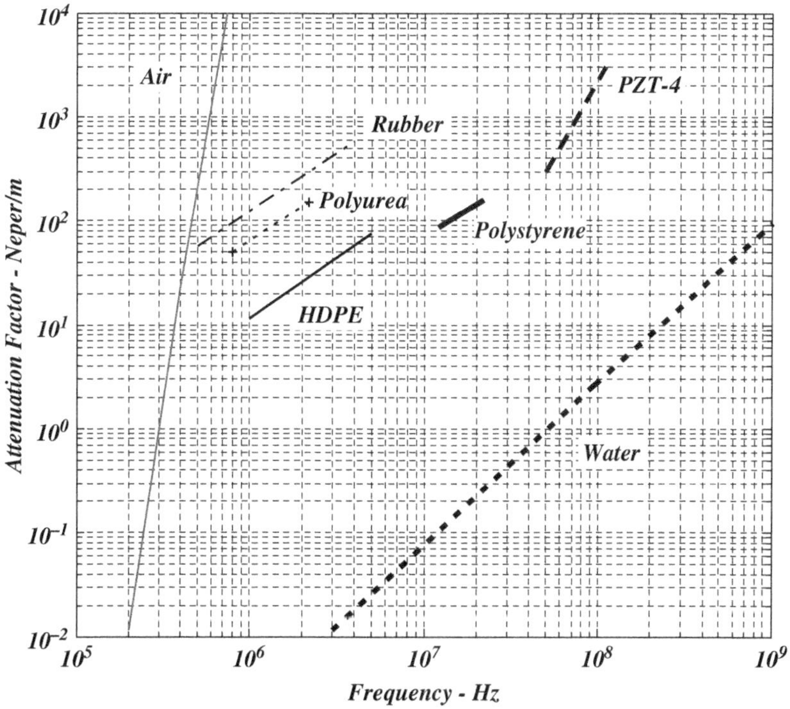

Figure 3.58 The attenuation factor as function of frequency for different polymers and other materials (HDPE = high density polyethylene).

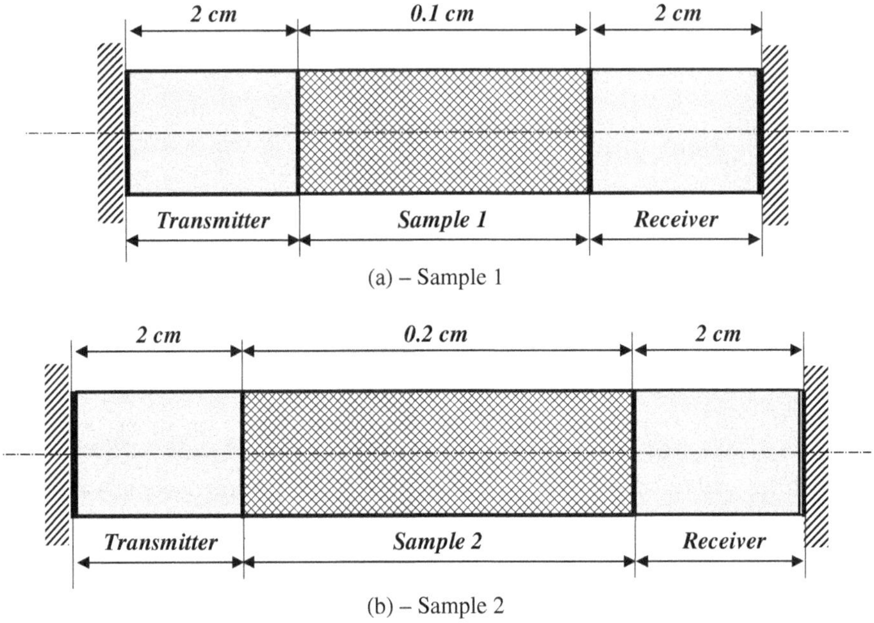

Figure 3.59 The ultrasonic set-up for characterizing the properties of a VEM using two test samples (a) and (b).

Table 3.7 Effect of the excitation frequency on the amplitudes and time delay of the two VEM samples.

	Frequency (MHz)							
Parameter	**0.5**	**0.75**	**1**	**1.5**	**2**	**3**	**4**	**5**
A_1 (mv)	50.00	54.30	46.50	20.5	11.4	3.80	1.80	1.10
A_2 (mv)	24.10	23.70	19.20	8.30	5.00	2.00	1.30	0.90
t_d (μs)	0.2446	0.2093	0.1840	0.1581	0.1416	0.1221	0.1082	0.0973

Example 3.12 Consider the ultrasonic material characterization set-up shown schematically in Figure 3.59. The set-up is used to characterize a VEM that has a density ρ = 1, 100 kg m^{-3} and a complex modulus to be described by the following expression:

$$E = E_0 \left[1 + \alpha \frac{s^2 + 2\zeta\omega_n s}{s^2 + 2\zeta\omega_n s + \omega_n^2} \right] \tag{3.110}$$

where E_0, α, ζ, and ω_n are parameters to be determined to best fit the experimental results. The main geometrical parameters of the considered test samples are displayed in Figure 3.59. A summary of the experimental results obtained by testing the two VEM samples over a frequency range between 0.5 and 5 MHz are listed in Table 3.7.

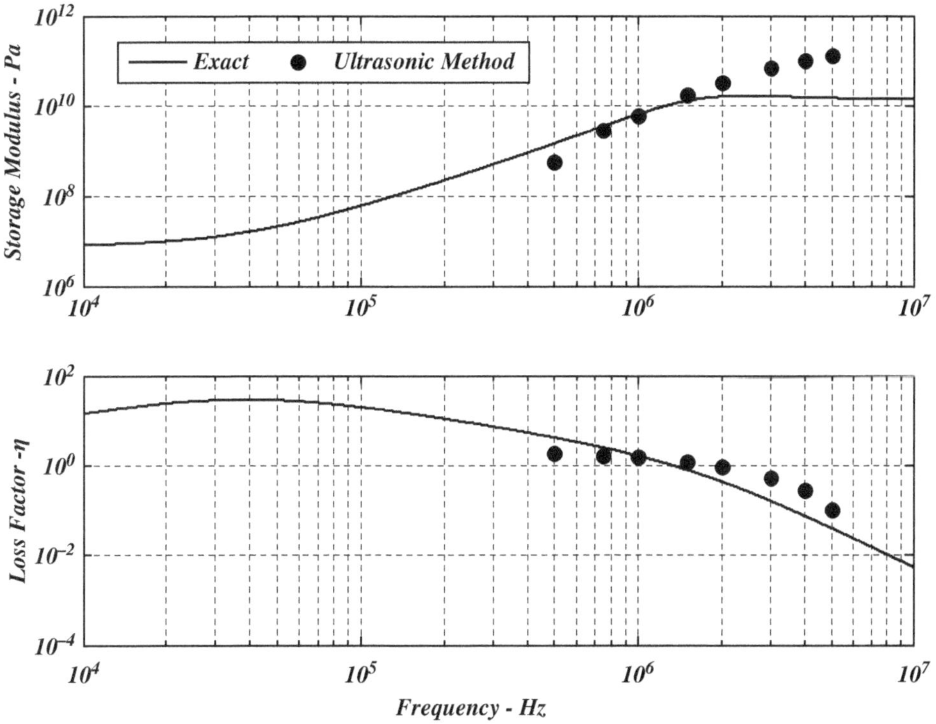

Figure 3.60 The extracted and exact complex moduli of the VEM using the ultrasonic method.

Determine the experimental complex modulus of the VEM and estimate the parameters E_0, α, ζ, and ω_n of the model given by Eq. (3.110) that best fit the experimental results.

Solution

The theory presented in Section 3.5.4.3 is used to extract the storage modulus and loss factor of the VEM from the experimental results listed in Table 3.7.

Figure 3.60 displays the experimental storage modulus and loss factor of the VEM as function of the excitation frequency.

The figure also shows a comparison with the storage modulus and loss factor as obtained from Eq. (3.110) with the optimally determined parameters E_0 = *1E7 Pa*, $\alpha = 1750$, $\zeta = 0.7$, and ω_n = 1E7 rad s^{-1}. It is evident that the model, as it stands as a single oscillator, is inadequate to represent the experimental results. Multi- oscillators are needed to capture the experimental results.

3.6 Summary

This chapter has presented a comprehensive coverage of the methods used to characterize the damping properties of VEM both in the frequency and the time-domain. The frequency domain methods generate the complex modulus of the VEM whereas the time-domain methods generate the creep compliance and relaxation modulus. The theoretical basics that govern the interconversion from the time-domain parameters to the equivalent frequency domain parameters are described in detail. Hence, it is sufficient to characterize the properties of the VEM either in the time or the frequency domain as the interconversion rules can be readily used to convert from one domain to the other.

Practical issues related to the implementation of the interconversion relationship have been addressed and efficient approaches that can ensure accuracy and speed of conversion are presented.

Special emphasis is placed on outlining the approaches to be adopted to generate the master curves of the VEM whereby narrow band test measurements can be used to predict the performance over a wide range of time and frequencies. Such extrapolation capabilities are made possible using the temperature-frequency superposition and temperature-time superposition principles. The bases for these two principles are traced back to molecular and kinetic theories of VEM.

The chapter has concluded with a presentation of two of the most common methods of characterization of VEM that rely on wave propagation principles. These methods are the SHPB method, the wave propagation method, and the ultrasonic method.

The presented characterization methods will enable the development of mathematical models of the VEM damping behavior that can be integrated with the dynamics of structures in order to predict their response to either time-domain or frequency domain excitations. Chapter 4 will discuss the basics of such an integration process.

References

Alig, I. and Lellinger, D. (2000). Ultrasonic methods for characterizing polymeric material. *Chemical Innovation* 30 (2): 12–18.

Anderssen, R.S., Davies, A.R., and de Hoog, F.R. (2008). On the Volterra integral equation relating creep and relaxation. *Inverse Problems* 24 (3): 035009.

Ashby, M.F. (1999). *Materials Selection Mechanical Design*, 2e. Oxford: Butterworth-Heinemann.

ASTM E 756-93, "Standard test method for measuring vibration damping properties of materials", ASTM Standard, American Society for Testing and Materials, West Conshohocken, PA, 1993.

Brinson, H.F. and Brinson, L.C. (2008). *Polymer Engineering Science and Viscoelasticity: An Introduction*. New York, NY: Springer.

Brown, R. and Read, B. (1984). *Measurement Techniques for Polymers Solids*. New York, NY: Elsevier Applied Science Publishers.

Cebon, D. and Ashby, M.F. (1994). Materials selection for precision instruments. *Measurement Science and Technology* 5: 296–306.

Chae, S.-H., Zhao, J.-H., Edwards, D.R., and Ho, P.S. (2010). Characterization of the viscoelasticity of molding compounds in the time domain. *Journal of Electronic Materials* 39 (4): 419–425.

Challis, R., Blarel, F., Unwin, M. et al. (2009). Models of ultrasonic wave propagation in epoxy materials. *IEEE Transactions on Ultrasonics, Ferroelectrics, and Frequency Control* 56 (6): 1225–1237.

Chen, H.Y., Stepanov, E.V., Chum, S.P. et al. (1999). Creep Behavior of Amorphous Ethylene–Styrene Interpolymers in the Glass Transition Region. *Journal of Polymer Science: Part B: Polymer Physics* 37: 2373–2382.

Christensen, R.M. (1982). *Theory of Viscoelasticity: An Introduction*, 2e. New York, NY: Academic Press Inc.

Cowans J., "The effects of viscoelastic behavior on the operation of a delayed resonator vibration absorber", Masters Thesis, Clemson University, 2006.

Dlubac, J., Lee, G., Duffy, J. et al. (1990). Comparison of the complex dynamic modulus as measured by three apparatus. In: *Sound and Vibration Damping with Polymers*, vol. ACS 424 (ed. R. Crosaro and L. Sperling), 49–62. Washington, DC: American Chemical Society.

Drake M. L. and Terborg G. E., "Polymeric material testing procedures to determine damping properties and the results of selected commercial material", Technical Report AFWAL-TR-80-4093, 1979.

Evans, R.M.L., Tassieri, M., Auhl, D., and Waigh, T.A. (2009). Direct conversion of rheological compliance measurements into storage and loss moduli. *Physical Review E* 80: 012501.

Ferry, J.D. (1980). *Viscoelastic Properties of Polymers*. New York, NY: Wiley.

Gama, B.A., Lopatnikov, S.L., and Gillespie, J.W. Jr. (2004). Hopkinson bar experimental technique: a critical review. *Applied Mechanics Review* 57 (4): 223–250.

Garibaldi, L. and Onah, H.N. (1996). *Viscoelastic Material Damping Technology*. Turin, Italy: Becchis Osiride.

Gray, G.T. III (2000). Classic split-Hopkinson pressure bar testing. In: *ASM Handbook Vol 8, Mechanical Testing and Evaluation*, 462–476. Materials Park, OH: ASM Intl.

Gross, B. (1953). *Mathematical Structure of the Theories of Viscoelasticity*. Paris, France: Hermann.

Haddad, Y.M. (1995). *Viscoelasticity of Engineering Materials*. London: Chapman & Hall.

He, P. (1999). Experimental verification of models for determining dispersion from attenuation. *IEEE Transacions on Ultrasonics, Ferroelectrics, and Frequency Control* 46 (3): 706–714.

He, P. and Zheng, J. (2001). Acoustic dispersion and attenuation measurement using both transmitted and reflected pulses. *Ultrasonics* 39: 27–32.

Hillstrom, L.M., Mossberg, M., and Lundberg, B. (2000). Identification of complex modulus from measured strains on an axially impacted bar using least squares. *Journal of Sound and Vibration* 230 (3): 689–707.

Kolsky, H. (1949). An investigation of the mechanical properties of materials at very high rates of loading. *Proceedings of the Physical Society London, Section B* 62 (II-B): 676–700.

Lakes, R.S. (1999). *Viscoelastic Solids*. Boca Raton, FL: CRC Press.

Lakes, R.S. (2004). Viscoelastic Measurement Techniques. *Review of Scientific Instruments* 75 (4): 797–810.

Lakes, R.S. (2009). *Viscoelastic Materials*. Cambridge: Cambridge University Press.

Madigosky, W. and Lee, G. (1983). Improved resonance technique for materials characterization. *Journal of Acoustic Society of America* 73: 1374–1377.

Magrab, E. (1984). Torqued cylinder apparatus. *Journal of Research of National Bureau of Standards* 89: 193–207.

Mark, J.E. (2007). *Physical Properties of Polymers Handbook*, 2e. New York, NY: Springer.

McHugh J., "Ultrasound technique for the dynamic mechanical analysis (DMA) of polymers", Ph.D. Thesis, The Technical University in Berlin, BAM (Bundesanstalt for Materialforschung und–prüfung) 2008.

Mousavi, S., Nicolas, D.F., and Lundberg, B. (2004). Identification of complex moduli and Poisson's ratio from measured strains on an impacted bar. *Journal of Sound and Vibration* 277: 971–986.

Muller, P. (2005). Are the Eigensolutions of a 1-D.O.F. system with viscoelastic damping oscillatory or not? *Journal of Sound and Vibration* 285: 501–509.

Nashif, A., Jones, D., and Henderson, J. (1985). *Vibration Damping.* New York, NY: Wiley.

Park, S.W. and Schapery, R.A. (1999). Methods of interconversion between linear viscoelastic material functions. Part I - a numerical method based on Prony series. *International Journal of Solids and Structures* 36: 1653–1675.

Parot, J.-M. and Duperray, B. (2008). Exact computation of creep compliance and relaxation modulus from complex modulus measurement data. *Mechanics of Materials* 40: 575–585.

Qiao, J., Amirkhizi, A., Schaaf, K. et al. (2011). Dynamic mechanical and ultrasonic properties of polyurea. *Mechanics of Materials* 43: 598–607.

Rose, J.L. (1999). *Ultrasonic Waves in Solid Media.* Cambridge, UK: Cambridge University Press.

Schapery, R.A. and Park, S.W. (1999). Methods of interconversion between linear viscoelastic material functions. Part II - an approximate analytical method. *International Journal of Solids and Structures* 36: 1677–1699.

Schwarzl, F.R. (1970). On the interconversion between viscoelastic material functions. *Pure and Applied Chemistry* 23 (2–3): 219–234.

Ward, I.M. (1983). *Mechanical Properties of Polymers*, 2e. Chichester: John Wiley & Sons, Ltd.

Zhang, R., Jiang, W.H., and Gao, W.W. (2005). Frequency dispersion of ultrasonic velocity and attenuation of longitudinal waves propagating in 0.68Pb $(Mg_{1/3}\ Nb_{2/3})O_3 - 0.32PbTiO_3$ single crystals poled along [001] and [110]. *Applied Physics Letters* 87: 182903.

3.A Convolution Theorem

Theorem

If $X(s)$ is the Laplace transform of a time-domain signal $x(t)$ such that:

$$X(s) = F_1(s)F_2(s) \tag{3.A.1}$$

where $F_1(s)$ and $F_2(s)$ are the Laplace transforms of the functions $f_1(t)$ and $f_2(t)$ respectively. Then, $x(t)$ is given by:

$$x(t) = \int_0^t f_1(\tau)f_2(t-\tau)d\tau \tag{3.A.2}$$

Proof

From the definition of the Laplace transform:

$$X(s) = F_1(s)F_2(s)$$

$$= \int_{\tau}^{\infty} f_1(v)e^{-sv}dv \int_{0}^{\infty} f_2(\tau)e^{-s\tau}d\tau = \int_{0}^{\infty}\left[\int_{0}^{\infty} f_1(v)e^{-s(v+\tau)}dv\right]f_2(\tau)d\tau \tag{3.A.3}$$

Let $t = v + \tau$. If τ is held constant, then $d\tau = dv$ and Eq. (3.A.3) reduces to:

$$X(s) = \int_{0}^{\infty}\left[\int_{\tau}^{\infty} f_1(t-\tau)e^{-st}dt\right]f_2(\tau)d\tau \, Rubber$$

Reversing the order of integration gives:

$$X(s) = \int_{0}^{\infty}\left[\int_{0}^{t} f_1(t-\tau)f_2(\tau)d\tau\right]e^{-st}dt \, Plexi-glass$$

Applying the inverse Laplace transform on $X(s)$ gives:

$$L^{-1}[X(s)] = x(t) = \int_{0}^{t} f_1(t-\tau)f_2(\tau)d\tau \, HDPE$$

Problems

3.1 Show that the viscous damping coefficient $c_{d_{T_0}}$, at a given temperature T_0, of the Maxwell model can be calculated from the integration of the relaxation modulus $E(t,T_0)$ with respect to time t as follows:

$$c_{d_{T_0}} = \int_{0}^{\infty} E(t, T_0)dt$$

If the temperature is changed to T and the time is shifted to a time t' such that:

$$t' = t/\alpha_T$$

where α_T is the temperature shift factor, show that:

$$c_{d_T} = \int_{0}^{\infty} E(t', T)\, dt' = c_{d_{T_0}}\alpha_T$$

Or that the temperature shift factor α_T is given by:

$$\alpha_T = c_{d_T}/c_{d_{T_0}}$$ [1]

1 Similar result can be obtained using Rouse equation that relates the ratio of the relaxation time τ at one temperature T to that at a reference temperature T_0 to the temperature shift factor as $\tau_T/\tau_{T_0} \cong c_{d_T}/c_{d_{T_0}} = \alpha_T$ (Brinson and Brinson, 2008).

3.2 The properties of the VEM *Dyad*609 (*Soundcoat, Inc.*) are listed here at different temperatures and frequencies.

Temp (°F)	Freq (Hz)	G (psi)	Loss factor
25	10	70,000	0.045 0
50	10	60,000	0.180 0
75	10	18,000	0.410 0
100	10	3200	0.750 0
125	10	1000	0.950 0
150	10	450	0.600 0
175	10	300	0.300 0
200	10	230	0.180 0
225	10	210	0.110 0
250	10	210	0.062 0
275	10	210	0.041 0
300	10	210	0.029 0
25	100	70,000	0.029 0
50	100	69,000	0.120 0
75	100	40,000	0.270 0
100	100	13,000	0.500 0
125	100	3200	0.750 0
150	100	1300	0.950 0
175	100	550	0.700 0
200	100	350	0.420 0
225	100	280	0.280 0
250	100	220	0.160 0
275	100	215	0.110 0
300	100	210	0.075 0
25	1000	70,000	0.018 0
50	1000	70,000	0.080 0
75	1000	60,000	0.170 0
100	1000	31,000	0.310 0
125	1000		0.500 0
150	1000	3900	0.720 0
175	1000	1700	0.930 0
200	1000	800	0.900 0
225	1000	500	0.600 0
250	1000	330	0.400 0
275	1000	290	0.270 0
300	1000	260	0.170 0

Generate the master curves for this material and plot the corresponding temperature shift factor using the WLF formula. Compare the obtained results with those generated by the Arrhenius formula.

3.3 For the Poynting–Thomas model, show that the creep compliance $J(t)$ and the relaxation modulus $E(t)$ are given by:

(a) $J(t) = \frac{1}{E_\infty}\left[1 - \left(1 - \frac{E_\infty}{E_s}\right)e^{-t/\lambda}\right]$, where $\lambda = \frac{c_d}{E_p}$ and $E_\infty = \frac{E_s E_p}{(E_s + E_p)}$

(b) $E(t) = E_\infty\left[1 + \left(\frac{E_s}{E_\infty} - 1\right)e^{-t/\alpha}\right]$, where $\alpha = \frac{c_d}{(E_s + E_p)}$.

Using $J(t)$ and $E(t)$ determine $J^*(\omega)$ and $E^*(\omega)$. Show that:

$$E' = E_\infty\left[\frac{1 + \alpha\lambda\omega^2}{1 + \alpha^2\omega^2}\right] \text{ and } \eta = (\lambda - \alpha)\omega/\left[1 + \alpha\lambda\omega^2\right]$$

Solve both by hand and by MATLAB "*fourier*" command. Check and comment on the results.

3.4 The relaxation modulus $E(t$ and) the creep compliance $J(t)$ of VEM are usually given by the GMM or the generalized Kelvin–Voigt model shown in Figure P3.1a,b respectively.

Mathematically, these models are represented by the following Prony series:

$$E(t) = E_0 + \sum_{i=1}^{n} E_i e^{-t/\rho_i} \quad \textit{Generalized Maxwell model}$$

$$\text{and } J(t) = J_0 + \sum_{j=1}^{m} J_j\left(1 - e^{-t/\tau_j}\right) \quad \textit{Generalized Kelvin – Voigt model}$$

where E_0 = the equilibrium modulus, E_i = relaxation strength, and ρ_i = relaxation times = η_i/E_i. Also, J_0 = the glassy compliance, J_j = retardation strength, and τ_j = retardation times such that $J_0 = 1/E_0$, $J_j = 1/E_j$, $\tau_j = \eta_j/E_j$. Note that E_0, E_i, ρ_i, J_0, J_j, and τ_j are all positive constants to be identified experimentally.

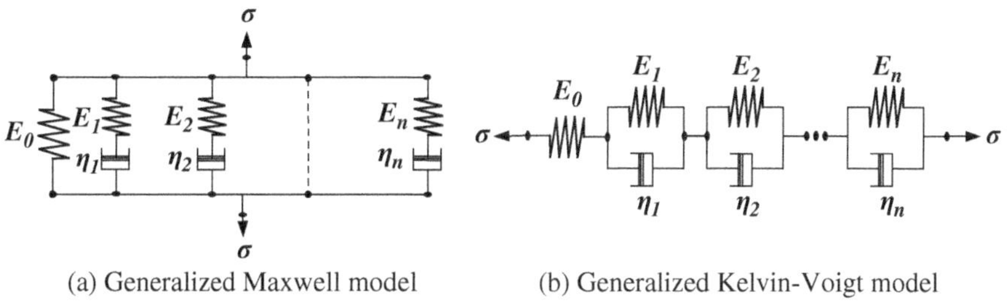

Figure P3.1 (a) Generalized Maxwell model and (b) generalized Kelvin–Voigt model.

Show that:

$$a.E^*(\omega) = E_0 + \sum_{k=1}^{n} E_k \frac{i\omega\rho_k}{i\omega\rho_k + 1} \quad \text{and} \quad b.J^*(\omega) = J_0 + \sum_{l=1}^{m} J_l \left(1 - \frac{i\omega\tau_l}{i\omega\tau_l + 1}\right)$$

3.5 Consider a VEM described by either of the following creep compliance $J(t)$ or the relaxation modulus $E(t)$:

$$E(t) = E_0 + \sum_{i=1}^{n} E_i e^{-t/\rho_i} \quad \text{or} \quad J(t) = J_0 + \sum_{j=1}^{n} J_j\left(1 - e^{-t/\tau_j}\right)$$

such that ρ_i and τ_i are selected a priori and set equal to each other. Then, when one set of constants either "E_0, E_i," or "J_0, J_j" is known, the other "*unknown*" set of constants can be determined by solving the following equation:

$$\int_0^t E(t-\tau)\cdot\frac{dJ(\tau)}{d\tau}d\tau = 1$$

In case "E_0, E_i, ρ_i" are known and it is desired to determine "J_0, J_j," show that this equation results in the following set of linear equations:

AD = B

where A, D, and B are matrices with dimensions $P \times n$, $n \times 1$, and $P \times 1$, respectively, with P denoting the number of time samples and n defines the number of terms of the Prony series.

The elements of matrices **A, B,** and **D** are given by:

$$A_{kj} = E_0\left(1 - e^{-t_k/\tau_j}\right) + \sum_{i=1}^{n} \frac{\rho_i E_i}{\rho_i - \tau_j}\left(e^{-t_k/\rho_i} - e^{-t_k/\tau_j}\right) \quad when\, \rho_i \neq \tau_j$$

$$= E_0\left(1 - e^{-t_k/\tau_j}\right) + \sum_{i=1}^{n} \frac{t_k E_i}{\tau_j} e^{-t_k/\rho_i} \quad when\, \rho_i = \tau_j$$

where $k = 1,\ldots,n$, $p = 1,\ldots,P$

$$B_k = 1 - \left(E_0 + \sum_{i=1}^{n} E_i e^{-t_k/\rho_i}\right) \Big/ \left(E_0 + \sum_{i=1}^{n} E_i\right),$$

$$D_k = [J_1\, J_2 \ldots J_N]^T,$$

and

$$J_0 = 1 \Big/ \left(E_0 + \sum_{i=1}^{n} E_i\right).$$

Consider a VEM that has $E_0 = 2.24E6\ \mathrm{N\,m^{-2}}$ and has the following relaxation modulus parameters E_i, ρ_i for a 12-term Prony series:

Term	1	2	3	4	5	6	7	8	9	10	11
E_i $(GN\,m^{-2})$	0.194	0.283	0.554	0.602	0.388	0.156	0.041	0.013 8	0.003 68	0.79E-3	0.96E-3
ρ_i (sec)	2E-2	2E-1	2E0	2E1	2e2	2E3	2E4	2E5	2E6	2E7	2E8

Determine:

(a) the complex modulus of the material and plot its storage modulus and loss factor as function of the frequency.

(b) the parameters J_0 and J_j $(j = 1,\ldots n)$ for an equivalent creep compliance as described by a generalized Kelvin–Voigt model.

(c) the complex modulus of the material as obtained from the creep compliance calculated in item (b). Plot the storage modulus and loss factor as function of the frequency. Compare with those obtained in item (a).

3.6 The relaxation modulus for a VEM is determined experimentally as listed in the following table:

Time – s	1E-6	1E-5	1E-4	1E-3	1E-2	1E-1	1E-0
$E(t)$ – $\mathrm{MN\,m^{-2}}$	382.9	155.6	30.3	6.5	3.93	3.27	3.26

Determine:

(a) the optimal parameters E_i $(i = 0,\ldots, n)$ of the Prony series:

$$E(t) = E_0 + \sum_{i=1}^{n} E_i e^{-t/\rho_i}$$

Using MATLAB "*optimtool*" GUI (Graphical User Interface). Assume that $n = 7$ and ρ_i is given by:

Term	1	2	3	4	5	6	7
ρ_i (s)	1E-6	1E-5	1E-4	1E-3	1E-2	1E-1	1E0

(b) the complex modulus of the material and plot its storage modulus and loss factor as function of the frequency by the following two methods:

i) Using the Laplace transform of the Prony series as outlined in Problem 3.5.

ii) Using the Fourier transform (MATLAB "*fft*" command) on the original data. Compare and comment on the results.

3.7 Consider a mass supported on a VEM as shown in Figure P3.2 that has a relaxation stiffness represented by the following GMM:

$$K(t) = K_0\left(1 + \alpha_1 e^{-t/\rho_1}\right)$$

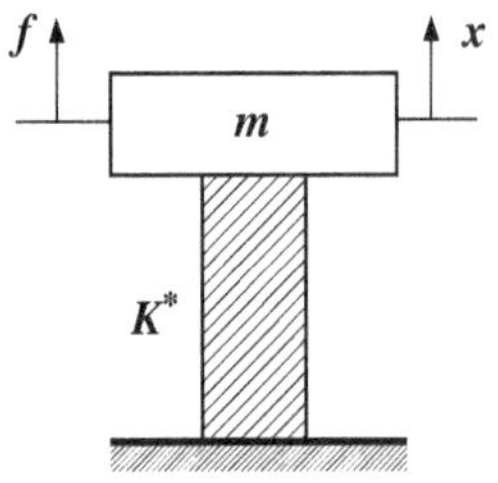

Figure P3.2 A mass supported on a VEM.

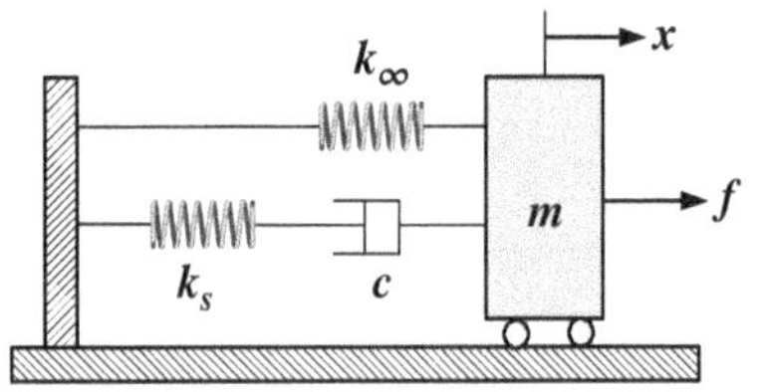

Figure P3.3 Mass supported on a Zener VEM model.

Show that the complex stiffness of the VEM is given by:

$$K^*(s) = K_0\left(1 + \alpha_1 \frac{s\rho_1}{s\rho_1 + 1}\right)$$

Using the following transformation equation:

$$z = \frac{1}{\rho_1 s + 1} x$$

where z is called an internal degree of freedom for describing the dynamics of the VEM.

Show that the dynamics of the mass/VEM system is given by:

$$\begin{Bmatrix} \dot{x} \\ \ddot{x} \\ \dot{z} \end{Bmatrix} = \begin{bmatrix} 0 & 1 & 0 \\ -\frac{K_0(1+\alpha_1)}{m} & 0 & \frac{K_0\alpha_1}{m} \\ \frac{1}{\rho_1} & 0 & -\frac{1}{\rho_1} \end{bmatrix} \begin{Bmatrix} x \\ \dot{x} \\ z \end{Bmatrix} + \begin{Bmatrix} 0 \\ \frac{1}{m} \\ 0 \end{Bmatrix} f$$

3.8 Consider a mass m that is supported on a VEM described by a Zener model as shown in Figure P3.3. Show that the equation of motion of the mass is given by:

$$m\ddot{x}(t) + k_s \int_{-\infty}^{t} e^{-(t-\tau)/\lambda}\dot{x}(\tau)d\tau + k_\infty x(t) = f(t)$$

where $\lambda = c/k_s$.

(*Hint*: Derive first an expression of the relaxation modulus (i.e., stiffness $k(t)$) of the Zener model, then use the convolution integral of Eq. (3.48) to account for the effect of the VEM on the dynamics of the mass m).

3.9 Investigate the eigensolutions of the mass/VEM system described in Problem 3.8 as described by the homogeneous equation:

$$m\ddot{x}(t) + k_s \int_{-\infty}^{t} e^{-(t-\tau)/\lambda}\dot{x}(\tau)d\tau + k_\infty x(t) = 0$$

Assume that the eigensolution takes the form $x(t) = e^{st}$ and that the parameters k_s and k_∞ are related by $\alpha\, k_\infty = \lambda\, k_0$ where $k_0 = k_s + k_\infty$. Note that the parameters α and λ are denoted as the relaxation and the retardation times respectively (Muller 2005).

Show that the characteristic equation of the system reduces to:

$$s^3 + \frac{1}{\lambda}s^2 + \omega'^2 s + \frac{1}{\alpha}\omega'^2 = 0 \,\text{where}\, \omega'^2 = \frac{k_0}{m}$$

3.10 Assume that the parameters of the mass/VEM system of Problem 3.8 $\lambda = 1$ and $\omega' = 1$. Then put the characteristics equation of the system in the following root locus form:

$$1 + \frac{\beta}{s(s^2 + s + 1)} = 0 \;\; where\; \beta = \frac{1}{\alpha}$$

Plot the root locus of the characteristics equation for different values of the parameter α. Determine the range of values of α for which the system is stable (i.e., when the real parts of all the roots of the characteristic equation are negative).

4

Viscoelastic Materials

4.1 Introduction

Classical models of viscoelastic materials (VEMs), such as those by Maxwell, Kelvin–Voigt, Poynting and Thomson, and Zener, have been shown in Chapter 2 to have serious limitations in modeling accurately the behavior of realistic VEMs. Furthermore, the use of the classical approach of the *complex modulus* to represent the dynamics of VEMs is limited to *frequency domain* analysis. To avoid such limitations of these classical models and approaches, several advanced models of the VEMs have been developed. These include: Fractional Derivative (FD) model (Bagley and Torvik 1983), Golla–Hughes–McTavish (GHM) model (Golla and Hughes 1985), and the Augmented Temperature Fields (ATF) (Lesieutre and Mingori 1990).

In this chapter, considerable effort is devoted to present the merits and limitations of these models because of their wide and ease of use in the analysis of damped structures. Particular emphasis is placed on integrating the models with finite element models simulating the dynamics of structures treated with VEMs. Such integration is essential to the prediction of the response of these structures both in the time and frequency domains. This enables the computation of the structural response to transient, shock, as well as sinusoidal loading.

4.2 Golla–Hughes–McTavish (GHM) Model

The GHM model was developed by Golla and Hughes in 1985. The model describes the shear modulus of VEMs with a second order differential equation unlike the first order differential equations used to describe the Maxwell, Kelvin–Voigt, Poynting-Thomas, and Zener models. Such a distinction makes it easy to incorporate the dynamics of the VEMs into finite element models of vibrating structures.

According to the GHM formulation, the shear modulus G of VEMs can be written in the Laplace domain as

$$G(s) = G_0\left(1 + \sum_{n=1}^{N} \alpha_n \frac{s^2 + 2\zeta_n\omega_n s}{s^2 + 2\zeta_n\omega_n s + \omega_n^2}\right) \tag{4.1}$$

where G_0 is the equilibrium value of the modulus, that is, the initial value of $G(\omega = 0)$, and s is the Laplace domain variable. The parameters α_n, ζ_n and ω_n are obtained from curve

Active and Passive Vibration Damping, First Edition. Amr M. Baz.

fitting the complex modulus data for a particular VEM at a given temperature. The summation may be thought of as representing the material modulus as a series of mini-oscillators (second order equations) as suggested by Golla and Hughes (1985). These terms are a representation of the internal variables necessary to describe the characteristics of the VEMs. The number of terms kept in the expansion is determined by accuracy needed to replicate the real behavior of the material. In many cases, only two to four terms are necessary.

4.2.1 Motivation of the *GHM* Model

Consider a VEM described by one mini-oscillator (i.e., $n = 1$), which is coupled with a mass such that the equation of motion in the Laplace domain is

$$Ms^2\mathbf{x}(s) + K(s)\mathbf{x}(s) = \mathbf{F}(s) \tag{4.2}$$

where M is the mass, K the complex stiffness of the VEM, and $\mathbf{F}$ the forcing function, so Eq. (4.2) can be written as

$$Ms^2\mathbf{x}(s) + K\left(1 + \alpha_n \frac{s^2 + 2\zeta_n\omega_n s}{s^2 + 2\zeta_n\omega_n s + \omega_n{}^2}\right)\mathbf{x}(s) = \mathbf{F}(s). \tag{4.3}$$

$$\text{Let } z = \frac{\omega_n^2}{s^2 + 2\zeta_n\omega_n s + \ \omega_n^2 x} \tag{4.4}$$

Then, in the time domain, Equation (4.3) reduces to

$$\ddot{z} + 2\zeta_n\omega_n\dot{z} = \omega_n^2(x - z) \tag{4.5}$$

Substituting Eq. (4.4) into Eq. (4.5) gives:

$$Ms^2x + Kx + \frac{K\alpha_n}{\omega_n{}^2}\left(s^2 + 2\zeta_n\omega_n s\right)z = F(s)$$

$$\textit{In the time domain: } M\ddot{x} + Kx + \frac{K\alpha_n}{\omega_n{}^2}\omega_n{}^2(x - z) = f \tag{4.6}$$

Equation (4.6) reduces to

$$M\ddot{x} + (K + K\alpha)x - K\alpha z = f \tag{4.7}$$

Rewriting Eq. (4.5) as follows

$$\ddot{z} - \omega_n{}^2x + 2\zeta_n\omega_n\dot{z} + \omega_n{}^2z = 0 \tag{4.8}$$

and combining Eqs. (4.7) and (4.8) in a matrix form, gives:

$$\begin{bmatrix} M & 0 \\ 0 & 1 \end{bmatrix}\begin{Bmatrix} \ddot{x} \\ \ddot{z} \end{Bmatrix} + \begin{bmatrix} 0 & 0 \\ 0 & 2\zeta_n\omega_n \end{bmatrix}\begin{Bmatrix} \dot{x} \\ \dot{z} \end{Bmatrix} + \begin{bmatrix} K + K\alpha_n & -K\alpha_n \\ -\omega_n{}^2 & \omega_n{}^2 \end{bmatrix}\begin{Bmatrix} x \\ z \end{Bmatrix} = \begin{Bmatrix} f \\ 0 \end{Bmatrix} \tag{4.9}$$

This system has asymmetric stiffness matrix. However, by multiply last row by $\frac{K\alpha_n}{\omega_n{}^2}$, makes the stiffness matrix symmetric and as follows

$$\begin{bmatrix} M & 0 \\ 0 & \dfrac{K\alpha_n}{\omega_n{}^2} \end{bmatrix} \begin{Bmatrix} \ddot{x} \\ \ddot{z} \end{Bmatrix} + \begin{bmatrix} 0 & 0 \\ 0 & \dfrac{2\zeta_n K\alpha_n}{\omega_n} \end{bmatrix} \begin{Bmatrix} \dot{x} \\ \dot{z} \end{Bmatrix} + \begin{bmatrix} K + K\alpha_n & -K\alpha_n \\ -K\alpha_n & K\alpha_n \end{bmatrix} \begin{Bmatrix} x \\ z \end{Bmatrix} = \begin{Bmatrix} f \\ 0 \end{Bmatrix} \tag{4.10}$$

Equation (4.10) governs the dynamics of the mechanical system shown in Figure 4.1. Hence, the VEM is represented by a spring-mass-damper assembly that is connected in parallel with another spring K. Note that K, from Eq. (4.3), represents the stiffness of the VEM under static conditions (i.e., at zero frequency ω).

It is important here to note that z defines an "internal degree of freedom" (IDOF), which describes the motion of a VEM modeled by a single mini-oscillator. More IDOFs would be added when the VEM is modeled by N mini-oscillators. The addition of these damping IDOFs increases the size of the equations of motion of the structure considerably. Application of classical model reduction techniques, such as Guyan Reduction, is essential to reduce the size of the structure/VEM model to include only the structural degrees of freedom (DOF) in order to enhance the computational efficiency.

For a VEM that is modeled with N mini-oscillators, the equivalent mechanical system will be as shown in Figure 4.2.

Figure 4.1 Equivalent system of the GHM model.

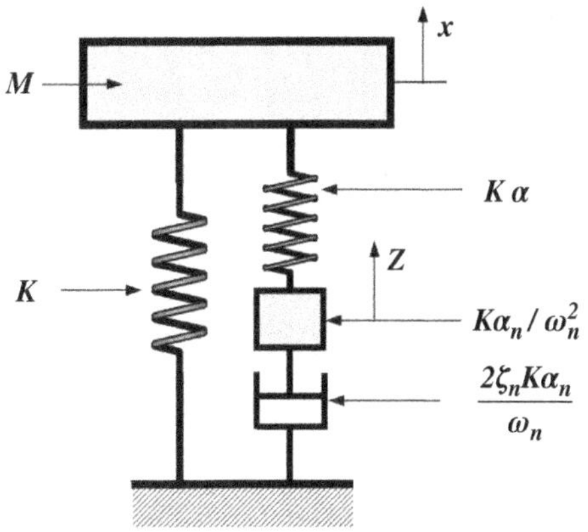

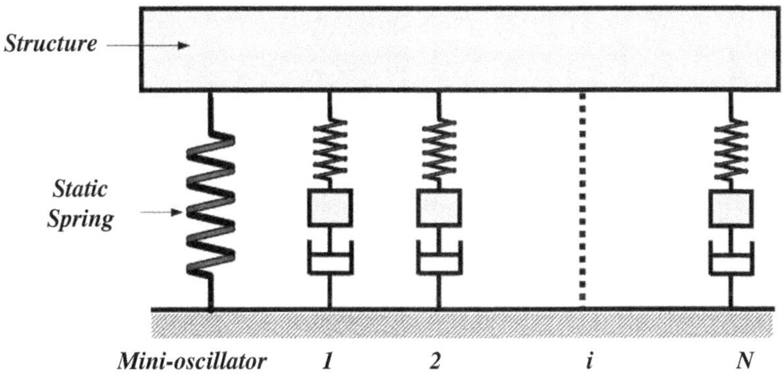

Figure 4.2 Equivalent system of the GHM model with N mini-oscillators.

Accordingly, the equations governing a model with N mini-oscillators are given by:

$$Ms^2x + K\left[1 + \alpha_1 \frac{s^2 + 2\zeta_1\omega_1 s}{s^2 + 2\zeta\omega_1 s + \omega_1^2} + \alpha_2 \frac{s^2 + 2\zeta_2\omega_2 s}{s^2 + 2\zeta_2\omega_2 s + \omega_2^2} + \ldots\ldots\ldots\ldots\right]x = F(s)$$

$$\text{Let}\, z_1 = \frac{\omega_1^2}{s^2 + 2\zeta_1\omega_1 s + \omega_1^2}x,\; z_2 = \frac{\omega_2^2}{s^2 + 2\zeta_2\omega_2 s + \omega_2^2}x,\; \ldots\ldots$$

Then,

$$\begin{bmatrix} M & 0 & 0 & \ldots & 0 \\ 0 & \frac{K\alpha_1}{\omega_1^2} & 0 & \ldots & 0 \\ 0 & 0 & \frac{K\alpha_2}{\omega_2^2} & \ldots & 0 \\ \ldots & \ldots & \ldots & \ldots & \ldots \\ 0 & 0 & 0 & \ldots & \frac{K\alpha_N}{\omega_N^2} \end{bmatrix} \begin{Bmatrix} \ddot{x} \\ \ddot{z}_1 \\ \ddot{z}_2 \\ \ldots \\ \ddot{z}_N \end{Bmatrix} + \begin{bmatrix} 0 & 0 & 0 & \ldots & 0 \\ 0 & \frac{2\zeta_1\alpha_1 K}{\omega_1} & 0 & \ldots & 0 \\ 0 & 0 & \frac{2\zeta_2\alpha_2 K}{\omega_2} & \ldots & 0 \\ \ldots & \ldots & \ldots & \ldots & \ldots \\ 0 & 0 & 0 & \ldots & \frac{2\zeta_N\alpha_N K}{\omega_N} \end{bmatrix} \begin{Bmatrix} \dot{x} \\ \dot{z}_1 \\ \dot{z}_2 \\ \ldots \\ \dot{z}_N \end{Bmatrix} +$$

$$\begin{bmatrix} K(1+\alpha_1+\alpha_2+\ldots) & -K\alpha_1 & -K\alpha_2 & \ldots & -K\alpha_N \\ -K\alpha_1 & K\alpha_1 & & \ldots & 0 \\ -K\alpha_2 & 0 & K\alpha_2 & \ldots & 0 \\ \ldots & \ldots & \ldots & \ldots & 0 \\ -K\alpha_N & 0 & 0 & 0 & K\alpha_N \end{bmatrix} \begin{Bmatrix} x \\ z_1 \\ z_2 \\ \ldots \\ z_N \end{Bmatrix} = \begin{Bmatrix} F \\ 0 \\ 0 \\ 0 \\ 0 \end{Bmatrix} \quad (4.11)$$

Example 4.1 Consider a mass = 100 kg supported on a VEM, which is modeled by the following GHM model with one mini-oscillator such that its stiffness is given by:

$$K = 100\left(1 + 7\frac{s^2 + 2000s}{s^2 + 2000s + 1E6}\right)$$

that is, $K = 100$, $\alpha_1 = 7$, $\zeta_1 = 1$, and $\omega_1 = 1000$. Determine the frequency response of the mass to a unit sinusoidal force when considering the full state vector $\{x\ z\}^T$. Compare the results with the response when the IDOF z of the VEM is eliminated.

Solution

Figure 4.3 shows the storage and the loss factor of the VEM under consideration.
The equations governing the dynamics of the mass/VEM system are given by:

$$\begin{bmatrix} 100 & 0 \\ 0 & 0.0007 \end{bmatrix} \begin{Bmatrix} \ddot{x} \\ \ddot{z} \end{Bmatrix} + \begin{bmatrix} 0 & 0 \\ 0 & 1.4 \end{bmatrix} \begin{Bmatrix} \dot{x} \\ \dot{z} \end{Bmatrix} + \begin{bmatrix} 800 & -700 \\ -700 & 700 \end{bmatrix} \begin{Bmatrix} x \\ z \end{Bmatrix} = \begin{Bmatrix} \sin(\omega t) \\ 0 \end{Bmatrix} \quad (4.12)$$

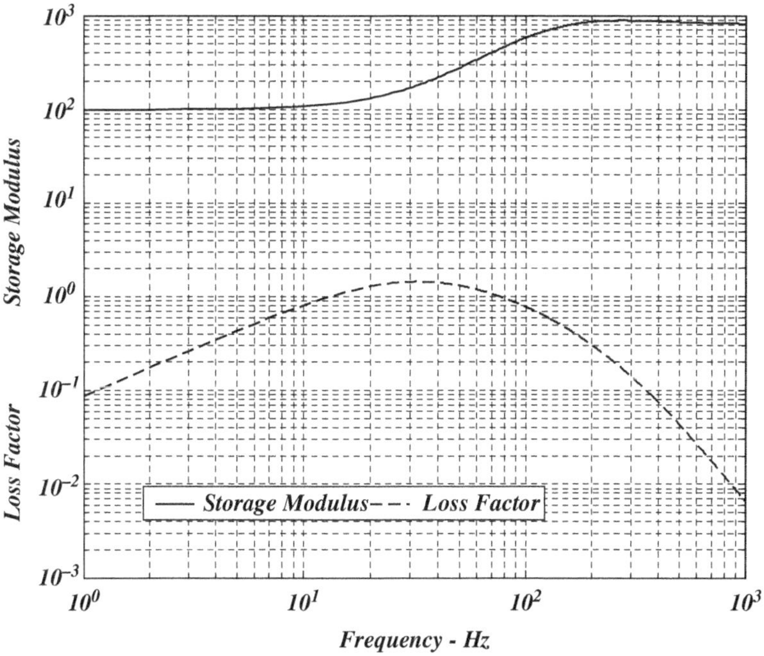

Figure 4.3 The storage and the loss factor of the VEM.

Hence, the steady-state response of the full state vector $\{x\ z\}^T$ will be given by:

$$\begin{Bmatrix} x \\ z \end{Bmatrix} = \left(-\begin{bmatrix} 100 & 0 \\ 0 & 0.0007 \end{bmatrix}\omega^2 + i\omega\begin{bmatrix} 0 & 0 \\ 0 & 1.4 \end{bmatrix} + \begin{bmatrix} 800 & -700 \\ -700 & 700 \end{bmatrix}\right)^{-1}\begin{Bmatrix} 1 \\ 0 \end{Bmatrix} \tag{4.13}$$

If one considers the static conditions only, then Eq. (4.12) reduces to:

$$\begin{bmatrix} 800 & -700 \\ -700 & 700 \end{bmatrix}\begin{Bmatrix} x \\ z \end{Bmatrix} = \begin{Bmatrix} F \\ 0 \end{Bmatrix}$$

which implies that $700x = 700z$ or $x = z$. Hence, one can write the full state vector $\{x\ \ z\}^T$ in terms of the structural DOF as follows:

$$\begin{Bmatrix} x \\ z \end{Bmatrix} = \begin{Bmatrix} 1 \\ 1 \end{Bmatrix} x = Tx \tag{4.14}$$

where T is a transformation matrix.

Using this transformation matrix, Eq. (4.12) can be *condensed* to be in terms of the structural DOF x as follows:

$$T^T\begin{bmatrix} 100 & 0 \\ 0 & 0.0007 \end{bmatrix}T\ddot{x} + T^T\begin{bmatrix} 0 & 0 \\ 0 & 1.4 \end{bmatrix}T\dot{x} + T^T\begin{bmatrix} 800 & -700 \\ -700 & 700 \end{bmatrix}Tx = \sin(\omega t)$$

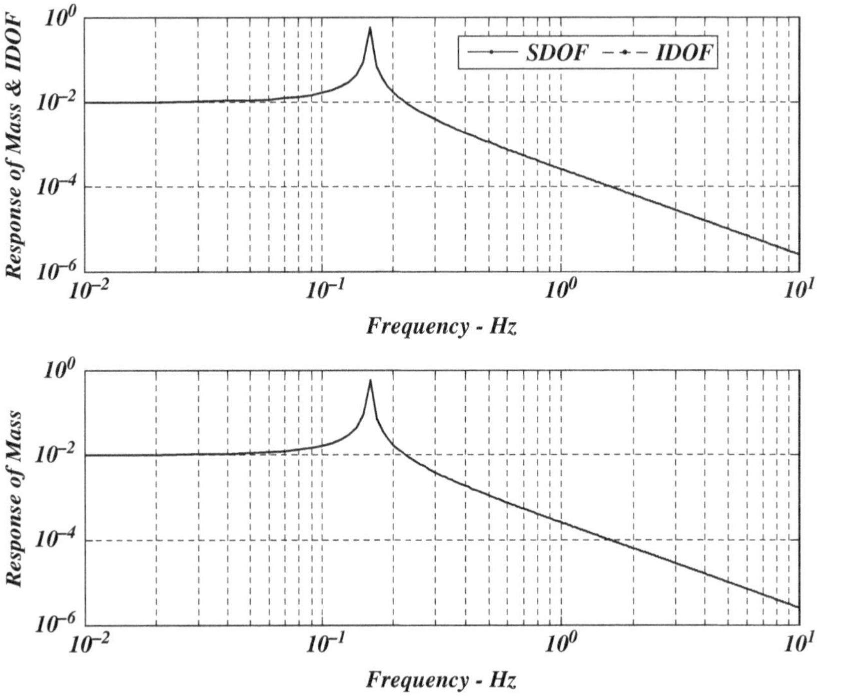

Figure 4.4 The response of the structure and IDOF of the VEM.

which has a the following steady-state solution:

$$x = \left(-T^T \begin{bmatrix} 100 & 0 \\ 0 & 0.0007 \end{bmatrix} T\omega^2 + i\omega T^T \begin{bmatrix} 0 & 0 \\ 0 & 1.4 \end{bmatrix} T + T^T \begin{bmatrix} 800 & -700 \\ -700 & 700 \end{bmatrix} T \right)^{-1} \tag{4.15}$$

Figure 4.4 shows a comparison between the responses of the mass as obtained by Eqs. (4.13) and (4.15), that is, based on the full order model or the reduced order model. The figure shows also the IDOF of the VEM as obtained by Eq. (4.13) and by Eq. (4.15) when coupled with the condensation Eq. (4.14).

It is evident that using the static condensation approach described here, which is also called the "Guyan Reduction," has resulted in a computationally efficient and accurate predictions of the structural response.

4.2.2 Computation of the Parameters of the *GHM* Mini-Oscillators

The parameters:G_0, α_n, ζ_n, and ω_n of the GHM mini-oscillators are determined in such a way that replicates closely the experimental behavior of actual VEM. The optimal values of these parameters are selected, on a rational basis, in order to minimize the deviations between the predictions of the GHM model and the experimental data. The optimization problem is formulated as follows:

$$\left\langle \begin{array}{l} \text{Determine } G_0,\ \alpha_n,\ \zeta_n, \text{ and } \omega_n \\ \text{To minimize } F = \sum_{Frequency} \left[\left(G'_{GHM} - \frac{G'_{Exp}}{G'_{Exp}} \right)^2 + \left(\eta_{GHM} - \frac{\eta_{Exp}}{\eta_{Exp}} \right)^2 \right] \\ \text{Such that } G_0,\ \alpha_n,\ \zeta_n, \text{ and } \omega_n > 0 \\ \text{and } G_0 \left(1 + \sum_{n=1}^{N} \alpha_n \right) = G_\infty \end{array} \right\rangle \tag{4.16}$$

where G_0 and G_∞ are the storage moduli of the VEM at $\omega = 0$ and ∞, respectively. Also, N denotes the number of the GHM mini-oscillators.

In Eq. (4.16), the objective function F is formulated in order to minimize the sum of the prediction error of the GHM model of both the storage modulus and the loss factor as computed over the entire experimental frequency range. The errors are cast in a normalized and quadratic form in order to make the optimization problem well-conditioned. Furthermore, constraints are imposed to guarantee that the parameters are all nonnegative and that the model can predict the storage modulus at large frequencies.

Figure 4.5 outlines the flow chart of the optimization process.

Example 4.2 Consider the experimental data listed in Table 4.1 for a VEM that is tested over a frequency range between 20 and 5000 Hz. Compute the optimal parameters of *GHM* models of the previous VEM using one, two, and three mini-oscillators. Compare the prediction accuracies of the three models.

Solution

For all the considered models, it is assumed that $\zeta_n = 1$

a) One mini-oscillator

From the experimental data, $G_0 = 750$ psi, and $G_\infty = 30{,}000$ psi, then $\alpha_1 = 39$. Then, starting with an initial guess of $\omega_1 = 10{,}000$ rad s^{-1}, a MATLAB solution of the optimization problem using the "fmincon" subroutine of the Optimization Toolbox gives an optimal $\omega_1 = 11{,}713.94$. The corresponding value of the objective function $F = 41.97$.

b) Two mini-oscillators

The parameters to be determined are: α_1, ω_1, α_2, and ω_2. An initial guess is assumed to be: 20, 500, 20, and 1000. The optimal values obtained are: 5.7, 1154.6, 33.3, and 30,515.55 with an optimal objective function value of 4.22. Hence, with two mini-oscillators, the prediction accuracy of the GHM has improved by a factor of 41.97/4.22 = 9.945.

c) Three mini-oscillators

The parameters to be determined are: α_1, ω_1, α_2, ω_2, α_3, and ω_3. An initial guess is assumed to be: 18, 1000, 15, 1000, 6, and 5000. The optimal values obtained are:

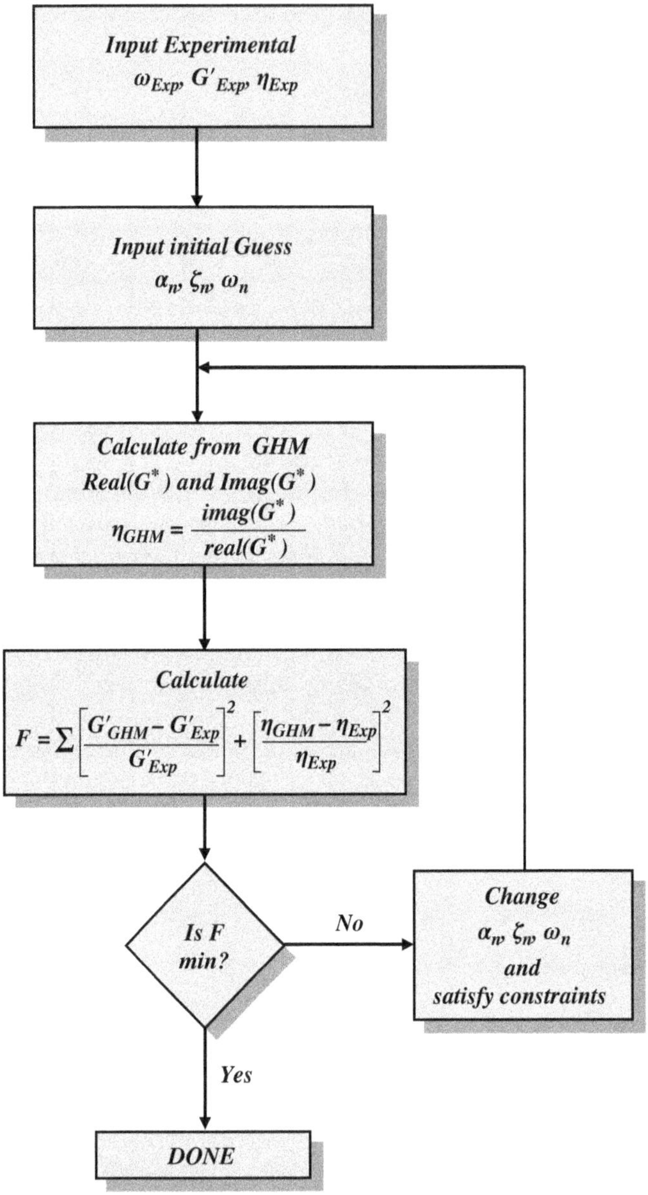

Figure 4.5 Flowchart of the optimal selection of the GHM parameters.

31.36, 31 840.6, 6.91, 2216.9, 0.728, and 0.001 with an optimal objective function value of 1.59. Hence, with three mini-oscillators, the prediction accuracy of the GHM has improved by a factor of 41.97/2.459 = 17.06.

Figure 4.6 shows comparisons between the predictions of the three GHM models and the experimental results.

Table 4.1 Storage and loss factor of a VEM.

Frequency (Hz)	20	30	40	50	60	70	80	90	100	200	300
G' (kpsi)	0.75	1.1	1.3	1.6	1.8	2	2.2	2.3	2.6	4	5.2
Loss factor η	0.95	0.96	0.98	0.99	1.	1.	1.	1.	1.	0.99	0.95
Frequency (Hz)	**400**	**500**	**600**	**700**	**800**	**900**	**1000**	**2000**	**3000**	**4000**	**5000**
G' (kpsi)	6.3	7.2	8.5	9.5	11	12	13	19	23	26	30
Loss factor η	0.91	0.88	0.85	0.82	0.79	0.77	0.75	0.63	0.56	0.51	0.48

4.2.3 On the Structure of the GHM Model

4.2.3.1 Other Forms of GHM Structures

There are several forms that are proposed for the GHM model other than the original three-parameter (α_n, ζ_n, ω_n) structure given by Eq. (4.1) as:

$$G(s) = G_0\left(1 + \alpha_n \frac{s^2 + 2\zeta_n\omega_n s}{s^2 + 2\zeta_n\omega_n s + \omega_n^2}\right) \tag{4.17}$$

Among these forms are:

a) *Friswell et al. Model* (1997): This model is a four-parameter model including (α_n, γ_n, β_n, δ_n) and takes the following form:

$$G(s) = G_0\left(1 + \frac{\alpha_n s^2 + \gamma_n s}{s^2 + \beta_n s + \delta_n}\right) \tag{4.18}$$

and

b) *Martin Model* (2011): This model is also a four-parameter model including (α_n, ζ_n, ω_n, ψ_n) and takes the following form:

$$G(s) = G_0\left(1 + \alpha_n \frac{s^2 + 2\zeta_n\omega_n\psi_n s}{s^2 + 2\zeta_n\omega_n s + \omega_n^2}\right) \tag{4.19}$$

These two modified forms provide additional DOF to the model in order to improve its curve fitting capabilities and reduce the number of mini-oscillators needed to replicate the experimental behavior of the VEM. However, these two forms maintain the basic nature of the GHM model whereby the VEM is modeled by a second order differential equation that can be directly coupled with the dynamical equations of motion of the structures.

4.2.3.2 Relaxation Modulus of the GHM Model

The relaxation modulus of the VEM can be extracted from the GHM model using the approaches presented in Chapter 3. Hence, the GHM model can be written as:

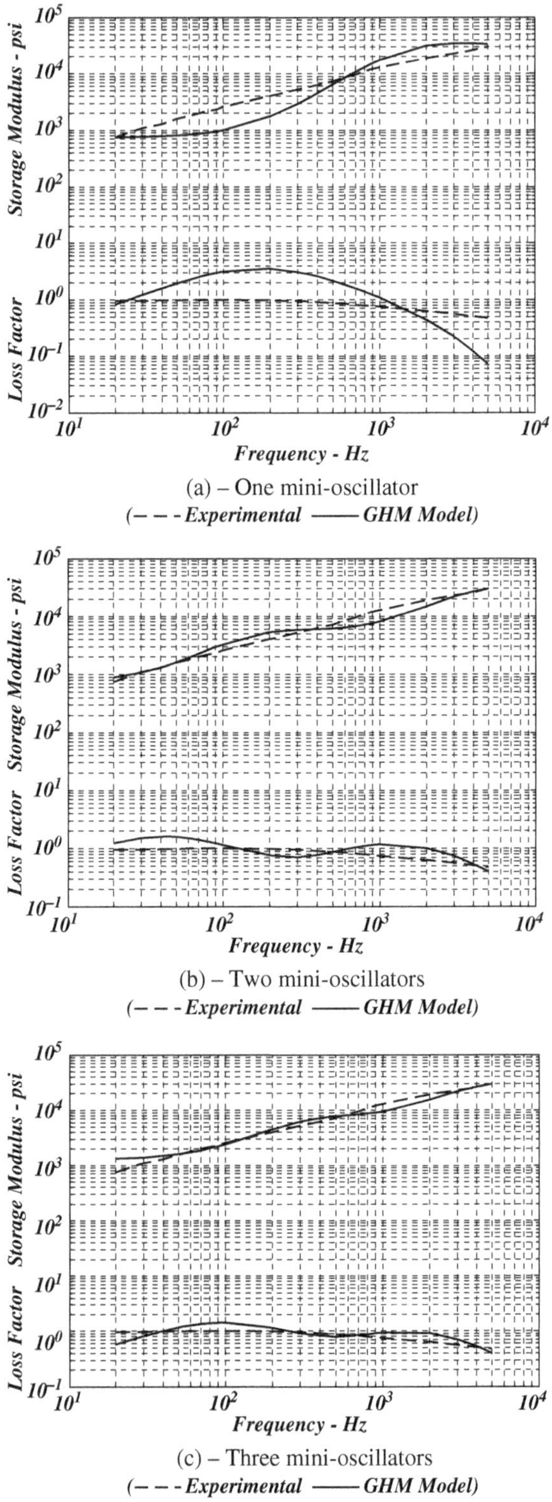

Figure 4.6 Prediction accuracy of GHM models with 1, 2, and 3 mini-oscillators. (a) One mini-oscillator, (b) Two mini-oscillators, and (c) Three mini-oscillators.

$$\frac{G(s)}{s} = \frac{G_0}{s} + \alpha_n \frac{s + 2\zeta_n \omega_n}{s^2 + 2\zeta_n \omega_n s + \omega_n^2}$$

Applying the inverse Laplace transform gives the shear relaxation modulus of the VEM as follows:

$$\begin{aligned} G(t) &= G_0 \left(1 + \alpha_n e^{-\zeta_n \omega_n t} \left[\cos\sqrt{1-\zeta_n^2}\,\omega_n t + \frac{\zeta_n}{\sqrt{1-\zeta_n^2}} \sin\sqrt{1-\zeta_n^2}\,\omega_n t \right] \right) \\ &= G_0 \left(1 + \alpha_n \frac{1}{\sqrt{1-\zeta_n^2}} e^{-\zeta_n \omega_n t} \cos\left(\omega_d t - \phi\right) \right) \end{aligned} \tag{4.20}$$

where $\omega_d = \sqrt{1-\zeta_n^2}\,\omega_n$ and $\phi = \tan^{-1}\left(\frac{\zeta_n}{\sqrt{1-\zeta_n^2}}\right)$.

This expression is rather complicated, but if $\zeta_n = 1$, that is, all the mini-oscillators are selected to be critically damped, then Eq. (4.20) reduces to:

$$G(t) = G_0[1 + \alpha_n e^{-\omega_n t}(1 + \omega_n t)] \tag{4.21}$$

For N mini-oscillators, Eq. (4.21) becomes:

$$G(t) = G_0 \left[1 + \sum_{n=1}^{N} \alpha_n e^{-\omega_n t}(1 + \omega_n t) \right] \tag{4.22}$$

Note that this form of the shear relaxation modulus is resembles the corresponding expression of the *Generalized Maxwell Model* (GMM) given by:

$$G(t) = G_0 \left[1 + \sum_{n=1}^{N} \alpha_n e^{-t/\rho_n} \right] \tag{4.23}$$

The only difference is the multiplier $(1 + \omega_n t)$ which appears in the GHM modulus and results from the fact that the mini-oscillators are selected to be critically damped. With this selection, the curve fitting of the GHM parameters becomes faster. However, more mini-oscillators may be needed to replicate the experimental behavior of the VEM.

4.2.4 Structural Finite Element Models of Rods Treated with VEM

In this section, the GHM model is integrated with the structural finite element model of rods which experience longitudinal vibrations only. The rods are treated with either unconstrained or constrained VEM layers in order to illustrate the importance of constraining the VEM as effective means for enhancing the damping characteristics of the rod/VEM assembly.

4.2.4.1 Unconstrained Layer Damping

Figure 4.7 illustrates a rod treated with an unconstrained VEM layer. The rod is modeled by N one-dimensional finite elements. Each element is bounded by two nodes and each node has a single DOF that is the longitudinal deflection u.

The VEM layer is bonded, from one side, to the rod surface while its other side is left unconstrained, that is, free to move. In this manner, the VEM will experience the same deflection u as the rod. Hence, one can write the potential and kinetic energies of a rod/VEM element as follows

Potential Energy:

$$P.E = P.E_{structure} + P.E_{VEM}$$
$$= \frac{1}{2}\int_0^L E_s A_s u_{,x}^2 dx + \frac{1}{2}\int_0^L E_v A_v u_{,x}^2 dx \tag{4.24}$$

where E_i and A_i are Young's modulus and the area of ith layer with subscript s and v denoting the structure and the VEM, respectively. Also, $u_{,x}$ denotes partial derivative of the deflection u with respect to x.

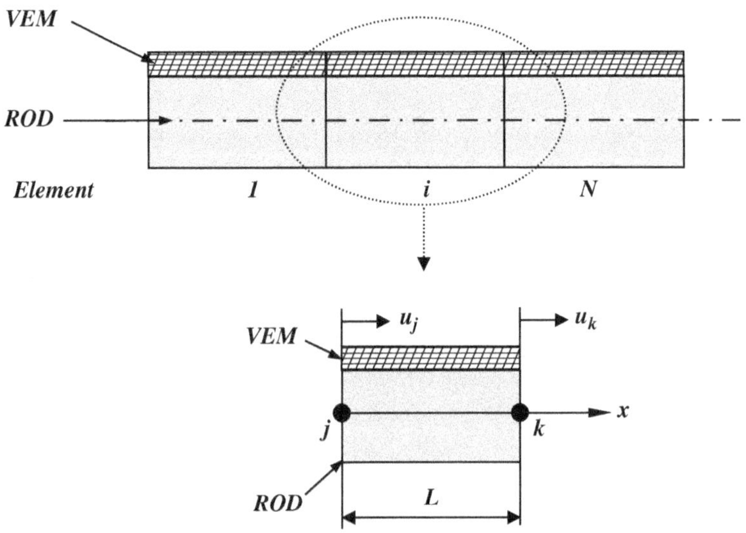

Figure 4.7 Finite element of the rod/unconstrained VEM assembly.

Using a linear shape function, one can write the deflection u at any location x, in terms of the nodal deflection as follows:

$$u = (1 - x/L)\,u_j + x/L\,u_k$$
$$= \{(1 - x/L) \;\; x/L\} \begin{Bmatrix} u_j \\ u_k \end{Bmatrix} = \{N\}\{\Delta_i\} \tag{4.25}$$

where $\{N\} = \{(1 - x/L) \;\; x/L\}$ = interpolating vector and $\{\Delta_i\} = \{u_j \;\; u_k\}^T$ = nodal deflection vector of the ith element.

From Eqs. (4.24) and (4.25), the potential energy (PE) reduces to

$$P.E = \frac{1}{2}\{\Delta_i\}^T \int_0^L E_s A_s \{N_{,x}\}^T \{N_{,x}\} dx \{\Delta_i\} + \frac{1}{2}\{\Delta_i\}^T \int_0^L E_v A_v \{N_{,x}\}^T \{N_{,x}\} dx \{\Delta_i\}$$
$$= \frac{1}{2}\{\Delta_i\}^T [E_s A_s + E_v A_v] \int_0^L \{N_{,x}\}^T \{N_{,x}\} dx \{\Delta_i\} \tag{4.26}$$
$$= \frac{1}{2}\{\Delta_i\}^T [K_s + K_v]\{\Delta_i\}$$

where $[K_s] = E_s A_s \int_0^L \{N_{,x}\}^T \{N_{,x}\} dx = \frac{E_s A_s}{L}\begin{bmatrix} 1 & -1 \\ -1 & 1 \end{bmatrix}$,

$$\text{and } [K_v] = E_v A_v \int_0^L \{N_{,x}\}^T \{N_{,x}\} dx = \frac{E_v A_v}{L}\begin{bmatrix} 1 & -1 \\ -1 & 1 \end{bmatrix} \tag{4.27}$$

where $[K_s]$ and $[K_v]$ denote structural stiffness matrix and VEM stiffness matrix. Note that $[K_s]$ is a real stiffness matrix whereas $[K_v]$ is complex matrix as represented by the *GHM* model. Also, in Eq. (4.26), $\{N_{,x}\}$ denotes the partial derivative of $\{N\}$ with respect to x.

Kinetic Energy:

$$K.E. = \frac{1}{2}\int_0^L (\rho_s A_s + \rho_v A_v)\dot{u}^2 dx \tag{4.28}$$

where ρ_s and ρ_v denote the density of the structure and the VEM, respectively.

But, as $u = \{N\}\{\Delta_i\}$, then $\dot{u} = \{N\}\{\dot{\Delta}_i\}$ and Eq. (4.28) reduces to:

$$
\begin{aligned}
K.E. &= \frac{1}{2}\{\dot{\Delta}_i\}^T(\rho_s A_s + \rho_v A_v)\int_0^L \{N\}^T\{N\}dx\{\dot{\Delta}_i\} \\
&= \frac{1}{2}\{\dot{\Delta}_i\}^T[M]\{\dot{\Delta}_i\}
\end{aligned}
\tag{4.29}
$$

where $[M] = \dfrac{(\rho_s A_s + \rho_v A_v)L}{6}\begin{bmatrix} 2 & 1 \\ 1 & 2 \end{bmatrix}$ = mass matrix of the rod/VEM assembly.

From Eqs. (4.26) and (4.29), the equation of motion of the element can be obtained using the Lagrangian dynamics as follow

$$
\frac{d}{dt}\frac{\partial KE}{\partial\{\dot{\Delta}_i\}} + \frac{\partial PE}{\partial\{\Delta_i\}} = \{\Gamma_i\}
$$

or

$$
[M]s^2\{\Delta_i\} + [K_s + K_v]\{\Delta_i\} = \{F_i\} \tag{4.30}
$$

where $\{F_i\}$ is the vector of the forces acting on the ith element.

The equation of motion of the entire rod/VEM system can then be determined by assembling the mass and stiffness matrices of the individual elements to yield:

$$
[M_o]s^2\{\Delta\} + [K_{s_o} + K_{v_o}]\{\Delta\} = \{F_o\} \tag{4.31}
$$

where $[M_o]$, $[K_{s_o}]$, and $[K_{v_o}]$ are the overall mass matrix, overall structural stiffness matrix, and overall VEM stiffness matrix. Also, $\{\Delta\}$ and $\{F_o\}$ denote the deflection and load vectors of the entire rod/VEM assembly.

Before describing the $[K_{v_o}]$ in terms of the GHM model, the boundary conditions are imposed on the structure to eliminate the rigid body modes. This is essential because these modes do to not contribute any damping to the flexible body modes.

If one GHM mini-oscillator is used, one can write $[K_{v_o}]$ for one element as follows:

$$
\begin{aligned}
[K_{v0}] &= \left[1 + \alpha_1 \frac{s^2 + 2\zeta_1\omega_1 s}{s^2 + 2\zeta\omega_1 s + \omega_1{}^2}\right]\frac{E_{v0}A_v}{L}\begin{bmatrix} 1 & -1 \\ -1 & 1 \end{bmatrix} \\
&= \left[1 + \alpha_1 \frac{s^2 + 2\zeta_1\omega_1 s}{s^2 + 2\zeta\omega_1 s + \omega_1{}^2}\right][K_v]
\end{aligned}
\tag{4.32}
$$

where $[K_v] = \dfrac{E_{v0}A_v}{L}\begin{bmatrix} 1 & -1 \\ -1 & 1 \end{bmatrix}$. Introducing the IDOF z to describe the dynamics of the VEM, and employing the approach outlined in Section 4.2.1, Eq. (4.31) reduces to:

$$
[M_T]\{\ddot{X}\} + [C_T]\{\dot{X}\} + [K_T]\{X\} = \{F_T\} \tag{4.33}
$$

where $[M_T] = \begin{bmatrix} [M_o] & 0 \\ 0 & \dfrac{\alpha_1 [K_{V_o}]}{\omega_1^2} \end{bmatrix}$, $[C_T] = \begin{bmatrix} 0 & 0 \\ 0 & \dfrac{2\zeta_1 \alpha_1 [K_{V_o}]}{\omega_1} \end{bmatrix}$,

$$[K_T] = \begin{bmatrix} [K_{so}] + (1+\alpha_1)[K_{V_o}] & -\alpha_1 [K_{V_o}] \\ -\alpha_1 [K_{V_o}] & \alpha_1 [K_{V_o}] \end{bmatrix}, \{X\} = \begin{Bmatrix} \Delta \\ z \end{Bmatrix} \text{ and}$$

$$\{F_T\} = \begin{Bmatrix} F_o \\ 0 \end{Bmatrix}.$$

If the "Static Condensation" is used to condense the IDOF z of the GHM model, then Eq. (4.33) reduces to:

$$\begin{bmatrix} [K_{so}] + (1+\alpha_1)[K_{V_o}] & -\alpha_1 [K_{V_o}] \\ -\alpha_1 [K_{V_o}] & \alpha_1 [K_{V_o}] \end{bmatrix} \begin{Bmatrix} \Delta \\ z \end{Bmatrix} = \begin{Bmatrix} F_o \\ 0 \end{Bmatrix}$$

which yields: $-\alpha_1 [K_{V_o}] \{\Delta\} + \alpha_1 [K_{V_o}] \{z\} = 0$ or $\{\Delta\} = \{z\}$.

Hence, the full order state vector $\{X\}$ can be written, in terms of $\{\Delta\}$ as

$$\{X\} = \{I I\}^T \{\Delta\} = T\{\Delta\} \tag{4.34}$$

Combining Eqs. (4.33) and (4.34) yields the following equation of motion of the reduced order model

$$[M_R]\{\ddot{\Delta}\} + [C_R]\{\dot{\Delta}\} + [K_R]\{\Delta\} = \{F_o\} \tag{4.35}$$

where the reduced mass, damping, and stiffness matrices are given by:

$$[M_R] = T^T [M_T] T, \ [C_R] = T^T [C_T] T, \ \text{and} \ [K_R] = T^T [K_T] T \tag{4.36}$$

The time and frequency responses of the full order and reduced order rod/VEM systems can be computed from Eqs. (4.33) and (4.36). Furthermore, the natural frequencies and corresponding damping ratios can be determined by casting the homogenous parts of the two equations in a state-space form as follows:

Full order system: $\{\dot{Y}_T\} = [A_T]\{Y_T\}$, and

Reduced order system: $\{\dot{Y}_R\} = [A_R]\{Y_R\}$

where $\{Y_T\} = \{X \ \dot{X}\}^T$ and $\{Y_R\} = \{\Delta \ \dot{\Delta}\}^T$. Also, the state matrices $[A_T]$ and $[A_R]$ are given by:

$$[A_T] = \begin{bmatrix} 0 & I \\ -[M_T]^{-1}[K_T] & -[M_T]^{-1}[C_T] \end{bmatrix} \text{ and}$$

$$[A_R] = \begin{bmatrix} 0 & I \\ -[M_R]^{-1}[K_R] & -[M_R]^{-1}[C_R] \end{bmatrix}$$

Using MATLAB commands "damp (A_T)" and "damp (A_R)" give directly the natural frequencies and corresponding damping ratios of the full order and reduced order rod/VEM systems.

Example 4.3 Consider the fixed-free rod/VEM system shown in Figure 4.8. The rod is made of aluminum with width of 0.025 m, thickness of 0.025 m, and length of 1 m. The VEM has width of 0.025 m, thickness = 0.025 m, and density of 1100 kg m^{-3}. The storage modulus and loss factor of the VEM are predicted by GHM model with one mini-oscillator ($E_0 = 15.3MPa, \alpha_1 = 39, \zeta_1 = 1, \omega_1 = 19,058 rad/s$). Determine the time and frequency response of the free end (Node 3) when the rod is subjected to a unit force acting at the same node. Determine also the natural frequencies and damping ratios of the full and reduced order models of the rod/VEM system assuming that the system is modeled by a two element finite element model. Assume that the time domain excitation is carried by a unit force of duration 100 μs.

Solution

Figure 4.9 shows the frequency and time response of the full and reduced order models.

It is evident that the response of the reduced order system matches closely that of the full order system. One should, however, note that the response suggests very clearly that the system seems to be very lightly damped in spite of the use of the VEM treatment. Such an observation is confirmed by computing the natural frequencies and corresponding damping ratios of the full and reduced order models using the "damp" command of MATLAB. These parameters are given in Table 4.2.

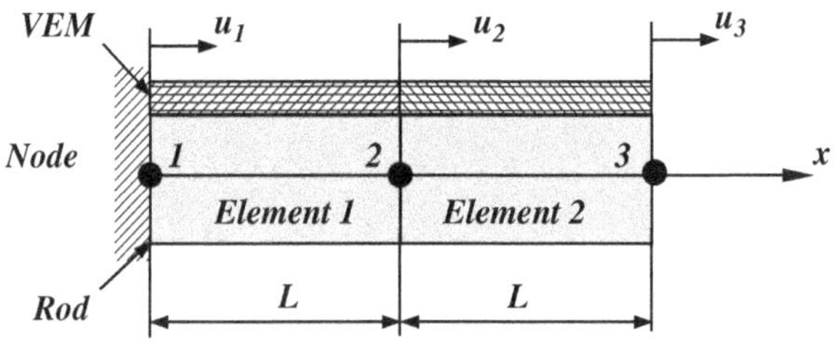

Figure 4.8 Two element model a rod/unconstrained VEM.

4.2.4.2 Constrained Layer Damping

Figure 4.10 displays a rod treated with a constrained VEM layer. The rod is modeled by N one-dimensional finite elements. Each element is bounded by two nodes and each node has a single DOF, which is the longitudinal deflection u.

The VEM layer is bonded, from one side, to the rod surface while its other side is constrained from motion by a cover sheet called "constraining layer." In this manner, the VEM will experience shear strain γ as shown in Figure 4.10c due to the relative motion between its bottom and top surfaces. Hence, one can write the potential and kinetic energies of a rod/VEM element as follows:

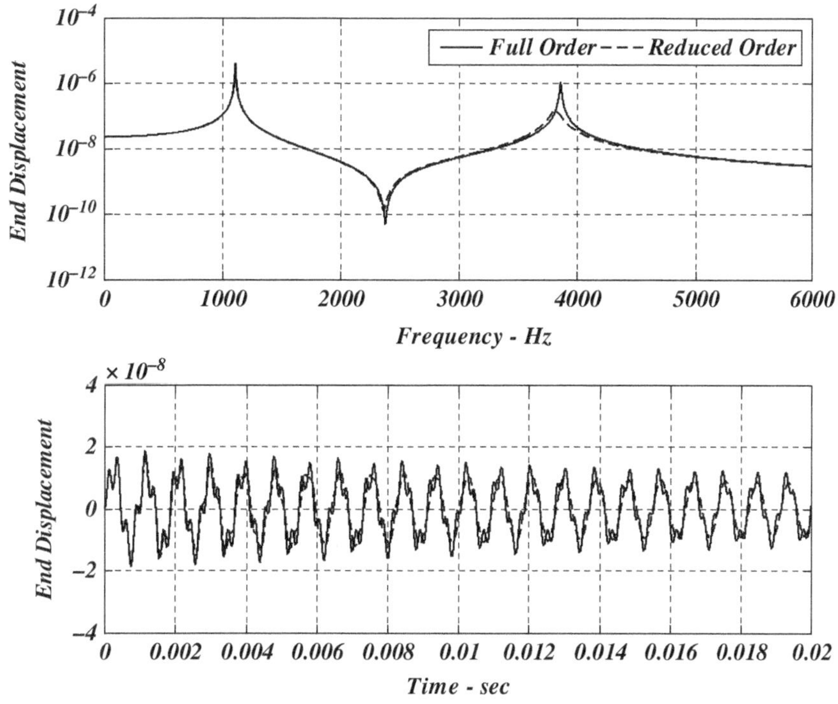

Figure 4.9 Frequency response of the rod/unconstrained VEM.

Table 4.2 Modal parameters of unconstrained layer damping.

	Full order model		Reduced order model	
Mode	Frequency (Hz)	Damping ratio	Frequency (Hz)	Damping ratio
1	1103.5	0.002 42	1100.3	0.003 10
2	3869.4	0.001 57	3821.6	0.010 70

Potential Energy:

$$P.E = P.E_{structure} + P.E_{Constraining\ Layer} + P.E_{VEM}$$
$$= \frac{1}{2}\int_0^L E_3 A_3 u_{3,x}^2\,dx + \frac{1}{2}\int_0^L E_1 A_1 u_{1,x}^2\,dx + \frac{1}{2}\int_0^L G_v A_v \gamma^2 dx \tag{4.37}$$

where E_i and A_i are Young's modulus and the area of ith layer with subscripts 1 and 3 denoting the constraining layer and the base structure, respectively. Also, G_v and A_v denote the shear modulus and shear area of the VEM. Considering the geometry

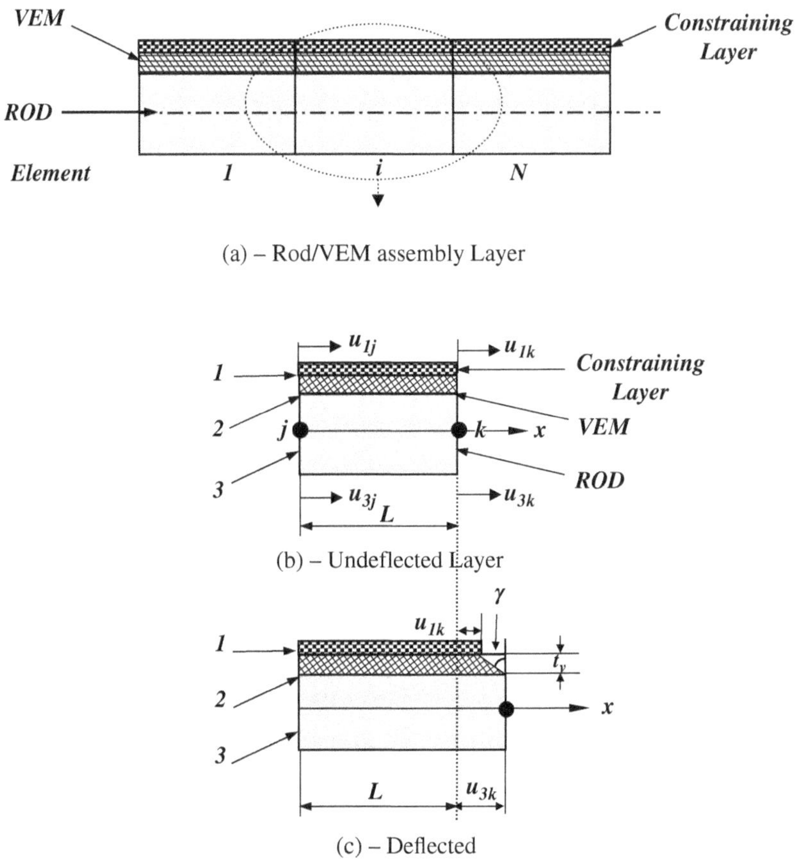

Figure 4.10 Finite element of model the rod/constrained VEM assembly. (a) Rod/VEM assembly layer, (b) undeflected layer, and (c) deflected.

of the deflected rod/VEM shown in Figure 4.10c, the shear strain γ can be determined from

$$\gamma = \frac{(u_3 - u_1)}{t_v} \tag{4.38}$$

Equation (4.38) is obtained by assuming small shear strain γ, which implies that $\gamma \cong \tan(\gamma)$.

Assuming that the deflections u_1 and u_3 are represented by linear shape functions, then

$$u_1 = \{N_1\}\{\Delta_i\} \text{ and } u_3 = \{N_3\}\{\Delta_i\} \tag{4.39}$$

where $\{N_1\} = \{(1 - x/L) \quad 0 \quad x/L \quad 0\}$ and $\{N_3\} = \{0 \quad (1 - x/L) \quad 0 \quad x/L\}$ with $\{\Delta_i\} = \{u_{1j} \; u_{3j} \; u_{1k} \; u_{3k}\}^T$ = nodal deflection vector of the ith element.

From Eqs. (4.37) through (4.39), the potential energy reduces to

$$
\begin{aligned}
P.E &= \frac{1}{2}\{\Delta_i\}^T \int_0^L E_1A_1\{N_{1,x}\}^T\{N_{1,x}\}dx\{\Delta_i\} \\
&+ \frac{1}{2}\{\Delta_i\}^T \int_0^L E_3A_3\{N_{3,x}\}^T\{N_{3,x}\}dx\{\Delta_i\} \\
&+ \frac{1}{2}\{\Delta_i\}^T G_vA_v/t_v^2 \int_0^L \{N_3-N_1\}^T\{N_3-N_1\}dx\{\Delta_i\} \\
&= \frac{1}{2}\{\Delta_i\}^T[K_1+K_3+K_{vG}]\{\Delta_i\}
\end{aligned}
\tag{4.40}
$$

where $[K_i] = E_iA_i\int_0^L\{N_{i,x}\}^T\{N_{i,x}\}dx, i = 1,3$

$$
\text{and } [K_{vG}] = G_vA_v/t_v^2 \int_0^L \{N_3-N_1\}^T\{N_3-N_1\}dx \tag{4.41}
$$

where $[K_1]$, $[K_3]$, and $[K_{vG}]$ denote the stiffness matrices of the constraining layer, base structures, and VEM, respectively. Note that $[K_1]$ and $[K_3]$ are real stiffness matrices whereas $[K_{vG}]$ is complex matrix as represented by the GHM model.

Kinetic Energy:

$$
K.E. = \frac{1}{2}\int_0^L \rho_1A_1\dot{u}_1^2dx + \frac{1}{2}\int_0^L \rho_3A_3\dot{u}_3^2dx + \frac{1}{2}\int_0^L \rho_vA_v\dot{u}_v^2dx \tag{4.42}
$$

where ρ_1, ρ_3, and ρ_v denote the density of the constraining layer, base structure, and the VEM, respectively. In Eq. (4.42), $\dot{u}_v$ is the velocity of the VEM that is assumed to be the average of the velocities of the constraining layer and the base structure, that is, it is given by

$$
\dot{u}_v = \frac{1}{2}(\dot{u}_1 + \dot{u}_3) \tag{4.43}
$$

But, as $u = \{N\}\{\Delta_i\}$, then $\dot{u} = \{N\}\{\dot{\Delta}_i\}$ and Eq. (4.42) reduces to:

$$
\begin{aligned}
K.E. &= \frac{1}{2}\{\dot{\Delta}_i\}^T \rho_1A_1\int_0^L\{N_1\}^T\{N_1\}dx\{\dot{\Delta}_i\} \\
&+ \frac{1}{2}\{\dot{\Delta}_i\}^T \rho_3A_3\int_0^L\{N_3\}^T\{N_3\}dx\{\dot{\Delta}_i\}
\end{aligned}
$$

$$+\frac{1}{2}\{\dot{\Delta}_i\}^T\frac{1}{4}\rho_v A_v\int_0^L\{N_1+N_3\}^T\{N_1+N_3\}dx\{\dot{\Delta}_i\}$$
$$=\frac{1}{2}\{\dot{\Delta}_i\}^T[M_1+M_3+M_v]\{\dot{\Delta}_i\} \tag{4.44}$$

where $[M_i]=\rho_i A_i\int_0^L\{N_i\}^T\{N_i\}dx, i=1,3$

$$\text{and } [M_v]=\frac{1}{4}\rho_v A_v\int_0^L\{N_1+N_3\}^T\{N_1+N_3\}dx \tag{4.45}$$

From Eqs. (4.40) and (4.44), the equation of motion of the individual elements can be obtained using the Lagrangian dynamics and assembly of these equations gives the equation of motion of the entire rod/VEM system as follows

$$[M_o]s^2\{\Delta\}+[K_{1_o}+K_{3_o}+K_{v_{Go}}]\{\Delta\}=\{F_o\} \tag{4.46}$$

where $[M_o],[K_{1_o}],[K_{3_o}]$, and $[K_{v_{Go}}]$ are the overall mass matrix, overall constraining layer stiffness matrix, base structure stiffness matrix, and overall VEM stiffness matrix. Also, $\{\Delta\}$ and $\{F_o\}$ denote the deflection and load vectors of the entire assembly.

Imposing the boundary conditions is essential before describing the $[K_{v_{Go}}]$ in terms of the GHM model in order to eliminate the rigid body modes because these modes do to not contribute any damping to the flexible body modes. In the case of the constraining layer damping, this process will not be enough to eliminate all the rigid body modes as $[K_{v_{Go}}]$ remains singular as will be outlined in Example 4.4. Further manipulations are necessary.

Introducing the IDOF z to describe the dynamics of the VEM with a single mini-oscillator, and employing the approach outlined in Section 4.2.1, Eq. (4.46) reduces to:

$$[M_T]\{\ddot{X}\}+[C_T]\{\dot{X}\}+[K_T]\{X\}=\{F_T\} \tag{4.47}$$

where $[M_T]=\begin{bmatrix}[M_o] & 0\\ 0 & \dfrac{\alpha_1[K_{v_{Go}}]}{\omega_1^2}\end{bmatrix}, [C_T]=\begin{bmatrix}0 & 0\\ 0 & \dfrac{2\zeta_1\alpha_1[K_{v_{Go}}]}{\omega_1}\end{bmatrix},$

$$[K_T]=\begin{bmatrix}[K_{1o}]+[K_{3_o}]+(1+\alpha_1)[K_{v_{Go}}] & -\alpha_1[K_{v_{Go}}]\\ -\alpha_1[K_{v_{Go}}] & \alpha_1[K_{v_{Go}}]\end{bmatrix}, \{X\}=\begin{Bmatrix}\Delta\\ z\end{Bmatrix}$$

and $\{F_T\}=\begin{Bmatrix}F_o\\ 0\end{Bmatrix}$.

If the matrix $[K_{v_{Go}}]$ remains singular, then introduce the following transformation of the IDOF z to $\bar{z}$ such that

$$z=R_n\bar{z} \tag{4.48}$$

where R_n is the eigenvector matrix of the non-zero eigenvalues Λ of $[K_{vGo}]$ such that

$$[K_{vGo}] = R_n \Lambda R_n^T \text{ with } R_n^T R_n = I \tag{4.49}$$

Note that the $[K_{vGo}]$ can also be written as:

$$[K_{vGo}] = R_T \Lambda_T R_T^T = [R_0 \;\; R_n] \begin{bmatrix} 0 & 0 \\ 0 & \Lambda \end{bmatrix} \begin{bmatrix} R_0^T \\ R_n^T \end{bmatrix} = R_n \Lambda R_n^T$$

where R_T is the total eigenvector matrix of $[K_{vGo}]$ which includes the eigenvector matrices for the zero and non-zero eigenvalues R_0 and R_n, respectively.

Substituting Eqs. (4.48) and (4.49) into Eq. (4.47), it reduces to

$$[M_t]\{\ddot{X}_t\} + [C_t]\{\dot{X}_t\} + [K_t]\{X_t\} = \{F_t\} \tag{4.50}$$

where $[M_t] = \begin{bmatrix} [M_o] & 0 \\ 0 & \dfrac{\alpha_1 \Lambda}{\omega_1^2} \end{bmatrix}, [C_t] = \begin{bmatrix} 0 & 0 \\ 0 & \dfrac{2\zeta_1 \alpha_1 \Lambda}{\omega_1} \end{bmatrix},$

$$[K_t] = \begin{bmatrix} [K_{1o}] + [K_{3_o}] + (1+\alpha_1)[K_{vGo}] & -\alpha_1 R_n \Lambda \\ -\alpha_1 \Lambda R_n^T & \alpha_1 \Lambda \end{bmatrix}, \{X_t\} = \begin{Bmatrix} \Delta \\ \bar{z} \end{Bmatrix} \text{ and}$$

$$\{F_t\} = \begin{Bmatrix} F_o \\ 0 \end{Bmatrix}.$$

In this form, the mass and stiffness matrices $[M_t]$ and $[K_t]$ are non-singular matrices with all the modes are flexible body modes.

If the Static Condensation is used to condense the IDOF z of the GHM model, then Eq. (4.33) reduces to:

$$\begin{bmatrix} [K_{1o}] + [K_{3_o}] + (1+\alpha_1)[K_{vGo}] & -\alpha_1 R_n \Lambda \\ -\alpha_1 \Lambda R_n^T & \alpha_1 \Lambda \end{bmatrix} = \begin{Bmatrix} \Delta \\ \bar{z} \end{Bmatrix} = \begin{Bmatrix} F_0 \\ 0 \end{Bmatrix}$$

which yields: $-\alpha_1 \Lambda R_n^T \{\Delta\} + \alpha_1 \Lambda \{\bar{z}\} = 0$ or $\{\bar{z}\} = R_n^T \{\Delta\}$

Hence, the full order state vector $\{X\}$ can be written, in terms of $\{\Delta\}$ as

$$\{X_t\} = \{I R_n^T\}^T \{\Delta\} = T_c \{\Delta\} \tag{4.51}$$

Combining Eqs. (4.50) and (4.51) yields the following equation of motion of the reduced order model

$$[M_r]\{\ddot{\Delta}\} + [C_r]\{\dot{\Delta}\} + [K_r]\{\Delta\} = \{F_o\} \tag{4.52}$$

where the reduced mass, damping, and stiffness matrices are given by:

$$[M_r] = T_c^T [M_t] T_c, \; [C_r] = T_c^T [C_t] T_c, \text{ and } [K_r] = T_c^T [K_t] T_c \tag{4.53}$$

The time and frequency responses of the full order and reduced order rod/VEM systems can be computed from Eqs. (4.50) and (4.52). Furthermore, the natural frequencies and corresponding damping ratios can be determined by casting the homogenous parts of the two equations in a state-space form as described in the case of the unconstrained layer damping.

For VEM represented by N mini-oscillators, Eq. (4.50) takes the following form:

$$[M_{t_N}]\{\ddot{X}_t\} + [C_{t_N}]\{\dot{X}_t\} + [K_{t_N}]\{X_t\} = \{F_t\} \tag{4.54}$$

where

$$[M_{t_N}] = \begin{bmatrix} [M_o] & 0 & 0 & 0 \\ 0 & \dfrac{\alpha_1\Lambda}{\omega_1^2} & 0 & \ldots\ldots \\ \cdots & 0 & \ldots\ldots & 0 \\ & & & \\ 0 & \ldots\ldots & 0 & \dfrac{\alpha_N\Lambda}{\omega_N^2} \end{bmatrix}, [C_{t_N}] = \begin{bmatrix} 0 & 0 & \ldots\ldots & 0 \\ 0 & \dfrac{2\zeta_1\alpha_1\Lambda}{\omega_1} & 0 & \ldots\ldots \\ \ldots\ldots & 0 & \ldots\ldots & 0 \\ & & & \\ 0 & \ldots\ldots & 0 & \dfrac{2\zeta_N\alpha_N\Lambda}{\omega_N} \end{bmatrix},$$

and

$$[K_{t_N}] = \begin{bmatrix} [K_{1o}] + [K_{3_o}] + \left(1 + \sum_{i=1}^{N}\alpha_i\right)[K_{V_{Go}}] & -\alpha_1 R_n\Lambda & \ldots\ldots & \ldots\ldots & -\alpha_N R_n\Lambda \\ -\alpha_1\Lambda R_n^T & \alpha_1\Lambda & 0 & 0 & 0 \\ \ldots\ldots & 0 & \alpha_2\Lambda & \ldots\ldots & 0 \\ \ldots\ldots & 0 & 0 & \ldots\ldots & 0 \\ -\alpha_N\Lambda R_n^T & 0 & 0 & 0 & \alpha_N\Lambda \end{bmatrix}$$

Example 4.4 Consider the fixed-free rod/VEM system shown in Figure 4.11. The rod is made of aluminum with width of 0.025 m, thickness of 0.025 m, and length of 1 m. The VEM has width of 0.025 m, thickness = 0.025 m, and density of 1100 kg m^{-3}. The VEM is constrained by an aluminum constraining layer which is 0.025 m wide and 0.0025 m thick. Using the GHM modeling approach of the VEM with one mini-oscillator ($E_0 = 15.3MPa$, $\alpha_1 = 39$, $\zeta_1 = 1$, $\omega_1 = 19,058 rad/s$), determine the time and frequency response of the free end (Node 3) when the rod is subjected to a unit load acting at the same node. Determine also the natural frequencies and damping ratios of the full and reduced order models of the rod/VEM system assuming that the system is modeled by a two element finite element model. Assume that the time domain excitation is carried by a unit force of duration 100 μs.

Solution

Figure 4.12 shows the frequency and time response of the full and reduced order models. It is clear from Figure 4.12 that the response of the reduced order system adequately matches that of the full order system. One should, however, note that the time response

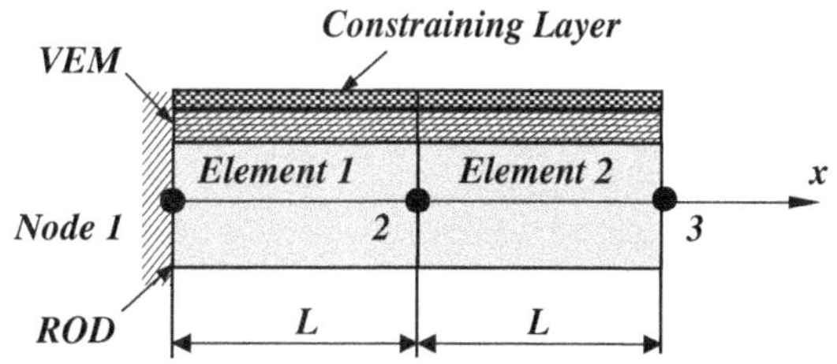

Figure 4.11 Two element model a rod/constrained VEM.

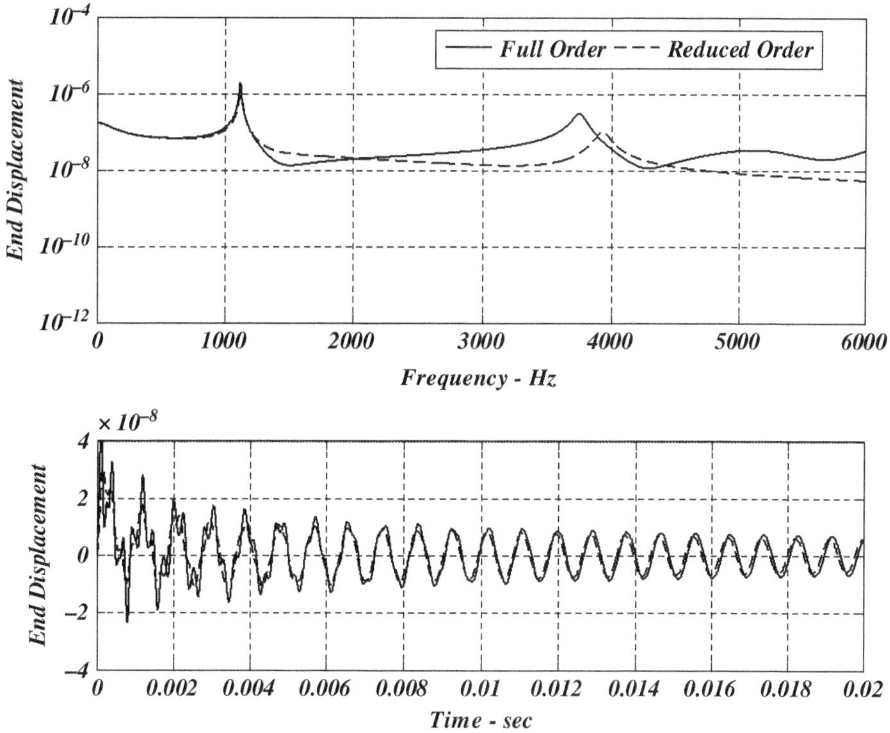

Figure 4.12 Frequency and time response of the rod/constrained VEM.

indicates clearly that the damping layer, in its constrained configuration, becomes effective in damping out the structural vibration of the rod. This is in spite of the fact that the constraining layer is a very thin layer that did not add weight or stiffness to the entire assembly. Such an observation is confirmed by computing the natural frequencies and corresponding damping ratios of the reduced order model using the "damp" command of MATLAB. These parameters are given in Table 4.3.

Comparing the results shown in Tables 4.2 and 4.3 indicates that constraining the VEM increases the damping ratios by two- and 10-fold for the first and second modes as compared to those when the VEM is left unconstrained. Such an enhancement of the damping characteristics is attributed to changing the damping mechanism from tension, in the case of the unconstrained layer damping, to shear in the case of the constrained

Table 4.3 Modal parameters of constrained layer damping.

Model	Full order model		Reduced order model	
Mode	Frequency (Hz)	Damping ratio	Frequency (Hz)	Damping ratio
1	1130.05	0.004 89	1116.24	0.005 1
2	3757.96	0.012 30	3933.12	0.012 6

layer damping. More detailed analysis of the energy dissipation in these two configurations will be presented in Chapter 6.

4.3 Structural Finite Element Models of Beams Treated with VEM

In this section, finite element models are developed for beams treated with passive constrained layer damping (PCLD) treatments. Validation of the predictions of these models against the predictions of the commercial software package ANSYS are also presented.

4.3.1 Degrees of Freedom

A schematic drawing of a finite element of a beam treated with a constrained VEM layer is shown in Figure 4.13.

The element has two nodes (*j* and *k*), each node has four DOF including: u_1, u_3, w, and $w_{,x}$ that are: axial displacement of the constraining layer, axial displacement of the beam, as well as the transverse and rotational deflections of beam, respectively.

The spatial distributions of these deflections, along the element, are described by the following shape functions:

$$u_1 = b_1 + b_2 x, \quad u_3 = b_3 + b_4 x,$$

and

$$w = a_1 + a_2 x + a_3 x^2 + a_4 x^3 \tag{4.55}$$

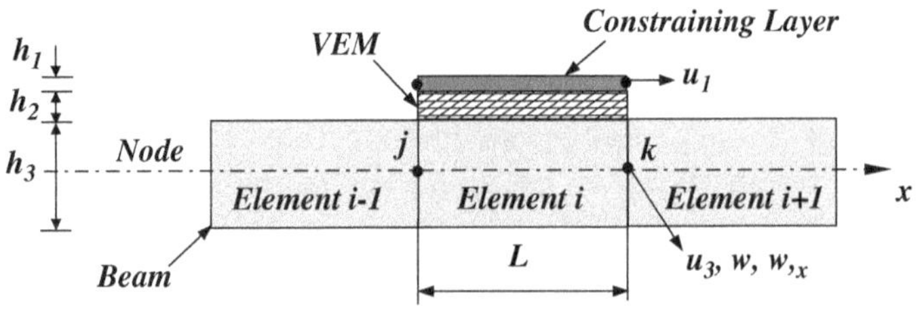

Figure 4.13 Beam element with a constrained VEM damping treatment.

where the $a_i's$ and $b_i's$ are constants that are determined in terms of the nodal deflection vector $\{\Delta_i\}$ as outlined in Section 4.2.4 for the case of rods.

Accordingly, Eq. (4.55) can be written as follows:

$$\begin{Bmatrix} u_1 \\ u_3 \\ w \end{Bmatrix} = \begin{Bmatrix} N_{u_1} \\ N_{u_3} \\ N_w \end{Bmatrix} \{\Delta_i\} = \begin{bmatrix} N_1 & 0 & 0 & 0 & N_2 & 0 & 0 & 0 \\ 0 & N_1 & 0 & 0 & 0 & N_2 & 0 & 0 \\ 0 & 0 & N_{1w} & N_{1w,x} & 0 & 0 & N_{2w} & N_{2w,x} \end{bmatrix} \{\Delta_i\} \tag{4.56}$$

where $\{\Delta_i\} = \{u_{1j}\ u_{3j}\ w_j\ w_{,xj}\ u_{1k}\ u_{3k}\ w_k\ w_{,xk}\}^T$ denotes the nodal deflection vector. Also, it can be easily shown that the interpolating functions $N_i's$ are given by: $N_1 = (1 - x/L)$, $N_2 = x/L$, $N_{1w} = 1-3\,(x/L)^2 + 2\,(x/L)^3$, $N_{1w,x} = x\,[1-2\,(x/L) + (x/L)^2]$, $N_{2w} = 3\,(x/L)^2 - 2\,(x/L)^3$, and $N_{2w,x} = x\,[-(x/L) + (x/L)^2]$. Note that x and L define the position along the element and the element length, respectively.

4.3.2 Basic Kinematic Relationships

Figure 4.14 shows a schematic drawing of the deflected beam/VEM assembly. The figure displays the deflections experienced by the different layers of the assembly.

From the geometry, it can be seen that:

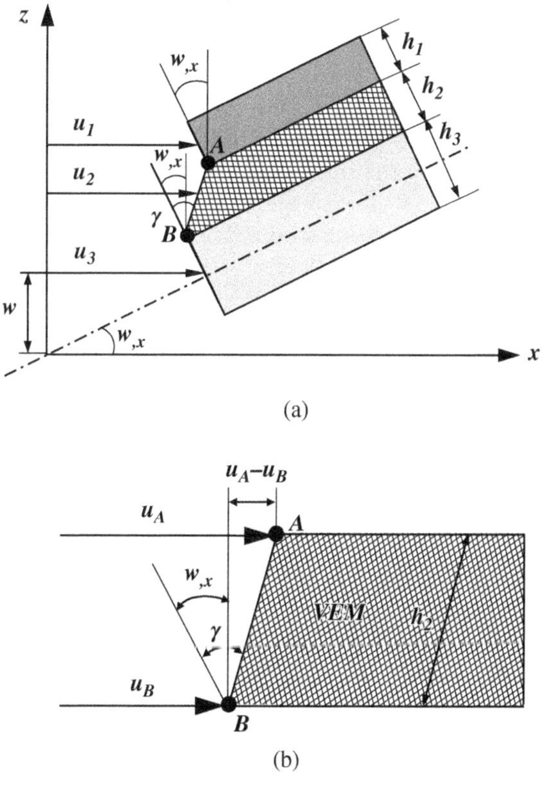

Figure 4.14 Deflected beam/VEM Assembly.

$$u_A = u_1 + \frac{h_1}{2} w_{,x} \text{ and } u_B = u_3 - \frac{h_3}{2} w_{,x} \tag{4.57}$$

Hence, the shear strain γ experienced by the VEM layer can be determined from:

$$\gamma \cong w_{,x} + \frac{1}{h_2}(u_A - u_B) = \frac{h}{h_2} w_{,x} + \frac{1}{h_2}(u_1 - u_3) \tag{4.58}$$

where $h = h_2 + \frac{1}{2}(h_1 + h_3)$. It is assumed here that the longitudinal deflection u_2 of the VEM can be determined from:

$$u_2 = \frac{1}{2}(u_A + u_B) = \frac{1}{2}\left[\frac{1}{2}(h_1 - h_3) w_{,x} + (u_1 + u_3)\right] \tag{4.59}$$

4.3.3 Stiffness and Mass Matrices of the Beam/VEM Element

In order to determine the stiffness and mass matrices of the Beam/VEM Element, it is essential to develop expressions for the potential and kinetic energies.

a) ***Potential Energy***: The PE of the element is a combination of potential energies of constraining layer, VEM layer, and beam layers. It is given by:

$$PE = \frac{1}{2}\int_0^L \left[\sum_{i=1}^{3} E_i A_i u_{i,x}^2 + (E_1 I_1 + E_3 I_3) w_{,xx}^2 + G_2 A_2 \gamma^2\right] dx \tag{4.60}$$

where $E_i A_i$ and $E_i I_i$ denote the longitudinal and flexural rigidity of the ith layer. Also, $G_2 A_2$ denotes the shear rigidity of the VEM with G_2 defining its shear modulus. The terms inside the integral of Eq. (4.60) define the potential energies due to extension, bending, and shear, respectively.

In Eq. (4.60), it is assumed that the three layers move equally in the transverse direction and that the flexural rigidity of the VEM is negligible. These assumptions are justifiable as the thicknesses of the VEM and the constraining layer are small.

Using the interpolating Eq. (4.56), Eq. (4.60) reduces to:

$$PE = \frac{1}{2}\{\Delta_i\}^T \left([K_u] + [K_w] + [K_\gamma]\right)\{\Delta_i\} = \frac{1}{2}\{\Delta_i\}^T [K_e]\{\Delta_i\} \tag{4.61}$$

where

$$[K_u] = \int_0^L \left[E_1 A_1 \left(N_{u_1,x}^T N_{u_1,x}\right) + E_3 A_3 \left(N_{u_3,x}^T N_{u_3,x}\right)\right] dx$$

$$+ \int_0^L \frac{1}{2} E_2 A_2 \left[\left(N_{u_1,x} + \frac{1}{2}(h_1 - h_3) N_{w,xx} + N_{u_3,x}\right)^T \left(N_{u_1,x} + \frac{1}{2}(h_1 - h_3) N_{w,xx} + N_{u_3,x}\right)\right] dx,$$

$$[K_w] = \int_0^L \left[(E_1 I_1 + E_3 I_3) \left(N_{w,xx}^T N_{w,xx} \right) \right] dx,$$

$$\left[K_\gamma\right] = \int_0^L \frac{G_2 A_2}{h_2^2} \left[(N_{u_1} - N_{u_2} + hN_{w,x})^T (N_{u_1} - N_{u_2} + hN_{w,x}) \right] dx,$$

and $[K_e] = [K_u] + [K_w] + \left[K_\gamma\right]$ = element stiffness matrix.

Note that $[K_u]$, $[K_w]$, and $[K_\gamma]$ are the stiffness matrices due to extension, bending, and shear. These stiffness matrices are of dimension 8 × 8.

b) ***Kinetic Energy***: The kinetic energy (KE) accounts for the energies of the different layers as follows:

$$KE = \frac{1}{2} \int_0^L \left[\sum_{i=1}^{3} \rho_i A_i \dot{u}_i^2 + \sum_{i=1}^{3} \rho_i A_i \dot{w}^2 \right] dx \tag{4.62}$$

Note that the terms inside the integral of Eq. (4.62) define the KEs due to extension and bending only. It is assumed that the KE due to rotary inertia is negligible with ρ_i denoting the density of the ith layer.

Using the interpolating Eq. (4.56), Eq. (4.62) reduces to:

$$KE = \frac{1}{2} \left\{ \dot{\Delta}_i \right\}^T ([M_u] + [M_w]) \left\{ \dot{\Delta}_i \right\} = \frac{1}{2} \left\{ \dot{\Delta}_i \right\}^T [M_e] \left\{ \dot{\Delta}_i \right\} \tag{4.63}$$

where

$$[M_u] = \int_0^L \left[\left(\rho_1 A_1 N_{u_1}^T N_{u_1} + \rho_3 A_3 N_{u_3}^T N_{u_3} \right) \right] dx$$

$$+ \int_0^L \left[\frac{1}{4} \rho_2 A_2 \left(\frac{1}{2} (h_1 - h_3) N_{w,x} + N_{u_1} + N_{u_3} \right)^T \left(\frac{1}{2} (h_1 - h_3) N_{w,x} + N_{u_1} + N_{u_3} \right) \right] dx$$

$$[M_w] = \int_0^L \left[\sum_{i=1}^{3} \rho_i A_i N_w^T N_w \right] dx, \quad \text{and} \quad [M_e] = [M_u] + [M_w] = \text{element mass matrix.}$$

Note that $[M_u]$ and $[M_w]$ are the mass matrices due to extension and bending. These mass matrices have also a dimension of 8 × 8.

4.3.4 Equations of Motion of the Beam/VEM Element

From Eqs. (4.61) and (4.63), the equation of motion of the individual elements can be obtained using the Lagrangian dynamics to yield the following equation:

$$[M_e] \left\{ \ddot{\Delta}_i \right\} + [K_e] \left\{ \Delta_i \right\} = \left\{ F_i \right\} \tag{4.64}$$

where $[M_e]$ and $[K_e]$ are the element mass and stiffness matrices with $\{\Delta_i\}$ and $\{F_i\}$ denoting the nodal deflection and load vectors of the beam/VEM element.

The equations of motion of the entire beam/VEM assembly can be derived by assembling the element matrices using the approach outlined in Section 4.2.4 for the case of rods. Then, the boundary conditions are imposed and the resulting system can be augmented with the dynamics of the VEM using one or more GHM mini-oscillators. Condensation of the IDOF and elimination of the rigid body modes are carried out in the same manner as described in Section 4.2.4 for the case of rods.

Example 4.5 Consider the fixed-free beam/VEM system shown in Figure 4.15. The physical and geometrical parameters of the beam/VEM system are listed in Table 4.4. The beam is divided into 10 finite elements and is subjected to a force F acting along the transverse direction. Determine the response of the beam/VEM system for:

a) Sinusoidal force of $F = 0.001 \sin(\omega t)$ where ω is the excitation frequency.
b) Impulse of $F = 0.001N$ for time $t \leq 0.02$ s and $F = 0$ for $t > 0.02$ s.

Compare the response, in the time and frequency domains, with those of plain and untreated beam. Assume that the complex shear modulus G_2 of the VEM is represented by a GHM model with four mini-oscillators that have the properties listed in Table 4.5 such that:

$$G_2(s) = G_0 \left[1 + \sum_{i=1}^{4} \alpha_i \frac{s^2 + 2\omega_i s}{s^2 + 2\omega_i s + \omega_i^2} \right] \quad \text{with } G_0 = 2.72\,\text{MPa}$$

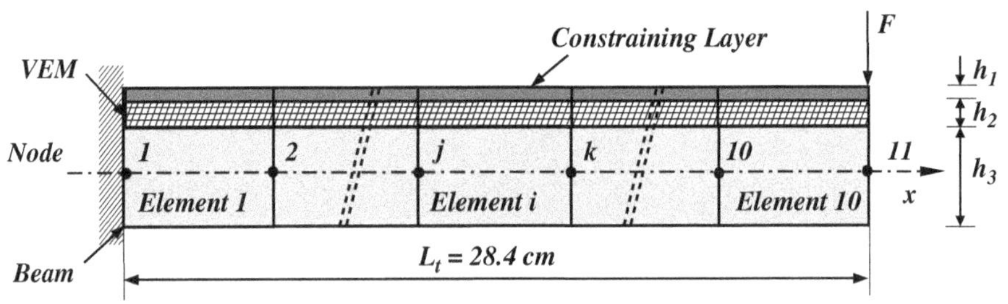

Figure 4.15 Ten-element cantilever beam with a constrained VEM damping treatment.

Table 4.4 Physical and geometrical parameters of beam/VEM system.

Layer	Thickness (m)	Width (m)	Young's modulus (GPa)	Density (kg m^{-3})	Poisson's ratio
Constraining layer	$h_1 = 0.0025$	0.025	70	2700	0.3
VEM	$h_2 = 0.0025$	0.025	GHM	1100	0.5
Beam	$h_3 = 0.0025$	0.025	70	2700	0.3

Table 4.5 Optimized parameters of the GHM four-mini-oscillator model of the VEM.

Oscillator	1	2	3	4
ω_i (rad s^{-1})	80,000	35,000	4000	500
α_i	15	55	18	4

Solution

The finite element formulation presented in Section 4.3 is employed to investigate the effect of viscoelastic damping treatment on the vibration response of the beam both in the time and frequency domains. Figure 4.16 shows comparisons between the response of the beam when treated with the constrained damping treatment (PCLD) and that of a plain beam. The benefits are very clear both in the time and frequency domains.

4.4 Generalized Maxwell Model (GMM)

4.4.1 Overview

The GMM presents an alternative approach to the GHM model that enables also the modeling of the behavior of VEM both in the time and frequency domains especially when it is integrated with a finite element model of a vibrating structure. The GMM is extensively used in numerous commercial finite element software packages such as ANSYS.

The relaxation modulus $E(t)$ of VEM is usually given by the GMM as shown in Figure 4.17.

Mathematically, this model is represented, in the time domain, by the following Prony series:

$$E(t) = E_0\left[1 + \sum_{i=1}^{n} \alpha_i e^{-t/\rho_i}\right] \tag{4.65}$$

where E_0 = the equilibrium modulus, α_i = the ith relative modulus, and ρ_i = the ith relaxation times = η_i/E_i. Note that E_0, α_i, and ρ_i are all positive constants. Also, the summation term $\sum_{i=1}^{n}\alpha_i e^{-t/\rho_i}$ is designated the "relaxation kernel."

In the frequency domain, Eq. (4.65) reduces to:

$$E^*(s) = E_0\left[1 + \sum_{i=1}^{n} \alpha_i \frac{\rho_i s}{\rho_i s + 1}\right] \tag{4.66}$$

where s is the Laplace complex variable. It can be seen that E_0 is the limiting value of $E(t)$, which is attained at $t = \infty$ when the VEM is totally relaxed. Equivalently, E_0 is attained when the VEM operates under static conditions at $\omega = 0$.

Also, at time $t = 0$ or $\omega = \infty$, the relaxation modulus assumes a value E_∞ given by:

$$E_\infty = E_0\left[1 + \sum_{i=1}^{n} \alpha_i\right] \tag{4.67}$$

where E_∞ denotes the instantaneous modulus.

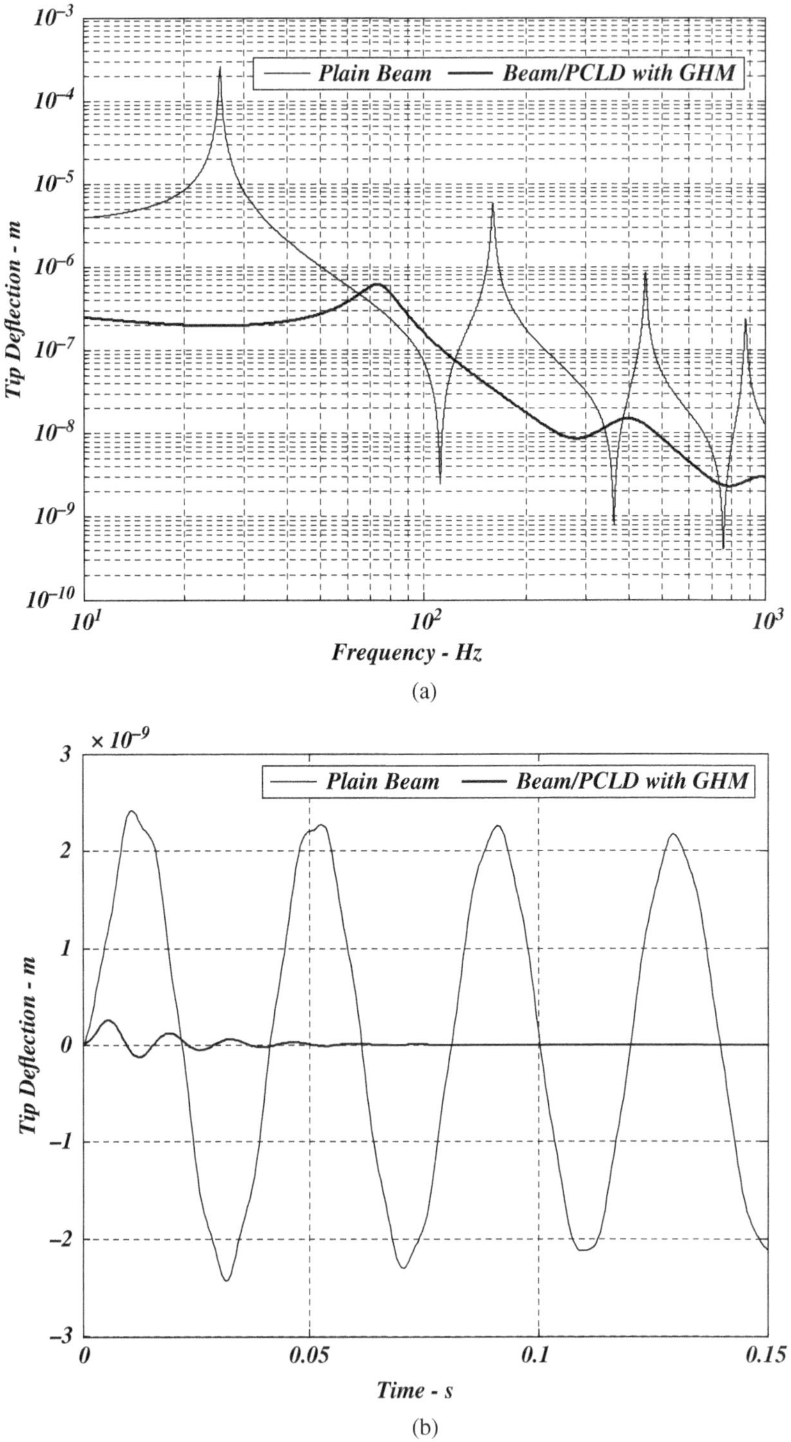

Figure 4.16 Response of the 10-element cantilever beam with and without VEM treatment in the frequency (a) and time domains (b).

Figure 4.17 Generalized Maxwell model.

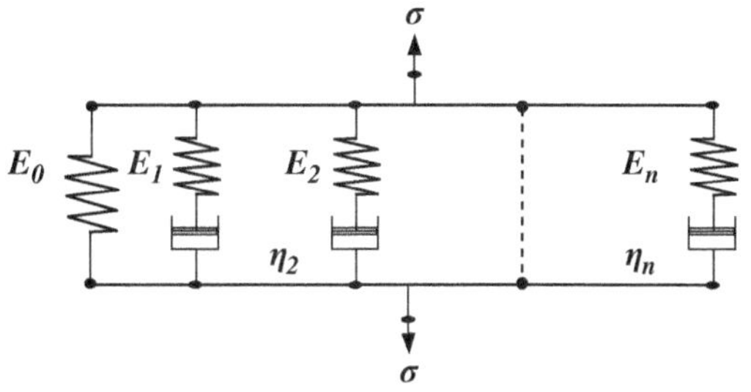

Alternative form of the *GMM*

Equations (4.65) and (4.66) can also be rewritten in the following forms:

$$E(t) = E_\infty \left[\beta_\infty + \sum_{i=1}^{n} \beta_i e^{-t/\rho_i}\right] \text{ and } E^*(s) = E_\infty \left[\beta_\infty + \sum_{i=1}^{n} \beta_i \frac{\rho_i s}{\rho_i s + 1}\right] \tag{4.68}$$

where E_∞ is the instantaneous modulus and β_i is the ith relative modulus.

Equation (4.68) suggests that E_∞ is the limiting value of $E(t)$, which is attained at $t = 0$ with $\beta_\infty + \sum_{i=1}^{n} \beta_i = 1$. Equivalently, E_∞ is attained when the VEM operates at a very high frequency such that $\omega = \infty$.

The relative moduli α_i and β_i are related by the following expressions:

$$\beta_\infty = E_0/E_\infty \text{ and } \beta_i = (E_0/E_\infty)\,\alpha_i \tag{4.69}$$

The alternative form of the GMM, given by Eq. (4.68), is used in most of the commercial finite element codes, such as ANSYS, to express the behavior of the VEM.

4.4.2 Internal Variable Representation of the GMM

4.4.2.1 Single-DOF System

Consider a VEM described by a single GMM mini-oscillator (i.e., $n = 1$), which is coupled with a mass such that the equation of motion in the Laplace domain is:

$$Ms^2X(s) + K(s)X(s) = F(s) \tag{4.70}$$

where M is the mass, K is the complex stiffness of the VEM, and F is the forcing function. Using Eq. (4.66), Eq. (4.70) can be written as

$$Ms^2X(s) + K\left(1 + \alpha_n \frac{\rho_n s}{\rho_n s + 1}\right)X(s) = F(s). \tag{4.71}$$

Now let the internal variable $Z(s)$ be defined as:

$$Z = \frac{1/\rho_n}{s + 1/\rho_n} X \tag{4.72}$$

$$In\ the\ time\ domain: \dot{z} = \frac{1}{\rho_n}(x - z) \tag{4.73}$$

Substituting Eq. (4.72) into Eq. (4.71) gives:

$$Ms^2X(s) + KX(s) + K\alpha_n\rho_n sZ(s) = F(s) \tag{4.74}$$

Transforming Eq. (4.74) into the time domain, and using Eq. (4.73), yields:

$$M\ddot{x} + Kx + K\alpha_n(x - z) = f \tag{4.75}$$

Combining Eqs. (4.75) and (4.73) in a matrix form, gives:

$$\begin{bmatrix} M & 0 \\ 0 & 0 \end{bmatrix}\begin{Bmatrix} \ddot{x} \\ \ddot{z} \end{Bmatrix} + \begin{bmatrix} 0 & 0 \\ 0 & 1 \end{bmatrix}\begin{Bmatrix} \dot{x} \\ \dot{z} \end{Bmatrix} + \begin{bmatrix} K(1+\alpha_n) & -K\alpha_n \\ -\dfrac{1}{\rho_n} & \dfrac{1}{\rho_n} \end{bmatrix}\begin{Bmatrix} x \\ z \end{Bmatrix} = \begin{Bmatrix} f \\ 0 \end{Bmatrix} \tag{4.76}$$

Multiplying the second row of Eq. (4.76) by $K\alpha_n\rho_n$ yields the following system that has symmetric matrices:

$$\begin{bmatrix} M & 0 \\ 0 & 0 \end{bmatrix}\begin{Bmatrix} \ddot{x} \\ \ddot{z} \end{Bmatrix} + \begin{bmatrix} 0 & 0 \\ 0 & K\alpha_n\rho_n \end{bmatrix}\begin{Bmatrix} \dot{x} \\ \dot{z} \end{Bmatrix} + \begin{bmatrix} K(1+\alpha_n) & -K\alpha_n \\ -K\alpha_n & K\alpha_n \end{bmatrix}\begin{Bmatrix} x \\ z \end{Bmatrix} = \begin{Bmatrix} f \\ 0 \end{Bmatrix} \tag{4.77}$$

Note that the mass matrix of the system given by Eq. (4.77) is singular. It is therefore essential to statically condense the IDOF z of the VEM.

Using the static condensation method, as outlined in Section 4.2, gives:

$$z = x \quad \text{and} \quad \begin{Bmatrix} x \\ z \end{Bmatrix} = \begin{Bmatrix} 1 \\ 1 \end{Bmatrix} x = Tx \tag{4.78}$$

Accordingly, Eq. (4.77) reduces to:

$$M\ddot{x} + K\alpha_n\rho_n\dot{x} + Kx = f \tag{4.79}$$

Equation (4.79) indicates that the reduced order system, which contains only the structural degree of freedom x, is now damped by virtue of the VEM damping parameters (α_n, ρ_n). Solution of Eq. (4.79), either in the time or frequency domain, yields the system response x from which the response of the IDOF z can be reconstructed using Eq. (4.78).

4.4.2.2 Multi-Degree of Freedom System

Equation (4.77) can be extended to describe the dynamics of a multi-degree of freedom system coupled with a VEM that is modeled by a multi-term GMM. The extension can be carried out in a manner similar to that of the multi-resonator GHM model described in Section 4.2. The corresponding equation is given by:

$$\begin{bmatrix} \mathbf{M}_s+\mathbf{M}_v & 0 & 0 & \dots & 0 \\ 0 & 0 & 0 & \dots & 0 \\ 0 & 0 & 0 & \dots & 0 \\ \dots & \dots & \dots & \dots & \dots \\ 0 & 0 & 0 & \dots & 0 \end{bmatrix} \begin{Bmatrix} \ddot{\mathbf{x}} \\ \ddot{\mathbf{z}}_1 \\ \ddot{\mathbf{z}}_2 \\ \dots \\ \ddot{\mathbf{z}}_n \end{Bmatrix} + \begin{bmatrix} 0 & 0 & 0 & \dots & 0 \\ 0 & \mathbf{K}_v\alpha_1\rho_1 & 0 & \dots & 0 \\ 0 & 0 & \mathbf{K}_v\alpha_2\rho_2 & \dots & 0 \\ \dots & \dots & \dots & \dots & \dots \\ 0 & 0 & 0 & \dots & \mathbf{K}_v\alpha_n\rho_n \end{bmatrix} \begin{Bmatrix} \dot{\mathbf{x}} \\ \dot{\mathbf{z}}_1 \\ \dot{\mathbf{z}}_2 \\ \dots \\ \dot{\mathbf{z}}_n \end{Bmatrix} +$$

$$\begin{bmatrix} \mathbf{K}_s+\mathbf{K}_v(1+\alpha_1+\alpha_2+\dots) & -\mathbf{K}_v\alpha_1 & -\mathbf{K}_v\alpha_2 & \dots & -\mathbf{K}_v\alpha_n \\ -\mathbf{K}_v\alpha_1 & \mathbf{K}_v\alpha_1 & & \dots & 0 \\ -\mathbf{K}_v\alpha_2 & 0 & \mathbf{K}_v\alpha_2 & \dots & 0 \\ \dots & \dots & \dots & \dots & 0 \\ -\mathbf{K}_v\alpha_n & 0 & 0 & 0 & \mathbf{K}_v\alpha_n \end{bmatrix} \begin{Bmatrix} \mathbf{x} \\ \mathbf{z}_1 \\ \mathbf{z}_2 \\ \dots \\ \mathbf{z}_n \end{Bmatrix} = \begin{Bmatrix} \mathbf{f} \\ 0 \\ 0 \\ 0 \\ 0 \end{Bmatrix} \tag{4.80}$$

where **M** and **K** are the mass and stiffness matrices of the base structure. Also, α_i and ρ_i denote the ith relative modulus and the ith relaxation time of the GMM description of the VEM.

In a compact form, Eq. (4.80) reduces to

$$\mathbf{M_o}\ddot{\mathbf{X}} + \mathbf{C_o}\dot{\mathbf{X}} + \mathbf{K_o}\mathbf{X} = \mathbf{f} \tag{4.81}$$

where $\mathbf{M_o}$, $\mathbf{C_o}$, and $\mathbf{K_o}$ are the overall mass, damping, and stiffness matrices of the structure/VEM assembly.

4.4.2.3 Condensation of the Internal Degrees of Freedom

The condensation of the IDOF vector ($\mathbf{z} = \{\mathbf{z}_1, \mathbf{z}_2, \dots, \mathbf{z}_n\}^T$) of the VEM is essential, at this stage, to avoid the singularity of the mass matrix of the combined structure/VEM assembly. The static condensation yields:

$$\mathbf{z} = \begin{bmatrix} I_{n\times n} \\ I_{n\times n} \\ \dots \\ I_{n\times n} \end{bmatrix} \mathbf{x} = I_{nn\times n}\mathbf{x} \quad \text{and} \quad \begin{Bmatrix} \mathbf{x} \\ \mathbf{z} \end{Bmatrix} = \begin{Bmatrix} I_{n\times n} \\ I_{nn\times n} \end{Bmatrix} \mathbf{x} = T\mathbf{x} \tag{4.82}$$

Hence, Eq. (4.80) reduces to:

$$\mathbf{M}_R\ddot{\mathbf{x}} + \mathbf{C}_R\dot{\mathbf{x}} + \mathbf{K}_R\mathbf{x} = \mathbf{f} \tag{4.83}$$

where $\mathbf{M}_R$, $\mathbf{C}_R$, and $\mathbf{K}_R$ are the reduced mass, damping, and stiffness matrices of the structure/VEM assembly, given by:

$$\mathbf{M}_R = \mathbf{T}^T\mathbf{M_o}\mathbf{T}, \mathbf{C}_R = \mathbf{T}^T\mathbf{C_o}\mathbf{T}, \quad \text{and} \quad \mathbf{K}_R = \mathbf{T}^T\mathbf{K_o}\mathbf{T}. \tag{4.84}$$

4.4.2.4 Direct Solution of Coupled Structural and Internal Degrees of Freedom

A direct solution of the coupled structural and IDOF may be necessary to achieve excellent numerical accuracy, which may not be obtained with the static condensation. For such a direct solution, Eq. (4.80) is rewritten in the following partitioned form:

$$\begin{bmatrix} \mathbf{M}_{xx} & 0 \\ 0 & 0 \end{bmatrix} \begin{Bmatrix} \ddot{\mathbf{x}} \\ \ddot{\mathbf{z}} \end{Bmatrix} + \begin{bmatrix} 0 & 0 \\ 0 & \mathbf{C}_{zz} \end{bmatrix} \begin{Bmatrix} \dot{\mathbf{x}} \\ \dot{\mathbf{z}} \end{Bmatrix} + \begin{bmatrix} \mathbf{K}_{xx} & -\mathbf{K}_{xz} \\ -\mathbf{K}_{xz} & \mathbf{K}_{zz} \end{bmatrix} \begin{Bmatrix} \mathbf{x} \\ \mathbf{z} \end{Bmatrix} = \begin{Bmatrix} \mathbf{f} \\ 0 \end{Bmatrix} \tag{4.85}$$

where $\mathbf{M}_{xx}$, $\mathbf{C}_{zz}$, $\mathbf{K}_{xx}$, $\mathbf{K}_{xz}$, **and** $\mathbf{K}_{zz}$ are partitioned mass, damping, and stiffness matrices of the structure/VEM assembly.

4.4.2.4.1 *Time Domain Analysis*

Equation (4.82) can be expanded and written as follows:

$$\mathbf{M}_{xx}\ddot{\mathbf{x}} + \mathbf{K}_{xx}\mathbf{x} - \mathbf{K}_{xz}\mathbf{z} = \mathbf{f} \tag{4.86}$$

and

$$\mathbf{C}_{zz}\dot{\mathbf{z}} + \mathbf{K}_{zz}\mathbf{z} = \mathbf{K}_{xz}\mathbf{x} \tag{4.87}$$

In state-space representation, Eqs. (4.86) and (4.87) can be cast in the following standard form:

$$\dot{\mathbf{X}} = \mathbf{A}\mathbf{X} + \mathbf{B}u$$

and

$$y = \mathbf{C}\mathbf{X} + \mathbf{D}u \tag{4.88}$$

where $\mathbf{A} = \begin{bmatrix} 0 & \mathbf{I} & 0 \\ -\mathbf{M}_{xx}^{-1}\mathbf{K}_{xx} & 0 & \mathbf{M}_{xx}^{-1}\mathbf{K}_{xz} \\ \mathbf{C}_{zz}^{-1}\mathbf{K}_{xz} & 0 & -\mathbf{C}_{zz}^{-1}\mathbf{K}_{zz} \end{bmatrix}$, $\mathbf{B} = \begin{Bmatrix} 0 \\ \mathbf{M}_{xx}^{-1}\mathbf{f} \\ 0 \end{Bmatrix}$, $\mathbf{C} = [\mathbf{C}_{xx} \;\; \mathbf{0}]$, and $\mathbf{D} = 0$ with $\mathbf{X} = \{\mathbf{x} \;\; \mathbf{z}\}^T$ and $\mathbf{C}_{xx}$ denoting the measurement matrix of the structural system with entries 0 and 1 depending on the location of the sensor.

The state-space system given by Eq. (4.88) can be solved by direct integration for any particular set of initial conditions or input excitations.

4.4.2.4.2 *Frequency Domain Analysis*

Taking the Laplace transform of the Eqs. (4.86) and (4.87) yields:

$$[\mathbf{M}_{xx}s^2 + \mathbf{K}_{xx}]\mathbf{x} - \mathbf{K}_{xz}\mathbf{z} = \mathbf{f} \tag{4.89}$$

and

$$[\mathbf{C}_{zz}s + \mathbf{K}_{zz}]\mathbf{z} = \mathbf{K}_{xz}\mathbf{x} \tag{4.90}$$

For sinusoidal input excitation $\mathbf{f}$ of magnitude $\mathbf{f}_0$ and frequency ω, these equations yield:

$$\mathbf{z} = [\mathbf{C}_{zz}\omega i + \mathbf{K}_{zz}]^{-1}\mathbf{K}_{xz}\mathbf{x}$$

and

$$\mathbf{x} = \left(\left[-\mathbf{M}_{xx}\omega^2 + \mathbf{K}_{xx}\right] - \mathbf{K}_{xz}\left[\mathbf{C}_{zz}\omega i + \mathbf{K}_{zz}\right]^{-1}\mathbf{K}_{xz}\right)^{-1}\mathbf{f}_0 \tag{4.91}$$

It is important to note that if $\mathbf{K}_{zz}$ and $\mathbf{C}_{zz}$ are singular because of the singularity of $\mathbf{K}_v$, then the zero eigenvalues of $\mathbf{K}_v$ should be eliminated using the approach outlined in the GHM method using Eqs. (4.48) and (4.49).

Example 4.6 Consider the fixed-free beam/VEM system of Example 4.5. Determine the response of the system assuming that the complex shear modulus G_2 of the VEM is represented by a five-term GMM model that have the properties listed in Table 4.6 such that:

$$G_2(s) = G_\infty\left[\beta_\infty + \sum_{i=1}^{5}\beta_i\frac{\rho_i s}{\rho_i s + 1}\right] \quad \text{with } G_\infty = 292.01\,\text{MPa and}$$

$\beta_\infty = 0.007.$

Compare the response with that predicted by ANSYS, which requires the use of the GMM form described before in terms of the final equilibrium modulus G_∞ and relative moduli β_i's.

Solution

The finite element formulation presented in Section 4.4 is employed to investigate the effect of viscoelastic damping treatment on the vibration response of the beam both in the time and frequency domains. Figure 4.18 shows comparisons between the response of the PCLD beam using the described finite element method and ANSYS. Perfect agreement is evident between the predictions of the two approaches.

The Prony series description of the *GMM* in *ANSYS* is given as follows:

```
TB,PRONY, 3, 1, 5, SHEAR
TBTEMP, 0
TBDATA, 1, 0.7026,4E-5, 0.2342, 3E-4, 0.0468, 3E-3
TBDATA, 7, 0.0047, 3E-2, 0.0047, 3E-1,,
```

The resulting mode shapes of the beam/PCLD assembly at the first two natural frequencies are displayed in Figure 4.19.

Table 4.6 Optimized parameters of GMM five-term model of the VEM.

Term	1	2	3	4	5
β_i	0.7026	0.2342	0.0468	0.0047	0.0047
ρ_i (s)	4E-5	3E-4	3E-3	3E-2	3E-1

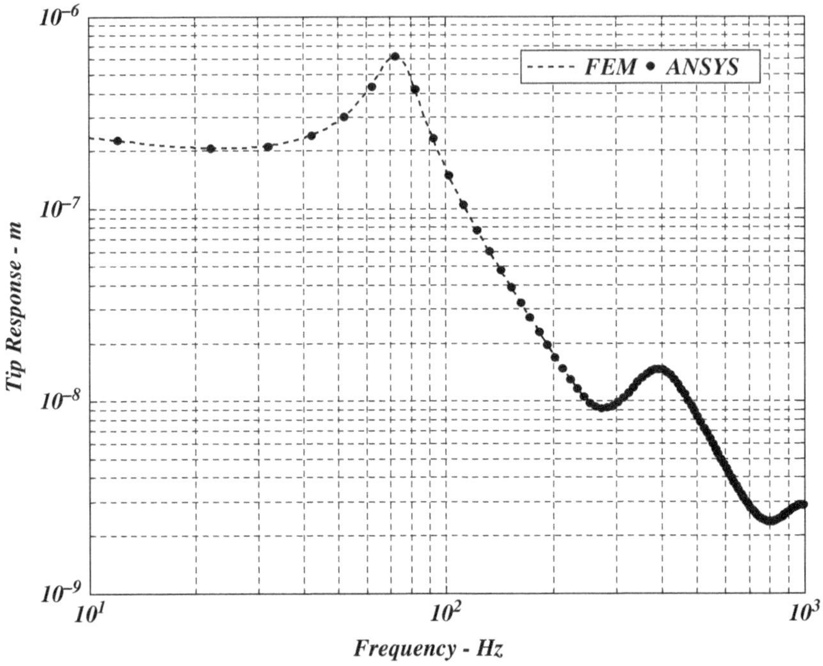

(a)

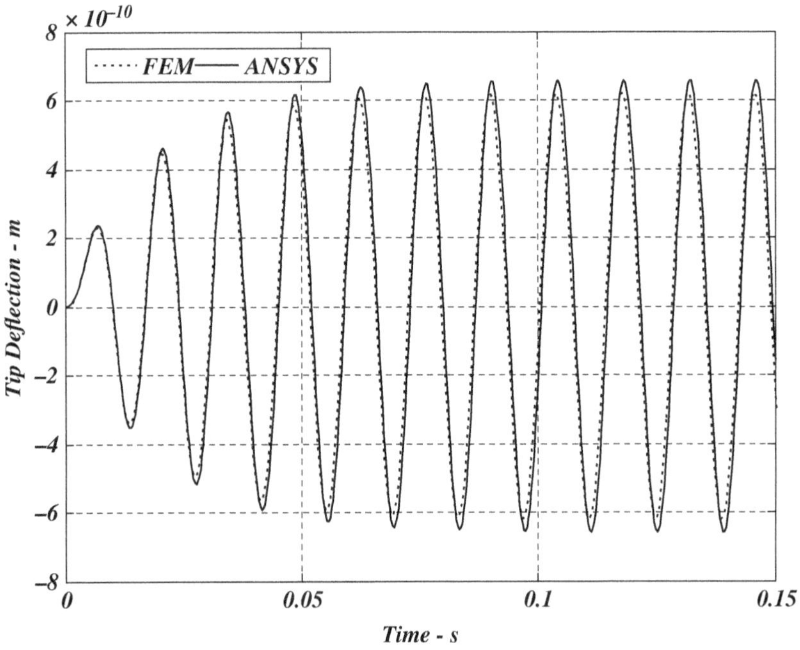

(b)

Figure 4.18 Comparison between the predictions of the finite element method and ANSYS using the GMM. (a) Frequency response and (b) time response at 72 Hz).

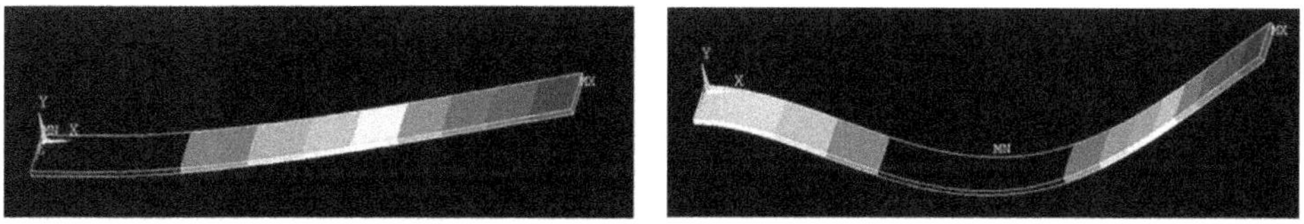

(a) – First mode at 72Hz (b) – Second mode at 392 Hz

Figure 4.19 Modes shapes of the beam/PCLD assembly at the first two natural frequencies using ANSYS. (a) First mode at 72 Hz and (b) second mode at 392 Hz.

4.5 Augmenting Thermodynamic Field (ATF) Model

4.5.1 Overview

The ATF was developed by Lesieutre in 1989 as a physics-based approach to modeling the behavior of VEM both in the time and frequency domains because it can be easily integrated with finite element models of vibrating structures as the GHM and GMM models. Several follow-up studies have been carried out to develop better understanding of the ATF approach and extend its applicability to various types of structures as reported, for example, by Lesieutre and Mingori (1990), Rusovici (1999), and Trindade et al. (2000).

The ATF approach is developed using the following Helmholtz free energy density function f_{ATF} in N m^{-2}:

$$f_{ATF} = \frac{1}{2}E_{\infty}\varepsilon^2 - \delta_{ATF}\varepsilon z + \frac{1}{2}\gamma_{ATF}z^2 \tag{4.92}$$

where ε is the strain, E_{∞} is the un-relaxed or high frequency modulus of elasticity, and z is a single ATF. Note that z is the IDOF z of the VEM, which is denoted as the "temperature field." It resembles the IDOFs described also for the GHM and GMM approaches. The coupling term between the mechanical displacement field and the ATF is δ_{ATF} while γ_{ATF} is an effective modulus of the ATF.

The resulting constitutive equations are given by:

$$\sigma = \frac{\partial f_{ATF}}{\partial \varepsilon} = E_{\infty}\varepsilon - \delta_{ATF}z, \tag{4.93}$$

and

$$A_{ATF} = -\frac{\partial f_{ATF}}{\partial z} = \delta_{ATF}\,\varepsilon - \gamma_{ATF}z. \tag{4.94}$$

where A_{ATF} denotes the affinity, which along with the z forms the thermodynamic pair that is equivalent to the mechanical pair: the stress σ and the strain ε.

The evolution equation of the IDOF z is obtained using the basic assumption of irreversible thermodynamics, where the rate of change of z is proportional to A_{ATF}. In other words, the rate of change of z is proportional to its deviation from an equilibrium value $\bar{z}$ such that:

$$\dot{z} = L_{ATF}A_{ATF} = -\frac{1}{\rho_{ATF}}(z - \bar{z}) \tag{4.95}$$

where L_{ATF} is a constant of proportionality and ρ_{ATF} denotes the relaxation time constant at constant strain. Also, $\bar{z}$ is given by:

$$\bar{z} = z|_{A=0} = (\delta_{ATF}/\gamma_{ATF})\varepsilon \tag{4.96}$$

Combining Eqs. (4.95) and (4.96) gives:

$$\dot{z} = \left(\frac{\delta_{ATF}}{\gamma_{ATF}\rho_{ATF}}\right)\varepsilon - \frac{1}{\rho_{ATF}}z \tag{4.97}$$

Hence, the constitutive equations of the VEM are given by:

$$\sigma = E_{\infty}\varepsilon - \delta_{ATF}z, \tag{4.98}$$

and

$$\dot{z} = \left(\frac{\delta_{ATF}}{\gamma_{ATF}\rho_{ATF}}\right)\varepsilon - \frac{1}{\rho_{ATF}}z \tag{4.99}$$

Taking the Laplace transform of these constitutive equations and eliminating z between them gives:

$$z = \frac{\delta_{ATF}}{\gamma_{ATF}(\rho_{ATF}s + 1)}\varepsilon \text{ and } \sigma = \frac{E_{\infty}\rho_{ATF}s + \left(E_{\infty} - \frac{\delta_{ATF}^2}{\gamma_{ATF}}\right)}{\rho_{ATF}s + 1}\varepsilon \tag{4.100}$$

$$\text{Let } E_0 = \left(E_{\infty} - \frac{\delta_{ATF}^2}{\gamma_{ATF}}\right),$$

$$\text{Then, } \sigma = E_0\left[1 + \left(\frac{E_{\infty} - E_0}{E_0}\right)\frac{\rho_{ATF}s}{\rho_{ATF}s + 1}\right]\varepsilon = E^*(s)\,\varepsilon \tag{4.101}$$

where $E^*(s)$ is the complex modulus of the VEM, which can be rewritten as follows:

$$E^*(s) = E_0\left[1 + \Delta_{ATF}\frac{\rho_{ATF}s}{\rho_{ATF}s + 1}\right] \tag{4.102}$$

where $\Delta_{ATF} = \left(\frac{E_{\infty} - E_0}{E_0}\right)$ is called the "Relaxation Resistance."

4.5.2 Equivalent Damping Ratio of the *ATF* Model

Consider a mass M supported on a VEM that has a stiffness $K^*(s)$ described by the following single term ATF model:

$$K^*(s) = K_0\left[1 + \Delta_{ATF}\frac{\rho_{ATF}s}{\rho_{ATF}s + 1}\right] \text{ with } \Delta_{ATF} = (K_{\infty} - K_0)/K_0 \tag{4.103}$$

Then, according to approaches presented in Sections 4.4.2 and 4.5.1, the equations of motion that govern the dynamics of the mass, subject to a force **f**, and VEM DOF (x and z) are given by:

$$\begin{bmatrix} M & 0 \\ 0 & 0 \end{bmatrix}\begin{Bmatrix} \ddot{x} \\ \ddot{z} \end{Bmatrix} + \begin{bmatrix} 0 & 0 \\ 0 & K\Delta_{ATF}\rho_{ATF} \end{bmatrix}\begin{Bmatrix} \dot{x} \\ \dot{z} \end{Bmatrix} + \begin{bmatrix} K(1+\Delta_{ATF}) & -K\Delta_{ATF} \\ -K\Delta_{ATF} & K\Delta_{ATF} \end{bmatrix}\begin{Bmatrix} x \\ z \end{Bmatrix} = \begin{Bmatrix} \mathbf{f} \\ 0 \end{Bmatrix} \tag{4.104}$$

Using the static equation yields the following reduced order system:

$$M\ddot{x} + K_0\Delta_{ATF}\rho_{ATF}\dot{x} + K_0 x = \mathbf{f}$$

Or

$$\ddot{x} + \frac{(K_\infty - K_0)\rho_{ATF}}{M}\dot{x} + \frac{K_0}{M}x = \frac{1}{M}\mathbf{f} \tag{4.105}$$

Equation (4.105) can be written as:

$$\ddot{x} + 2\zeta\omega_n x + \omega_n^2 x = F \tag{4.106}$$

where $\omega_n = \sqrt{\frac{K_0}{M}}, \zeta_{ATF} = \frac{\rho_{ATF}(K_\infty - K_0)}{2\sqrt{K_0/M}} = \frac{1}{2}\rho_{ATF}\Delta_{ATF}\omega_n, F = \frac{1}{M}\mathbf{f}.$

Hence, the equivalent damping ratio of a single term ATF model is $\zeta_{ATF} = \frac{1}{2}\rho_{ATF}\Delta_{ATF}\omega_n$.

4.5.3 Multi-degree of Freedom ATF Model

Equation (4.77) can be extended to describe the dynamics of a multi-degree of freedom system coupled with a VEM, which is modeled by a multi-term ATF as outlined in the GHM and GMM approaches to yield:

$$E^*(s) = E_0\left[1 + \sum_{i=1}^{n}\Delta_{ATF_i}\frac{\rho_{ATF_i}s}{\rho_{ATF_i}s + 1}\right] \tag{4.107}$$

Note that Eq. (4.103) resembles Eq. (4.66) for the GMM. The correspondence between the two equations suggests that E_0 is equivalent to the equilibrium or relaxed modulus of the VEM at $\omega = 0$ and $t = \infty$. Also, Δ_{ATF_i} and ρ_{ATF_i} are equivalent to the relative moduli and relaxation time for the GMM model.

4.5.4 Integration with a Finite Element Model

Equation (4.77) can be extended to describe the dynamics of a multi-degree of freedom system coupled with a VEM that is modeled by a multi-term ATF. The extension can be carried out in a manner similar to that of the multi-resonator GHM model described in Section 4.2. The corresponding equation is given by:

$$\begin{bmatrix} \mathbf{M}_s + \mathbf{M}_v & 0 & 0 & \dots & 0 \\ 0 & 0 & 0 & \dots & 0 \\ 0 & 0 & 0 & \dots & 0 \\ \dots & \dots & \dots & \dots & \dots \\ 0 & 0 & 0 & \dots & 0 \end{bmatrix} \begin{Bmatrix} \ddot{\mathbf{x}} \\ \ddot{\mathbf{z}}_1 \\ \ddot{\mathbf{z}}_2 \\ \dots \\ \ddot{\mathbf{z}}_n \end{Bmatrix} + \begin{bmatrix} 0 & 0 & 0 & \dots & 0 \\ 0 & \mathbf{K}_v\Delta_{ATF_1}\rho_{ATF_1} & 0 & \dots & 0 \\ 0 & 0 & \dots & \dots & 0 \\ \dots & \dots & \dots & \dots & \dots \\ 0 & 0 & 0 & \dots & \mathbf{K}_v\Delta_{ATF_n}\rho_{ATF_n} \end{bmatrix} \begin{Bmatrix} \dot{\mathbf{x}} \\ \dot{\mathbf{z}}_1 \\ \dot{\mathbf{z}}_2 \\ \dots \\ \dot{\mathbf{z}}_n \end{Bmatrix} +$$

$$\begin{bmatrix} \mathbf{K}_s + \mathbf{K}_v(1 + \Delta_{ATF_1} + \Delta_{ATF_2} + \dots) & -\mathbf{K}_v\Delta_{ATF_1} & \dots & \dots & -\mathbf{K}_v\Delta_{ATF_n} \\ -\mathbf{K}_v\Delta_{ATF_1} & \mathbf{K}_v\Delta_{ATF1} & 0 & \dots & 0 \\ -\mathbf{K}_v\Delta_{ATF_2} & 0 & \dots & \dots & 0 \\ \dots & \dots & 0 & \dots & 0 \\ -\mathbf{K}_v\Delta_{ATF_n} & 0 & 0 & 0 & \mathbf{K}_v\Delta_{ATF_n} \end{bmatrix} \begin{Bmatrix} \mathbf{x} \\ \mathbf{z}_1 \\ \mathbf{z}_2 \\ \dots \\ \mathbf{z}_n \end{Bmatrix} = \begin{Bmatrix} \mathbf{f} \\ 0 \\ 0 \\ 0 \\ 0 \end{Bmatrix} \tag{4.108}$$

where **M** and **K** are the mass and stiffness matrices of the base structure. Also, Δ_{ATF_i} and ρ_{ATF_i} denote the ith relative modulus and the ith relaxation time of the ATF description of the VEM.

In a compact form, Eq. (4.108) reduces to:

$$\mathbf{M_o}\ddot{\mathbf{X}} + \mathbf{C_o}\dot{\mathbf{X}} + \mathbf{K_o}\mathbf{X} = \mathbf{f} \tag{4.109}$$

where $\mathbf{M_o}$, $\mathbf{C_o}$, and $\mathbf{K_o}$ are the overall mass, damping, and stiffness matrices of the structure/VEM assembly. Solution of Eq. (4.105) can be obtained using the same approaches considered for solving Eq. (4.81) for the case of the GMM. These approaches can be carried out either by condensing the VEM DOF or manipulating the directly coupled structural and VEM DOF.

Example 4.7 Consider the fixed-free beam/VEM system of Example 4.5. Determine the response of the system assuming that the complex shear modulus G_2 of the VEM is represented by a three-term ATF model and its equivalent three-term GHM that have the properties listed in Table 4.7 such that:

$$\text{ATF model: } G_2(s) = G_0\left[1 + \sum_{i=1}^{3} \Delta_{ATF_i} \frac{\rho_{ATF_i} s}{\rho_{ATF_i} s + 1}\right] \quad \text{with } G_0 = 0.5\,\text{MPa}$$

Compare the responses predicted by using the ATF and GHM models of the VEM.

Solution

The finite element formulation presented in Section 4.4 is employed to investigate the effect of viscoelastic damping treatment on the vibration response of the beam both in

Table 4.7 Optimized parameters of three-term ATF and GHM models of the VEM.

Model	Term	1	2	3
ATF	Δ_{ATF_i}	0.746	3.265	43.284
	$\rho_{ATF_i} - s$	2.1E-3	2.11E-4	1.40E-5
GHM	ω_i (rad s^{-1})	6502.9	50,618.8	352,782
	α_i	0.742	3.237	41.654
	ζ_i	6.97	5.38	2.56

the time and frequency domains. Figure 4.20 shows comparisons between the responses of the PCLD beam with a VEM core modeled by the ATF and GHM methods. Exact agreement is evident between the predictions of the two approaches.

4.6 Fractional Derivative (FD) Models

4.6.1 Overview

In this section, the fractional derivative approach for modeling VEM is integrated with finite element formulations of basic structures. In this context, extensive efforts have been exerted to approximate the fractional order derivative of the stress–strain relationships using appropriate time discretization schemes. Distinct among these efforts is the pioneering work of Padovan (1987) that introduced several implicit, explicit, and predictor-corrector type algorithms. In 1998, Escobedo-Torres and Ricles (1998) employed a central difference method along with a stability analysis. Enelund and Josefson (1997) utilized convolution integral description with a singular kernel function of Mittag–Leffler type. Galerkin projections have also been considered for approximating fractional derivatives of linear and nonlinear systems by Singh and Chatterjee (2006). However, all these efforts were limited to simple vibrating systems such as single-DOF systems and rod-type structures.

Recently, the emphasis has been placed on extending these efforts to develop finite element algorithms for handling more complex structural systems such as plain beams (Sorrentino and Fasana 2007), beams treated with unconstrained layer damping (Cortes and Elejabarrieta 2007a,b), beams treated with constrained layer damping (Galucio et al. 2004), and plates (Schmidt and Gaul 2002).

In 2009, Bekuit et al. (2009) extended these efforts and presented a quasi-2D finite element formulation for modeling actively constrained layer beams with viscoelastic cores described by fractional derivative constitutive equations, which are represented by the Grunwald approximation.

Other more recent advanced efforts for integrating fractional derivative approach for modeling VEM with finite element formulations include modeling of large deformation of viscoelastic beams (Bahraini et al. 2013), modeling of the dynamics of curved beams by Piovan et al. (2009), and condensation of finite element models (Catania et al. 2008).

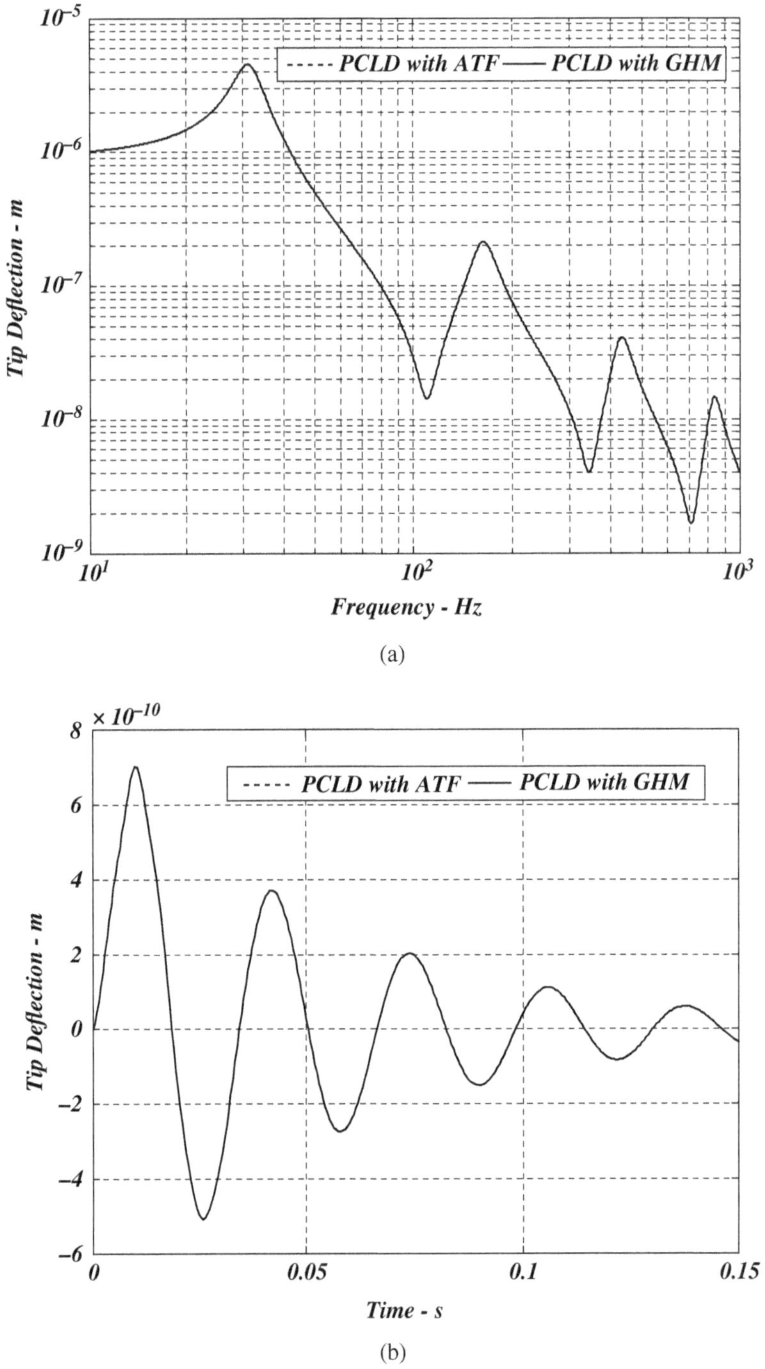

Figure 4.20 Response of the 10-element cantilever beam with and without VEM treatment in the (a) frequency and (b) time domains.

In this section, the approach considered is parallel to those adopted for handling GHM, GMM, and ATF models. IDOF are introduced to simplify the constitutive equations of the VEM. Then, the most common Grunwald approximation is used to discretize the resulting equations in the time domain to enable the coupling of the VEM DOF with those of the host structure.

4.6.2 Internal Degrees of Freedom of Fractional Derivative Models

Consider the following four-parameter fractional derivative constitutive model by Bagley and Torvik (1983) that relates the stress σ to its strain ε of a VEM as follows:

$$\sigma(t)+\tau^{\alpha}D^{\alpha}\sigma(t)=E_0\varepsilon(t)+E_{\infty}\tau^{\alpha}D^{\alpha}\varepsilon(t) \tag{4.110}$$

where D^{α} is a fractional derivative of order α such that $0<\alpha<1$. Also, E_0 is the relaxed elastic modulus, E_{∞} is the un-relaxed elastic modulus, and τ is the relaxation time.

Let a strain function $\bar{\varepsilon}$ denote the internal variable of the VEM such that:

$$\bar{\varepsilon}=\varepsilon-\frac{\sigma}{E_{\infty}} \tag{4.111}$$

Then, Eq. (4.110) can be rewritten as follows:

$$\frac{\sigma(t)}{E_{\infty}}-\frac{E_0}{E_{\infty}}\varepsilon(t)-\tau^{\alpha}\left[D^{\alpha}\varepsilon(t)-\frac{1}{E_{\infty}}D^{\alpha}\sigma(t)\right]=0$$

Add and subtract $\varepsilon(t)$ to the left-hand side of this equation, which gives:

$$\frac{\sigma(t)}{E_{\infty}}-\varepsilon(t)+\varepsilon(t)-\frac{E_0}{E_{\infty}}\varepsilon(t)-\tau^{\alpha}\left[D^{\alpha}\varepsilon(t)-\frac{1}{E_{\infty}}D^{\alpha}\sigma(t)\right]=0 \tag{4.112}$$

Substituting Eq. (4.111) into Eq. (4.112) yields:

$$\tau^{\alpha}D^{\alpha}\bar{\varepsilon}+\bar{\varepsilon}=\left(\frac{E_{\infty}-E_0}{E_{\infty}}\right)\varepsilon \tag{4.113}$$

Note that Eq. (4.112) contains the fractional derivatives of only the internal variable $\bar{\varepsilon}$ unlike Eq. (4.110) that depends on the fractional derivatives of both the stress and strain. Such a transformation and simplification of the constitutive equation enables the development and implementation of the integrated structural/VEM finite element formulation.

4.6.3 Grunwald Approximation of Fractional Derivative

Using the Grunwald approximation, described in Chapter 2 and given by Eq. (2B.14), the fractional derivative $\tau^{\alpha}D^{\alpha}\bar{\varepsilon}$ of the internal variable $\bar{\varepsilon}$ can be written as follows:

$$\tau^{\alpha}D^{\alpha}\bar{\varepsilon}=\left(\frac{\tau}{\Delta t}\right)^{\alpha}\bar{\varepsilon}+\left(\frac{\tau}{\Delta t}\right)^{\alpha}\sum_{j=1}^{N-1}A_{j+1}\bar{\varepsilon}(t-jt/N) \tag{4.114}$$

where $\Delta t=\frac{t}{N}$ and $A_{j+1}=[(j-1-\alpha)/j]A_j$ for $0<j<\infty$.

Accordingly, Eq. (4.113) reduces to:

$$\left(\frac{\tau}{\Delta t}\right)^{\alpha}\bar{\varepsilon}+\left(\frac{\tau}{\Delta t}\right)^{\alpha}\sum_{j=1}^{N-1}A_{j+1}\bar{\varepsilon}(t-jt/N)+\bar{\varepsilon}=\left(\frac{E_{\infty}-E_{0}}{E_{\infty}}\right)\varepsilon \tag{4.115}$$

Let $c=\dfrac{\tau^{\alpha}}{\tau^{\alpha}+\Delta t^{\alpha}}$, then Eq. (4.115) yields the following equation that describes the "fading memory phenomenon" of the VEM as a function of the time step n:

$$\text{or } \bar{\varepsilon}_{n+1}=(1-c)\left(\frac{E_{\infty}-E_{0}}{E_{\infty}}\right)\varepsilon_{n+1}-c\sum_{j=1}^{N-1}A_{j+1}\bar{\varepsilon}_{n+1-j} \tag{4.116}$$

From Eqs. (4.111) and (4.116), the stress σ_{n+1}, at time step $n+1$, can be written as:

$$\begin{aligned}\sigma_{n+1}&=E_{\infty}(\varepsilon_{n+1}-\bar{\varepsilon}_{n+1})\\&=E_{\infty}\left(\varepsilon_{n+1}-(1-c)\left(\frac{E_{\infty}-E_{0}}{E_{\infty}}\right)\varepsilon_{n+1}+c\sum_{j=1}^{N-1}A_{j+1}\bar{\varepsilon}_{n+1-j}\right)\\ \text{i.e. } \sigma_{n+1}&=E_{0}\left(\left[1+c\left(\frac{E_{\infty}-E_{0}}{E_{0}}\right)\right]\varepsilon_{n+1}+c\frac{E_{\infty}}{E_{0}}\sum_{j=1}^{N-1}A_{j+1}\bar{\varepsilon}_{n+1-j}\right)\end{aligned} \tag{4.117}$$

Equation (4.117) constitutes the basic expression for extracting the potential energy of a VEM described by a fractional derivative constitutive relationship. The potential energy can then be used, along with appropriate finite element shape functions, to compute stiffness matrix of the VEM as will be outlined in Section 4.6.4.

4.6.4 Integration Fractional Derivative Approximation with Finite Element

The Grunwald approximation of fractional derivative constitutive equations of the VEM, as described by Eq. (4.117), is coupled with the structural description of a plain rod and then a beam treated with PCLD treatment.

4.6.4.1 Viscoelastic Rod

As shown in Section 4.2.4, the *PE* of the VEM rod element is given by:

$$P.E=\frac{1}{2}A\int_{0}^{L}\sigma\varepsilon\,dx=\frac{1}{2}\{\Delta\}^{T}[K_{v}]\{\Delta\} \tag{4.118}$$

where A is the cross sectional area of the VEM. Also, σ and ε are the stress and strain experienced by the VEM. Furthermore, $\{\Delta\}$ and $[K_v]$ denote nodal deflection vector and the VEM stiffness matrix, respectively. Note that the strain ε and the nodal deflection vector $\{\Delta\}$ are related, according to Eq. (4.25), by the following interpolating equation:

$$\varepsilon=[N_{,x}]\{\Delta\} \tag{4.119}$$

where $\{N_{,x}\}$ denotes the partial derivative of $\{N\}$ with respect to x.

Combining Eq. (4.119) and (4.118) gives:

$$P.E. = \frac{1}{2}\{\Delta\}^T A \int_0^L [N_{,x}]^T \sigma \, dx = \frac{1}{2}\{\Delta\}^T [K_v]\{\Delta\}$$

$$\text{that is, } A \int_0^L [N_{,x}]^T \sigma \, dx = [K_v]\{\Delta\} \tag{4.120}$$

Using Eq. (4.117) to substitute for σ in terms of ε and $\bar{\varepsilon}$ while assuming that $\bar{\varepsilon}$ can be represented in a form similar to ε such that: $\bar{\varepsilon} = [N_{,x}]\{\bar{\Delta}\}$with $\{\bar{\Delta}\}$ denoting the internal deflection vector of the VEM, then Eq. (4.120) at time step $n + 1$ reduces to:

$$\begin{aligned}[K_v]\{\Delta_{n+1}\} &= E_0 A \int_0^L [N_{,x}]^T \left[1 + c\left(\frac{E_\infty - E_0}{E_0}\right)\right][N_{,x}]\{\Delta_{n+1}\} dx \\ &+ E_0 A \int_0^L c\frac{E_\infty}{E_0}[N_{,x}]^T \sum_{j=1}^{N-1} A_{j+1}[N_{,x}]\{\bar{\Delta}_{n+1-j}\} dx\end{aligned} \tag{4.121}$$

$$\text{Let } [K_c] = E_0 A \int_0^L [N_{,x}]^T [N_{,x}] dx, \; [\bar{K}_c] = \left[1 + c\left(\frac{E_\infty - E_0}{E_0}\right)\right][K_c],$$

$$\text{and } \bar{F}_{n+1} = -c\frac{E_\infty}{E_0}[K_c]\sum_{j=1}^{N-1} A_{j+1}\{\bar{\Delta}_{n+1-j}\}. \tag{4.122}$$

Then, Eq. (4.121) reduces to:

$$[K_v]\{\Delta_{n+1}\} = [\bar{K}_c]\{\Delta_{n+1}\} - \bar{F}_{n+1} \tag{4.123}$$

Now, one can write the expressions for the kinetic energy KE of the VEM and the work done W_{nc} by the non-conservative external loads F_{n+1} as follows:

$$K.E. = \frac{1}{2}\{\dot{\Delta}_{n+1}\}^T [M_v]\{\dot{\Delta}_{n+1}\} \tag{4.124}$$

and

$$W_{nc} = F_{n+1}\{\Delta_{n+1}\} \tag{4.125}$$

Using the Lagrangian dynamics, the equation of motion of the VEM rod with fractional derivative constitutive equation can be extracted from:

$$\frac{d}{dt}\frac{\partial K.E.}{\partial\{\dot{\Delta}_{n+1}\}} - \frac{\partial P.E.}{\partial\{\Delta_{n+1}\}} = F_{n+1} \tag{4.126}$$

Substituting from Eqs. (4.118), (4.123), and (4.124) into Eq. (4.126) gives:

$$[M_v]\{\ddot{\Delta}_{n+1}\} + [\bar{K}_c]\{\Delta_{n+1}\} = F_{n+1} + \bar{F}_{n+1} \tag{4.127}$$

$$\text{that is}, [M_v]\{\ddot{\Delta}_{n+1}\} + \left[1 + c\left(\frac{E_\infty - E_0}{E_0}\right)\right][K_c]\{\Delta_{n+1}\} = F_{n+1}$$
$$- c\frac{E_\infty}{E_0}[K_c]\sum_{j=1}^{N-1} A_{j+1}\{\bar{\Delta}_{n+1-j}\} \tag{4.128}$$

Note that the left-hand side of Eq. (4.128) includes the inertia force of the rod with mass matrix $[M_v]$, the restoring force resulting from its static structural stiffness matrix $[K_c]$, and the additional stiffness generated by the fractional derivative effect $c(E_\infty/E_0 - 1)[K_c]$. The right hand side of Eq. (4.128) includes the external forces F_{n+1} as well as the loading vector $\bar{F}_{n+1}$ arising from the viscoelastic damping that manifests itself in the form of the fading memory of the time history of the internal variable deflection vector $\{\bar{\Delta}\}$.

Example 4.8 Consider a fixed-free VEM rod that has a length $L = 0.5$ m, width $b = 0.05$ m, and thickness $h = 0.05$ m. The rod is divided into 10 finite elements. The VEM of the rod has a density of 1000 kg m^{-3} and is described by the following fractional derivative model:

$$E^* = \frac{\sigma(s)}{\varepsilon(s)} = \frac{E_0 + E_\infty(\tau s)^\alpha}{1 + (\tau s)^\alpha}$$

where $E_0 = 7MPa$, $E_\infty = 10MPa$, $\tau = 0.02s$, and $\alpha = 0.5$.

Determine the time response of the rod when it is subjected to a unit step load $F = 1$ N at its free end.

Solution

A finite element model is developed to model the dynamics of the viscoelastic rod using the approach outlined in Section 4.6.4. In particular, the model capitalizes on the formulation described by Eq. (4.128).

Figure 4.21a displays a dimensionless time response of the rod where the tip displacement is normalized with respect to a nominal displacement $u_0 = FL/(bhE_0)$ and the time is normalized with respect to the relaxation time constant τ. The obtained results conform exactly with the results reported by Enelund and Josefson (1997) and Galucio et al. (2004).

Figure 4.21b shows the frequency content of the rod time response indicating a dominant frequency at 48.84 Hz.

4.6.4.2 Beam with Passive Constrained Layer Damping (PCLD) Treatment

In this section, the expression of the PE of a beam treated with PCLD treatment, which is presented in Section 4.3.3 by Eq. (4.61), will be reexamined to account for the fractional derivative description of the behavior of the viscoelastic core. In particular, the stiffness matrices $[K_{u_v}]$ and $[K_\gamma]$ of the VEM associated with axial and shear deformations will be modified to include the contributions to the additional stiffening and fading memory effects produced by the fractional derivative formulation of the VEM.

As the potential energy of the beam/PCLD assembly is given by:

$$P.E. = \frac{1}{2}\{\Delta_i\}^T\left(\left[K_{u_{b,c}}\right] + \left[K_{u_v}\right] + [K_w] + \left[K_\gamma\right]\right)\{\Delta_i\} = \frac{1}{2}\{\Delta_i\}^T[K_e]\{\Delta_i\} \tag{4.129}$$

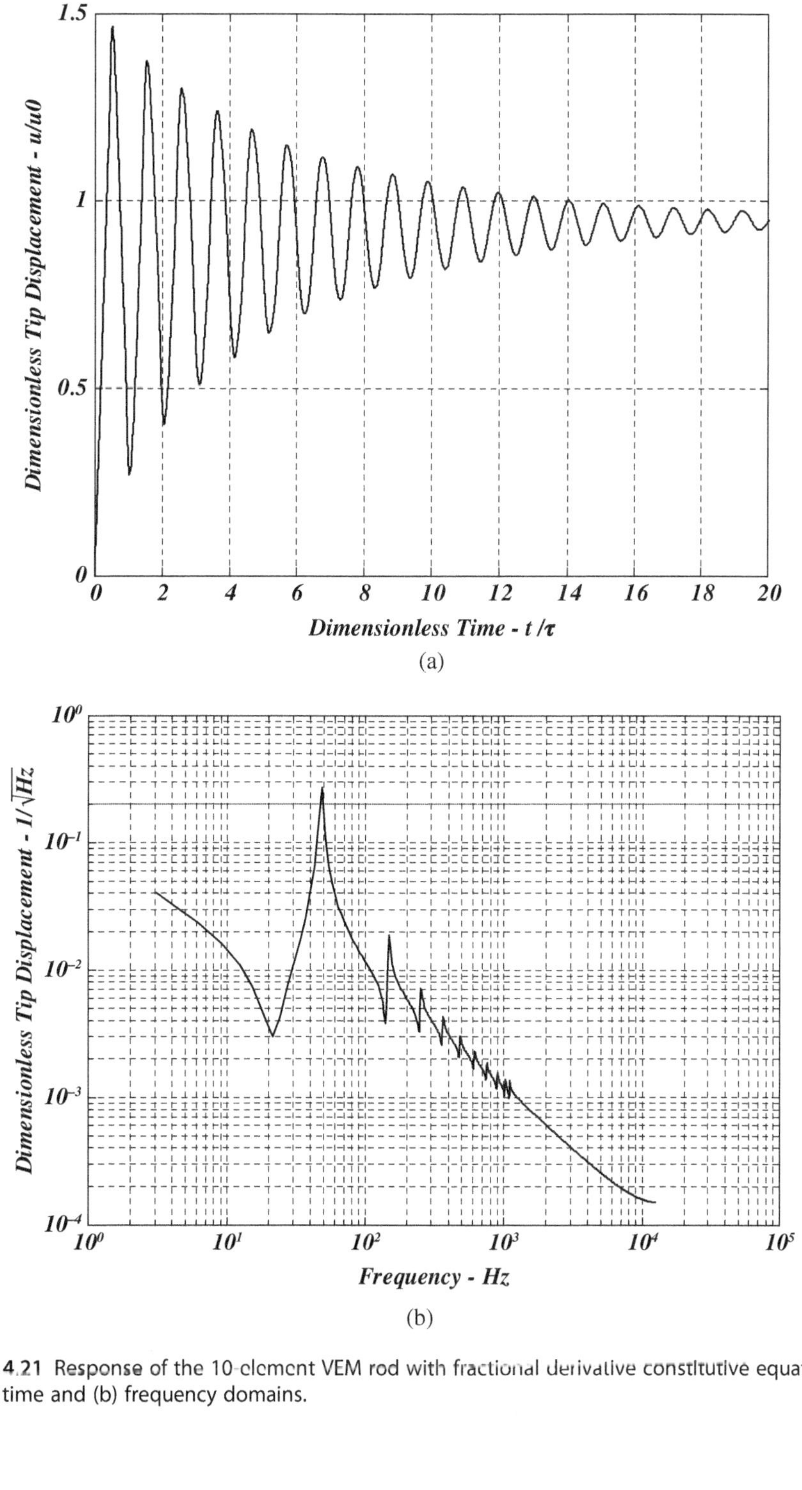

Figure 4.21 Response of the 10-element VEM rod with fractional derivative constitutive equations in the (a) time and (b) frequency domains.

where $[K_{u_{b,c}}]$, $[K_{u_v}]$, $[K_w]$, and $[K_\gamma]$ denote the combined axial stiffness of the beam and constraining layers, the axial stiffness of the VEM layer, the combined bending stiffness of the beam and constraining layers, and the shear stiffness of the VEM layer, respectively. These stiffness matrices are given by:

$$[K_{u_{b,c}}] = \int_0^L \left[E_1A_1\left(N_{u_1,x}^T N_{u_1,x}\right) + E_3A_3\left(N_{u_2,x}^T N_{u_2,x}\right)\right] dx,$$

$$[K_{u_v}] = \int_0^L \frac{1}{2}E_2A_2\left[\left(N_{u_1,x} + \frac{1}{2}(h_1 - h_3)N_{w,xx} + N_{u_2,x}\right)^T \left(N_{u_1,x} + \frac{1}{2}(h_1 - h_3)N_{w,xx} + N_{u_2,x}\right)\right] dx,$$

$$[K_w] = \int_0^L \left[(E_1I_1 + E_3I_3)\left(N_{w,xx}^T N_{w,xx}\right)\right] dx,$$

$$[K_\gamma] = \int_0^L \frac{G_2A_2}{h_2^2}\left[(N_{u_1} - N_{u_2} + hN_{w,x})^T (N_{u_1} - N_{u_2} + hN_{w,x})\right] dx.$$

and

$$[K_e] = [K_{u_{b,c}}] + [K_{u_v}] + [K_w] + [K_\gamma] = \text{element stiffness matrix.} \tag{4.130}$$

Then, it is necessary to modify E_2 and G_2 of the VEM, in the same manner as was presented in Section 4.6.4.1 for the case of viscoelastic rods, in order to account for the fractional derivative description of the behavior of the viscoelastic core. Such modifications will be first reflected in modifying the stiffness matrices $[K_{u_v}]$ and $[K_\gamma]$ of the VEM associated with axial and shear deformations such that:

$$[\bar{K}_{u_v}] = \left[1 + c\left(\frac{E_\infty - E_0}{E_0}\right)\right][K_{u_v}] \tag{4.131}$$

and

$$[\bar{K}_\gamma] = \left[1 + c\left(\frac{G_\infty - G_0}{G_0}\right)\right][K_\gamma] \tag{4.132}$$

where $E_2 = \dfrac{E_0 + E_\infty(\tau s)^\alpha}{1 + (\tau s)^\alpha}$ and $G_2 = \dfrac{G_0 + G_\infty(\tau s)^\alpha}{1 + (\tau s)^\alpha}$.

Also, the modifications generate loading vectors $\bar{F}_{n+1}^u$ and $\bar{F}_{n+1}^\gamma$ by virtue of the fading memory effect, such that:

$$\bar{F}_{n+1}^u = -c\frac{E_\infty}{E_0}[K_{u_v}]\sum_{j=1}^{N-1} A_{j+1}\{\bar{\Delta}_{n+1-j}\} \text{ and } \bar{F}_{n+1}^\gamma = -c\frac{G_\infty}{G_0}[K_\gamma]\sum_{j=1}^{N-1} A_{j+1}\{\bar{\Delta}_{n+1-j}\} \tag{4.133}$$

Then, the resulting equations of motion of the beam/PCLD assembly are given by:

$$[M_v]\{\ddot{\Delta}_{n+1}\} + \left([K_{u_{b,c}}] + [\bar{K}_{u_v}] + [K_w] + [\bar{K}_\gamma]\right)\{\Delta_{n+1}\} = F_{n+1} + \bar{F}_{n+1}^u + \bar{F}_{n+1}^\gamma \tag{4.134}$$

Equation (4.134) can be used to predict the time response of beam/PCLD assemblies when subjected to external loading conditions F_{n+1}.

Example 4.9 Consider the fixed-free beam/PCLD assembly shown in Figure 4.22a. The beam has a length $L = 0.2$ m and width $b = 0.01$ m. The thicknesses of the constraining layer, VEM, and beam are $h_1 = 0.0025$ m, $h_2 = 0.0025$ m, and $h_3 = 0.0025$ m, respectively. The beam/PCLD assembly is divided into five finite elements. The beam and the constraining layers are made of aluminum while the VEM has a density of 1600 kg m^{-3}, Poisson's ratio $\nu = 0.49$, and is described by the following fractional derivative model:

$$E^* = \frac{\sigma(s)}{\varepsilon(s)} = \frac{E_0 + E_\infty(\tau s)^\alpha}{1 + (\tau s)^\alpha}$$

where $E_0 = 1.5MPa$, $E_\infty = 69.9495MPa$, $\tau = 0.014052s$, and $\alpha = 0.7915$.

Determine the time response of the beam/PCLD assembly when it is subjected to the transverse pulse loading, shown in Figure 4.22b, at its free end.

Validate the prediction of the fractional derivative finite element model against that of an equivalent ANSYS model that uses a GMM description of the VEM, which is given by:

$$E_2(s) = E_\infty\left[\beta_\infty + \sum_{i=1}^{3}\beta_i\frac{\rho_i s}{\rho_i s + 1}\right] \quad \text{with } E_\infty = 69.9495\,\text{MPa} \text{ and } \beta_\infty = 0.007.$$

The values of the relative moduli and relaxation time constants of the three-term GMM model are listed in Table 4.8.

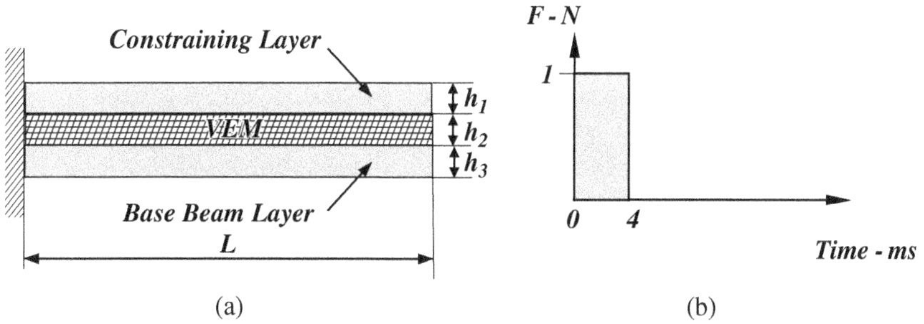

Figure 4.22 Schematic drawing of a (a) sandwich beam and (b) corresponding loading.

Table 4.8 Parameters of the three-term model GMM of the VEM.

Term	1	2	3
β_i	0.003	0.060	0.930
ρ_i (s)	2.1E-3	2.11E-4	1.4E-5

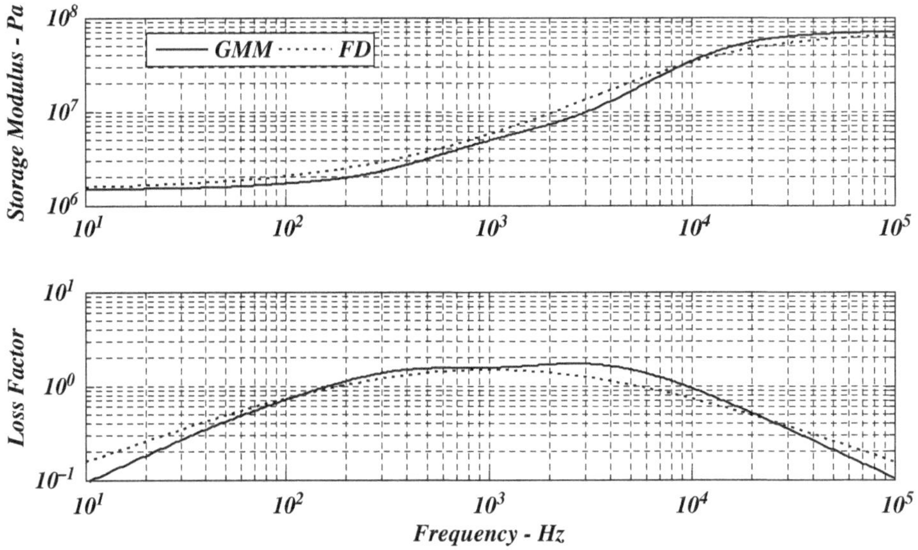

Figure 4.23 Comparisons between the storage modulus and loss factor of the fractional derivative (FD) and the GMM models.

Comparisons between the storage modulus and loss factor of the FD model and the GMM are shown in Figure 4.23. It is evident that the two models are fairly equivalent to each other.

Solution

A finite element model is developed to model the dynamics of the beam/PCLD assembly using the approach outlined in Section 4.6.4.2. In particular, the model capitalizes on the formulation described by Eq. (4.134).

Figure 4.24a displays the time response of the tip displacement of the sandwich beam. The obtained results are in excellent agreement exactly with ANSYS results. Figure 4.24b shows the frequency content of the response indicating a dominant frequency at 27.45 Hz.

4.7 Finite Element Modeling of Plates Treated with Passive Constrained Layer Damping

4.7.1 Overview

Consider the quadrilateral plate element shown in Figure 4.25. The element consists of a base plate treated with a constrained damping treatment. The sides of the element are a and b long, respectively. The thicknesses of the plate, VEM, and constraining layers are assumed to be h_1, h_2, and h_3 respectively.

Each node of the plate/VEM/constraining element has seven DOF resulting in a total of 28 DOF per element. These DOF are namely the transverse and angular deflections

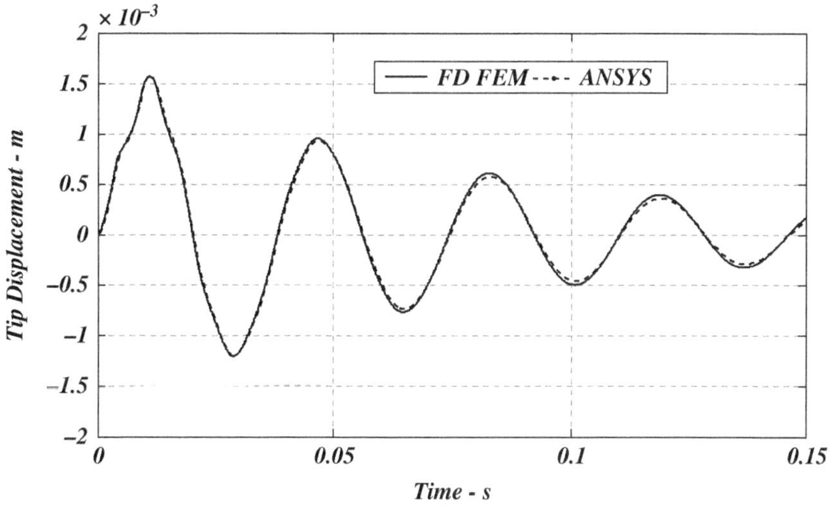

(a)

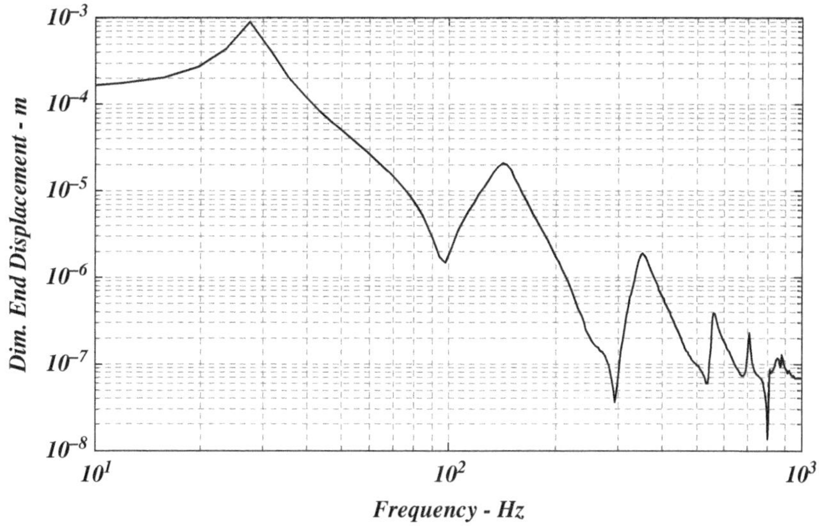

(b)

Figure 4.24 Response of the five-element beam/PCLD with fractional derivative constitutive equations in the time domains.

(w, $w_{,x}$, and $w_{,y}$) of the plate as well as the axial displacements of the plate (u_3 and v_3) and the constraining layer (u_1 and v_1) along the x and y directions.

In here, the plate, VEM, and constraining layers are assumed to have the same transverse deflection, that is, the compression of the VEM is negligible. Such an assumption is justifiable for VEM with a small thickness.

The nodal deflection vector $\{\Delta_e\}$ for the eth element is expressed accordingly as follows:

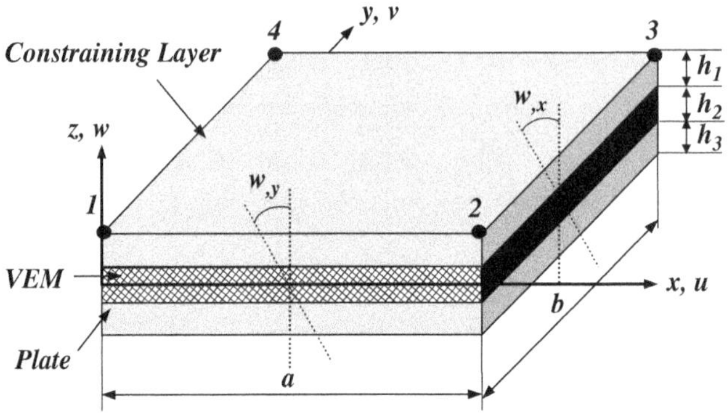

Figure 4.25 Plate treated with constrained damping treatment.

$$\{\Delta_e\} = \left\{ \begin{matrix} u_{1_1}\ u_{3_1}\ v_{1_1}\ v_{3_1}\ w_1\ w_{,x_1}\ w_{,y_1}\ u_{1_2}\ u_{3_2}\ v_{1_2}\ v_{3_2}\ w_2\ w_{,x_2}\ w_{,y_2} \\ u_{1_3}\ u_{3_3}\ v_{1_3}\ v_{3_3}\ w_3\ w_{,x_3}\ w_{,y_3}\ u_{1_4}\ u_{3_4}\ v_{1_4}\ v_{3_4}\ w_4\ w_{,x_4}\ w_{,y_4} \end{matrix} \right\}^T \tag{4.135}$$

where sub-subscripts 1 through 4 denote the numbers designated to the element nodes as indicated in Figure 4.25.

4.7.2 The Stress and Strain Characteristics

4.7.2.1 The Plate and the Constraining Layers

The strain–displacement relationships are expressed as follows:

$$\begin{Bmatrix} \varepsilon_{xi} \\ \varepsilon_{yi} \\ \gamma_{xyi} \end{Bmatrix} = \begin{bmatrix} \dfrac{\partial}{\partial x} & 0 & z_i\dfrac{\partial^2}{\partial x^2} \\ 0 & \dfrac{\partial}{\partial y} & z_i\dfrac{\partial^2}{\partial y^2} \\ \dfrac{\partial}{\partial y} & \dfrac{\partial}{\partial x} & 2z_i\dfrac{\partial^2}{\partial x \partial y} \end{bmatrix} \begin{Bmatrix} u_i \\ v_i \\ w \end{Bmatrix} i = 1,3 \tag{4.136}$$

Also, these strains are related to the stresses by the following constitutive equation:

$$\{\sigma_{xi}\ \sigma_{yi}\ \tau_{xyi}\}^T = [D_i]\{\varepsilon_{xi}\ \varepsilon_{yi}\ \gamma_{xyi}\}^T \tag{4.137}$$

where $[D_i] = \dfrac{E_i}{1-\nu_i^2}\begin{bmatrix} 1 & \nu_i & 0 \\ \nu_i & 1 & 0 \\ 0 & 0 & \frac{1}{2}(1-\nu_i) \end{bmatrix}$ with E_i and ν_i denoting Young's modulus and Poisson's ratio, respectively.

4.7.2.2 The VEM Layer

The shear strains γ_{x_2} and γ_{y_2} of the VEM layer can be written as

$$\gamma_{x2} = \frac{h}{h_2}\frac{\partial w}{\partial x} + \frac{u_1 - u_3}{h_2} \quad \text{and} \quad \gamma_{y2} = \frac{h}{h_2}\frac{\partial w}{\partial y} + \frac{v_1 - v_3}{h_2} \tag{4.138}$$

where $h = h_2 + \frac{1}{2}(h_1 + h_3)$.

These strains are related to the shear stresses by:

$$\tau_{x2} = G_2\gamma_{x2} \quad \text{and} \quad \tau_{y2} = G_2\gamma_{y2} \tag{4.139}$$

where G_2 denotes the shear modulus of the VEM that can be described by the GHM, GMM, or ATF approaches.

4.7.3 The Potential and Kinetic Energies

The PE and the KE associated with the plate, VEM, and constraining layers are given by:

$$\begin{aligned} P.E. &= \sum_{i=1,3}\frac{1}{2}\int_V [D_i]\left(\varepsilon_{xi}^2 + \varepsilon_{yi}^2\right)dV + \int_V G_2\left(\gamma_{x2}^2 + \gamma_{y2}^2\right)dV \\ &= \frac{1}{2}\{\Delta_e\}^T[K_e]\{\Delta_e\} \end{aligned} \tag{4.140}$$

and

$$\begin{aligned} K.E. &= \sum_{i=1,3}\frac{1}{2}\int_A \rho_i h_i\left(u_i^2 + v_i^2\right)dA + \sum_{i=1}^{3}\frac{1}{2}\int_A \rho_i h_i \dot{w}^2 dA \\ &= \frac{1}{2}\{\dot{\Delta}_e\}^T[M_e]\{\dot{\Delta}_e\} \end{aligned} \tag{4.141}$$

where $[K_e]$ and $[M_e]$ are the element stiffness and mass matrices, respectively.

4.7.4 The Shape Functions

The transverse and longitudinal displacements can be expressed in terms of the following shape functions:

$$\begin{aligned} w(x,y) &= c_1 + c_2x_i + c_3y_i + c_4x_i^2 + c_5x_iy_i + c_6y_i^2 \\ &\quad + c_7x_i^3 + c_8x_i^2y_i + c_9x_iy_i^2 + c_{10}y_i^3 + c_{11}x_i^3y_i + c_{12}x_iy_i^3, \\ u_1(x,y) &= c_{13} + c_{14}x + c_{15}y + c_{16}xy, \\ v_1(x,y) &= c_{17} + c_{18}x + c_{19}y + c_{20}xy, \\ u_3(x,y) &= c_{21} + c_{22}x + c_{23}y + c_{24}xy, \text{ and} \\ v_3(x,y) &= c_{25} + c_{26}x + c_{27}y + c_{28}xy. \end{aligned} \tag{4.142}$$

where the constant coefficients c_1 through c_{28} are determined in terms of the nodal deflection vector $\{\Delta_e\}$, that is, the 28 DOF of the element, resulting in the following interpolating equations:

$$w = [N_w]\{\Delta_e\}, \quad u_i = [N_{ui}]\{\Delta_e\}, \quad i = 1,3, \quad \text{and}$$
$$v_i = [N_{vi}]\{\Delta_e\}, \quad i = 1,3 \tag{4.143}$$

where $[N_w]$, $[N_{ui}]$, and $[N_{vi}]$ are the transverse and axial shape vectors. Note also that $[N_w] = [N_{w1} \;\; N_{w2} \;\; N_{w3} \;\; N_{w4}]$.

If $\bar{x} = x/a$ and $\bar{y} = y/b$, then $[N_{w1}]$, $[N_{w2}]$, $[N_{w3}]$, and $[N_{w4}]$ can be easily shown to take the following forms as reported by Yeh and Chen (2007):

$$[N_{w1}] = \left[0,0,0,0,\frac{1}{8}(1-\bar{x})(1-\bar{y})\left(2-\bar{x}-\bar{x}^2-\bar{y}-\bar{y}^2\right),\right.$$
$$\left.\frac{b}{8}(1-\bar{x})(1-\bar{y})\left(1-\bar{y}^2\right),\frac{a}{8}(1-\bar{x})(1-\bar{y})\left(1-\bar{x}^2\right)\right],$$

$$[N_{w2}] = \left[0,0,0,0,\frac{1}{8}(1+\bar{x})(1-\bar{y})\left(2+\bar{x}-\bar{x}^2-\bar{y}-\bar{y}^2\right),\right.$$
$$\left.\frac{b}{8}(1+\bar{x})(1+\bar{y})\left(1-\bar{y}^2\right),\frac{a}{8}(1+\bar{x})(1+\bar{y})\left(1-\bar{x}^2\right)\right],$$

$$[N_{w3}] = \left[0,0,0,0,\frac{1}{8}(1+\bar{x})(1+\bar{y})\left(2+\bar{x}-\bar{x}^2+\bar{y}-\bar{y}^2\right),\right.$$
$$\left.\frac{b}{8}(1+\bar{x})(1+\bar{y})\left(1-\bar{y}^2\right),\frac{a}{8}(1+\bar{x})(1+\bar{y})\left(1-\bar{x}^2\right)\right],$$

and

$$[N_{w4}] = \left[0,0,0,0,\frac{1}{8}(1-\bar{x})(1+\bar{y})\left(2-\bar{x}-\bar{x}^2+\bar{y}-\bar{y}^2\right),\right.$$
$$\left.\frac{b}{8}(1-\bar{x})(1+\bar{y})\left(1-\bar{y}^2\right),\frac{a}{8}(1-\bar{x})(1+\bar{y})\left(1-\bar{x}^2\right)\right]. \tag{4.144}$$

Also, the axial shape functions $[N_{ui}]$ and $[N_{vi}]$ are given by

$$[N_{u1}] = [N_{u11} \; N_{u12} \; N_{u13} \; N_{u14}],$$
$$[N_{u3}] = [N_{u31} \; N_{u32} \; N_{u33} \; N_{u34}] \tag{4.145}$$

$$[N_{v1}] = [N_{v11} \; N_{v12} \; N_{v13} \; N_{v14}] \quad \text{and}$$
$$[N_{v3}] = [N_{v31} \; N_{v32} \; N_{v33} \; N_{v34}]. \tag{4.146}$$

where

$$N_{u11} = \left[\tfrac{1}{4}(1-\bar{x})(1-\bar{y}),0,0,0,0,0,0,0\right],\; N_{u12} = \left[\tfrac{1}{4}(1+\bar{x})(1-\bar{y}),0,0,0,0,0,0,0\right],$$
$$N_{u13} = \left[\tfrac{1}{4}(1+\bar{x})(1+\bar{y}),0,0,0,0,0,0,0\right],\; N_{u14} = \left[\tfrac{1}{4}(1-\bar{x})(1+\bar{y}),0,0,0,0,0,0,0\right],$$
$$N_{u31} = \left[0,0,\tfrac{1}{4}(1-\bar{x})(1-\bar{y}),0,0,0,0,0\right],\; N_{u32} = \left[0,0,\tfrac{1}{4}(1+\bar{x})(1-\bar{y}),0,0,0,0,0\right],$$
$$N_{u33} = \left[0,0,\tfrac{1}{4}(1+\bar{x})(1+\bar{y}),0,0,0,0,0\right],\; N_{u34} = \left[0,0,\tfrac{1}{4}(1-\bar{x})(1+\bar{y}),0,0,0,0,0\right],$$
$$N_{v11} = \left[0,\tfrac{1}{4}(1-\bar{x})(1-\bar{y}),0,0,0,0,0,0\right],\; N_{v12} = \left[0,\tfrac{1}{4}(1+\bar{x})(1-\bar{y}),0,0,0,0,0,0\right],$$
$$N_{v13} = \left[0,\tfrac{1}{4}(1+\bar{x})(1+\bar{y}),0,0,0,0,0,0\right],\; N_{v14} = \left[0,\tfrac{1}{4}(1-\bar{x})(1+\bar{y}),0,0,0,0,0,0\right],$$
$$N_{v31} = \left[0,0,0,\tfrac{1}{4}(1-\bar{x})(1-\bar{y}),0,0,0,0\right],\; N_{v32} = \left[0,0,0,\tfrac{1}{4}(1+\bar{x})(1-\bar{y}),0,0,0,0\right],$$
$$N_{v33} = \left[0,0,0,\tfrac{1}{4}(1+\bar{x})(1+\bar{y}),0,0,0,0\right],\; N_{v34} = \left[0,0,0,\tfrac{1}{4}(1-\bar{x})(1+\bar{y}),0,0,0,0\right].$$

4.7.5 The Stiffness Matrices

Substituting Eqs. (4.142) through (4.146) into Eq. (4.140) yields the following expressions for the stiffness matrices of the assembly:

i) Axial Stiffness Matrices

$$[K_a] = \sum_{i=1,3} h_i \int_A [N_i]^T [D_{ui}][N_i] dA \tag{4.147}$$

where $[N_i] = \begin{bmatrix} N_{ui,x} \\ N_{vl,x} \\ N_{ui,y} + N_{vi,x} \end{bmatrix}$ and $[D_{ui}] = \dfrac{E_i}{1-\nu_i^2}\begin{bmatrix} 1 & \nu_i & 0 \\ \nu_i & 1 & 0 \\ 0 & 0 & \frac{1}{2}(1-\nu_i) \end{bmatrix}$ for $i = 1,3$

ii) Bending Stiffness Matrices

$$[K_b] = \sum_{i=1,3} \int_A [N_{wb}]^T [D_{wi}][N_{wb}] dA \tag{4.148}$$

where $[N_{wb}] = \begin{bmatrix} N_{w,xx} \\ N_{w,yy} \\ 2N_{w,xy} \end{bmatrix}$ and $[D_{wi}] = \dfrac{E_i I_i}{1-\nu_i^2}\begin{bmatrix} 1 & \nu_i & 0 \\ \nu_i & 1 & 0 \\ 0 & 0 & \frac{1}{2}(1-\nu_i) \end{bmatrix}$ for $i = 1,3$

with I_i denoting the area moment of inertia of the ith layer.

iii) Shear Stiffness Matrix

$$[K_v] = G_2 h_2 \int_A [N_v]^T [N_v] dA \tag{4.149}$$

where $[N_v] = \dfrac{1}{h_2}\left[(N_{u1} - N_{u3}) + hN_{w,x} \quad (N_{v1} - N_{v3}) + hN_{w,y}\right]^T$.

Hence, the element total stiffness matrix $[K_e]$ is:

$$[K_e] = [K_a] + [K_b] + [K_v] \tag{4.150}$$

4.7.6 The Mass Matrices

Substituting Eqs. (4.142) through (4.146) into Eq. (4.141) yields the following expressions:

i) Axial Mass Matrices

$$[M_a] = \sum_{i=1,3} \int_A \rho_i h_i \left([N_{ui}]^T [N_{ui}] + [N_{vi}]^T [N_{vi}] \right) dA \tag{4.151}$$

ii) Bending Mass Matrices

$$[M_b] = \sum_{i=1}^{3} \int_A \rho_i h_i [N_w]^T [N_w] dA \tag{4.152}$$

Then, the element total mass matrix $[M_e]$ is:

$$[M_e] = [M_a] + [M_b] \tag{4.153}$$

4.7.7 The Element and Overall Equations of Motion

Using the Lagrangian dynamics approach, the element equations of motion can be written as:

$$[M_e]\{\ddot{\Delta}_e\} + [K_e]\{\Delta_e\} = \{F_e\} \tag{4.154}$$

The equations of motion of the entire plate/VEM/constraining layer assembly can be derived by assembling the element matrices. The boundary conditions are then imposed and the resulting system can be augmented with the dynamics of the VEM using the GHM, GMM, ATF, or the fractional derivative approach.

Example 4.10 Consider the cantilevered plate/VEM/constraining layer assembly shown in Figure 4.26. The main physical and geometrical parameters of the different layers are listed in Table 4.9. The VEM is described by GHM and ATF models such that:

$$ATF \text{ model: } G_2(s) = G_0\left[1 + \sum_{i=1}^{3} \Delta_i \frac{\rho_i s}{\rho_i s + 1}\right] \text{ with } G_0 = 0.5MPa$$

$$GHM \text{ model: } G_2(s) = G_0\left[1 + \sum_{i=1}^{3} \alpha_i \frac{s^2 + 2\zeta_i \omega_i s}{s^2 + 2\zeta_i \omega_i s + \omega_i^2}\right] \text{ with } G_0 = 0.5MPa$$

The parameters Δ_i, ρ_i, α_i, ζ_i, and ω_i are listed in Table 4.7.

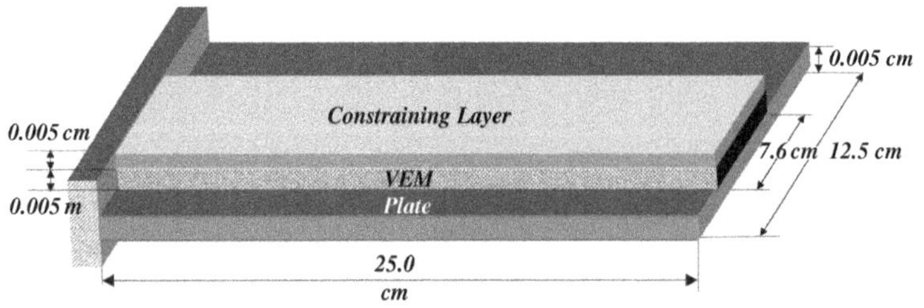

Figure 4.26 Geometry of a cantilevered plate/VEM/constraining layer assembly.

Table 4.9 Parameters of the plate/VEM/constraining layer assembly.

Parameter	Length (cm)	Thickness (m)	Width (cm)	Density (kg km^{-3})	Young's modulus (GPa)
Plate	25	0.005	12.0	2700	70.00
VEM	25	0.005	6.0	1104	[a]
Constraining layer	25	0.005	6.0	2700	70.00

[a] From the ATF and GHM models with $E* = 3G_2$.

Determine the frequency response of the system when excited at the free end by a unit force. Determine also the time response of system when a unit pulse force of duration 0.10 ms is applied at the free end. Compare the response when the VEM is described by the ATF and GHM models.

Determine the time and frequency response when the VEM and constraining layer cover the entire plate surface.

Solution

The approach outlined in Section 4.7 is used to extract the time and frequency response of the plate/VEM/constraining layer assembly. Figure 4.27a,b display the time and

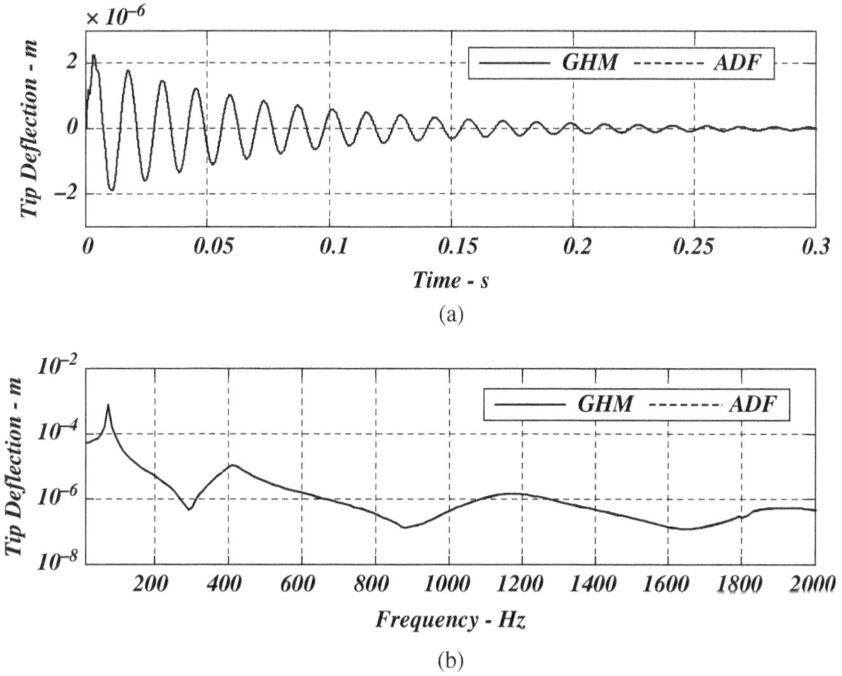

Figure 4.27 Response of the partially covered plate/PCLD with ATF and GHM constitutive equations.

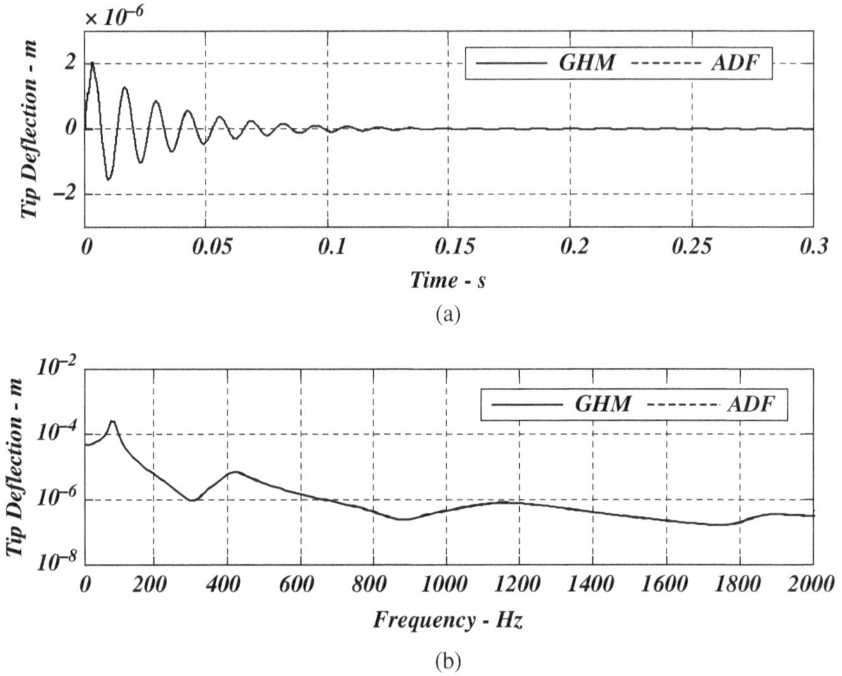

Figure 4.28 Response of the fully covered plate/PCLD with ATF and GHM constitutive equations.

frequency response of the partially covered plate. The obtained results emphasize the excellent agreement between the predictions using the ATF and GHM models.

Figure 4.28 shows the corresponding results when the plate is fully covered by the PCLD treatment. It is evident that better damping characteristics can be achieved.

Example 4.11 Consider the cantilevered plate/VEM/constraining layer assembly shown in Figure 4.26 and described in Example 4.9. Validate the prediction of the finite element model against that of an equivalent ANSYS model that uses a GMM description of the VEM that has the relative moduli and relaxation time constants listed in Table 4.8.

Solution

Figure 4.29 displays the *ANSYS* finite element model of the cantilevered plate/VEM/constraining layer assembly.

Figure 4.30 shows the mode shapes of the assembly at the first four natural frequencies as predicted by ANSYS.

Figure 4.31 shows a comparison between the frequency responses as obtained by the finite element approach outlined in Section 4.7 and the ANSYS software package. Adequate agreement is achieved particularly for the first and second modes of vibration.

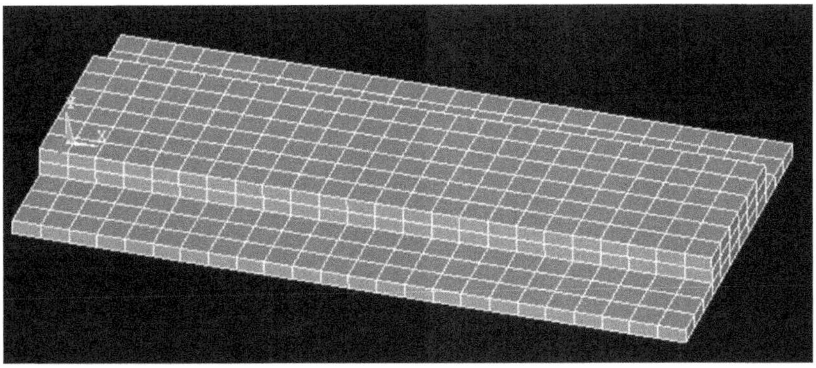

Figure 4.29 The ANSYS finite element model of the cantilevered plate/VEM/constraining layer assembly.

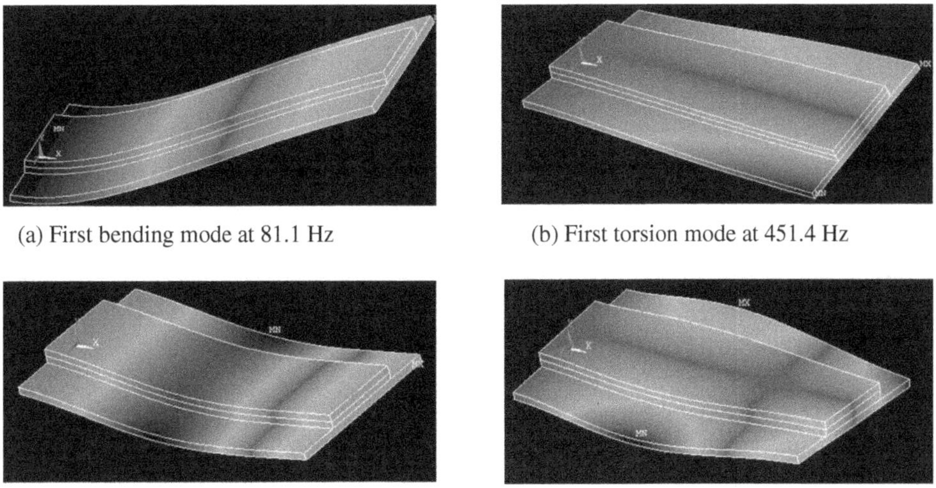

(a) First bending mode at 81.1 Hz

(b) First torsion mode at 451.4 Hz

(c) Second bending mode at 1,272. Hz

(d) Second torsion mode at 1,931 Hz

Figure 4.30 The mode shapes of the assembly at the first four natural frequencies. (a) First bending mode at 81.1 Hz, (b) first torsion mode at 451.4 Hz, (c) second bending mode at 1272 Hz, and (d) second torsion mode at 1931 Hz.

4.8 Finite Element Modeling of Shells Treated with Passive Constrained Layer Damping

4.8.1 Overview

Extensive efforts have been exerted to analyze of the dynamics and damping characteristics of shells treated with PCLD treatments. Such interest is driven by the importance of this class of damped structures in various critical applications such as aircraft cabins, automobiles, and rockets. Distinct among these efforts is the pioneering work of Markuš

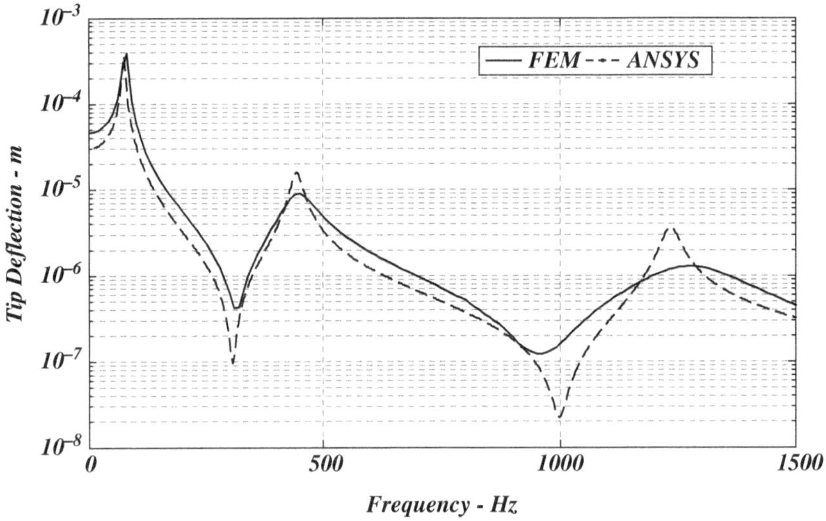

Figure 4.31 Frequency response of the plate/PCLD assembly.

(1976) whereby the damping properties of layered cylindrical shells are developed for axially-symmetric modes of vibration. In Ramesh and Ganesan (1993, 1994), determined the natural frequencies and loss factors for isotropic cylindrical shell treated with PCLD using the finite element approach coupled with the first order shear deformation theory (FSDT). In 1999, Chen and Huang (1999) employed the assumed mode method to study the response of cylindrical shells which are partially treated with PCLD using the Donnell–Mushtari–Vlasov thin shell theory. Sainsbury and Masti (2007) employed the finite element model by Sivadas and Ganesan (1993) along with the strain energy method to optimize the placement of partial PCLD treatments over the surface of cylindrical shells. In 2012, Mohammadi and Sedaghati presented (2012) a general approach to determining the damping characteristics of three-layered sandwich cylindrical shells with thin or thick viscoelastic core using semi-analytical finite element method.

4.8.2 Stress–Strain Relationships

Figure 4.32 shows a schematic drawing of a quadrilateral element of a cylindrical shell treated with a PCLD treatment. The shell has a radius R and thickness h_1. Also, the thicknesses of the VEM and constraining layers are assumed to be h_2, and h_3, respectively. Associated with the shell/PCLD assembly is the coordinate system x, y, z. The stress–strain relationships for the different layers of the assembly are as follows:

4.8.2.1 Shell and Constraining Layer

The constitutive relationships for the shell and constraining layer are given by:

$$\left\{\sigma_{xi}\ \sigma_{yi}\ \tau_{xyi}\right\}^T = [D_i]\left\{\varepsilon_{xi}\ \varepsilon_{yi}\ \gamma_{xyi}\right\}^T \quad i = 1,3 \tag{4.155}$$

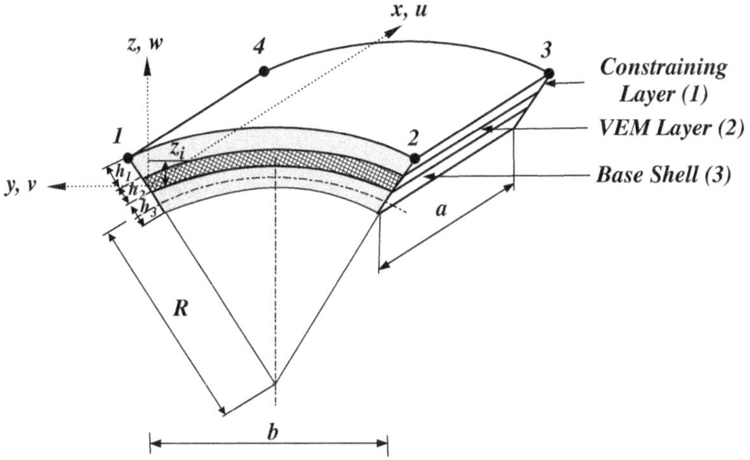

Figure 4.32 Cylindrical shell element treated with passive constrained layer damping (PCLD) treatment.

where $[D_i] = \frac{E_i}{1-\nu_i^2}\begin{bmatrix} 1 & \nu_i & 0 \\ \nu_i & 10 & \\ 0 & 0 & \frac{1}{2}(1-\nu_i) \end{bmatrix}$ with E_i and ν_i denoting Young's modulus and Poisson's ratio, respectively.

According to the Donnell–Mushtari–Vlasov thin shell theory (Soedel 2004), the strain–displacement relationships are expressed as follows:

$$\begin{Bmatrix} \varepsilon_{xi} \\ \varepsilon_{yi} \\ \varepsilon_{xyi} \\ \varepsilon_{xzi} \\ \varepsilon_{yzi} \end{Bmatrix} = \begin{bmatrix} \frac{\partial}{\partial x} & 0 & z_i \frac{\partial^2}{\partial x^2} \\ 0 & \frac{\partial}{\partial y} & \left(\frac{1}{R} + z_i \frac{\partial^2}{\partial y^2}\right) \\ \frac{\partial}{\partial y} & \frac{\partial}{\partial x} & 2 z_i \frac{\partial^2}{\partial x \partial y} \\ 0 & 0 & \frac{\partial}{\partial x} \\ 0 & -\frac{1}{R} & \frac{\partial}{\partial y} \end{bmatrix} \begin{Bmatrix} u_i \\ v_i \\ w \end{Bmatrix} \quad i = 1,3 \tag{4.156}$$

where z_i denotes the distance from the neutral surface along the transverse direction.

4.8.2.2 Viscoelastic Layer

The constitutive relationships for the VEM layer are given by:

$$\sigma_{xz} = G_2 \gamma_{xz} \text{ and } \sigma_{yz} = G_2 \gamma_{yz} \tag{4.157}$$

where G_2 denotes the shear modulus of the VEM. Also, γ_{xz} and γ_{yz} denote the shear strain in the VEM, in the x–z and y–z planes, respectively, which are shown in Figure 4.33.

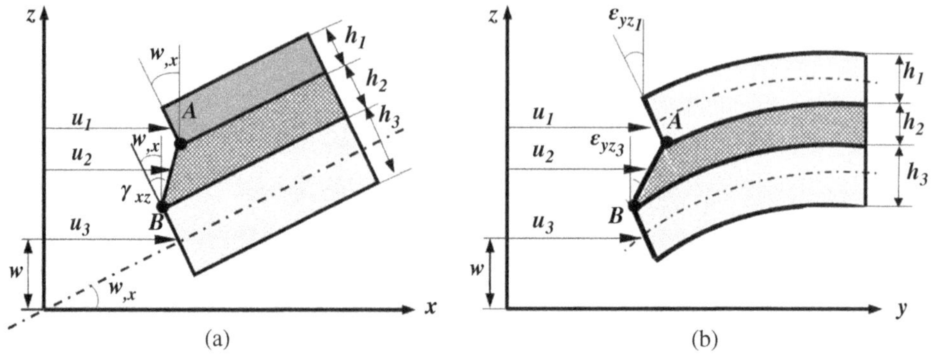

Figure 4.33 Shear strains in the VEM treatment of a shell/PCLD assembly.

The shear strain γ_{xz}, in the x–z plane, is as given in the case of flat plates by Eq. (4.58) as follows:

$$\gamma_{xz} = \frac{h}{h_2} w_{,x} + \frac{1}{h_2}(u_1 - u_3) \tag{4.158}$$

where $h = h_2 + \frac{1}{2}(h_1 + h_3)$.

However, the shear strain γ_{yz}, in the y–z plane, can be determined as follows:

$$\gamma_{yz} = \left(\frac{\partial w}{\partial y} - \frac{v_B}{R}\right) + \frac{1}{h_2}(v_A - v_B) \tag{4.159}$$

Equation (4.158) is derived using Eq. (4.156) and the geometrical configuration of the shell/PCLD assembly shown in Figure 4.33b. It is also assumed the shell radius R is much larger than the thicknesses of the shell, VEM, and the constraining layer. Hence, the radii of the different layers of the assembly are nearly equal to R.

Furthermore, the in-plane deflections v_A and v_B are expressed in terms of the mid-plane deflections v_1 and v_3 as follows:

$$v_A = v_1 + \frac{h_1}{2}\left(\frac{\partial w}{\partial y} - \frac{v_1}{R}\right) \tag{4.160}$$

and

$$v_B = v_3 - \frac{h_3}{2}\left(\frac{\partial w}{\partial y} - \frac{v_3}{R}\right) \tag{4.161}$$

Substituting Eqs. (4.160) and (4.161) into Eq. (4.158) yields:

$$\begin{aligned} \gamma_{yz} = {} & \left(\frac{1}{h_2} - \frac{h_1}{2h_2R}\right)v_1 - \left[\left(\frac{1}{R} + \frac{1}{h_2}\right)\left(1 + \frac{h_3}{2R}\right)\right]v_3 \\ & + \left[1 + \left(\frac{1}{R} + \frac{1}{h_2}\right)\frac{h_3}{2} + \frac{h_1}{2h_2}\right]\frac{\partial w}{\partial y} \end{aligned} \tag{4.162}$$

Note that when R tends to ∞, Eq. (4.162) reduces to Eq. (4.138) for a flat plate.

4.8.3 Kinetic and Potential Energies

The PE and the KE associated with the shell/PCLD assembly are given by:

$$P.E. = \sum_{i=1,3}\frac{1}{2}\int_V [D_i]\left(\varepsilon_{xi}^2+\varepsilon_{yi}^2\right)dV + \int_V G_2\left(\gamma_{xz}^2+\gamma_{yz}^2\right)dV$$
$$= \frac{1}{2}\{\Delta_e\}^T[K_e]\{\Delta_e\} \tag{4.163}$$

and

$$K.E. = \sum_{i=1,3}\frac{1}{2}\int_A \rho_i h_i\left(u_i^2+v_i^2\right)dA + \sum_{l=1}^{3}\frac{1}{2}\int_A \rho_i h_i \dot{w}^2 dA$$
$$= \frac{1}{2}\{\dot{\Delta}_e\}^T[M_e]\{\dot{\Delta}_e\} \tag{4.164}$$

where $[K_e]$ and $[M_e]$ are the element stiffness and mass matrices, respectively. Also, $\{\Delta_e\}$ denotes the nodal deflection vector for the eth element that is given by:

$$\{\Delta_e\} = \begin{Bmatrix} u_{1_1}\ u_{3_1}\ v_{1_1}\ v_{3_1}\ w_1\ w_{,x_1}\ w_{,y_1}\ u_{1_2}\ u_{3_2}\ v_{1_2}\ v_{3_2}\ w_2\ w_{,x_2}\ w_{,y_2} \\ u_{1_3}\ u_{3_3}\ v_{1_3}\ v_{3_3}\ w_3\ w_{,x_3}\ w_{,y_3}\ u_{1_4}\ u_{3_4}\ v_{1_4}\ v_{3_4}\ w_4\ w_{,x_4}\ w_{,y_4} \end{Bmatrix}^T \tag{4.165}$$

where sub-subscripts 1 through 4 denote the numbers designated to the element nodes as indicated in Figure 4.32.

4.8.4 The Shape Functions

The transverse and longitudinal displacements can be expressed in terms of the nodal deflection vector and appropriate shape function vectors as follows

$$w = [N_w]\{\Delta_e\},\quad u_i = [N_{ui}]\{\Delta_e\},\quad i = 1,3,\ \text{and}$$
$$v_i = [N_{vi}]\{\Delta_e\},\quad i = 1,3 \tag{4.166}$$

Note that $[N_w]$ denotes the bending shape function] whereas $[N_{ui}]$ and $[N_{vi}]$ define the axial shape functions, respectively. These functions can be extracted as outlined in Section 4.7.4.

4.8.5 The Stiffness Matrices

Substituting Eq. (4.166) into Eq. (4.163) yields the following expressions for the stiffness matrices of the assembly:

i) Axial Stiffness Matrices

$$[K_a] = \sum_{i=1,3} h_i \int_A [N_i]^T[D_{ui}][N_i]dA \tag{4.167}$$

where $[N_i] = \begin{bmatrix} [N_{ui,x}] \\ [N_{vi,x}] \\ [N_{ui,y} + N_{vi,x}] \end{bmatrix}$ and $[D_{ui}] = \frac{E_i}{1-\nu_i^2}\begin{bmatrix} 1 & \nu_i & 0 \\ \nu_i & 1 & 0 \\ 0 & 0 & \frac{1}{2}(1-\nu_i) \end{bmatrix}$ For $i = 1,3$

ii) Bending Stiffness Matrices

$$[K_b] = \sum_{i=1,3} \int_A [N_{wb}]^T [D_{wi}][N_{wb}] dA \tag{4.168}$$

where $[N_{wb}] = \begin{bmatrix} [N_{w,xx}] \\ [N_{w,yy}] \\ [2N_{w,xy}] \end{bmatrix}$ and $[D_{wi}] = \frac{E_i I_i}{1-\nu_i^2}\begin{bmatrix} 1 & \nu_i & 0 \\ \nu_i & 1 & 0 \\ 0 & 0 & \frac{1}{2}(1-\nu_i) \end{bmatrix}$ For $i = 1,3$

with I_i denoting the area moment of inertia of the ith layer.

iii) Shear Stiffness Matrix

$$[K_v] = G_2 h_2 \int_A \left([N_v]^T [N_v]\right) dA \tag{4.169}$$

where $[N_v] = \frac{1}{h_2}\left[([N_{u1}] - [N_{u3}]) + h\,[N_{w,x}] \quad (n_1[N_{v1}] - n_2[N_{v3}]) + n_3\,[N_{w,y}]\right]^T$

with $n_1 = \left(1 - \frac{h_1}{2R}\right)$, $n_2 = \left[\left(\frac{h_2}{R} + 1\right)\left(1 + \frac{h_3}{2R}\right)\right]$, and $n_3 = \frac{1}{2}\left(h + \frac{h_2 h_3}{R}\right)$.

Hence, the element total stiffness matrix $[K_e]$ is:

$$[K_e] = [K_a] + [K_b] + [K_v] \tag{4.170}$$

4.8.6 The Mass Matrices

Substituting Eq. (4.166) into Eq. (4.164) yields the following expressions:

i) Axial Mass Matrices

$$[M_a] = \sum_{i=1,3} \int_A \rho_i h_i \left([N_{ui}]^T [N_{ui}] + [N_{vi}]^T [N_{vi}]\right) dA \tag{4.171}$$

ii) Bending Mass Matrices

$$[M_b] = \sum_{i=1}^{3} \int_A \rho_i h_i [N_w]^T [N_w] dA \tag{4.172}$$

Then, the element total mass matrix $[M_e]$ is:

$$[M_e] = [M_a] + [M_b] \tag{4.173}$$

4.8.7 The Element and Overall Equations of Motion

Using the Lagrangian dynamics approach, the element equations of motion can be written as:

$$[M_e]\{\ddot{\Delta}_e\} + [K_e]\{\Delta_e\} = \{F_e\} \tag{4.174}$$

The equations of motion of the entire shell/PCLD assembly can be derived by assembling the element matrices. The boundary conditions are then imposed and the resulting system can be augmented with the dynamics of the VEM using the GHM, GMM, ATF, or the fractional derivative approach.

Example 4.12 Consider the clamped-free shell/PCLD assembly shown in Figure 4.34. The main physical and geometrical parameters of the shell and PCLD are listed in Table 4.10. The shell has an internal radius R that is 0.1016 m. The PCLD treatment consists of two patches as displayed in Figure 4.34. The patches are bonded 180° apart on the outer surface of the cylinder with each of which subtending an angle of 90° angle at the center of the shell as shown in Figure 4.34.

Determine the frequency response of the system when excited at the free end by a unit force. Determine also the time response of system when a unit pulse force of duration 0.10 ms is applied at the free end. Compare the response when the response is predicted using ANSYS and the finite element approach outlined in Section 4.7.

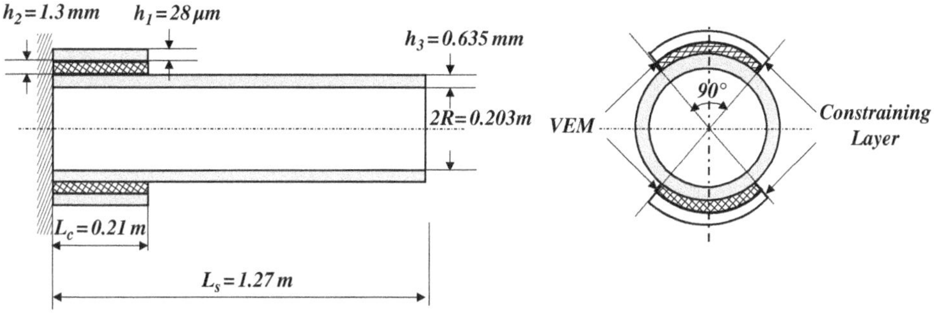

Figure 4.34 Configuration of the shell/PCLD assembly.

Table 4.10 Parameters of the shell/PCLD assembly.

Parameter	Length (m)	Thickness (mm)	Density (kg m^{-3})	Young's modulus (GPa)
Shell	1.270	0.635	7800	210
VEM	0.212	1.300	1140	[a]
Constraining layer	0.212	0.028	7800	210

[a] $G_\infty = 292.01$ MPa and $\beta_\infty = 0.007$ for a five-term GMM of the VEM as described in Example 4.6 and Table 4.6.

Solution

a) The characteristics of the plain shell

The natural frequencies and the corresponding mode shapes of the plain shell, shown in Figure 4.35, as predicted by using ANSYS and the finite element approach outlined in Section 4.7 are listed in Table 4.11.

b) The characteristics of the shell/PCLD

The natural frequencies and the corresponding mode shapes of the plain shell, shown in Figure 4.36, as predicted by using ANSYS and the finite element approach outlined in Section 4.7 are listed in Table 4.12.

Figure 4.37 shows a comparison between the frequency responses of the plain shell as predicted by the finite element model and ANSYS. Adequate agreement is indicated in the figure particularly for the first three modes of vibrations.

Figure 4.38 shows a comparison between the frequency responses of the shell/PCLD as predicted by the finite element model and ANSYS. Adequate agreement is indicated in the figure, particularly for the first two modes of vibrations.

4.9 Summary

This chapter has presented the Golla–Hughes–McTavish model as an effective means for modeling and integrating the dynamics of VEM into the finite element models of structures to be damped by the VEM. The physical meaning of the model is presented indicating that it is a combination of the Maxwell and Kelvin–Voigt models with an added mass inserted between the components of the Maxwell model. The use of the

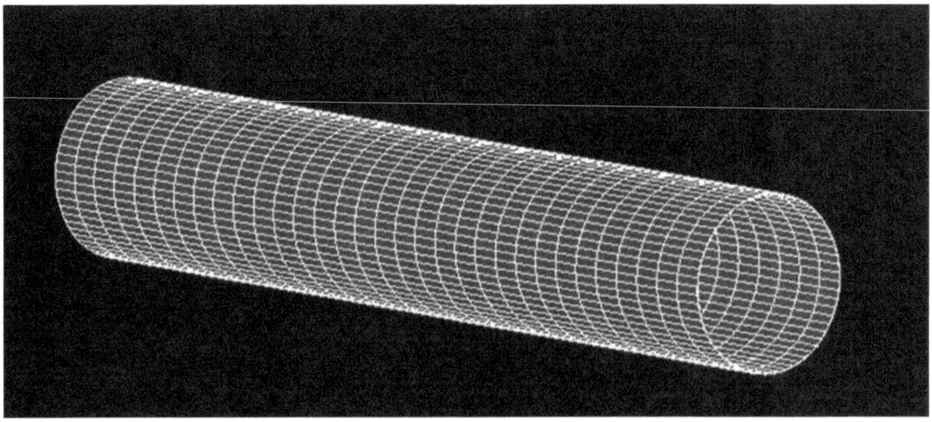

Figure 4.35 Finite element model of the plain shell.

Table 4.11 The natural frequencies and mode shapes of the plain shell.

Finite Element	*ANSYS*
58 Hz 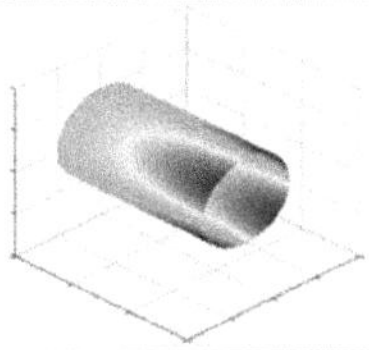	**58 Hz** 
59 Hz 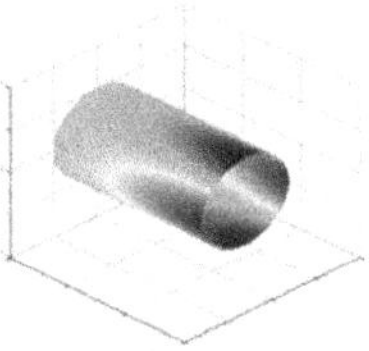	**58 Hz** 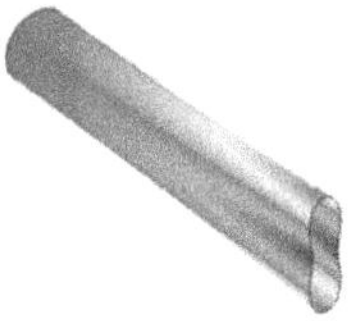
119 Hz 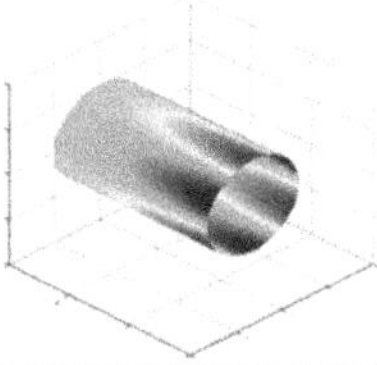	**118 Hz** 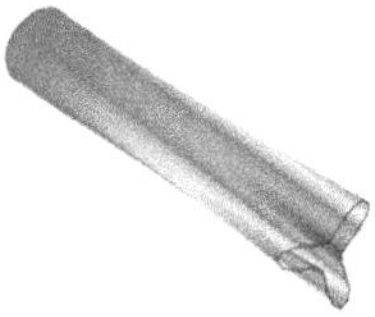
119 Hz 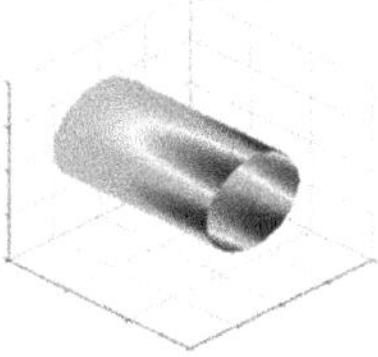	**118 Hz** 
124 Hz 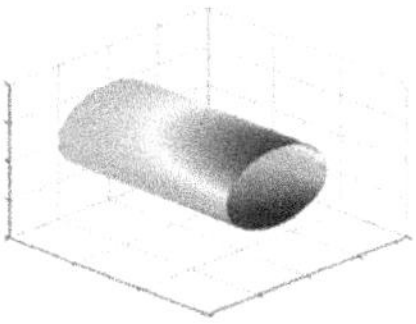	**124 Hz** 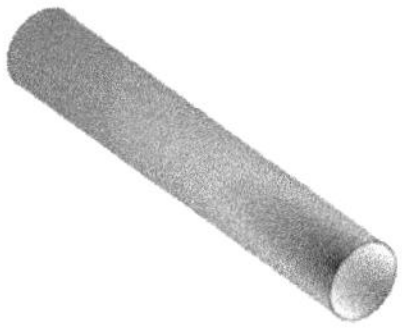
124 Hz 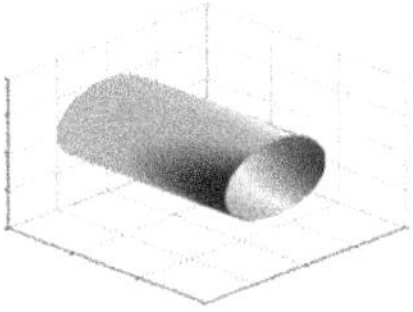	**124 Hz** 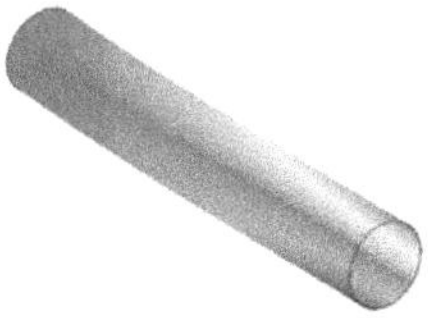

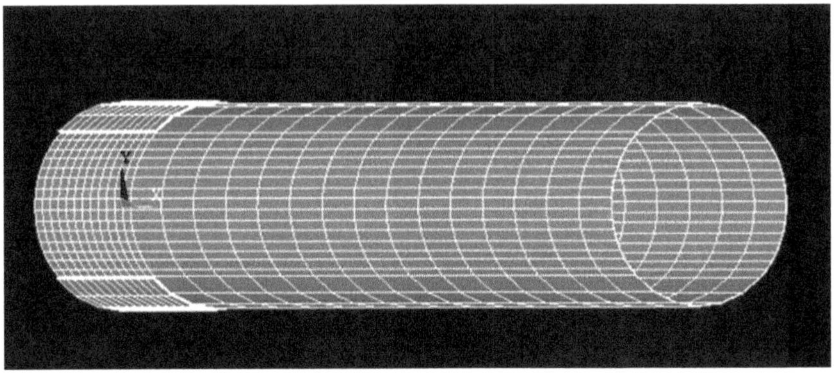

Figure 4.36 Finite element model of the shell/PCLD system.

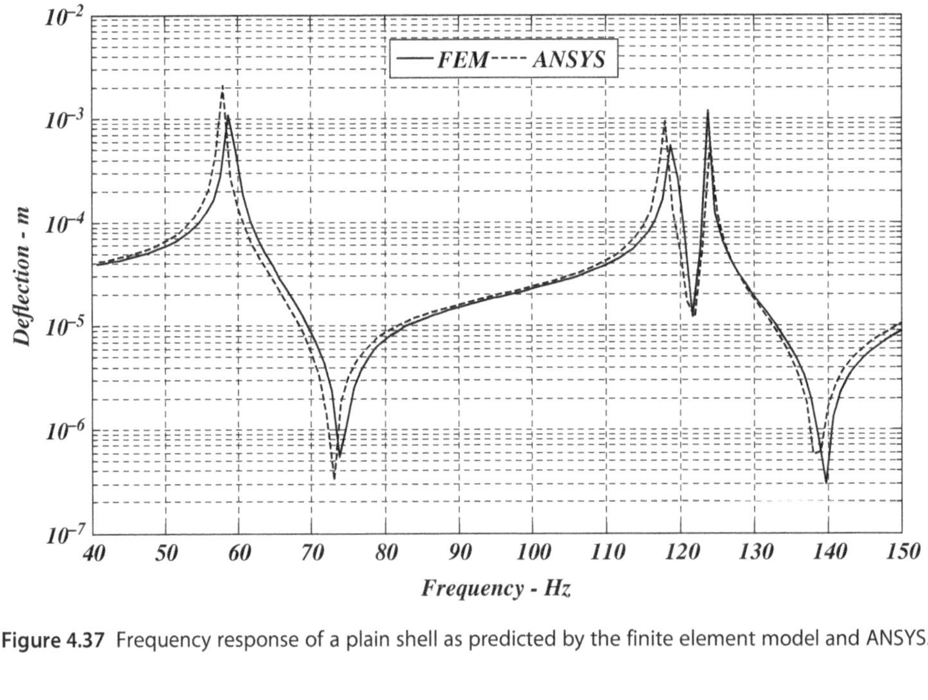

Figure 4.37 Frequency response of a plain shell as predicted by the finite element model and ANSYS.

GHM model to describe the dynamics of rods treated with VEM in constrained and unconstrained configurations is discussed in details. It is concluded that using the VEM in its constrained configuration enhances the damping characteristics considerably as compared to the unconstrained configuration.

Table 4.12 The natural frequencies and mode shapes of the shell/PCLD.

Finite Element	*ANSYS*
60 Hz 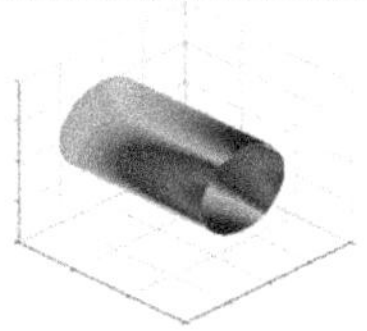	60 Hz 
62 Hz 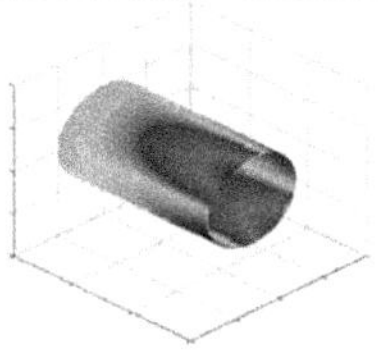	60 Hz 
119 Hz 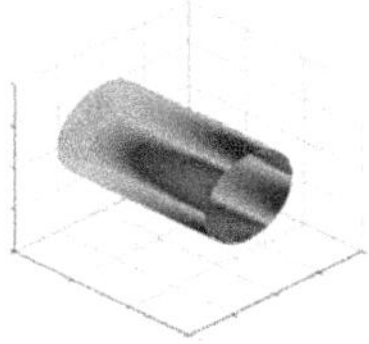	118 Hz 
120 Hz 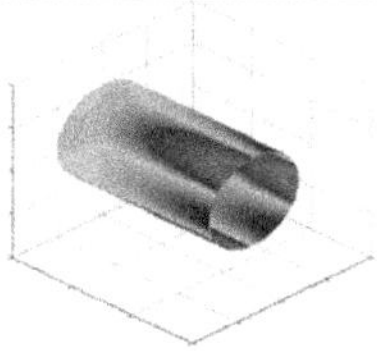	118 Hz 
125 Hz 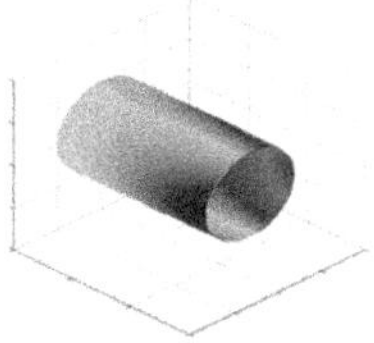	128 Hz 
129 Hz 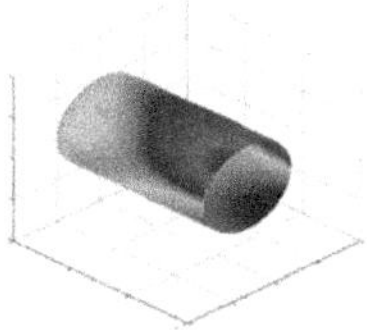	138 Hz 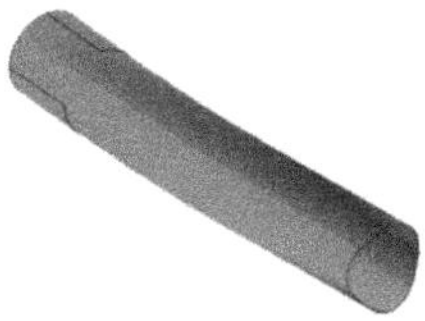

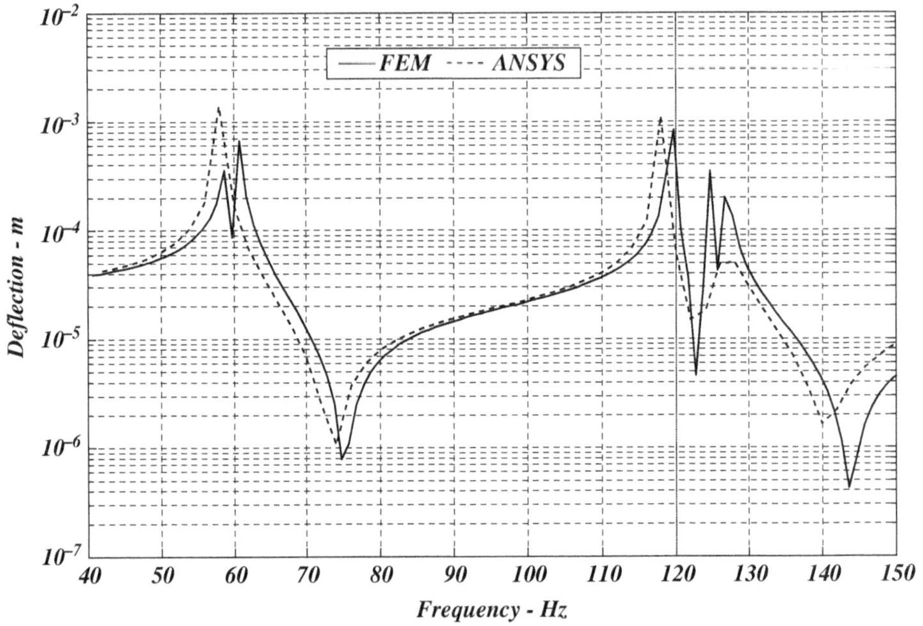

Figure 4.38 Frequency response of a shell/PCLD as predicted by the finite element model and ANSYS.

References

Bagley, R.L. and Torvik, P.J. (1983). Fractional calculus – a different approach to the analysis of viscoelastically damped structures. *AIAA Journal* 21: 741–749.

Bahraini, S.M.S., Eghtesad, M., Farid, M., and Ghavanloo, E. (2013). Large deflection of viscoelastic beams using fractional derivative model. *Journal of Mechanical Science and Technology* 27 (4): 1063–1070.

Bekuit, J.-J.R.B., Oguamanam, D.C.D., and Damisa, O. (2009). Quasi-2D finite element formulation of active-constrained layer beams. *Journal of Smart Materials and Structures* 18: 095003.

Catania, G., Sorrentino, S., and Fasana, A. (2008). A condensation technique for finite element dynamic analysis using fractional derivative viscoelastic models. *Journal of Vibration and Control* 14 (9–10): 1573–1586.

Chen, L.H. and Huang, S.C. (1999). Vibrations of a cylindrical shell with partially constrained layer damping (CLD) treatment. *International Journal of Mechanical Sciences* 41: 1485–1498.

Cortes, F. and Elejabarrieta, M.J. (2007a). Finite element formulations for transient dynamic analysis in structural systems with viscoelastic treatments containing fractional derivative models. *International Journal of Numerical Methods in Engineering* 69: 2173–2195.

Cortes, F. and Elejabarrieta, M.J. (2007b). Homogenized finite element formulations for transient dynamic analysis of unconstrained layer damping beams involving fractional derivative models. *Computational Mechanics* 40: 313–324.

Enelund, M. and Josefson, B.L. (1997). Time-domain finite element analysis of viscoelastic structures with fractional derivatives constitutive relations. *AIAA Journal* 35 (10): 1630–1637.

Escobedo-Torres, J. and Ricles, J.M. (1998). The fractional order elastic-viscoelastic equations of motion: formulation and solution methods. *Journal of Intelligent Material Systems and Structures* 9: 489–502.

Friswell, M.I., Inman, D.J., and Lam, M.J. (1997). On the realization of GHM models in viscoelasticity. *Journal of Intelligent Material Systems and Structures* 8 (11): 986–993.

Galucio, A.C., Deü, J.F., and Ohayon, R. (2004). Finite element formulation of viscoelastic sandwich beams using fractional derivative operators. *Computational Mechanics* 33: 282–291.

Golla, D.F. and Hughes, P.C. (1985). Dynamics of viscolelastic structures – a time domain finite element formulation. *ASME Journal of Applied Mechanics* 52: 897–600.

Lesieutre G. A., "Finite element modeling of frequency-dependent material damping using augmenting thermodynamic fields", Ph.D. Dissertation, University of California Los Angeles (UCLA), CA, 1989.

Lesieutre, G.A. and Mingori, D.L. (1990). Finite element modeling of frequency-dependent material damping using augmenting thermodynamic fields. *Journal of Guidance, Control and Dynamics* 13 (6): 1040–1050.

Markuš, Š. (1976). Damping properties of layered cylindrical shells vibrating in axially symmetric mode. *Journal of Sound and Vibration* 48 (4): 511–524.

Martin L. A., "A novel material modulus function for modeling viscoelastic materials", Ph.D. Dissertation, Virginia Polytechnic Institute and State University, 2011.

Mohammadi, F. and Sedaghati, R. (2012). Linear and nonlinear vibration analysis of sandwich cylindrical shell with constrained viscoelastic core layer. *International Journal of Mechanical Sciences* 54 (1): 156–171.

Padovan, J. (1987). Computational algorithms for FE formulations involving fractional operators. *Computational Mechanics* 2: 271–287.

Piovan, M.T., Sampaiob, R., and Deu, J.-F. (2009). Dynamics of sandwich curved beams with viscoelastic core described by fractional derivative operators. *Mecánica Computacional* XXVIII: 691–710, Edited by: C. G. Bauza, P. Lotito, L. Parente, and M. Vénere, Tandil, Argentina, .

Ramesh, T.C. and Ganesan, N. (1993). Vibration and damping analysis of cylindrical shells with a constrained damping layer. *Computers and Structures.* 46 (4): 751–758.

Ramesh, T.C. and Ganesan, N. (1994). Finite element analysis of cylindrical shells with a constrained viscoelastic layer. *Journal of Sound and Vibration* 172 (3): 359–370.

Rusovici R., "Modeling of shock wave propagation and attenuation in viscoelastic structures", Ph.D. Dissertation, Virginia Polytechnic Institute and State University, 1999.

Sainsbury, M.G. and Masti, R.S. (2007). Vibration damping of cylindrical shells using strain-energy-based distribution of an add-on viscoelastic treatment. *Finite Element in Analysis and Design* 43: 175–192.

Schmidt, A. and Gaul, L. (2002). Finite element formulation of viscoelastic constitutive equations using fractional time derivatives. *Nonlinear Dynamics* 29: 37–55.

Singh, S.J. and Chatterjee, A. (2006). Galerkin projections and finite elements for fractional order derivatives. *Nonlinear Dynamics* 45 (1–2): 183–206.

Sivadas, K.R. and Ganesan, N. (1993). Axisymmetric vibration analysis of thick cylindrical shell with variable thickness. *Journal of Sound and Vibration* 160: 387–400.

Soedel, W. (2004). *Vibrations of Shells and Plates*, 3rde. CRC Press.
Sorrentino, S. and Fasana, A. (2007). Finite element analysis of vibrating linear systems with fractional derivative viscoelastic models. *Journal of Sound and Vibration* 299: 839–853.
Trindade, M.A., Benjeddou, A., and Ohayon, R. (2000). Modeling of frequency-dependent viscoelastic materials for active-passive vibration damping. *Journal of Vibration and Acoustics* 122 (2): 169–174.
Yeh, J.Y. and Chen, L.W. (2007). Finite element dynamic analysis of orthotropic sandwich plates with an electrorheological fluid core layer. *Composite Structures* 78 (3): 368–376.

Problems

4.1 For a *Dyad*606 VEM of *Soundcoat* operating at 100°F (38°C), determine the optimal parameters of a GHM model with four mini-oscillators. Compare the predictions of the model with the actual experimental results given in the table next.
Develop the GHM for the storage modulus E' by converting G' to E' (by multiplying by $2(1+\nu) \approx 3$, where is ν Poisson's ratio ≈ 0.5) and then converting kpsi to Pa.

Frequency (Hz)	20	30	40	50	60	70	80	90	100	200	300
G' (kpsi)	0.75	1.1	1.3	1.6	1.8	2	2.2	2.3	2.6	4	5.2
Loss factor η	0.95	0.96	0.98	0.99	1	1	1	1	1	0.99	0.95
Frequency (Hz)	**400**	**500**	**600**	**700**	**800**	**900**	**1000**	**2000**	**3000**	**4000**	**5000**
G' (kpsi)	6.3	7.2	8.5	9.5	11	12	13	19	23	26	30
Loss factor η	0.91	0.88	0.85	0.82	0.79	0.77	0.75	0.63	0.56	0.51	0.48

4.2 Consider a mass M (=10 kg mass) is supported on a *DYAD*606 VEM. Consider the *Dyad*606 to be modeled with four mini-oscillators. Compute the response of the mass to a unit sinusoidal force as follows:
a) using the full dynamic equations of the mass and the four mini-oscillators.
b) using the reduced dynamic equations that are obtained by applying the Static Condensation to eliminate the IDOF corresponding to the mini-oscillators.

Compare the response of the full and reduced order models. Assume that the *Dyad* VEM has a cross sectional area of 0.01 m^2 and thickness of 0.1 m.

4.3 Consider the dynamic system shown in Figure P4.1 that simulates the dynamics of VEMs. Derive the following:
a) relationship between the force F and the net deflection (x_1) in the following Laplace domain form:

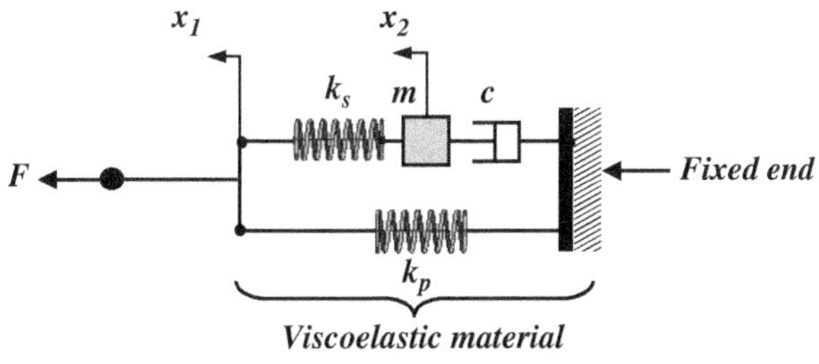

Figure P4.1 Dynamic system simulating a viscoelastic material.

$F = K^* x_1$

where K^* = complex stiffness = $K'(1 + i\,\eta)$ with K' and η denoting the storage and loss factor, respectively.

b) storage modulus of the VEM K' in terms of the frequency.

c) loss factor η of the VEM in terms of the frequency.

4.4 Consider the GHM model of a VEM given by:

$$\sigma = E_0\left(1 + \hat{\alpha}\frac{s^2 + 2\hat{\zeta}\hat{\omega}s}{s^2 + 2\hat{\zeta}\hat{\omega}s + \hat{\omega}^2}\right)\varepsilon$$

where $\hat{\alpha}, \hat{\zeta}\, and\, \hat{\omega}$ are the *GHM* parameters. Also, $\sigma\, and\, \varepsilon$ are the stress and the strain of the VEM. In the above equation s denotes the Laplace complex number. Determine expressions of the storage modulus E' and the loss factor η of the *GHM* as function of the frequency ω such that the complex modulus $E = E'(1 + \eta i)$.

Discuss why the GHM is structured as given before and not as follows:

a) $\sigma = E_0\left(1 + \hat{\alpha}\frac{s^2}{s^2 + \hat{\omega}^2}\right)\varepsilon,$

b) $\sigma = E_0\left(1 + \hat{\alpha}\frac{2\hat{\zeta}\hat{\omega}s}{s^2 + 2\hat{\zeta}\hat{\omega}s + \hat{\omega}^2}\right)\varepsilon,$

or

a) $\sigma = E_0\left(1 + \hat{\alpha}\frac{s^2}{s^2 + 2\hat{\zeta}\hat{\omega}s + \hat{\omega}^2}\right)\varepsilon.$

4.5 Consider the equation of motion of a mass M supported by VEM that has a stiffness described by the four-parameter GHM model of Friswell et al. (1997) that includes $(\alpha_n, \gamma_n, \beta_n, \delta_n)$ such that:

$$Ms^2X + K_0\left(1 + \alpha_n\frac{s^2 + \gamma_n s}{s^2 + \beta_n s + \delta_n}\right)X = F$$

where s and X denote the Laplace complex number and Laplace transform of the displacement x of the mass M under the influence of the force F.

Extract the internal DOF z of the VEM and then cast the final equation of motion of the mass/VEM combination in the form:

$$[M_T]\begin{Bmatrix}\ddot{x}\\ \ddot{z}\end{Bmatrix}+[C_T]\begin{Bmatrix}\dot{x}\\ \dot{z}\end{Bmatrix}+[K_T]\begin{Bmatrix}x\\ z\end{Bmatrix}=\begin{Bmatrix}F\\ 0\end{Bmatrix}$$

Determine the elements of the *symmetric* mass, damping, and stiffness matrices $[M_T]$, $[C_T]$, and $[K_T]$.

From the final equation of motion, identify the physical realization of the mass/VEM system as an assembly of springs, masses, and dampers.

4.6 Consider the following ATF model of a VEM given by:

$$\sigma = E_u\varepsilon - \delta\zeta$$
$$\dot{\zeta} + B\zeta = (B\delta/\alpha)\,\varepsilon$$

where the σ, ε, and ζ are the stress, strain, and temperature field. Also, E_u, B, δ *and* α are the ATF parameters.

Put the model in the Laplace domain (s) to show that the stress–strain relationship is given by:

$$\sigma = E_r\left(\frac{E_u/E_r s + B}{s+B}\right)\varepsilon \text{ where } E_r = \left(E_u - \delta^2/\alpha\right)$$

For a mass M supported by a VEM such that:

$$K_r\left(\frac{K_u/K_r s + B}{s+B}\right)$$

Put this equation of motion in the state-space such that $X = \{q,\dot{q},z\}^{\mathrm{T}}$ is the state-space vector with q and z denoting the DOF of the mass and the VEM, respectively. Show how to use the resulting equation to determine the system natural frequency and damping ratio.

4.7 Consider the fixed-free beam/VEM system shown in Figure P4.2. The beam is made of aluminum with a width of 0.025 m, thickness 0.0025 m, and length 1 m. The VEM has width of 0.025 m, thickness = 0.0025 m, and density of 1100 kg m^{-3}. The storage modulus and loss factor of the VEM are as given in Example 4.2. Using the GHM modeling approach of the VEM, determine the frequency

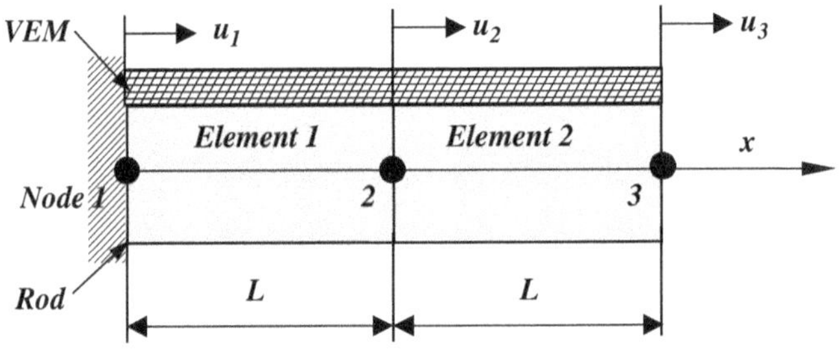

Figure P4.2 A fixed-free beam/VEM system.

and time response of the transverse vibration of the free end (Node 3) when the beam is subjected to a unit transverse load at the same end. Determine also the natural frequencies and damping ratios of the full and reduced order models of the beam/VEM system assuming that the system is modeled by a two-element finite element model.

Note that the beam element is a one-dimensional element that has two nodes and each node has two DOF. These DOF are the linear and angular deflection w and w_x. The shape function of the beam is assumed cubic as follows

$$w(x) = a_1 + a_2 x + a_3 x^2 + a_4 x^3$$

and the PE and KE of the beam element are given by:

$$PE = \frac{1}{2} EI \int_0^L w_{xx}^2 dx \text{ and } KE = \frac{1}{2} m \int_0^L \dot{w}^2 \, dx$$

where EI is the flexural rigidity of the beam and m is the beam mass per unit length.

Use a *GHM* model with a single mini-oscillator with $\alpha_1 = 39$, $\omega_1 = 19,058 rad/s$, and $\zeta_1 = 1$.

4.8 Consider the fixed-free rod/VEM system shown in Figure P4.3. The rod is made of aluminum with a width of 0.025 m, thickness 0.025 m, and length 0.30 m. The VEM has a width of 0.025 m, thickness = 0.025 m, and density 1100 kg m^{-3}. The storage modulus and loss factor of the VEM are predicted by GHM model with one mini-oscillator ($E_0 = 15.3 MPa, \alpha_1 = 39, \zeta_1 = 1, \omega_1 = 19,058 rad/s$). Determine the frequency and time response of the free end (Node 4) when the rod is subjected to a unit load $\boldsymbol{F}$ at the same end. Determine also the natural frequencies and damping ratios of the full and reduced order models of the rod/VEM system assuming that the system is modeled by a three element finite element model.

4.9 Consider the fixed-free beam/VEM system shown in Figure P4.4. The beam is made of aluminum with a width of 0.025 m, thickness 0.0025 m, and length 1 m. The VEM has a width of 0.025 m, thickness = 0.0025 m, and density 1100 kg m^{-3}. The storage modulus and loss factor of the VEM are as given in Example 4.2. The VEM is constrained by an aluminum constraining layer that is 0.025 m wide and 0.0025 m thick. Using the GHM modeling approach of the VEM, determine the frequency and time response of the transverse vibration of the free end

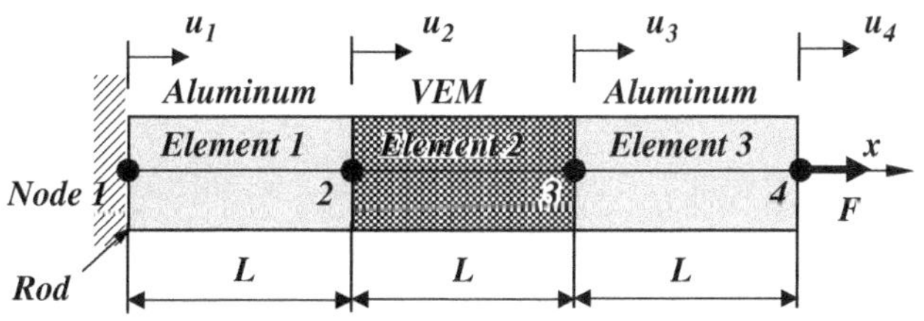

Figure P4.3 A fixed-free rod/VEM system.

(Node 3) when the beam is subjected to a unit transverse load at the same end. Determine also the natural frequencies and damping ratios of the full and reduced order models of the beam/VEM system assuming that the system is modeled by a two element finite element model.

Use a GHM model with a single mini-oscillator with $\alpha_1 = 39, \omega_1 = 19,058 rad/s$, and $\zeta_1 = 1$.

4.10 Consider the aluminum cantilever plate shown in Figure P4.5. The plate is treated with one PCLD patch placed at the fixed end of the plate. The dimensions of the plate and the PCLD are shown in the figure with $h_1 = h_2 = h_3 = 0.005m$. The VEM is modeled using a GHM model with three mini-oscillators such that:

$$G_2(s) = G_0\left[1 + \sum_{i=1}^{3} \alpha_i \frac{s^2 + 2\zeta_i\omega_i s}{s^2 + 2\zeta_i\omega_i s + \omega_i^2}\right] \quad \text{with} \quad G_0 = 0.5MPa$$

The parameters α_i, ζ_i, and ω_i are listed in Table 4.7. Assume that the constraining layer is made of aluminum.

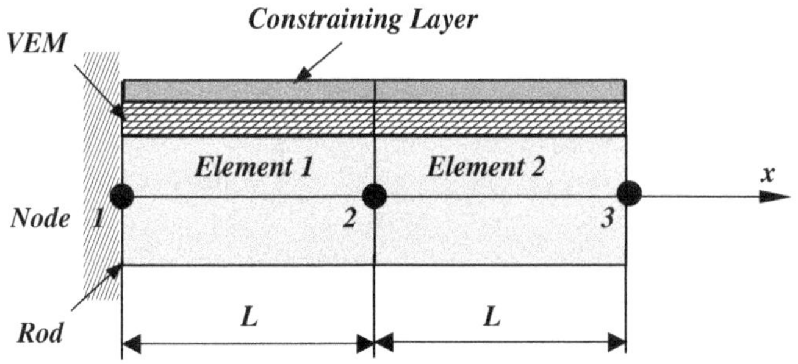

Figure P4.4 A fixed-free beam/constrained VEM system.

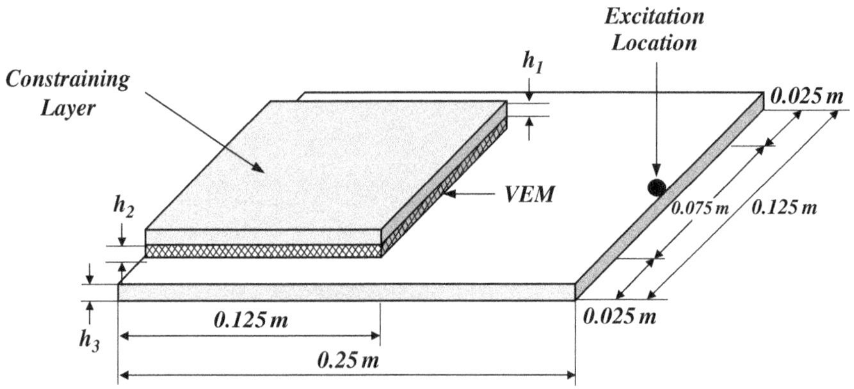

Figure P4.5 A cantilever plate with partial PCLD treatment.

Determine the frequency and time response of the plate when is excited by a unit force acting at the middle of the free end.

Compare the obtained responses when an ATF model is used to model the VEM such that:

ATF model: $G_2(s) = G_0\left[1 + \sum_{i=1}^{3} \Delta_{ATF_i} \frac{\rho_{ATF_i} s}{\rho_{ATF_i} s + 1}\right]$ with $G_0 = 0.5\,\text{MPa}$. The parameters Δ_{ATF_i} and ρ_{ATF_i} of the ATF model are listed in Table 4.7.

5

Finite Element Modeling of Viscoelastic Damping by Modal Strain Energy Method

5.1 Introduction

The modal strain energy (MSE) method has been widely accepted as an effective and practical means for predicting the modal parameters of complex structures treated with viscoelastic damping treatments. The method is based on estimating the modal strain energies of the structure and the viscoelastic material (VEM) by using the undamped (real) mode shapes of the structure/VEM assembly instead of the exact damped (complex) mode shape. Such an approximation makes it easy to integrate the MSE with commercial finite element codes that generally do not use complex eigenvalue problem solvers. The theoretical basis of the original MSE method and several of its modified versions are presented. Application of the method to various types of viscoelastic damping treatments is discussed and compared with the predictions obtained by using the exact complex eigenvalue problem solvers and the *Golla–Hughes–McTavish Model* (GHM) approach discussed in Chapter 4.

5.2 Modal Strain Energy (MSE) Method

The MSE method was originally introduced by Kerwin and Ungar (1962) and then extended by Johnson and Kienholz (1982) as an approximate means for predicting the modal parameters of complex structures treated with viscoelastic damping treatments. The method is based on describing the viscoelastic material by the "complex modulus" approach and therefore, it is limited to frequency domain analysis. To account for the variation of the VEM properties with frequency, the MSE method becomes iterative in nature. In spite of these two limitations, the MSE method has been widely accepted because of its practicality and ease of integration with finite element models of structures treated with VEM without increasing the size of models as in the case of the GHM method (Golla and Hughes 1985).

The theory behind the MSE method starts by describing the dynamics of the structure/VEM system by the following finite element equation:

$$[M]\{\ddot{X}\} + [K]\{X\} = \{0\} \tag{5.1}$$

where $\{X\}$ is the nodal deflection vector of the structure, $[M]$ its mass matrix (real), and $[K]$ is its stiffness matrix, which is complex to account for the VEM.

Active and Passive Vibration Damping, First Edition. Amr M. Baz.

The stiffness matrix can be written as

$$\begin{aligned}[K] &= [K_e] + [K_v] \\ &= [K_e] + \{[K_{v_r}] + i\,[K_{v_i}]\} \\ &= [K_e] + [K_{v_r}] + i\eta_v[K_{v_r}] \\ &= [K_R] + i\,[K_I]\end{aligned} \tag{5.2}$$

where $[K_e]$ and $[K_v]$ are the stiffness matrices of the elastic structure and VEM, respectively. Also, $[K_{v_r}]$ and $[K_{v_i}]$ are the stiffness matrices corresponding to the storage and the dissipative components of the VEM stiffness, respectively. In Eq. (5.2), η_v denotes the loss factor of the VEM. Also, $[K_R]$ and $[K_I]$ denote the total elastic stiffness matrix of the structure/VEM and the dissipative stiffness matrix of the VEM. These matrices are given by

$$[K_R] = [K_e] + [K_{v_r}] \quad \text{and} \quad [K_I] = \eta_v[K_{v_r}] \tag{5.3}$$

To cast the finite element model of Eq. (5.1) as an eigenvalue problem, the solution $\{X\}$ can be written as

$$\{X\} = \phi_n^* e^{i\omega_n^* t} \tag{5.4}$$

where ϕ_n^* and ω_n^* denote the eigenvector (mode shape) and eigenvalue (natural frequency) of the structure/VEM at mode n. These two quantities are complex and can be described as follows

$$\phi_n^* = \phi_{n_r} + i\,\phi_{n_i} \quad \text{and} \quad \omega_n^* = \omega_n\sqrt{(1 + i\eta_n)} \tag{5.5}$$

where ϕ_{n_r} and ϕ_{n_i} are the real and imaginary components of the eigenvector ϕ_n^*. Also, ω_n and η_n are the natural frequency and the loss factor of the nth mode.

Substituting Eq. (5.3) into Eq. (5.1) gives the following complex eigenvalue problem

$$[K]\{\phi_n^*\} = \omega_n^{*^2}[M]\{\phi_n^*\} \tag{5.6}$$

Solution of this problem requires an iterative approach to account for the variation of both the storage and dissipative components of the stiffness matrix $[K]$ with the frequency.

Substituting Eqs. (5.3) and (5.5) into Eq. (5.6) yields the following

$$[[K_R] + i[K_I]]\{\varphi_n^*\} = \omega_n^2(1 + i\eta_n)[M]\{\phi_n^*\} \tag{5.7}$$

Equating the real and imaginary parts on the two sides of Eq. (5.5), gives

$$[K_R]\{\phi_n^*\} = \omega_n^2[M]\{\phi_n^*\} \tag{5.8}$$

and

$$[K_I]\{\phi_n^*\} = \omega_n^2\eta_n[M]\{\phi_n^*\} \tag{5.9}$$

Pre-multiplying Eqs. (5.8) and (5.9) by $\{\phi_n^*\}^T$ gives

$$\omega_n^2 = \frac{\{\phi_n^*\}^T[K_R]\{\phi_n^*\}}{\{\phi_n^*\}^T[M]\{\phi_n^*\}}, \tag{5.10}$$

and

$$\omega_n^2 \eta_n = \frac{\{\phi_n^*\}^T [K_I] \{\phi_n^*\}}{\{\phi_n^*\}^T [M] \{\phi_n^*\}} \tag{5.11}$$

Dividing Eq. (5.11) by Eq. (5.10) and substituting for $[K_I]$ using Eq. (5.3) gives the modal loss factor η_n as follows

$$\eta_n = \frac{\{\phi_n^*\}^T [K_I] \{\phi_n^*\}}{\{\phi_n^*\}^T [K_R] \{\phi_n^*\}} = \eta_v \frac{\{\phi_n^*\}^T [K_{v_r}] \{\phi_n^*\}}{\{\phi_n^*\}^T [K_R] \{\phi_n^*\}} = \eta_v \frac{\{\phi_n^*\}^T [K_{v_r}] \{\phi_n^*\}}{\{\phi_n^*\}^T [[K_e] + [K_{v_r}]] \{\phi_n^*\}} \tag{5.12}$$

The *MSE* method simplifies Eq. (5.12) by replacing the exact damped (complex) mode shape $\{\phi_n^*\}$ by the undamped (real) mode shapes of the structure/VEM assembly $\{\phi_{n_r}\}$. Such simplification makes it easy to integrate the *MSE* with commercial finite element codes that generally do not use complex eigenvalue problem solvers.

This yields the following expression

$$\eta_n = \eta_v \frac{\{\phi_{n_r}\}^T [K_{v_r}] \{\phi_{n_r}\}}{\{\phi_{n_r}\}^T [[K_e] + [K_{v_r}]] \{\phi_{n_r}\}} \tag{5.13}$$

Physically, Eq. (5.13) means

$$\eta_n = \eta_v \frac{MSE\ of\ VEM}{MSE\ of\ \ Stucture + MSE\ of\ VEM} \tag{5.14}$$

that is, the loss factor of a structure/VEM assembly at the *n*th mode is equal to the loss factor of the VEM at the same frequency multiplied by the ratio of the MSE of the VEM to that of the structure/VEM assembly.

Implementation of MSE method is carried out according to the iterative scheme shown in Figure 5.1 in order to account for the variation of the VEM properties with frequency, the MSE method becomes in nature.

Example 5.1 Use the MSE method to calculate the natural frequencies and the corresponding modal loss factor (or damping ratios) for the rod/unconstrained VEM system given in Example 4.3. Compare the results with the predictions obtained by using GHM approach discussed in Chapter 4.

Solution

Table 5.1 lists the natural frequencies and the corresponding modal damping ratios (MDRs) for the rod/unconstrained VEM system as obtained by the GHM, and the MSE methods. Table 5.2 lists the results of the iterative solution and its convergence when using the MSE method.

The displayed results suggest that the MSE method accurately predicts the modal parameters of the considered case. Furthermore, the MSE method converges after three iterations to the final modal parameters.

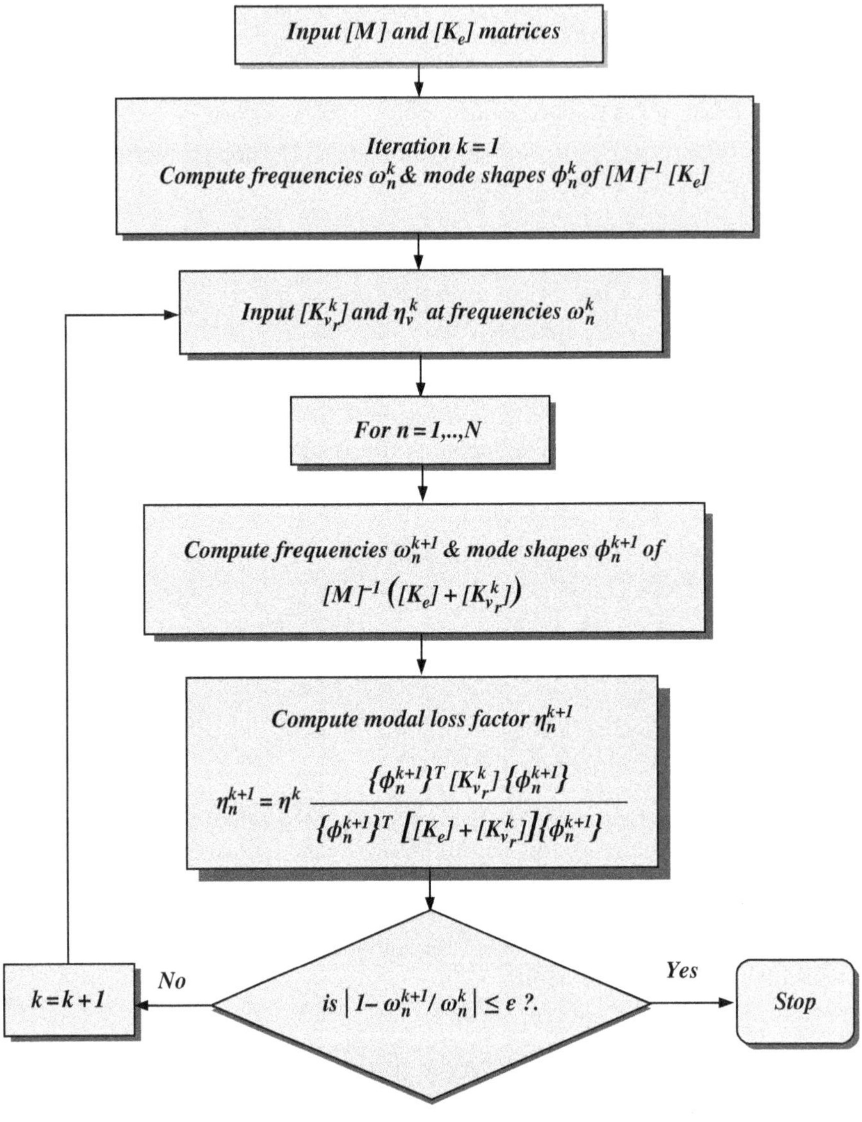

Figure 5.1 Implementation of the MSE method.

Example 5.2 Use the MSE method to calculate the natural frequencies and the corresponding modal loss factor (or damping ratios) for the rod/constrained VEM system given in Example 4.4 with VEM that has the following complex modulus

$$G^* = G_0\left(1 + \alpha\frac{s^2 + 2\omega_1 s}{s^2 + 2\omega_1 s + \omega_1^2}\right)$$

where G_0 = 15.3 MN m^{-2}, α_1 = 39, and ω_1 = 1.905 80E4 rad s^{-1}.

Compare the results with the predictions obtained by using the exact complex eigenvalue problem solvers and by applying the GHM approach discussed in Chapter 4.

Table 5.1 Comparison between the predictions of GHM and MSE methods for the rod/unconstrained VEM system.

Method	GHM		MSE	
Mode	Freq. (Hz)	Damping ratio	Freq. (Hz)	Damping ratio
1	1103.50	0.002 42	1102.93	0.00 241
2	3869.42	0.001 57	3865.52	0.001 56

Table 5.2 Convergence of the iterative solution of the MSE method for the rod/unconstrained VEM system.

Iteration	Mode 1 (Hz)	Mode 2 (Hz)	ζ_1	ζ_2
1	1101.41	3847.69	0.000 17	0.000 036
2	1102.93	3865.51	0.002 40	0.001 56
3	1102.93	3865.52	0.002 41	0.001 56

Solution

Table 5.3 lists the natural frequencies and the corresponding MDRs for the rod/constrained VEM system as obtained by the GHM and the MSE methods.

The displayed results show that the predictions of the modal parameters by the MSE are inaccurate for the considered case as compared to the predictions of the exact or the GHM methods. Such inaccuracy is attributed to the fact that approximating the exact (complex) eigenvectors by the undamped (real) eigenvectors is far from accurate because the loss factor is high in this case.

For this reason, several modified versions of the original MSE method have been developed to improve the estimates of the eigenvectors.

Table 5.3 Comparison between the predictions of the GHM and MSE methods for the rod/constrained VEM system.

Method	*GHM*		*MSE*	
Mode	Frequency (Hz)	Damping ratio	Frequency (Hz)	Damping ratio
1	1113.06	0.0489	1103.81	0.0179
2	3757.96	0.0123	3740.42	0.0136
3	5191.08	0.0869	5071.87	0.0959
4	6417.20	0.0341	6395.56	0.0353

5.3 Modified Modal Strain Energy (MSE) Methods

Four modified versions of the original MSE method will be discussed in this section. These methods aim at developing improved eigenvectors to account for the imaginary component that has been neglected in the original MSE method. These methods range from heuristic methods, as the Weighted Stiffness Matrix (WSM) method (Hu et al. 1995) and the Weighted STorage Modulus method (WSTM) (Xu et al. 2002) to the more rigorous methods as the Improved Reduction System method (O'Callahan 1989; Scarpa et al. 2002) and the low frequency approximation (LFA) method (Scarpa et al. 2002).

5.3.1 Weighted Stiffness Matrix Method (WSM)

This method is based on substituting Eq. (5.5) into Eq. (5.5) to give

$$[[K_R] + i[K_I]]\left\{\phi_{n_r} + i\phi_{n_i}\right\} = \omega_n^2(1 + i\eta_n)[M]\left\{\phi_{n_r} + i\phi_{n_i}\right\}$$

Equating the real part on both sides of this equation gives

$$[K_R]\left\{\phi_{n_r}\right\} - [K_I]\left\{\phi_{n_i}\right\} = \omega_n^2[M]\left\{\phi_{n_r}\right\} - \omega_n^2\eta_n[M]\left\{\phi_{n_i}\right\} \tag{5.15}$$

Assume a vector $\{\bar{\phi}\}$ to be defined such that:

$$\{\bar{\phi}\} = a\left\{\phi_{n_r}\right\} \text{ and } \{\bar{\phi}\} = -b\left\{\phi_{n_i}\right\}$$

Then, Eq. (5.15) reduces to

$$([K_R] + \beta[K_I])\{\bar{\phi}\} = \omega_n^2(1 + \beta\eta_n)[M]\{\bar{\phi}\}$$

or

$$[K_M]\{\bar{\phi}\} = \bar{\omega}_n^2[M]\{\bar{\phi}\} \tag{5.16}$$

where $[K_M] = [K_R] + \beta[K_I]$, $\beta = a/b$, and $\bar{\omega}_n^2 = \omega_n^2(1 + \beta\eta_n)$.

Eq. (5.16) represents a modified eigenvalue problem that has real eigenvalue $\bar{\omega}_n^2$ and eigenvector $\{\bar{\phi}\}$. Note that $[K_M]$ is a modified stiffness matrix that augments the elastic stiffness matrix $[K_R]$ with a weighted contribution of the imaginary component of the stiffness $[K_I]$. The weighting parameter β is calculated from the following empirical formula proposed by Hu et al. (1995)

$$\beta = \frac{trace[K_I]}{trace[K_R]} \tag{5.17}$$

If $\beta = 0$, the modified eigenvalue problem reduces to that used in computing the real eigenvectors employed by the *MSE* method. If $\beta \neq 0$, the modified eigenvalue problem attempts to heuristically account for the imaginary component of the stiffness in order to generate a better estimate $\{\bar{\phi}\}$ of the real eigenvector. This estimate is used to compute the loss factor η_n of the nth mode as follows

$$\eta_n = \eta_v \frac{\{\bar{\phi}\}^T[K_{v_r}]\{\bar{\phi}\}}{\{\bar{\phi}\}^T[[K_e] + [K_{v_r}]]\{\bar{\phi}\}} \tag{5.18}$$

5.3.2 Weighted Storage Modulus Method (WSTM)

In this method, the shear modulus of the VEM as described by

$$G^* = G'(1 + i\eta_v)$$

has a storage modulus G' and loss factor η_v. Hence, to generate the real eigenvectors only G' is used because it directly affects the real stiffness matrix $[K_{v_r}]$ of the VEM. However, a better estimate of the real eigenvectors can be obtained if the storage modulus is modified to account for the dissipative part. Xu et al. (2002) proposed to modify the storage modulus as follows

$$G'_{\text{modified}} = G'\sqrt{1 + \eta_v^2} \tag{5.19}$$

In this manner, the modified storage modulus is the magnitude of the shear modulus that augments the storage modulus by the contribution of the loss modulus. Such a heuristic modification was motivated by the fact that increasing the storage modulus increases the natural frequencies and the observations that the natural frequencies increase with increasing the loss factor of the VEM as reported by Xu and Chen (2000) when using the exact complex eigenvalue problem solvers.

The loss factor of the structure/VEM system is then determined from

$$\eta_n = \eta_v \frac{\left\{\tilde{\phi}\right\}^T \left[\tilde{K}_{v_r}\right] \left\{\tilde{\phi}\right\}}{\left\{\tilde{\phi}\right\}^T \left[[K_e]\sqrt{1 + \eta_v^2} + \left[\tilde{K}_{v_r}\right]\right] \left\{\tilde{\phi}\right\}} \tag{5.20}$$

where $\left[\tilde{K}_{v_r}\right]$ is the elastic component of the stiffness matrix of the VEM as modified by the weighted storage modulus, that is, $\left[\tilde{K}_{v_r}\right] = \sqrt{1 + \eta_v^2}[K_{v_r}]$.

Also, $\left\{\tilde{\phi}\right\}$ is the eigenvector of the following eigenvalue problem:

$$\left([K_e] + \left[\tilde{K}_{v_r}\right]\right)\left\{\tilde{\phi}\right\} = \omega_n^2 [M]\left\{\tilde{\phi}\right\} \tag{5.21}$$

5.3.3 Improved Reduction System Method (IRS)

This method is based on assuming the following modal transformation

$$\{X\} = [\Phi]\{q\} \tag{5.22}$$

where $[\Phi]$ is the eigenvector matrix for the undamped part of Eq. (5.1), that is,

$$[M]\{\ddot{X}\} + [K_R]\{X\} = 0$$

such that

$$[\Phi]^T [M][\Phi] = [I] \quad \text{and} \quad [\Phi]^T [K_R][\Phi] = [\Lambda] \tag{5.23}$$

Also, in Eq. (5.22), $\{q\}$ denotes the modal displacement vector, which is given by

$$\{q\} = [\{q_r\} + i\{q_i\}]e^{i\omega t} \tag{5.24}$$

But, for the damped system we have

$$[M]\{\ddot{X}\} + ([K_R] + i[K_I])\{X\} = 0 \tag{5.25}$$

Substituting Eqs. (5.22) and (5.24) into Eq. (5.25) gives

$$([K_R] + i[K_I])\,[\Phi]\{q_r + iq_i\} - \omega^2[M][\Phi]\{q_r + iq_i\} = \{0\} \tag{5.26}$$

Pre-multiplying Eq. (5.26) by $[\Phi]^T$and using Eq. (5.23) gives

$$\left([\Lambda] + i\left[\tilde{K}_I\right]\right)\{q_r + iq_i\} - \omega^2 I\{q_r + iq_i\} = \{0\} \tag{5.27}$$

where $\left[\tilde{K}_I\right] = [\Phi]^T[K_I][\Phi]$.

Equating the real and imaginary parts on both sides of Eq. (5.23), gives the following matrix equation

$$\left\{\begin{bmatrix}[\Lambda] & -\left[\tilde{K}_I\right]\\ \left[\tilde{K}_I\right] & [\Lambda]\end{bmatrix} - \omega^2\begin{bmatrix}[I] & 0\\ 0 & [I]\end{bmatrix}\right\}\begin{Bmatrix}q_r\\ q_i\end{Bmatrix} = \begin{Bmatrix}0\\ 0\end{Bmatrix}$$

or

$$\{[K_T] - \omega^2[M_T]\}\begin{Bmatrix}q_r\\ q_i\end{Bmatrix} = 0 \tag{5.28}$$

Using static condensation, the second row of Eq. (5.28) gives

$$\{q_i\} = -[\Lambda]^{-1}\left[\tilde{K}_I\right]\{q_r\} = [S]\{q_r\} \tag{5.29}$$

and

$$\begin{Bmatrix}q_r\\ q_i\end{Bmatrix} = \begin{Bmatrix}[I]\\ [S]\end{Bmatrix}\{q_r\} = [T]\{q_r\} \tag{5.30}$$

Hence, the condensed system can be obtained by combining Eqs. (5.28) and (5.30) to yield the following

$$\{[K_c] - \omega^2[M_c]\}\{q_r\} = 0 \tag{5.31}$$

where

$$[K_c] = [T]^T[K_T][T] = [\Lambda] + [S]^T\left[\tilde{K}_I\right] - \left[\tilde{K}_I\right][S] + [S]^T[\Lambda][S]$$

and

$$[M_c] = [T]^T[M_T][T] = [I] + [S]^T[S].$$

Solution of the eigenvalue problem given by Eq. (5.31) yields the eigenvalue ω and the eigenvector $\{q_r\}$. The full complex eigenvector can be extracted as follows

$$\{\phi^*\} = [\Phi][[I] + i[S]]\{q_r\} \tag{5.32}$$

This eigenvector can be used to compute the modal loss factor of the structure/VEM assembly as follows

$$\eta_n = \eta_v \frac{\{\phi^*\}^T [K_{v_r}] \{\phi^*\}}{\{\phi^*\}^T [[K_e] + [K_{v_r}]] \{\phi^*\}} \tag{5.33}$$

5.3.4 Low Frequency Approximation Method (LFA)

This method is based on expanding the second row of Eq. (5.28) to give

$$\left[\tilde{K}_I\right]\{q_r\} + \left([\Lambda] - \omega^2[I]\right)\{q_i\} = \{0\}$$

or

$$\begin{aligned}\{q_i\} &= -\left([\Lambda] - \omega^2[I]\right)^{-1}\left[\tilde{K}_I\right]\{q_r\} \\ &= -[\Lambda]^{-1}\left([I] - \omega^2[\Lambda]^{-1}[I]\right)^{-1}\left[\tilde{K}_I\right]\{q_r\}\end{aligned} \tag{5.34}$$

For low frequencies, the Taylor series expansion of Eq. (5.29) in terms of ω is given by

$$\{q_i\} = -[\Lambda]^{-1}\left[\tilde{K}_I\right]\{q_r\} - [\Lambda]^{-2}\left[\tilde{K}_I\right]\omega^2\{q_r\} \tag{5.35}$$

Note that the first term of Eq. (5.35) is corresponding to the static condensation Eq. (5.29). In this manner, Eq. (5.35) includes the contribution of the inertia terms and accordingly presents a dynamic condensation of $\{q_i)$ in terms of $\{q_r\}$.

Expanding the first row of Eq. (5.28) gives:

$$\left([\Lambda] + \beta\left[\tilde{K}_I\right]\right)\{q_r\} = \omega^2[I]\{q_r\} \tag{5.36}$$

Combining Eqs. (5.35) and (5.36) gives

$$\{q_i\} = -\left[[\Lambda]^{-1}\left[\tilde{K}_I\right] + [\Lambda]^{-2}\left[\tilde{K}_I\right]\left[[\Lambda] + \beta\left[\tilde{K}_I\right]\right]\right]\{q_r\} = \left[\tilde{S}\right]\{q_r\} \tag{5.37}$$

Note that β is given by Eq. (5.17) as suggested by Hu et al. (1995).

Equation (5.37) presents a condensation equation with the transformation

$$\begin{Bmatrix} q_r \\ q_i \end{Bmatrix} = \begin{Bmatrix} [I] \\ \left[\tilde{S}\right] \end{Bmatrix}\{q_r\} = \left[\tilde{T}\right]\{q_r\} \tag{5.38}$$

Then, solving the original system given by Eq. (5.28) reduces to the following condensed system:

$$\left\{\left[\tilde{K}_c\right] - \omega^2\left[\tilde{M}_c\right]\right\}\{q_r\} = 0 \tag{5.39}$$

where

$$\left[\tilde{K}_c\right] = \left[\tilde{T}\right]^T [K_T]\left[\tilde{T}\right] = [\Lambda] + \left[\tilde{S}\right]^T\left[\tilde{K}_I\right] - \left[\tilde{K}_I\right]\left[\tilde{S}\right] + \left[\tilde{S}\right]^T[\Lambda]\left[\tilde{S}\right]$$

and

$$\left[\widetilde{M}_c\right] = \left[\widetilde{T}\right]^T [M_T]\left[\widetilde{T}\right] = [I] + \left[\widetilde{S}\right]^T\left[\widetilde{S}\right].$$

Solution of the eigenvalue problem given by Eq. (5.39) yields the eigenvalue ω and the eigenvector $\{q_r\}$. The full complex eigenvector can be reconstructed as follows

$$\left\{\widetilde{\phi}^*\right\} = [\Phi]\left[[I] + i\left[\widetilde{S}\right]\right]\{q_r\} \tag{5.40}$$

This eigenvector can be used to compute the modal loss factor of the structure/VEM assembly as follows

$$\eta_n = \eta_v \frac{\left\{\widetilde{\phi}^*\right\}^T [K_{v_r}]\left\{\widetilde{\phi}^*\right\}}{\left\{\widetilde{\phi}^*\right\}^T [[K_e] + [K_{v_r}]]\left\{\widetilde{\phi}^*\right\}} \tag{5.41}$$

Example 5.3 Use the different modified MSE methods to calculate the natural frequencies and the corresponding modal loss factor (or damping ratios) for the rod/constrained VEM system given in Example 4.4 with VEM that has the following complex modulus

$$G^* = G_0\left(1 + \alpha\frac{s^2 + 2\omega_1 s}{s^2 + 2\omega_1 s + \omega_1^2}\right)$$

where G_0 = 15.3 MN m^{-2}, α_1 = 39, and ω_1 = 19,058 rad s^{-1}.

Compare the results with the predictions obtained by using the original MSE method and the different modified MSE methods.

Solution

Tables 5.4 and 5.5 list the natural frequencies and the corresponding MDRs for the rod/constrained VEM system as obtained by the exact eigenvalue problem solver of MATLAB, the MSE method, and four modified MSE methods.

It is clear that the four modified methods have improved the accuracy of the MSE. All the methods provide adequate predictions of the natural frequencies. Also, all the methods have predicted accurately the damping ratios except for the first mode with the exception of the LFA method.

Table 5.4 Comparison between the natural frequency predictions of the different modified MSE methods for the rod/constrained VEM system.

Mode	*GHM*	*MSE*	Weighted stiffness	Weighted storage	*IRS*	*LFA*
1	1113.06	1103.81	1105.26	1109.67	1111.89	1112.33
2	3757.96	3740.42	3740.81	3748.69	3746.69	3744.90
3	5191.08	5071.87	5072.19	5118.93	5059.11	5055.82
4	6417.20	6395.56	6395.32	6409.19	6391.55	6396.75

Table 5.5 Comparison between the damping ratio predictions of the different modified MSE methods for the rod/constrained VEM system.

Mode	Exact	MSE	Weighted stiffness	Weighted storage	IRS	LFA
1	0.0049	0.0179	0.0078	0.0049	0.0062	0.0346
2	0.0123	0.0136	0.0124	0.0124	0.0132	0.0135
3	0.0869	0.0959	0.0943	0.0942	0.0964	0.0952
4	0.0341	0.0353	0.0352	0.0352	0.0354	0.0361

5.4 Summary of Modal Strain Energy Methods

Table 5.6 summarizes the basic equations that are used to compute the modal loss factor using the conventional or the modified MSE methods. The table also lists the different forms of the eigenvectors needed to predict the modal loss factors for the considered MSE methods. Note that the MSE, WSM, and WSTM methods all use real eigenvectors while the IRS and LFA methods use imaginary eigenvectors.

5.5 Modal Strain Energy as a Metric for Design of Damping Treatments

The MSE, as described in Sections 5.1 through 5.4, serves as an important design metric for selecting the optimal design parameters (Lepoittevin and Kress 2009; Sainsbury and Masti 2007); location (Ro and Baz 2002); and topology of damping treatments (Ling et al. 2010).

Figure 5.2 summarizes the basic concept behind using the MSE as a design metric. For a given base structure (i.e., known $[K_e]$ and $[M]$), an initial guess of the design parameters and/or topology (i.e., $[K_v]$), of the VEM is input to the MSE module to determine the modal loss factors η_n for the first N modes. These factors can be maximized for a particular mode or a group of critical modes by adjusting the design parameters and/or topology (i.e., $[K_v]$), of the VEM, in a rational manner, using available optimization tools such as the MATLAB Optimization Toolbox. This process is repeated until an optimal configuration of the VEM is attained while satisfying a set of design constraints.

In this section, the MSE will be utilized in selecting the optimal thickness of unconstrained damping layers which are used to treat rods undergoing longitudinal vibrations.

In order to illustrate the utility of the MSE as a design metric, consider the following example:

Example 5.4 Consider 1D fixed-free rod shown in Figure 5.3. The rod is treated by an unconstrained VEM and is divided into N finite elements. It is desired to determine the optimal distribution of the thickness $[t_{vi}(x)$ for $i = 1, \ldots, N]$ of the VEM along the rod in order to maximize the modal loss factor over the first five natural frequencies while minimizing the weight of the treatment.

Table 5.6 The basic equations used to determine the modal loss factor using the conventional or the modified MSE methods.

Method		Modal loss factor	Eigenvectors
MSE		$\eta_n = \eta_v \frac{\{\phi_{nr}\}^T[K_{vr}]\{\phi_{nr}\}}{\{\phi_{nr}\}^T[[K_e]+[K_{vr}]]\{\phi_{nr}\}}$	The real eigenvector $\{\phi_n\}$ is solution of: $[[K_e]+[K_{v_r}]]\{\phi_n\} = \omega_n^2[M]\{\phi_n\}$
MODIFIED MSE	Weighted Stiffness Matrix Method	$\eta_n = \eta_v \frac{\{\bar{\phi}\}^T[K_{vr}]\{\bar{\phi}\}}{\{\bar{\phi}\}^T[[K_e]+[K_{vr}]]\{\bar{\phi}\}}$	The real eigenvector $\{\bar{\phi}\}$ is solution of: $[[K_R]+\beta[K_I]]\{\bar{\phi}\} = \bar{\omega}_n^2[M]\{\bar{\phi}\}$ where $\beta = trace[K_I]/trace[K_R]$, $\bar{\omega}_n^2 = \omega_n^2(1+\beta\eta_n)$
	Weighted Storage Modulus Method	$\eta_n = \eta_v \frac{\{\tilde{\phi}\}^T[\tilde{K}_{v_r}]\{\tilde{\phi}\}}{\{\tilde{\phi}\}^T\{[K_e]\sqrt{1+\eta_v^2}+[\tilde{K}_{v_r}]\}\{\tilde{\phi}\}}$	The real eigenvector $\{\tilde{\phi}\}$ is solution of: $[[K_e]+[\tilde{K}_{v_r}]]\{\tilde{\phi}\} = \omega^2[M]\{\tilde{\phi}\}$ where $[\tilde{K}_{v_r}] = \sqrt{1+\eta_v^2}[K_{v_r}]$
	Improved Reduction System Method (IRS)	$\eta_n = \eta_v \frac{\{\phi^*\}^T[K_{vr}]\{\phi^*\}}{\{\phi^*\}^T[[K_e]+[K_{vr}]]\{\phi^*\}}$	The imaginary eigenvector $\{\phi^*\}$ is given by: $\{\phi^*\} = [\Phi][[I]+i[S]]\{q_r\}$ where $[S] = -[\Lambda]^{-1}[\tilde{K}_I]$, $[\tilde{K}_I] = [\Phi]^T[K_I][\Phi]$, with $[\Lambda]$and $[\Phi]$ are the eigenvalues and vectors of $[M]\{\ddot{X}\}+[K_R]\{X\}=0$ and $\{q_r\}$ = eigenvector of $\{[K_c]-\omega^2[M_c]\}\{q_r\} = 0$ with $[M_c] = [I]+[S]^T[S]$, $[K_c] = [\Lambda]+[S]^T[\tilde{K}_I]-[\tilde{K}_I][S]+[S]^T[\Lambda][S]$
	Low Frequency Approximation Method (LFA)	$\eta_n = \eta_v \frac{\{\tilde{\phi}^*\}^T[K_{vr}]\{\tilde{\phi}^*\}}{\{\tilde{\phi}^*\}^T[[K_e]+[K_{vr}]]\{\tilde{\phi}^*\}}$	The imaginary eigenvector $\{\tilde{\phi}^*\}$ is given by: $\{\tilde{\phi}^*\} = [\Phi][[I]+i[\tilde{S}]]\{q_r\}$ where $\{\tilde{S}\} = -[[\Lambda]^{-1}[\tilde{K}_I]+[\Lambda]^{-2}[\tilde{K}_I][[\Lambda]-\beta[\tilde{K}_I]]]$, $[\tilde{K}_I] = [\Phi]^T[K_I][\Phi]$, $\beta = trace[K_I]/trace[K_R]$, and $\{q_r\}$ = eigenvector of $\{[\tilde{K}_c]-\omega^2[\tilde{M}_c]\}\{q_r\}=0$ with $[\tilde{M}_c] = [I]+[\tilde{S}]^T[\tilde{S}]$, $[\tilde{K}_c] = [\Lambda]+[\tilde{S}]^T[\tilde{K}_I]-[\tilde{K}_I][\tilde{S}]+[\tilde{S}]^T[\Lambda][\tilde{S}]$

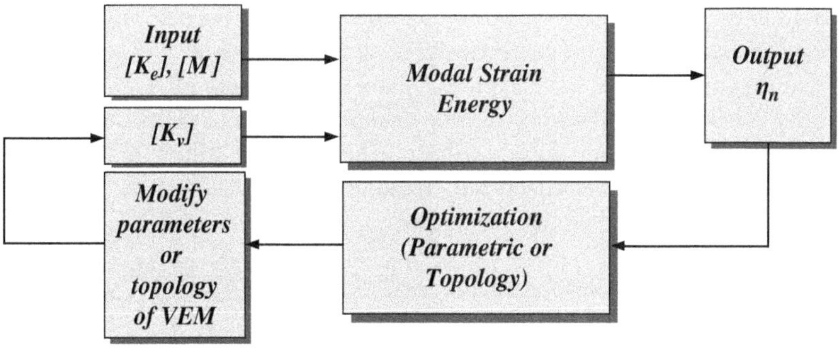

Figure 5.2 MSE as a design metric of VEM.

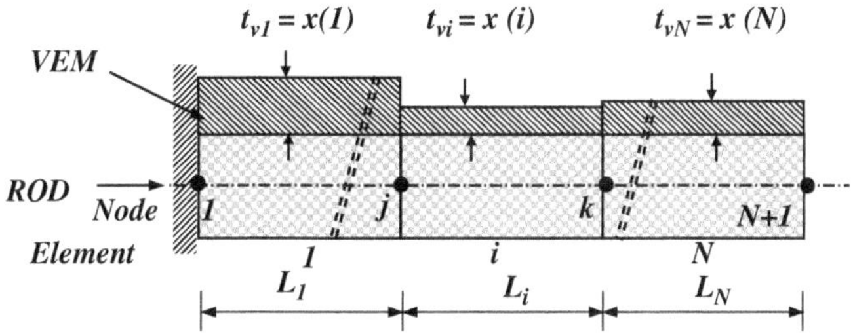

Figure 5.3 Finite element model of a rod treated with unconstrained VEM.

Assume that the rod is made of aluminum with width of 0.025 m, thickness of 0.025 m, and length of 1 m. Assume also that the rod is divided into five finite elements and that the VEM has a width of 0.025 m and a density of 1100 kg m^{-3}. The storage modulus and loss factor of the VEM are predicted by GHM model with one mini-oscillator ($E_0 = 15.3MPa$, $\alpha_1 = 39$, $\zeta_1 = 1$, $\omega_1 = 19,058 rad/s$).

Solution

The design problem is formulated mathematically as follows:

$$\left\langle \begin{array}{l} \text{Determine the thickness distribution} \\ \text{of the damping layer, } i.e.\, t_{v1}, t_{v2}, \ldots, t_{vN} \\ \text{To maximize } F = \left[\sum_{n=1}^{5} \eta_n\right] \Big/ \left[\rho_v b_v \sum_{n=1}^{5} t_{v_n}\right] \\ \text{Such that: } t_{v1}, t_{v2}, \ldots, t_{vN} > 0 \\ \text{and } t_{vmin} < t_{vi} < t_{vmax} \quad \text{for } i = 1, .., N \end{array} \right\rangle \tag{5.42}$$

In this optimum design problem, the objective function is written as the ratio of the sum of the modal loss factor for the first five modes to the total weight of the treatment. In this manner, maximizing F will simultaneously ensure maximizing the modal loss factor and minimizing the total weight. The constraints imposed on the design problem ensure that all the design variables t_{vi} are positive and each is bounded from below and from above by t_{vmin} and t_{vmax}, respectively.

Note that the lower bound is selected to ensure adequate damping and avoid reaching the unrealistic trivial solution where all the design variables t_{vi} vanish, which makes the mass of the treatment minimum (= zero) and the objective function maximum (= ∞). However, the upper bound is selected to avoid an impractically thick VEM.

Consider the following objective functions:

i) ***F_1 = sum of the modal loss factor for the first five modes*** $F_1 = \left[\sum_{n=1}^{5} \eta_n\right]$

Two sets of constraints are imposed:

***Set 1:* t_{vmin} = 0.001 and t_{vmax} = 0.01 m**

Let the initial guess of the thickness distribution is $[t_{v1}, t_{v2}, ..., t_{v5}]$ = [0.005 0.005 0.005 0.005 0.005]. The MATLAB solution of the optimization problem using the "fmincon" subroutine of the Optimization Toolbox gives optimal thicknesses = [0.01 0.01 0.01 0.01 0.01] as displayed in Figure 5.4. The corresponding value of the objective function F_1 = 0.002 043.

***Set 2:* t_{vmin} = 0.001 and t_{vmax} = 0.025 m**

If the initial guess of the thickness distribution is maintained at $[t_{v1}, t_{v2}, ..., t_{v5}]$ = [0.005 0.005 0.005 0.005 0.005]. The MATLAB solution of the optimization problem using the "fmincon" subroutine of the Optimization Toolbox gives optimal thicknesses = [0.025 0.025 0.025 0.025 0.025] as displayed in Figure 5.5. The corresponding value of the objective function F_1 = 0.005 44.

In both cases, the optimum is attained when the thickness of each element attains the allowable upper bound in order to maximize the sum of the modal loss factor for the first five modes. Accordingly, the optimization algorithm pushes the VEM thickness to its maximum limit without any regard to the weight.

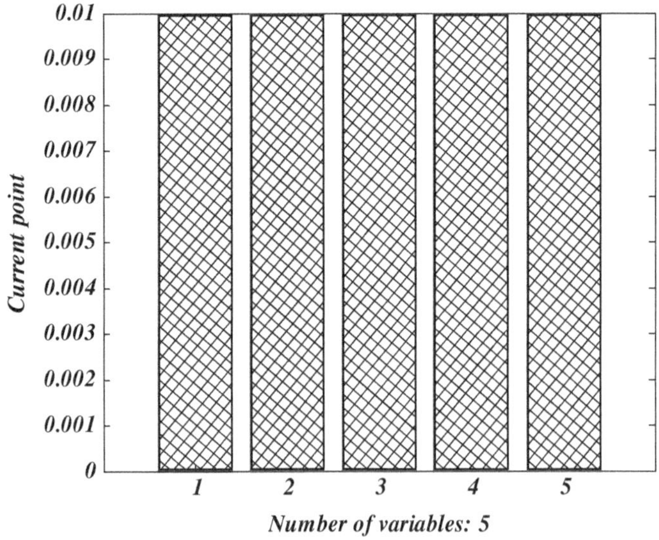

Figure 5.4 Optimal thickness distribution of VEM for constraint set 1 and objective function F_1.

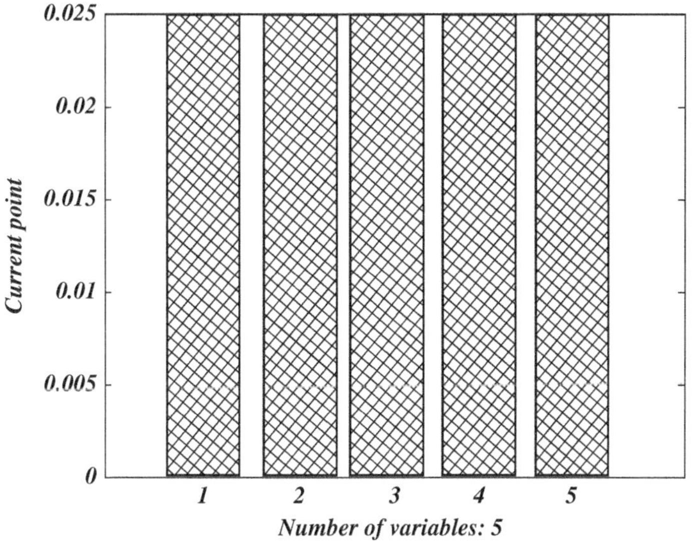

Figure 5.5 Optimal thickness distribution of VEM for constraint set 2 and objective function F_1.

ii) ***F_2 = sum of the modal loss factors/total weight*** $F_2 = \left[\sum_{n=1}^{5} \eta_n\right] / \left[\rho_v b_v \sum_{n=1}^{5} t_{v_n}\right]$

In this case, the objective function F_2 puts a penalty on the weight of the *VEM*. Two sets of constraints are imposed:

Set 1: t_{vmin} **= 0.001 and** t_{vmax} **= 0.025 m**

If the initial guess of the thickness distribution is $[t_{v1}, t_{v2}, \ldots, t_{v5}]$ = [0.005 0.005 0.005 0.005 0.005]. *MATLAB* Optimization Toolbox gives optimal thicknesses = [0.025 0.001 0.001 0.001 0.001] as displayed in Figure 5.6. The corresponding value of the objective function F_2 = 0.002 28.

Set 2: t_{vmin} **= 0.01 and** t_{vmax} **= 0.025 m**

If the initial guess of the thickness distribution is $[t_{v1}, t_{v2}, \ldots, t_{v5}]$ = [0.01 0.01 0.01 0.01 0.01]. MATLAB Optimization Toolbox gives optimal thicknesses = [0.025 0.012 0.01 0.01 0.01] as displayed in Figure 5.7. The corresponding value of the objective function F_2 = 0.001 73.

Figure 5.8a,b show the time response of the rod when treated with the optimal damping treatments corresponding to constraint sets 1 and 2, which are imposed on objective function F_2. The rod in both cases is subjected to a unit impulse at its free end.

It is important to note that although the optimal objective function for the first set of constraints is F_2 = 0.002 28 and that for the second set is F_2 = 0.001 73, the vibration damping characteristics for the second set is better than the first. This is attributed to the fact that the weight of the VEM for the second set is 2.39 times that of the first set. Hence, the sum of the modal loss factors for the second set is almost twice that of the first set. Accordingly, more damping is achieved with the second set of constraints in spite of the lower value of the objective function.

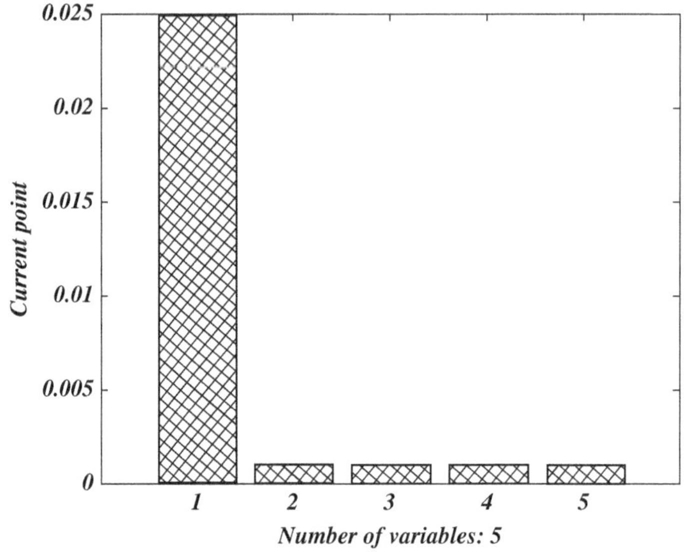

Figure 5.6 Optimal thickness distribution of VEM for constraint set 1 and objective function F_2.

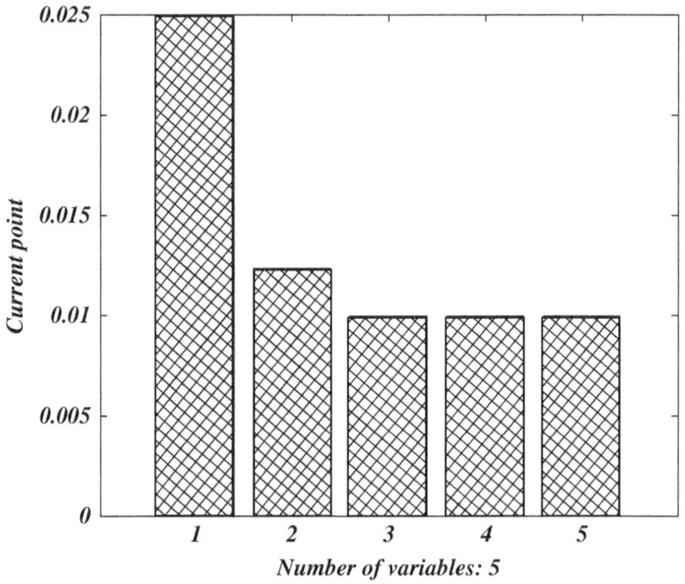

Figure 5.7 Optimal thickness distribution of VEM for constraint set 2 and objective function F_2.

5.6 Perforated Damping Treatments

5.6.1 Overview

Engineered Damping Treatments (EDTs) that have high damping characteristics per unit volume are presented in this section. The EDTs under consideration consist of cellular viscoelastic damping matrices with optimally selected cell configuration, size, and

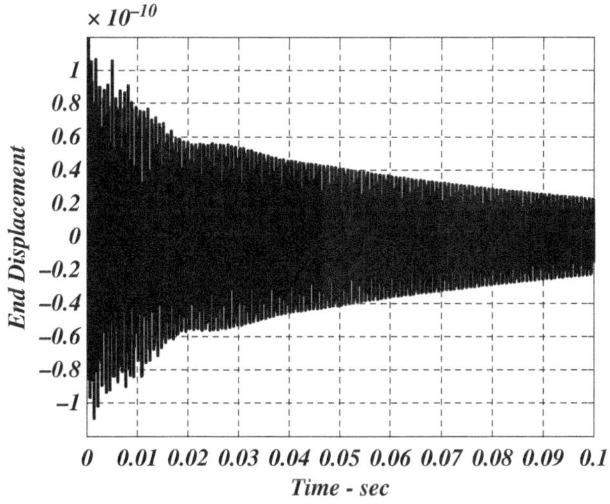

(a) – Constraint set 1

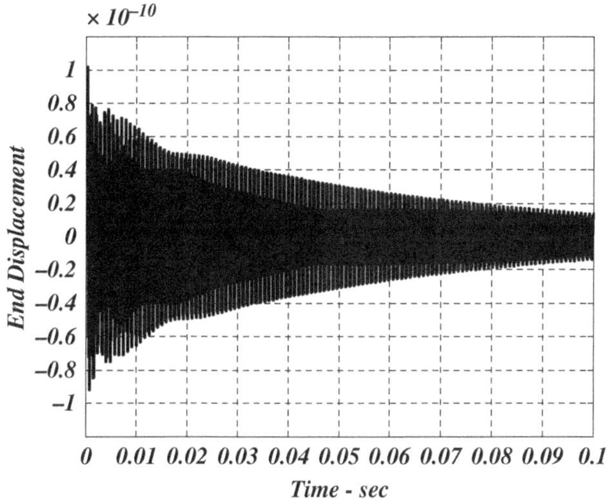

(b) – Constraint set 2

Figure 5.8 Time response of optimal damping treatment designs based on objective function F_2. (a) Constraint set 1 and (b) constraint set 2.

(a) Conventional Treatment | (b) Treatment with square holes | (c) Treatment with reentrant hexagon holes

Figure 5.9 Perforated damping treatment. (a) Conventional treatment. (b) Treatment with square holes. (c) Treatment with reentrant hexagonal holes.

distribution. These perforated EDTs are intended to replace conventional viscoelastic damping treatments to improve their damping characteristics and reduce their weight at the same time. Examples of such perforated EDTs are shown in Figure 5.9. The number, shape, and spacing of the perforations are critical to the effective damping characteristics of the treatment and to the minimization of its weight. Figure 5.9a–c shows a conventional damping treatment, treatment with square holes that has a positive Poisson's ratio, and treatment with reentrant hexagon holes that has a negative Poisson's ratio.

The cellular topologies of the EDTs are modeled using the finite element method in an attempt to determine the optimal topologies that maximize the strain energy, maximize the damping characteristics, and minimize the total weight. The damping characteristics of the manufactured EDTs are evaluated and compared with the corresponding characteristics obtained by conventional solid damping treatments in order to emphasize the importance of using optimally configured damping treatment to achieve high damping characteristics.

5.6.2 Finite Element Modeling

Consider the configuration shown in Figure 5.10. It consists of a base plate covered from one side with a layer of VEM. The base plate is isotropic and linearly elastic with density, elasticity modulus, and Poisson's ratio of ρ_p, E_p, ν_p, respectively. The VEM layer properties are denoted by ρ_v, E_v, ν_v where the complex elasticity modulus $E_v = E_0(1 + j\eta)$ is used to describe the viscoelastic properties of the layer. The viscoelastic layer is conventionally solid with constant thickness or it can have a variable thickness to maximize the damping behavior.

Figure 5.10b shows a finite element of this composite. The element is a four-noded rectangular element with dimensions $2a \times 2b$ while the thicknesses of the base layer and the treatment are h_p and h_v, respectively. The plate element is aligned to the x–y-plane. The displacement components of the base plate are u, v, and w in the x, y, and z directions. Also, θ_x and θ_y are the rotational components about the x and y directions.

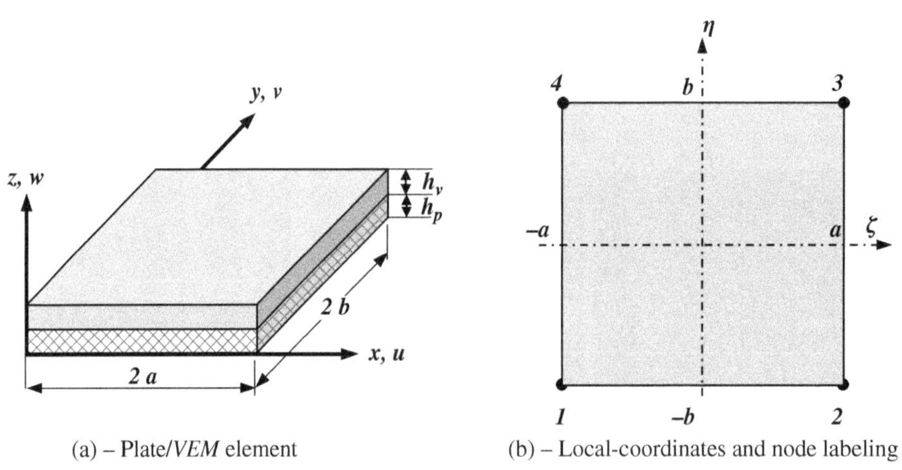

Figure 5.10 Plate treated with unconstrained damping treatment. (a) Plate/VEM element. (b) Local coordinates and node labeling.

The displacement vector **u** of the base plate is given by: $u = \{u \;\; v \;\; w \;\; \theta_x \;\; \theta_y\}^T$. Accordingly, each node has five degrees of freedom corresponding to the five displacement components. Furthermore, the base plate is assumed to be thin and hence the strain components ε_{xz}, ε_{yz}, and ε_{zz} vanish. According to this assumption, the rotation degrees of freedom can be expressed in terms of the gradients of the lateral deflection such that

$$\theta_x = \frac{\partial w}{\partial y}, \theta_y = -\frac{\partial w}{\partial x} \tag{5.43}$$

The other strain components are

$$\varepsilon_{xx} = \frac{\partial u}{\partial x} - z\frac{\partial^2 w}{\partial x^2} \tag{5.44}$$

$$\varepsilon_{yy} = \frac{\partial v}{\partial y} - z\frac{\partial^2 w}{\partial y^2} \tag{5.45}$$

$$\varepsilon_{xy} = \frac{1}{2}\left(\frac{\partial u}{\partial y} + \frac{\partial v}{\partial x}\right) - z\frac{\partial^2 w}{\partial x \partial y} \tag{5.46}$$

Within the finite element, the displacement vector is approximated as

$$\mathbf{u} = t\begin{Bmatrix} u \\ v \\ w \\ \theta_x \\ \theta_y \end{Bmatrix} = \mathbf{N}\mathbf{q} \tag{5.47}$$

where $q = \{p_1 \;\; p_2 \;\; p_3 \;\; p_4\}^T$= nodal deflection vector of the four nodes and N represents the appropriate shape function. Using bilinear interpolation functions ϕ_1 for the variation of in-plane displacement components (u and v) and bi-cubic interpolation functions ϕ_2 for the variation of lateral displacement and rotation components (w, θ_x, or θ_y) such that:

$$\phi_1(\zeta,\eta) = \{1 \;\; \zeta \;\; \eta \;\; \zeta\eta\}$$

and

$$\phi_2(\zeta,\eta) = \{1 \;\; \zeta \;\; \eta \;\; \zeta^2 \;\; \zeta\eta \;\; \eta^2 \;\; \zeta^3 \;\; \zeta^2\eta \;\; \zeta\eta^2 \;\; \eta^3 \;\; \zeta^3\eta \;\; \zeta\eta^3\} \tag{5.48}$$

Then, **C** is a 5 × 20 matrix that includes the basis functions for all the five variables, that is:

$$\mathbf{C}(\zeta,\eta) = \begin{bmatrix} \phi_1 & & 0 \\ & \phi_1 & \\ 0 & & \phi_2 \\ & & \dfrac{\partial \phi_2}{\partial y} \\ & & -\dfrac{\partial \phi_2}{\partial x} \end{bmatrix} \tag{5.49}$$

Hence, the shape function **N** is also a 5 × 20 matrix, which can be expressed as:

$$\mathbf{N} = \mathbf{C}(\zeta,\eta)\begin{bmatrix} \mathbf{C}(-a,-b) \\ \mathbf{C}(a,-b) \\ \mathbf{C}(a,b) \\ \mathbf{C}(-a,b) \end{bmatrix}^{-1} \tag{5.50}$$

For the VEM layer, the five degrees of freedom are as follows:

$$\begin{Bmatrix} u_v \\ v_v \\ w_v \\ \theta_{x_v} \\ \theta_{y_v} \end{Bmatrix} = \begin{bmatrix} 1 & 0 & 0 & 0 & (h_p+h_v)/2 \\ 0 & 1 & 0 & -(h_p+h_v)/2 & 0 \\ 0 & 0 & 1 & 0 & 0 \\ 0 & 0 & 0 & 1 & 0 \\ 0 & 0 & 0 & 0 & 1 \end{bmatrix} \begin{Bmatrix} u \\ v \\ w \\ \theta_x \\ \theta_y \end{Bmatrix} = \mathbf{T}\mathbf{N}\mathbf{q} \tag{5.51}$$

where $\mathbf{T}$ is the transformation matrix relating the degrees of freedom of the VEM to those of the base plate.

5.6.2.1 Element Energies

The total kinetic energy T of the composite plate is the summation of the kinetic energies of the plate and VEM, which are denoted by T_p and T_v respectively. T is given by:

$$T = T_p + T_v \tag{5.52}$$

For the ith layer in the eth element:

$$T^e = \frac{1}{2}\rho_i \iiint_V \left[\left(\frac{\partial u}{\partial t}\right)^2 + \left(\frac{\partial v}{\partial t}\right)^2 + \left(\frac{\partial w}{\partial t}\right)^2\right] dV = \frac{1}{2}\dot{\mathbf{q}}^{e^T}\mathbf{M}_i^e\dot{\mathbf{q}}^e. \tag{5.53}$$

Similarly, the total potential energy of the composite plate is the summation of the plate and VEM elastic energies.

$$V = V_p + V_v \tag{5.54}$$

For the ith layer in the eth element:

$$E_i^e = \frac{1}{2}\rho_i \iiint_V \left[\varepsilon_i^{*^T}\sigma_i\right] dV = \frac{1}{2}\mathbf{q}^{e^T}\mathbf{K}_i^e\mathbf{q}^e \tag{5.55}$$

In Eq. (5.55), the stress–strain relationship is given by:

$$\sigma_i = \frac{E_i}{1-\nu_i^2}\begin{bmatrix} 1 & \nu_i & 0 \\ \nu_i & 1 & 0 \\ 0 & 0 & \frac{1-\nu_i}{2} \end{bmatrix}\varepsilon_i \tag{5.56}$$

The equation of motion for plate/*VEM* system can then be reduced to:

$$M\ddot{X} + (K_R + jK_I)X = 0 \tag{5.57}$$

where K_R is the real part of global stiffness matrix and K_I is the imaginary part of global stiffness matrix. Therefore, the nth MDR can be written:

$$\zeta_n = \frac{1}{2}\frac{\phi_n^T K_I \phi_n}{\phi_n^T K_R \phi_n} \tag{5.58}$$

where ϕ_n is the nth eigenvector.

Example 5.5 Consider the three viscoelastic treatments, shown in Figure 5.11, which are manufactured from *Flexane* 80 (ITW Devcon, Danvers, MA). The VEM has a Young's modulus = 3.6 MPa, density = 2300 kg m^{-3}, and Poisson's ratio = 0.49. The treatments are 0.080″ thick and bonded, in unconstrained manner, to an aluminum base structure that is 0.040″ thick.

Determine the strain energy distribution for each of the three VEM/plate assemblies. Assume that the damping treatments are clamped from one side and loaded under axial loading from the opposite side. The remaining two sides of each assembly are maintained free.

Solution

Figure 5.12 displays the finite element models of the considered three damping treatment configurations.

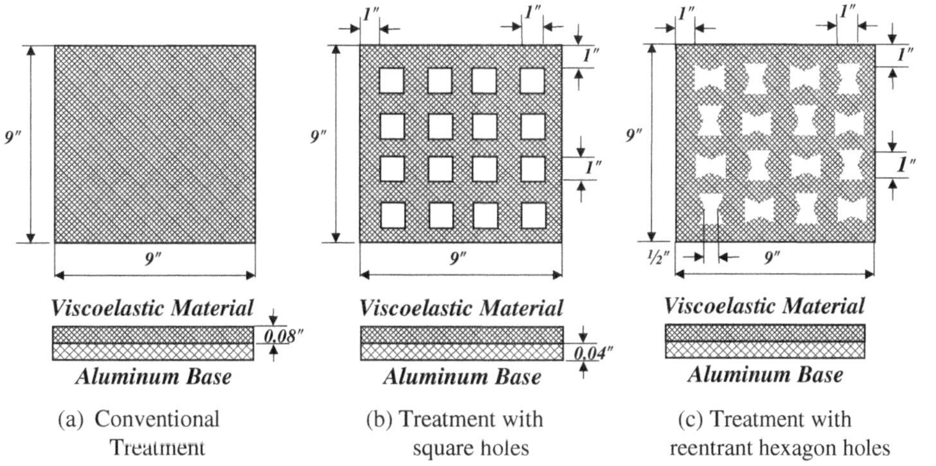

Figure 5.11 Three types of damping treatment. (a) Conventional treatment. (b) Treatment with square holes. (c) Treatment with reentrant hexagonal holes.

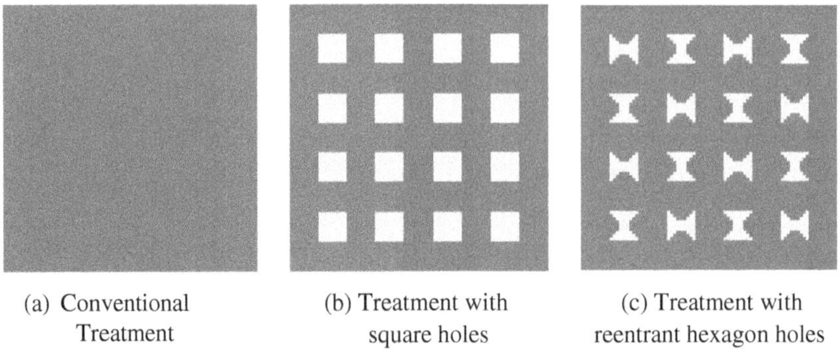

Figure 5.12 Finite element models of the three viscoelastic treatments. (a) Conventional treatment. (b) Treatment with square holes. (c) Treatment with reentrant hexagonal holes.

The results in Table 5.7 indicate that the natural frequencies and MSE in the VEM of the three treatments are nearly the same. However, the MDRs of the conventional treatment are slightly higher than those of treatments with square or reentrant hexagon holes. But, the results reveal that the treatments with square perforations has the highest MSE in VEM (i.e. dissipated energy) per unit volume of VEM followed by the treatments with reentrant hexagon holes and then the conventional treatments came last. The percentage

Table 5.7 Comparison between the natural frequencies, modal damping ratios (MDR), modal strain energy (MSE) in VEM and modal strain energy MSE per unit volume for the different viscoelastic treatments.

	Natural frequencies (Hz)		
Mode number	**Conventional**	**Squares**	**Reentrant hexagons**
1	24	23	23
2	52	50	51
3	114	113	113
4	145	143	143
	MDR (%)		
Mode number	**Conventional**	**Squares**	**Reentrant hexagons**
1	0.006015	0.005551	0.005734
2	0.006070	0.005350	0.005627
3	0.008950	0.007724	0.008188
4	0.009249	0.007556	0.008196

	MSE in VEM (mJ cycle^{-1})		
Mode Number	**Conventional**	**Squares**	**Reentrant hexagons**
1	0.48	0.45	0.46
2	1.03	0.99	1.01
3	2.26	2.22	2.24
4	2.86	2.82	2.83
	MSE in VEM/Treatment Volume (GJ m^{-3})		
Mode Number	**Conventional**	**Squares**	**Reentrant hexagons**
1	4.74	5.54	5.33
2	11.02	12.19	11.71
3	22.32	27.32	25.97
4	28.25	34.71	32.81
Average	**16.58**	**19.94**	**18.96**
%Gain	**0**	**20.26%**	**14.32%**

gains in the dissipated energy relative to the conventional treatments are 20.26 and 14.32% for treatments with square holes and treatments with reentrant hexagon holes, respectively.

5.6.2.2 Topology Optimization of Unconstrained Layer Damping

The energy dissipation of plate/constrained layer damping comes from the shearing deformation in VEM layer, it is reasonable to consider the MDR or modal loss factor of damping structures as the objective function of topology optimization (El-Sabbagh and Baz 2014). Therefore, the objective function is written:

$$f = \sum_{r=1}^{m} \xi_r \tag{5.59}$$

where f denotes the objective function of the optimization problem in present study. m represents the number of the considered MDR. Define the relative density of each VEM element as the design variable vector:

$$\rho = \{\rho_1, \rho_2, \ldots, \rho_N\}^T \tag{5.60}$$

In addition, the constraint is considered to limit the consumption of VEM, the volume fraction is considered. The optimization problem is formulated as follows:

$$\left\langle \begin{array}{l} find: \rho = \{\rho_1, \rho_2, \ldots, \rho_N\}^T \in R \\ to \quad min: f = \sum_{r=1}^{m} \xi_r \\ such \quad that: \begin{cases} \sum_{i=1}^{n} \rho_e - V_0\alpha \le 0 \\ \left(K - \omega_j^2 M\right)\Phi_j = 0 \\ 0 \le \rho_e \le 1, e = 1, 2, \ldots N \end{cases} \end{array} \right\rangle \tag{5.61}$$

where N is the number of elements, Φ_j is the eigenvector, M and K are the global mass and stiffness matrices, and $V/V_0 = \alpha =$ volume fraction of viscoelastic material on the plate.

According to the *Solid Isotropic Material with Penalization* (SIMP) topology optimization method, the element mass and stiffness matrices can be expressed as the product of variables density and the entity element mass and stiffness matrices. The penalty factors p, q; $p, q \ge 1$ are put in to accelerate the convergence of iteration results, that is:

$$M_v(\rho_e) = \rho_e^p M_v^{(e)}, K_v(\rho_e) = \rho_e^q K_v^{(e)} \tag{5.62}$$

where ρ_e is variable density of each VEM element; it is a relative quantity and $0 \le \rho_e \le 1$. Note that if, $\rho_e = 0$ then there is no VEM treatment for this element or equivalently, the thickness of the VEM is equal to zero. Similarly, if $\rho_e = 1$, then the thickness of VEM in this element is equal to the assigned thickness. Also, $M_v^{(e)}$, $K_v^{(e)}$ are the mass and stiffness matrices of VEM element. Keep the base plate and constrained layer unchanged, the global mass, and stiffness matrices can be calculated as follows:

$$M = \sum_{e=1}^{N} \left(M_p^{(e)} + \rho_e^p M_v^{(e)} + M_c^{(e)}\right) \tag{5.63}$$

$$K = \sum_{e=1}^{N} \left(K_p^{(e)} + \rho_e^p \left[K_v^{(e)} + K_{\beta v}^{(e)}\right] + K_c^{(e)}\right) \tag{5.64}$$

where p and q are penalty factors, $p = 1$, $q = 3$. The layout of VEM on the plate can be determined by searching the optimal relative density of each VEM element. In order to solve the presented optimum problem, *Method of Moving Asymptote* (MMA) method is employed.

5.6.2.3 Sensitivity Analysis

According to MSE method, approximate expressions of the ith MDR can be obtained. The equation of motion for plate/VEM system is given:

$$M\ddot{X} + (K_R + iK_I)X = 0 \tag{5.65}$$

where K_R is the real part of global stiffness matrix, K_I is the image part of global stiffness matrix. Therefore, the nth MDR can be written:

$$\zeta_n = \frac{1}{2}\frac{\phi_n^T K_I \phi_n}{\phi_n^T K_R \phi_n} \tag{5.66}$$

where φ_r is the nth eigenvector, the derivatives of Eq. (5.66) to design variables are:

$$\frac{\partial \zeta_n}{\partial \rho_i} = \frac{1}{2}\frac{\left(\phi_n^T \frac{\partial K_I}{\partial \rho_i}\phi_n\right)\left(\phi_n^T K_R \phi_n\right) - \left(\phi_n^T K_I \phi_n\right)\left(\phi_n^T \frac{\partial K_R}{\partial \rho_i}\phi_n\right)}{\left(\phi_n^T K_R \phi_n\right)^2} \tag{5.67}$$

The derivatives of the stiffness matrix are obtained by solving the following sensitivity equations:

$$\frac{\partial K_I}{\partial \rho_i} = imag\sum_{i=1}^{N} q\rho_i^{(q-1)} Kv$$

and

$$\frac{\partial K_R}{\partial \rho_i} = real\sum_{i=1}^{N} q\rho_i^{(q-1)} Kv \tag{5.68}$$

when $p = 1$, $q = 3$, the derivative of MDR can be expressed as:

$$\begin{aligned}\frac{\partial \zeta_n}{\partial \rho_i} &= \frac{1}{2}\frac{\left[\phi_n^T imag(3\rho_i^3 Kv)\phi_n\right]\left[\phi_n^T real(\rho_i^3 Kv + Kp + Kc)\phi_n\right]}{\left[\phi_n^T real(\rho_i^3 Kv + Kp + Kc)\phi_n\right]^2} \\ &\quad - \frac{1}{2}\frac{\left[\phi_n^T imag(\rho_i^3 Kv)\phi_n\right]\left[\phi_n^T real(3\rho_i^3 Kv)\phi_n\right]}{\left[\phi_n^T real(\rho_i^3 Kv + Kp + Kc)\phi_n\right]^2}\end{aligned} \tag{5.69}$$

The dynamic constraint function is

$$f_j = \phi_j^T\left(K - \omega_j^2 M\right)\phi_j \tag{5.70}$$

The derivative of the dynamic constraint function is calculated as follows:

$$\frac{\partial f_j}{\partial \rho_i} = \phi_j^T\left(\frac{\partial K}{\partial \rho_i} - \omega_j^2\frac{\partial M}{\partial \rho_i}\right)\phi_j \tag{5.71}$$

and

$$\frac{\partial M}{\partial \rho_i} = \sum_{i=1}^{n} p\rho_i^{(q-1)} Mv \tag{5.72}$$

When $p = 1$, $q = 3$, the sensitivity of constrained functions become:

$$\frac{\partial f_j}{\partial \rho_i} = \phi_j^T\left(\sum_{i=1}^{N} 3\rho_i^2 Kv - \omega_j^2\sum_{i=1}^{N} Mv\right)\phi_j \tag{5.73}$$

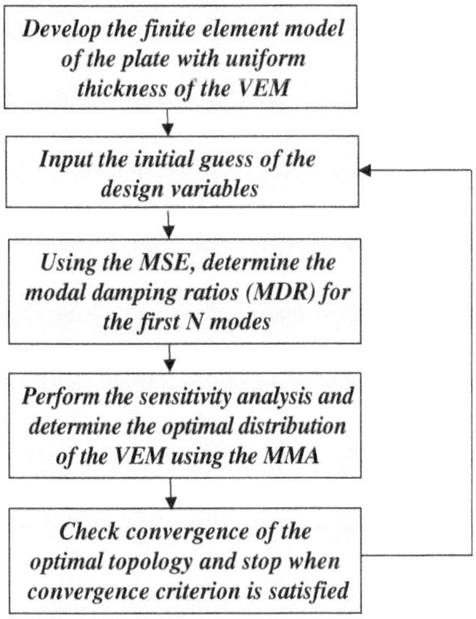

Figure 5.13 Block diagram for the MMA optimization procedures.

Figure 5.13 presents the flow chart of the topology optimization using the MMAs proposed by Svanberg (1987, 2002) to determine the optimal distribution of damping material over structures. It is noted that the computation time is mainly determined by building/solving the finite element model with the complex stiffness matrices and using the MMA approach.

Example 5.6 Consider a plate that is treated initially with a conventional viscoelastic treatment as shown in Figure 5.14. The VEM has a Young's modulus = 3.6 Mpa, density = 2300 kg m^{-3}, and Poisson's ratio = 0.49. The treatments are 0.080″ thick and bonded, in unconstrained manner, to aluminum base structure that is 0.040″ thick. The plate is fixed at its left edge and all its other three edges are free.

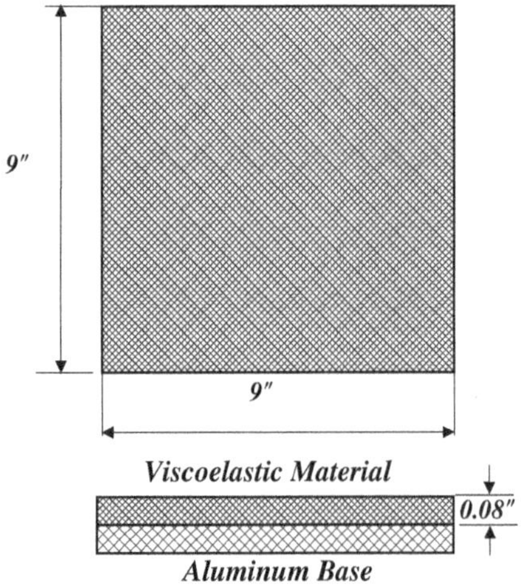

Figure 5.14 Initial configuration of a plate treated with an unconstrained VEM damping treatment.

Determine the optimal topologies of the VEM for volume ratios of 0.25, 0.5, and 0.75 in order to maximize the MDR of the first mode.

Plot also the effect of optimization iteration number on the MDRs of the first four modes.

Solution

A finite element code is developed to describe the dynamics of a plate treated with VEM as outlined in Section 5.6.2. The finite element model is used to extract the strain energy in order serve as a quantitative measure for computing the MDR. Then, the MMA topology optimization method outlined in Sections 5.6.2.2 and 5.6.2.3 is employed to determine the optimal topology of surface damping treatments.

Figures 5.15–5.17 present the optimal topologies of the VEM for volume ratios of 0.25, 0.5, and 0.75 in order to maximize the MDR of the first mode. The figures display also the effect of optimization iteration number on the MDRs of the first four modes.

The figures indicate that the optimum MDRs for the first mode attain their peak values nearly after 2000 iterations. These optimal damping ratios are 0.0002, 0.00026, and 0.00024 for volume ratios of 0.25, 0.5, and 0.75 respectively. These damping ratios as well as those for higher order modes increase with increasing volume ratios. The figures indicate also that the obtained optimal topologies tend to concentrate the VEM near the fixed end of the plate with additional treatments placed at the mid-section of the plate near its free end.

Furthermore, it is observed that the MMA algorithm converges faster as the volume ratio of the VEM is increased.

Example 5.7 Consider a plate that is treated initially with a conventional viscoelastic treatment as shown in Figure 5.14. The VEM has a Young's modulus = 3.6 Mpa, density = 2300 kg m^{-3}, and Poisson's ratio = 0.49. The treatments are 0.080″ thick and

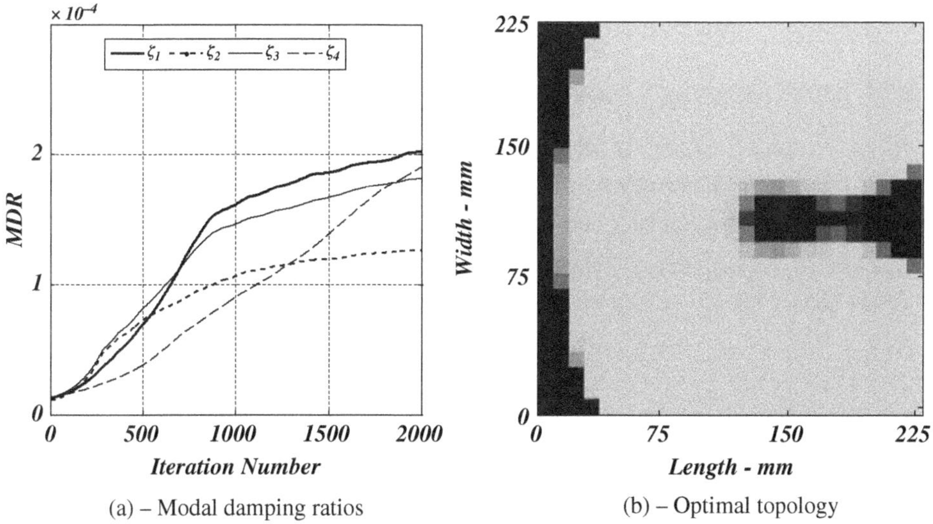

Figure 5.15 Optimal topology of a VEM treatment of a cantilevered plate with volume ratio of 0.25. (a) Modal damping ratios. (b) Optimal topology.

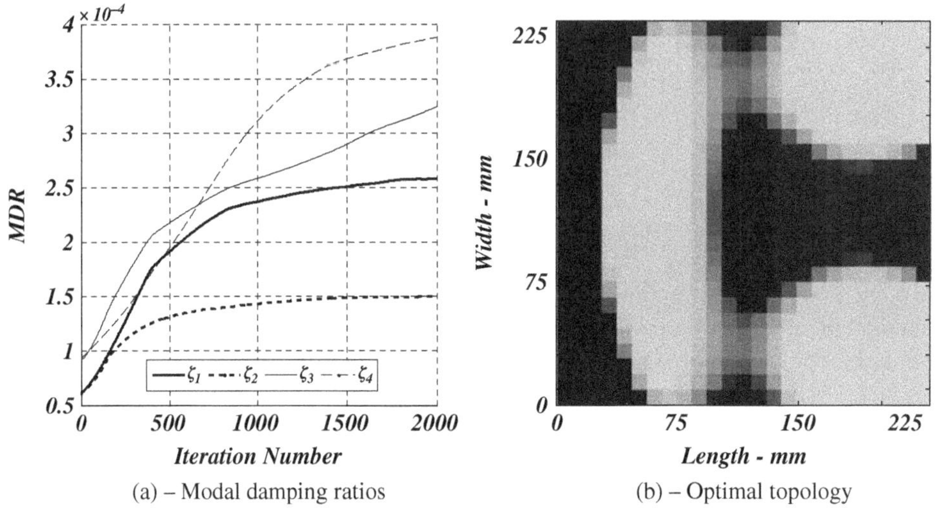

Figure 5.16 Optimal topology of a VEM treatment of a cantilevered plate with volume ratio of 0.5. (a) Modal damping ratios. (b) Optimal topology.

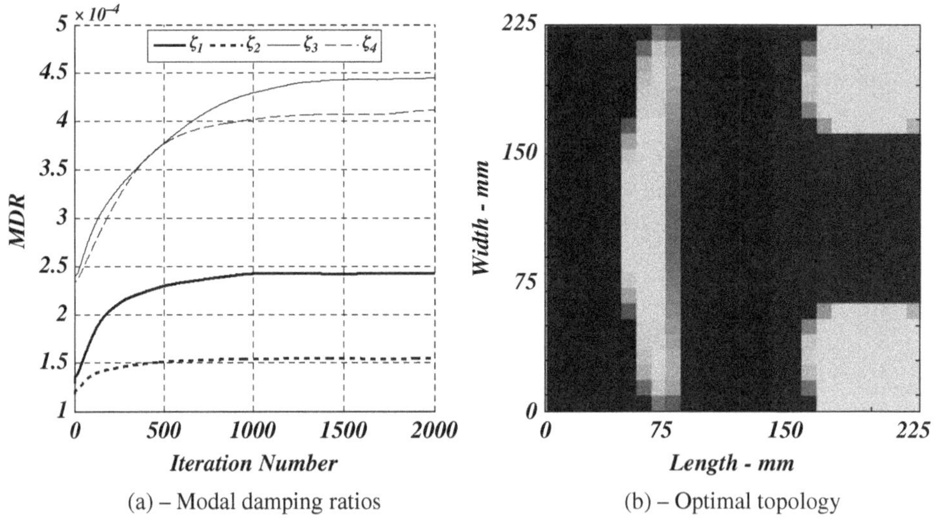

Figure 5.17 Optimal topology of a VEM treatment of a cantilevered plate with volume ratio of 0.75. (a) Modal damping ratios. (b) Optimal topology.

bonded, in an unconstrained manner, to an aluminum base structure that is 0.040″ thick. The plate is simply supported at all its edges.

Determine the optimal topologies of the VEM for volume ratios of 0.25, 0.5, and 0.75 in order to maximize the MDR of the first mode.

Also plot the effect of optimization iteration number on the MDRs of the first four modes.

Solution

The approach adopted in solving Problem 5.6 is used. However, the simply supported boundary condition is imposed on the finite element model. Solving the topology optimization problem of the VEM, for volume ratios of 0.25, 0.5, and 0.75, yields the optimal topologies shown in Figures 5.18–5.20, respectively. The figures display also the effect of optimization iteration number on the MDRs of the first four modes.

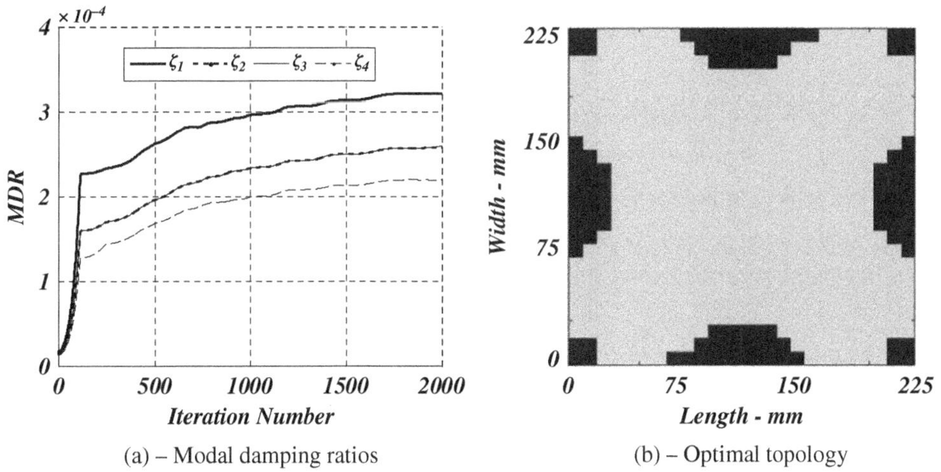

Figure 5.18 Optimal topology of a VEM treatment of a simply supported plate with volume ratio of 0.25. (a) Modal damping ratios. (b) Optimal topology.

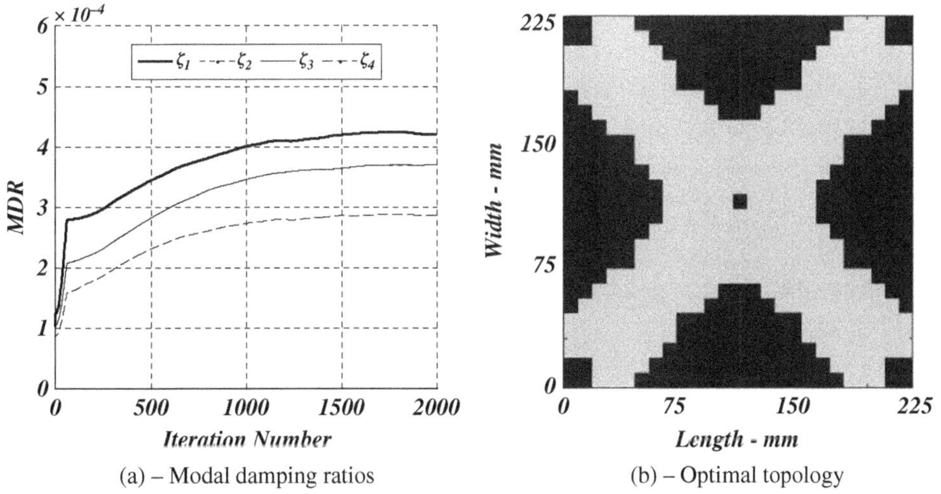

Figure 5.19 Optimal topology of a VEM treatment of a simply supported plate with volume ratio of 0.50. (a) Modal damping ratios. (b) Optimal topology.

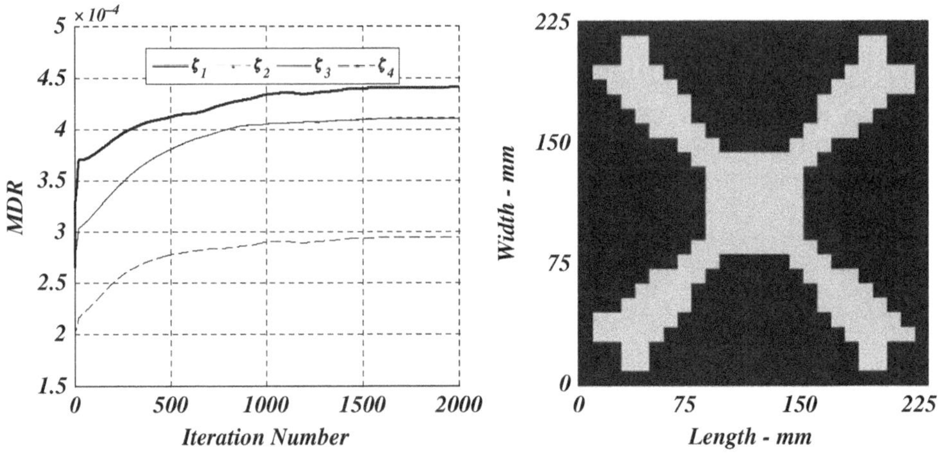

(a) – Modal damping ratios (b) – Optimal topology

Figure 5.20 Optimal topology of a VEM treatment of a simply supported plate with volume ratio of 0.75. (a) Modal damping ratios. (b) Optimal topology.

The figures indicate that the optimum MDRs for the first mode attain their peak values nearly after 2000 iterations. These optimal damping ratios are 0.00032, 0.00042, and 0.00044 for volume ratios of 0.25, 0.5, and 0.75 respectively. These damping ratios as well as those for higher order modes increase with increasing volume ratios. The figures indicate also that the obtained optimal topologies tend to concentrate the VEM near the fixed end of the plate with additional treatments placed at the mid-section of the plate near its free end.

5.7 Summary

This chapter has presented the MSE method with its original and modified forms. The theories behind the MSE method and its modifications are presented. Numerical examples are given to quantify the accuracy of the different method in predicting the modal parameters of rods treated with VEM.

The MSE has also been shown to serve as an important design metric for selecting the optimal design parameters and/or determining the optimal topologies of damping treatments based on rational design objectives.

References

Curà, F., Mura, A., and Scarpa, F. (2011). Modal strain energy based methods for the analysis of complex patterned free layer damped plates. *Journal of Vibration and Control* 18 (9): 1291–1302.

El-Sabbagh, A. and Baz, A. (2014). Topology optimization of unconstrained damping treatments for plates. *Engineering Optimization* 46 (9): 1153–1168.

Golla, D.F. and Hughes, P.C. (1985). Dynamics of viscolelastic structures – a time domain finite element formulation. *ASME Journal of Applied Mechanics* 52: 897–600.

Hu, B.-G., Dokainish, M.A., and Mansour, W.M. (1995). Modified MSE method for viscoelastic systems: a weighted stiffness matrix approach. *Journal of Vibration and Acoustics, Transactions of the ASME* 117 (2): 226–231.

Johnson, C.D. and Kienholz, D.A. (1982). Finite element prediction of damping in structures with constrained viscoelastic layers. *AIAA Journal* 20: 1284–1290.

Kerwin, E.M. and Ungar, E.E. (1962). Loss factors of viscoelastic systems in terms of energy concepts. *Journal of Acoustical Society of America* 34: 954–957.

Lepoittevin, G. and Kress, G. (2009). Optimization of segmented constrained layer damping with mathematical programming using strain energy analysis and modal data. *Materials & Design* 31 (1): 14–24.

Ling, Z., Ronglu, X., Yi, W., and El-Sabbagh, A. (2010). Topology optimization of constrained layer damping on plates using method of moving asymptote (MMA) approach. *Shock and Vibration* doi: 10.3233/SAV-2010-0583.

O'Callahan J., "A procedure for an improved reduced system", Proceedings of the International Modal Analysis Conference (IMAC), pp. 17–21, 1989.

Ro, J. and Baz, A. (2002). Optimal placement and control of active constrained layer damping using modal strain energy approach. *Journal of Vibration and Control* 8 (8): 861–876.

Sainsbury, M.G. and Masti, R.S. (2007). Vibration damping of cylindrical shells using strain-energy-based distribution of an add-on viscoelastic treatment. *Finite Elements in Analysis and Design* 43 (3): 175–192.

Scarpa F., Landi F. P., Rongong J. A., DeWitt L., and Tomlinson G., "Improving the MSE method for viscoelastic damped structures", Proceedings of SPIE – The International Society for Optical Engineering, Vol. 4697, pp. 25–34, 2002.

Svanberg, K. (1987). The method of moving asymptotes: a new method for structural optimization. *International Journal for Numerical Methods in Engineering* 24: 359–373.

Svanberg, K. (2002). A class of globally convergent optimization methods based on conservative convex separable approximations. *SIAM Journal on Optimization* 12 (2): 555–573.

Xu Y. and Chen D., "Finite element modeling for the flexural vibration of damped sandwich beams considering complex modulus of the adhesive layer", Proceedings of SPIE – Damping and Isolation, Vol. 3989, pp. 121–129, 2000.

Xu Y., Liu Y., and Wang B., "Revised modal strain energy method for finite element analysis of viscoelastic damping treated structures", Proceedings of SPIE – The International Society for Optical Engineering, Vol. 4697, pp. 35–42, 2002.

Problems

5.1 Consider the fixed-free rods shown in Figure P5.1. Determine the natural frequencies and the damping ratios of the two rod/VEM systems using the MSE method and its four modified versions discussed in Section 5.3.

Assume the dimensions and the material properties of the two systems are as given in Example 4.3.

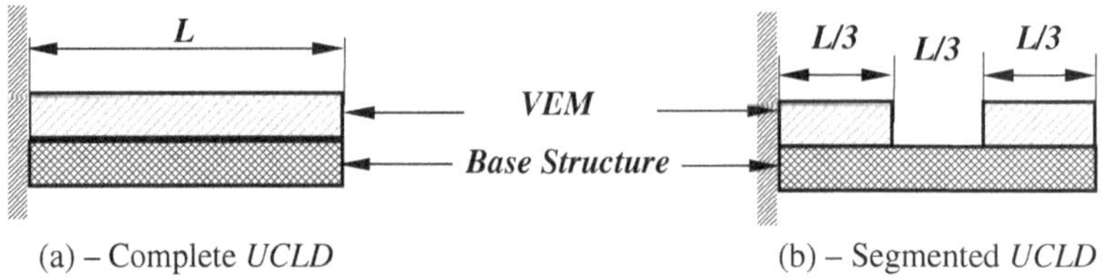

Figure P5.1 Complete and segmented UCLD treatments. (a) Complete UCLD. (b) Segmented UCLD.

Assume also that the two systems are modeled using three finite elements.

5.2 Consider the fixed-free rods shown in Figure P5.2. Determine the natural frequencies and the damping ratios of the two rod/VEM systems using the MSE method and its four modified versions discussed in Section 5.3.

Assume the dimensions and the material properties of the two systems are as given in Example 4.4.

Assume also that the two systems are modeled using three finite elements.

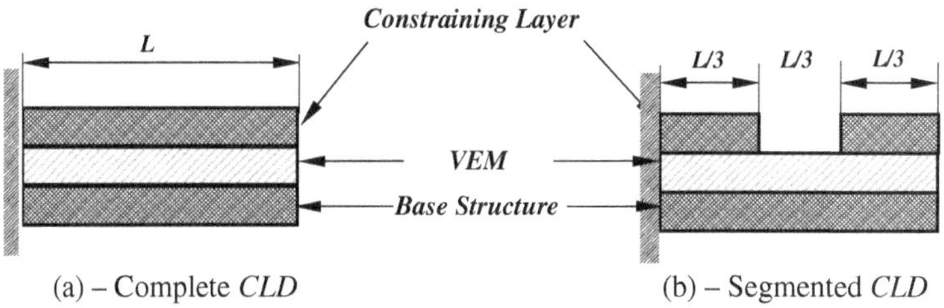

Figure P5.2 Complete and segmented CLD treatments. (a) Complete CLD. (b) Segmented CLD.

5.3 Consider the dynamical system shown in Figure P5.3a. The mass m is supported on two springs. One of these springs is an elastic spring with real stiffness k_1 while the third spring is viscoelastic with complex stiffness $k_2^* = k_2(1 + \eta i)$ where η is the loss factor of the VEM. Show that:

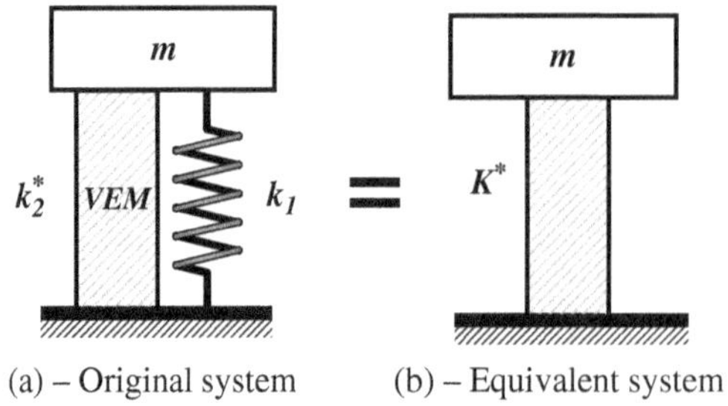

Figure P5.3 Mass supported on a VEM and a spring. (a) Original system. (b) Equivalent system.

a) the equivalent system shown in Figure P5.3b has:

$$K^* = K_R(1 + \eta_s i)$$

where

$$\eta_s = \eta \frac{k_2}{k_1 + k_2}$$

b) the loss factor η_{MSE} of this system, as calculated by using the MSE concept, is given by:

$$\eta_{MSE} = \eta \frac{k_2}{k_1 + k_2}$$

5.4 Consider the dynamical system shown in Figure P5.4a. The mass m is supported on three springs. Two are elastic springs with real stiffness k_1 and k_3, respectively, while the third spring is viscoelastic with complex stiffness $k_2^* = k_2(1 + \eta i)$ where η is the loss factor of the VEM. Show:

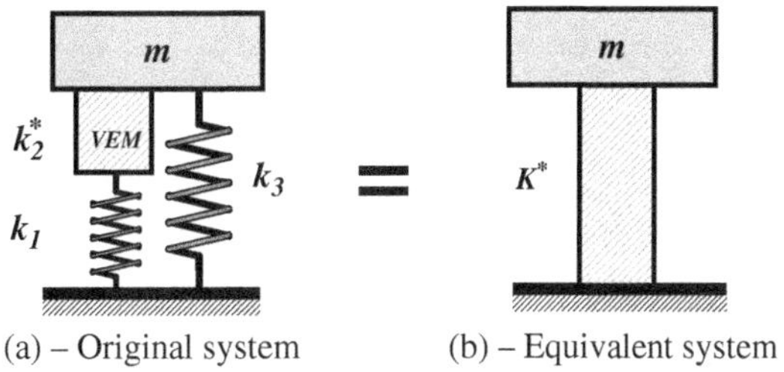

Figure P5.4 Mass supported on a VEM and two springs (a) Original system. (b) Equivalent system.

a) that this system is equivalent to the system shown in Figure P5.4b such that:

where $\eta_s = \dfrac{\eta}{[1 + K_{21}(1 + \eta^2)] + \frac{1}{K_{23}}\left[(1 + K_{21})^2 + (\eta K_{21})^2\right]}$, with $K_{21} = k_2/k_1$ and $K_{23} = k_2/k_3$.

b) that the loss factor η_{MSE} of this system as calculated by using the MSE concept is given by:

$$\eta_{MSE} = \frac{\eta}{1 + \frac{k_2}{k_1} + \frac{k_3}{k_2}\left(1 + \frac{k_2}{k_1}\right)^2}$$

c) that the error between the exact loss factor η_s and that η_{MSE} estimated by MSE is:

$$\frac{\Delta\eta_s}{\eta_s} = \frac{\eta_{MSE} - \eta_s}{\eta_s} = \eta^2 \left[\frac{k_2}{k_1} + \frac{k_3}{k_2}\left(\frac{k_2}{k_1}\right)^2\right] \Big/ \left[1 + \frac{k_2}{k_1} + \frac{k_3}{k_2}\left(1 + \frac{k_2}{k_1}\right)^2\right] > 0$$

That is, MSE over estimates the exact loss factor.

5.5 Consider the dynamical system shown in Figure P5.5. The vibration of the primary system (m_1–k_1) is controlled by the secondary system (i.e., a dynamic damper system). Assume that $m_1 = m_2 = 1kg$, $k_1 = 1\ \mathrm{N\,m^{-1}}$, and $k_2^* = (1 + i)$, then:

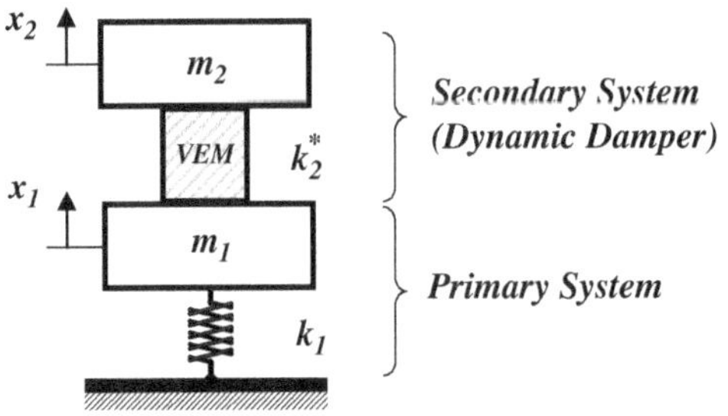

Figure P5.5 Primary system with a dynamic damper.

a) Derive the equations of motion of the system and cast in the following form:

$$[M]\{\ddot{X}\} + [[K_R] + i[K_I]]\{X\} = 0$$

where $[M]$ = mass matrix, $[K_R]$ = real stiffness matrix, $[K_I]$ = imaginary stiffness matrix, and

b) Determine the MDRs of the two modes of vibration of the system using the *Original Modal Strain Method* (MSE).

5.6 Consider the dynamical system shown in Figure P5.6. The mass m is supported on four springs. Three of these springs is an elastic spring with real stiffness k_1 while the fourth spring is viscoelastic with complex stiffness $k_2^* = k_2(1 + \eta i)$ where η is the loss factor of the VEM. Determine the loss factor of the system using:

a) the equivalent system shown in Figure P5.6 such that $K^* = K_R(1 + \eta_s i)$
b) the *MSE* concept such that the loss factor = η_{MSE}
c) Compare the results obtained in *a* and *b*.

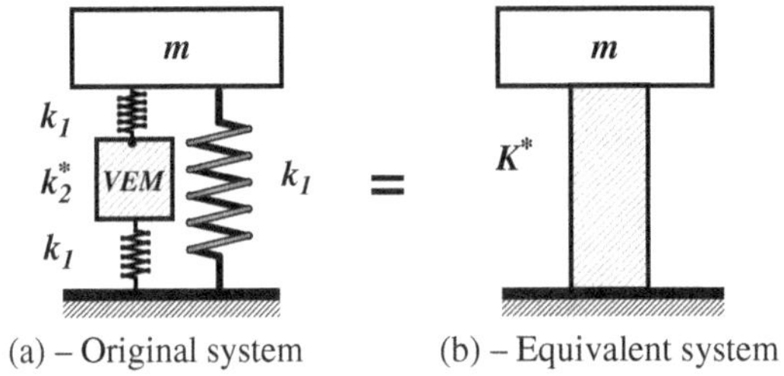

Figure P5.6 Mass supported on a VEM and three springs (a) Original system. (b) Equivalent system.

5.7 Consider the fixed-free rod/VEM system shown in Figure P5.7. The rod is made of aluminum with width of 0.025 m, thickness of 0.025 m, and length of 1 m. The VEM has a width of 0.025 m, thickness = t_v m, and density 1100 kg m^{-3}. The

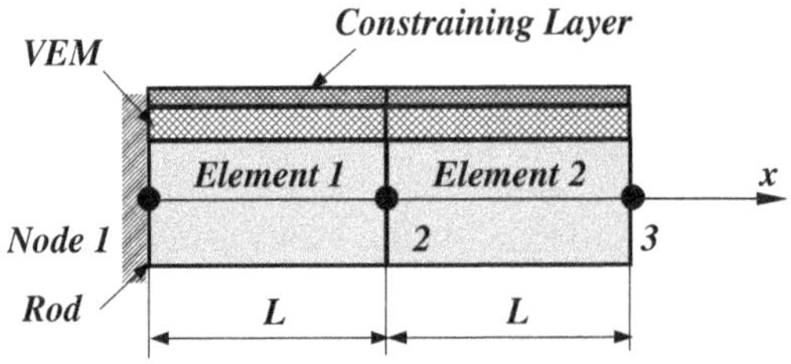

Figure P5.7 A fixed-free rod treated with a constrained VEM layer.

VEM is constrained by an aluminum constraining layer that is 0.025 m wide and 0.0025 m thick. Using the GHM modeling approach of the VEM with one mini-oscillator ($E_0 = 15.3MPa, \alpha_1 = 39, \zeta_1 = 1,\ \omega_1 = 19,058 rad/s$), determine:

a) the optimum thickness t_v of the VEM that maximizes the sum of the MDRs for the two modes of vibration the end/VEM (use the MSE as a design criterion).
b) the optimum thickness t_v of the VEM that maximizes the sum of the MDRs for the two modes of vibration the end/VEM per unit weight of the VEM treatment (use the MSE as a design criterion).

Assume that the system is modeled by a two element finite element model and that t_v is constrained such that: $0.005 \le t_v \le 0.05\ m$.

Comment on the optimum results obtained in items a and b.

5.8 Consider the cantilever beam/passive constrained layer damping (PCLD) system shown as Figure P5.8. The main physical and geometrical parameters of the base beam, the viscoelastic layer and constrained layer are listed in Table P5.1. The beam is made of aluminum and is mounted in a cantilever configuration. The beam is treated with segmented arrangements of viscoelastic material that are constrained by an aluminum layer as shown in Figure P5.8 (Curà et al. 2011).

Determine the sum of the MDRs of the beam/stripped VEM assembly for the first four modes of vibration.

Determine such sum for three arrangements shown in Figure P5.8.

Table P5.1 Physical and geometrical parameters of a plate/PCLD system.

Layer	Length (m)	Width (m)	Thickness (mm)	Density (kg m^{-3})	Modulus (MPa)
Base Plate	0.32	0.125	0.50	2700	7100[a]
Viscoelastic	stripped	0.125	0.50	1140	20[b]
Constraining Layer	stripped	0.125	0.25	2700	7100[a]

[a] Young's modulus.
[b] Shear modulus, $\eta = 0.5$.

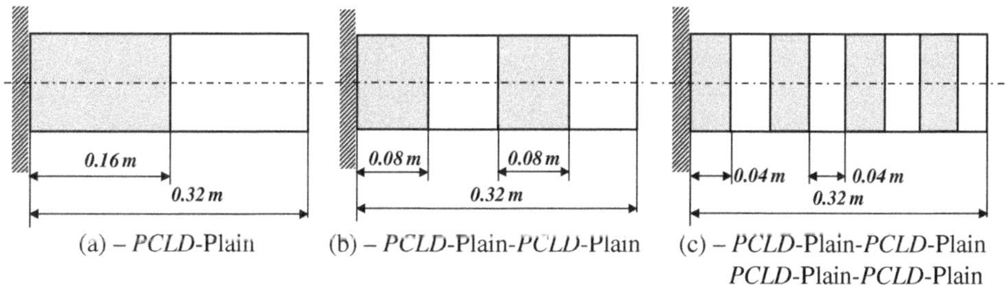

Figure P5.8 Cantilever beam treated with segmented PCLD treatments (a) PCLD-Plain. (b) PCLD-Plain-PCLD-Plain. (c) PCLD-Plain-PCLD-Plain PCLD-Plain-PCLD-Plain.

5.9 Consider the cantilever plate/PCLD system shown as Figure P5.9. The main physical and geometrical parameters of the base plate, the viscoelastic layer and constrained layer are listed in Table P5.2. The plate is made of aluminum and is mounted in a cantilever configuration. The plate is treated with viscoelastic material, which is constrained by a continuous Polyvinylidene Fluoride (PVDF) piezoelectric layer as shown in Figure P5.9.

Using the topology optimization approach, determine the optimal distribution of the VEM over the plate surface in order to maximize the sum of the MDRs of the plate/VEM assembly for the first four modes of vibration.

Determine such optimal topologies for volume fractions of the VEM of 0.2, 0.4, 0.6, and 0.8 (Ling et al. 2010).

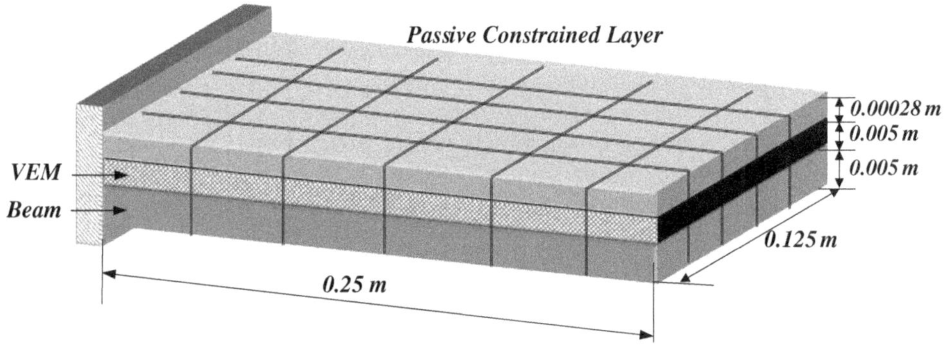

(a) – Plate with *PCLD* treatment

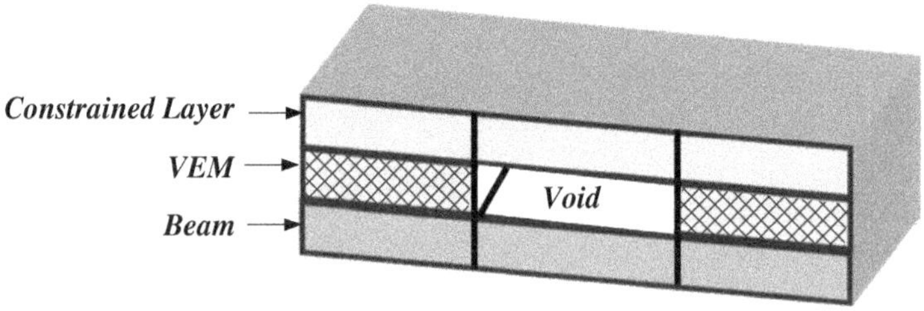

(b) – *VEM* treatment with optimally placed voids

Figure P5.9 A cantilever plate treated with PCLD treatments (a) Plate with PCLD treatment. (b) VEM treatment with optimally placed voids.

5.10 Using the topology optimization approach, determine the optimal distribution of the VEM over the plate surface of Problem 5.9 when the plate is anchored in a simply-supported configuration as shown in Figure P5.10. The topology optimization aims at maximizing the sum of the MDRs of the plate/VEM assembly for the first four modes of vibration.

Table P5.2 Physical and geometrical parameters of a plate/PCLD system.

Layer	Length (m)	Width (m)	Thickness (mm)	Density (kg m^{-3})	Modulus (MPa)
Base Plate	0.25	0.125	0.5	2700	7,100[a]
Viscoelastic	0.25	0.125	0.5	1140	20[b]
PVDF	0.25	0.125	0.028	1800	2,250[a]

[a] Young's modulus.
[b] Shear modulus, $\eta = 0.5$.

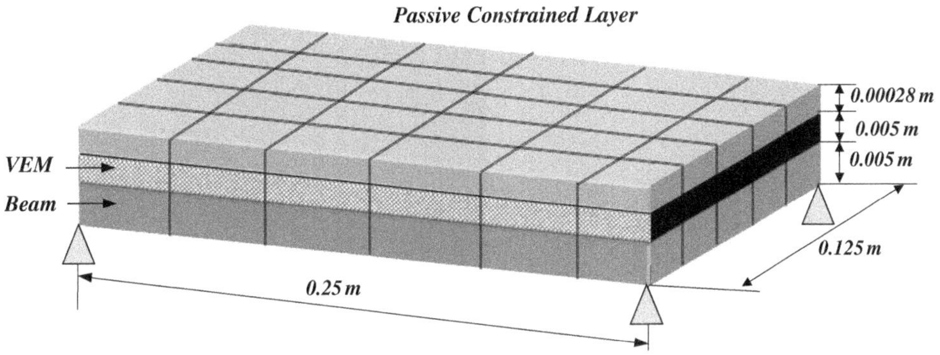

Figure P5.10 A simply supported plate treated with PCLD treatment.

Determine such optimal topologies for volume fractions of the VEM of 0.2, 0.4, 0.6, and 0.8.

6

Energy Dissipation in Damping Treatments

6.1 Introduction

The energy dissipation characteristics of various types of viscoelastic damping treatments are presented for rods, beams, and plates. Passive damping treatments in constrained and unconstrained configurations are considered. Also, Active Constrained Layer Damping (ACLD) treatments, consisting of viscoelastic cores constrained by active piezoelectric layers, are presented as effective means for enhancing the damping characteristics and compensating for the performance degradation of passive treatments.

6.2 Passive Damping Treatments of Rods

The energy dissipation characteristics of structures treated with passive damping treatments in constrained and unconstrained configurations are considered in this section.

6.2.1 Passive Constrained Layer Damping

Figures 6.1 and 6.2 show drawings of the passive constrained layer damping (PCLD) treatment. It is assumed that the thicknesses of the constraining and viscoelastic layers are very small compared to that of the base structure. Hence, the bending effects are negligible, the constraining layer is subjected to longitudinal strains only, and the viscoelastic core is subjected to shear only. It is also assumed that the longitudinal stresses in the viscoelastic core are negligible. Furthermore, the constraining layer is assumed to be elastic and dissipates no energy, whereas the core is assumed to be linearly viscoelastic. In addition, the base structure is subjected to an axial strain ε_0 that is assumed to be spatially uniform over the interface between the base structure and the viscoelastic layer. The strain ε_0 is also assumed to be temporally varying in a sinusoidal fashion at a frequency ω due to cyclic vibration of the base structure.

6.2.1.1 Equation of Motion

From the geometry of Figure 6.2, the shear strain γ in the viscoelastic core is given by:

$$\gamma = (u - u_o)/h_1 \tag{6.1}$$

where u and u_o are the longitudinal deflections of the constraining layer and base structure, respectively. Also, h_1 denotes the thickness of the viscoelastic layer.

Active and Passive Vibration Damping, First Edition. Amr M. Baz.

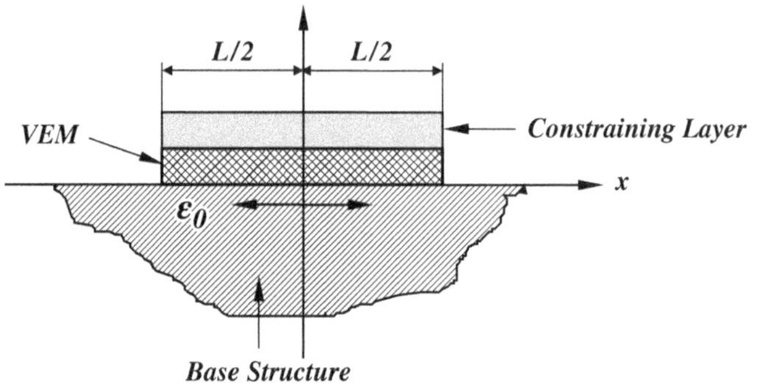

Figure 6.1 Structure/constrained VEM assembly.

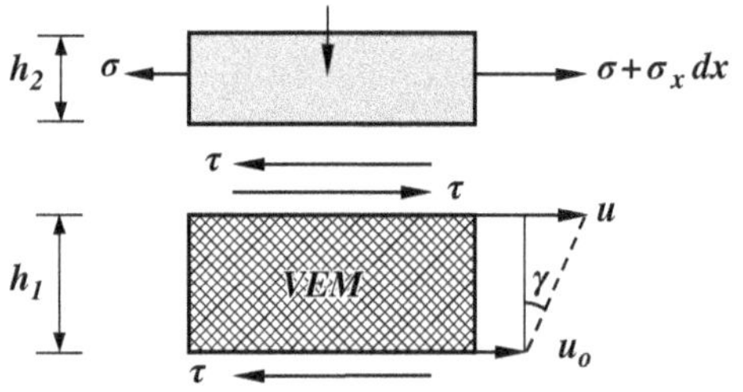

Figure 6.2 Free body diagram of constrained VEM.

The potential energy PE associated with the extension of the constraining layer and the shearing of the viscoelastic layer are:

$$PE = \frac{1}{2} E_2 h_2 b \int_{-L/2}^{L/2} u_{,x}^2 \, dx + \frac{1}{2} G' h_1 b \int_{-L/2}^{L/2} \gamma^2 \, dx \tag{6.2}$$

where E_2, h_2, and b denote Young's modulus, thickness, and width of the constraining layer, respectively. Also, subscript x denotes partial differentiation with respect to x. In Eq. (6.1), it is assumed that the behavior of the viscoelastic layer is linear and described in terms of the complex modulus $G^* = G'(1 + i\eta_g)$ with G', η_g, and i denoting the storage shear modulus, loss factor, and $i = \sqrt{-1}$.

The kinetic energy KE associated with the longitudinal deflection u is given by:

$$KE = \frac{1}{2} mb \int_{-L/2}^{L/2} \dot{u}^2 dx \tag{6.3}$$

where m is the mass/unit width and unit length of the constraining layer. In Eq. (6.3), the rotary inertia of the viscoelastic layer is neglected and also the inertia of the base structure is not considered.

The work W_d dissipated in the viscoelastic core is given by:

$$W_d = -h_1 b \int_{-L/2}^{L/2} \tau_d \gamma \, dx \tag{6.4}$$

where τ_d is the dissipative shear stress developed by the viscoelastic core. It is given by:

$$\tau_d = \left(G'\eta_g/\omega\right)\dot{\gamma} = \left(G'\eta_g\right)\gamma i \tag{6.5}$$

where ω denotes the excitation frequency of the base structure. In Eq. (6.5), $(G'\eta_g/\omega)$ quantifies the equivalent viscous damping of the ViscoElastic Material (VEM) (Nashif et al. 1985).

The equations and boundary conditions governing the operation of the PCLD system are obtained by applying Hamilton's principle (Meirovitch 1967):

$$\int_{t_1}^{t_2} \delta(KE - PE)dt + \int_{t_1}^{t_2} \delta(W_d)\, dt = 0 \tag{6.6}$$

where $\delta(.)$ denotes the first variation in the quantity inside the parentheses. Also, t denotes time with t_1 and t_2 defining the integration time limits.

From Eqs. (6.1) through (6.6), we have

$$\begin{aligned} &\int_{t_1}^{t_2} \left[-mb \int_{-L/2}^{L/2} \ddot{u}\delta u \, dx - E_2 h_2 b [u_{,x}\delta u]_{-L/2}^{L/2} + E_2 h_2 b \int_{-L/2}^{L/2} u_{,xx}\delta u dx \right] dt \\ &- \int_{t_1}^{t_2} \left[G'b/h_1 \int_{-L/2}^{L/2} (u-u_0)\delta u \, dx + G'b/h_1 \eta_g i \int_{-L/2}^{L/2} (u-u_0)\delta u dx \right] dt = 0 \end{aligned} \tag{6.7}$$

Hence, the resulting equation of the Constrained Layer Damping (CLD) system is:

$$mh_1/G^* \ddot{u} = B^{*^2} u_{,xx} - (u - u_0) \tag{6.8}$$

with the following boundary conditions:

$$u_x = 0 \, at \, x = \pm L/2 \tag{6.9}$$

where $B^* = \sqrt{h_1 h_2 E_2 / G^*}$ is a characteristic complex length of the passive treatment. It is important here to note that the second order partial differential Eq. (6.8) describing the PCLD system is the same as that obtained by Plunkett and Lee (1970) if the inertia of the constraining layer is set to zero.

Neglecting the inertia term in Eq. (6.8) gives the following quasi-static equilibrium equation:

$$B^{*2} u_{,xx} - u = -u_0 \quad \text{or} \quad B^{*^2} u_{,xx} - u = -\varepsilon_0 x \tag{6.10}$$

which is subjected to the boundary conditions given by Eq. (6.9). This equation has the following general solution:

$$u = a_1 e^{-x/B^*} + a_2 e^{x/B^*} + \varepsilon_0 x$$

where a_1 and a_2 can be determined from the boundary conditions giving:

$$a_1 = -a_2 = \frac{1}{2}\varepsilon_0 B^* / \cosh\left(\frac{L}{2B^*}\right)$$

Hence, u and γ are given by

$$u = \varepsilon_0 \left[x - B^* \sinh\left(\frac{x}{B^*}\right) / \cosh\left(\frac{L}{2B^*}\right)\right]$$

and

$$\gamma = -\frac{\varepsilon_0 B^*}{h_1} \sinh\left(\frac{x}{B^*}\right) / \cosh\left(\frac{L}{2B^*}\right) \tag{6.11}$$

Equation (6.11) indicates that when $x = 0$, that is, at the middle of the VEM, both the deflection u and the shear strain γ is zero. Hence, at and near this location; the VEM is ineffective in energy dissipation. Most of the energy dissipation occurs near the edges of the VEM where the shear strain is a maximum.

Example 6.1 Determine the normalized shear strain $(\gamma h_1/\varepsilon_0 L)$ distribution over the length of the VEM using:

a) The closed-form Eq. (6.11).
b) A numerical MATLAB solution of Eq. (6.10) subject to the boundary conditions expressed by Eq. (6.9).

Assume that $L/B_0 = 3.28$ where $B_0 = \sqrt{h_1 h_2 E_2 / G}$ and G is given by:

$$G^* = G(\cos\theta + i\sin\theta) = G\cos\theta[1 + i\tan\theta] = G'\left[1 + i\eta_g\right]$$

Note that $\tan(\theta) = \eta_g$ = loss factor of *VEM* and $B^* = B_0[\cos(\theta/2) - i\sin(\theta/2)]$. Assume that VEM loss factor $\eta_g = 1$.

Solution

Figure 6.3 displays a comparison between the shear strain distributions as obtained by the closed-form Eq. (6.11) and by MATLAB "*bvp4c*" command that require the definition of the following two functions:

```
% ******************************
function dydx=lee1(x,y)
w=3.28;
dydx=[y(2);w^2*(y(1)-x)];
% ******************************
function res=lee2(ya,yb)
res= [ya(2); yb(2)];
% ******************************
```

➡

```
% ********PCLD-BVP***********
% ******************************
xint=linspace(-0.5,0.5,101);
solinit=bvpinit(xint,[1 0]);
sol=bvp4c(@lee1,@lee2,solinit)
yint=deval(sol,xint);
strain=abs(yint(1,:)-xint);
% ******************************
```

The first function defines the differential Eq. (6.10) as a set of first order differential equations and the second defines the boundary conditions.

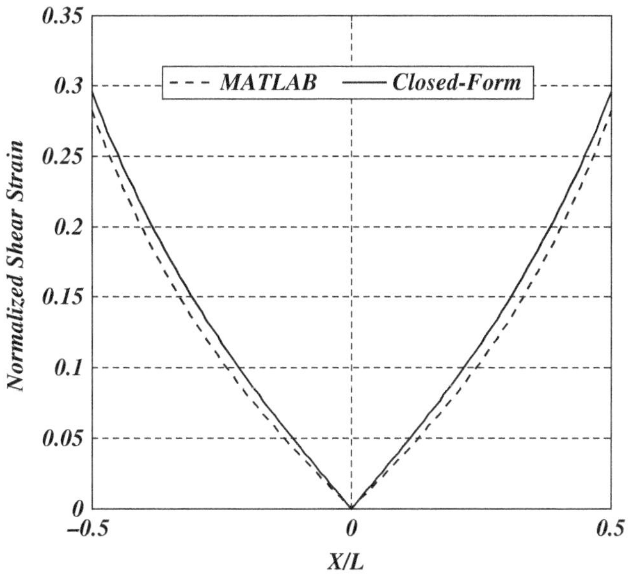

Figure 6.3 Comparison between the shear strain distributions obtained by closed-form Eq. (6.11) and by MATLAB *"bvp4c"* command.

Note the close agreement between the exact and MATLAB methods in predicting the strain distribution. It is important to note here that the shear strain is highest near the edges of the VEM/constraining layer and vanishes at the middle of the treatment. This observation has significant implications in the design of effective damping treatment where the energy dissipation can be maintained high while reducing the weight of the treatment. For example, cutting a hole at the center of the treatment can reduce the weight of the VEM while not significantly affecting the energy dissipation. Alternatively, a functionally graded VEM with its shear modulus and/or loss factor varying along the length of the treatment can result in higher energy dissipation, even for perforated damping treatments.

6.2.1.2 Energy Dissipation

The energy dissipation in the VEM/unit length/cycle is given by

$$\Delta W_p = \pi b h_1 G'' \int_{-L/2}^{L/2} |\gamma|^2 dx \tag{6.12}$$

Substituting Eq. (6.11) into Eq. (6.12) gives:

$$\Delta W_p = \pi b h_1 G'' \int_{-L/2}^{L/2} \left(\frac{\varepsilon_0 B_0}{h_1}\right)^2 \sinh^2\left(\frac{x}{B^*}\right)/\cosh^2\left(\frac{L}{2B^*}\right) dx \tag{6.13}$$

Closed-form integration of Eq. (6.13) requires the use of the complex variable identities given in Appendix 6.A. This gives

$$\Delta W_p = 2\pi\varepsilon_0^2 E_2 h_2 Lb/\omega^* \frac{[\sinh[\omega^* \cos(\theta/2)]\sin(\theta/2) - \sin[\omega^* \sin(\theta/2)]\cos(\theta/2)]}{[\cosh[\omega^* \cos(\theta/2)] + \cos[\omega^* \sin(\theta/2)]]} \tag{6.14}$$

where $\theta = \tan^{-1}(\eta_g)$ with η_g= loss factor of *VEM*.

Now, let the nominal energy W_n be as defined by Plunkett and Lee (1970) for the PCLD treatments:

$$W_n = \frac{1}{2}\varepsilon_0^2 E_2 h_2 Lb \tag{6.15}$$

The nominal energy W_n denotes the maximum strain energy of the constraining layer if the entire layer is strained by ε_0. Then, Eq. (6.14) and (6.15) can be normalized with respect to the nominal energy W_n to give the equivalent loss coefficient η_p of the PCLD that quantifies the energy by the treatment as follows:

$$\eta_p = 4\pi/\omega^* \frac{[\sinh[\omega^* \cos(\theta/2)]\sin(\theta/2) - \sin[\omega^* \sin(\theta/2)]\cos(\theta/2)]}{[\cosh[\omega^* \cos(\theta/2)] + \cos[\omega^* \sin(\theta/2)]]} \tag{6.16}$$

where $\omega^* = L/B_0$.

Example 6.2 Calculate the equivalent coefficient η_p of the PCLD treatment for different values of the dimensionless length $\omega^* = L/B_0$ and loss factor η_g of the VEM. Determine also the optimal values of the dimensionless length $\omega^* = L/B_0$ that maximize the equivalent coefficient η_p for different values of the loss factor η_g of the VEM.

Solution

Figure 6.4 shows the effect of the dimensionless length $\omega^* = L/B_0$ and the loss factor η_g of the *VEM* on the equivalent loss coefficient η_p as calculated from Eq. (6.16). Note that there is an optimal length of the PCLD treatment at which the equivalent loss coefficient η_p attains a maximum value. The value of the optimal dimensionless length is found to be nearly independent of the loss factor η_g of the viscoelastic layer and is equal approximately to 3.28.

The results displayed in Figure 6.4 have other important practical implications as far as the application PCLD treatments to large structures. For such structures, complete PCLD treatments, as shown in Figure 6.5a, will be ineffective as the equivalent loss coefficient η_p will be low. For more effective damping, the PCLD treatments should be "segmented," as shown in Figure 6.5b, with the length of each segment approximately equal to the optimal length of 3.28 B_0.

6.2.2 Passive Unconstrained Layer Damping

Figure 6.6 shows a schematic drawing of the passive unconstrained layer damping (PUCLD) treatment.

In this case, the VEM will experience the same strain ε_0 as the base structure. Hence, the energy dissipation $\Delta W_{unconstrained}$ in the VEM/unit length/unit width/cycle is given by

$$\Delta W_{unconstrained} = \pi b h_1 E_1'' \int_{-L/2}^{L/2} \varepsilon_0^2 dx = \pi b h_1 E_1'' L \varepsilon_0^2 \tag{6.17}$$

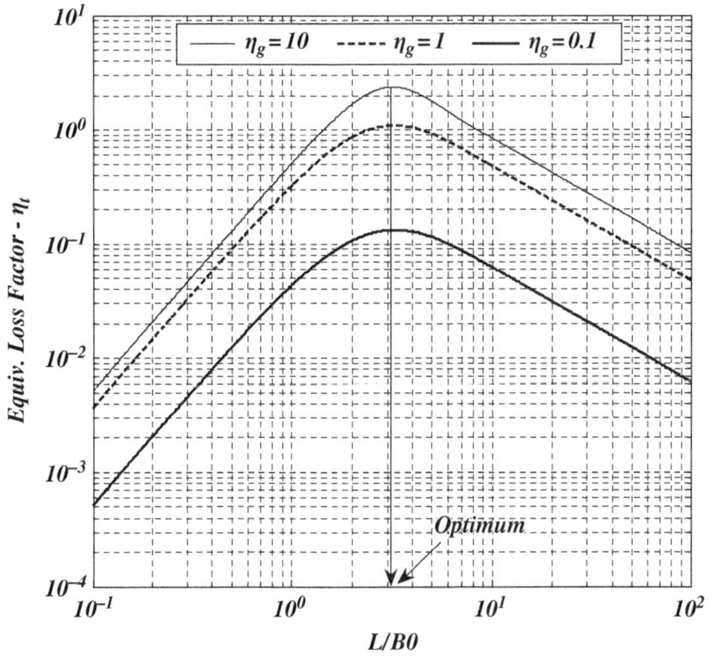

Figure 6.4 Effect of length (L/B_0) on total loss coefficient (η_t) for different viscoelastic material loss factors η_g.

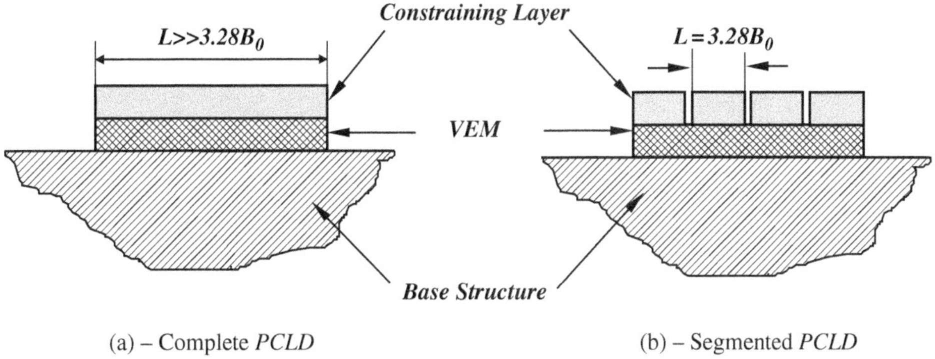

Figure 6.5 Complete and segmented PCLD treatments. (a) Complete PCLD. (b) Segmented PCLD.

Accordingly, the equivalent loss coefficient $\eta_{unconstrained}$ of the PUCLD normalized with respect to the nominal energy W_n can be obtained from

$$\eta_{unconstrained} = 2\pi h_1 \frac{E_1'}{h_2 E_2} \eta_g \tag{6.18}$$

The ratio between the energy dissipated in the PCLD and PUCLD can be determined by dividing Eq. (6.15) by Eq. (6.17) or Eq. (6.16) by Eq. (6.18) to give

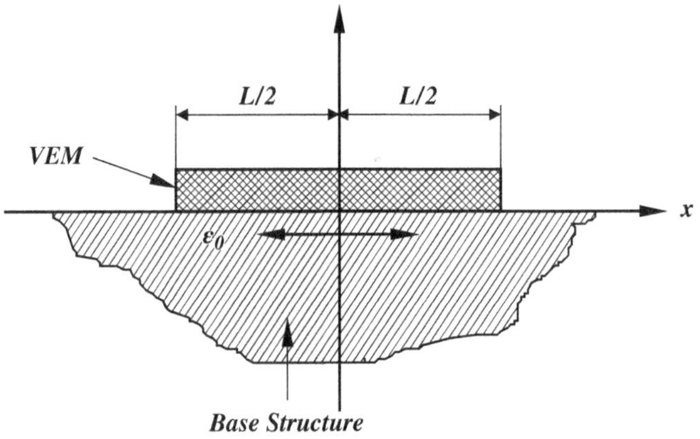

Figure 6.6 Structure/unconstrained VEM assembly.

$$\frac{\Delta W_p}{\Delta W_{unconstrained}} = \frac{2h_2E_2}{3h_1G\omega^*\sin\theta}4\pi \bigg/ \frac{[\sinh[\omega^*\cos(\theta/2)]\sin(\theta/2) - \sin[\omega^*\sin(\theta/2)]\cos(\theta/2)]}{[\cosh[\omega^*\cos(\theta/2)] + \cos[\omega^*\sin(\theta/2)]]} \tag{6.19}$$

Example 6.3 Consider a PCLD with constraining layer to VEM thickness ratio $h_2/h_1 = 1$, loss factor η_g of the VEM = 1, and dimensionless length $\omega^* = L/B_0 = 3.28$. Determine the equivalent coefficient η_p of the PCLD treatment assuming that $E_2 = 70$ Gpa and $G' = 10$ Mpa. If the constraining layer is removed and the treatment is reduced to PUCLD, determine:

a) the equivalent coefficient $\eta_{unconstrained}$ of the PUCLD treatment, and
b) the ratio between the energy dissipated in the PCLD and PUCLD.

Solution

As the loss factor η_g of the VEM = 1, then $\theta = 45°$ and Eq. (6.16) gives

$$\eta_p = 1.104$$

Also, Eqs. (6.18) and (6.19) give

$$\eta_{unconstrained} = 0.0027, \text{and} \frac{\Delta W_p}{\Delta W_{unconstrained}} = 409.98$$

This example clearly emphasizes the importance of using the VEM in a constrained configuration in order to considerably enhance its damping characteristics as compared to the unconstrained configuration. The example indicates that the loss factor of the PCLD is three orders of magnitude higher than that of the PUCLD.

6.3 Active Constrained Layer Damping Treatments of Rods

ACLD treatments have been recognized as an effective means for damping out the vibration of various structural members (Baz 1996, 1997a,b,c). In this class of damping treatments, viscoelastic damping layers are constrained by active piezoelectric layers whose longitudinal strains are controlled in response to the structural vibrations in order to enhance the energy dissipation characteristics as shown in Figures 6.7 and 6.8.

Figure 6.8 indicates that when the base structure experiences a longitudinal displacement u_O, the inactive constraining layer deflects longitudinally u and the viscoelastic layer is subjected to shear strain γ_p as shown in Figure 6.7b. Under these conditions, the ACLD acts as a conventional PCLD treatment. But, when the constraining layer is properly activated by the controller, an additional deflection u_p is generated by the piezoelectric effect to increase the shear strain of the viscoelastic core to γ_a as shown in Figure 6.7c. Such an increase in the shear strain (γ_a-γ_p) enhances the energy dissipation characteristics of the ACLD and results in effective damping of the structural vibrations.

6.3.1 Equation of Motion

In order to quantify the performance of the ACLD, the work done W_{piezo} by the non-conservative piezoelectric control forces must be considered. Appendix 6.B includes a brief summary of the basics of one-dimensional piezoelectricity indicating that the work done W_{piezo} can be determined from

$$W_{piezo} = E_2 h_2 b \int_{-L/2}^{L/2} \varepsilon_p u_x dx \tag{6.20}$$

where ε_p is the strain induced in the piezoelectric constraining layer. In this study, ε_p is assumed constant over the entire length of the constraining layer in order to maintain and emphasize the simplicity and practicality of the ACLD treatment.

Then, the equations and boundary conditions governing the operation of the ACLD system are obtained by applying Hamilton's principle (Meirovitch 1967):

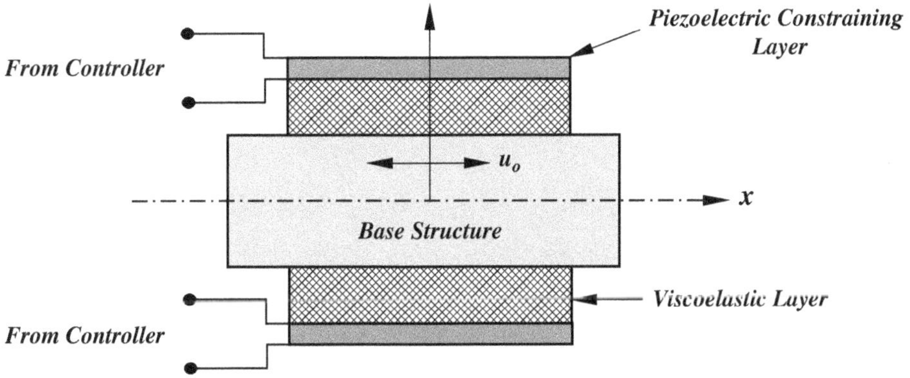

Figure 6.7 Active constrained layer damping (ACLD).

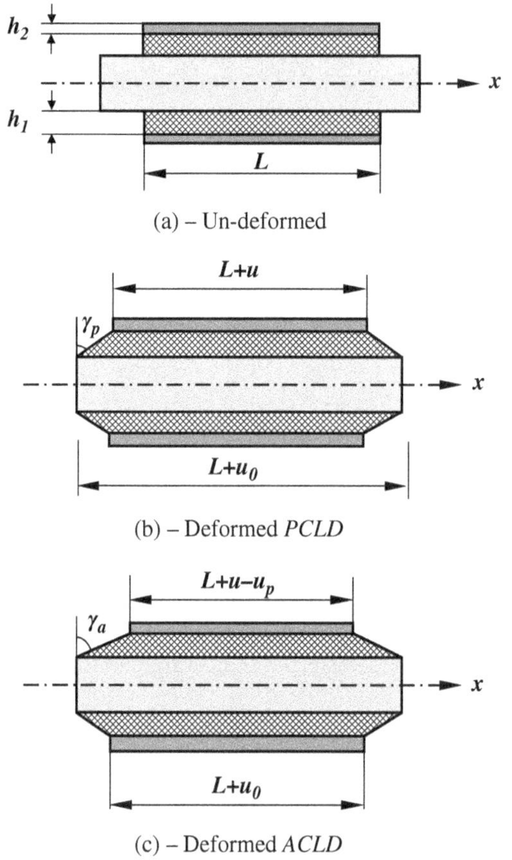

Figure 6.8 Operating principles of the PCLD and ACLD treatments. (a) Un-deformed. (b) Deformed PCLD. (c) Deformed ACLD.

$$\int_{t_1}^{t_2} \delta(KE - PE)dt - \int_{t_1}^{t_2} \delta\left(W_d - W_{piezo}\right) dt = 0 \tag{6.21}$$

where $\delta(.)$ denotes the first variation in the quantity inside the parentheses. Also, t denotes time with t_1 and t_2 defining the integration time limits.

From Eqs. (6.1) through (6.6) and (6.20), we have:

$$\begin{aligned}&\int_{t_1}^{t_2}\left[-mb\int_{-L/2}^{L/2}\ddot{u}\delta u\,dx - E_2h_2b[u_{,x}\delta u]_{-L/2}^{L/2} + E_2h_2b\int_{-L/2}^{L/2}u_{,xx}\delta u dx\right]dt\\&-\int_{t_1}^{t_2}\left[G'b/h_1\int_{-L/2}^{L/2}(u-u_0)\delta u\,dx + G'b/h_1\eta_g i\int_{-L/2}^{L/2}(u-u_0)\delta u dx\right]dt\\&+\int_{t_1}^{t_2}\left[E_2h_2b\left[\varepsilon_p\delta u\right]_{-L/2}^{L/2}\right]dt = 0\end{aligned} \tag{6.22}$$

Hence, the resulting equation of the ACLD system is:

$$mh_1/G^*\ddot{u} = B^{*^2} u_{,xx} - (u - u_0) \tag{6.23}$$

with the following boundary conditions:

$$u_x = \varepsilon_p \;\; at \;\; x = \pm L/2 \tag{6.24}$$

Note that the second order partial differential Eq. (6.23) describing the ACLD system is the same as that describing conventional PCLD as given by Eq. (6.8). However, the boundary conditions given by Eq. (6.24) are modified to account for the control action generated by the strain ε_p induced by the active constraining layer at the free end of the constraining layer (i.e., at $x = \pm L/2$). Therefore, the particular nature of operation of the ACLD system implies the existence of boundary control action ε_p.

6.3.2 Boundary Control Strategy

A boundary control strategy is devised to capitalize on this inherent operating nature of the ACLD system in such a manner that global stability of all the vibration modes of the system is ensured. The distributed-parameter control theory (Butkovskiy 1969) is used to devise the boundary control strategy that generates the boundary control action ε_p in order to ensure global stability of all the vibration modes of the ACLD-treated structure. The control strategy is devised to ensure that the total energy of the ACLD system is a strictly non-increasing function of time.

The total energy E_n of the ACLD system is obtained using Eqs. (6.2) and (6.3) as follows:

$$E_n = PE + KE$$

or

$$E_n/b = \frac{1}{2}\left(E_2 h_2 \int_{-L/2}^{L/2} u_x^2 \, dx + G' h_1 \int_{-L/2}^{L/2} \gamma^2 \, dx + m \int_{-L/2}^{L/2} \dot{u}^2 \, dx \right) \tag{6.25}$$

Equation (6.25) gives the energy norm of the ACLD system, which is quadratic and strictly positive. This norm is equal to zero if and only if u and $\dot{u}$ are zeros for all the points along the constraining layer between $[-L/2, L/2]$. This condition is ensured only when the ACLD system reverts back to its original undeflected equilibrium position.

Differentiating Eq. (6.25) with respect to time, integrating by parts and using Eqs. (6.23) and (6.24) gives:

$$\dot{E}_n/b = E_2 h_2 [\dot{u}(L/2) - \dot{u}(-L/2)]\varepsilon_p - \left(G'\eta_g h_1/\omega\right) \int_{-L/2}^{L/2} \dot{\gamma}^2 \, dx \tag{6.26}$$

As the second term is strictly negative, hence a globally stable boundary controller with a continuously decreasing energy norm (i.e., $\dot{E}_n < 0$) is obtained when the control action ε_p takes the following form:

$$\varepsilon_p = -K_g \left[\dot{u}(L/2) - \dot{u}(-L/2)\right] \tag{6.27}$$

where K_g is the gain of the boundary controller.

Equation (6.27) indicates that the control action is a velocity feedback of the differential longitudinal displacement of the ends of the piezoelectric constraining layer.

It is important also here to note that when the active control action ε_p ceases or fails to operate for one reason or another (i.e., when $\varepsilon_p = 0$), the rod system remains globally stable as indicated by Eq. (6.26). Such inherent stability is attributed to the second term in the equation that quantifies the contribution of the PCLD. Hence, the two terms in Eq. (6.26) provide quantitative means for weighing the individual contributions of the ACLD and the PCLD to the total rate of energy dissipation of the base structure.

The globally stable boundary controller can be easily implemented by solving the quasi-static form of the partial differential Eq. (6.23) subject to its boundary conditions in Eq. (6.24). This gives the following closed-form solution:

$$u - u_0 = (\varepsilon_p - \varepsilon_0) B^* \sinh(x/B^*)/\cosh(L/2B^*) \tag{6.28}$$

Substituting Eq. (6.28) into (6.27) gives:

$$\varepsilon_p = \varepsilon_0 (2K_g s B^*)\{[\tanh(L/2B^*) - (L/2B^*)]/[1 + (2K_g s B^*)\tanh(L/2B^*)]\} \tag{6.29}$$

where s is the Laplace complex number.

Implementation of the control strategy requires that the actuator be designed with self-sensing capabilities using the approaches suggested by Dosch et al. (1992). It is also important to note that the temporal derivatives of u can be determined by monitoring the current of the piezo-sensor rather than its voltage as described, for example, by Miller and Hubbard (1987).

6.3.3 Energy Dissipation

The energy characteristics of the ACLD are quantified by calculating the energies ΔW_p and ΔW_a, which are dissipated per vibration cycle by the passive and active components of the ACLD treatment as follows:

$$\Delta W_p = \int_0^{2\pi/\omega} (G' \eta_g h_1 b/\omega) \int_{-L/2}^{L/2} \dot{\gamma}^2 dx\, dt \tag{6.30}$$

and

$$\Delta W_a = \int_0^{2\pi/\omega} (E_2 h_2 [\dot{u}(L/2) - \dot{u}(-L/2)] \varepsilon_p)\, dt \tag{6.31}$$

where $2\pi/\omega$ is the period of a vibration cycle of frequency ω.

Using Eqs. (6.1), (6.27)–(6.29), reduces Eqs. (6.30) and (6.31) to:

$$\Delta W_p = \pi G' \eta_g b/h_1 \varepsilon_0^2 (\varepsilon_p/\varepsilon_0 - 1)^2 B_0^2 \int_{-L/2}^{L/2} [\sinh(x/B^*)/\cosh(L/2B^*)]^2 dx \tag{6.32}$$

and

$$\Delta W_a = 4\pi E_2 h_2 \omega K_g \varepsilon_0 B_0^2 \left\{ [\tanh(A) - A] / \left[1 + \left(2K_g \omega B^* i\right)\tanh(A)\right] \right\}^2 \tag{6.33}$$

where B_o is the magnitude of the complex characteristic length B^* as obtained from Eq. (6.A.2) in Appendix 6.A. Also, $A = \omega^*/2[\cos(\theta/2) - i\sin(\theta/2)]$.

Equations (6.32) and (6.33) can be normalized with respect to the nominal energy W_n, Eq. (6.15), to give the dimensionless loss factor η_p and η_a that quantify the energies dissipated by the passive and active components of the ACLD treatment as follows:

$$\eta_p = 4\pi/\omega^* \left(\varepsilon_p/\varepsilon_0 - 1\right)^2 \frac{[\sinh[\omega^* \cos(\theta/2)] \sin(\theta/2) - \sin[\omega^* \sin(\theta/2)] \cos(\theta/2)]}{[\cosh[\omega^* \cos(\theta/2)] + \cos[\omega^* \sin(\theta/2)]]} \tag{6.34}$$

and

$$\eta_a = 8\pi\left(K_g \omega B_0\right) \frac{[\tanh(A) - A]}{[1 + \left(2K_g \omega B_0\right)[i\cos(\theta/2) + \sin(\theta/2)]\tanh(A)]^2} \tag{6.35}$$

where $\omega^* = L/B_0$.

Therefore, Eqs. (6.34) and (6.35) provide closed-form expressions of the loss coefficients η_p and η_a as functions of three dimensionless parameters θ, ω^*, and $(K_g\omega B_0)$. These parameters define the loss factor of the viscoelastic layer ($\tan(\theta)$), a dimensionless length of the constraining layer (L/B_0) and a dimensionless control gain ($K_g\omega B_0$). The sum of these two equations gives the total loss coefficient η_t of the ACLD treatment due to the combined passive and active components as:

$$\eta_t = \eta_p + \eta_a \tag{6.36}$$

Example 6.4 Calculate the total loss coefficient η_t of the ACLD treatment for different values of the control gain ($K_g\omega B_0$) and loss factor η_g of the VEM when the dimensionless length $\omega^* = L/B_0 = 3.28$.

Solution

Figure 6.9 shows the effect of the dimensionless control gain ($K_g\omega B_0$) and the loss factor η_g of the *VEM* on the loss coefficient η_t as calculated from Eq. (6.32). The displayed results suggest that there is an optimal control gain for each VEM. Such an optimal gain increases as the loss factor η_g of the viscoelastic core is increased. The figure indicates also that as $K_g\omega B_0$ approaches zero, the loss coefficient tends to a low limiting value, which equals that of the PCLD treatment. However, as the gain is increased to infinity, the loss coefficient decreases to another limiting value equal to η_a, which is generated only by the active component of the ACLD treatment.

Example 6.5 Calculate the individual contribution of the passive and the active components of the ACLD to the total loss coefficient η_t of the treatment when the control gain ($K_g\omega B_0$) = 1 and loss factor η_g = 1.

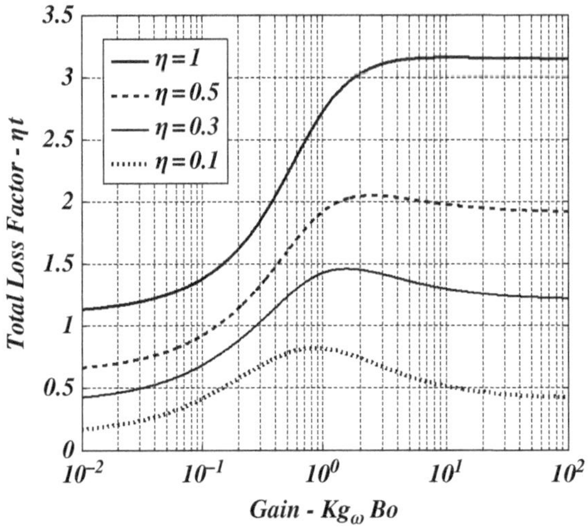

Figure 6.9 Effect of control gain ($K_g\omega B_0$) on total loss coefficient (η_t) for different values of the viscoelastic material loss factor η_g.

Solution

Figure 6.10 shows the effect of the dimensionless length ((L/B_0) on the passive and the active components of the loss coefficients of the ACLD treatment for $K_g\omega B_0 = 1$ and $\eta_g = 1$ as calculated from Eqs. (6.32) through (6.36). Plotted also on the figure are the characteristics of the PCLD for comparison purposes.

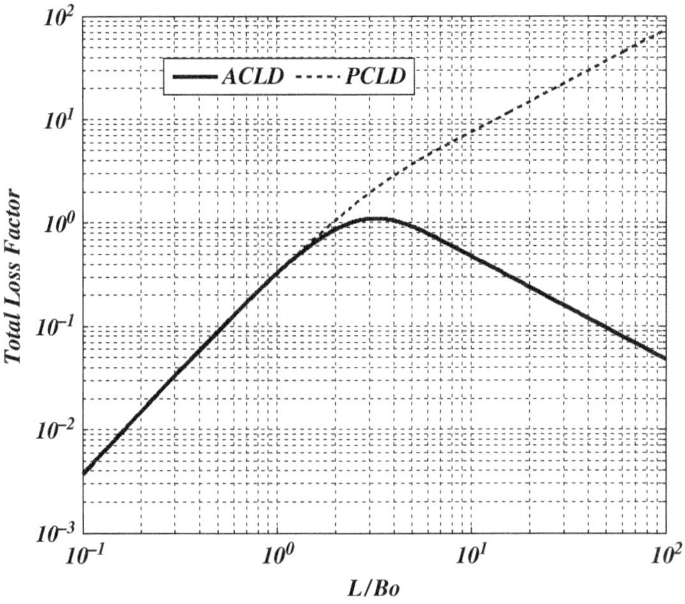

Figure 6.10 Loss factor of the ACLD and PCLD treatments for $\eta_g = 1$ and gain $K_g\omega B_0 = 1$.

Example 6.6 Calculate the shear strain distribution inside the viscoelastic core due to the use of the ACLD when the control gain ($K_g\omega B_0$) = 1 and loss factor η_g = 1. Assume that when the dimensionless length ω^* = L/B_0 = 3.28. Compare the results with the shear distribution for the PCLD.

Solution

The shear strain distribution inside the viscoelastic core is calculated from Eqs. (6.11) to (6.28). Figure 6.11 displays the distributions for both the ACLD and PCLD for comparison purposes.

The figure shows the improvement in the shear strain distribution inside the viscoelastic core due to the use of the ACLD configuration. It is evident also that the shear strain γ increases over the entire length of the treatment, which in turn results in increasing the passive energy dissipation per cycle according to Eqs. (6.14) and (6.30).

6.4 Passive Constrained Layer Damping Treatments of Beams

6.4.1 Basic Equations of Damped Beams

The equation of motion of a damped beam can be given by (Kerwin 1959):

$$D_t^* w_{,xxxx} - m\omega^2 w = 0 \tag{6.37}$$

where $D_t^* = D_t(1 + i\eta_B)$ = complex bending stiffness, m = mass per unit length, w = transverse deflection, and ω = frequency of oscillation. Note that η_B = loss factor of the damped beam.

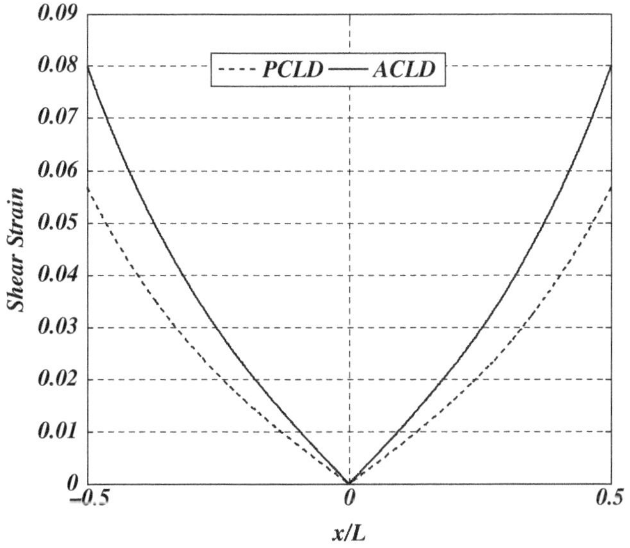

Figure 6.11 Comparison between the shear strain distributions of the ACLD and *PCLD* treatments for η_g = 1 and gain ($K_g\omega B_0$) = 1.

Equation (6.37) can be rewritten as:

$$w_{,xxxx} - k_B^{*4} w = 0 \tag{6.38}$$

where $k_B^* = \left(m\omega^2/D_t^*\right)^{1/4} \cong k_B(1 - i\eta_B/4) \cong k_B$ bending wave number for small values of η_B with $k_B = (m\omega^2/D_t)^{1/4}$.

A solution for Eq. (6.38) may take the following form:

$$w = w_0 e^{-i(k_B x - \omega t)} \tag{6.39}$$

where w_0 is an initial deflection.

6.4.2 Bending Energy of Beams

The elastic energy W_e associated with bending of the beam can be determined as follows (Mandal and Biswas 2005):

$$W_e = \frac{1}{2}\operatorname{Re}\left(\int_0^{2\pi/\omega} \left(F\dot{w}^\dagger + M\dot{w}^\dagger_{,x}\right) dt\right) \tag{6.40}$$

where F = the shear force = $-D_t^* w_{,xxx}$ and $M = D_t^* w_{,xx}$ are the bending moment respectively. Also "*Re*" denotes real part and $(.)^\dagger$ defines the complex conjugate of (.).

Then, using Eq. (6.39) to substitute for w as well as its spatial and temporal derivatives, reduces Eq. (6.40) to:

$$W_e = \frac{1}{2}\operatorname{Re}\left(\int_0^{2\pi/\omega} \left(-D_t^* w_{,xxx}\dot{w}^\dagger + D_t^* w_{,xx}\dot{w}^\dagger_{,x}\right) dt\right) = \pi D_t k_B^3 w_0^2 \tag{6.41}$$

6.4.3 Energy Dissipated in Beams with Passive Constrained Layer Damping

Consider the beam/constrained VEM system shown in Figure 6.12. The base beam and the constraining layer experience longitudinal vibrations u_1 and u_3, respectively. The VEM undergoes shear strain γ while the entire assembly is subjected to transverse

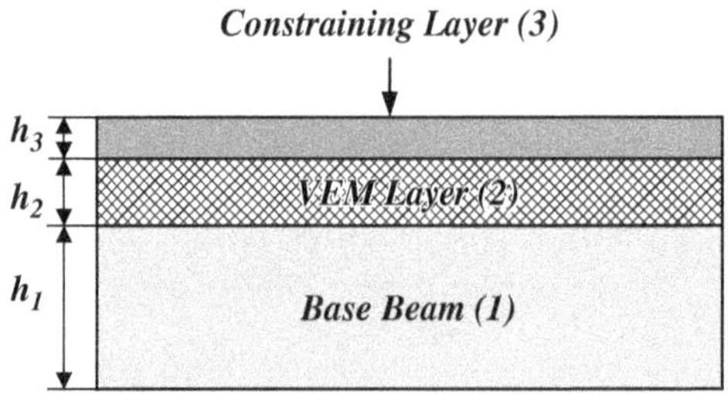

Figure 6.12 A schematic drawing of an undeflected beam/constrained VEM.

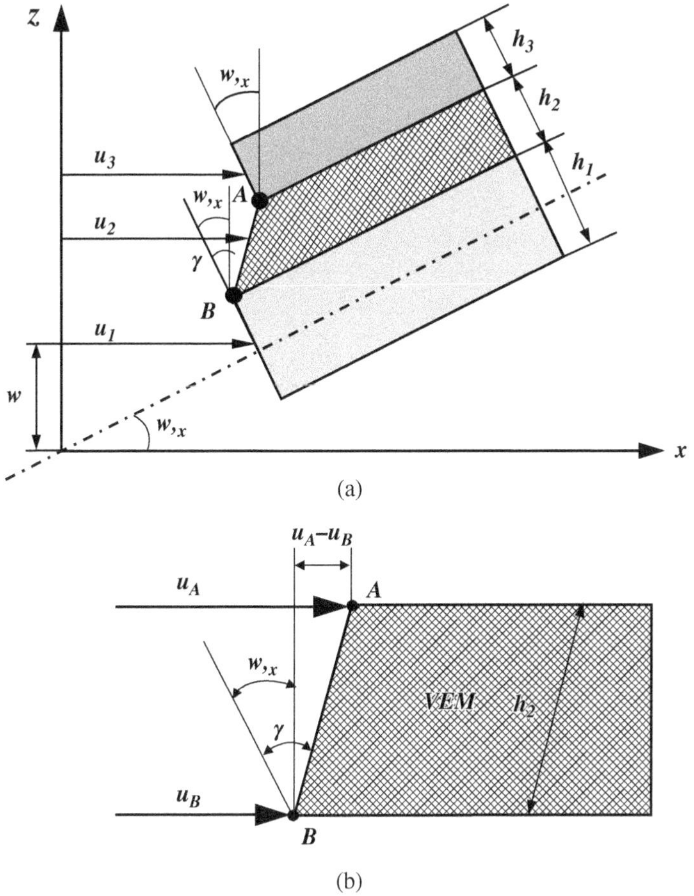

Figure 6.13 Deflected beam/VEM assembly.

deflection w and angular deflection $\partial w/\partial x= w_{,x}$. Figure 6.13 displays the deflected beam/VEM assembly along with the associated deflections relative to the fixed coordinate system x-z.

From the geometry of Figure 6.13a, the shear strain γ of the viscoelastic core is determined as follows:

$$u_A = u_3 + \frac{h_3}{2}\frac{\partial w}{\partial x} \quad \text{and} \quad u_B = u_1 - \frac{h_1}{2}\frac{\partial w}{\partial x} \tag{6.42}$$

Also, from Figure 6.13b

$$\gamma - w_{,x} = \sin(\gamma - w_{,x}) \cong \frac{1}{h_2}(u_A - u_B) \tag{6.43}$$

Combining Eqs. (6.42) and (6.43) gives:

$$h_2\gamma = h_2\frac{\partial w}{\partial x} + u_A - u_B$$
$$= h_2\frac{\partial w}{\partial x} + u_3 + \frac{h_3}{2}\frac{\partial w}{\partial x} - u_1 + \frac{h_1}{2}\frac{\partial w}{\partial x},$$
$$= \left(h_2 + \frac{h_1}{2} + \frac{h_3}{2}\right)\frac{\partial w}{\partial x} + u_3 - u_1$$

or

$$\gamma = \frac{h}{h_2}\frac{\partial w}{\partial x} + \frac{(u_3 - u_1)}{h_2} \tag{6.44}$$

where $h = h_2 + (h_1 + h_3)/2$.

Note that h_1, h_2, and h_3 denote the thickness of the constraining layer, the viscoelastic layer, and the base structure, respectively.
Consider the following two cases:

a) $u_1 = 0$, that is, the base beam is very stiff compared to the constraining layer.

Eq. (6.44) gives:

$$u_3 = \gamma h_2 - h\frac{\partial w}{\partial x} \tag{6.45}$$

Consider the free body diagrams of the constraining and the VEM layers shown in Figure 6.14.

$$\text{Then,}\, \Delta\sigma h_3 b = \tau b dx \quad \text{or} \quad \sigma_{,x} = G^*\gamma/h_3 \tag{6.46}$$

where

$$\sigma = E_3\, u_{3,x} = \frac{k_3}{h_3}u_{3,x} \tag{6.47}$$

with $k_3 = E_3h_3$ = longitudinal rigidity of constraining layer per unit width and $G^* = G(1 + i\eta_v)$.

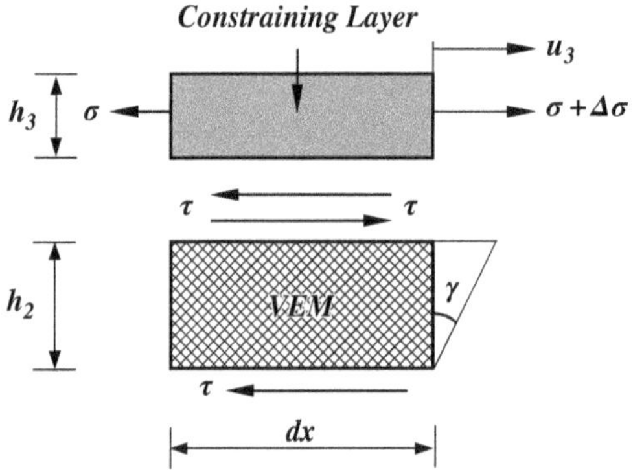

Figure 6.14 Free body diagrams of the constraining and VEM layers.

Combining Eqs. (6.45) through (6.47) yields:

$$\gamma_{,xx} - \frac{G^*}{k_3 h_2}\gamma = \frac{h}{h_2} w_{,xxx} \quad (6.48)$$

Using Eq. (6.39), a solution for Eq. (6.48) can be obtained and it takes the following form:

$$\gamma = -\frac{ihk_B}{h_2\left[1 + \frac{G^*}{k_B^2 k_3 h_2}\right]} w \quad (6.49)$$

It can be easily seen, by direct substitution, that this solution satisfies the differential Eq. (6.48).

Then, the energy dissipated in the VEM can be determined in a manner similar to Eq. (2.49) to assume the following form:

$$W_D = \pi G' \eta_v h_2 \gamma^2 \quad (6.50)$$

Accordingly, the loss factor of a beam treated with a passive constrained damping layer can be determined as follows:

$$\eta = \frac{W_D}{W_e} \quad (6.51)$$

Using Eqs. (6.41) and (6.50), the loss factor η reduces to:

$$\eta = \frac{\pi G' \eta_v h_2 \gamma^2}{\pi D_t\, k_B^3\, w_0^2} \quad (6.52)$$

Substituting Eq. (6.49) into Eq. (6.52) and assuming that $G^* \simeq G'$ for $\eta_v < 1$ (Kerwin 1959), then:

$$\frac{\eta}{\eta_v} = \frac{(h^2 k_3 / D_t)\, g}{(1+g)^2} \quad (6.53)$$

where $g = \dfrac{G'}{h_2 k_3 k_B}$ = dimensionless shear parameter.

Example 6.7 Determine the value of the shear parameter g for which the dimensionless loss factor η/η_v, given by Eqs. (6.53), attains a maximum for a given value of the dimensionless parameter $(h^2 k_3 / D_t)$.

Solution

Consider the following function ***f*** such that:

$f = \dfrac{\eta}{\eta_v (h^2 k_3 / D_t)}$, then Eq. (6.53) yields:

$$f = \frac{g}{(1+g)^2}$$

f attains a maximum when the dimensionless loss factor η/η_v is also maximum. Both maxima occur when:

$$\frac{df}{dg} = 0 \text{ or when } \frac{(1+g)^2 - 2g(1+g)}{(1+g)^4} = 0 \text{ that is, at } g = 1$$

This yields f_{max} = 0.25 as shown in Figure 6.15.

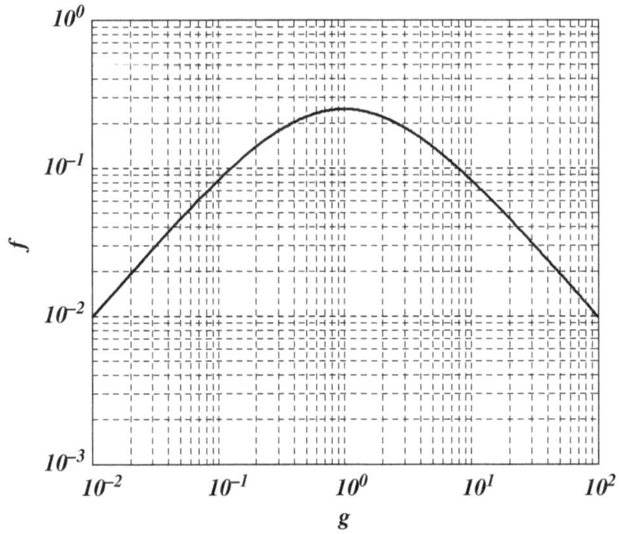

Figure 6.15 Effect of shear parameter g on the dimensionless loss factor.

Example 6.8 Consider the CLD treatment shown in Figure 6.16. The treatment is subjected to the following constraints: $u_1 = 0$ and $w = 0$. The constraining layer is given a static displacement $u_3 = u_{30}$ at $x = 0$. Determine the distance x_e along the VEM when $u_{3e} = u_{30}/e$ where e is the exponential function.

Discuss the physical meaning of the obtained results.

Solution

For the constraints under consideration, Eqs. (6.45) and (6.48) reduce to:

$$\gamma = u_3/h_2 \text{ and } \gamma_{,xx} - \frac{G^*}{k_3 h_2}\gamma = 0$$

$$\text{or } u_{3,xx} - \frac{G^*}{k_3 h_2} u_3 = 0$$

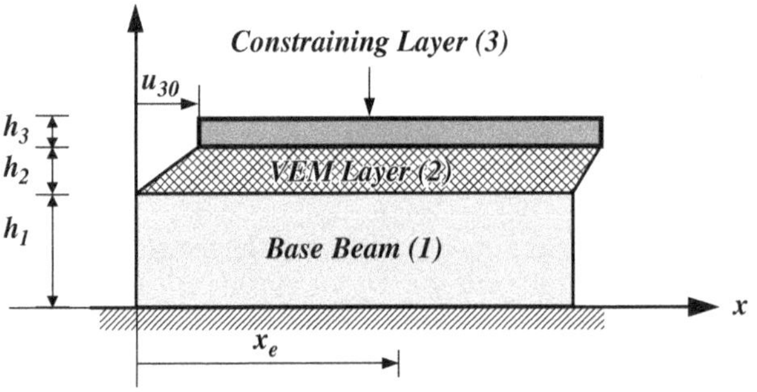

Figure 6.16 A special constrained layer damping treatment.

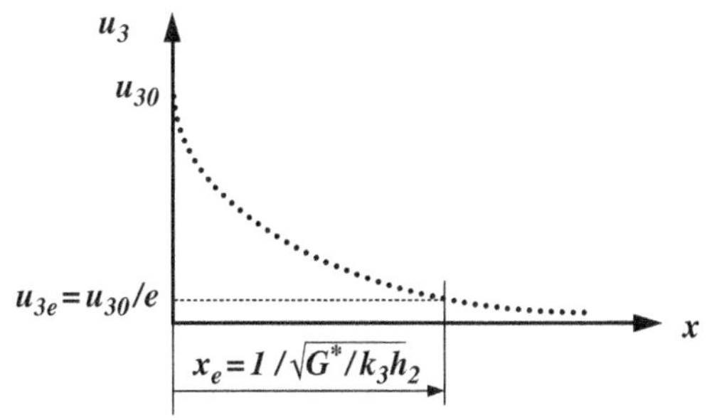

Figure 6.17 The $u_3 - x$ characteristics.

This equation has a possible solution given by:

$$u_3 = u_{30}\, e^{-\sqrt{G^*/k_3h_2}\,x}$$

At $x = x_e$, $u_{3e} = u_{30}/e$, that is,

$$u_{3e} = u_{30}\, e^{-\sqrt{G^*/k_3h_2}\,x_e} = u_{30}/e$$

This yields: $x_e = 1/\sqrt{G^*/k_3h_2}$.

Note that x_e is called "shear length" as it quantifies the distance along the constraining layer where a localized shear disturbance can "still" be felt after its magnitude has dropped by a factor of 2.718 (i.e., e).

Figure 6.17 displays the $u_3 - x$ characteristics and shows the u_{30}, u_{3e}, and x_e. Consider the following ratio $1/\left(k_B^2x_e^2\right)$, gives:

$$1/\left(k_B^2x_e^2\right) = G^*/k_B^2k_3h_2 = g$$

Note that this ratio can be written as:

$$1/\left(k_B^2x_e^2\right) = \lambda_B^2/\left(4\pi^2x_e^2\right) = g$$

where $\lambda_B = 2\pi/k_B$ = bending wavelength.

Hence, the "shear parameter" g is a direct measure of the ratio of the "bending wavelength" λ_B to the "shear length" x_e.

b) $u_1 \neq 0$, that is, the base beam and constraining layer have comparable stiffness.

In this case, the equilibrium of forces in the x direction requires that:

$$k_1 u_1 = -k_3 u_3 \tag{6.54}$$

where $k_1 = E_1h_1$ = longitudinal rigidity of the base beam per unit width.

Hence, Eq. (6.44) reduces to:

$$\gamma = \frac{h}{h_2}\frac{\partial w}{\partial x} + \frac{(u_3 + (k_3/k_1)u_3)}{h_2}$$

or

$$u_3 = \frac{k_1h_2}{k_1 + k_3}\gamma - \frac{k_1h}{k_1 + k_3}\frac{\partial w}{\partial x} \tag{6.55}$$

Following an approach similar to that used in case (a), this yields the following differential equation for the shear strain:

$$\gamma_{,xx} - \frac{G^*(k_1 + k_3)}{k_1 k_3 h_2}\gamma = \frac{h}{h_2} w_{,xxx} \tag{6.56}$$

which has the following solution:

$$\gamma = -\frac{ihk_B}{h_2\left[1 + \dfrac{G^*}{k_B^2 \frac{k_1 k_3}{k_1 + k_3} h_2}\right]} w \tag{6.57}$$

This yields the dimensionless loss factor of the beam/constraining layer assembly as follows:

$$\frac{\eta}{\eta_v} = \frac{(h^2 k_3/D_t)\, g}{[1 + (1 + k_3/k_1) g]^2} \tag{6.58}$$

Example 6.9 Determine the value of the shear parameter g for which the dimensionless loss factor η/η_v, given by Eq. (6.58), attains a maximum for a given value of the dimensionless parameter $(h^2 k_3/D_t)$.

Solution

Consider the following function f such that:

$$f = \frac{\eta}{\eta_v (h^2 k_3/D_t)}, \text{ then Eq. (6.58) yields:}$$

$$f = \frac{g}{[1 + (1 + k_3/k_1) g]^2}$$

f attains a maximum when the dimensionless loss factor η/η_v is also maximum. Both maxima occur when:

$$\frac{df}{dg} = 0 \text{ yielding } g = 1/(1 + K_r)$$

where $K_r = k_3/k_1$. This yields $f_{max} = 1/[4(1 + K_r)]$ as shown in Figure 6.18.

Note that maximum loss factor is obtained when the base beam is very stiff compared to the constraining layer, that is, $K_r = k_3/k_1 = 0$.

6.5 Active Constrained Layer Damping Treatments of Beams

Consider the beam/constraining layer damping treatment assembly shown in Figure 6.19. The beam is acted upon by an external moment M_0. Using Eq. (6.39), this moment results in a transverse deflection w_0 such that:

$$M_0 e^{-i(k_B x - \omega t)} = -D_t w_{,xx} = -w_0 D_t k_B^2 e^{-i(k_B x - \omega t)} \tag{6.59}$$

or

$$w_0 = -M_0/\left(D_t k_B^2\right) \tag{6.60}$$

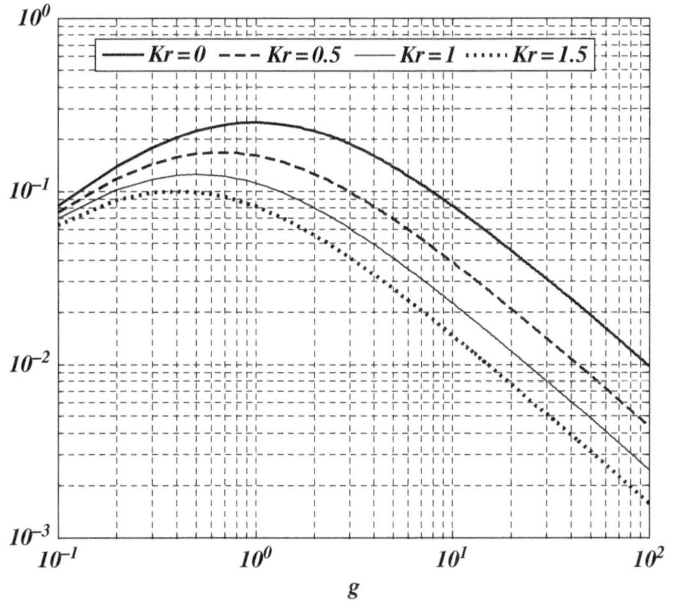

Figure 6.18 Effect of shear parameter g and ratio of longitudinal rigidity $K_r = k_3/k_1$ on the dimensionless loss factor.

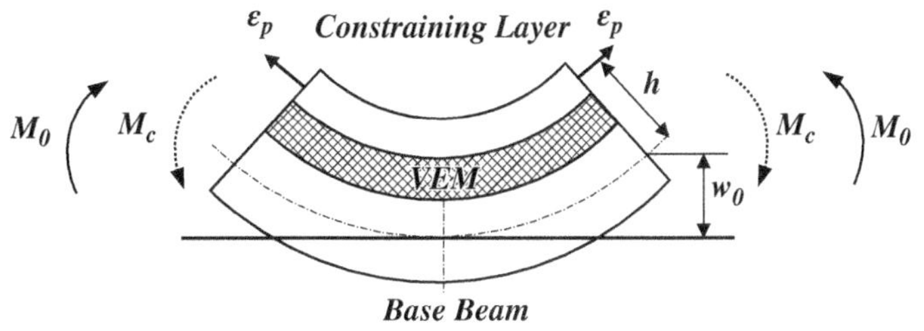

Figure 6.19 Actively controlled beam.

If the passive constraining layer is replaced by an active piezoelectric layer that generates a piezoelectric strain ε_p, then, a restoring control moment M_c is developed such that:

$$M_c = k_3 h \varepsilon_p \tag{6.61}$$

Accordingly, the net moment acting on the beam becomes:

$$\left(M_0 - k_3 h \varepsilon_p\right) e^{-i(k_B x - \omega t)} = -D_t\, w_{,xx} = -w_c D_t k_B^2\, e^{-i(k_B x - \omega t)} \tag{6.62}$$

where w_c is the resulting transverse deflection of the beam with the active piezoelectric layer. This deflection is given by:

$$w_c = -\left(\frac{M_0 - k_3 h \varepsilon_p}{D_t\, k_B^2}\right) \tag{6.63}$$

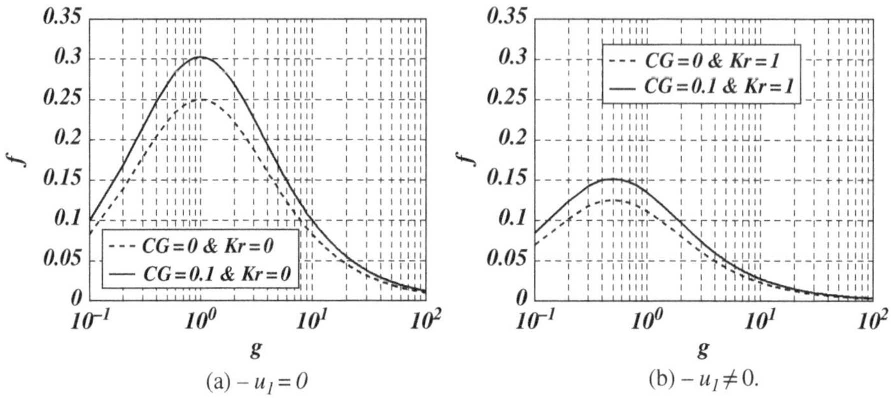

Figure 6.20 The dimensionless loss factors for actively controlled beams. (a) $u_1 = 0$. (b) $u_1 \neq 0$.

Substituting Eq. (6.60) into Eq. (6.63), yields:

$$w_c = w_0\left(1 + \frac{k_3 h \varepsilon_p}{w_0 D_t\, k_B^2}\right) \tag{6.64}$$

If the piezoelectric strain is generated according to a simple proportional control law, such that:

$$\varepsilon_p = K_G w_0 \tag{6.65}$$

where K_G = control gain, then Eq. (6.64) reduces to:

$$w_c = w_0(1 + C_G) \tag{6.66}$$

where $C_G = \dfrac{k_3 h K_G}{D_t\, k_B^2}$ = Control parameter.

Example 6.10 Determine the dimensionless loss factor η/η_v for a beam with ACLD treatment when:

Case (a). $u_1 = 0$ and Case (b). $u_1 \neq 0$.

In Case (a), plot η/η_v as a function of the value of the shear parameter g and the control parameter C_G for a given value of the dimensionless parameter($h^2 k_3/D_t$). In Case (b), plot η/η_v as a function of the value of the shear parameter g, the longitudinal rigidity $K_r = k_3/k_1$, and the control parameter C_G for a given value of the dimensionless parameter($h^2 k_3/D_t$).

Solution

In the two cases under consideration, the loss factor is calculated using Eq. (6.52) such that:

$$\eta = \frac{\pi G' \eta_v h_2 \gamma^2}{\pi D_t\, k_B^3\, w_0^2}$$

where the shear strain γ is given by:

$$\text{Case (a): } \gamma = -\frac{ihk_B}{h_2\left[1+\frac{G^*}{k_B^2 k_3 h_2}\right]}w$$

with $w = w_c e^{-i(k_B x - \omega t)}$ and $w_c = w_0(1 + C_G)$. This yields the following dimensionless loss factor η/η_v:

$$\frac{\eta}{\eta_v} = \frac{(h^2 k_3/D_t)g}{(1+g)^2}(1+C_G)^2 \tag{6.67}$$

$$\text{Case (b): } \gamma = -\frac{ihk_B}{h_2\left[1+\frac{G^*}{k_B^2 \frac{k_1 k_3}{k_1+k_3} h_2}\right]}w$$

with $w = w_c e^{-i(k_B x - \omega t)}$ and $w_c = w_0(1 + C_G)$. This yields the following dimensionless loss factor η/η_v:

$$\frac{\eta}{\eta_v} = \frac{(h^2 k_3/D_t)g}{[1+(1+k_3/k_1)g]^2}(1+C_G)^2 \tag{6.68}$$

Equations (6.67) and (6.68) are used to plot the scaled dimensionless loss factor f given by:

$$f = \frac{\eta}{\eta_v(h^2 k_3/D_t)} \tag{6.69}$$

The displayed results clearly indicate that adding active control capabilities to the constraining layers has resulted in an increase in the loss factor of the treatments even with a small gain ($C_G = 0.1$).

6.6 Passive and Active Constrained Layer Damping Treatments of Plates

In this section, the emphasis is placed on extending the work presented in Sections 6.6.2 and 6.3 to flat plates treated with PCLD or ACLD treatments. Details of this extension are presented by Ray and Baz (1997).

A typical configuration of the ACLD treatment of a flat plate is shown in Figures 6.21 and 6.22. Assume that the base structure experiences longitudinal displacements, u_0 and

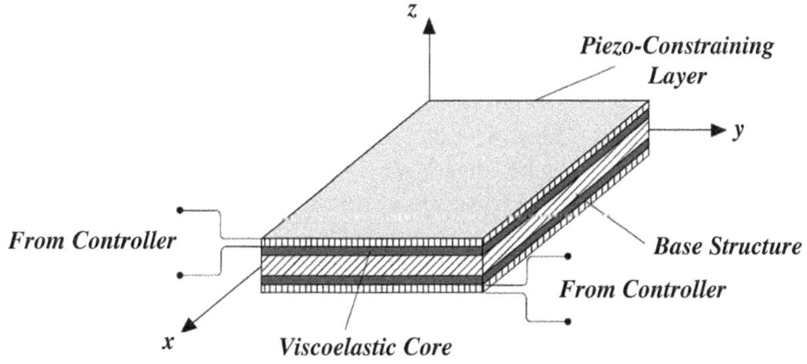

Figure 6.21 A plate treated with an active constrained layer damping.

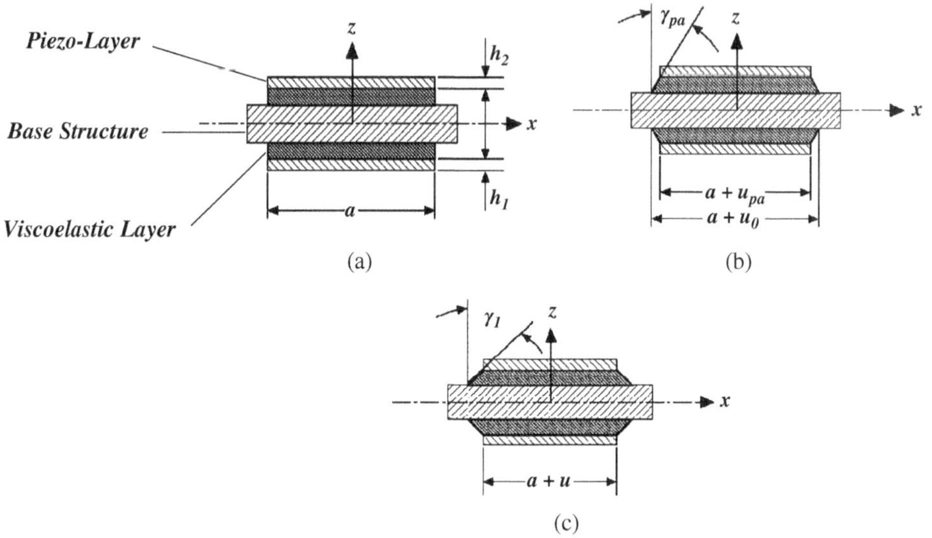

Figure 6.22 Operating principles of the PCLD and the ACLD treatments: (a) Undeformed. (b) PCLD. (c) ACLD.

v_0 at the interface with the viscoelastic core in the x and y directions, respectively. This makes the active constraining layer undergo displacements u_{pa} and v_{pa} at the interface with the viscoelastic core. Consequently, the viscoelastic layer is subjected to a passive shear stain γ_{pa} in the x-z plane as shown in Figure 6.22b. Under these conditions, the ACLD acts as a conventional PCLD.

But, when the constraining layer is activated properly by the controller, the passive displacements u_{pa} and v_{pa} change to u and v, respectively. Thus, an additional displacement (u-u_p) is generated by the piezoelectric effect to increase the shear strain of the viscoelastic core to γ_1 as shown in Figure 6.22c in the x-z plane. The corresponding increase in the shear strain $(\gamma_1 - \gamma_{pa})$ enhances the energy dissipation characteristics of the ACLD and results in effective damping of the structural vibrations. Similarly, the activated piezo-layer also undergoes an additional deflection (v-v_{pa}) in the y direction resulting in increasing the shear strain of the viscoelastic core in the y-z plane to γ_2.

6.6.1 Kinematic Relationships

Here, it is assumed that the bending effects in the plate are negligible, the constraining layer is subjected to in-plane strains only and the viscoelastic core is subjected to shear only. The piezoelectric constraining layer is assumed to be elastic and dissipate no energy whereas the core is assumed to be linearly viscoelastic.

From the geometry of Figure 6.22, the shear strains γ_1 and γ_2 in the viscoelastic core, in the x and y directions, respectively, are expressed as:

$$\gamma_1 = (u - u_0)/h_1, \quad \gamma_2 = (v - v_0)/h_1 \tag{6.70}$$

where h_1 denotes the thickness of the viscoelastic layer

6.6.2 Energies of the PCLD and ACLD Treatments

6.6.2.1 The Potential Energies

U_1 and U_2 of the plane stress deformations of the constraining layer and the shearing of the VEM core are given by:

$$U_1 = \frac{1}{2}h_2 \int_{-b/2}^{b/2}\int_{-a/2}^{a/2} \left[C_{11}u_{,x}^2 + 2C_{12}u_{,x}v_{,y} + C_{22}v_{,y}^2 + C_{66}\left(u_{,y} + v_{,x}\right)^2\right] dxdy,$$

and

$$U_2 = \frac{1}{2}G' h_1 \int_{-b/2}^{b/2}\int_{-a/2}^{a/2} \left[\gamma_1^2 + \gamma_2^2\right] dxdy, \tag{6.71}$$

where C_{ij} are the elastic constants of the constraining layer. Also, h_2, a, and b are its thickness, length, and width, respectively. Note also that the subscripts x and y denote partial differentiation with respect to x and y.

In Eq. (6.71), the complex modulus of the VEM is given by $G^* = G'(1 + \eta_g i)$ with G' and η_g denoting the storage modulus and the loss factor. Also, $i = \sqrt{-1}$.

6.6.2.2 The Kinetic Energy

T associated with the longitudinal displacements u and v is given by:

$$T = \frac{1}{2}m \int_{-b/2}^{b/2}\int_{-a/2}^{a/2} \left[u_{,t}^2 + v_{,t}^2\right] dxdy \tag{6.72}$$

where m is the mass per unit area and t denotes partial differentiation with respect to time.

6.6.2.3 Work Done

For the case of the ACLD treatment, the work done by the piezoelectric control forces is given by:

$$W_1 = -h_2 \int_{-b/2}^{b/2}\int_{-a/2}^{a/2} \left[\left(C_{11}\varepsilon_{px} + C_{12}\varepsilon_{py}\right)u_{,x} + \left(C_{12}\varepsilon_{px} + C_{22}\varepsilon_{py}\right)u_{,y}\right] dxdy \tag{6.73}$$

where ε_{px} and ε_{py} are the piezoelectric strains induced in the x and y directions, respectively. Note that for the case of the PCLD treatment, the work done $W_1 = 0$.

The work W_2 that is dissipated in the VEM is given by:

$$W_2 = -h_1 \int_{-b/2}^{b/2}\int_{-a/2}^{a/2} \left[\tau_{dx}\gamma_1 + \tau_{dy}\gamma_2\right] dxdy, \tag{6.74}$$

where τ_{dx} and τ_{dy} are the dissipative shear stresses developed in the VEM core. These stresses are given by:

$$\tau_{dx} = \left(G'\eta_g/\omega\right)\gamma_{1,t} = \left(G'\eta_g i\right)\gamma_1,$$
$$\tau_{dy} = \left(G'\eta_g/\omega\right)\gamma_{2,t} = \left(G'\eta_g i\right)\gamma_2. \tag{6.75}$$

Note that Eq. (6.75) is developed in the same manner as Eq. (6.5).

6.6.3 The Models of the PCLD and ACLD Treatments

The equations and the boundary conditions for the PCLD and ACLD treatments are developed using Hamilton's principle:

$$\int_{t_1}^{t_2} \delta\left(T - \sum_{i=1}^{2} U_i + \sum_{j=1}^{2} W_j\right) dt = 0 \tag{6.76}$$

where $\delta(.)$ denotes the first variation in the quantity inside the parentheses.

The resulting equations of motion are:

$$mh_1/G^* u_{,tt} = B_x^{*^2} u_{,xx} + B_{xy}^{*^2} v_{,xy} + B_z^{*^2} u_{,yy} - (u - u_0),$$

and

$$mh_1/G^* v_{,tt} = B_z^{*^2} v_{,xx} + B_{xy}^{*^2} u_{,xy} + B_y^{*^2} v_{,yy} - (v - v_0) \tag{6.77}$$

The associated boundary conditions are:

$$PCLD\, u_{,x} = 0 \;\; at\; x = \pm a/2, \quad v_{,y} = 0 \;\; at\; y = \pm b/2,$$

and

$$u_{,y} + v_{,x} = 0 \;\; at\; x = \pm a/2, y = \pm b/2. \tag{6.78}$$

$$ACLD\, u_{,x} = \varepsilon_{px} \;\; at\; x = \pm a/2, \quad v_{,y} = \varepsilon_{py} \;\; at\; y = \pm b/2,$$

and

$$u_{,y} + v_{,x} = 0 \;\; at\; x = \pm a/2, y = \pm b/2. \tag{6.79}$$

with $B_x^* = \sqrt{h_1 h_2 C_{11}/G^*}$, $B_y^* = \sqrt{h_1 h_2 C_{22}/G^*}$, $B_z^* = \sqrt{h_1 h_2 C_{66}/G^*}$,
$B_{xy}^* = \sqrt{h_1 h_2 (C_{12} + C_{66})/G^*}$.

Note that Eqs. (6.77) through (6.79) are the equations that represent the *2D* dynamics of a plate with PCLD/ACLD treatments. These equations have exactly the same structure as Eqs. (6.8) and (6.9), described in Section 6.6.2, to simulate the 1D dynamics of rods treated with PCLD treatments and Eqs. (6.23) and (6.24), outlined in Section 6.6.3, to model the 1D dynamics of rods treated with ACLD treatments.

6.6.4 Boundary Control of Plates with ACLD Treatments

The adopted control strategy is an energy-based approach that similar to that outlined in Section 6.6.3. The control structure is extracted to ensure that the rate of decay of the total energy E_n of the system is strictly negative at each instant of time.

The total energy is given by:

$$E_n = U_1 + U_2 + T$$

or

$$E_n = \frac{1}{2}\int_{-b/2}^{b/2}\int_{-a/2}^{a/2}\begin{bmatrix} h_2\left[C_{11}u_{,x}^2 + 2C_{12}u_{,x}v_{,y} + C_{22}v_{,y}^2 + C_{66}\left(u_{,y}+v_{,x}\right)^2\right] + \\ \left[G'h_1\left(\gamma_1^2+\gamma_2^2\right)\right] + \left[m\left(u_{,t}^2 + v_{,t}^2\right)\right]\end{bmatrix} dxdy \tag{6.80}$$

Differentiating Eq. (6.80) with respect to time, integrating by parts, and using Eqs. (6.77) and (6.79), yields:

$$\begin{aligned}\dot{E}_n = &\int_{-b/2}^{b/2} h_2\left(C_{11}\varepsilon_{px} + C_{12}\varepsilon_{py}\right)\left[u_{,t}(a/2) - u_{,t}(-a/2)\right]dy \\ &+ \int_{-a/2}^{a/2} h_2\left(C_{12}\varepsilon_{px} + C_{22}\varepsilon_{py}\right)\left[v_{,t}(a/2) - v_{,t}(-a/2)\right]dx \\ &- G'\eta_g h_1 \int_{-b/2}^{b/2}\int_{-a/2}^{a/2}\left[\gamma_1^2 + \gamma_2^2\right] dxdy.\end{aligned} \tag{6.81}$$

Note that the control action is coupled in nature and the piezo-strains, ε_{px} and ε_{py} are not independent but are related by:

$$\varepsilon_{py} = \left(d_{32}/d_{31}\right)\varepsilon_{px} \tag{6.82}$$

where d_{31} and d_{32} are the piezo-strain constants in the x and y directions (1 and 2) due to the applied electric field in the z direction (3). Hence, in order to guarantee that the energy norm will be continuously decreasing, the control action ε_{px} should take the following form:

$$\varepsilon_{px} = -K_x b\left[u_{,t}(a/2) - u_{,t}(-a/2)\right] - K_y a\left[v_{,t}(b/2) - v_{,t}(-b/2)\right] \tag{6.83}$$

provided that:

$$K_y/K_x = \left(C_{12} + [d_{32}/d_{31}]C_{22}\right)/\left(C_{11} + [d_{32}/d_{31}]C_{12}\right) \tag{6.84}$$

where K_x and K_y are the control gains of the globally stable boundary controller.

6.6.5 Energy Dissipation and Loss Factors of Plates with PCLD and ACLD Treatments

The energy dissipated (ΔW_{pa}, ΔW_a) of plates treated with PCLD and ACLD treatments can be determined as follows:

$$\Delta W_{pa} = \left(G'\eta_g h_1/\omega\right)\int_0^{2\pi/\omega}\int_{-b/2}^{b/2}\int_{-a/2}^{a/2}\left[\gamma_1^2 + \gamma_2^2\right] dxdydt \tag{6.85}$$

and

$$\Delta W_a = \int_0^{2\pi/\omega}\int_{-b/2}^{b/2} h_2(C_{11}\varepsilon_{px} + C_{12}\varepsilon_{py})[u_{,t}(a/2) - u_{,t}(-a/2)]dydt + \int_0^{2\pi/\omega}\int_{-a/2}^{a/2} h_2(C_{12}\varepsilon_{px} + C_{22}\varepsilon_{py})[v_{,t}(a/2) - v_{,t}(-a/2)]dxdt \tag{6.86}$$

Now, defining a nominal energy W_n to denote the maximum strain energy of the constraining layer if the whole layer is subjected to uniform longitudinal strains ε_{0x} and ε_{0y} only in the x and y directions, then:

$$W_n = \frac{1}{2}abh_2\,\varepsilon_{0x}^2\,C_{11}\left[1 + 2(C_{12}/C_{11})(\varepsilon_{0y}/\varepsilon_{0x}) + (C_{22}/C_{11})(\varepsilon_{0y}/\varepsilon_{0x})^2\right] \tag{6.87}$$

Hence, the loss factors (η_{pa}, η_p) of plates treated with PCLD and ACLD treatments can be determined as follows:

$$\eta_{pa} = \Delta W_{pa}/W_n \tag{6.88}$$

and

$$\eta_a = \Delta W_a/W_n \tag{6.89}$$

Substituting from Eqs. (6.77) through (6.89), the loss factors (η_{pa}, η_p) of plates treated with PCLD and ACLD treatments can be determined from:

$$\eta_{pa} = \frac{4\pi}{\left[1 + 2(C_{12}/C_{11})(\varepsilon_{0y}/\varepsilon_{0x}) + (C_{22}/C_{11})(\varepsilon_{0y}/\varepsilon_{0x})^2\right]} \times \left[\frac{D\sinh(Cw_x) - C\sin(Dw_x)}{w_x[\cosh(Cw_x) - \cos(Dw_x)]} + \frac{C_{22}}{C_{11}}\left(\frac{\varepsilon_{0y}}{\varepsilon_{0x}}\right)^2 \frac{D\sinh(Cw_y) - C\sin(Dw_y)}{w_y[\cosh(Cw_y) - \cos(Dw_y)]}\right] \tag{6.90}$$

and

$$\eta_a = \frac{8\pi[1 + (d_{32}/d_{31})(C_{12}/C_{11})]}{\left[1 + 2(C_{12}/C_{11})(\varepsilon_{0y}/\varepsilon_{0x}) + (C_{22}/C_{11})(\varepsilon_{0y}/\varepsilon_{0x})^2\right]} \times \left[\frac{K_x\omega B_x^* b\left[\tanh(A) - A + (a/b)(\varepsilon_{0y}/\varepsilon_{0x})(K_y/K_x)(B_y^*/B_x^*)(\tanh(B) - B)\right]}{w_x\left[1 + 2K_x\omega B_x^* b\left[\tanh(A) + (a/b)(K_y/K_x)(B_y^*/B_x^*)(d_{32}/d_{31})\tanh(B)\right]\right]}\right] \times \left[\begin{array}{l}(\varepsilon_{0y}/\varepsilon_{0x} - 1)\tanh(A) - A + (a/b)(\varepsilon_{0y}/\varepsilon_{0x})(K_y/K_x)(B_y^*/B_x^*) \\ \times [(\varepsilon_{py}/\varepsilon_{px} - 1)\tanh(B) + B]\end{array}\right] \tag{6.91}$$

where $w_x = a/B_x$, $w_y = b/B_y$, $C = \cos(\theta/2)$ and $D = \sin(\theta/2)$ with $\theta = \tan^{-1}(\eta_g)$. Also, B_x and B_y denote the magnitude of the B_x^* and B_y^*, respectively.

Example 6.11 Determine the loss factor η_{pa} for a square plate ($a/b = 1$) treated with a PCLD treatment as a function of the dimensionless length $w_x = w_y$ for different values of the VEM loss factor η_g. Assume that: $C_{22}/C_{11} = 1$, $C_{12}/C_{11} = 0.33$ = Poisson's ratio, and $\varepsilon_{0x}/\varepsilon_{0y} = 1$.

Solution

Equation (6.90) is used to determine the loss factor η_{pa} for the square plate/PCLD treatment as a function of the dimensionless length $w_x = w_y$ for the considered values of the parameters C_{22}/C_{11}, C_{12}/C_{11} and $\varepsilon_{0x}/\varepsilon_{0y}$. The obtained results are displayed in Figure 6.23 for values of the VEM loss factor η_g = 1, 0.5, and 0.1.

Note that the maximum loss factor of the PCLD treatment of a plate occurs at an optimum dimensionless length ($w_x = a/B_x$) of 3.26. This is exactly the same value which is obtained for the case of a rod as outlined in Section 6.6.2.

Example 6.12 Determine the loss factor η_{pa} for a square plate ($a/b = 1$) treated with an ACLD treatment as a function of the dimensionless control effort ($K_x\omega B_x b$) for different values of the VEM loss factor η_g. Assume that: $w_x = w_y = 3.26$, $C_{22}/C_{11} = 1$, $C_{12}/C_{11} = .33$ = Poisson's ratio, and $\varepsilon_{0x}/\varepsilon_{0y} = 1$. Assume also that the constraining layer is made of a piezo-ceramic material with $d_{31}/d_{32} = 1$.

Solution

Equation (6.91) is used to determine the loss factor η_{pa} for the square plate/ACLD treatment as a function of the dimensionless control effort ($K_x\omega B_x b$) for the considered values

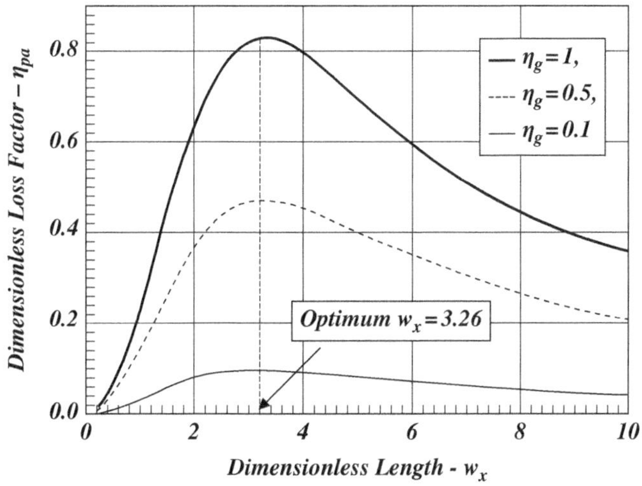

Figure 6.23 Effect of length w_x on the passive loss factor for a different *VEM* loss factor η_g.

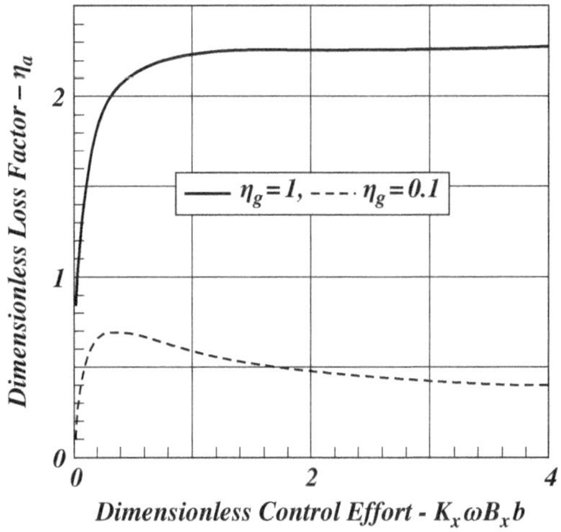

Figure 6.24 Effect of control effort ($K_x\omega B_x b$) on the active loss factor for a different VEM loss factor η_g.

of the parameters d_{31}/d_{32}, w_x, C_{22}/C_{11}, C_{12}/C_{11} and $\varepsilon_{0x}/\varepsilon_{0y}$. The obtained results are displayed in Figure 6.24 for values of the VEM loss factor η_g = 1 and 0.1.

The displayed results indicate that the addition of the active control capabilities has resulted in significant enhancement of the loss factor for the ACLD treatment as compared to the corresponding loss factor of the PCLD treatment displayed in Figure 6.23.

6.7 Passive and Active Constrained Layer Damping Treatments of Axi-Symmetric Shells

Distributed-parameter modeling of thin cylindrical shells that are fully treated with ACLD is presented. Hamilton's principle is utilized to develop the shell/ACLD model as well as the associated boundary conditions. A globally stable boundary control strategy is developed to damp out the vibration of the shell/ACLD system. The devised boundary controller is compatible with the operating nature of the ACLD treatments where the strain induced, in the active constraining layer, generates a control force acting at the boundary of the treated shell. As the boundary control strategy is based on a distributed-parameter model of the shell/ACLD system, the classical spillover problems resulting from using "truncated" finite element models is eliminated. Also, such an approach makes the boundary controller capable of controlling all the modes of vibration of the shell/ACLD and guarantees that the total energy norm of the system is continuously decreasing with time. Numerical examples are presented to demonstrate the effectiveness of the ACLD in damping out the vibration of cylindrical shells. Such effectiveness is determined for different control gains and compared with the performance of conventional PCLD. The results obtained demonstrate the high damping characteristics of the boundary controller particularly over broad frequency bands.

6.7.1 Background

Extensive efforts have been exerted to control the vibration of cylindrical shells using either passive or active control means. For example, Markus (1976, 1979) used unconstrained passive damping layer treatments to suppress the axi-symmetric vibrations of thin cylindrical shells. However, for higher damping characteristics; the *PCLD* treatments have been successfully employed to various types of cylindrical shells (Jones and Salerno 1966; Pan 1969; DiTaranto 1972; Lu et al. 1973; Leissa and Iyer 1981; Alma and Asnani 1984; Ramesh and Ganesan 1994, 1995). Recently, several attempts have been made to actively control the vibration of shells using discrete piezoelectric actuators (Forward 1981; Lester and Lefebvre 1993; Zhou et al. 1993; Chaudhry et al. 1994; Banks et al. 1995; Sonti and Jones 1996) bonded to the shell surfaces or distributed piezoelectric actuators embedded in the composite fabric of the shell (Tzou 1993).

In all these studies, the emphasis is placed on using separately the passive or the active vibration control actions. In the present section, a radically different approach is adopted whereby the passive and active control strategies are combined to operate in unison. In the proposed hybrid configuration, an optimal balance is achieved between the simplicity of the passive damping and the efficiency of the active control. A preferred embodiment of such hybrid configuration is the ACLD treatment, which has been recognized as an effective means for damping the vibration of beams and plates (Baz 1996, 1997a,b,c; Baz and Ro 1994, 1995, 1996). The ACLD treatments have been controlled using simple proportional and/or derivative feedback of the transverse deflection or the slope of the deflection line. The control gains have generally been arbitrarily selected to be small enough to avoid instability problems. In Shen 1994, developed the stability bounds for full ACLD treatments and Baz and Ro (1995) devised optimal control strategies for selecting the gains. In 1996, the control gains are selected by Baz using the theory of robust controls to ensure stability in the presence of parameter uncertainty and to reject the effect of external disturbances (Baz 1998).

In this section, the focus is placed on using the ACLD treatments to control the vibration of cylindrical shells with particular emphasis on developing a distributed-parameter model using Hamilton's principle to describe the axi-symmetric vibrations of shells that are fully treated with ACLD treatments. The variational formulation, being energy-based, is much simpler than the force equilibrium-based shear model of Pan (1969), which is used to analyze the dynamics of circular sandwiched shells treated with PCLD treatments. Also, it directly provides the boundary conditions associated with the ACLD treatment. The present model is an extension of the boundary control model developed by Deng (1995) to control the vibrations of plain and untreated shells. The variational model is utilized to devise a globally stable boundary control strategy that is compatible with the operating nature of the ACLD treatments. In this manner, the instability problems associated with the simple proportional and/or derivative controllers are completely avoided. Furthermore, as the control strategy is based on a distributed-parameter model, hence the classical spillover problems resulting from using "truncated" finite element models are eliminated. Accordingly, the devised boundary controller will be able to control all the modes of vibration of the ACLD-treated structures.

6.7.2 The Concept of the Active Constrained Layer Damping

The ACLD treatment consists of a conventional PCLD that is augmented with efficient active control means to control the strain of the constraining layer, in response to the shell vibrations as shown in Figure 6.25. The shear deformation of the viscoelastic damping layer is controlled by an active piezoelectric constraining layer that is energized by a control voltage V_c. This control voltage is generated based on the boundary control strategy devised in this study. In this manner, the ACLD when bonded to the shell acts as a smart constraining layer damping treatment with built-in actuation capabilities. With appropriate strain control, through proper manipulation of V_c, all the structural modes of vibration can be damped out.

Also, the ACLD provides a practical means for controlling the vibration of massive structures with the currently available piezoelectric actuators without the need for excessively large actuation voltages. This is due to the fact that the ACLD properly utilizes the piezoelectric actuator to control the shear in the soft viscoelastic core, which is a task compatible with the low control authority capabilities of the currently available piezoelectric materials.

6.7.3 Variational Modeling of the Shell/ACLD System

6.7.3.1 Main Assumptions of the Model

Figure 6.26 shows a schematic drawing of the ACLD treatment of a sandwiched cylindrical shell. It is assumed that the shear strains in the piezoelectric actuator layer and in the base shell are negligible. It is also assumed that the longitudinal and tangential stresses in the viscoelastic core are negligible. The transverse displacements w of all points on any cross section of the sandwiched shell are considered to be equal. Furthermore, the piezoelectric actuator layer and the base shell are assumed to be elastic and dissipate no energy, whereas the core is assumed to be linearly viscoelastic.

In addition, it is assumed that the thickness and modulus of elasticity of the sensor are negligible as compared to those of the base shell.

6.7.3.2 Kinematic Relationships

From the geometry of Figure 6.26, the shear strain γ in the core is:

$$\gamma = \frac{[hw_{,x} + (u_1 - u_3)]}{h_2}, \tag{6.92}$$

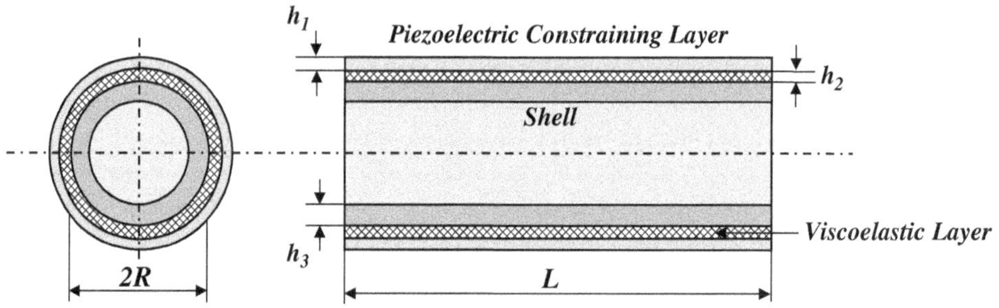

Figure 6.25 The shell/ACLD system.

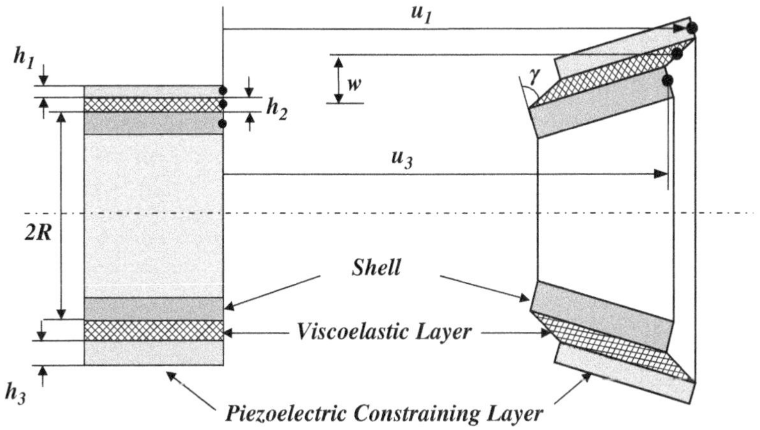

Figure 6.26 Schematic drawing of the structure and geometry of shell/ACLD system.

where

$$h = h_2 + \frac{1}{2}(h_1 + h_3). \tag{6.93}$$

In these equations, u_1 and u_3 are the longitudinal deflections of the piezo-actuator layer and shell layer, respectively; and w denotes the transverse deflection of the shell system. Subscript x denotes partial differentiation with respect to x and h_1, h_2, and h_3 define the thicknesses of the piezo-actuator, the viscoelastic layer, the base shell system, respectively.

6.7.3.3 Stress-Strain Relationships

a) Cylindrical shell

Using the Donnell–Mushtari theory of thin cylindrical shells (Leissa 1973), one can write the longitudinal strains ε_{ix} and the tangential strains $\varepsilon_{i\theta}$ in the ith layer as follows:

$$\varepsilon_{1x} = u^0_{1,x} - \left[z + \frac{1}{2}(h_1 + h_2)\right] w_{,xx}, \ \varepsilon_{3x} = u^0_{3,x} - \left[z + \frac{1}{2}(h_2 + h_3)\right] w_{,xx}, \tag{6.94}$$

and

$$\varepsilon_\theta = -\frac{w}{R}. \tag{6.95}$$

where $u^0_{i,x}$, w, and R denote the longitudinal strain at the middle plane of the ith layer, the transverse deflection, and the radius of the mid-surface of the core layer, respectively. Subscript i is set equal 1 for the base shell and 3 for the piezoelectric constraining layer.

Hence, the corresponding longitudinal and tangential stresses σ_{ix} and $\sigma_{i\theta}$ in the ith layer are given by:

$$\sigma_{ix} = \frac{E_i}{1 - \nu_i^2}\left[\varepsilon_{ix} + \nu_i \varepsilon_{i\theta}\right], \tag{6.96}$$

and

$$\sigma_{i\theta} = \frac{E_i}{1-\nu_i^2}\left[\varepsilon_{i\theta} + \varepsilon_{ix}\nu_i\right], \tag{6.97}$$

where E_i and ν_i are Young's modulus and Poisson's ratio for the ith layer, respectively.

The longitudinal and tangential forces, N_{ix} and $N_{i\theta}$, acting on the ith layer can be obtained by integrating the stresses over the cross section of that layer as follows:

$$N_{ix} = \int_{b_i}^{a_i} \sigma_{ix}\, dz \quad \text{and} \quad N_{i\theta} = \int_{b_i}^{a_i} \sigma_{i\theta}\, dz \tag{6.98}$$

Note that when $i = 1$, $a_i = -h_2/2$ and $b_i = -(h_2/2 + h_1)$, and when $i = 3$, $a_i = h_2/2 + h_3$ and $b_i = h_2/2$.

Substituting Eqs. (6.94) through (6.98) give:

$$N_x = N_{1x} + N_{3x} \quad \text{and} \quad N_\theta = N_{1\theta} + N_{3\theta} \tag{6.99}$$

where

$$N_{ix} = K_i\left[\varepsilon_{ix}^0 + \nu_i\varepsilon_\theta\right] \quad \text{and} \quad N_{i\theta} = K_i\left[\varepsilon_\theta + \nu_i\,\varepsilon_{ix}^0\right]. \tag{6.100}$$

where $K_i = \dfrac{E_i h_i}{1-\nu_i^2}$ with $i = 1$ and 3.

b) Piezoelectric constraining layer

The strain ε_p induced by the piezoelectric layer due to the application of a control voltage V_c is:

$$\varepsilon_p = \left[d_{31}\ d_{32}\ 0\right]^T \frac{V_c}{h_1}, \tag{6.101}$$

where d_{31}, d_{32} are the piezoelectric strain coefficients.

Hence, the corresponding induced stress σ_p is obtained from:

$$\sigma_p = \frac{E_p}{1-\nu_p^2}\begin{bmatrix} 1 & \nu_p & 0 \\ \nu_p & 1 & 0 \\ 0 & 0 & \dfrac{1-\nu_p}{2} \end{bmatrix} \varepsilon_p \tag{6.102}$$

where E_p and ν_p are Young's modulus and Poisson's ratio for the piezoelectric actuator layer, respectively.

Integrating the piezo-stresses over the cross section of the actuator gives the control forces and moments generated by the actuator. It is important to note here that the control forces $N_{p\theta}$ along the tangential direction and the associated control moments vanish because of the axi-symmetric nature of the vibration of the shell. Only the longitudinal control forces, N_{px} generated along the x-axis exist and are given by:

$$N_{px} = \int_{b_p}^{a_p} \sigma_{px}\, dz, \tag{6.103}$$

where $a_p = -h_2/2$ and $b_p = -(h_2/2 + h_1)$. Also, σ_{px} is the x component of the piezo-stress σ_p.

Equations (6.101) through (6.103) yield the following expression for the control force N_{x_p}:

$$N_{px} = K_1\left(d_{31} + \nu_p d_{32}\right)\frac{V_c}{h_1} \tag{6.104}$$

The expressions given by Eq. (6.100) for the longitudinal and tangential forces, N_{ix} and $N_{i\theta}$, acting on the different layers of the shell, as well as the piezo-actuator control force given by Eq. (6.103), are used to compute the potential and control energies of the shell/ACLD system.

6.7.3.4 Energies of Shell/ACLD System

a) Potential energies

The potential energies associated with the extension U_1, bending U_2, and shearing U_3 of the different layers of the shell/ACLD system are given by:

$$U_1 = \frac{1}{2}\pi R\sum_{i=1,3}\left[\int_0^L K_i\left(\varepsilon_{ix}^0 + \nu_i\varepsilon_\theta\right)\varepsilon_{ix}^0 dx + \int_0^L K_i\left(\nu_i\varepsilon_{ix}^0 + \varepsilon_\theta\right)\varepsilon_\theta dx\right], \tag{6.105}$$

$$U_2 = \frac{1}{2}\pi R D_t\int_0^L w_{,xx}^2\, dx, \tag{6.106}$$

and

$$U_3 = \frac{1}{2}\pi R G_2' h_2\int_0^L \gamma^2\, dx. \tag{6.107}$$

where $D_t = \sum_{i=1}^{3} E_i h_i^3/\left(1-\nu_i^2\right)$ with $E_i h_i^3$ denoting the flexural rigidity of the ith layer and G_2' is the storage shear modulus of the viscoelastic layer.

b) Kinetic energies

The KE associated only with the transverse deflection w of the shell/ACLD system is given by:

$$KE = \frac{1}{2}m\int_0^L \dot{w}^2\, dx. \tag{6.108}$$

where m is the mass per unit length of the sandwiched shell system.

c) Work done on shell/ACLD system

The work done W_1 by the external transverse loads q acting on the shell/ACLD system per unit perimeter length of the sandwiched shell is given by:

$$W_1 = \pi R \int_0^L q w dx, \tag{6.109}$$

and the work done W_2 by the piezoelectric control forces is given by:

$$W_2 = \pi R \int_0^L N_{px} \varepsilon_{1x}^0 dx. \tag{6.110}$$

In this study, N_{px} is assumed constant over the entire length of the constraining layer in order to maintain and emphasize the simplicity and practicality of the ACLD treatment.

The work W_3 dissipated in the viscoelastic core is given by:

$$W_3 = -\pi R h_2 \int_0^L \tau_d \gamma \, dx \tag{6.111}$$

where τ_d is the dissipative shear stress developed by the viscoelastic core. It is given by:

$$\tau_d = \left(\frac{G_2' \eta_v}{\omega} \right) \gamma_t = G_2' \eta_v \gamma i. \tag{6.112}$$

where η_v, ω, and i denote the loss factor of the viscoelastic core, the frequency, and $\sqrt{-1}$, respectively.

In Eq. (6.112), the behavior of the viscoelastic core is modeled using the common complex modulus approach, which is a frequency domain-based method (Nashif et al. 1985). Adoption of this approach results in a variational model of the ACLD than can easily be reduced to the classical model by Pan (1969) when the piezoelectric strain is set equal to zero.

6.7.3.5 The Model

The equations and boundary conditions governing the operation of the shell/ACLD system are obtained by applying Hamilton's principle as follows (Meirovitch 1967):

$$\int_{t_1}^{t_2} \delta \left(KE - \sum_{i=1}^{3} U_i \right) dt + \int_{t_1}^{t_2} \delta \left(\sum_{i=1}^{3} W_i \right) dt = 0. \tag{6.113}$$

where δ (.) denotes the first variation in the quantity inside the parentheses, t denotes the time and t_1 and t_2 define the bounds of the time interval where the shell/ACLD dynamics are considered.

The resulting equations of the shell/ACLD system are:

$$-K_1 u_{1,_{xx}} + \nu_1 K_1 \frac{w_{,x}}{R} + \frac{G_2^*}{h_2} [h w_{,x} + (u_1 - u_3)] = 0, \tag{6.114}$$

$$-K_3 u_{3,_{xx}} + \nu_3 K_3 \frac{w_{,x}}{R} - \frac{G_2^*}{h_2} [h w_{,x} + (u_1 - u_3)] = 0, \tag{6.115}$$

and

$$\frac{m}{\pi R}w_{,tt} + D_t w_{,xxxx} + (K_1 + K_3)\frac{w}{R^2} - \frac{1}{R}\left(\nu_1 K_1 u_{1,x} + \nu_3 K_3 u_{3,x}\right)$$
$$- \frac{G_2^* h}{h_2}\left[h w_{,xx} + \left(u_{1,x} - u_{3,x}\right)\right] = 0. \tag{6.116}$$

where $G_2^* = G_2'(1 + \eta_v i)$ is the complex modulus of the VEM.

These equations are subject to the following boundary conditions:

$$K_1\left[u_{1,x} - \nu_1\frac{w}{R}\right]\delta u_1\Big|_0^L = (d_{31} + \nu_1 d_{32})\frac{V_C}{h_1}\Big|_0^L, \tag{6.117}$$

$$K_3\left[u_{3,x} - \nu_3\frac{w}{R}\right]\delta u_3\Big|_0^L = 0, \tag{6.118}$$

$$D_t w_{,xx}\,\delta w_x\big|_0^L = 0, \tag{6.119}$$

and

$$\left[D_t w_{,xxx} + \frac{G_2^* h}{h_2}(h w_{,x} + u_1 - u_3)\right]\delta w\big|_0^L = 0. \tag{6.120}$$

Eliminating u_1 and u_3 from Eqs. (6.114) to (6.116) yields the following sixth order partial differential equation in the transverse deflection w of the shell/ACLD system:

$$D_t w_{,xxxxxx} - D_t g(1 + Y)w_{,xxxx}$$
$$+ \left[\frac{1-\nu_1^2}{R^2}K_1 + \frac{1-\nu_3^2}{R^2}K_3 - \frac{2(\nu_1 - \nu_3)}{Rh_2}G_2^* h\right]w_{,xx}$$
$$- \left[g\left(\frac{1-\nu_1^2}{R^2}K_1 + \frac{1-\nu_3^2}{R^2}K_3\right) - \frac{(\nu_1 - \nu_3)^2}{R^2}G_2^* h\right]w \tag{6.121}$$
$$- \frac{m}{\pi R}g w_{,tt} + \frac{m}{\pi R}w_{,xxtt} = 0$$

where

$$g = \frac{G_2^*}{h_2}\left[\frac{K_1 + K_3}{K_1 K_3}\right] \quad \text{and} \quad Y = \frac{h^2}{D_t}\left[\frac{K_1 K_3}{K_1 + K_3}\right]. \tag{6.122}$$

For a simply supported shell/ACLD system, the eight boundary conditions given by Eqs. (6.117) and (6.120) reduce to the following six boundary conditions:

At $x = 0$ and L

$$\frac{m}{\pi R}w_{,tt} + D_t w_{,xxxx} + \left(\frac{K_1\nu_1}{R} - \frac{G_2^* h}{h_2}\right)\left(1 + \frac{d_{32}}{d_{31}}\right)\frac{d_{31}V_C}{h_1} = q, \tag{6.123}$$

$$w = 0, \tag{6.124}$$

and

$$w_{,xx} = 0. \tag{6.125}$$

Similar expressions can be easily obtained for other boundary conditions.

It is important here to note that the sixth order partial differential equation describing the shell/ACLD system (Eq. (6.121)) is the same as that describing a shell treated with conventional PCLD as obtained by Pan (1969). However, the boundary condition given by Eq. (6.123) is modified to account for the control action generated by the control voltage V_C applied to the Active Constraining Layer at the free ends of the shell (i.e., at $x = 0$ and L).

Therefore, the particular nature of operation of the shell/ACLD system implies the existence of boundary control action at $x = 0$ and L. In Section 6.6.4, a boundary control strategy is devised to capitalize on this inherent operating nature of the shell/ACLD system in such a manner that ensures global stability of all the vibration modes of the system.

6.7.4 Boundary Control Strategy

6.7.4.1 Overview

Distributed-parameter control theory is used to devise a boundary control strategy that generates the boundary control action in order to ensure global stability of all the vibration modes of the shell/ACLD system. The control strategy is devised to ensure that the total energy of the shell/ACLD system is a strictly non-increasing function of time.

6.7.4.2 Control Strategy

The total energy E_n of the shell/*ACLD* system is obtained using Eqs. (6.105) through (6.108) as follows:

$$E_n = U_1 + U_2 + U_3 + T, \tag{6.126}$$

or

$$E_n = \frac{1}{2}\pi R \sum_{i=1,3}\left[\int_0^L K_i\left(\varepsilon_{ix}^0 - \nu_i\frac{w}{R}\right)\varepsilon_{ix}^0 dx - \int_0^L K_i\left(\nu_i\varepsilon_{ix}^0 - \frac{w}{R}\right)\frac{w}{R}dx\right]$$
$$+\frac{1}{2}\pi R D_t\int_0^L w_{,xx}^2\,dx + \frac{1}{2}\pi R G_2' h_2\int_0^L \gamma^2\,dx + \frac{1}{2}m\int_0^L \dot{w}^2\,dx. \tag{6.127}$$

Equation (6.127) quantifies the energy norm of the shell/ACLD system, which is quadratic and strictly positive. This norm is equal to zero if, and only if, u_1, u_3, w, w_x, w_{xx}, and w_t are all zero for all the points on the shell between $[0, L]$. This condition is ensured only when the shell/ACLD system reverts back to its original undeflected equilibrium position.

Differentiating the different components of Eq. (6.127) with respect to time, integrating by parts and imposing the boundary conditions, gives:

$$\dot{E}_n = N_{px}\left[\dot{u}_1(L) - \dot{u}_1(0)\right] - \frac{G_2'\eta_v h_2}{\omega}\int_0^L \dot{\gamma}^2 dx. \tag{6.128}$$

As the second term is strictly negative, hence a globally stable boundary controller with a continuously decreasing energy norm (i.e., $\dot{E}_n < 0$) is obtained when the control action N_{px} takes the following form:

$$N_{px} = -K_g[\dot{u}_1(L) - \dot{u}_1(0)]. \tag{6.129}$$

where K_g is the gain of the boundary controller. Equation (6.130) indicates that the control action is a velocity feedback of the longitudinal displacement of the piezoelectric constraining layer.

It is also important to note that when the active control action N_{px} ceases or fails to operate for one reason or another (i.e., when the control voltage $V_c = 0$ as indicated in Eq. (6.123)), the shell system remains globally stable as indicated by Eq. (6.128). Such inherent stability is attributed to the second term in the equation that quantifies the contribution of the PCLD. Hence, the two terms of Eqs. (6.128) provide quantitative means for weighing the individual contributions of the ACLD and the PCLD to the total rate of energy dissipation of the shell system.

6.7.4.3 Implementation of the Boundary Control Strategy

The globally stable boundary controller can be implemented using Eqs. (6.104) and (6.129) to generate the control voltage V_c as follows:

$$\begin{aligned} V_C &= \left[\frac{h_1}{K_1(d_{31} + \nu_1 d_{32})}\right] N_{px} = -\left(K_g\left[\frac{h_1}{K_1(d_{31} + \nu_1 d_{32})}\right]\right)[\dot{u}_1(L) - \dot{u}_1(0)] \\ &= K_G[\dot{u}_1(L) - \dot{u}_1(0)] \end{aligned} \tag{6.130}$$

where K_G denotes the equivalent gain of the boundary controller such that $K_G = -K_g h_1/(K_1(d_{31} + \nu_1 d_{32}))$. Such an equivalent gain combines the control gain K_g and the piezo-actuator parameters (h_1, K_1, d_{31}, d_{32}, and ν_1), which are generally unknown constants.

Implementation of this control strategy requires that the actuator must be designed as an actuator with self-sensing capabilities using the approaches suggested by Dosch et al. (1992) to measure u_1. It is important to note that the temporal derivatives of u_1 can be determined by monitoring the current of the piezo-sensor rather than its voltage as described, for example, by Miller and Hubbard (1987).

The effectiveness of the boundary controller, given by Eq. (6.130), in suppressing the vibration of a shell treated with ACLD treatment is determined in Example 6.14 when the shell system is subjected to axi-symmetric sinusoidal transverse load acting uniformly over the entire span of the shell.

6.7.4.4 Transverse Compliance and Longitudinal Deflection

The transverse compliance is obtained by solving the partial differential equation of the shell/ACLD system, Eq. (6.121) is obtained using the classical separation of variables approach. In this approach, the transverse deflection w is given by:

$$w = W(x)T(t) \tag{6.131}$$

where $W(x)$ is a spatial function in x and $T(t)$ is a temporal function in t such that $\ddot{T}/T = -\omega^2$. From Eqs. (6.131) and (6.121), the following characteristic equation for the shell/ACLD system is obtained:

$$\lambda^6 - \alpha_1\lambda^4 + \alpha_2\lambda^2 + \alpha_3 = 0 \tag{6.132}$$

where

$$\alpha_1 = g(1+Y),$$

$$\alpha_2 = \frac{1}{D_t}\left[\frac{1-\nu_1^2}{R^2}K_1 + \frac{1-\nu_3^2}{R^2}K_3 - \frac{2(\nu_1-\nu_3)}{Rh_2}G_2^* h\right] - \frac{m}{\pi R D_t}\omega^2,$$

and

$$\alpha_3 = -\frac{1}{D_t}\left[g\left(\frac{1-\nu_1^2}{R^2}K_1 + \frac{1-\nu_3^2}{R^2}K_3\right) - \frac{(\nu_1-\nu_3)^2}{R^2}G_2^* h\right] + \frac{mg\omega^2}{\pi R D_t}. \tag{6.133}$$

where λ is the differential operator with respect to x. The roots of these characteristics equations are: $\pm\delta_1$, $\pm\delta_2$, and $\pm\delta_3$.

Hence, the spatial function $W(x)$ is given by:

$$W(x) = \sum_{i=1}^{3} C_i e^{\delta_i x} + \sum_{j=4}^{6} C_j e^{-\delta_{j-3} x} \tag{6.134}$$

The six coefficients C_i's and C_j's are determined from the six boundary conditions given by Eqs. (6.123) through (6.125) as follows:

$$\begin{bmatrix} R_1 & R_2 & R_3 & R_4 & R_5 & R_6 \\ 1 & 1 & 1 & 1 & 1 & 1 \\ \delta_1^2 & \delta_2^2 & \delta_3^2 & \delta_1^2 & \delta_2^2 & \delta_3^2 \\ S_1 & S_2 & S_3 & S_4 & S_5 & S^6 \\ e^{\delta_1 L} & e^{\delta_2 L} & e^{\delta_3 L} & e^{-\delta_1 L} & e^{-\delta_2 L} & e^{-\delta_3 L} \\ \delta_1^2 e^{\delta_1 L} & \delta_2^2 e^{\delta_2 L} & \delta_3^2 e^{\delta_3 L} & \delta_1^2 e^{-\delta_1 L} & \delta_2^2 e^{-\delta_2 L} & \delta_3^2 e^{-\delta_3 L} \end{bmatrix} \begin{Bmatrix} C_1 \\ C_2 \\ C_3 \\ C_4 \\ C_5 \\ C_6 \end{Bmatrix} = \begin{Bmatrix} q \\ 0 \\ 0 \\ q \\ 0 \\ 0 \end{Bmatrix} \tag{6.135}$$

where

$$\begin{aligned} R_i &= \delta_i^4 D_t - \frac{m}{\pi R}\omega^2 + K_G\,\Delta\dot{u}_i \quad For\ i = 1,2,3 \\ &= \delta_{i-3}^4 D_t - \frac{m}{\pi R}\omega^2 + K_G\,\Delta\dot{u}_i \quad For\ i = 4,5,6 \end{aligned}, \tag{6.136}$$

with

$$K_G = \left(\frac{K_1\nu_1}{R} - \frac{G_2^* h}{h_2}\right)\left(1 + \frac{d_{32}}{d_{31}}\right)\frac{d_{31}K_g}{h_1}. \tag{6.137}$$

and

$$\Delta\dot{u}_i = i\omega g\left[\frac{Ag-B}{2g(\sqrt{g}+\delta_i)}\left(1-e^{-\sqrt{g}L}\right) + \frac{Ag-B}{2g(\sqrt{g}-\delta_i)}\left(e^{\sqrt{g}L}-1\right) + \frac{A\,\delta_i^2-B}{\delta_i(\delta_i^2-g)}\left(e^{\delta_i L}-1\right)\right] \text{ for}$$

$i = 1,\ldots,3$

$$\Delta\dot{u}_i = i\omega g\left[\frac{Ag-B}{2g(\sqrt{g}-\delta_i)}\left(1-e^{-\sqrt{g}L}\right) + \frac{Ag-B}{2g(\sqrt{g}+\delta_i)}\left(e^{\sqrt{g}L}-1\right) + \frac{A\,\delta_i^2-B}{\delta_i(\delta_i^2-g)}\left(e^{-\delta_i L}-1\right)\right] \text{ for}$$

$i = 4,\ldots,6$

where $A = \left(\frac{\nu_1}{gL^2\frac{R}{L}} - \frac{1}{\frac{K_1}{K_3}+1}\frac{h}{L}\right)$ and $B = \left(\frac{\nu_1}{\left[\frac{K_3}{K_1}+1\right]\frac{R}{L}} - \frac{\nu_3}{\left[\frac{K_1}{K_3}+1\right]\frac{R}{L}}\right)$

Also,

$$S_i = \left(\delta_i^4 D_t - \frac{m}{\pi R}\omega^2\right)e^{\delta_i L} + K_G\,\Delta\dot{u}_i \;\; For\; i = 1,2,3$$
$$= \left(\delta_{i-3}^4 D_t - \frac{m}{\pi R}\omega^2\right)e^{-\delta_{i-3} L} + K_G\,\Delta\dot{u}_i \;\; For\; i = 4,5,6 \qquad (6.138)$$

Hence, the transverse deflection w at any x and t can then be determined from Eqs. (6.131) and (6.134). For a unit transverse load q, the resulting deflection is the compliance of the shell.

Example 6.13 If the constrained layer and base shell are manufactured from the same material, and the thickness of the viscoelastic beam h_2 is set equal to zero, that is, the shell is un-damped, show that Eq. (6.121) reduces to the equation of motion describing the dynamics of a homogeneous axi-symmetrical circular shell of thickness $h_1 + h_3$.

Solution

Setting $E_1 = E_3 = E$, and $\nu_1 = \nu_3 = \nu$, then $K_1 = \frac{Eh_1}{1-\nu^2}$ and $K_3 = \frac{Eh_3}{1-\nu^2}$. Also, dividing Eq. (6.121) by g, it reduces to:

$$\frac{1}{g}D_t w_{,xxxxxx} - D_t(1+Y)w_{,xxxx} + \frac{1}{g}\frac{E(h_1+h_3)}{R^2}w_{,xx}$$
$$-g\frac{E(h_1+h_3)}{R^2}w - \frac{m}{\pi R}w_{,tt} + \frac{1}{g}\frac{m}{\pi R}w_{,xxtt} = 0$$

But, as g tends to ∞ as $h_2 = 0$, then:

$$D_t(1+Y)w_{,xxxx} + \frac{E(h_1+h_3)}{R^2}w + \frac{m}{\pi R}w_{,tt} = 0 \qquad (6.139)$$

Also as $Y = \frac{K_1K_3}{K_1+K_3}\frac{h^2}{D_t}$, $D_t = \frac{E(h_1^3+h_3^3)}{12(1-\nu^2)}$ and $h = \frac{1}{2}(h_1+h_3)$ when $h_2 = 0$, then:

$$D_t(1+Y) = D_t\left(1 + \frac{E(h_1h_3)}{(1-\nu^2)(h_1+h_3)}\frac{(h_1+h_3)^2}{4D_t}\right)$$
$$= \left(\frac{E(h_1^3+h_3^3)}{12(1-\nu^2)} + \frac{E(h_1h_3)}{(1-\nu^2)(h_1+h_3)}\frac{(h_1+h_3)^2}{4}\right) = \frac{E(h_1+h_3)^3}{12(1-\nu^2)}$$

Then, Eq. (6.139) reduces to:

$$\frac{E(h_1+h_3)^3}{12(1-\nu^2)}w_{,xxxx} + \frac{E(h_1+h_3)}{R^2}w + \frac{m}{\pi R}w_{,tt} = 0 \qquad (6.140)$$

which is the same as equation of motion of describing the dynamics of a homogeneous axi-symmetrical circular shell of thickness $h_1 + h_3$.

Example 6.14 Consider an ACLD treatment used to fully treat a simply supported aluminum shell that is 0.3048 m long, 0.005 m thick, and 0.60 m in the outside radius. The damping treatment consists of an acrylic-based VEM, which is 0.005 m thick and has a complex shear modulus $G_2 = 20\ (1 + 0.5i)\ \mathrm{MN\,m^{-2}}$.

Assume that the viscoelastic core is constrained by an active polymeric piezoelectric film (PolyVinyliDene Fluoride, PVDF) whose thickness, h_1, and Young's modulus, E_1, are 0.005 m and 2.25 $\mathrm{GN\,m^{-2}}$, respectively. The piezoelectric strain constants d_{31} and d_{32} are 23×10^{-12} and $3 \times 10^{-12}\ \mathrm{m\,V^{-1}}$, respectively. The density of the piezoelectric film (PVDF) is 1800 $\mathrm{kg\,m^{-3}}$.

Determine the compliance at the mid-span using the mechanical compliance approach when the ACLD treatment is controlled by the boundary control approach described in Section 6.7.4 with control gain K_g of $10^6\ \mathrm{N\,(m\,s)^{-1}}$. Compare the performance with that of a PCLD treatment, that is, when $K_g = 0$.

Note that the aluminum shell under consideration is subject to sinusoidal transverse loading that is uniformly distributed over its entire span.

Solution

Figure 6.27a shows the compliance of the shell/ACLD system for the gain K_g of the boundary control set to $10^6\ \mathrm{N\,(m\,s)^{-1}}$. Also shown in the figure is the compliance of the uncontrolled shell that is treated with the PCLD treatment. In that case, the control loop that regulates the interaction between the piezo-sensor and the piezo-actuator is maintained open, that is, $K_g = 0$. It is evident that the ACLD treatment has effectively attenuated the vibration of the shell over the frequency band under consideration as compared to the conventional PCLD treatment.

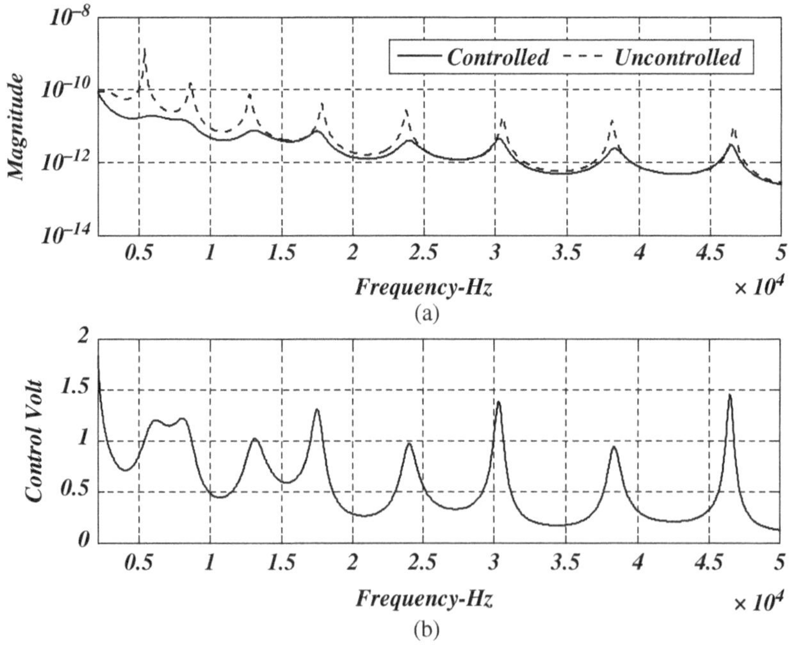

Figure 6.27 Performance of the ACLD with the boundary controller. (a) Compliance. (b) Control voltage.

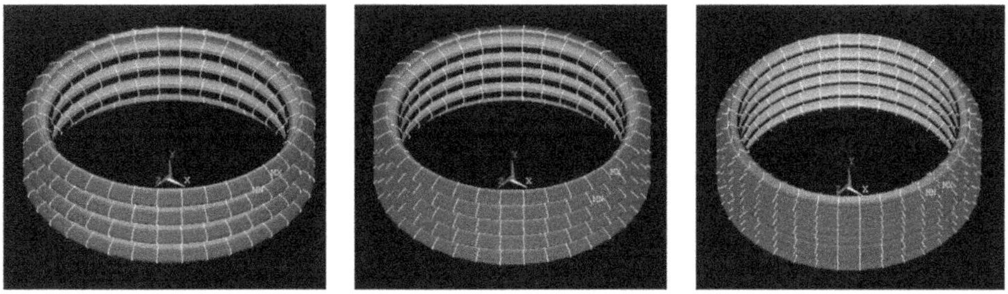

Figure 6.28 Shape of the dominant modes of vibration of the shell/ACLD system.

The corresponding control voltage used to activate the piezo-constraining layer is shown in Figure 6.27b for different levels of the control gains. Note that effective vibration attenuation can be achieved by the devised boundary control strategy without the need for excessively high control voltages. A maximum voltage of 1.81 V is used to reduce the amplitude of vibration from 1.42E-9m to 1.79E-11m; that is, an attenuation of 98.74% is achieved.

Figure 6.27 shows also the dominant modes of vibration in the considered frequency range. The first three modes occur at 5377, 8585, and 12,740 Hz. The corresponding mode shapes of these three modes are displayed in Figure 6.28 as obtained by using a commercial finite element method (ANSYS). Note that the predictions of ANSYS for the shell/ACLD natural frequencies are 5667, 8912, and 12,985 Hz.

Close agreement is evident between the predictions of the system natural frequencies by ANSYS and the mechanical compliance approach described in Section 6.7.4.5. Prediction errors of 5.1, 3.67, and 1.88% are obtained at the considered three modes.

6.7.5 Energy Dissipated in the ACLD Treatment of an Axi-Symmetric Shell

The energy dissipated in the ACLD treatment of the shell can be determined by combining Eqs. (6.111) and (6.112) to yield the following expression:

$$E_D = \pi R h_2 G_2' \eta_v \int_0^L \gamma^2 \, dx \tag{6.141}$$

In order to determine the shear strain γ, Eq. (6.114) is divided by K_1 and Eq. (6.115) K_3. The resulting two equations are subtracted from each other to yield:

$$z_{,xx} - gz = \left[\frac{\nu_1 - \nu_3}{R} + gh\right] w_{,x} \tag{6.142}$$

where $z = u_1 - u_3$. Differentiating Eq. (6.134) with respect to x and substituting into Eq. (6.142) gives the spatial component $Z(x)$ of $z(x,t)$ as:

$$Z_{,xx} - gZ = \left[\frac{\nu_1 - \nu_3}{R} + gh\right]\left(\sum_{i=1}^{3} \delta_i C_i e^{\delta_i x} - \sum_{j=4}^{6} \delta_{j-3} C_j e^{-\delta_{j-3} x}\right) = f(x) \tag{6.143}$$

Equation (6.143) is solved to yield the spatial distribution of z along the shell longitudinal axis using the convolution integral as follows:

$$Z(x) = \int_0^x \bar{h}(x-y) f(y) dy \tag{6.144}$$

where $\bar{h}(x)$ is the unit impulse response of the system $(Z_{,xx} - gZ)$, which is given by:

$$\bar{h}(x) = sinh\left(\sqrt{g}x\right) / \sqrt{g} \tag{6.145}$$

Once the spatial distribution of $z(x)$ along the shell longitudinal axis is determined, then the shear strain spatial distribution can be determined from:

$$\gamma = hW_{,x} + Z \tag{6.146}$$

Equation (6.141) can then be used for computing the energy dissipated by the ACLD treatment.

Example 6.15 Consider the ACLD treatment used to fully treat a simply supported aluminum shell described in Example 6.14. Determine the shape and the energy dissipation in the ACLD treatment when it is controlled by a boundary controller with control gain K_g of 10^6 N (m s)$^{-1}$. Compare the performance with that of a PCLD treatment, that is, when $K_g = 0$. Assume that the shell/ACLD assembly is excited sinusoidally at a frequency of 5377 Hz, which corresponds to mode (4,0) as shown in Figure 6.28.

Solution

Figure 6.29 displays the shape and the energy dissipation in the ACLD treatment when it is controlled by a boundary controller with control gain K_g of 10^6 N (m s)$^{-1}$.

Note that the predicted theoretical shape of the shell/ACLD, shown in Figure 6.29a, matches that obtained by the ANSYS finite element model. The corresponding energy dissipated due to the active control is shown in Figure 6.29b.

6.8 Summary

This chapter presented the energy dissipation characteristics of various types of viscoelastic damping treatments as applied to vibrating rods, beams, plates, and shells. Passive damping treatments in constrained and unconstrained configurations were considered. The energy dissipation metrics for these two configurations were derived indicating that the use of the constrained VEM configuration results in significant improvement of the

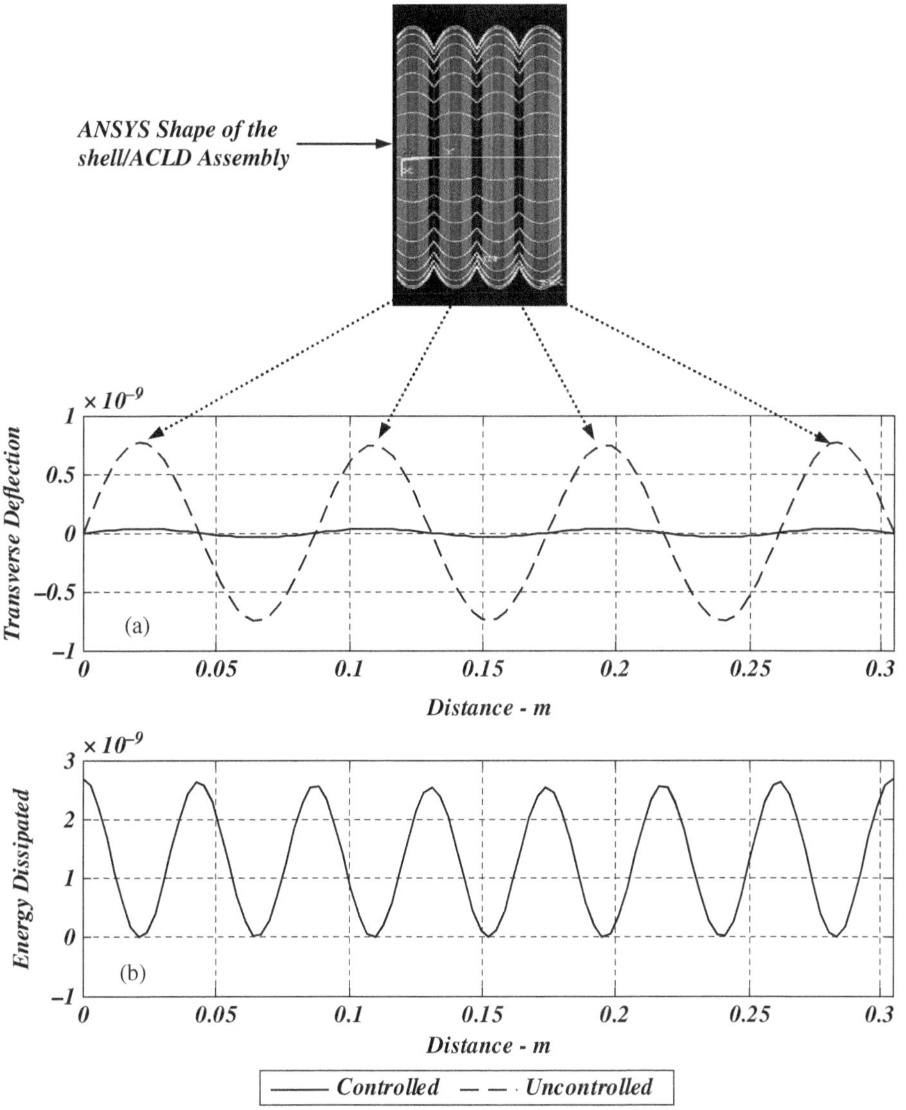

Figure 6.29 Shape and energy dissipation in the shell/ACLD system at excitation frequency 5377 Hz corresponding to Mode (4,0).

damping characteristics compared to those obtained with the unconstrained configuration. Also, ACLD treatments, consisting of viscoelastic cores constrained by active piezoelectric layers, were presented as effective means for enhancing the damping characteristics and compensating for the performance degradation of passive treatments. The boundary control strategy was presented, which ensures global stability of the ACLD treatment. The individual contributions of the passive and active components of the ACLD were isolated indicating that failure of the active component will not result in unstable operation of the damping treatment.

References

Alam, N. and Asnani, N.T. (1984). Vibration and damping analysis of a multi-layered cylindrical shell, part I: theoretical analysis. *AIAA Journal* 22 (6): 803–810.

Alberts, T.E. and Xia, H.C. (1995). Design and analysis of fiber enhanced viscoelastic damping polymers. *Journal of Vibration and Acoustics* 117 (4): 398–404.

Banks, H.T., Smith, R.C., and Wang, Y. (1995). The modeling of piezoceramic patch interactions with shells, plates, and beams. *Quarterly of Applied Mathematics* LIII (2): 353–381.

Baz A., Active Constrained Layer Damping, US Patent 5,485,053, filed October 15 1993 and issued January 16 1996.

Baz, A. (1997a). Boundary control of beams using active constrained layer damping. *ASME Journal of Vibration and Acoustics* 119 (2): 166–172.

Baz, A. (1997b). Dynamic boundary control of beams using active constrained layer damping. *Journal of Mechanical Systems & Signal Processing* 11 (6): 811–825.

Baz, A. (1997c). Optimization of energy dissipation characteristics of active constrained layer damping. *Journal of Smart Materials & Structures* 6: 360–368.

Baz, A. (1998). Robust control of active constrained layer damping. *Journal of Sound & Vibration* 211 (3): 467–480.

Baz, A. and Ro, J. (1994). Actively-controlled constrained layer damping. *Sound and Vibration Magazine* 28 (3): 18–21.

Baz, A. and Ro, J. (1995). Optimum design and control of active constrained layer damping. *ASME Journal of Vibration and Acoustics* 117: 135–144.

Baz, A. and Ro, J. (1996). Vibration control of plates with active constrained layer damping. *Journal of Smart Materials and Structures* 5: 261–271.

Butkovskiy, A.G. (1969). *Distributed Control Systems*. New York, NY: Elsevier Publishing Co., Inc.

Chaudhry Z., Lalande F. and Rogers C. A., "Special considerations in the modeling of induced strain actuator patches bonded to shell structures", Proceedings of the SPIE Conference on Smart Structures, Vol. 2190, Orlando, FL, pp. 563–570, 1994.

Demoret K. and Torvik P., "Optimal length of constrained layers on a substrate with linearly varying strains", Proceedings of the ASME Design Engineering Technology Conference, DE-Vol. 84–3, Boston, MA, pp. 719–726, 1995.

Deng, Y. (1995). Boundary stabilization of a thin circular cylindrical shell subject to axisymmetric deformation. *Dynamics and Control* 5: 205–218.

DiTaranto, R.A. (1972). Free and forced response of a laminated ring. *The Journal of the Acoustical Society of America* 53 (3): 748–757.

Dosch, J.J., Inman, D.J., and Garcia, E. (1992). A self-sensing piezoelectric actuator for collocated control. *Journal of Intelligent Material Systems and Structures* 3: 166–184.

Forward, R.L. (1981). Electronic damping of orthogonal bending modes in a cylindrical mast experiment. *Journal of Spacecraft* 18: 11–17.

ANSI/IEEE Std. 176-1987. IEEE Standard on Piezoelectricity, IEEE Standards Organization, Piscataway, NJ, 1987.

Jones, I.W. and Salerno, V.L. (1966). The effect of structural damping on the forced vibration of cylindrical sandwich shells", Transactions of the American Society of Mechanical Engineers. *Journal of Engineering for Industry* 88: 318–324.

Kerwin, E.M. (1959). Damping of flexural waves by a constrained viscoelastic layer. *The Journal of the Acoustical Society of America* 31 (7): 952–962.

Leissa, A.W. (1973). *Vibration of Shells (NASA SP-288)*. Washington, DC: Government Printing Office.

Leissa, A.W. and Iyer, K.M. (1981). Modal response of circular cylindrical shells with structural damping. *Journal of Sound and Vibration* 77 (1): 1–10.

Lester, H.C. and Lefebvre, S. (1993). Piezoelectric actuator model for active sound and vibration control cylinders. *Journal of Intelligent Material System and Structures* 4: 295–306.

Lu, Y.P., Douglas, B.E., and Thomas, E.V. (1973). Mechanical impedance of damped three-layered sandwich rings. *AIAA Journal* 11 (3).

Mandal, N.K. and Biswas, S. (2005). Vibration power flow: a critical review. *The Shock and Vibration Digest* 37 (1): 3–11.

Markus, S. (1976). Damping properties of layered cylindrical shells vibrating in axially symmetric modes. *Journal of Sound and Vibration* 48 (4): 511–524.

Markus, S. (1979). Refined theory of damped axisymmetric vibration of double-layered cylindrical shells. *Journal of Mechanical Engineering Science* 21 (1): 33–37.

Meirovitch, L. (1967). *Analytical Methods in Vibrations.* New York, NY: Macmillan Publishing Co., Inc.

Miller S. and Hubbard Jr. J., "Observability of a bernoulli–euler beam using pvf2 as a distributed sensor", Seventh Conference on Dynamics & Control of Large Structures, VPI & SU, Blacksburg, VA, pp. 375–930, 1987.

Nashif, A., Jones, D.I., and Henderson, J.P. (1985). *Vibration Damping.* New York, NY: John Wiley & Sons, Inc.

Pan, H.H. (1969). Axisymmetric vibrations of a circular sandwich shell with a viscoelastic core layer. *Journal of Sound and Vibration* 9 (2): 338–348.

Plunkett, R. and Lee, C.T. (1970). Length optimization for constrained viscoelastic layer damping. *Journal of Acoustical Society of America* 48 (1 (Part 2)): 150–161.

Ramesh, T.C. and Ganesan, N. (1994). Finite element analysis of cylindrical shells with a constrained viscoelastic layer. *Journal of Sound and Vibration* 172 (3): 359–370.

Ramesh, T.C. and Ganesan, N. (1995). Vibration and damping analysis of cylindrical shells with constrained damping treatment – a comparison of three theories. *Journal of Vibration and Acoustics* 117: 213–219.

Ray, M.C. and Baz, A. (1997). Optimization of energy dissipation of active constrained layer damping treatments of plates. *Journal of Sound and Vibration* 208 (3): 391–406.

Shen, I.Y. (1994). Hybrid damping through intelligent constrained layer treatments. *ASME Journal of Vibration and Acoustic* 116 (3): 341–348.

Sonti, V.R. and Jones, J.D. (1996). Curved piezo-actuator model for active vibration control of cylindrical shells. *AIAA Journal* 34 (5): 1034–1040.

Tzou, H.S. (1993). *Piezoelectric Shells: Distributed Sensing and Control of Continua.* Dordrecht, The Netherlands: Kluwer Academic Publishers.

Zhou, S., Liang, C., and Rogers, C.A. (1993). Impedance modeling of two-dimensional piezoelectric actuators bonded on a cylinder. *Adaptive Structures and Material Systems ASME* 35: 247–255.

6.A Basic Identities

Let $G^* = G(\cos\theta + i\sin\theta) = G\cos\theta[1 + i\tan\theta] = G'[1 + i\eta_g]$
that is,

$$G' = G\cos\theta \text{ and } \eta_g = \tan\theta. \tag{6.A.1}$$

$$B^* = \sqrt{\frac{h_1 h_2 E_2}{G^*}} = \sqrt{\frac{h_1 h_2 E_2}{G(\cos\theta + i\sin\theta)}} = \sqrt{\frac{h_1 h_2 E_2}{G}(\cos\theta - i\sin\theta)} = \sqrt{\frac{h_1 h_2 E_2}{G}e^{-i\theta}} = B_0 e^{-i\frac{\theta}{2}}$$

or,

$$B^* = B_0[c(\theta/2) - is(\theta/2)] \text{ and } B_0 = \sqrt{\frac{h_1 h_2 E_2}{G}} \tag{6.A.2}$$

where $c(\theta/2) = \cos(\theta/2)$ and $s(\theta/2) = \sin(\theta/2)$.

$$\frac{x}{B^*} = \frac{x}{B_0[c(\theta/2) - is(\theta/2)]} = \frac{x}{B_0}[c(\theta/2) + is(\theta/2)] = u + iv \tag{6.A.3}$$

$$\frac{L}{B^*} = \frac{L}{B_0}[c(\theta/2) + is(\theta/2)] \tag{6.A.4}$$

$$\sinh(x/B^*) = \sinh(u + iv) = \sinh u \cos v + i\cosh u \sin v \tag{6.A.5}$$

$$\begin{aligned}\sinh^2(x/B^*) &= \sinh^2 u\cos^2 v + \cosh^2 u\sin^2 v \\ &= \sinh^2 u + \sin^2 v = \sinh^2\left[\frac{x}{B_o}c(\theta/2)\right] + \sin^2\left[\frac{x}{B_o}s(\theta/2)\right]\end{aligned} \tag{6.A.6}$$

$$\int_0^{L/2} \sinh^2(x/B^*)\,dx = \frac{B_0}{2c(\theta/2)s(\theta/2)}\left[\sinh\left[\frac{L}{B_0}c(\theta/2)\right]s(\theta/2) - \sin\left[\frac{L}{B_0}s(\theta/2)\right]c(\theta/2)\right] \tag{6.A.7}$$

$$\begin{aligned}\cosh^2\left(\frac{L}{2B^*}\right) &= \cosh^2\left(\frac{L}{2B_0}c(\theta/2)\right)\cos^2\left(\frac{L}{2B_0}s(\theta/2)\right) \\ &\quad + \sinh^2\left(\frac{L}{2B_0}c(\theta/2)\right)\sin^2\left(\frac{L}{2B_0}s(\theta/2)\right) \\ &= \frac{1}{2}\left[\cosh\left(\frac{L}{2B_0}c(\theta/2)\right) + \cos\left(\frac{L}{2B_0}s(\theta/2)\right)\right]\end{aligned} \tag{6.A.8}$$

6.B Piezoelectricity*

6.B.1 Piezoelectric Effects

Figure 6.B.1 displays the two piezoelectric effects: the "Forward" and the "Reverse" effects. In the forward effect, a piezoelectric film generates an output electric voltage when it is subjected to an input mechanical stress as shown in Figure 6B.1a. In the reverse effect, the film develops an output mechanical deformation when an input electric voltage is applied across its surfaces as shown in Figure 6B.1b. Hence, the forward effect makes the film act as a sensor whereas in the reverse effect, the film acts as an actuator.

6.B.2 Basic Constitutive Equations

The constitutive equations describe the interaction between the mechanical and electrical characteristics of piezoelectric films. In the one-dimensional case, these equations, as given in the IEEE Standard on Piezoelectricity (1987), are:

$$S_1 = s_{11}^E T_1 + d_{31} E_3, \tag{6.B.1}$$

and

$$D_3 = d_{31} T_1 + e_{33}^T E_3 \tag{6.B.2}$$

Equation (6.B.1) describes the mechanical strain S_1, in direction 1, due to a mechanical stress T_1, in direction 1, and an electrical field E_3, in direction 3. In Eq. (6.B.2), the electric displacement D_3, in direction 3, is generated by the mechanical stress T_1 and the electric field E_3. The directions assigned in the constitutive equations are defined in Figure 6.B.2 in relation to the poling axis P of the film.

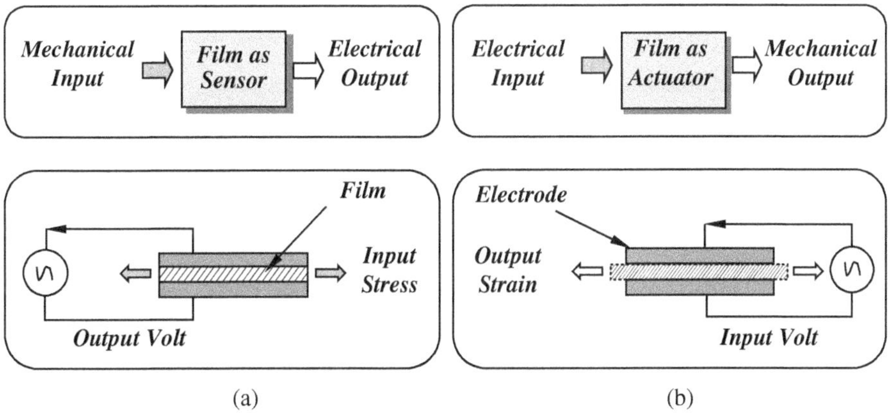

Figure 6.B.1 Forward and reverse piezoelectric effects.

* The word "*piezo*" is derived from the Greek word "*press*" and the discovery of the piezoelectric phenomenon, in 1880, is credited to Pierre and Jacque Curie.

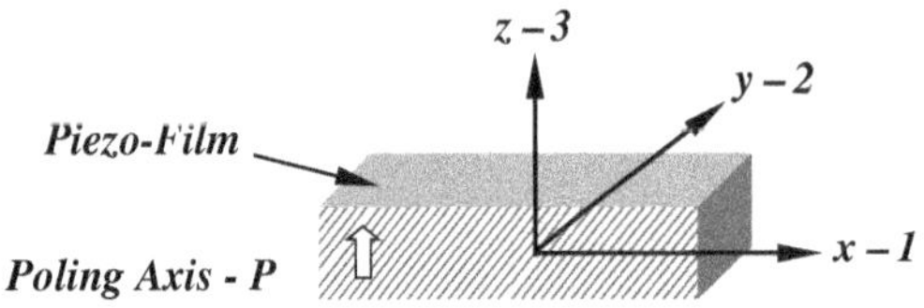

Figure 6.B.2 Coordinate system of piezoelectric films.

Table 6.B.1 Piezoelectric variables: definition and units.

Symbol	Definition	Units
S_1	Strain in direction 1	dimensionless
T_1	Stress in direction 1	$\mathrm{N\,m^{-2}}$
s_{11}^{E}	Compliance, in direction 1, at constant electric field intensity E	$\mathrm{m^2\,N^{-1}}$
d_{31}	Piezoelectric strain constant, measured in direction, due to an electric field applied in direction 3	$\mathrm{m\,V^{-1}}$ or Coulomb $\mathrm{N^{-1}}$
E_3	Electric field intensity (voltage across film per unit thickness)	$\mathrm{V\,m^{-1}}$
D_3	Electric displacement (charge per unit area or electric flux density)	Coulomb $\mathrm{m^{-2}}$
ε_{33}^{T} [a]	Permittivity (or dielectric constant) in direction 3	Farad $\mathrm{m^{-1}}$

[a] $\varepsilon_{33}^{T} = \varepsilon_r.\varepsilon_o$, with ε_r = relative permittivity and ε_o = permittivity of free space = 8.85E – 12 Farad/m.

The definition and units of the different terms in Eqs. (6.B.1) and (6.B.2) are given in Table 6.B.1. Typical values of the piezoelectric constants for different types of piezoelectric materials are given in Table 6.B.2.

Table 6.B.2 Physical parameters of typical piezoelectric materials.

Material	s_{11}^{3} ($\mathrm{m^2\,N^{-1}}$)	d_{31} ($\mathrm{m\,V^{-1}}$)	ε_{33}^{T} (Farad $\mathrm{m^{-1}}$)	Density ($\mathrm{kg\,m^{-3}}$)
PZT-4	0.159E-10	−180 E-12	1.50E-8	7600
PZT-5H	0.165E-10	−274 E-12	2.49E-8	7300
PVDF	0.500E-9	23 E-12	110E-12	1780

PZT, Lead zirconate titanate (ceramic) and *PVDF*, PolyVinyliDene Fluoride (Polymer).

Problems

6.1 Consider the PCLD treatment shown in Figure 6.1, derive the equation of motion and boundary conditions when the VEM is assumed to be made of "Functionally Graded Material" (FGM) such that its shear modulus G^* is linearly varying as follows

$$G^* = G\left(1 + \eta_g i\right).[1 + \alpha x/(L/2)] \text{ for } 0 \le x \le L/2$$

$$\text{and } G^* = G\left(1 + \eta_g i\right).[1 - \alpha x/(L/2)] \text{ for } -L/2 \le x \le 0$$

where α is a constant.

Figure P6.1 Shear modulus distribution along a VEM.

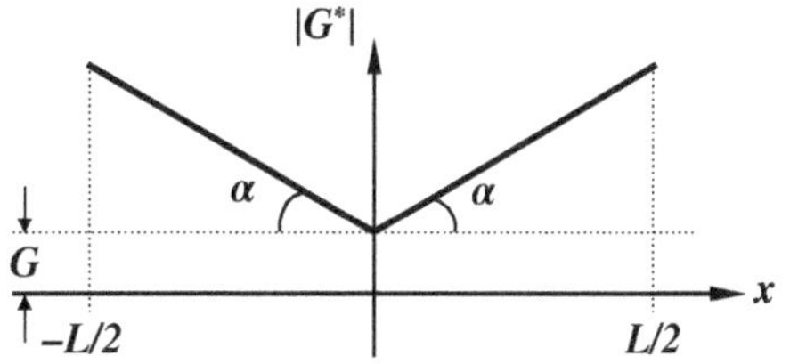

Determine:

a) the longitudinal deflection u and shear strain γ.
b) the energy dissipated per cycle.
c) the loss factor.

Compare the performance of the PCLD/FGM with that of a conventional PCLD with VEM that has constant shear modulus. Assume $\alpha = 1$ and 10.

6.2 Consider the PCLD treatment shown in Figure 6.1, derive the equation of motion and boundary conditions when the base structure is assumed to be subject to linearly varying strain as follows

Figure P6.2 Strain distribution along a VEM.

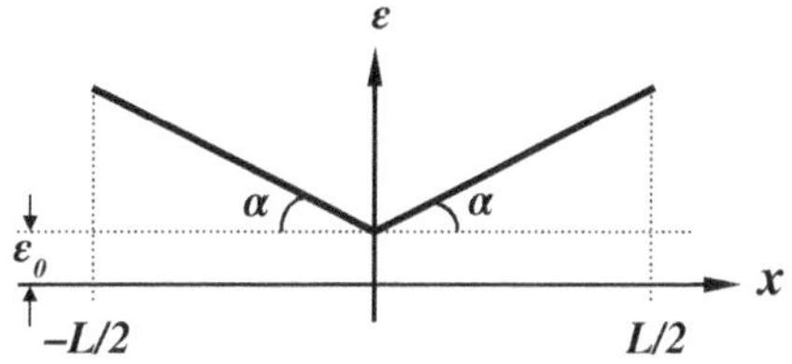

$\varepsilon = \varepsilon_0.[1 + \alpha x/(L/2)]\,\text{for}\,0 \le x \le L/2$

and $\varepsilon = \varepsilon_0.\ [1 - \alpha x/(L/2)]$ for $-L/2 \le x \le 0$ (α is a constant).

Determine (Demoret and Torvik 1995):

a) the longitudinal deflection u and shear strain γ.
b) the energy dissipated per cycle.
c) the loss factor.

Compare the performance with that obtained when the base structure is subjected to constant strain ε_0. Assume $\alpha = 1$ and 10.

6.3 Consider the ACLD treatment shown in Figure 6.7, derive the equation of motion and boundary conditions when the VEM is assumed to be made of FGM such that its shear modulus G^* linearly varies as follows

Figure P6.3 Shear modulus distribution along VEM.

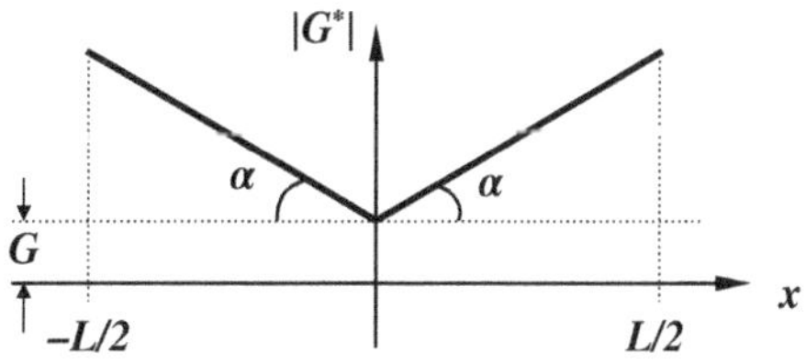

$$G^* = G\left(1+\eta_g i\right).[1+\alpha x/(L/2)] \text{ for } 0 \le x \le L/2$$

$$\text{and } G^* = G\left(1+\eta_g i\right).[1-\alpha x/(L/2)] \text{ for } -L/2 \le x \le 0$$

where α is a constant.

Determine:

a) the longitudinal deflection u and shear strain γ
b) the energy dissipated per cycle
c) the loss factor

Compare the performance of the ACLD/FGM with that of a conventional ACLD with VEM that has constant shear modulus. Assume α = 1 and 10.

6.4 Consider the fiber-reinforced VEM shown in Figure P6.4, which has the representative volume shown in Figure P6.5.

Determine the equation that governs the quasi-static equilibrium of the unit volume in terms of the geometrical and physical properties of the fiber and VEM. Determine the longitudinal deflection of the fibers and shear strain of the VEM.

Use this information to show that the average normal stress σ_a and average strain ε_a are given by (Alberts and Xia 1995)

$$\sigma_a = \frac{\sigma}{2}\frac{d_f^2}{\left(t_v+d_f\right)^2}$$

$$\text{and } \varepsilon_a = u\left(L_f/4, t_v/2\right)/\left(L_f/4\right) = \frac{\sigma}{2E_f}\left(1+\frac{1}{\beta}\coth\beta\right)$$

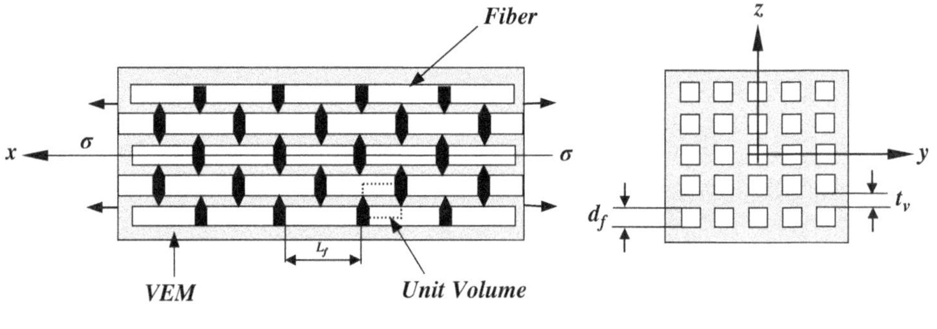

Figure P6.4 Fiber-reinforced composite.

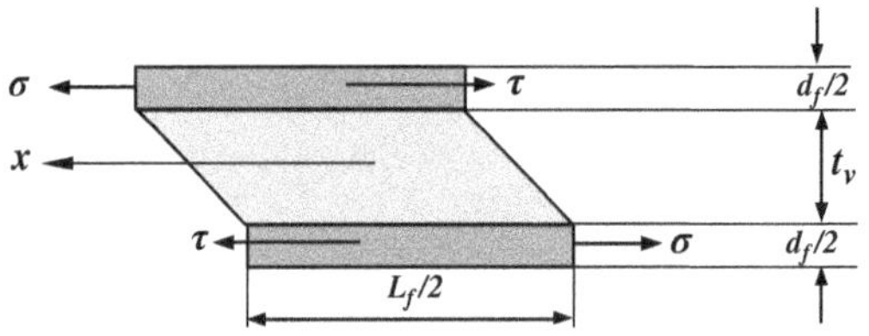

Figure P6.5 Unit volume.

where $\beta = \dfrac{L_f\sqrt{G_v}}{\sqrt{2t_v d_f E_f}}$.

with E_f and G_v denoting the elastic modulus of the fibers and the shear modulus of the VEM, respectively.

6.5 Consider the free-free rod/*VEM* system shown in Figure P6.6. The rod is made of aluminum with width of 0.025 m, thickness of 0.0025 m, and length of 1 m. The VEM has width of 0.025 m, thickness = 0.0025 m, and density of 1100 kg m^{-3}. The storage modulus and loss factor of the VEM are 15.3 MPa and 1, respectively. The VEM is constrained by an aluminum constraining layer that is 0.025 m wide and 0.0025 m thick.

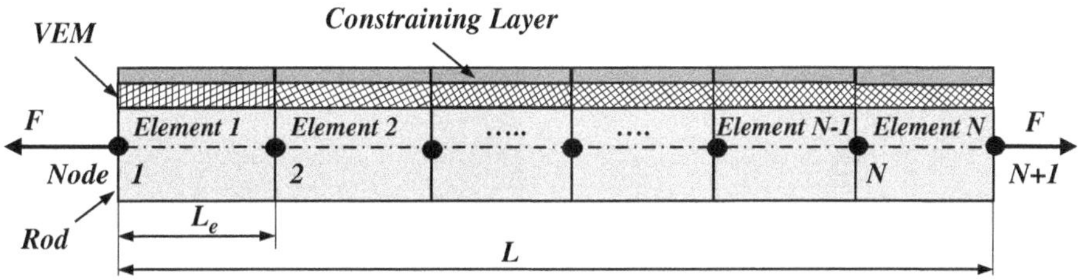

Figure P6.6 A free-free rod treated with a constrained VEM treatment.

Determine the shear strain γ and the energy dissipation E_d distributions along the VEM (i.e., plot γ-x and E_d-x).

Assume the rod/VEM system is modeled by a ten-element finite element model ($N = 10$) with the two ends of the rod (nodes 1 and 11) subjected to sinusoidal excitations at 1000 Hz of magnitudes 10 and −10 KN.

6.6 Consider the free-free rod/VEM system shown in Figure P6.7. The geometrical and physical properties of the rod/VEM/constraining layers are described in Problem 6.5.

Derive the distributed-parameter equations of motion that govern the longitudinal deflections u_3 of the rod and u_1 of the constraining layer using Hamilton's principle.

Derive also the associated boundary conditions corresponding to free-free constraining layer and to the loading of the two ends of the rod by sinusoidal excitations at 1000 Hz of magnitudes 10 and −10 KN.

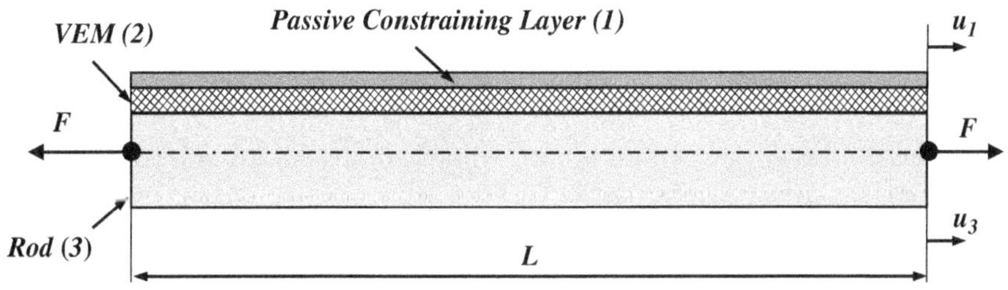

Figure P6.7 A free-free rod treated with an unconstrained VEM treatment.

Determine the shear strain γ and the energy dissipation E_d distributions along the VEM (i.e., plot γ-x and E_d-x) using the MATLAB subroutine ***bvp4c***.

Compare the predictions of the distributed-parameter model with the predictions of the finite element model developed in Problem 6.6.

6.7 Consider the free-free rod/VEM system shown in Figure P6.8. The geometrical and physical properties of the rod/VEM layers are as described in Problem 6.5.

Assume that the constraining layer is active and manufactured from piezoelectric material that is 0.025 m wide and 0.0025 m thick. The piezo-layer has Young's modulus of 60 GPa and density of 7800 kg m^{-3}.

Derive the distributed-parameter equations of motion that govern the longitudinal deflections u_3 of the rod and u_1 of the active constraining layer using Hamilton's principle.

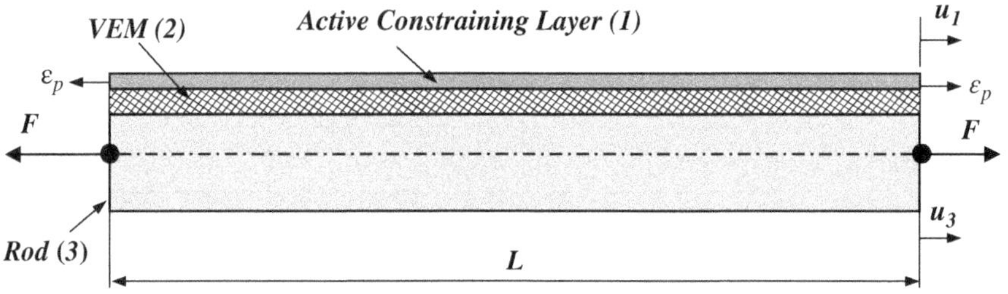

Figure P6.8 A free-free rod treated with an active constrained VEM treatment.

Derive also the associated boundary conditions corresponding to free-free active constraining layer and to the loading of the two ends of the rod by sinusoidal excitations at 1000 Hz of magnitudes 10 and −10 KN.

Determine an expression for the control law necessary to predict the piezo-strain ε_p in order to ensure that the time rate of change of the kinetic and potential energies of the rod/VEM/constraining layer system is strictly negative.

Determine the shear strain γ and the energy dissipation E_d distributions along the VEM (i.e., plot γ-x and E_d-x) for different values of the control gain K_g.

Compare the predictions of the distributed-parameter model of the assembly when $K_g = 0$.

6.8 Consider the idealized ACLD system shown in Figure 6.7, which is governed by Eqs. (6.23) and (6.24). Show that the transfer function of the system is given by:

$$\frac{\bar{u}}{\varepsilon_p} = \frac{sinh(\lambda\bar{x})}{2\lambda\ cosh(\lambda/2)}$$

where $\bar{u} = u/L, \bar{x} = x/L, \lambda^2 = \bar{\omega}^{*2}(\alpha\bar{s}^2 + 1), \bar{\omega}^* = L/B^*, \alpha = mh_1\omega_0^2/G^*, \bar{s} = s/\omega_0$, and ω_0 is a characteristic frequency.

If the piezoelectric control strain ε_p is generated according to the globally stable control law given by Eq. (6.27), show that the transfer function of the system reduces to:

$$\frac{\bar{s}[\bar{u}(1/2)-\bar{u}(-1/2)]}{\varepsilon_p}=\frac{\bar{s}\,sinh(\lambda/2)}{\lambda\,cosh(\lambda/2)}$$

leading to the following block diagram of the closed-loop ACLD system shown in Figure P6.9:

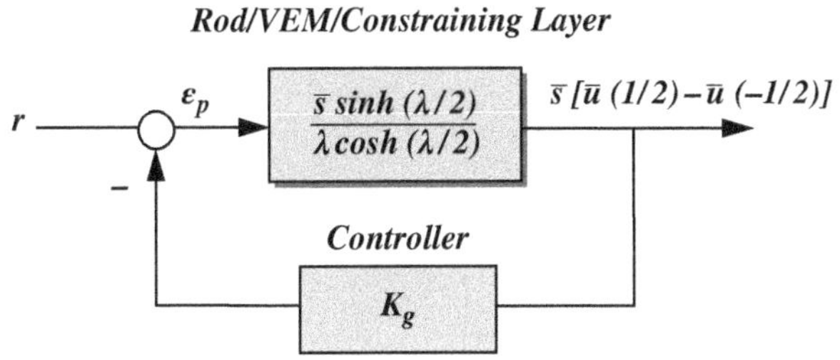

Figure P6.9 Block diagram of the closed-loop active constrained layer damping system.

6.9 Plot the root locus of the idealized ACLD system shown in Figure 6.7 and described by the open-loop transfer function given in Problem 6.8. Discuss the stability of the closed-loop system for different values of the control gain K_g.

Investigate the effect of the parameters of $\bar{\omega}^*$, and α on the stability characteristics of the system.

6.10 Consider the idealized ACLD system shown in Figure 6.7. Assume that the control action ε_p is generated using the following dynamic controller:

$$\dot{v}=av+bF$$

$$\varepsilon_p=-K_g c v$$

where $F=[\dot{u}(L/2)-\dot{u}(-L/2)]$, K_g denotes the control gain, and the parameters of the control law *a, b,* and *c* satisfy the following relationships called the Kalman–Yakubovitch lemma (Baz 1997a):

$$a^T P+Pa=-Q$$

$$Pb=c^T$$

with P and Q are symmetric positive definite matrices. If the total energy E_n/b of the system, as described by Eq. (6.25), is modified to account for the energy of the dynamic controller $v^T Pv$ such that:

$$E_n/b=\frac{1}{2}\left(E_2h_2\int_{-L/2}^{L/2}u_x^2\,dx+G'h_1\int_{-L/2}^{L/2}\gamma^2\,dx+m\int_{-L/2}^{L/2}\dot{u}^2\,dx\right)+v^T Pv$$

Show that if the piezoelectric strain ε_p is generated according to the dynamic boundary controller, then the controller ensures global stability of the system with $\dot{E}_n/b\le 0$.

Part II
Advanced Damping Treatments

7

Vibration Damping of Structures Using Active Constrained Layer Damping

7.1 Introduction

The performance characteristics of Active Constrained Layer Damping (ACLD) are presented. The ACLD consists of a viscoelastic damping layer that is sandwiched between two layers of piezoelectric sensor and actuator. The composite ACLD when bonded to a vibrating structure acts as a smart treatment whose shear deformation can be controlled and tuned to the structural response in order to enhance the energy dissipation mechanism and improve the vibration damping characteristics.

Particular emphasis is placed on studying the performance of ACLD treatments that are provided with sensing layers of different spatial distributions. The effect of varying the controller gains and the operating temperature on the ACLD performance is determined for beams, plates, and shells. Comparisons with the performance of conventional passive constrained layer damping (PCLD) are also presented.

7.2 Motivation for Using Passive and Active Constrained Layer Damping

In this section, a control theory perspective is presented to motivate the need for ACLD as an effective means for controlling the structural vibrations. In order to illustrate such effectiveness, several structural configurations are considered as displayed in Figure 7.1. The considered configurations include: an undamped structure, a structure treated with an unconstrained damping layer, a structure treated with a constrained damping layer, and a structure treated with ACLD. For the purpose of illustrating the basic features of these configurations, the base structure is assumed without a loss of generality to be a rod undergoing longitudinal vibrations.

The considered rod is assumed to be mounted in a fixed-free configuration and made of aluminum with width of 0.025 m, thickness of 0.025 m, and length of 1 m. Also, it is assumed that the viscoelastic material (VEM) treatment is a full treatment with a width of 0.025 m, thickness = 0.025 m, and density of 1100 kg m^{-3}. The storage modulus and loss factor of the VEM are predicted by Golla–Hughes–McTavish model (GHM) model with one mini-oscillator with $E_0 = 15.3MPa, \alpha_1 = 39, \zeta_1 = 1, \omega_1 = 19,058 rad/s$. Furthermore, the VEM is constrained by an aluminum constraining layer that is 0.025 m wide and

Active and Passive Vibration Damping, First Edition. Amr M. Baz.

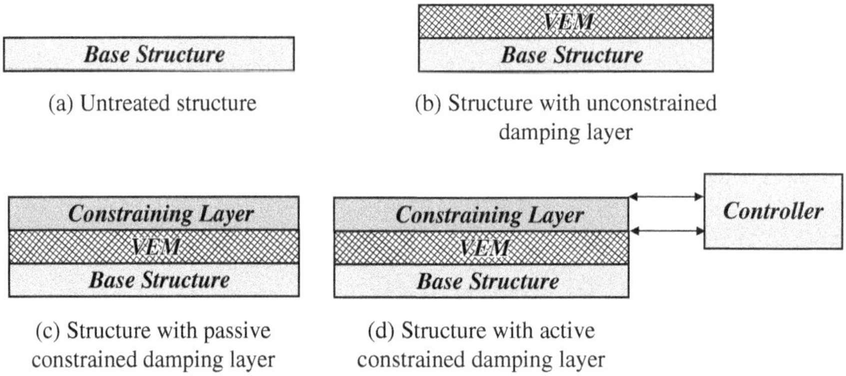

Figure 7.1 Configurations of structures treated with various passive and active constrained layer damping treatments. (a) untreated structure, (b) structure with unconstrained damping layer, (c) structure with passive constrained damping layer, and (d) structure with active constrained damping layer

0.0025 m thick. For simplicity purposes, the structural system is modeled by one finite element model.

7.2.1 Base Structure

Consider the dynamics of the untreated plain rod shown in Figure 7.2. The equation of motion of the single element undamped rod is given by:

$$[M_T]\{\ddot{u}_1\} + [K_T]\{u_1\} = \{F\} \tag{7.1}$$

where $[M_T] = \dfrac{\rho_s A_s L}{3}, [K_T] = \dfrac{E_s A_s}{L}$, and $\{F_T\} = F$. Note that ρ_s, A_s, L, E_s and are the density, area, length, and Young's modulus of the rod.

In the Laplace domain, Eq. (7.1) leads to the transfer function between deflection u_1 and force F as follows:

$$\frac{u_1}{F} = \frac{1}{\left(\frac{\rho_s A_s L}{3}\right)s^2 + \left(\frac{E_s A_s}{L}\right)} \tag{7.2}$$

The transfer function has no zeros, that is, the roots of the numerator, but has two poles, that is, the roots of the denominator, which are located at:

$$s = \pm\sqrt{\left(\frac{3E_s}{\rho_s L^2}\right)}i \tag{7.3}$$

where s denotes the Laplace complex number.

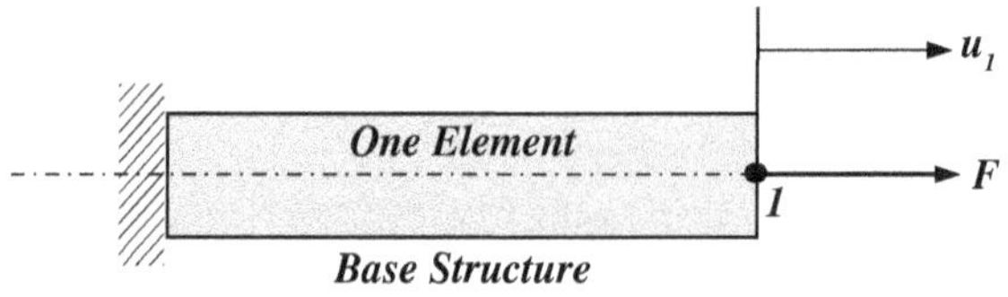

Figure 7.2 Untreated plain rod.

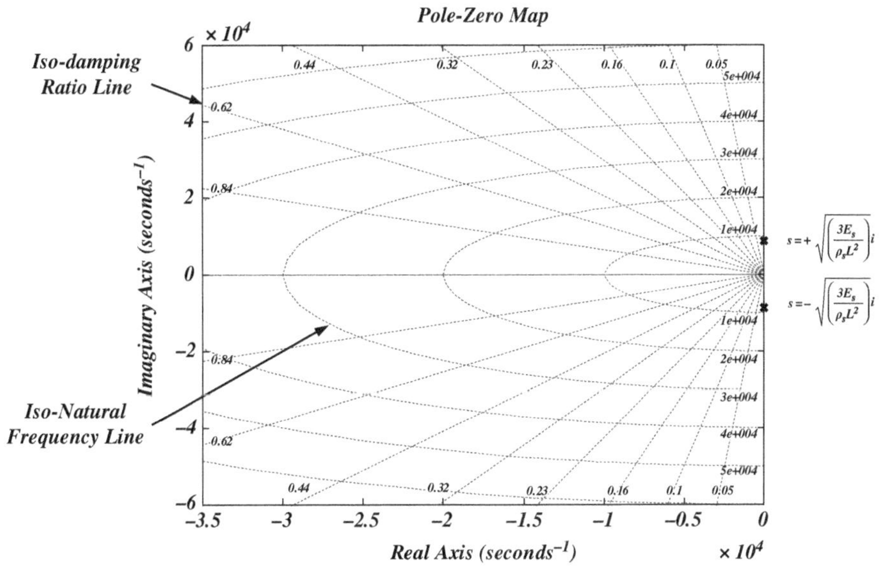

Figure 7.3 Pole-zero map of a plain untreated rod.

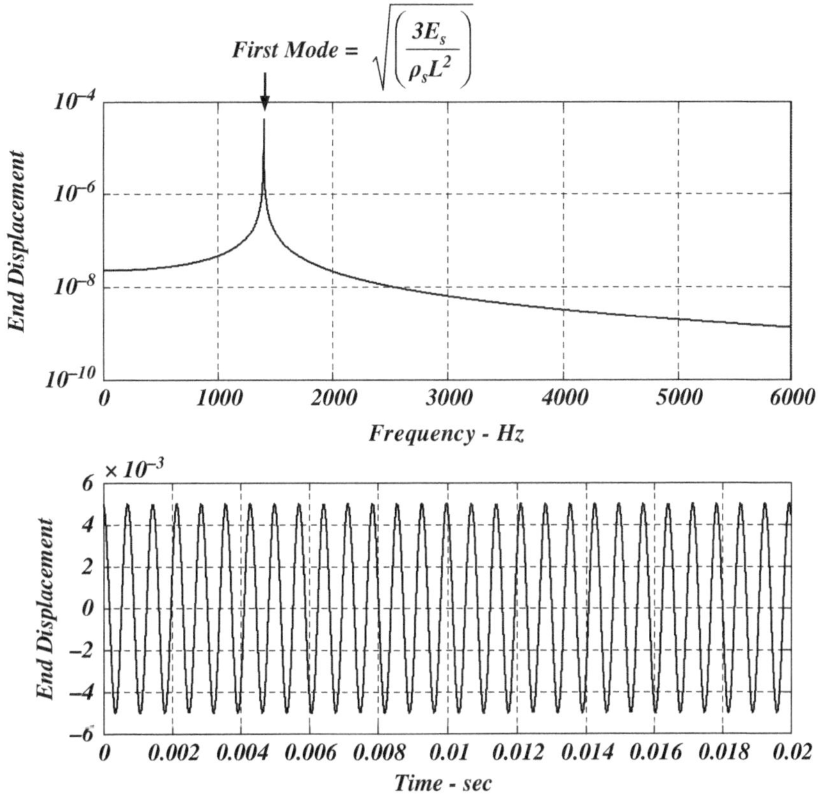

Figure 7.4 Frequency and time response of a plain untreated rod.

Figure 7.3 shows a display of the pole-zero map of the undamped rod. The poles are designated by × and are located on the imaginary axis where the damping ratio is zero.

For this undamped rod, the frequency and time response characteristics are shown in Figure 7.4.

7.2.2 Structure Treated with Unconstrained Passive Layer Damping

Consider the dynamics of a rod treated with an unconstrained layer damping as shown in Figure 7.5. The equation of motion of the single element of the rod is given in Chapter 4 by:

$$[M_T]\{\ddot{X}\} + [C_T]\{\dot{X}\} + [K_T]\{X\} = \{F_T\} \tag{7.4}$$

$$[K_T] = \begin{bmatrix} [K_{so}] + E_{v_o}(1+\alpha_1)[\bar{K}_{v_o}] & -E_{v_o}\alpha_1[\bar{K}_{v_o}] \\ -E_{v_o}\alpha_1[\bar{K}_{v_o}] & E_{v_o}\alpha_1[\bar{K}_{v_o}] \end{bmatrix}, \{X\} = \begin{Bmatrix} u_1 \\ z \end{Bmatrix} \text{and} \{F_T\} = \begin{Bmatrix} F \\ 0 \end{Bmatrix}. \tag{7.5}$$

where $[M_o] = \frac{(\rho_s A_s + \rho_v A_v)L}{3}$, $[K_{s_o}] = \frac{(E_s A_s)}{L}$, and $[K_v] = \frac{(E_v A_v)}{L}$. Also, u_1 and z denote the deflection of the rod and internal degree of freedom of the VEM, respectively. Note that ρ_i, A_i, L, and E_i define the density, area, length, and Young's modulus of the rod ($i = s$) and VEM ($i = v$). Also, $[K_v] = E_{v0}[\bar{K}_{v_o}]$ and the GHM model of the VEM is given by:

$$[K_{v_o}] = E_{v0}\left[1 + \alpha_1 \frac{s^2 + 2\zeta_1\omega_1 s}{s^2 + 2\zeta\omega_1 s + \omega_1{}^2}\right][\bar{K}_{v_o}] \tag{7.6}$$

As in the Laplace domain, Eq. (7.1) leads to the transfer function between deflection u_1 and force F as follows:

$$\frac{u_1}{F} = \frac{s^2 + 2\zeta\omega_1 s + \omega_1{}^2}{M_o s^4 + 2\zeta\omega_1 M_o s^3 + [M_o\omega_1{}^2 + (K_{s_o} + K_v)]s^2 + 2\zeta\omega_1(K_{s_o} + K_v)s + (K_{s_o} + K_v)\omega_1{}^2} \tag{7.7}$$

The transfer function has two zeros, that is, the roots of the numerator, at $-\omega_1$. These two repeated roots depend only on the parameters of the GHM model of the VEM. However, the transfer function has four poles, that is, the roots of the denominator. The locations of these poles depend on the interaction between the parameters of the rod and the VEM.

Figure 7.6a shows a display of the pole-zero map of the rod with unconstrained layer damping. The poles are designated "×" and the zeros are designated "o." The figure displays clearly two sets of pairs of poles. The first set of poles appears to be located on the imaginary axis defining the poles of the structure. The second set of poles consists of

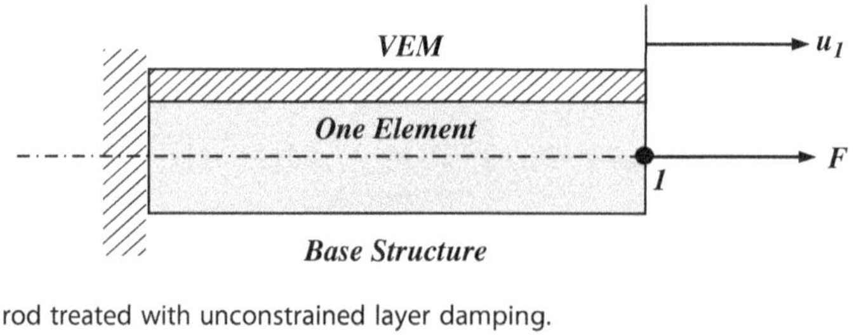

Figure 7.5 A rod treated with unconstrained layer damping.

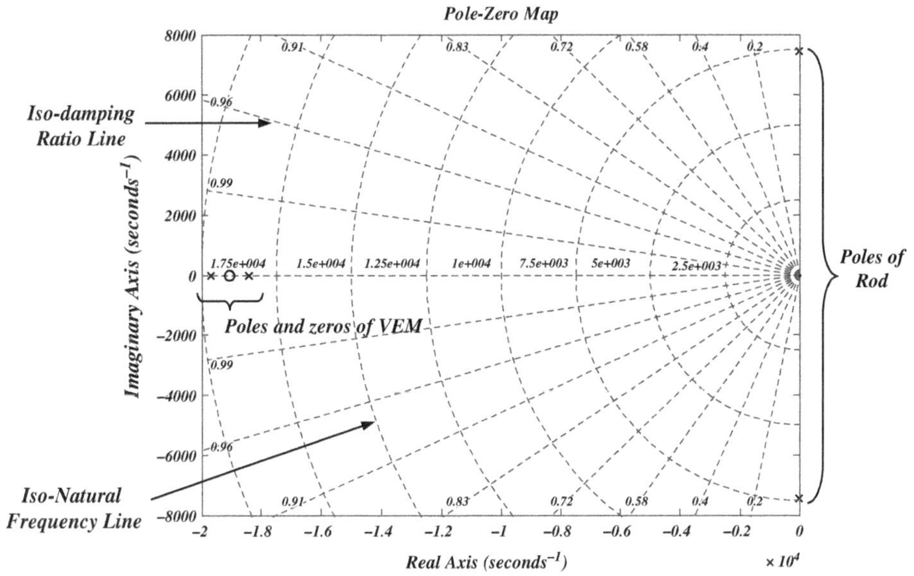

(a) – Full view

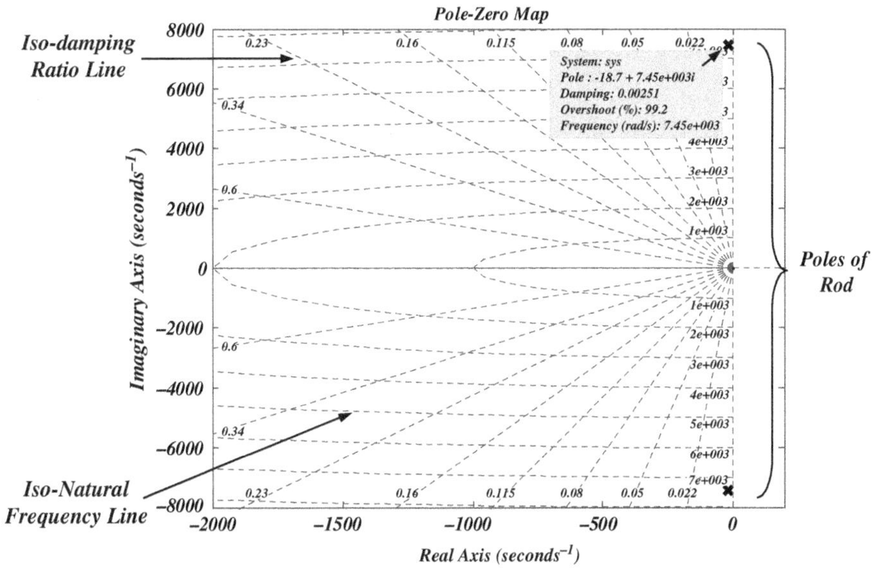

(b) – Close-up view

Figure 7.6 Pole-zero map of a rod treated with unconstrained layer damping. (a) Full view, (b) Close-up view.

distinct negative real numbers defining the poles of the VEM. These VEM poles sandwich the two repeated zeros of the VEM as indicated clearly in Figure 7.6a.

Note that the poles and the zeros of the VEM appear on the real axis as the damping ratio of the GHM mini-oscillator model is selected to be equal to 1, that is, $\zeta = 1$ or critically damped.

A closer look at the two poles of the rod, as displayed in Figure 7.6b, indicates that these two poles are actually not lying on the imaginary axis, but are located inside the left hand side of pole-zero map. The migration of these poles from the imaginary axis has resulted from the interaction of the parameters of the rod and the VEM. However, the extent of migration is very limited as indicated by the resulting damping ratio, which is $\zeta = 0.002\,51$. Such a very low damping ratio suggests the ineffectiveness of unconstrained layer damping in enhancing the damping characteristics of structures. This fact emphasizes the findings reported in Chapters 4 and 6. The natural frequency of the pole is 7450 rad s^{-1} (i.e., 1186.3 Hz), which matches that predicted by the frequency response characteristics shown in Figure 7.7.

For this configuration of a rod treated with unconstrained layer damping, the corresponding frequency and time response characteristics are shown in Figure 7.7. The results displayed in Figure 7.7 indicate that use of the unconstrained layer damping has introduced light damping of the rod vibration in comparison with the characteristics displayed in Figure 7.4 for the undamped rod.

7.2.3 Structure Treated with Constrained Passive Layer Damping

Consider the dynamics of a rod treated with a constrained layer damping as shown in Figure 7.8. The equation of motion of the single element of the rod is given in Chapter 4 by Eq. (4.50):

$$[M_t]\{\ddot{X}_t\} + [C_t]\{\dot{X}_t\} + [K_t]\{X_t\} = \{F_t\} \tag{7.8}$$

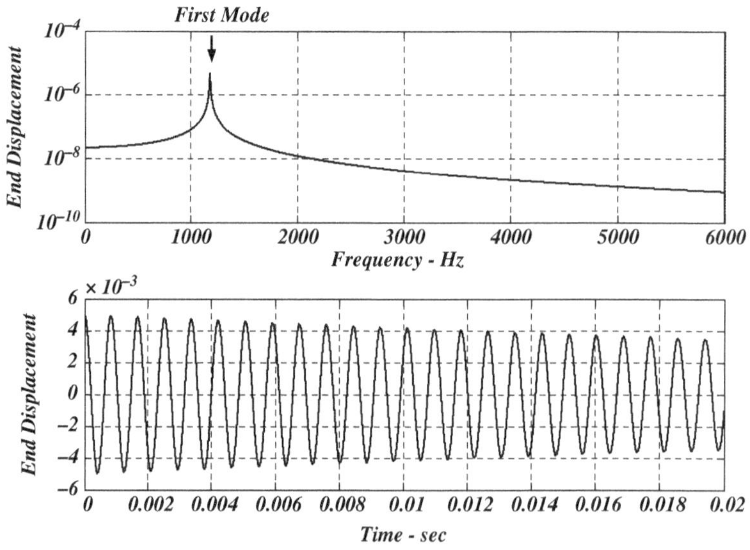

Figure 7.7 Frequency and time response of a rod treated with unconstrained layer damping.

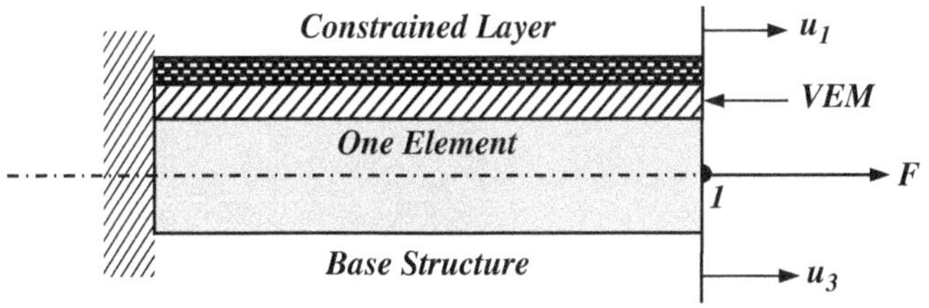

Figure 7.8 Rod treated with a passive constrained layer damping.

where $\{X_t\} = \{u_1\ u_3\ \bar{z}\}^T$ and $\{F_t\} = \{0\ \ F\ \ 0\}^T$ with $u_1, u_3, \bar{z}$ denoting the deflection of the constraining layer, deflection of the rod, and internal degree of freedom of the VEM, respectively. The matrices $[M_t]$, $[C_t]$, and $[K_t]$ are described in Section 4.2.4.2 of Chapter 4. Equation (7.8) can be written in the following state-space representation as:

$$\dot{\mathbf{X}} = A\mathbf{X} + Bu \tag{7.9}$$

where $A = \begin{bmatrix} 0 & I \\ -[M_t]^{-1}[K_t] & -[M_t]^{-1}[C_t] \end{bmatrix}$, $B = \begin{Bmatrix} 0 \\ [M_t]^{-1}\{0\ 1\ 0\}^T \end{Bmatrix}$ and $u = F$.

In the Laplace domain, Eq. (7.9) reduces to:

$$\mathbf{X} = (sI - A)^{-1}Bu \tag{7.10}$$

Then, the transfer function between the rod deflection u_3 and the applied force F is given by:

$$\frac{u_3}{F} = C(sI - A)^{-1}B = \frac{N(s)}{D(s)} \tag{7.11}$$

where C = measurement matrix = [0 1 0], $N(s)$ = numerator, and $D(s)$ = denominator.

Figure 7.9 shows a display of the pole-zero map of the rod with constrained layer damping. The poles are designated by "×" and the zeros are designated by "**o**." The figure displays clearly two sets of poles and zeros. The first set of four complex poles is located near the imaginary axis and sandwiching two closely placed complex zeros. The second set of negative real poles sandwiching two negative real zeros. The first set of poles/zeros are related to the rod structural system whereas the second set of poles/zeros defines the dynamics of the VEM with its internal degrees of freedom (DOF).

In Figure 7.9b, a close-up view of the pole-zero map indicates clearly that the dominant pole of the rod system is located such that the damping ratio ζ = 0.00524 and a natural frequency of 7510 rad s^{-1} (i.e., 1196 Hz). This pole coincides with the first natural frequency of the rod system as is confirmed by the frequency and time response characteristics shown in Figure 7.10.

Furthermore, the pole-zero map of Figure 7.9b also displays complex poles, which are located in the s-plane such that the damping ratio ζ = 0.0857 and a natural frequency of 32,700 rad s^{-1} (i.e., 5220 Hz). These poles coincide with the second natural frequency of the rod system as is confirmed by the frequency and time response characteristics shown in Figure 7.10.

It is important to note that the low damping ratio (ζ = 0.00524) of the first mode is evident in the sharpness of the resonance of the frequency response and in the low decay

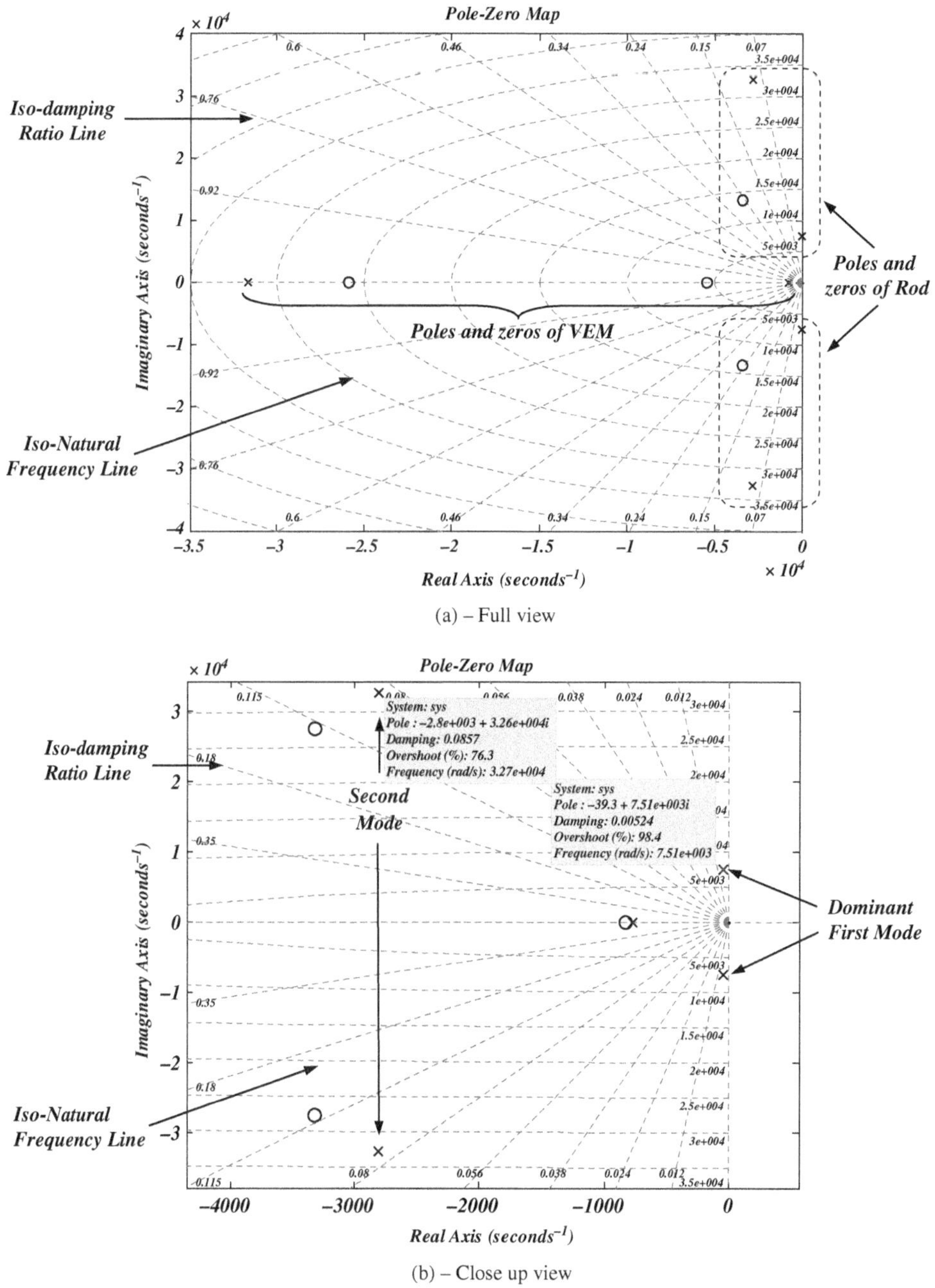

Figure 7.9 Pole-zero map of a rod treated with a passive constrained layer damping. (a) Full view and (b) close-up view.

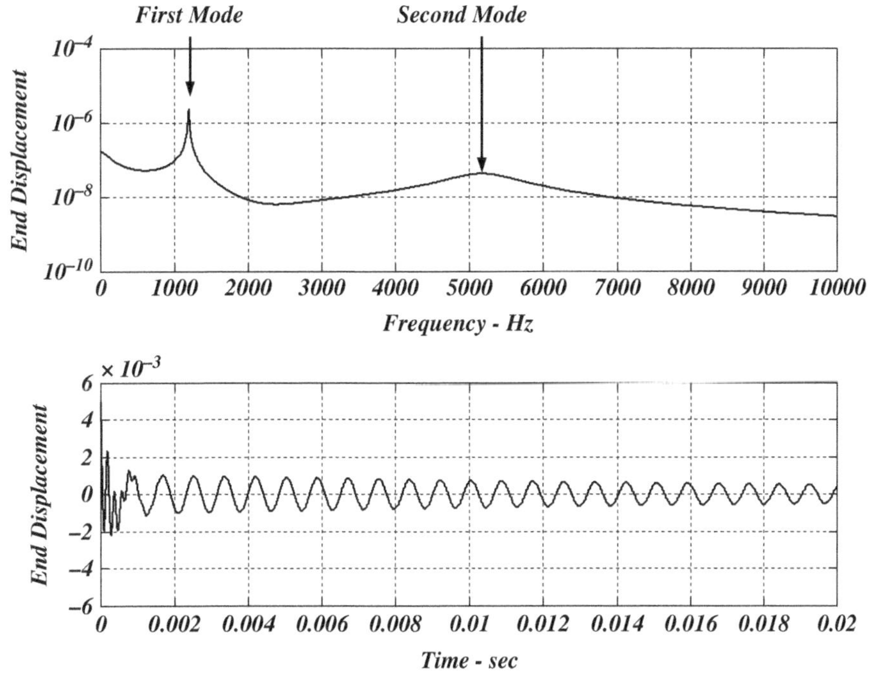

Figure 7.10 Frequency and time response characteristics of a rod treated with a passive constrained layer damping.

of the vibration in the time response displayed in Figure 7.10. Also, the higher damping ratio ($\zeta = 0.0857$) of the second mode is manifested clearly by the flatness of the resonance of the frequency response and by the very fast decay of vibration in the time response displayed in Figure 7.10.

Comparison between the obtained characteristics of rods treated with unconstrained and constrained layer damping indicates that constraining the VEM has significantly enhanced the damping ratio. For the considered examples, the damping ratio of the first mode has increased from 0.002 51 with unconstrained layer damping to almost double the ($\zeta = 0.005\,24$) when using constrained layer damping. Also, with constrained layer damping, high damping ratios are obtained for higher order modes as it is predicted that the damping ratio for the second mode reaches $\zeta = 0.0857$, which is 20-fold that of the first mode.

7.2.4 Structure Treated with Active Constrained Passive Layer Damping

Consider the dynamics of a rod treated with an active constrained layer damping as shown in Figure 7.11. The equation of motion of the single element of the rod is given in Chapter 4:

$$[M_t]\{\ddot{X}_t\} + [C_t]\{\dot{X}_t\} + [K_t]\{X_t\} = \{C_e\}F_e + \{C_c\}F_c \tag{7.12}$$

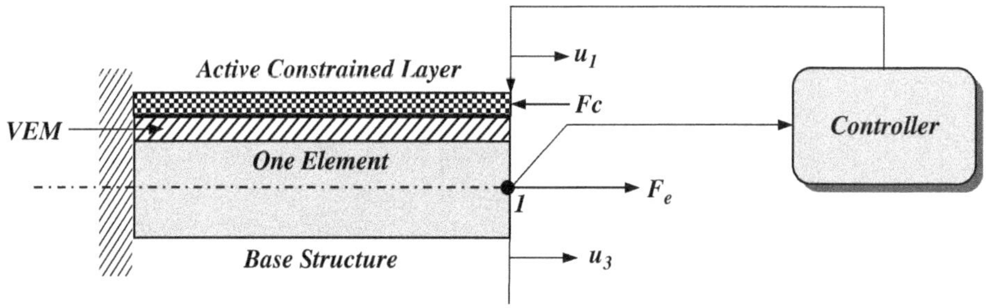

Figure 7.11 Rod treated with active constrained layer damping.

where $\{X_t\} = \{u_1\ u_3\ \bar{z}\}^T$, $\{C_e\} = [0\ \ 1\ \ 0]^T$, and $\{C_c\} = [1\ \ 0\ \ 0]^T$. Also, F_e and F_c denote the external and control forces, respectively. The matrices $[M_t]$, $[C_t]$, and $[K_t]$ are as described in Section 4.2.4.2 of Chapter 4.

The control force F_c can be generated using the following control law:

$$F_c = -K_G\{C_e\}^T\{X_t\} - K_G k_r\{C_e\}^T\{\dot{X}_t\} \tag{7.13}$$

where K_G is the proportional control gain and $K_G k_r$ is the derivative control gain. In this form, k_r is the ratio between the derivative and proportional control gains.

Putting Eq. (7.12) into the state-space representation, gives:

$$\dot{\mathbf{X}} = A\mathbf{X} + BF_c + B_eF_e \tag{7.14}$$

where $A = \begin{bmatrix} \mathbf{0} & \mathbf{I} \\ -[M_t]^{-1}[K_t] & -[M_t]^{-1}[C_t] \end{bmatrix}$, $B = \begin{Bmatrix} 0 \\ [M_t]^{-1}[C_c] \end{Bmatrix}$ and $B_e = \begin{Bmatrix} 0 \\ [M_t]^{-1}[C_e] \end{Bmatrix}$.

In the Laplace domain, Eq. (7.14) reduces to:

$$\mathbf{X} = (sI - A)^{-1}(BF_c + B_eF_e) \tag{7.15}$$

Figure 7.12 shows a display of the block diagram of the ACLD/rod system that translates the interactions between Eqs. (7.12) and (7.15).

Then, the transfer function between the rod deflection u_3 and the applied force F_e is given by:

$$\frac{u_3}{F_e} = \frac{N_c(s)}{D_c(s)} \tag{7.16}$$

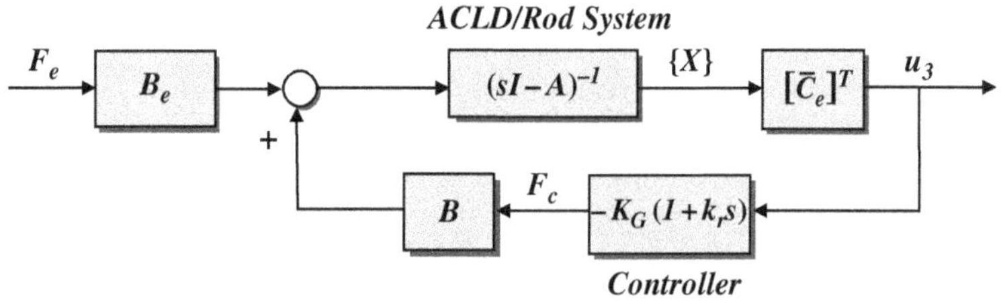

Figure 7.12 Block diagram of a rod treated with active constrained layer damping.

where $N_c(s)$ and $D_c(s)$ are the numerator and denominator of the closed-loop system. Also, $[\bar{C}_e]$ is given by: $[\bar{C}_e] = [C_e\, \mathbf{0}_{1\times 3}]$.

The performance of the closed-loop system is governed by the denominator of the transfer function u_3/F, that is, by the roots of its characteristics equation:

$$D_c(s) = 0 \tag{7.17}$$

For a given value of k_r, Eq. (7.17) can be reduced to the following form:

$$D_c(s) = 1 + Gain\frac{N}{D} = 0 \tag{7.18}$$

where N/D is denoted as the open-loop transfer function.

Equation (7.18) can be rewritten to take one of the following two forms:

a. $$D + Gain \ \ N = 0 \tag{7.19}$$

when the $Gain = 0$, the roots of the characteristics equation has roots equal to the roots of the denominator $D = 0$; that is, poles of the open-loop transfer function.

b. $$\frac{1}{Gain}D + N = 0 \tag{7.20}$$

when the $Gain = \infty$, the roots of the characteristics equation has roots equal to the roots of the denominator $N = 0$, that is, zeros of the open-loop transfer function.

Hence, by varying the gain from 0 to ∞, the roots of the characteristics equation of the closed-loop system start from the poles of the open-loop transfer function and end at the zeros of the open-loop transfer function. A plot of the loci of the roots of the characteristics equation of the closed-loop system is denoted the "root locus" plot.

Figure 7.13 displays the root locus plot of the roots of the characteristics equation of the closed-loop system of the ACLD/rod.

Note that, when the $Gain = 0$, the roots of the characteristics equation of the closed-loop system of the ACLD/rod coincide with those of a rod treated with PCLD, which is shown in Figure 7.9.

The importance of the "roots locus" plot lies in its display of a map of all the roots of the closed-loop system of the rod/ACLD for different values of the gain. Hence, by varying the gain, it is possible to place the roots of the system at desirable locations that may have favorable damping characteristics. These locations can be different from the poles of the rod/PCLD system. As a matter of fact, these locations can extend over a wider range between the poles and zeros of the rod/PCLD system. This enables placement of the roots of the rod/ACLD at locations where higher damping and stability of the closed-loop system can be achieved. In this manner, the performance of the rod/ACLD can be significantly enhanced over that of the rod/PCLD.

For example, the observed low damping ratio of the first mode of the rod treated with PCLD $\zeta = 0.005\,24$ can be improved to 0.0841, as indicated in the close-up view of the root locus plot shown in Figure 7.14, by selecting the gain to be 9.5E4 N m^{-1}. Higher values of the gain can still result in better performance but increasing the gain beyond 2.85E5 N m^{-1} can destabilize the second mode of vibration as the roots start crossing the imaginary axis into the positive side of the s-plane.

The resulting performance characteristics in the frequency and time domains are shown in Figure 7.15. The displayed performance of the rod/ACLD system is compared with the corresponding performance of rod/PCLD system and plain rod.

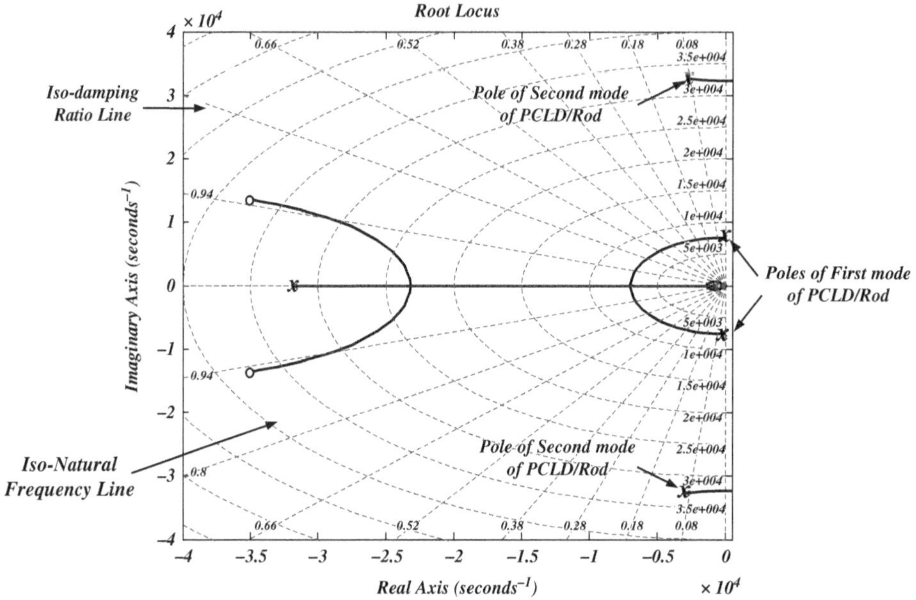

Figure 7.13 Root locus of a rod treated with active constrained layer damping with $k_r = 0.01$.

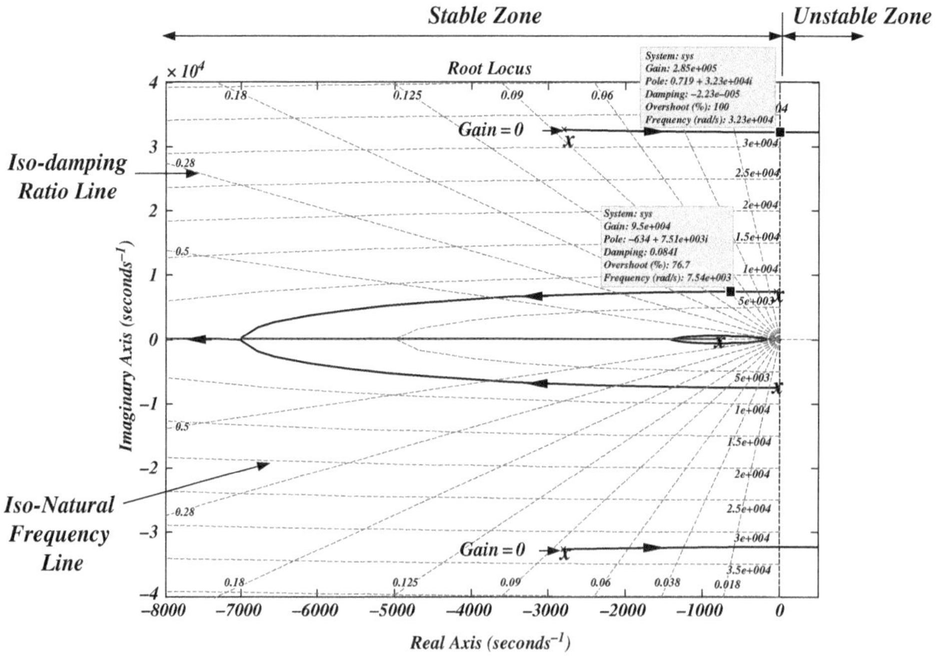

Figure 7.14 Close-up view of the root locus of a rod treated with active constrained layer damping with $k_r = 0.01$.

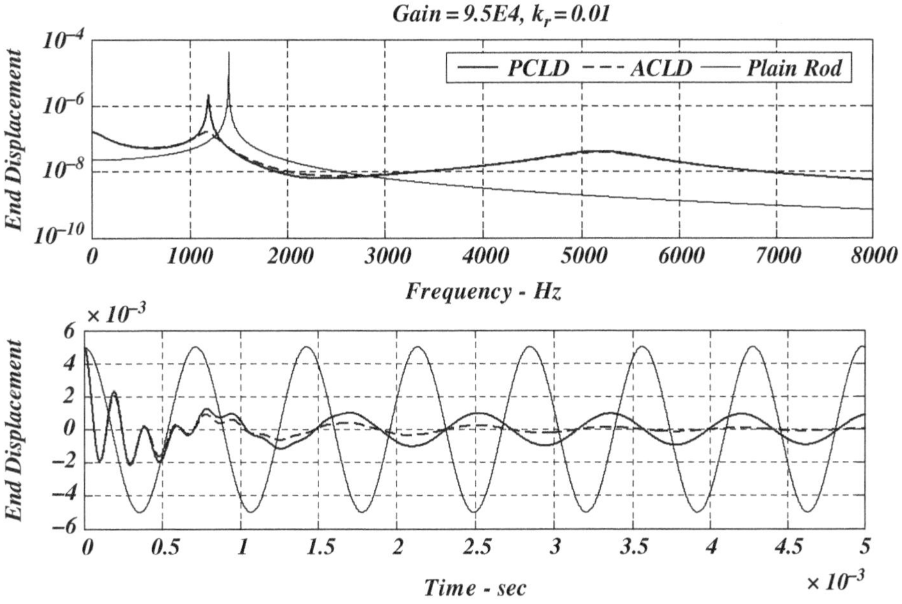

Figure 7.15 Frequency and time response of a rod treated with active constrained layer damping with Gain = 9.5E4 N m^{-1} and k_r = 0.01.

The figure indicates that the use of the ACLD treatment has significantly improved the damping characteristics of the first mode but has not affected the second mode. This is manifested clearly, in the frequency domain, as the behavior with the ACLD coincides with that with the PCLD treatment. Also, in the time domain, the high frequency components lie on top of each other between 0 and 0.75 ms, whereas the low frequency component with the ACLD is seen to die out very fast as the damping ratio is increased dramatically with the addition of the active control.

Table 7.1 summarizes the main features of the considered basic configurations of the undamped structure, a structure treated with an unconstrained damping layer, a structure treated with a constrained damping layer, and a structure treated with ACLD.

Table 7.1 Features of damping configurations.

Features	Plain rod	Rod with unconstrained layer damping	Rod/PCLD	Rod/ACLD
Poles	Imaginary	Complex with small real part	Complex with larger real part	Complex with much higher real part
zeros	None	Sandwiched between the poles	Sandwiched between the poles	Sandwiched between the poles
Damping Ratio	zero	Very Low	larger	Much higher
Stability	Limited stability	Always stable	Always stable	Can be unstable at high control gains

7.3 Active Constrained Layer Damping for Beams

7.3.1 Introduction

PCLD treatments have been successfully utilized, as a simple and reliable means, for damping out the vibration of a wide variety of flexible structures (Cremer, et al., 1988). However, for effective performance over a broad range of temperatures and frequencies, the weight of PCLD treatments can pose serious limitations to their use in applications where weight is critically important.

It is therefore the purpose of this section to elaborate on the concepts presented in Section 7.2 by considering ACLD treatments (Baz, 1993, 1996; Baz and Ro 1993a,b, 1994) as a viable alternative to PCLD treatments. ACLD combines the attractive attributes of both passive and active controls to achieve optimal vibration damping. In particular, it provides an effective means for augmenting the simplicity and reliability of passive damping with the low weight and high efficiency of active controls to attain high damping characteristics over broad frequency bands. Such characteristics are particularly suitable for damping the vibration of critical systems such as rotorcraft blades where damping-to-weight ratio is very important.

In this chapter, the emphasis is placed on developing distributed-parameter and finite element models to describe the behavior of structures treated with ACLD treatment. The finite element method (FEM) models, in particular, are intended to enhance the practicality of predicting the behavior of structures such as beams, plates, and shells that are subject to a wide variety of boundary conditions and are partially treated with multi-patches of ACLD treatments. The models will also allow the prediction of the ACLD performance when specific modes are targeted with proper spatial shaping of the sensing layer.

7.3.2 Concept of Active Constrained Layer Damping

The ACLD, as shown in Figure 7.16, consists of a conventional PCLD, which is augmented with active control means to control the strain of the constrained layer, in response to the structural vibrations. The viscoelastic damping layer is sandwiched between two piezoelectric layers. The three-layer composite ACLD when bonded to the beam acts as a "smart" constraining layer damping treatment with built-in sensing and actuation capabilities. The effect of interaction between the sensor and the actuator on the operation of the ACLD can best be understood by considering the motion experienced by the beam during a typical vibration cycle.

In Figure 7.16a, as the beam moves downward under the action of an external moment M_e, the sensor is subjected to tensile stresses that generate a positive voltage Vs by the direct piezoelectric effect. This voltage is amplified, its polarity is reversed, and the resulting voltage Vc is fed back to activate the piezoelectric constraining layer, which shrinks by virtue of the reverse piezoelectric effect. The shrinkage results in a shear deformation angle γ_a, in the viscoelastic layer, which is larger than the angle γ_p developed by a conventional passive constraining.

Similarly, Figure 7.16b describes the operation of the ACLD during the upward motion of the beam. During this part of the vibration cycle, the top part of the beam as well as the piezoelectric sensor experiences compressive stresses and a negative voltage is generated

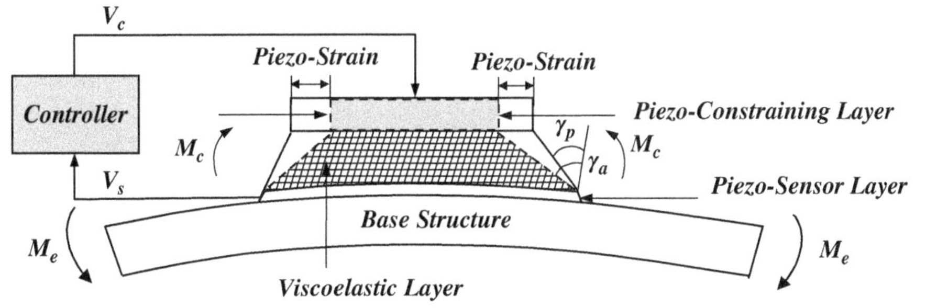

(a)

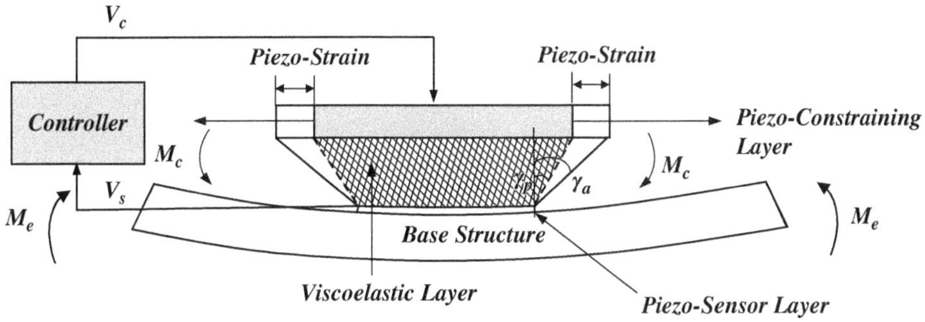

(b)

Figure 7.16 Concept of active constrained layer damping treatment.

by the sensor. Direct feedback of the sensor signal to the active constraining layer makes it extend and increase the shear deformation angle to γ_a compared to γ_p for the conventional constraining layer. The increase of the shear deformation of the viscoelastic layer, during the entire vibration cycle, is accompanied with an increase in the energy dissipated.

Furthermore, the shrinkage (or expansion) of the piezoelectric layer during the upward motion (or during the downward motion) produces a bending moment M_c on the beam, which tends to bring the beam back to its equilibrium position and counterbalance the effect of the external moment M_e. Therefore, the dual effect of the enhanced energy dissipation and the additional restoring bending moment will quickly damp out the vibration of the flexible beam. This dual effect, which does not exist in conventional constrained damping layers, significantly contributes to the damping effectiveness of the smart ACLD. In this manner, the smart ACLD consists of a conventional PCLD, which is augmented with the described dual effect to actively control the strain of the constrained layer, in response to the structural vibrations. With appropriate strain control strategy, the shear deformation of the viscoelastic damping layer can be increased, the energy dissipation mechanism can be enhanced and the vibration can be damped out. One possible strategy is the direct feedback of the sensor voltage to power the active constraining layer. Another strategy will rely on feeding back both the sensor voltage and its derivative to obtain proportional and derivative control action. With such a strategy additional damping can be imparted to the vibrating beam system and the versatility of active controls can be utilized to considerably improve the damping characteristics of the ACLD.

In this manner, the ACLD provides a practical means for controlling the vibration of massive structures with currently available piezoelectric actuators without the need for excessively large actuation voltages. This is due to the fact that the ACLD properly utilizes the piezoelectric actuator to control the shear in the soft viscoelastic core, which is a task compatible with the low control authority capabilities of the currently available piezoelectric materials.

Figure 7.17 shows different configurations of ACLD treatment that differ in the arrangement of the sensors. The first configuration shown in Figure 7.17a, is the classical three-layer ACLD treatment that includes a VEM sandwiched between a piezo-constraining layer and a piezo-sensor layer.

Figure 7.17b shows a two-layer ACLD treatment with the piezo-constraining layer acting simultaneously as an actuator and a sensor. This layer is called a "self-sensing piezo-layer," which ensures collocated sensor/actuator arrangement and when controlled properly guarantees global stability as discussed in Chapter 6 (Ro and Baz 2002).

In the third configuration of Figure 7.17c, the two-layer ACLD is provided with a discrete sensor to complete the control loop. The discrete sensor can be an accelerometer or a strain gage that is placed as close as possible to the treatment in order to approximate the case of collocated sensor/actuator arrangement.

7.3.3 Finite Element Modeling of a Beam/ACLD Assembly

A finite element model is developed in this section to describe the behavior of beams with ACLD treatments. The model extends the studies of Trompette et al. (1978), Rao (1976) used to analyze the dynamics of PCLD treatments. It accounts for the behavior of the distributed and spatially shaped piezoelectric sensor (Miller and Hubbard 1987) and the distributed piezoelectric actuator (Crawley and de Luis 1987). Appropriate control laws are considered to control the interaction between the piezo-sensor and actuator in order to achieve enhanced vibration control characteristics.

The emphasis is placed, in this section, on the development of a model for Bernoulli–Euler beams that are treated with multi-patches of ACLD layers in order to demonstrate the feasibility and merits of the ACLD concept. ACLD treatments with uniform and spatially shaped sensors are considered in following analyses in order to investigate the potential of targeting specific modes with the shaped sensors.

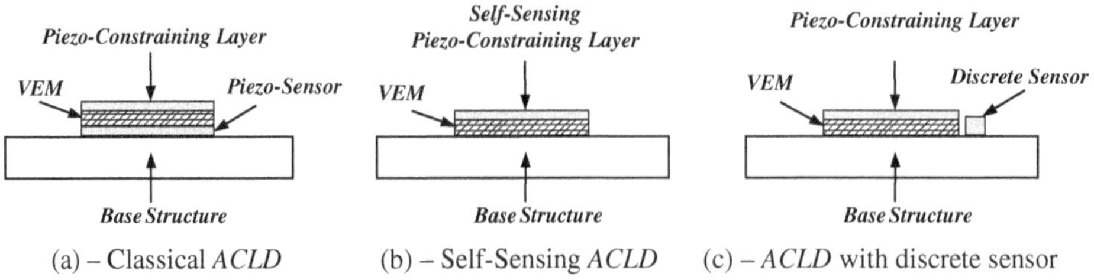

Figure 7.17 Different configurations of active constrained layer damping treatment. (a) Classical ACLD, (b) self-sensing ACLD, and (c) ACLD with discrete sensor.

7.3.3.1 The Model

Figure 7.18 shows a schematic drawing of the ACLD treatment of a sandwiched beam divided into N finite elements. It is assumed that the shear strains, in the piezoelectric sensor/actuator layers and in the base beam, are negligible.

It is also assumed that the transverse displacements w of all points on any cross section of the sandwiched beam are considered to be equal. Furthermore, the piezoelectric sensor/actuator layers and the base beam are assumed to be elastic and to dissipate no energy, whereas the core is assumed to be linearly viscoelastic. In addition, the piezoelectric sensor and the base beam are considered to be perfectly bonded together such that they can be reduced to a single equivalent layer. Accordingly, the original four-layer sandwiched beam reduces to an equivalent three-layer beam. Hence, the modeling approach presented in Section 4.3 for beams treated with PCLD can be extended to account for the effect of the piezo-sensing and actuation.

The piezo-sensor can take the general, uniform, or linear shapes shown in Figures 7.19a–c, respectively.

The complete description of the beam/ACLD assembly model includes:

Potential Energy

This remains as described by Eq. (4.60) as follows:

$$PE = \frac{1}{2}\int_0^L \left[\sum_{i=1}^{3} E_i A_i u_{i,x}^2 + (E_1 I_1 + E_3 I_3) w_{,xx}^2 + G_2 A_2 \gamma^2\right] dx \tag{7.21}$$

where u_1, u_3, w, and $w_{,x}$ denote the axial displacement of the constraining layer, axial displacement of the beam, as well as the transverse and rotational deflections of beam, respectively. Also, γ defines the shear strain of the VEM given by Eq. (4.58). Note that $E_i A_i$ and $E_i I_i$ denote the longitudinal and flexural rigidity of the ith layer. Also, $G_2 A_2$ denotes the shear rigidity of the VEM with G_2 defining its shear modulus.

Kinetic Energy

This is also as given by Eq. (4.62) as follows:

$$KE = \frac{1}{2}\int_0^L \left[\sum_{i=1}^{3} \rho_i A_i \dot{u}_i^2 + \sum_{i=1}^{3} \rho_i A_i \dot{w}^2\right] dx \tag{7.22}$$

with ρ_i denoting the density of the ith layer.

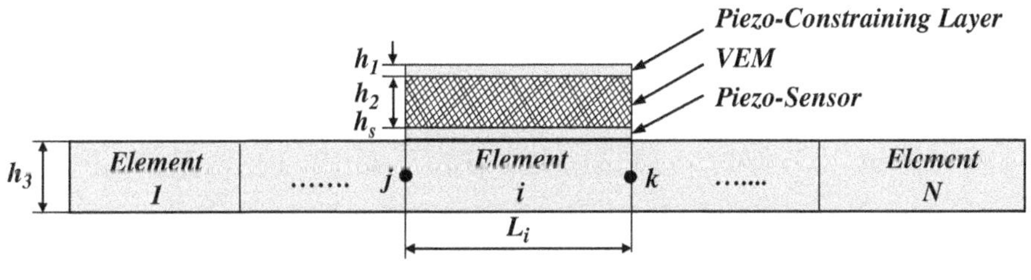

Figure 7.18 Schematic drawing of a deflected beam treated with ACLD treatment.

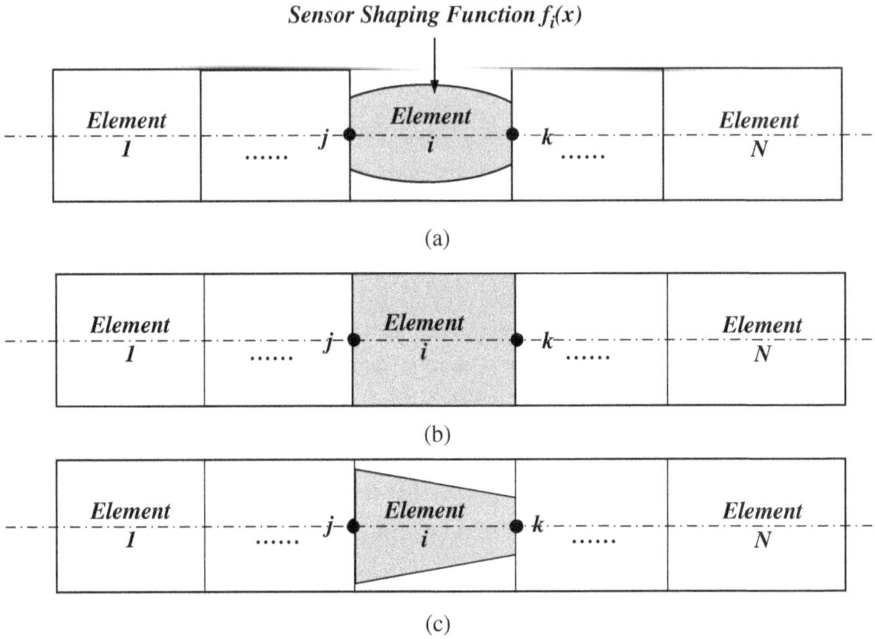

Figure 7.19 Shaped, uniform, and linearly varying sensors.

Piezo-Control Forces and Moments

1) *Piezo-Actuator.* The strain ε_p induced in the piezoelectric actuator is given by (Crawley and de Luis 1987):

$$\varepsilon_p = \left(\frac{d_{31}}{h_1}\right) V_c \tag{7.23}$$

where d_{31} is the piezoelectric strain constant resulting from the application of the voltage V_c across the piezo-actuator layer. In Eq. (7.23), V_c is assumed constant over the length of the beam element. The voltage V_c is generated from the proper manipulation of the piezo-sensor voltage V_s.

2) *Piezo-Sensor.* The strain ε_s induced in the piezo-sensor is proportional to the beam curvature ($w_{,xx}$) and is given by:

$$\varepsilon_s = -hw_{,xx} \tag{7.24}$$

where h is the distance from the beam neutral axis to the sensor surface.

The induced strain ε_s integrated over the entire length of the sensor due to its distributed nature, generates an output voltage V_s given by (Appendix 7.A):

$$V_s = -\left[\frac{k_{31}^2 hb}{g_{31}C}\right] \sum_{i=i_s}^{i_f} \int_0^{L_i} f_i(x)\, w_{,xx} dx \tag{7.25}$$

where $f_i(x)$ is a spatial distribution function that defines the shaping of the sensor over the ith element. For uniform sensor $f_i(x) = 1$ and for a linearly shaped sensor $f_i(x) = 1 - x/L$ as shown in Figure 7.19c. In Eq. (7.25), the sensor is extended between elements i_s and i_f. Also, k_{32}^1 is the electromechanical coupling factor, g_{31} is the piezoelectric voltage constant and C is the capacitance of the sensor, which is given by:

$$C = 8.854\left(10^{-12}\right)AK_{3t}/h_s \tag{7.26}$$

where A is the sensor surface area and K_{3t} is the dimensionless dielectric constant.

3) *Control Law.* The manipulation of the piezo-sensor voltage V_s to generate the actuator voltage V_c is governed by the following proportional and derivative control law:

$$V_c = -\left(K_P V_s + K_d \frac{dV_s}{dt}\right) \tag{7.27}$$

where K_p and K_d are the proportional and derivative control gains, respectively.

4) *Control Forces and Moments.* The vector $\{F_c\}$ of the control forces and moments generated by the piezo-constraining layer on the treated beam element can be expressed in the following matrix form:

$$\{F_c\} = \left\{F_{pj}, 0, 0, M_{pj}, F_{pk}, 0, 0, M_{pk}\right\}^T \tag{7.28}$$

where F_{pj}, F_{pk}, M_{pj}, and M_{pk} denote the control forces and moments generated at nodes j and k, which are given by:

For a uniform sensor

$$F_{pj} = -F_{pk} = \frac{-g}{2} G_c\left(w_{,x_{i_s}} - w_{,x_{i_f+1}}\right),$$

$$F_{wj} = F_{wk} = 0,$$

and

$$M_{pj} = -M_{pk} = -g\,G_c\left[\frac{1}{2}\left(u_{1i_s} - u_{1i_f+1}\right) - D_a\left(w_{,x_{i_s}} - w_{,x_{i_f+1}}\right)\right]. \tag{7.29}$$

For a shaped sensor

$$F_{pj} = -F_{pk} = \frac{-g}{2L} G_c\left[\left(w_{i_s} - w_{i_f+1}\right) + Lw_{,x_{i_s}}\right],$$

$$F_{wj} = F_{wk} = \frac{-g}{2L} G_c\left[\left(u_{1i_s} - u_{1i_f+1}\right) - D_a\left(w_{,x_{i_s}} - w_{,x_{i_f+1}}\right)\right],$$

$$M_{pj} = \frac{-g}{2} G_c\left[\left(u_{1i_s} - u_{1i_f+1}\right) - D_a\left(w_{,x_{i_s}} - w_{,x_{i_f+1}}\right) - \frac{D_a}{L}\left(w_{i_s} - w_{i_f+1}\right)\right],$$

and

$$M_{pk} = \frac{-D_a g}{2} G_c\left[w_{,x_{i_s}} + \frac{1}{L}\left(w_{i_s} - w_{i_f+1}\right)\right]. \tag{7.30}$$

where $g = E_1 b^2 d_{31}\left[k_{31}^2 D_s / g_{31} C\right]$, $D_a = \left(\frac{h_1 + h_3}{2} + h_2\right)$ is the distance between a neutral axis of entire sandwiched beam and piezo-actuator, and p is the d/dt operator. Also, $G_c = (K_P + K_d\, p)$ denotes the controller gain.

7.3.3.2 Equations of Motion

The stiffness matrix $[K_i]$, the mass matrix $[M_i]$, and the control force vector $\{F_c\}$ are combined to describe the dynamics of the ACLD-treated beam element as follows:

$$[M_i]\{\ddot{\Delta}_i\} + [K_i]\{\Delta_i\} = \{F_c\} \tag{7.31}$$

The effect of the proportional and derivative control actions on the performance of the assembled closed-loop system, given by Eq. (7.31), is determined by computing the eigenvalues (i.e., natural frequencies and damping ratios) of the closed-loop system and comparing these eigenvalues with those of the open-loop system. Note that $\{\Delta_i\}$ denotes the nodal deflection vector of the ith element that is bounded between nodes j and k. The nodal deflection vector $\{\Delta_i\}$ is given by:

$$\{\Delta_i\} = \left\{u_{1j}, u_{3j}, w_j, w_{,x_i}, u_{1k}, u_{3k}, w_k, w_{,x_k}\right\}^T \tag{7.32}$$

Also, stiffness matrix $[K_i]$ and the mass matrix $[M_i]$ are extracted from the potential energy (PE) and the kinetic energy (KE) such that:

$$PE = \frac{1}{2}\{\Delta_i\}^T[K_i]\{\Delta_i\} \text{ and } KE = \frac{1}{2}\{\dot{\Delta}_i\}^T[M_i]\{\dot{\Delta}_i\} \tag{7.33}$$

Example 7.1 Consider a cantilever beam/ACLD system with a VEM that has shear modulus $G' = 20$ MPa, loss factor $\eta = 1$, and thickness $h_2 = 0.0025$ m. The physical and geometrical parameters of the different layers of the beam/ACLD system are listed in Tables 7.2 and 7.3.

Design a velocity feedback controller to control the beam vibration using the root locus approach.

Table 7.2 Physical properties of the base beam and the viscoelastic material.

Material	Length (m)	Thickness (m)	Width (m)	Density (kg m^{-3})	Young's modulus (MPa)
Steel	0.5	0.0125	0.05	7800	210,000
VEM	0.5	0.006 25	0.05	1104	60

Table 7.3 Physical properties of the piezoelectric constraining layer.

Length (m)	Thickness (m)	Width (m)	Density (kg m^{-3})	Young's modulus (GPa)	d_{31} (m V^{-1})	k_{31}	g_{31} (mV N^{-1})	k_{3t}
0.5	0.0025	0.05	7600	63	186E-12	0.34	116E-2	1950

Determine the frequency response of the free end of the beam and control voltage characteristics of the beam/ACLD system and compare these characteristics with those of the beam/PCLD system. Assume that the beam is excited by a 1 N sinusoidal load at its free end.

Solution

The equation of motion of the beam/ACLD system, which is divided into N finite elements, is as given in Sections 4.3 and 7.3.3 as follows:

$$[M_t]\{\ddot{X}_t\} + [C_t]\{\dot{X}_t\} + [K_t]\{X_t\} = \{C_e\}F_e + \{C_c\}F_c \tag{7.34}$$

where $\{X_t\}_{1\times 4N} = \{u_{1_1}, u_{3_1}, w_1, w_{,x_1}, \ldots\ldots u_{1_N}, u_{3_N}, w_N, w_{,x_N}\}^T$, $\{C_e\}_{1\times 4N} = [0\ \ 0\ \ \ldots\ldots\ldots 0\ \ 1\ \ 0]^T$, and $\{C_C\}_{1\times 4N} = [0\ \ 0\ \ ..\ \ \ldots\ \ 0\ \ 1]^T$.

Also, F_e and F_c denote the external and control forces, respectively. The matrices $[M_t]$, $[C_t]$, and $[K_t]$ are as described in Section 4.3.3 of Chapter 4.

The first natural frequencies of the open-loop beam/ACLD system are listed in Table 7.4 along with comparisons with the predictions obtained by ANSYS. Close agreement is evident between the predictions of the FEM and ANSYS models.

The control force F_c is generated using the following velocity feedback control law:

$$F_c = -K_v\{C_e\}^T\{\dot{X}_t\} \tag{7.35}$$

where K_v is the derivative control gain.

Combining Eqs. (7.34) and (7.35) yields the following state-space representation:

State Equations:

$$\dot{\mathbf{X}} = A\mathbf{X} + Bu \tag{7.36}$$

and

Output:

$$y = C\mathbf{X} + Du$$

where $A = \begin{bmatrix} \mathbf{0} & \mathbf{I} \\ -[M_t]^{-1}[K_t] & -[M_t]^{-1}[C_t] \end{bmatrix}$, $B = \begin{Bmatrix} 0 \\ [M_t]^{-1}\{C_e\} \end{Bmatrix}$, $C = [0_{1\times 4N}\ \ 0_{1\times(4N-1)}1\ ]$, $D = 0$, and $u = F_c$.

The root locus of the system can be plotted using the following MATLAB command:

$$rlocus\,(A, B, C, D)$$

Table 7.4 Comparison between the natural frequencies of the beam/ACLD system as predicted by FEM and ANSYS.

Mode	1	2	3
FEM (Hz)	48.88	248.4	668.78
ANSYS (Hz)	46.35	253.52	677.49

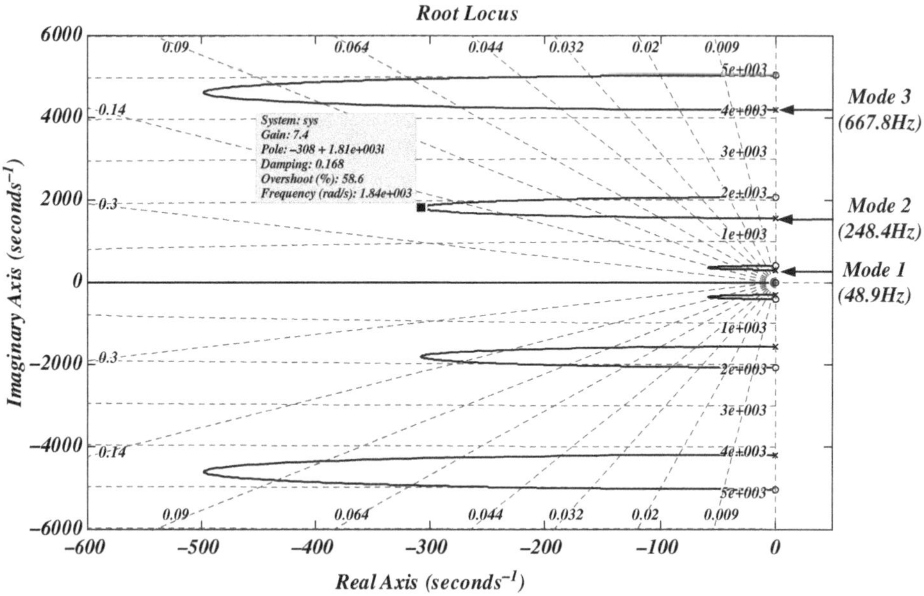

Figure 7.20 Root locus plot of beam/ACLD system.

Figure 7.20 displays the root locus of the system. The figure indicates the locations of the first three poles of the system that also coincide with the predictions of the FEM.

For maximum closed-loop damping, the velocity feedback gain is selected to be 7.4 as displayed in Figure 7.20.

The frequency response of the beam/ACLD system is shown in Figure 7.21a. The corresponding control voltage is also displayed in Figure 7.21b.

The displayed results indicate that the amplitude of vibration at the first mode is attenuated from 3.22E-5m for the uncontrolled case to 9.93E-6m when the system is controlled by a 38.15 control volts.

Example 7.2 Consider the cantilevered beam shown in Figure 7.22. The main geometrical and physical parameters of the beam and ACLD treatment are listed in Tables 7.5 and 7.6.

The beam is divided into 10 finite elements and is subjected to a force F acting along the transverse direction. Determine the response of the beam/VEM system:

1) In the frequency domain assuming $F = 0.1\, sin(\omega\, t)$ where ω is the excitation frequency.
2) In the time domain assuming $F = 0.1$ N for time $t \leq 2$ μs and $F = 0$ for $t > 2$ μs.

Compare the response, in the time and frequency domains, with those of a plain and untreated beam. Assume that the complex shear modulus G_2 of the VEM is represented by a GHM model with four mini-oscillators that have the properties listed in Table 4.5.

Also establish some comparisons between the response of the uncontrolled beam and that of the beam when it is controlled using a proportional and derivative controller that relies on either uniform or linearly shaped sensors.

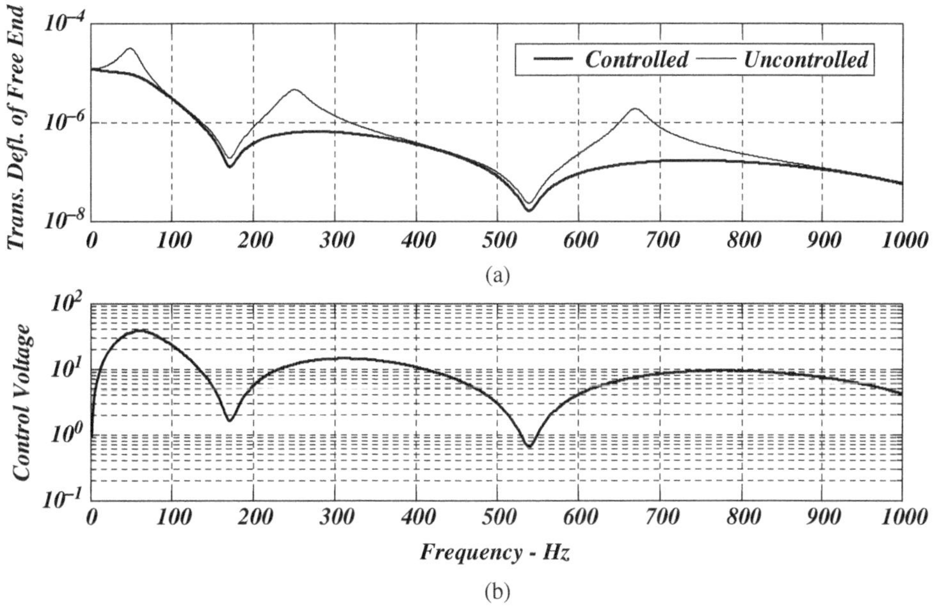

Figure 7.21 Frequency response and control voltage of the beam/ACLD system.

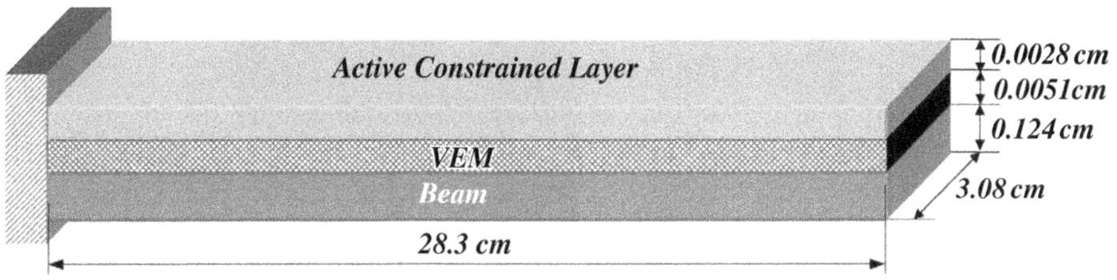

Figure 7.22 Beam with ACLD treatment.

Table 7.5 Beam system.

Parameter	Length (cm)	Thickness (m)	Width (m)	Density (kg m^{-3})	Young's modulus (MPa)
Beam	28.4	0.0025	3.08	2700	70,000
VEM	28.4	0.0025	3.08	1100	[a]
Piezo-film	28.4	28E-6	3.08	1780	4780

[a] Dyad 606 described GHM model with four mini-oscillators.

Table 7.6 Piezo-constraining layer – PVDF.

Parameter	d_{31} (m V^{-1})	K_{31}	g_{31} (Vm N^{-1})	k_{3i}
Value	23E-12	0.15	0.216	12

Solution

Figure 7.23a displays the finite element mesh of the beam/ACLD assembly. Also, Figures 7.23b through 7.23d show the modes and mode shapes of the first three natural frequencies of the assembly. Table 7.7 lists a comparison between the natural frequencies of the beam/ACLD assembly as obtained by the finite element model and ANSYS predictions.

Figure 7.24 displays the frequency response characteristics of the beam/ACLD with a uniform sensor when the proportional (P) and derivative (D) control gains K_p = 700 N m^{-1} and $k_r = 0.15$ where $k_r = K_d/K_p$. The figure displays also the response of the beam/PCLD for comparison purposes.

The displayed response indicates that the use of the ACLD has resulted in reducing the maximum amplitude of vibration at the first mode from 0.0724 to 0.002 42 m and required a control voltage of 78.87 V.

In Figure 7.25, the time response characteristics of the beam/ACLD with uniform sensor are displayed for proportional-derivative (PD) control gains K_p = 700 N m^{-1} and

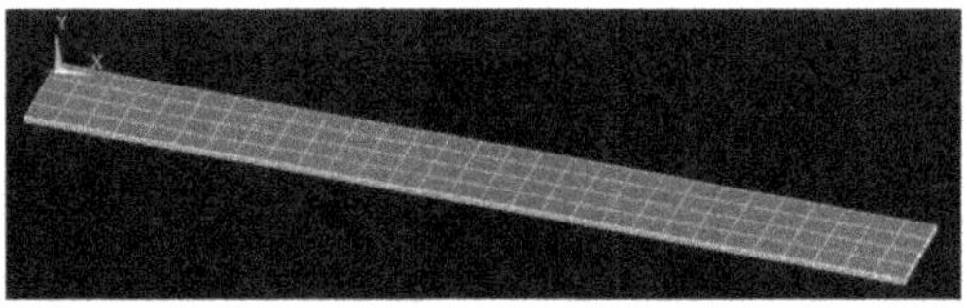

(a) – Finitc Element Mesh

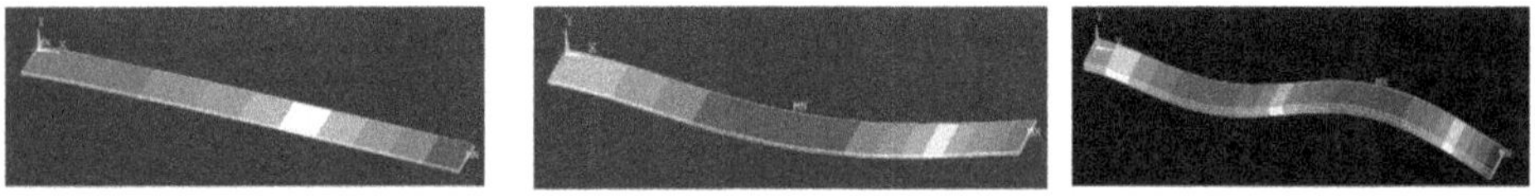

(b) – Mode 1 – 22.85 *Hz* (c) – Mode 2 – 143.10 *Hz* (d) – Mode 3 – 400.53 *Hz*

Figure 7.23 Finite element model and mode shapes of the beam/ACLD assembly. (a) Finite Element Mesh, (b) Mode 1 – 22.85 Hz, (c) Mode 2 – 143.10 Hz, and (d) Mode 3 – 400.53 Hz.

Table 7.7 Comparison between the natural frequencies of the beam/ACLD assembly as obtained by the finite element model and ANSYS predictions.

Method	Mode 1	Mode 2	Mode 3	Mode 4
ANSYS	22.85 Hz	143.10 Hz	400.53 Hz	784.70 Hz
MATLAB	21.80 Hz	136.50 Hz	383.30 Hz	754.90 Hz

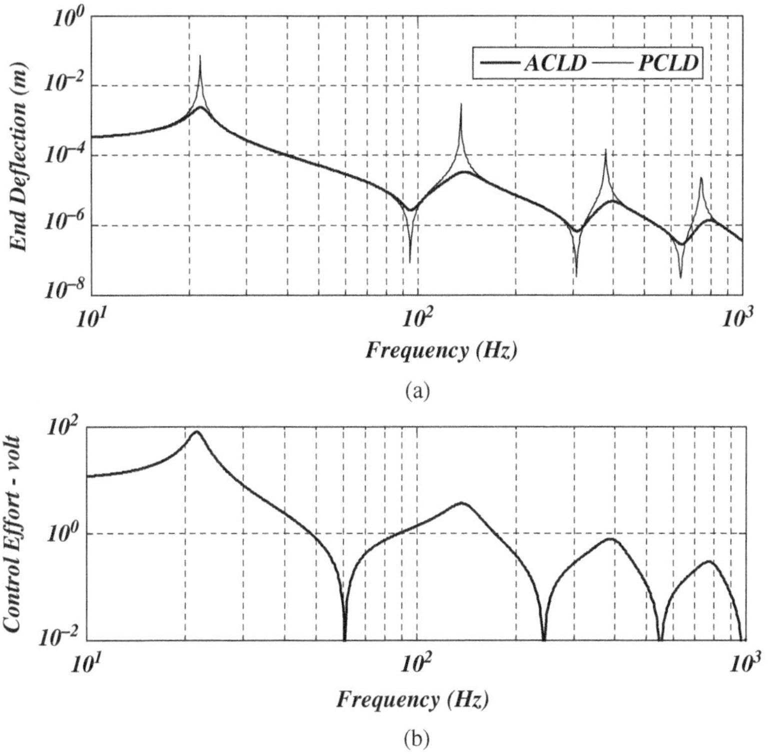

Figure 7.24 Frequency response of beam/ACLD system with uniform sensor.

$k_r = 0.15$ where $k_r = K_d/K_p$. The figure displays also the response of the beam/PCLD for comparison purposes.

The displayed response indicates that the use of the ACLD has also resulted in reducing the maximum amplitude of vibration to 0.0298 μm in about 0.1 s. Such a fast attenuation is achieved with a maximum control voltage of only 0.5 V.

Figures 7.26 and 7.27 display the frequency and time response characteristics of the beam/ACLD with a linearly shaped sensor using the same proportional (*P*) and derivative (*D*) control gains as those used with the uniform sensor. The figures also display the response of the beam/PCLD for comparison purposes.

The displayed frequency response emphasizes that the use of the ACLD has attenuated the maximum amplitude of vibration at the first mode to 0.0178 m with a control voltage of 336.3 V. Also, the displayed time response suggests that using the ACLD has reduced the maximum amplitude of vibration to 0.476 μm in about 0.1 s. Such a fast attenuation is achieved with a maximum control voltage of only 1.0 V.

Table 7.8 summarizes comparisons between the performance characteristics of the ACLD with uniform and linearly shaped sensors at the first mode of vibration. It can be seen that the use of ACLD with a shaped sensor has resulted in significant attenuation in the amplitudes of vibration and required lower voltage to achieve such attenuation as compared to the case of uniform shaped sensor.

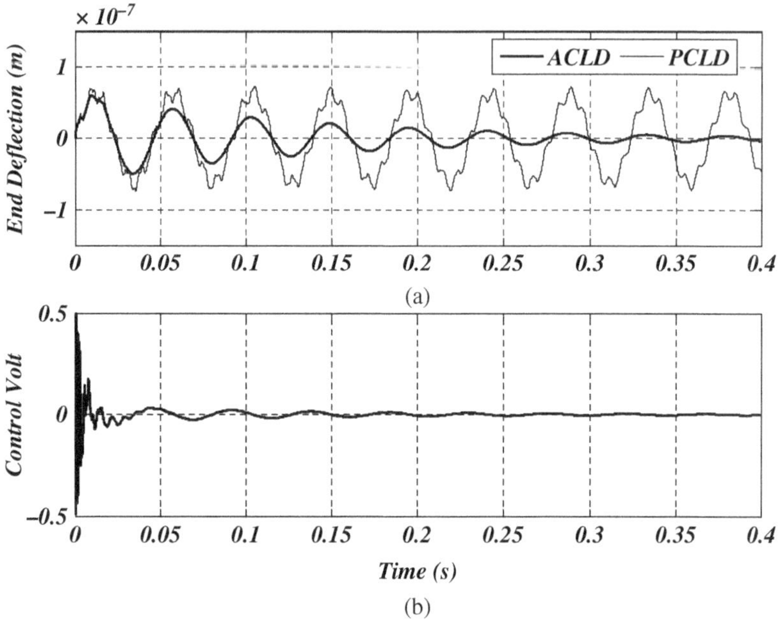

Figure 7.25 Time response of beam/ACLD system with uniform sensor.

7.3.4 Distributed-Parameter Modeling of a Beam/ACLD Assembly

7.3.4.1 Overview

This section focuses on developing a distributed-parameter model (DPM) using Hamilton's principle to describe the dynamics of beams that are fully treated with ACLD (Baz 1997a,b). The variational formulation, being energy-based, is much simpler to employ than the classical force equilibrium-based shear models (Mead and Markus 1969; DiTaranto 1965). Furthermore, the variational formulation provides directly the boundary conditions associated with the ACLD treatment.

For the active control purposes, the variational model provides also direct means for devising a globally stable boundary control strategy that is compatible with the operating nature of the ACLD treatments. In this manner, the instability problems associated with the simple proportional and/or derivative controllers are completely avoided. More importantly, as the control strategy is based on a DPM, the classical spillover problems resulting from using "truncated" finite element models are eliminated. Accordingly, the devised boundary controller enables the control of all the modes of vibration of the ACLD-treated structures.

7.3.4.2 The Energies and Work Done on the Beam/ACLD Assembly

The potential and kinetic energies as well as the external work done by the piezoelectric actuator and external loads, as described in Section 7.2.3, are recast as follows:

Potential Energy

Equation (7.21) is rewritten to include only the conservative component of the energy associated with the VEM as follows:

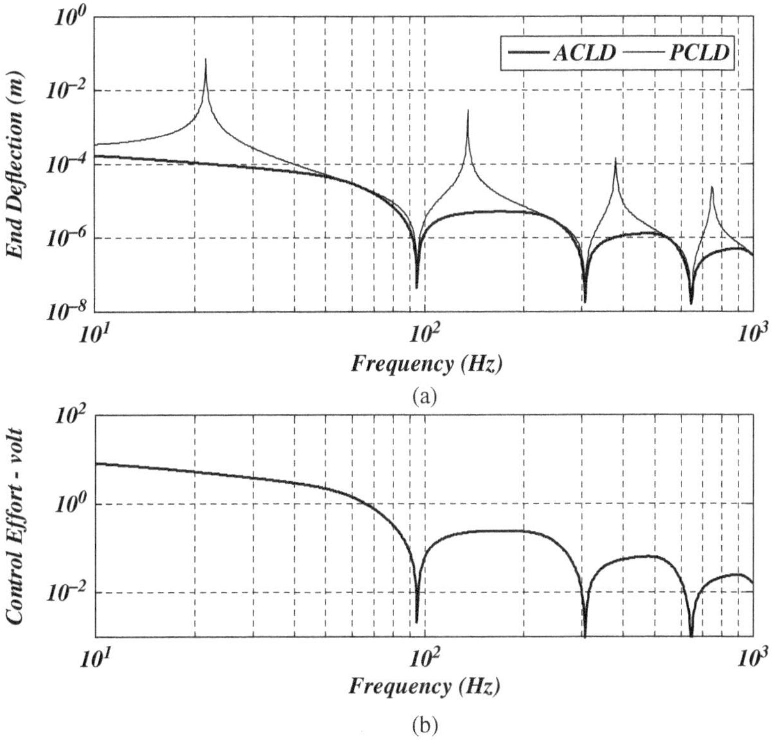

Figure 7.26 Frequency response of beam/ACLD system with linearly shaped sensor.

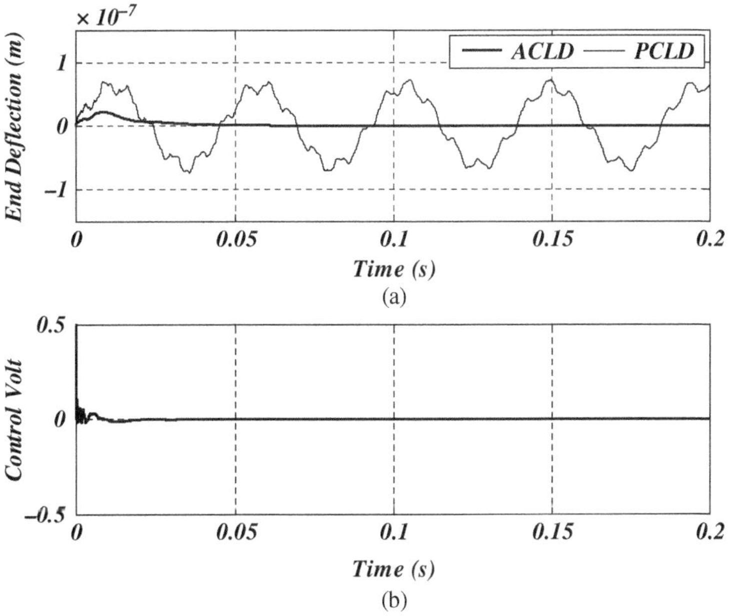

Figure 7.27 Time response of beam with ACLD treatment with linearly shaped sensor.

Table 7.8 Comparison between ACLD with uniform and linearly shaped sensors.

			Frequency response		Time response	
Sensor type	**Sensor shape**	**Measured parameter**	**Max. amplitude (m)**	**Max. volt**	**Max. amplitude (μm)**	**Max. volt**
Uniform	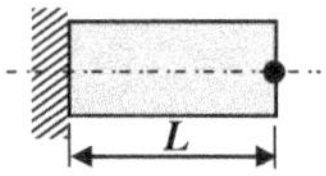	$w_{,x}(L)$	0.002 42	78.873	0.0585	0.5
Linearly Shaped	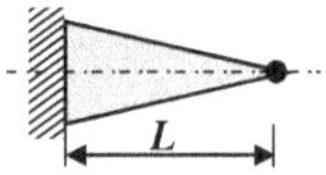	$w(L)/L$	0.001E-3	4.76	0.0217	0.5

$$PE = \frac{1}{2}\int_0^L \left[\left(K_1 u_{,x1}^2 + K_3 u_{,x3}^2\right) + D_t w_{,xx}^2 + G_2' A_2 \gamma^2\right] dx \tag{7.37}$$

where G_2' denotes the shear storage modulus of the VEM. Also, $K_1 = E_1A_1$, $K_3 = E_3A_3$, and $D_t = (E_1I_1 + E_3I_3)$.

Kinetic Energy

Equation (7.22) is simplified to only include the contribution of the transverse motion to the kinetic energy as the components associated with the longitudinal motions can be assumed negligible. Hence, Eq. (7.22) reduces to:

$$KE = \frac{1}{2}\int_0^L m_t \dot{w}^2 dx \tag{7.38}$$

where $m_t = \sum_{i=1}^{3} \rho_i A_i$.

External Work Done

The different components that account for the work done by the non-conservative loads include:

1) Work done W_1 by piezoelectric control forces. This is given by

$$W_1 = K_1 \int_0^L \varepsilon_p u_{,x1}\, dx \tag{7.39}$$

where ε_p denotes the piezo-strain that is induced in the piezoelectric actuator and given by Eq. (7.23). This strain ε_p is assumed constant over the entire length of the constraining layer in order to maintain and emphasize the simplicity and practicality of the ACLD treatment.

2) ***Work W_2 dissipated in the viscoelastic core.*** This is given by

$$W_2 = -A_1 \int_0^L \tau_d \gamma \, dx \tag{7.40}$$

where τ_d is the dissipative shear stress developed by the viscoelastic core. It is given by:

$$\tau_d = \left(G_2' \eta / \omega\right) \dot{\gamma} = G_2' \eta \gamma i \tag{7.41}$$

where η, ω, and i denote the loss factor of the viscoelastic core, the frequency and $\sqrt{-1}$, respectively. In Eq. (7.41), the behavior of the viscoelastic core is modeled using the common complex modulus approach, which is a frequency domain-based method.

7.3.4.3 The Distributed-Parameter Model

The equations and boundary conditions governing the operation of the beam/ACLD assembly are obtained by applying Hamilton's principle (Meirovitch 2010):

$$\int_{t_1}^{t_2} \left(\delta(KE - PE) + \delta \sum_{i=1}^{2} W_i \right) dt = 0 \tag{7.42}$$

where $\delta(.)$ denotes the first variation in the quantity inside the parentheses.

The resulting equations of the beam/ACLD assembly are:

$$-K_1 u_{,xx1} + G_2/h_2(u_1 - u_3 + hw_{,x}) = 0, \tag{7.43}$$

$$-K_3 u_{,xx3} - G_2/h_2(u_1 - u_3 + hw_{,x}) = 0, \tag{7.44}$$

and

$$D_t w_{,xxxx} + m_t w_{,tt} - G_2 h/h_2(u_{,x1} - u_{,x3} + hw_{,xx}) = 0 \tag{7.45}$$

where $G_2 = G_2'(1 + \eta i)$ is the complex modulus of the VEM.

For a cantilevered beam, Hamilton's principle yields the following boundary conditions:

At $x = L$:

$$u_1 = 0, u_3 = 0, \; w = 0, \; w_{,x} = 0 \tag{7.46}$$

and

At $x = L$:

$$u_{,x1} = \varepsilon_p, u_{,x3} = 0, \; w_{,xx} = 0, \; D_t w_{,xxx} - G_2 h/h_2(u_1 - u_3 + hw_{,x}) = 0 \tag{7.47}$$

Equations (7.43) and (7.44) give:

$$K_1 u_{,xx1} + K_3 u_{,xx3} = 0, \tag{7.48}$$

and

$$z_{,xx} = g(z + h w_{,x}) \tag{7.49}$$

where $z = (u_1 - u_3)$ and $g = G_2/h_2[(K_1 + K_3)/(K_1 K_3)]$.

Combining Eqs. (7.45) and (7.49) yields:

$$z_{,xx} = h\left[\frac{1}{gY}w_{,xxxxx} - w_{,xxx} + \frac{m_t}{gYD_t}w_{,xtt}\right] = gz + ghw_{,x} \tag{7.50}$$

or,

$$z = -hw_{,x} + \frac{h}{g}\left[\frac{1}{gY}w_{,xxxxx} - w_{,xxx} + \frac{m}{gYD_t}w_{,xtt}\right] \tag{7.51}$$

where $g = \frac{G_2}{h_2}\frac{K_1 + K_3}{K_1K_3}$ and $Y = \frac{h^2}{D_t}\frac{K_1K_3}{(K_1 + K_3)}$. Note that g and Y are defined as the shear parameter and geometrical factors, respectively, as introduced by DiTaranto (1973).

Differentiating Eq. (7.51) partially with respect to x, substituting into Eq. (7.22), and rearranging the terms yields a sixth order partial differential equation describing the dynamics of the beam/ACLD assembly as follows:

$$w_{,xxxxxx} - g(1 + Y)w_{,xxxx} + \frac{m}{D_t}w_{,xxtt} - \frac{gm}{D_t}w_{,tt} = 0 \tag{7.52}$$

The associated boundary conditions given by Eqs. (7.46) and (7.47) simplify to:

At x = 0:

$$z = u_1 - u_3 = 0,\ w = 0,\ w_{,x} = 0 \tag{7.53}$$

and

At x = L:

$$z_{,x} = u_{,x1} - u_{,x3} = \varepsilon_p, w_{,xx} = 0,\ D_t w_{,xxx} - G_2 h/h_2(u_1 - u_3 + hw_{,x}) = 0 \tag{7.54}$$

Further simplification of the boundary conditions can be obtained by using Eq. (7.51) and its spatial derivative with respect to x to yield:

At x = 0:

$$w_{,xxxxx} - gYw_{,xxx} = 0,\ w = 0,\ w_{,x} = 0 \tag{7.55}$$

and

At x = L:

$$\frac{D_t}{mg}w_{,xxxx} + \frac{1}{g}w_{,tt} - \frac{YD_t}{mh}\varepsilon_p = 0, w_{,xx} = 0,$$

$$\frac{D_t}{mg}w_{,xxxxx} - \frac{D_t}{m}(1 + Y)w_{,xxx} + \frac{1}{g}w_{,xtt} = 0 \tag{7.56}$$

It is important here to note that the sixth order partial differential equation describing the beam/ACLD system, given by Eq. (7.52) is the same as that describing a beam treated with conventional PCLD as obtained by Mead and Markus (1969). However, the boundary condition given by Eq. (7.56) is modified to account for the control action generated by the strain ε_p induced by the active constraining layer at the free end of the beam (i.e., at x = L). Therefore, the particular nature of operation of the beam/ACLD system implies the existence of boundary control action ε_p.

7.3.4.4 Globally Stable Boundary Control Strategy

The globally stable control strategy presented in this section aims at ensuring that the total energy $E_n = PE + KE$ of the beam/ACLD system is maintained as a strictly non-increasing function of time.

The total energy E_n is given by:

$$E_n = \frac{1}{2}\int_0^L \left[\left(K_1 u_{,x1}^2 + K_3 u_{,x3}^2\right) + D_t w_{,xx}^2 + G_2' A_2 \gamma^2\right] dx + \frac{1}{2}\int_0^L m_t \dot{w}^2 dx \tag{7.57}$$

Differentiating the total energy expression given by Eq. (7.57) with respect to the time and using Eqs. (7.52) through (7.54), yielding:

$$\dot{E}_n = K_1 u_{,t1}(L)\,\varepsilon_p - \left(G_2' \eta A_2/\omega\right)\int_0^L \gamma_{,t}^2\, dx \tag{7.58}$$

Equation (7.58) indicates that the second term is strictly negative, hence a globally stable boundary controller with a continuously decreasing energy norm (i.e., $\dot{E}_n < 0$) is obtained when the control action ε_p takes the following form:

$$\varepsilon_p = -K_g u_{,t1}(L) \tag{7.59}$$

where K_g is the gain of the boundary controller. This form of a controller makes the rate of change of the total energy of the entire assembly strictly negative.

Equation (7.59) indicates that the control action is a velocity feedback of the longitudinal displacement of the piezoelectric constraining layer. It is important also here to note that when the active control action ε_p ceases or fails to operate for one reason or another (i.e., when $\varepsilon_p = 0$), the beam system remains globally stable as indicated by Eq. (7.58). Such inherent stability is attributed to the second term in the equation that quantifies the contribution of the Passive Constrained Layer Treatment (PCLD).

7.3.4.5 Implementation of the Globally Stable Boundary Control Strategy

The boundary control strategy can be implemented in one of the following two ways:

i) in terms of the longitudinal displacement u_1 of the piezo-actuator by using Eq. (7.57) as follows:

$$\varepsilon_p = -K_g s u_1(L) \tag{7.60}$$

where s is the Laplace complex number.

ii) in terms of the longitudinal displacement u_3 of the base beam by using Eqs. (7.47) and (7.48), in order to relate u_3 to u_1 as follows:

$$u_3(L) = -\frac{K_1}{K_3} u_1(L) + \frac{K_1}{K_3 s}\varepsilon_p \tag{7.61}$$

Eliminating u_1 between Eqs. (7.60) and (7.61) gives the control action ε_p in terms of u_3 as follows:

$$\varepsilon_p = \frac{K_g K_3 s}{K_1 (1 + K_g)} u_3(L) \tag{7.62}$$

7.3.4.6 Response of the Beam/ACLD Assembly

The response w of the beam/ACLD assembly can be determined by solving Eq. (7.52) using the classical separation of variables method such that:

$$w = W(x) T(t) \tag{7.63}$$

where $W(x)$ and $T(t)$ denote spatial and temporal functions. Substituting Eq. (7.63) into Eq. (7.52) and setting $\ddot{T}/T = -\omega^2$ yields the following characteristic equation:

$$\lambda^6 - g(1+Y)\lambda^4 - \left(\frac{\omega^2 m_t}{D_t}\right)\lambda^2 + \left(\frac{\omega^2 m_t g}{D_t}\right) = 0 \tag{7.64}$$

where λ is a differential operator with respect to x. Let $\pm\delta_i$ with $i = 1,\ldots,3$ be the roots of the characteristics of the equation, then the spatial function $W(x)$ can be written as:

$$W(x) = C_1 e^{\delta_1 x} + C_2 e^{-\delta_1 x} + C_3 e^{\delta_2 x} + C_4 e^{-\delta_2 x} + C_5 e^{\delta_3 x} + C_6 e^{-\delta_3 x} \tag{7.65}$$

where the coefficients C_i's are to be determined from the boundary conditions given by Eqs. (7.55) and (7.56).

Example 7.3 Consider the cantilever beam/ACLD system with a VEM that is described in Example 7.1.

Determine the compliance and control voltage characteristics of the beam/ACLD system using a velocity feedback controller with a control gains $K_D = 7.4$. Compare these characteristics with those of the beam/PCLD system using the DPM approach outlined in Section 7.3.4. Also, compare the predictions of the DPM against the predictions of the finite element model when the beam is excited by a unit sinusoidal load at its free end.

Also determine the energy dissipation characteristics of the beam/ACLD system compared with that of the beam/PCLD system.

Solution

Note that when the beam/ACLD system is excited by a transverse end load F, the boundary condition at $x = L$, Eq. (7.56), must be modified to:

At $x = L$:

$$\frac{D_t}{mg} w_{,xxxxx} - \frac{D_t}{m}(1+Y) w_{,xxx} + \frac{1}{g} w_{,xtt} = F/(bm) \tag{7.66}$$

Accordingly, the transverse deflection $W(x)$ of beam can be determined using Eq. (7.65) subjected to the boundary conditions (7.55) and (7.56) after including the modifications imposed by Eq. (7.66).

Figure 7.28 displays a comparison between the frequency response characteristics of the open-loop beam/*ACLD* system (i.e., $K_D = 0$) as predicted by the FEM and DPM approaches.

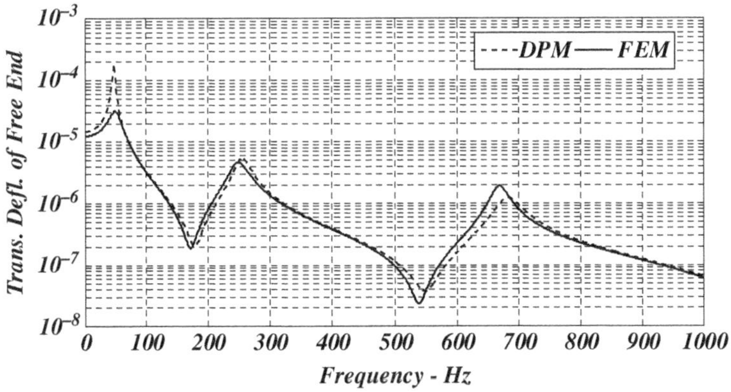

Figure 7.28 Frequency response and control voltage of beam with PCLD treatment when $K_D = 0$.

Table 7.9 Comparison between the natural frequencies of the beam/ACLD system as predicted by FEM, ANSYS, and DPM.

Mode	1	2	3
FEM (Hz)	48.88	248.4	668.78
ANSYS (Hz)	46.35	253.52	677.49
DPM –(Hz)	47.00	256.00	683.00

It can be seen that there is close agreement between the two approaches.

Table 7.9 lists also a comparison between the first three natural frequencies of the beam/ACLD system as computed using the FEM, ANSYS, and DPM approaches when $K_D = 0$.

It is evident that the predictions of the DPM are very close to those of the ANSYS model.

Figure 7.29 displays the compliance and control voltage characteristics of the beam/ACLD system for control gain $K_D = 7.4$ along with a comparison of the corresponding characteristics of the beam/PCLD system, that is, $K_D = 0$.

Note that there are discrepancies between the predictions of the DPM and the FEM of the closed-loop response characteristics of the bean/ACLD system. These discrepancies can be attributed to the fact that in the DPM, the control action is a linear force due to the strain ε_p applied to the constraining layer whereas in the FEM, the control action is a moment due to the piezo-strain applied to the entire assembly.

The energy dissipated in the beam/ACLD system is given by Eqs. (7.40) and (7.41) as:

$$D = G_2' \eta (bh_2) \int_0^L \gamma^2 \, dx \tag{7.67}$$

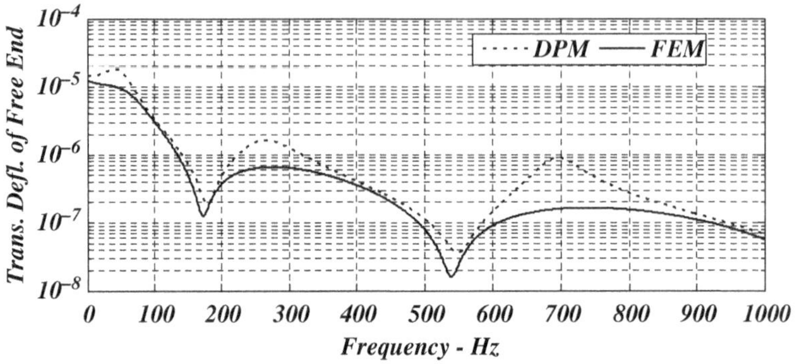

(a)

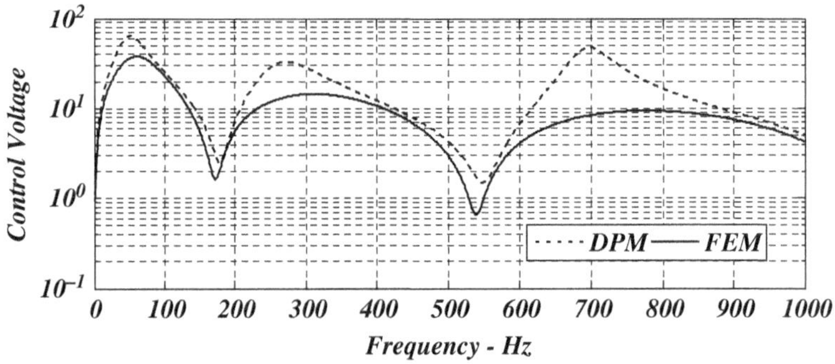

(b)

Figure 7.29 Frequency response (a) and control voltage (b) of beam with ACLD treatment when $K_D = 7.5$.

where the shear strain γ can be determined from Eq. (7.51) as follows:

$$\gamma = \frac{1}{h_2}(z + hw_{,x}) = \frac{h}{h_2 g}\left[\frac{1}{gY}w_{,xxxxx} - w_{,xxx} + \frac{m}{gYD_t}w_{,xtt}\right] \tag{7.68}$$

Figure 7.30 shows the corresponding energy dissipation of the beam/ACLD system in comparison with the corresponding characteristics of the beam/PCLD system, that is, $K_P = 0$. It is evident that the energy dissipation with the ACLD becomes higher than that of the PCLD particularly at high frequencies.

Table 7.10 summarizes the performance characteristics of the beam/ACLD as compared with that of the beam/PCLD system.

7.4 Active Constrained Layer Damping for Plates

A finite element model is developed in Section 4.7 to describe the behavior of plates with PCLD treatments. The effect of the control forces and moments generated by the active constrained layer are presented in this section.

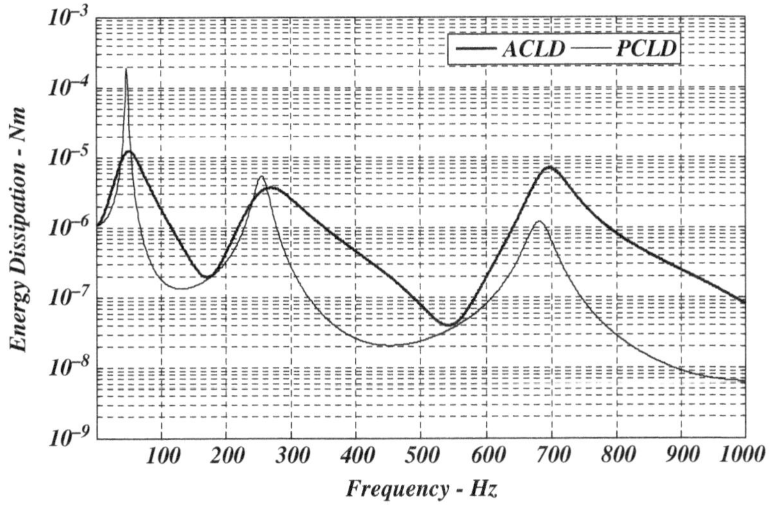

Figure 7.30 Energy dissipation of beam/ACLD system when $K_p = 7.4$.

Table 7.10 Performance of beam/ACLD assembly at different control gains.

	Uncontrolled			$K_D = 7.4$			
Mode	Frequency (Hz)	Amplitude (μm)	Energy dissipated (μJ)	Frequency (Hz)	Amplitude (μm)	Energy dissipated (μJ)	Volts (V)
1	47	187.4	187.8	42	0.1823	0.124	64.28
2	256	5.366	5.231	262	1.626	3.729	32.98
3	683	1.181	1.121	702	0.903	7.030	47.21

7.4.1 Control Forces and Moments Generated by the Active Constraining Layer

7.4.1.1 The In-Plane Piezoelectric Forces

These in-plane forces ($\boldsymbol{F_{px}}$, $\boldsymbol{F_{py}}$, and $\boldsymbol{F_{pxy}}$) at the jth node, are given by:

$$\begin{Bmatrix} F_{px_j} \\ F_{py_j} \\ F_{pxy_j} \end{Bmatrix} = (K_p + K_d p) V_s \iint_{a_i\, b_i} \left([B_{1p}]^T [D_{1p}] \begin{Bmatrix} d_{31} \\ d_{32} \\ 0 \end{Bmatrix} \right) dxdy \text{ For } j = 1, \ldots, 4 \quad (7.69)$$

with K_p and K_d denoting the proportional and derivative control gains. Also, p and V_s denote the operator d/dt and the sensor voltage. The constants d_{31} and d_{32} define the piezoelectric strain constants in the x and y directions. Also, $[B_{1p}]$ and $[D_{1p}]$ are given by (Section 4.8):

where

$$[B_{1p}] = \begin{bmatrix} N_{u1,x} \\ N_{v1,x} \\ N_{u1,y} + N_{v1,x} \end{bmatrix} \text{ and } [D_{1p}] = \frac{E_1}{1-\nu_1^2}\begin{bmatrix} 1 & \nu_1 & 0 \\ \nu_1 & 1 & 0 \\ 0 & 0 & \frac{1}{2}(1-\nu_1) \end{bmatrix} \quad (7.70)$$

where N_{u1} and N_{v1} define the axial shape functions, respectively. Also, E_1 and ν_1 denoting Young's modulus and Poisson's ratio of the piezoelectric constraining layer, respectively.

7.4.1.2 The Piezoelectric Moments

These piezoelectric moments $\{M_{pi}\}$ due to the bending of the piezoelectric constraining layer are given by:

$$\begin{Bmatrix} M_{px_j} \\ M_{py_j} \\ M_{pxy_j} \end{Bmatrix} = (K_p + K_d p) V_s h \int_{a_i}\int_{b_i} \left([B_{1b}]^T [D_{1b}] \begin{Bmatrix} d_{31} \\ d_{32} \\ 0 \end{Bmatrix} \right) dxdy \quad \text{for } j = 1,\ldots,4 \quad (7.71)$$

where $h = h_2 + \frac{1}{2}(h_1 + h_3)$.

7.4.1.3 Piezoelectric Sensor

The voltage, V_s, developed by the piezo-sensor is obtained from (Appendix 7.A, and Lee, 1987):

$$V_s = \left[\frac{k_{31}^2 b}{g_{31} C}\right] \sum_{i_{sx}}^{i_{fx}} \sum_{i_{sy}}^{i_{fy}} \int_{a_i}\int_{b_i} b(x,y) \left[(u_{1,x} + v_{1,y}) - h(w_{,xx} + w_{,yy}) \right] dxdy = [B_s]\{\Delta_i\} \quad (7.72)$$

where

$$[B_s] = \left[\frac{k_{31}^2 b}{g_{31} C}\right] \sum_{i_{sx}}^{i_{fx}} \sum_{i_{sy}}^{i_{fy}} \int_{a_i}\int_{b_i} b(x,y) \left[\left([N_{u_1}]_{,x} + [N_{v_1}]_{,y}\right) - h\left([N_w]_{,xx} + [N_w]_{,yy}\right) \right] dxdy \quad (7.73)$$

Also, h is the distance from the plate neutral plane to the sensor surface and $b(x, y)$ is a distribution shape function of the sensor [$b(x,y) = 1$ for a uniform sensor]. The sensor is extended between elements i_{sx} and i_{fx} in the x direction and i_{sy} and i_{fy} in the y direction. Also, k_{31}^2 is the electromechanical coupling factor, g_{31} is the piezoelectric voltage constant and C is the capacitance of the sensor that is given by:

$$C = 8.854 \times 10^{-12} A\, k_{3t}/h_1 \quad (7.74)$$

where A is the sensor surface area and k_{3t} is the dielectric constant.

Note also that $[N_{u_1}]$, $[N_{v_1}]$, and $[N_w]$ define the in-plane and transverse shape functions, respectively, as defined in Section 4.8.

7.4.1.4 Control Voltage to Piezoelectric Constraining Layer

The voltage V_A applied to the piezoelectric constraining layer is given by:

$$V_A = \left(K_p + K_d p\right) V_s \tag{7.75}$$

7.4.2 Equations of Motion

The dynamics of the ACLD-treated plate element is described by the following equation of motion:

$$[M_i]\{\ddot{\Delta}_i\} + [K_i]\{\Delta_i\} = \{F_c\} \tag{7.76}$$

where $[M_i]$ and $[K_i]$ denote the mass and stiffness matrices of the plate/ACLD element given in Appendix 7.A. The vector $\{F_c\}$ is the vector of control forces and moments generated by the piezo-constraining layers on the treated plate element. It is expressed as follows:

$$\{F_c\} = \{F_1, F_2, F_3, F_4\}^T \tag{7.77}$$

where

$$\{F_i\} = \left\{F_{pxi}, F_{pyi}, 0, 0, -F_{pxi}, -F_{pyi}, 0, M_{pxi}, M_{pyi}\right\}^T \; for\; i = 1, \ldots, 4 \tag{7.78}$$

Example 7.4 Consider the aluminum cantilever plate treated with two ACLD patches placed at the fixed end of the plate as shown in Figure 7.31. The dimensions of the plate and the PCLD patches are shown in the figure with $h_1 = h_2 = h_3 = 0.005m$.
The VEM is modeled using a GHM model with three mini-oscillators such that:

$$G_2(s) = G_0\left[1 + \sum_{i=1}^{3} \alpha_i \frac{s^2 + 2\zeta_i\omega_i s}{s^2 + 2\zeta_i\omega_i s + \omega_i^2}\right] \text{ with } G_0 = 0.5\,MPa$$

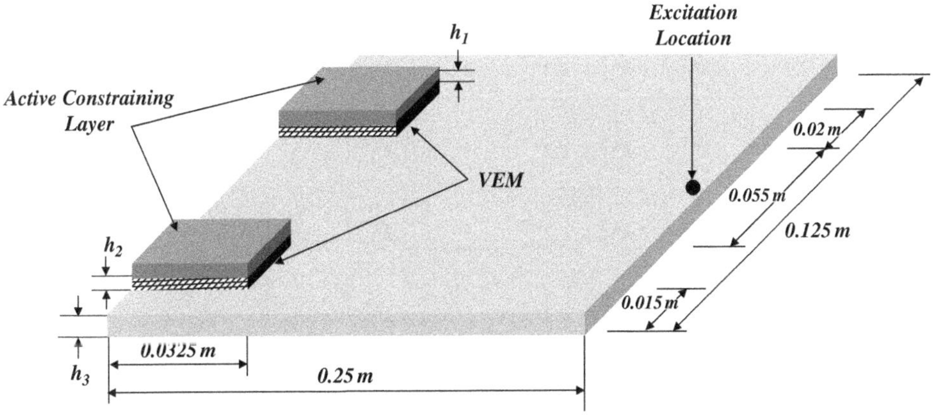

Figure 7.31 Schematic drawing of a cantilever plate treated with two ACLD Patches.

The parameters α_i, ζ_i, and ω_i are listed in Table 4.7. Assume that the constraining layer is made of piezoelectric material lead zirconate titanate (PZT-4) with $d_{31} = d_{32} = -123 \times 10^{-12}$ m V^{-1} and $C_{31}^E = C_{32}^E = 78.3GPa$. Determine the time and frequency response of the plate, at different control gains, when it is excited by a unit force acting at the middle of the free end.

Solution

Figure 7.32 displays the modes and mode shapes of the plate/ACLD patches.

The frequency response of the plate tip and the corresponding control voltage are displayed in Figure 7.33 when the ACLD patches are controlled with a velocity feedback that has a gain $K = 1$ Ns m^{-1}. Also displayed in the figure, for comparison purposes, is the response of the plate when $K = 0$ that corresponds to the open-loop condition. In this case, it is evident that the controller effectiveness is only limited to higher order modes.

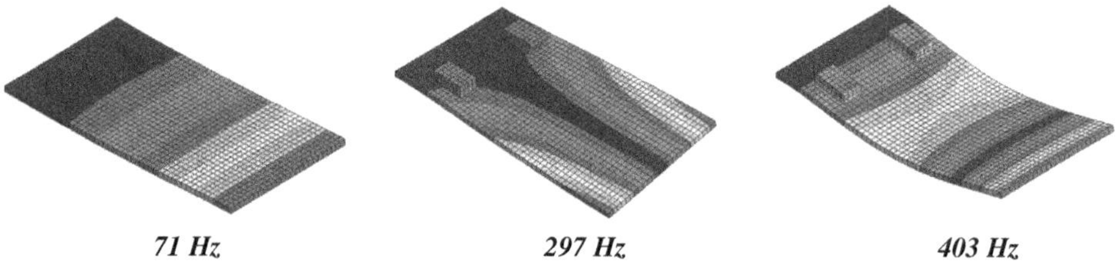

Figure 7.32 Modes and modes shapes of a cantilever plate treated with two ACLD Patches.

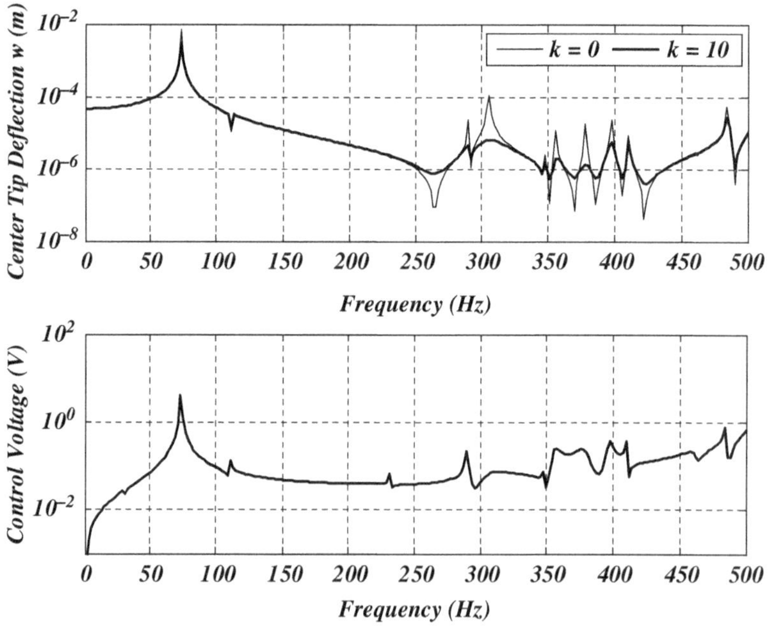

Figure 7.33 Frequency response and control voltage of the plate with ACLD patches when $K = 0$ (open-loop) and 10 Ns m^{-1} (closed-loop).

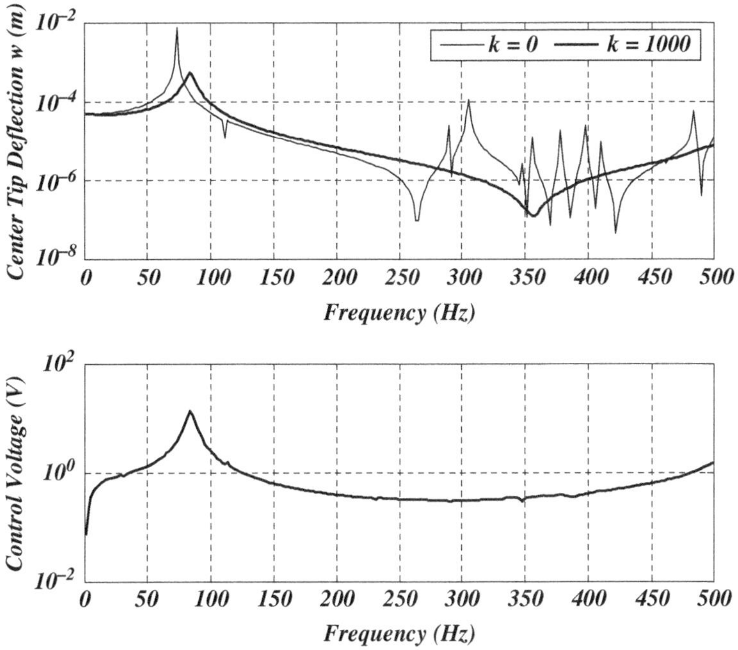

Figure 7.34 Frequency response and control voltage of the plate with ACLD patches when $K = 0$ (open-loop) and 1000 Ns m^{-1} (closed-loop).

When the controller gain is increased to $K = 1000$ Ns m^{-1}, the resulting plate response and the control voltage are shown in Figure 7.34. The effectiveness of the controller, in this case, in attenuating the vibration is clear over a wide frequency band and is extended to the low frequency modes. Note that the maximum control voltage required in this case is only 13.37 V.

Table 7.11 summarizes the performance characteristics of the plate/ACLD patches, at the first mode of vibration, for different control gains.

The time response of the plate tip and the corresponding control voltage are displayed in Figure 7.35 when the ACLD patches are controlled with a velocity feedback that has a gain $K = 1$ Ns m^{-1}. The displayed characteristics are computed when the plate is subjected, at its mid tip point, to a transient force of 1 N for a duration of 0.01 s. The closed-loop response is compared also with that of the open-loop, that is, when $K = 0$.

Table 7.11 Performance of plate/ACLD patches for different control gains in the frequency domain.

Control gain	Maximum deflection (m)	Maximum control voltage (V)
0	0.007 47	0
10	0.003 27	4.047
1000	0.000 54	13.37

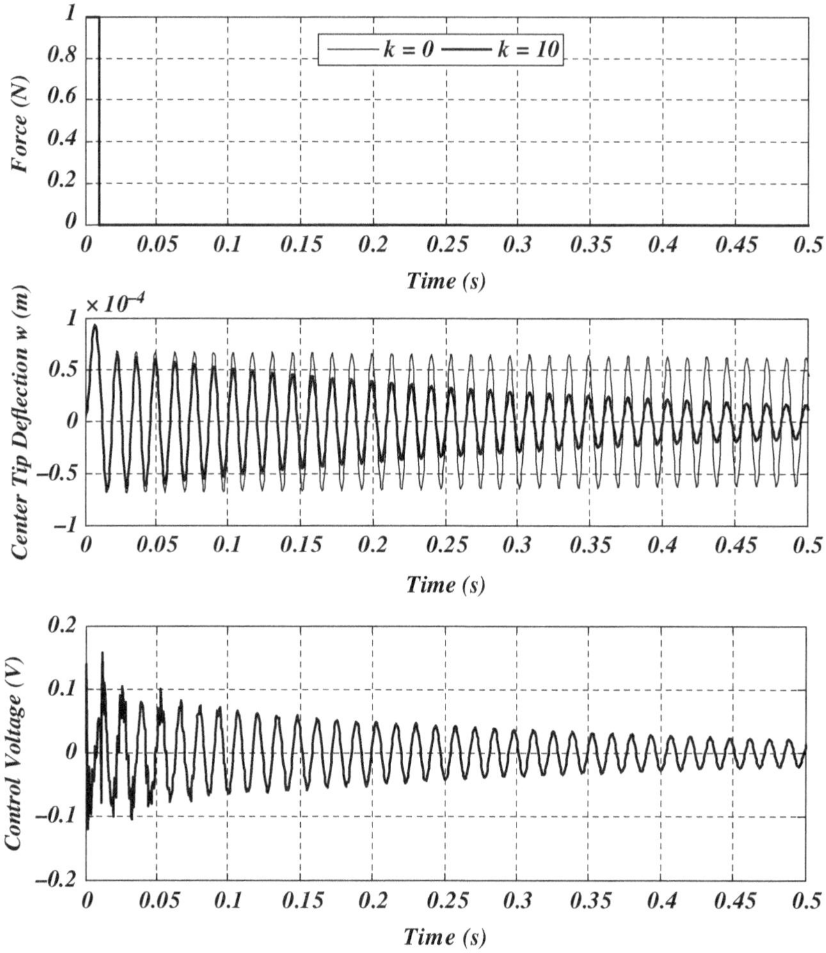

Figure 7.35 Time response and control voltage of the plate with ACLD patches when $K = 0$ (open-loop) and 10 Ns m^{-1} (closed-loop).

The obtained comparison indicates that the damping ratio is increased from 0.00196 for the open loop to 0.00710 for the closed-loop condition. Note that this 3.5-times increase in the damping ratio is achieved with only 0.158 V.

When the controller gain is increased to K = 1000 Ns m^{-1}, the resulting plate time response and the control voltage are shown in Figure 7.36. Such an increase in the control gain results in increasing the damping ratio from 0.001 96 for the open loop to 0.051 for the closed-loop condition. Note that this 25-times increase of the damping ratio is achieved with only 1.424 V.

Table 7.12 summarizes the closed-loop damping ratio and control voltage of the plate/ACLD patches, for different control gains in comparison with the corresponding characteristics of the open-loop system.

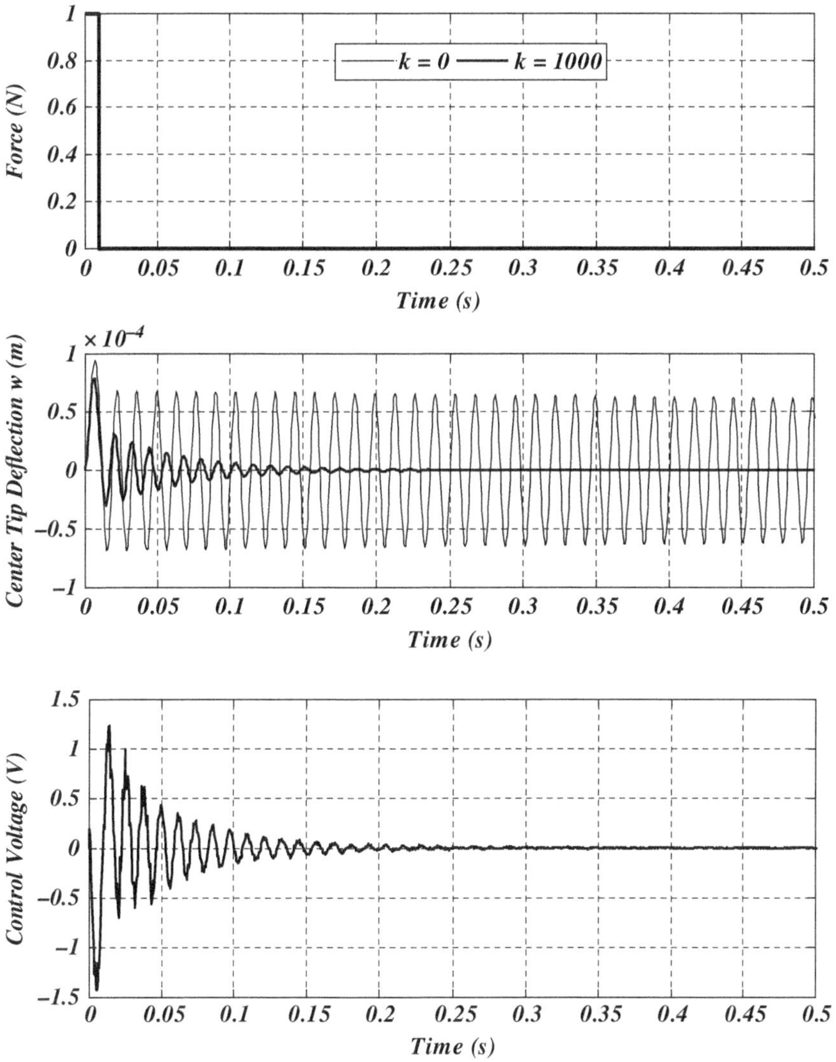

Figure 7.36 Time response and control voltage of the plate with ACLD patches when $K = 0$ (open-loop) and 1000 Ns m^{-1} (closed-loop).

Table 7.12 Performance of plate/ACLD patches for different control gains in the time domain.

Control gain	Closed-loop damping ratio (ζ)	Maximum control voltage (V)
0	0.001 96	0
10	0.007 10	0.158
1000	0.051 00	1.424

7.5 Active Constrained Layer Damping for Shells

A finite element model is developed in Section 4.8 to describe the behavior of shells with PCLD treatments. The effect of the control forces and moments generated by the active constrained layer are presented in this section.

7.5.1 Control Forces and Moments Generated by the Active Constraining Layer

Figure 7.37 shows a schematic drawing of a quadrilateral element of a cylindrical shell treated with an ACLD treatment.

The in-plane piezoelectric forces (F_{px}, F_{py}, and F_{pxy}) and the associated control moments ($\{M_{px}\}$, $\{M_{py}\}$, and $\{M_{pxy}\}$) due to the bending of the piezoelectric constraining layer are given by Eqs. (7.69) and (7.71), respectively, as in the case of the quadrilateral plate element described in Section 7.4.

Similarly, the voltage generated by the piezoelectric sensor and the voltage applied to the piezoelectric constraining layer are given by Eqs. (7.72) and (7.75), respectively.

7.5.2 Equations of Motion

The dynamics of the ACLD-treated plate element is described by the following equation of motion:

$$[M_i]\{\ddot{\Delta}_i\} + [K_i]\{\Delta_i\} = \{F_c\} \tag{7.79}$$

where $[M_i]$ and $[K_i]$ denote the mass and stiffness matrices of the plate/ACLD element given in Appendix 7.A. The vector $\{F_c\}$ is the vector of control forces and moments

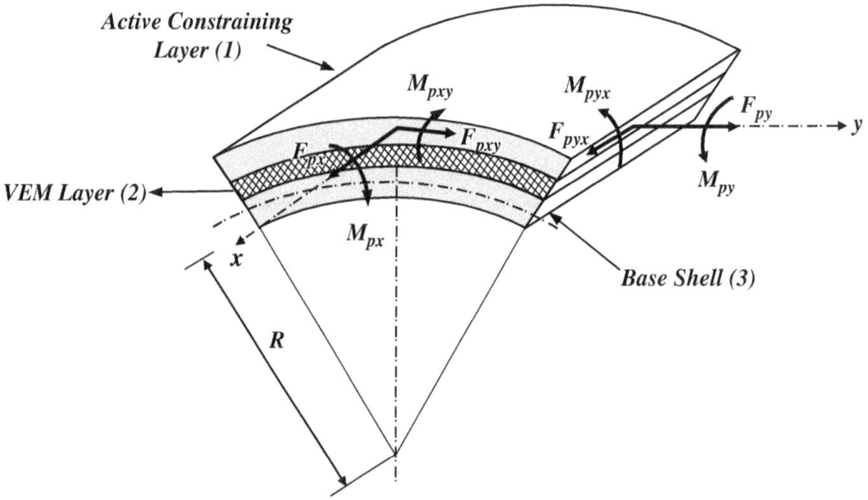

Figure 7.37 Piezoelectric forces and moments on a cylindrical shell element treated with active constrained layer damping treatment (ACLD).

generated by the piezo-constraining layers on the treated shell element are as given by Eqs. (7.77) and (7.78).

Example 7.5 Consider the clamped-free shell/ACLD assembly shown in Figure 7.38a. The main physical and geometrical parameters of the shell and PCLD are listed in Table 7.13. The shell has an internal radius R is 0.1016 m. The ACLD treatment consists of two patches as displayed in Figure 7.38b. The patches are bonded 180° apart on the outer surface of the cylinder with each of which subtending an angle of 90° angle at the center of the shell.

Determine the frequency response of the system when excited at the free end by a unit force. Also determine the time response of the system when a unit pulse force of duration 0.10 ms is applied at the free end. Compare the response when the response is predicted using ANSYS and the finite element approach outlined in Sections 4.7 and 7.5.

Solution

Figure 7.39a displays the frequency response of the shell/ACLD assembly, as predicted by the FEM approach from Sections 4.7 and 7.5, when using a proportional control gain K_P = 1.5E6 Nm m^{-1}. The corresponding control voltage is shown in Figure 7.39b. The effectiveness of the controller is very limited at the first and second modes (60 and 62 Hz). These modes and corresponding mode shapes are displayed in Table 4.12.

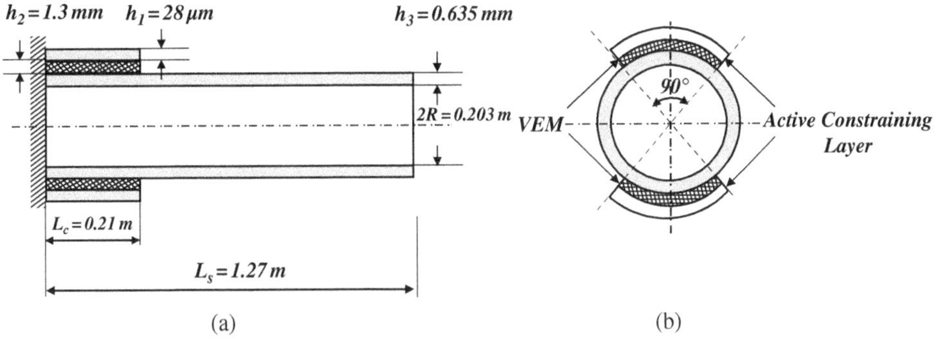

Figure 7.38 Configuration of the shell/ACLD assembly.

Table 7.13 Parameters of the shell/ACLD assembly.

Parameter	Length (m)	Thickness (mm)	Density (kg m^{-3})	Young's modulus (GPa)
Shell	1.270	0.635	7800	210
VEM	0.212	1.300	1140	[a]
PZT Constraining Layer	0.212	0.028	7600	66

[a] G_∞ = 292.01 MPa and β_∞ = 0.007 for a five-term GMM of the VEM as described in Example 4.6 and Table 4.6.

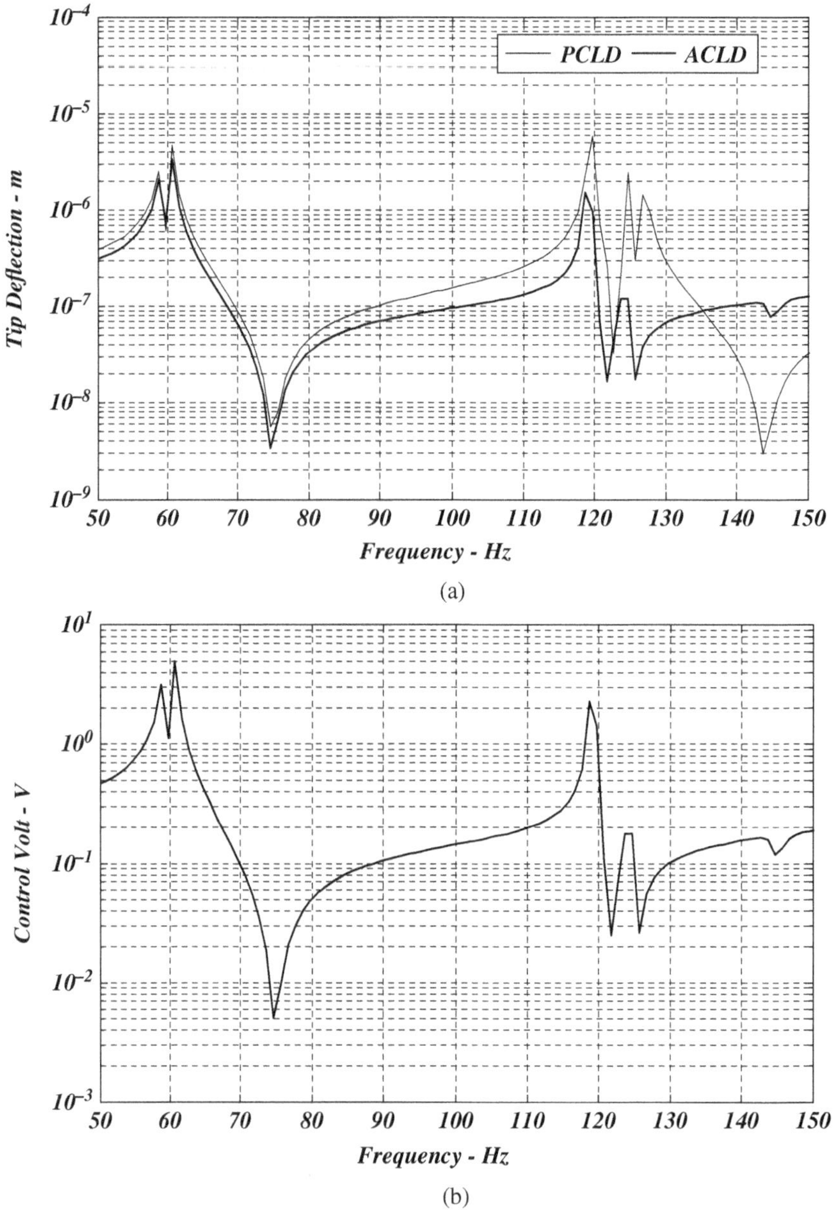

Figure 7.39 Frequency response (a) and control voltage (b) of the shell/ACLD assembly when K_P = 1.5E6 Nm m^{-1}.

The effectiveness of the controller becomes more apparent at the 125 Hz mode and higher.

The associated control voltage reaches 11 V at the first mode (60 Hz) and attains a maximum value of 24 V at the 119 Hz mode.

Increasing the control gain K_P to 3E6 Nm m^{-1} considerably improves the performance of the controller as can be seen in the frequency response displayed in Figure 7.40a. The effectiveness of the controller becomes evident at the 62 Hz mode as well at the higher order modes.

In this case, the control voltage increases by 30 V at the first mode (60 Hz) and attains a maximum value of 120 V at the 119 Hz mode.

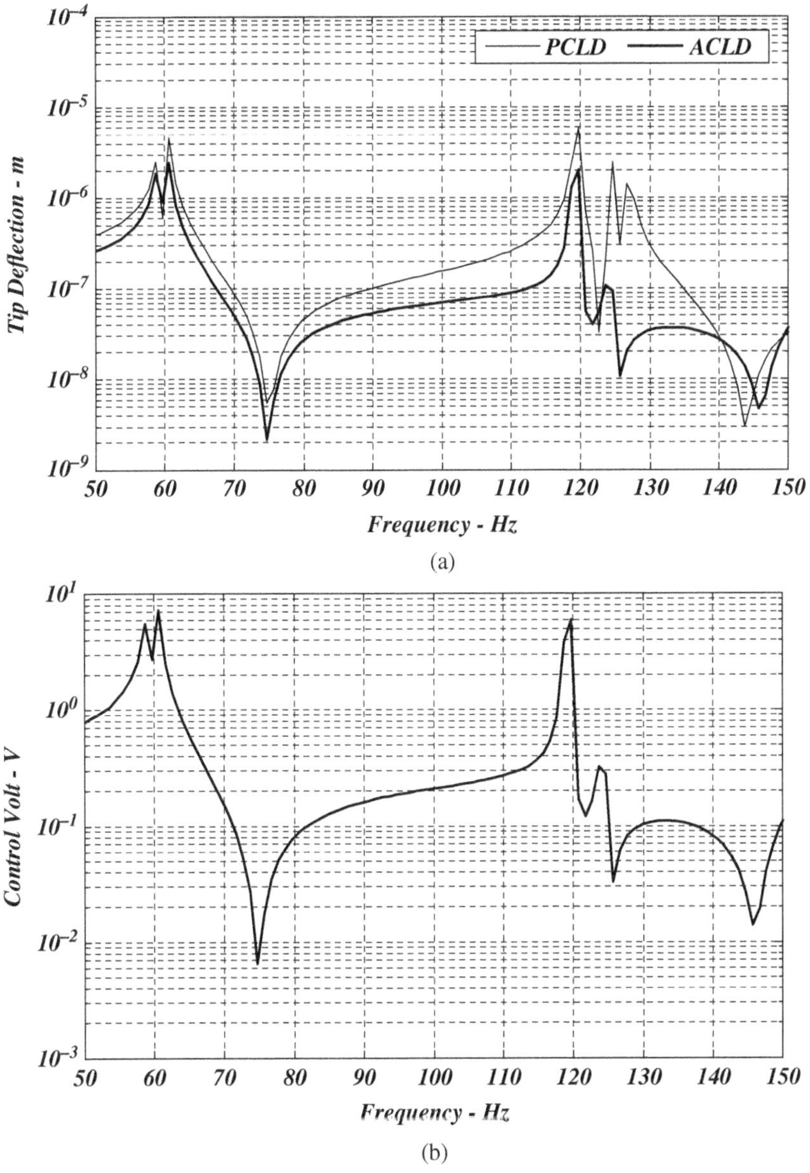

Figure 7.40 Frequency response (a) and control voltage (b) of the shell/ACLD assembly when K_P = 3E6 Nm m^{-1}.

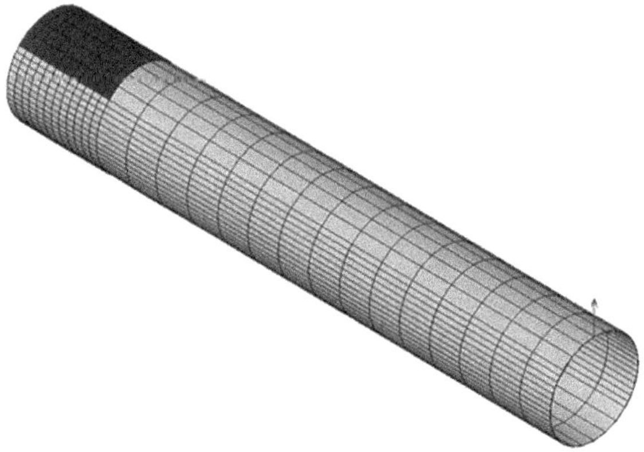

Figure 7.41 ANSYS finite element model of the shell/ACLD assembly.

Figure 7.41 displays the ANSYS finite element model of the shell/ACLD assembly. The ANSYS model is used to predict the time response characteristics of the assembly at different control gains.

Figure 7.42a shows the time response of the shell/ACLD assembly when using a proportional control gain $K_P = 1.5E6$ Nm m^{-1}. The corresponding control voltage is shown in Figure 7.42b. The effectiveness of the controller is limited as it has only attenuated the first and second modes (60 and 62 Hz) but the 119 and 120 Hz modes are still dominating the response.

A peak control voltage of 18 V is observed with this control gain.

Figure 7.43a shows the time response of the shell/ACLD assembly when the proportional control gain K_P is increased to 3E6 Nm m^{-1}. The corresponding control voltage is shown in Figure 7.43b. The effectiveness of the controller has improved considerably as demonstrated by the significant attenuation of the 119 and 120 Hz modes of vibrations. In this case, the peak control voltage required reaches 40 V.

Table 7.14 summarizes the performance of the shell/ACLD assembly for different control gains as compared with that of the shell/PCLD assembly. The performance is quantified by the maximum deflection and control voltage at the different modes of vibration.

7.6 Summary

This chapter has presented the utilization of active control as an effective means for enhancing the damping characteristics of conventional PCLD treatments. The resulting ACLD treatment is applied to attenuate the vibration of various structural members including: rods, beams, plates, and shells. The performance characteristics of the ACLD are predicted using the finite element models developed in Chapters 4 and 6 after integrating these with the appropriate equations for piezoelectric sensors and control actuators. The obtained predictions are validated also against the predictions of the commercial finite element package ANSYS.

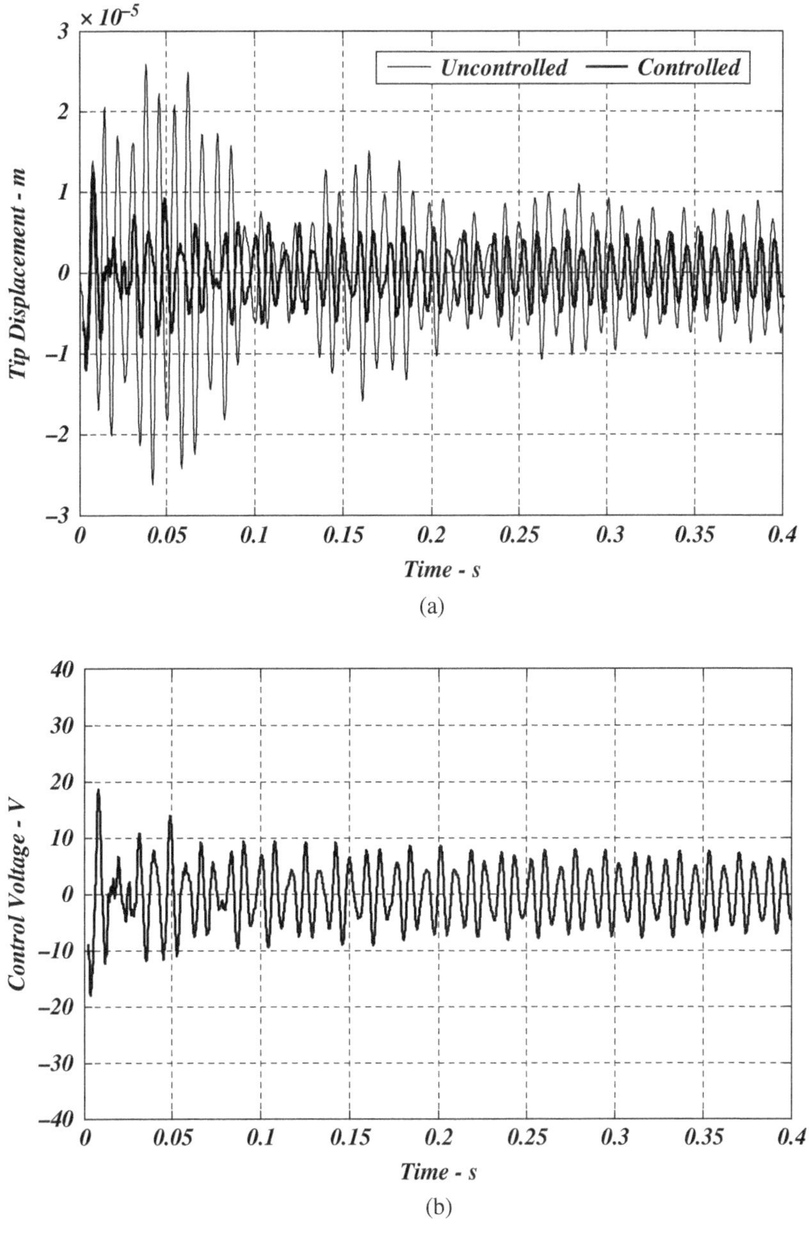

Figure 7.42 Time response and control voltage of the shell/ACLD assembly when $K_P = 1.5E6$ Nm m^{-1}.

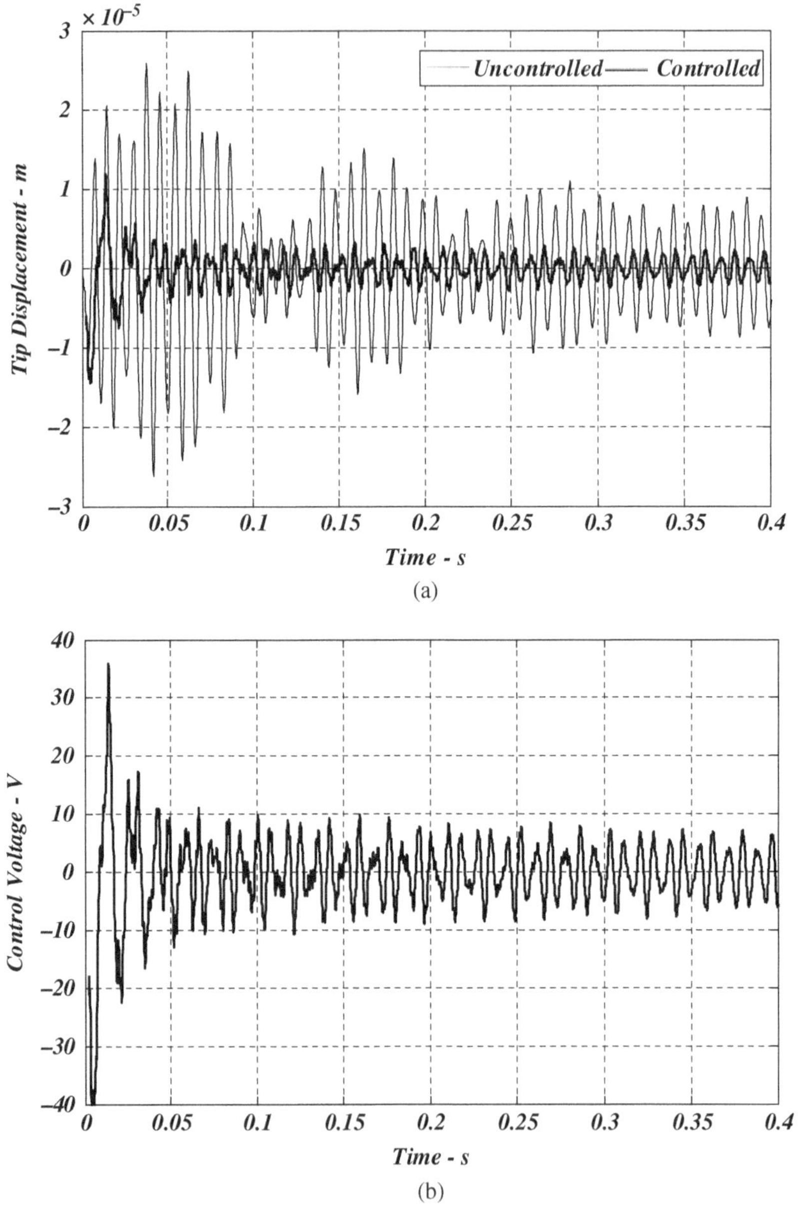

Figure 7.43 Time response (a) and control voltage (b) of the shell/ACLD assembly when K_P = 3E6 Nm m^{-1}.

Table 7.14 Performance of shell/ACLD patches for different control gains in the frequency domain.

	Maximum deflection (μm)				Maximum control voltage (V)			
Mode (Hz)	60	62	120	125	60	62	120	125
Gain = 0	2.51	4.66	5.86	2.44	0	0	0	0
Gain = 1.5E6	2.11	3.29	1.51	0.12	3.16	4.95	2.27	0.18
Gain = 3.0E6	1.85	2.45	2.00	0.11	5.55	7.34	6.01	0.32

References

Baz A., "Active constrained laer damping", Proceedings of DAMPING '93 Conference, San Francisco, CA, Wright Laboratory Document no. WL-TR-93-3105, pp. IBB 1–23, 1993.

Baz A., Active Constrained Layer Damping, US Patent 5,485,053, filed October 15 1993 and issued January 16 1996.

Baz, A. (1997a). Dynamic boundary control of beams using active constrained layer damping. *Mechanical Systems and Signal Processing* 88 (6): 811–825.

Baz, A. (1997b). Boundary control of beams using active constrained layer damping. *ASME Journal of Vibration and Acoustics* 119 (2): 166–172.

Baz A. and Ro J., "Partial treatment of flexible beams with active constrained layer damping", Proceedings of the American Society of Mechanical Engineers, No. AMD- Vol. 167, pp. 61–80, 1993a.

Baz A. and Ro J., "Finite element modeling and performance of active constrained layer damping" (ed. L. Meirovitch), Proceedings of the Ninth VPI & SU Conference on Dynamics & Control of Large Structures, Blacksburg, VA, pp. 345–358, 1993b.

Baz, A. and Ro, J. (1994). Actively-controlled constrained layer damping. *Sound & Vibration Magazine* 26 (3): 18–21.

Crawley, E. and De Luis, J. (1987). Use of piezoelectric actuators as elements in intelligent structures. *Journal of AIAA* 25: 1373–1385.

Cremer, L., Heckel, M., and Ungar, E. (1988). *Structure-Borne Sound: Structural Vibrations and Sound Radiation at Audio Frequencies*, 2e. Berlin: Springer-Verlag.

DiTaranto, R.A. (1965). Theory of vibratory bending for elastic and viscoelastic layered finite length beams. *ASME Journal of Applied Mechanics* 87: 881–886.

DiTaranto, R.A. (1973). Static analysis of a laminated beam. *Journal of Engineering For Industry, Transactions on ASME, Series B* 95 (3): 755–761.

Lee C. K., "Piezoelectric laminates for torsional and bending modal control: theory and experiment," Ph.D. dissertation, Cornell University, Ithaca, NY, 1987.

Mead, D. and Markus, S. (1969). The forced vibration of a three layer damped sandwich beam with arbitrary boundary conditions. *Journal of Sound and Vibration* 10: 163–175.

Meirovitch, L. (2010). *Fundamentals of Vibrations*, 1e. Long Grove, IL: Waveland Press Inc.

Miller S. and Hubbard J. Jr., "Observability of a Bernoulli–Euler Beam using PVF2 as a distributed sensor" (ed. L. Meirovitch), Proceedings of the Seventh Conference on Dynamics & Control of Large Structures, VPI & SU, Blacksburg, VA, pp. 375–390, 1987.

Rao, D.K. (1976). Static response of stiff-cored Unsymmetric sandwich beams. *ASME Journal of Engineering for Industry* 98: 391–396.

Ro, J. and Baz, A. (2002). Vibration control of plates using self-sensing active constrained layer damping networks. *Journal of Vibration and Control* 8 (8): 833–845.
Trompette, P., Boillot, D., and Ravanel, M.A. (1978). The effect of boundary conditions on the vibration of a Viscoelastically damped cantilever beam. *Journal of Sound and Vibration* 60 (3): 345–350.

7.A Piezoelectric Sensor Basic Equations

7.A.1 Basic Equations

Piezoelectric films bonded to vibrating structures are used to monitor its vibration. Figure 7.A.1 shows a typical arrangement of the sensor when bonded to a beam.

The constitutive equations of the piezo-sensor are:

$$S_1 = s_{11}^E T_1 + d_{31} E_3, \tag{7.A.1}$$

and

$$D_3 = d_{31} T_1 + \varepsilon_{33}^T E_3 \tag{7.A.2}$$

where S_1, T_1, E_3, and D_3 are the mechanical strain, mechanical stress, electrical field, and electric displacement, respectively. Also, s_{11}^E, d_{31}, and ε_{33}^T denote the compliance, piezo-strain constant, and permittivity.

Eliminating the stress T_1 from Eqs. (7.A.1) and (7.A.2) yields:

$$D_3 = \frac{d_{31}}{s_{11}^E} S_1 + \varepsilon_{33}^T \left(1 - k_{31}^2\right) E_3 \tag{7.A.3}$$

where k_{31}^2 is the electromechanical coupling factor $= \dfrac{d_{31}^2}{\left(s_{11}^E \, \varepsilon_{33}^T\right)}$.

For short-circuit conditions ($E_3 = 0$), then:

$$D_3 = \frac{d_{31}}{s_{11}^E} S_1 = e_{31} S_1 \tag{7.A.4}$$

where e_{31} is the piezo-stress constant and S_1 is given by:

$$S_1 = z\, w_{,xx} \tag{7.A.5}$$

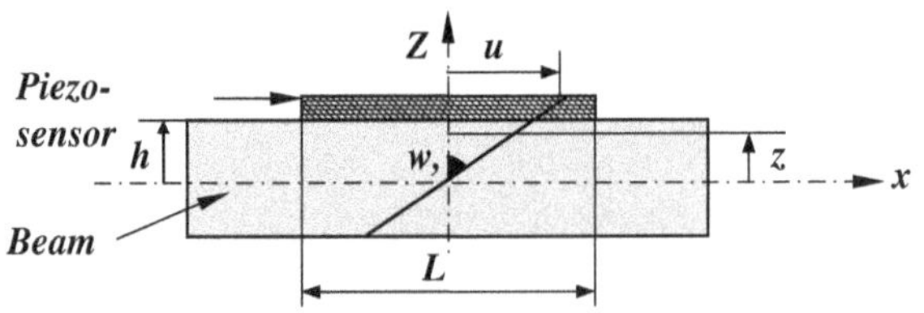

Figure 7.A.1 Piezoelectric sensor.

From Eqs. (7.A.4) and (7.A.5), the electric displacement D_3 at the sensor/beam interface ($z = h$) is given by:

$$D_3 = e_{31}\, h\, w_{,xx} \tag{7.A.6}$$

and the associated charge Q is:

$$Q = \int_0^L D_3 b(x)dx = e_{31}h \int_0^L w_{,xx} b(x)dx \tag{7.A.7}$$

7.A.2 Basic Sensor Configurations

Two configurations are considered here. These include the uniform and linearly shaped configurations shown in Figure 7.A.2.

For the uniform sensor, the width $b(x)$ is constant = b and the developed charge Q is:

$$Q = e_{31} h\, b[w_{,x}(L) - w_{,x}(0)] \tag{7.A.8}$$

For the linearly shaped sensor, the width $b(x) = b(1 - x/L)$ and the corresponding charge Q is:

$$Q = e_{31} h\, b/L[w(L) - w(0) - w_{,x}(0)L] \tag{7.A.9}$$

For a cantilevered beam, $w(0) = 0$ and $w_{,x}(0) = 0$, then:

$$Q_{uniform} = e_{31} h\, b\, w_{,x}(L) \text{ and } Q_{shaped} = e_{31} h\, b\, w(L)/L \tag{7.A.10}$$

Hence, the uniform sensor will monitor the angular deflection of the free end of the beam while the shaped sensor will generate a charge proportional to the transverse deflection of the free end.

7.A.3 Output Voltage of Sensor

The output voltage V of the sensor can be computed as follows:

$$V = \frac{1}{C} Q \tag{7.A.11}$$

where C denotes the capacitance of the sensor given by:

$$C = \frac{\varepsilon_{33} A_s}{h_s} \tag{7.A.12}$$

where ε_{33}, A_s, and h_s denote the permittivity, sensor surface area, and sensor thickness, respectively.

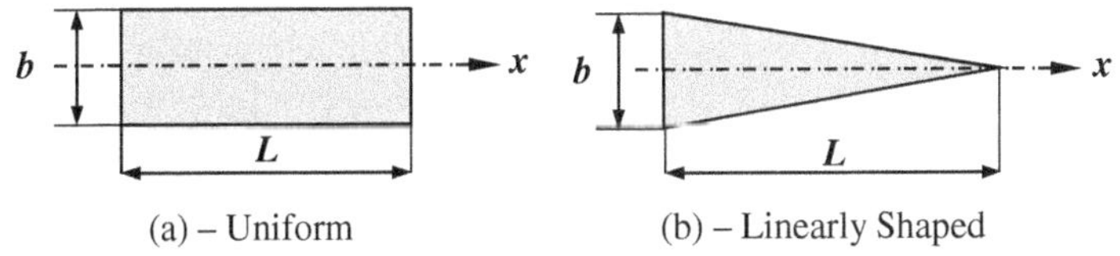

Figure 7.A.2 Uniform and linearly shaped sensors. (a) Uniform and (b) linearly shaped.

Combining Eqs. (7.A.7), (7.A.11), and (7.A.12) gives:

$$V_s = -\left[\frac{k_{31}^2 h b}{g_{31} C}\right] \int_0^L b(x)\, w_{,xx} dx \tag{7.A.13}$$

where $g_{31} = d_{31}/\varepsilon_{33}$= piezoelectric voltage constant. Also, $b(x)$ is a spatial distribution function that defines the width shaping of the sensor over the beam. For uniform sensor $b(x) = 1$ and for a linearly shaped sensor $b(x) = (1 - x/L)$.

Problems

7.1 Consider a mass = 100 kg supported on a VEM that is modeled on the following GHM model with one mini-oscillator such that its stiffness is given by:

$$K = 100\left(1 + 7\frac{s^2 + 2000s}{s^2 + 2000s + 1E6}\right)$$

that is, $K_o = 100$, $\alpha_1 = 7$, $\zeta_1 = 1$, and $\omega_1 = 1000$ (see Problem 4.1). Determine the state-space representation of the system. Plot the root locus of the system when the control law relies on:

a) a position sensor with $F_c = -k_g y$, where $y = [1\ 0\ 0\ 0]\{\Delta\}$.
b) a velocity sensor with $F_c = -k_g y$, where $y = [0\ 0\ 1\ 0]\{\Delta\}$.

with kg = control gain and $\{\Delta\} = \{x\ z\ \dot{x}\ \dot{z}\}^T$ where x and z are the DOF of the mass and the internal DOF of the VEM. Note that F_c is the control force as shown in Figure P7.1.

Identify the poles and zeros of the structure and of the VEM. Also determine the maximum possible damping ratio of the closed-loop system when the controller is provided with position or velocity sensor.

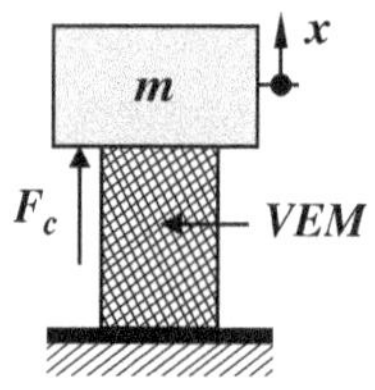

Figure P7.1 Mass supported on a VEM.

7.2 Consider the three-layer composite rod that is mounted in a cantilevered configuration as shown in Figure P7.2. The rod consists of a base structure treated with a VEM layer that is constrained by an active piezoelectric constraining layer. The assembly experiences longitudinal vibration along the x direction.

Derive expression for a globally stable boundary controller that ensures that the piezoelectric strain ε_p is generated to keep the rate of the total energy of the assembly $\dot{E}_n$ strictly negative where E_n is given by:

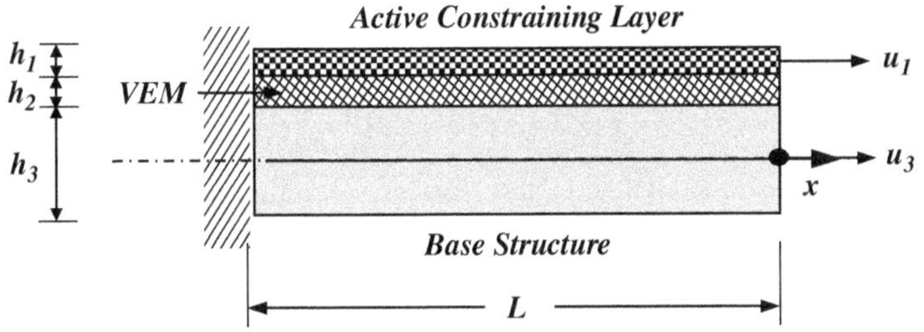

Figure P7.2 Cantilever rod treated with ACLD.

$E_n = T + U$

where T and U denote the kinetic and PE of the assembly such that:

$$T = \frac{1}{2} b \int_0^L \left(m_1 \dot{u}_1^2 + m_3 \dot{u}_3^2 \right) dx$$

and

$$U = \frac{1}{2} b \int_0^L \left(E_1 h_1 u_{1,x}^2 + E_3 h_3 u_{3,x}^2 + G' h_2 \gamma^2 \right) dx$$

where h_i, E_i, m_i, and u_i denote the thickness, Young's modulus, mass per unit length, and deflection of the ith layer (i = 1,...,3, such that 1 = piezo-layer, 2 = VEM layer, and 3 = base structure). Also, b and L denote the width and length of the rod, respectively. The shear storage modulus and shear strain of the *VEM* are G' and γ, respectively, with $\gamma = (u_1 - u_3)/h_2$.

Using Hamilton's principle show first that the equations of motion and the boundary conditions of the assembly are given by:

$$m_1 u_{1,tt} = E_1 h_1 u_{1,xx} - G^*/h_2 (u_1 - u_3) = 0,$$

and

$$m_3 u_{3,tt} = E_3 h_3 u_{3,xx} + G^*/h_2 (u_1 - u_3) = 0.$$

subject to the boundary conditions:

at $x = 0$: $u_{1,x} = \varepsilon_P$ and $u_{3,x} = 0$
at $x = L$: $u_1 = 0$ and $u_3 = 0$.

where $G^* = G'(1 + \eta i)$ with η denoting the loss factor of the VEM.

Using the equations of motion and the boundary conditions show that:

$$\dot{E}_n = b \int_0^L \left(E_1 h_1 \varepsilon_P u_{1,t} - (G' \eta / h_2 \omega) \gamma_{,t}^2 \right) dx$$

Hence, for $\dot{E}_n < 0$, show that the globally stable boundary control law is given by:
$\varepsilon_P = - K_G u_{1,t}$ with K_G = control gain > 0

7.3 Derive the finite element model for a single element of the cantilevered three-layer composite rod of Figure P7.2 such that:

$$[M_T]\{\ddot{\Delta}\} + [C_T]\{\dot{\Delta}_T\} + [K_T]\{\Delta_T\} = \{F_T\}$$

Assuming that the VEM is described by a single GHM mini-oscillator using the modeling approach given by Eq. (7.5) such that:

where $[M_T] = \begin{bmatrix} [M_o] & 0 \\ 0 & \dfrac{\alpha_1 E_{v_o}[\bar{K}_{v_o}]}{\omega_1^2} \end{bmatrix}$, $[C_T] = \begin{bmatrix} 0 & 0 \\ 0 & \dfrac{2\zeta_1\alpha_1 E_{v_o}[\bar{K}_{v_o}]}{\omega_1} \end{bmatrix}$, $[K_T] =$

$\begin{bmatrix} [K_{so}] + E_{v_o}(1+\alpha_1)[\bar{K}_{v_o}] & -E_{v_o}\alpha_1[\bar{K}_{v_o}] \\ -E_{v_o}\alpha_1[\bar{K}_{v_o}] & E_{v_o}\alpha_1[\bar{K}_{v_o}] \end{bmatrix}$, $\{\Delta_T\} = \begin{Bmatrix} u_1 \\ u_3 \\ z_1 \\ z_3 \end{Bmatrix}$, and $\{F_T\} = \begin{Bmatrix} 0 \\ F \\ 0 \\ 0 \end{Bmatrix}$.

with $[M_0]$ and $[K_{s_0}]$ are given by:

$$T = \frac{1}{2} b \int_0^L \left(m_1 \dot{u}_1^2 + m_3 \dot{u}_3^2\right) dx = \frac{1}{2}\{\dot{\Delta}\}^T [M_0]\{\dot{\Delta}\}$$

and $U = \dfrac{1}{2} b \displaystyle\int_0^L \left(E_1 h_1 u_{1,x}^2 + E_3 h_3 u_{3,x}^2 + G' h_2 \gamma^2\right) dx = \dfrac{1}{2}\{\Delta\}^T [K_{s0}]\{\Delta\}$

where $\{\Delta\} = \begin{Bmatrix} u_1 \\ u_3 \end{Bmatrix}$ = nodal deflection vector. The spatial distribution of $u_1(x)$ and $u_3(x)$ are assumed to be given by the following shape functions:

$$u_1(x) = a_1 e^{-ik_1 x} + a_2 e^{ik_1 x}$$

where $k_1 = \omega/\sqrt{E_1/\rho_1}$ = wave number of piezo-layer and

$$u_3(x) = b_1 e^{-ik_3 x} + b_2 e^{ik_3 x}$$

where $k_3 = \omega/\sqrt{E_3/\rho_3}$ = wave number of base structure with ω and ρ_i denote the excitation frequency and density of the ith layer.

Assume that the rod is made of aluminum with width of 0.025 m, thickness of 0.025 m, and length of 1 m. The rod is treated with a constrained damping treatment that has a width of 0.025 m, thickness = 0.025 m, and density of 1100 kg m^{-3}. The storage modulus and loss factor of the VEM are predicted by the GHM model with one mini-oscillator with $E_0 = 15.3 MPa$, $\alpha_1 = 39$, $\zeta_1 = 1$, $\omega_1 = 19,058 rad/s$. The physical and geometrical properties of the piezoelectric constraining layer are listed in Table 7.15 Physical and geometrical properties of the piezoelectric constraining layer.

Determine the frequency response of the treated rod when subjected to a longitudinal force $F = 10$ N at its free end. Compare the results with the response of the rod assembly when its finite element is extracted using classical linear shape functions.

Table 7.15 Physical and geometrical properties of the piezoelectric constraining layer.

Length (m)	Thickness (m)	Width (m)	Density (kg m^{-3})	Young's modulus (GPa)	d_{31} (m V^{-1})	k_{31}	g_{31} (mV N^{-1})	k_{3t}
1.0	0.0025	0.025	7600	63	186E-12	0.34	116E-2	1950

7.4 Consider a beam mounted in a fixed-free configuration as shown in Figure P7.3. The beam is made of aluminum with width of 0.025 m, thickness of 0.025 m, and length of 1 m. The beam is treated with an unconstrained damping treatment that has a width of 0.025 m, thickness = 0.025 m, and density of 1100 kg m^{-3}. The storage modulus and loss factor of the VEM are predicted by GHM model with one mini-oscillator with $E_0 = 15.3 MPa, \alpha_1 = 39, \zeta_1 = 1, \omega_1 = 19,058 rad/s$. For simplicity purposes, the structural system is modeled by one finite element model.

Plot the pole-zero map for the treated beam identifying the structural modes and VEM modes. Compare the results with the pole-zero map of the untreated beam.

Also determine the frequency response of the treated beam when subjected to a transverse force $F = 10$ N at its free end for a duration of 0.01 s. Compare the results with the response of the untreated beam.

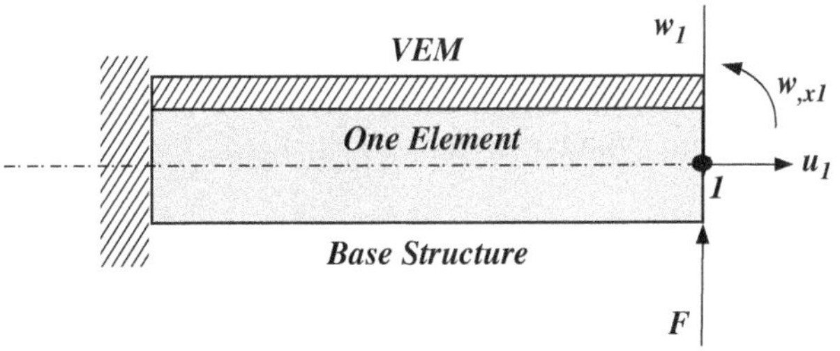

Figure P7.3 Cantilever beam treated with unconstrained VEM treatment.

7.5 For the beam considered in Problem 7.3, the VEM is constrained by an aluminum constraining layer that is 0.025 m wide and 0.0025 m thick as shown in Figure P7.4.

Plot the pole-zero map for the treated beam identifying the structural modes and VEM modes. Compare the results with the pole-zero map of the untreated beam.

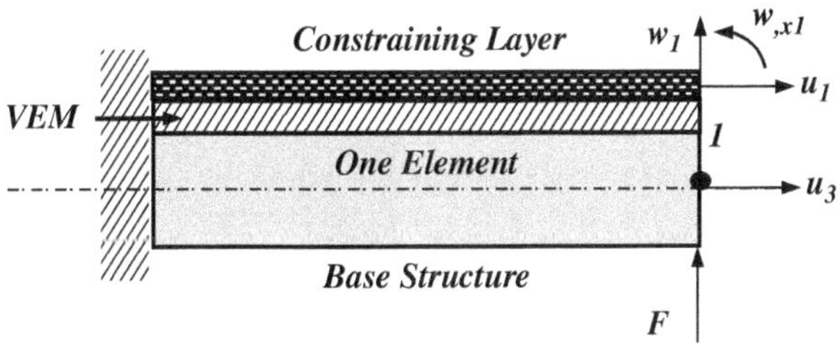

Figure P7.4 Cantilever beam treated with constrained VEM treatment.

Also determine the time response of the treated beam when subjected to a transverse force $F = 10$ N at its free end for a duration of 0.01 s. Compare the results with the response of the untreated beam.

7.6 Consider the beam system of Example 7.1 that has the physical and geometrical properties listed in Tables 7.2 and 7.3. The beam system is modeled by a single finite element as shown in Figure P7.5. The beam is controlled using a proportional-derivative (PD) controller such that $K_G(1 + 0.01s)$ where K_G is the control gain and s is the Laplace complex number. The input to the controller is the transverse deflection of the free end of the beam w_I.

Plot the root locus map for the actively treated beam identifying the structural modes and VEM modes for different control gains K_G. Select the control gain that ensures the maximum closed-loop damping ratio.

Also determine the time response of the controlled beam when subjected to a transverse force $F = 10$ N at its free end for a duration of 0.01 s. Compare the results with the response of the uncontrolled beam. Also plot the associated time history of the control effort.

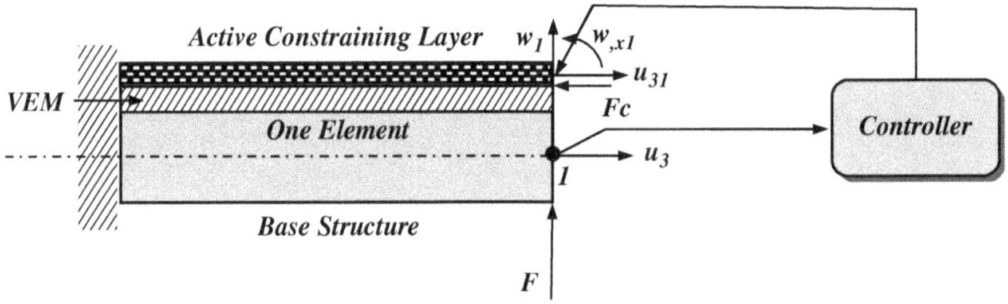

Figure P7.5 Cantilever beam treated with ACLD treatment with a controller.

7.7 For the distributed-parameter modeling of the beam/ACLD assembly described in Section 7.3.4, develop a globally stable boundary control law if the beam is mounted in a simply-supported configuration at both of its ends.

Assume the physical and geometrical parameters of the beam/ACLD system with the VEM are the same as those described in Example 7.1.

Determine the compliance and control voltage characteristics of the beam/ACLD system using a velocity feedback controller with a control gain $K_D = 7.4$. Compare these characteristics with those of the beam/PCLD system using the DPM approach outlined in Section 7.3.4. Also, compare the predictions of the DPM against the predictions of the finite element model when the beam is excited by a unit sinusoidal load at its free end.

Also determine the energy dissipation characteristics of the beam/ACLD system as compared with that of the beam/PCLD system.

7.8 Consider the aluminum cantilever plate that is fully treated with an ACLD patch placed as shown in Figure P7.6. The dimensions of the plate and the PCLD patches are shown in the figure with $h_1 = h_2 = h_3 = 0.005m$.

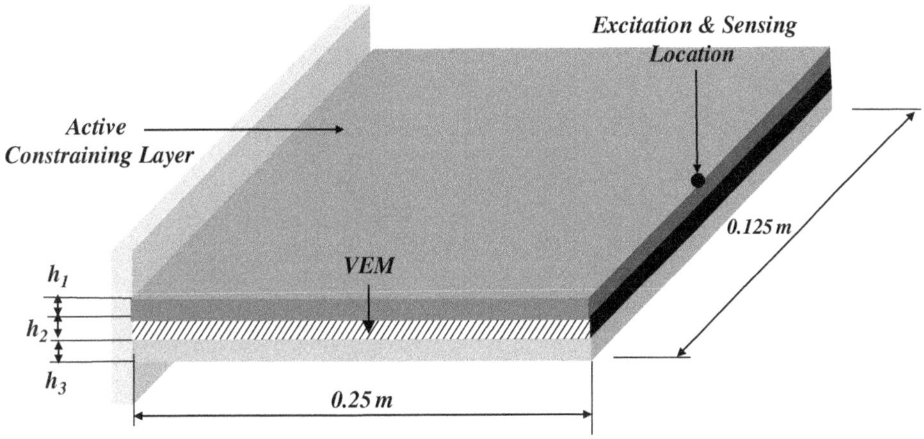

Figure P7.6 Cantilever plate treated with full constrained VEM treatment.

The VEM is modeled using a GHM model with three mini-oscillators such that:

$$G_2(s) = G_0\left[1 + \sum_{i=1}^{3} \alpha_i \frac{s^2 + 2\zeta_i\omega_i s}{s^2 + 2\zeta_i\omega_i s + \omega_i^2}\right] \text{ with } G_0 = 0.5MPa$$

The parameters α_i, ζ_i, and ω_i are listed in Table 4.7. Assume that the constraining layer is made of piezoelectric material (*PZT*-4) with $d_{31} = d_{32} = -123 \times 10^{-12}$ m V^{-1} and $C_{31}^E = C_{32}^E = 78.3GPa$. Determine the frequency response of the plate, at different control gains, when it is excited by a unit force acting at the middle of the free end. Assume a proportional controller that feeds the transverse deflection of the free end back to control the piezoelectric constraining layer.

Also determine the corresponding control effort.

7.9 Consider the aluminum cantilever plate that is partially treated with an ACLD patch placed as shown in Figure P7.7. The dimensions of the plate and the PCLD patches are shown in the figure with $h_1 = h_2 = h_3 = 0.005m$. The physical properties of the VEM and the piezoelectric constraining layers are the same as listed in Problem 7.6.

Determine the frequency response of the plate, at different control gains, when it is excited by a unit force acting at the middle of the free end. Assume a proportional controller that feeds the transverse deflection of the free end back to control the piezoelectric constraining layer.

Also determine the corresponding control effort. Compare the results with the corresponding results obtained when the plate is fully treated with the ACLD treatment as in Problem 7.8.

7.10 Consider the clamped-free shell/ACLD assembly shown in Figure P7.8a. The main physical and geometrical parameters of the shell and PCLD are listed in Table 7.13. The shell has an internal radius R is 0.1016 m. The ACLD treatment consists of two patches as displayed in Figure P7.8b. The patches are bonded 180°

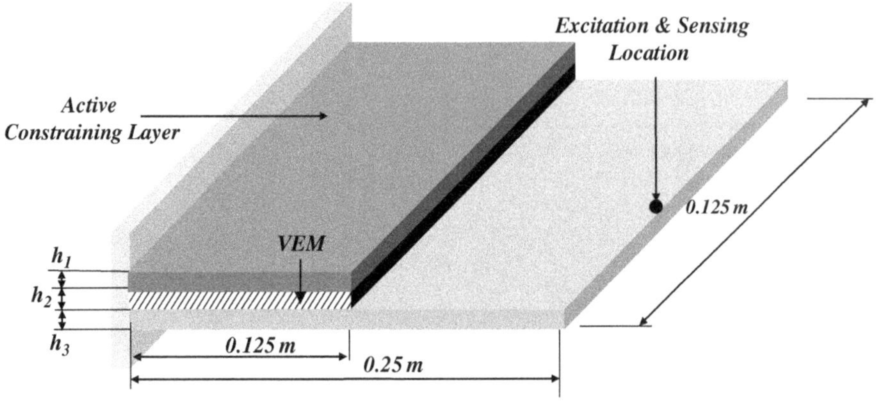

Figure P7.7 Cantilever plate treated with full constrained VEM treatment.

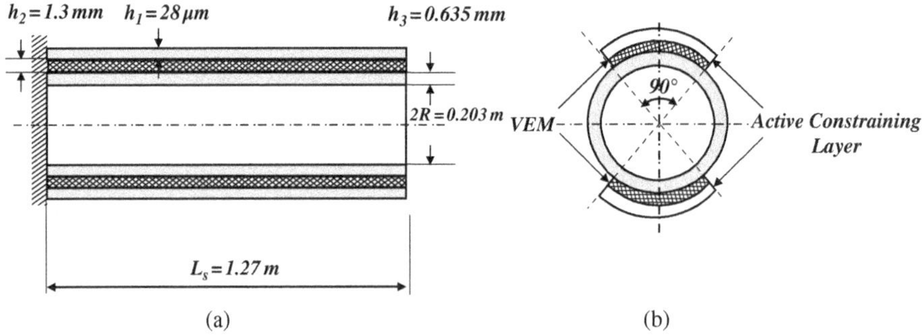

Figure P7.8 Cantilever shell treated with partial constrained VEM treatment.

apart on the outer surface of the cylinder with each of which subtending an angle of 90° angle at the center of the shell.

Determine the frequency response of the system when excited at the free end by a unit force. Also determine the time response of system when a unit pulse force of duration 0.10 ms is applied at the free end. Compare the response when the response is predicted using ANSYS and the finite element approach outlined in Sections 4.7 and 7.5.

Compare the results with those displayed in Example 7.5 for the partially treated shell/ACLD system.

8

Advanced Damping Treatments

8.1 Introduction

This chapter presents a brief summary of the basic characteristics of some of the important advanced damping treatments with attractive characteristics that potentially can present effective means for enhancing the structural damping either passively or actively.

Distinct among the considered treatments are: the stand-off damping treatment, functionally graded damping treatment, active piezoelectric damping composites (APDCs), passive magnetic composites (PMCs), and negative stiffness composites (NSCs). All five classes of damping treatments aim to provide novel and innovative means for improving the damping behavior over that of conventional constrained damping layer treatments.

In stand-off damping treatments, discussed in Section 8.2, the enhancement is achieved passively by using a stand-off layer (SOL) to move the viscoelastic layer away from the neutral axis of the base structure and thereby magnifying the shear strain and the energy dissipated by the viscoelastic matrix (VEM) (Whittier 1959).

The improvement in the damping characteristics by using the functionally graded damping treatment, in Section 8.3, takes place by optimally grading the shear modulus distribution over the length of the treatments. Optimal grading aims to place stiffer damping treatments in the high shear strain zones and softer damping treatment in the zones of low shear strain in order to maximize the total energy dissipation over the entire length of the treatment (Venkataraman and Sankar 2001).

In the APDCs, in Section 8.4, an array of piezoelectric rods is obliquely embedded across the thickness of a VEM in order to augment the classical shear damping component with a compression component. In this manner, the combined effect of shear and compression damping increases the total damping of the composite (Baz and Tempia 2004).

The PMCs, in Section 8.5, belong to the class of constrained damping treatments, where a viscoelastic core is constrained by magnetic constraining layer that interacts with neighboring layers in order to enhance the damping characteristics of the viscoelastic (Baz and Poh 2000; Omer and Baz 2000; Oh et al. 1999, 2000a, b; Ruzzene et al. 2000).

NSCs, in Section 8.6, have attracted the attention because of their ability to potentially exhibit enhanced vibrational damping characteristics. Such improved damping characteristics are attributed to the fact that negative stiffness elements tend to assist rather than resist the deformation of the composite under load. This increased deformation significantly contributes to enhancing the energy stored internally in the NSC. The pioneering work on NSC is attributed to Lakes (2001), Lakes et al. (2001), and Wang and Lakes (2004) as well as Platus (1999).

Active and Passive Vibration Damping, First Edition. Amr M. Baz.

8.2 Stand-Off Damping Treatments

Damping treatments with SOL have been widely accepted as an attractive alternative to conventional passive constrained layer damping (PCLD) treatments. Such an acceptance stems from the fact that the SOL, which is simply a slotted spacer layer sandwiched between the viscoelastic layer and the base structure, acts as a strain magnifier that considerably amplifies the shear strain and hence the energy dissipation characteristics of the viscoelastic layer. Accordingly, more effective vibration suppression can be achieved by using SOL as compared to employing PCLD.

In this section, a comprehensive finite element model of the SOL constrained damping treatment is presented. The model accounts for the geometrical and physical parameters of the slotted SOL, the viscoelastic, layer the constraining layer, and the base structure. The predictions of the model are validated against the predictions of a distributed transfer function (DTF) model and a model developed using the commercial finite element package ANSYS.

8.2.1 Background of Stand-Off Damping Treatments

The SOL concept was first introduced by Whittier (1959), who suggested that a spacer layer be introduced between the viscoelastic layer and the vibrating structure. The spacer layer moves the viscoelastic layer away from the neutral axis of the base structure. The shear strain is magnified by virtue of the displacement of the viscoelastic layer away from the neutral axis where the strain is minimal.

Most of the studies on passive stand-off layer (PSOL) damping assume ideal conditions of the SOL, that is, the SOL has infinite shear stiffness and hence passes the shear stresses to the viscoelastic layer. Furthermore, the SOL has been assumed also to have zero stiffness when bending in order not to contribute to the flexural rigidity of the base beam. Distinct among these studies are those of Rogers and Parin (1995), Falugi (1991), Falugi et al. (1989), and Parin et al. (1989) on slotted SOL damping treatments for aircraft applications. Several other analytical and finite element models have been developed, for example, by Yellin, et al. (2000), Tao et al. (1999), Mead (1998), and Garrison et al. (1994).

In this section, a finite element model is presented to enable studying the effect of the geometry of the slots on the SOL performance characteristics. The developed finite element model accounts for the bending stiffness and shear strength of SOL.

The predictions of the finite element model are validated and compared against the predictions of models developed using the DTF method and the commercial finite element software ANSYS.

8.2.2 The Stand-Off Damping Treatments

The SOL is one of the techniques used to enhance the damping characteristics of classical constrained layer damping treatments by increasing the distance between the viscoelastic layer and the neutral axis of the base structure. To achieve this goal, the SOL is sandwiched between the base structure and the viscoelastic layer forming a four-layer composite as shown in Figure 8.1. With a proper design of SOL, it can act as a strain magnifier that amplifies the shear strain of the viscoelastic layer and thereby enhancing the energy dissipation characteristics of the entire composite assembly.

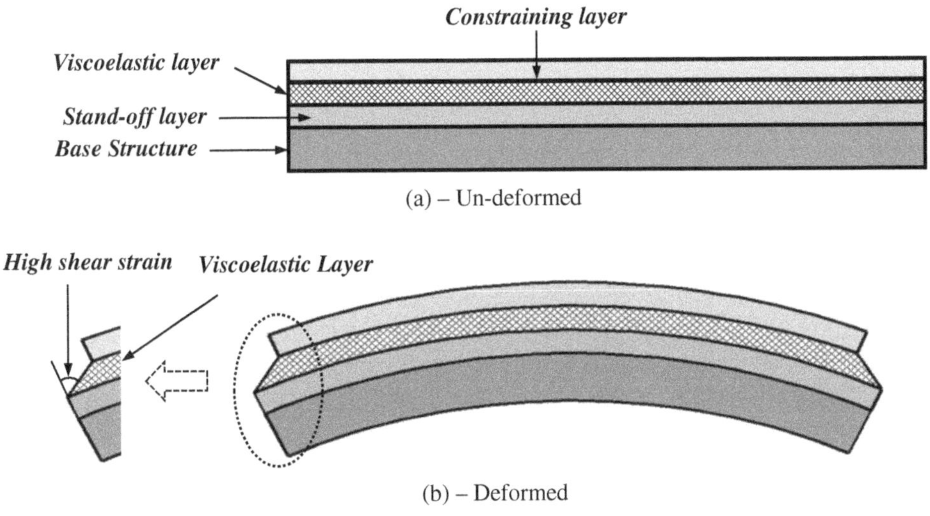

Figure 8.1 Passive stand-off layer damping treatment. (a) Undeformed. (b) Deformed.

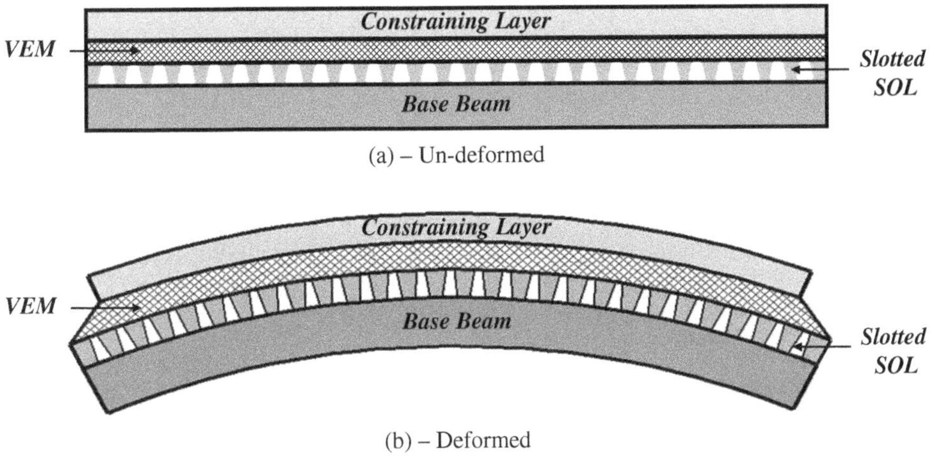

Figure 8.2 Slotted passive stand-off layer damping treatment. (a) Undeformed. (b) Deformed.

For effective operation, the SOL must have high shear stiffness and at the same time must not significantly affect the bending stiffness of the composite beam. If the SOL has high bending stiffness it will have no other effect than increasing the flexural rigidity of the constrained layer beam and thus will negatively affect its damping performance.

An effective method for avoiding this increase in the flexural rigidity of the treated beam is to use a slotted SOL of very high shear strength as compared to the viscoelastic material shown in Figure 8.2.

In the slotted SOL treatment, the geometry of the slots should be such that the SOL does not significantly contribute to the flexural rigidity of the base structure but also has enough shear strength such that it does not absorb shear deformation that is desired to be passed on to the viscoelastic layer, that is, it should pass on the shear stress to the viscoelastic layer.

8.2.3 Distributed-Parameter Model of the Stand-Off Layer Damping Treatment

The model presented here is summarized in this section based on the work of Yellin et al. (2000). This model is used to develop the finite element model of the PSOL, which is presented in Section 8.2.4.

The developed model is based on assuming small displacements with the lateral deflections of all the layers are equal. Also, it is assumed that there is no shear deformation experienced by neither the base beam nor the constraining layer. Furthermore, the viscoelastic layer is assumed to have no bending stiffness and therefore deforms only in pure shear. As for the PSOL, it is considered as a continuous and solid layer with the SOL and beam assembly modeled as an asymmetric composite so that planes remain planes under combined bending and axial loads. Also, SOL has finite shear stiffness, with the combined bending and shear stiffness of SOL less than that of the base beam and constraining layer to limit the shear deformation to the SOL and the VEM. Finally, it is assumed that the treatment is subject to steady-state harmonic excitation that permits the use of the complex modulus to describe the constitutive relationship of the VEM.

8.2.3.1 Kinematic Equations

The deformations in the stand-off and viscoelastic layers are generated by the axial deformation of the base beam and the constraining layer as well as by the lateral deflection of the beam. Hence, the total axial deflection, δ, as shown in Figure 8.3, is given by:

$$\delta = (u_c - u_b) + \frac{\partial w}{\partial x}\left(\frac{1}{2}h_c + h_s + h_v + \frac{1}{2}h_b\right) \tag{8.1}$$

where u_b and u_c are the axial deflections of the beam and the constraining layers, respectively. Also w and x denote the transverse deflection of the composite beam and the axial coordinate along the beam. Furthermore, h_i denotes the thickness of the ith layer where

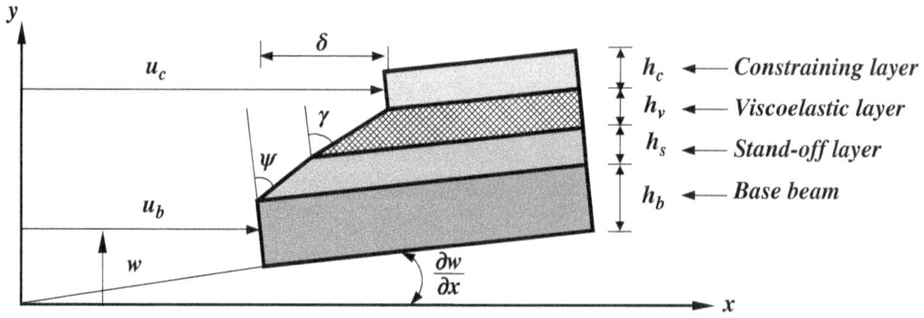

Figure 8.3 Deflections and geometry of the stand-off damping treatment.

the subscript $i = c$ = constraining layer, $i = v$ = viscoelastic layer, $i = s$ = stand-off layer, and $i = b$ = beam.

For small displacements, the deformation δ can also be expressed in terms of shear strains γ and ψ in the viscoelastic and SOLs as

$$\delta = h_s\psi + h_v\gamma \tag{8.2}$$

From Eqs. (8.1) and (8.2):

$$\psi = \frac{u_c - u_b}{h_s} + \left(\frac{2h_s + h_c + 2h_v + h_b}{2h_s}\right)\frac{\delta w}{\delta x} - \frac{h_v}{h_2}\gamma \tag{8.3}$$

8.2.3.2 Constitutive Equations

8.2.3.2.1 Beam and SOL

The axial strain at the neutral axis of the base beam ε_o and its curvature κ are given by

$$\varepsilon_o = \frac{\delta u_b}{\delta x} \quad \text{and} \quad \kappa = \frac{\delta^2 w}{\delta x^2} \tag{8.4}$$

Therefore, the axial strain across the composite beam (i.e. the base beam and SOL) is given by:

$$\varepsilon_x = \varepsilon_o - y\kappa \tag{8.5}$$

Hence, the stresses in axial direction for the base beam and the SOL (σ_{x_b} and σ_{x_s}) are given by:

$$\sigma_{x_b} = E_b(\varepsilon_o - y\kappa) \quad \text{and} \quad \sigma_{x_s} = E_s(\varepsilon_o - y\kappa) \tag{8.6}$$

Accordingly, the total axial force T_{bs} across the base beam and *SOL* is given by

$$\mathrm{T}_{bs} = \int_{b,s} \sigma_x \, dA = (EA)_{bs}\,\varepsilon_o - (EQ)_{bs}\,\kappa \tag{8.7}$$

where $(EA)_{bs}$ is axial rigidity of the composite base beam/*SOL* and $(EQ)_{bs}$ is product of elastic modulus of each layer and its first moment of area. These parameters are given by

$$(EA)_{bs} = E_bA_b + E_sA_s, \quad \text{and} \quad (EQ)_{bs} = E_bQ_b + E_sQ_s \tag{8.8}$$

Similarly, the total moment acting on the composite of the base beam/SOLs and taken around the beam neutral axis is

$$\mathrm{M}_{b,s} = -\int_{b,s} \sigma_x y \, dA = -(\mathrm{EA})_{bs}\varepsilon_o + (EI)_{bs}\kappa \tag{8.9}$$

where $(EI)_{bs}$ denotes the flexural rigidity $= E_bI_b + E_sI_s$

Also, the constitutive equation for the shear in the SOL is given by:

$$\tau_s = G_s\psi \tag{8.10}$$

where G_s is the shear modulus and Ψ is the shear strain of the SOL.

8.2.3.2.2 Constraining Layer

The internal axial tension T_c and bending moment M_c acting on the constraining layer are given by

$$\mathrm{T}_c = \mathrm{E}_c h_c b\frac{\partial u_c}{\partial x} \text{ and } \mathrm{M}_c = \frac{\mathrm{E}_c b h_c{}^3}{12}\frac{\partial^2 w}{\partial x^2} \tag{8.11}$$

8.2.3.2.3 Viscoelastic Layer

The constitutive equations for the viscoelastic layer is given by

$$\tau_v = G_v \gamma \tag{8.12}$$

where G_v is the shear modulus of the viscoelastic layer.

8.2.3.2.4 Equations of Motion

Consider the free body diagram of the PSOL composite beam shown in Figure 8.4, which gives the following equilibrium equations:

Force Equilibrium in Axial Direction

Base beam and SOL

$$\frac{d\mathrm{T}_{bs}}{dx} + \tau_s b + f_{bs} = (\rho_b + \rho_s)\ddot{u}_b. \tag{8.13}$$

Constraining layer

$$\frac{d\mathrm{T}_c}{dx} + \tau_v b + f_c = \rho_c \ddot{u}_c. \tag{8.14}$$

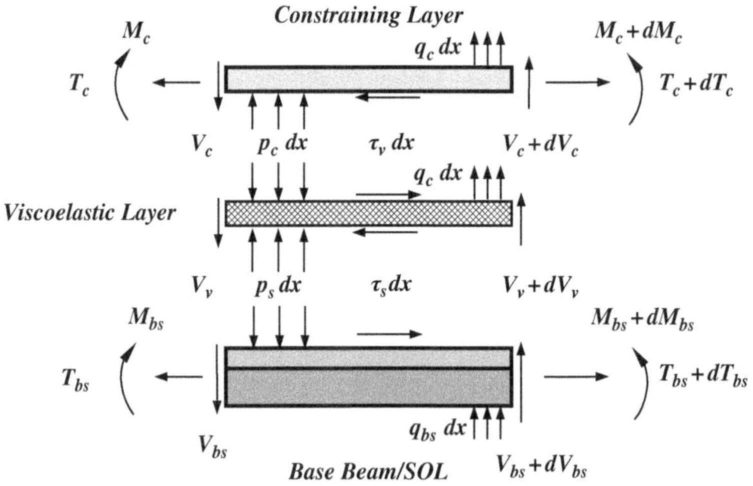

Figure 8.4 Free body diagram of the beam/passive stand-off layer assembly.

Force Equilibrium in Lateral Direction

Base beam and SOL

$$\frac{dV_{bs}}{dx} - p_s + q_{bs} = (\rho_b + \rho_s)\ddot{w} \tag{8.15}$$

Viscoelastic Layer

$$\frac{dV_v}{dx} - p_c + p_s + q_v = \rho_v \ddot{w} \tag{8.16}$$

Constraining layer

$$\frac{dV_c}{dx} + p_c + q_c = \rho_c \ddot{w} \tag{8.17}$$

Moment Equilibrium

Base beam and SOL

$$\frac{dM_{bs}}{dx} + V_{bs} - \tau_s b\left(h_s + \frac{1}{2}h_b\right) = 0 \tag{8.18}$$

Viscoelastic Layers

$$V_v = \tau_v b h_v \tag{8.19}$$

Constraining layer

$$\frac{dM_c}{dx} + V_c + \tau_v b \frac{h_c}{2} = 0 \tag{8.20}$$

In these equations, b denotes the beam width, τ_i is the shear stress, p_i defines the internal normal forces per unit length, q_i is the externally applied body forces per unit length, V_i and M_i denote the shear force and moment acting on ith layer, also ρ_i denotes the density of the ith layer with ($i = c, v, s, b$).

Note that for axial equilibrium of the viscoelastic layer $\tau_v = \tau_s$. Then, Eqs. (8.10) and (8.12) give

$$\psi = \frac{G_v}{G_s}\gamma \tag{8.21}$$

Combining Eqs. (8.3) and (8.21) gives

$$\psi = \left(\frac{G_v}{h_s G_v + h_v G_s}\right)(u_c - u_b) + \left(\frac{G_v}{h_s G_v + h_v G_s}\right)\left(\frac{2h_s + h_c + 2h_v + h_b}{2}\right)\frac{\delta w}{\delta x} \tag{8.22}$$

Let, $\alpha \equiv \frac{h_c}{2} + h_v + h_s + \frac{h_b}{2}$, $\rho = \rho_b + \rho_s + \rho_v + \rho_c$, $q = q_{bs} + q_v + q_c$, and

$$D_t = (\mathrm{EI})_{bs} + (EI)_c, \tag{8.23}$$

where ρ_i and h_i are the density and thickness of the ith layer.

Also, define the following normalized variables:

$$\bar{w}(\bar{x},\bar{s}) \equiv \frac{w(x,s)}{h_b},\ \bar{u}_c(\bar{x},\bar{s}) \equiv \frac{u_c(x,s)}{h_b}, \bar{u}_b(\bar{x},\bar{s}) \equiv \frac{u_b(x,s)}{h_b}, \bar{x} = \frac{x}{l},\ \bar{s} = \sqrt{\frac{\rho l^4}{D_t}}s,$$

$$\bar{q} \equiv \frac{q\,l^4}{D_t h_b}, \bar{f}_c \equiv \frac{f_c l^4}{D_t h_b}, \bar{f}_{bs} \equiv \frac{f_{bs} l^4}{D_t h_b}, \varepsilon(\bar{s}) \equiv \frac{b\alpha^2 G_v G_s l^2}{D_t(G_v h_s + G_s h_v)}, c_1^2 \equiv \frac{\rho_c}{\rho},$$

$$c_2{}^2 \equiv \frac{\rho_b + \rho_S}{\rho}, \beta \equiv \frac{l}{\alpha}, a_1 = \frac{(EQ_{bs} l)}{D_t}, a_2 \equiv \frac{E_c A_c l^2}{D_t}, \text{ and } a_3 \equiv \frac{(EA)_{bs} l^2}{D_t} \tag{8.24}$$

Then, by eliminating T, V, M, p, τ_v, and τ_s between Eqs. (8.13) through (8.20) knowing that $\tau_v = \tau_s$, and using Eqs. (8.23) and (8.24), results in the following normalized equations of motion

$$\left\{ \begin{bmatrix} D_t \frac{\partial^4}{\partial \bar{x}^4} - \frac{b\alpha^2 G_v G_s}{G_v h_s + G_s h_v} \frac{\partial^2}{\partial \bar{x}^2} & -\frac{b\alpha G_v G_s}{G_v h_s + G_s h_v} \frac{\partial}{\partial \bar{x}} & -(EQ)_{bs} \frac{\partial^3}{\partial \bar{x}^3} + \frac{b\alpha G_v G_s}{G_v h_s + G_s h_v} \frac{\partial}{\partial \bar{x}} \\ \frac{b\alpha G_v G_s}{G_v h_s + G_s h_v} \frac{\partial}{\partial \bar{x}} & -\mathrm{E}_c \mathrm{A}_c \frac{\partial^2}{\partial \bar{x}^2} + \frac{b G_v G_s}{G_v h_s + G_s h_v} & -\frac{b G_v G_s}{G_v h_s + G_s h_v} \\ (\mathrm{EQ})_{bs} \frac{\partial^3}{\partial \bar{x}^3} - \frac{b\alpha G_v G_s}{G_v h_s + G_s h_v} \frac{\partial}{\partial \bar{x}} & -\frac{b G_v G_s}{G_v h_s + G_s h_v} & -(EA)_{bs} \frac{\partial^2}{\partial \bar{x}^2} + \frac{b G_v G_s}{G_v h_s + G_s h_v} \end{bmatrix} \right.$$

$$\left. + \bar{s}^2 \begin{bmatrix} \rho & 0 & 0 \\ 0 & \rho_c & 0 \\ 0 & 0 & \rho_b + \rho_s \end{bmatrix} \right\} \begin{pmatrix} \bar{w}(\bar{x},\bar{s}) \\ \bar{u}_c(\bar{x},\bar{s}) \\ \bar{u}_b(\bar{x},\bar{s}) \end{pmatrix} = -\begin{Bmatrix} \bar{q} \\ \bar{f}_c \\ \bar{f}_{bs} \end{Bmatrix} \tag{8.25}$$

8.2.3.2.5 Boundary Conditions

The normalized boundary conditions for a cantilever beam excited at its base ($\bar{x} = 0$) are as follows:

Fixed end

$$\bar{w}(0,\bar{s}) = \mathrm{P}(\bar{s}), \text{ and } \frac{d\bar{w}(0,\bar{s})}{d\bar{x}} = \bar{u}_b(0,\bar{s}) = \frac{d\bar{u}_c(0,\bar{s})}{d\bar{x}} = 0$$

Free end

$$\frac{d^2\bar{w}(1,\bar{s})}{d\bar{x}^2} = 0, \frac{d\bar{u}_c(1,\bar{s})}{d\bar{x}} = \frac{d\bar{u}_b(1,\bar{s})}{d\bar{x}} = 0,$$

and

$$-\frac{d^3\bar{w}(1,\bar{s})}{d\bar{x}^3} + a_1 \frac{d^2\bar{u}_b(1,\bar{s})}{d\bar{x}^2} + \varepsilon(\bar{s}) \left\{ \frac{d\bar{w}(1,\bar{s})}{d\bar{x}} + \beta\bar{u}_c(1,\bar{s}) - \beta\bar{u}_b(1,\bar{s}) \right\} = 0 \tag{8.26}$$

These equations and boundary conditions will be solved using the "Distributed Transfer Function" method in Section 8.2.4 and the finite element method in Section 8.2.5: the predictions of the two methods will then be compared.

8.2.4 Distributed Transfer Function Method

The frequency response of the PSOL/beam assembly is determined analytically by using the DTF method developed by Yang and Tan (1992). Then, with all the initial boundary conditions assumed to be zero for sake of simplicity, Eqs. (8.25) and (8.26) reduce to:

$$\left\{\begin{bmatrix} \left(1-\frac{a_1{}^2}{a_3}\right)\frac{\partial^4}{\partial\bar{x}^4}+\bar{s}^2 & 0 & -\frac{a_1}{a_3}c_2{}^2\bar{s}^2\frac{\partial}{\partial\bar{x}} \\ 0 & -a_2\frac{\partial^2}{\partial\bar{x}^2}+c_1{}^2\bar{s}^2 & 0 \\ a_1\frac{\partial^3}{\partial\bar{x}^3} & 0 & -a_3\frac{\partial^2}{\partial\bar{x}^2}+c_2{}^2\bar{s}^2 \end{bmatrix} + \varepsilon(\bar{s})\bar{x}\begin{bmatrix} \left(\beta\frac{a_1}{a_3}-1\right)\frac{\partial^2}{\partial\bar{x}^2} & \left(\frac{\beta^2 a_1}{a_3}-\beta\right)\frac{\partial}{\partial\bar{x}} & \left(\beta-\frac{\beta^2 a_1}{a_3}\right)\frac{\partial}{\partial\bar{x}} \\ \beta\frac{\partial}{\partial\bar{x}} & \beta^2 & -\beta^2 \\ -\beta\frac{\partial}{\partial\bar{x}} & -\beta^2 & \beta^2 \end{bmatrix}\right\}\begin{pmatrix} \bar{w}(\bar{x},\bar{s}) \\ \bar{u}_c(\bar{x},\bar{s}) \\ \bar{u}_b(\bar{x},\bar{s}) \end{pmatrix} = \begin{pmatrix} -\bar{q}+\partial\frac{\bar{f}_{bs}}{\partial\bar{x}} \\ -\bar{f}_c \\ -\bar{f}_{bs} \end{pmatrix} \tag{8.27}$$

The equations of motion can be written in the state space form as follows

$$\frac{\partial}{\partial\bar{x}}\bar{y}(\bar{x},\bar{s}) = F(\bar{s})\,\bar{y}(\bar{x},\bar{s}) + \bar{q}(\bar{x},\bar{s}),\quad \bar{x}\in(0,1) \tag{8.28}$$

where $\bar{y}$ and $\bar{q}$ are given by:

$$\bar{y}(\bar{x},\bar{s}) = \begin{bmatrix} \bar{w} & \frac{\partial\bar{w}}{\partial\bar{x}} & \frac{\partial^2\bar{w}}{\partial\bar{x}^2} & \frac{\partial^3\bar{w}}{\partial\bar{x}^3} & \bar{u}_c & \frac{\partial\bar{u}_c}{\partial\bar{x}} & \bar{u}_b & \frac{\partial\bar{u}_b}{\partial\bar{x}} \end{bmatrix}^T,$$

and

$$\bar{q}(\bar{x},\bar{s}) = \{1\ 0\ 0\ 0\ 0\ 0\ 0\ 0\}^T.$$

$F(s)$ is given in Appendix 8.A and the boundary conditions formulated as follows:

$$\mathrm{M}(\bar{s})\bar{y}(0,\bar{s}) + \mathrm{N}(\bar{s})\bar{y}(1,\bar{s}) = \gamma(\bar{s}) \tag{8.29}$$

where $M(s)$, $N(s)$, and $\gamma(s)$ are given in Appendix 8.A.

The solution to Eqs. (8.28) and (8.29) is given by

$$\bar{y}(\bar{x},\bar{s}) = \int_0^1 \mathrm{G}(\bar{x},\varepsilon,\bar{s})\,\bar{q}(\varepsilon,\bar{s})d\varepsilon + \mathrm{H}(\bar{x},\bar{s})\gamma(\bar{s}),\quad \bar{x}\in(0,1)$$

where

$$G(\bar{x},\varepsilon,\bar{s}) = \begin{cases} e^{F(\bar{s})\bar{x}}\left(\mathrm{M}(\bar{s}) + \mathrm{N}(\bar{s})e^{F(\bar{s})}\right)^{-1}\mathrm{M}(\bar{s})e^{-F(\bar{s})\varepsilon}, & \varepsilon < \bar{x} \\ -e^{F(\bar{s})\bar{x}}\left(\mathrm{M}(\bar{s}) + \mathrm{N}(\bar{s})e^{F(\bar{s})}\right)^{-1}\mathrm{N}(\bar{s})e^{F(\bar{s})(1-\varepsilon)}, & \varepsilon > \bar{x} \end{cases},$$

and

$$\mathrm{H}(\bar{x},\bar{s}) = e^{F(\bar{s})\bar{x}}\left(\mathrm{M}(\bar{s}) + \mathrm{N}(\bar{s})e^{F(\bar{s})}\right)^{-1}$$

With all the external forces are set to zero, the solution reduces to

$$\bar{y}(\bar{x},\bar{s}) = H(\bar{x},\bar{s})\gamma(\bar{s}),\ \bar{x} \in (0,1) \ \text{ or } \ \bar{w}(\bar{x},\bar{s}) = \sum_{j=1}^{n} h_{1j}(\bar{x},\bar{s})\ \gamma_j(\bar{s}) \tag{8.30}$$

8.2.5 Finite Element Model

A finite element of the PSOL can be developed using the partial differential equations of the treatment as given by Eqs. (8.25) and (8.26) by employing the classical approaches of the assumed mode method or Galerkin's method as described by Meirovitch (2010). However, in this section, the approach presented in Chapters 4 through 6 will be adopted instead.

Figure 8.5a shows a schematic drawing of a finite element model of the PSOL. The model consists of two types of elements as displayed in Figure 8.5b. The first type of elements is represented by the ith element, which has a solid SOL, whereas the second type of element is illustrated by the neighboring $i + 1$th element which has a slot in the *SOL*.

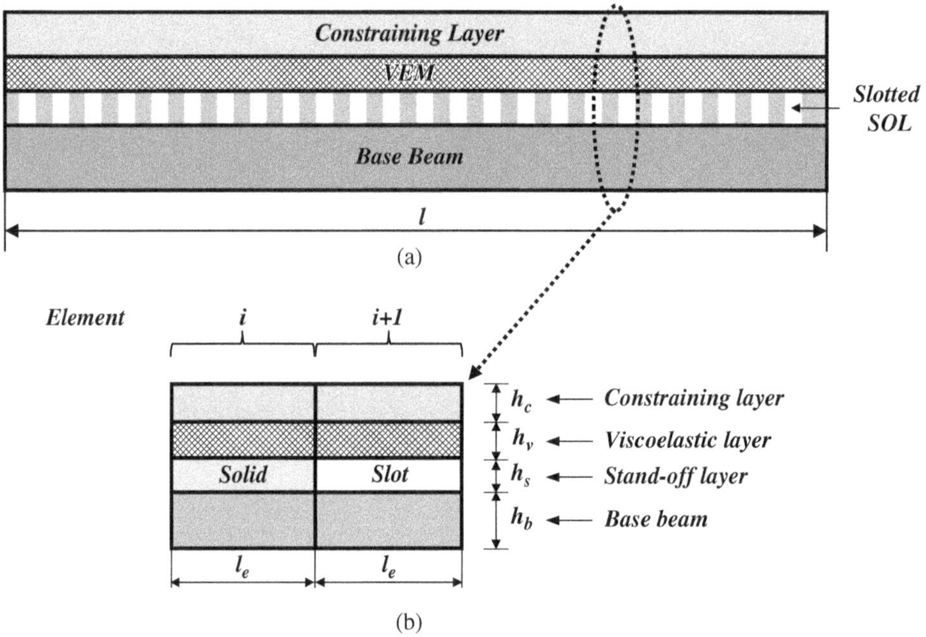

Figure 8.5 (a and b) Finite elements of the passive stand-off layer/composite beam.

These two types of elements are treated in a similar manner and incorporated in the general formulation of the finite element of the beam/PSOL assembly by appropriate modifications of the potential and kinetic energies as follows:

The potential energy PE of each element includes the contributions of the potential energies of beam, SOL, VEM, and constraining layers as follows:

$$PE = \frac{1}{2}\int_0^L \left[\sum_{j=b,s,v,c} E_j A_j u_{j,x}^2 + \sum_{j=b,s,c} E_j I_j w_{,xx}^2 + G_v A_v \gamma^2 \right] dx = \frac{1}{2}\{\Delta_e\}^T [K_e]\{\Delta_e\} \tag{8.31}$$

where E_jA_j and E_jI_j denote the longitudinal and flexural rigidity of the jth layer. Also, G_vA_v denotes the shear rigidity of the VEM with G_v defining its shear modulus and $\{\Delta_e\}$ defines the nodal deflection vector. In Eq. (8.31), E_s is set equal to Young's modulus of the SOL material for the elements that have solid SOL while E_s is set equal to zero for the elements that have slot in the SOL. This yields the appropriate stiffness matrix $[K_e]$ of the element.

Similarly, the kinetic energy (KE) of any general element of the beam/PSOL is given by:

$$KE = \frac{1}{2}\int_0^L \left[\sum_{j=b,s,v,c} \rho_j A_j \dot{u}_j^2 + \sum_{j=b,s,v,c} \rho_j A_j \dot{w}^2 \right] dx = \frac{1}{2}\{\dot{\Delta}_e\}^T [M_e]\{\dot{\Delta}_e\} \tag{8.32}$$

Note that A_j and ρ_j denote the area and density of the jth layer, respectively. In Eq. (8.32), ρ_s is set equal to density of the SOL material for the elements that have solid *SOL* while ρ_s is set equal to zero for the elements that have slot in the SOL. This yields the appropriate mass matrix $[M_e]$ of the element.

From Eqs. (8.31) and (8.32), the equation of motion of the individual elements can be obtained using the Lagrangian dynamics to yield the following equation:

$$[M_e]\{\ddot{\Delta}_e\} + [K_e]\{\Delta_e\} = \{F_e\} \tag{8.33}$$

where $[M_e]$ and $[K_e]$ are the element mass and stiffness matrices with $\{\Delta_e\}$ and $\{F_e\}$ denoting the nodal deflection and load vectors of the beam/PSOL element.

The equations of motion of the entire beam/PSOL assembly can be derived by assembling the element matrices using the approach outlined in Section 4.2.4 for the case of rods. Then, the boundary conditions are imposed enabling the computation of the natural frequencies, mode shapes, and response of the assembly.

Example 8.1 Consider a beam/PSOL assembly shown in Figure 8.6. The main geometrical and physical parameters of the beam/PSOL assembly are listed in Table 8.1. The assembly is mounted in a cantilevered manner at its end *A* and is excited at its free end *B* by a sinusoidal excitation in the transverse direction with amplitude of 1 mm.

Determine:

a) the natural frequencies and mode shapes of the beam/PSOL assembly.
b) the frequency response of the free end of the assembly at *B* using the DTF method, the finite element method (FEM), and the commercial software package ANSYS.

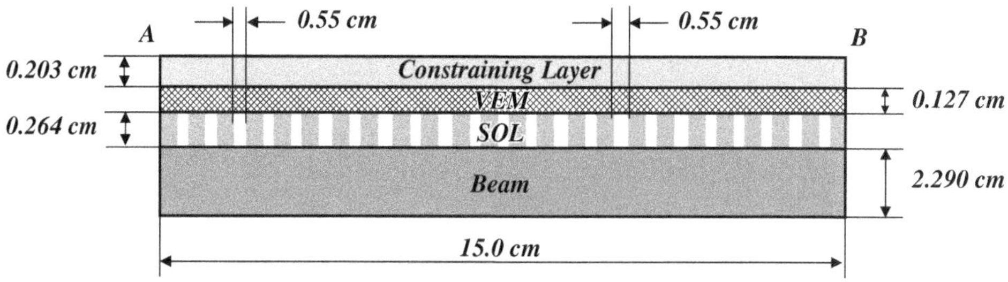

Figure 8.6 A passive stand-off layer/composite beam.

Table 8.1 The main parameters of the beam/PSOL treatment.

Parameter	Value
Beam material	*Aluminum*
SOL material	$E_s = 0.5$ GPa, $\rho_s = 1100$ kg m^{-3}
VEM material	$G_v = 1E5(1 + i)$ *Pa*, $\rho_v = 1100$ kg m^{-3}
Constraining layer material	*Aluminum*
Beam thickness (h_b)	2.290 mm
SOL thickness (h_s)	0.264 mm
VEM thickness (h_v)	0.127 mm
Constraining layer (h_c)	0.203 mm
Width (b)	11.75 mm
Length (l)	150 mm

Solution

Figure 8.7 shows a comparison between the predictions of the frequency response of the free end of a beam/PSOL assembly as obtained by the developed FEM and the method DTF. Excellent agreement is evident between the two methods.

Figure 8.8 shows another comparison between the predictions of the developed DTF method and ANSYS. It is clear that these predictions are in close agreement with each other.

Figure 8.9 displays the mode shapes of the first four modes of vibration of the assembly. The displayed mode shapes are corresponding to natural frequencies of 70.99, 440.9, 1248.1, and 2491.5 Hz, respectively.

Example 8.2 Consider a beam/PSOL assembly shown in Figure 8.6 that has a SOL with 25 slots. The main geometrical and physical parameters are listed in Table 8.1.

Determine:

a) the natural frequencies and mode shapes of the beam/PSOL assembly.
b) the frequency response of the free end of the assembly at B using the DTF method, the FEM, and the commercial software package ANSYS.

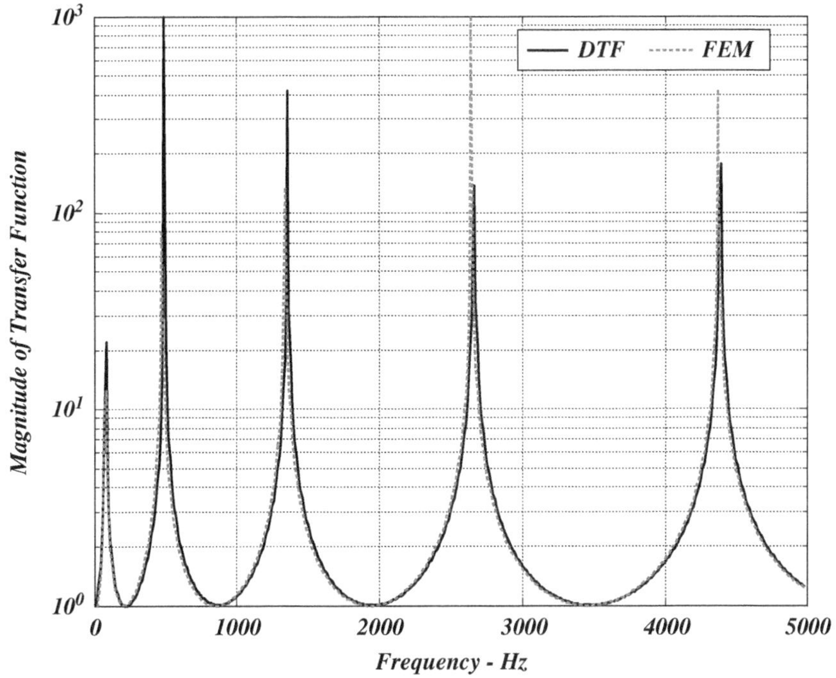

Figure 8.7 Comparisons between the frequency response of the beam/PSOL as obtained by the DTF and FEM methods

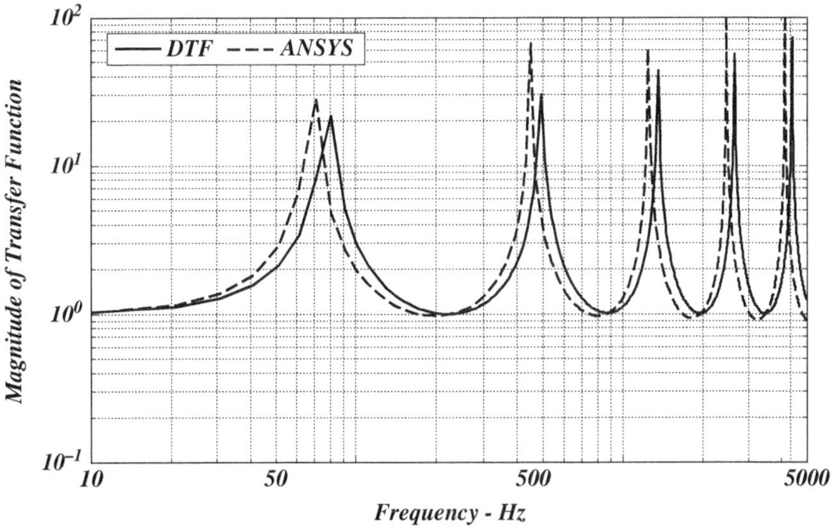

Figure 8.8 Comparisons between the frequency response of the beam/PSOL assembly as obtained by the DTF method and ANSYS.

Solution

Figure 8.10 shows the FEM of the beam/PSOL assembly as obtained by ANSYS.

Figure 8.11 shows a comparison between the predictions of the frequency response of the free end of a beam/PSOL assembly as obtained by the developed FEM, the method

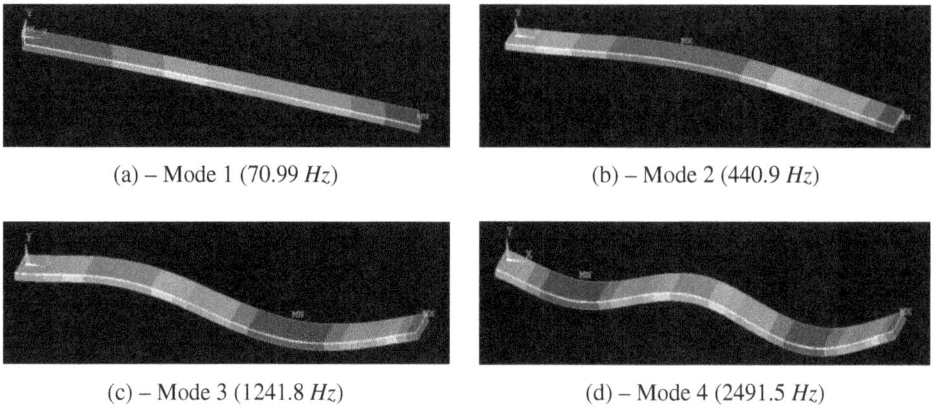

Figure 8.9 The mode shapes of the first four modes of vibration of the beam/PSOL assembly. (a) Mode 1 (70.99 Hz). (b) Mode 2 (440.9 Hz). (c) Mode 3 (1241.8 Hz). (d) Mode 4 (2491.5 Hz).

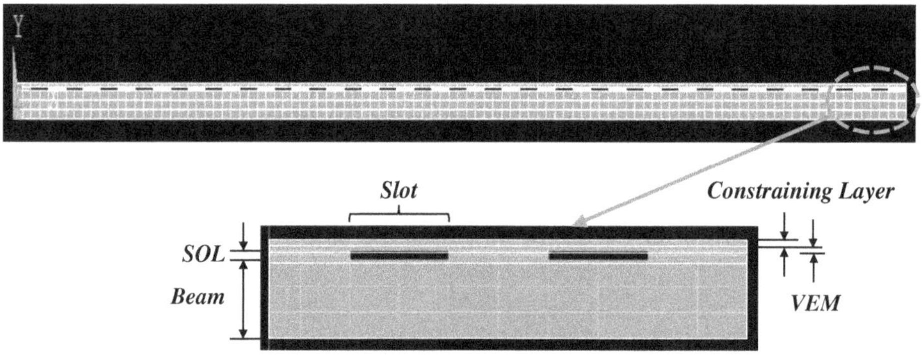

Figure 8.10 The finite element model of the beam/PSOL assembly.

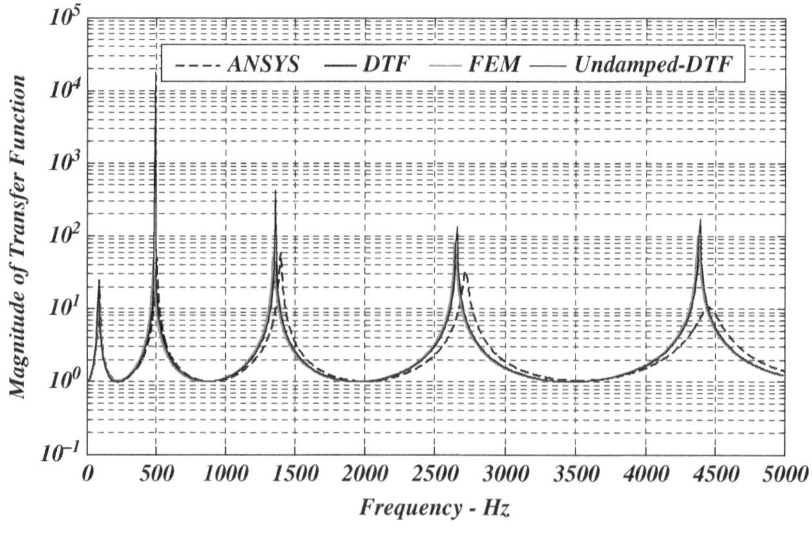

Figure 8.11 The mode shapes of the first four modes of vibration of the beam/slotted PSOL assembly.

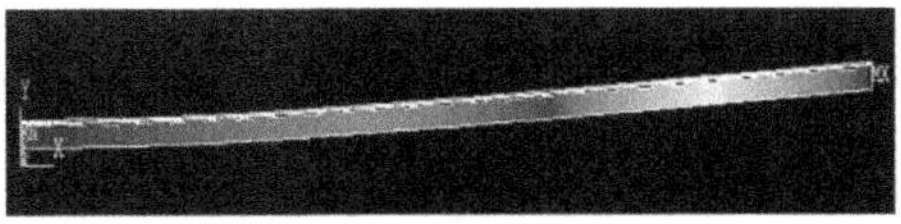

(a) – Mode 1 (81 Hz)

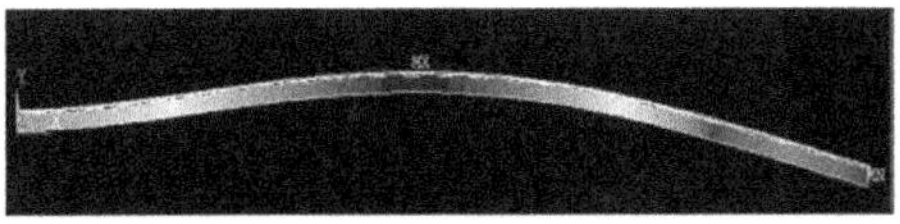

(b) – Mode 2 (501 Hz)

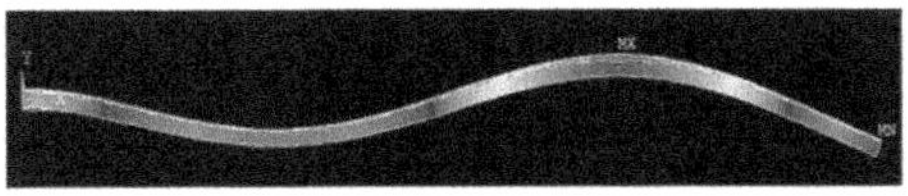

(c) – Mode 3 (1398 Hz)

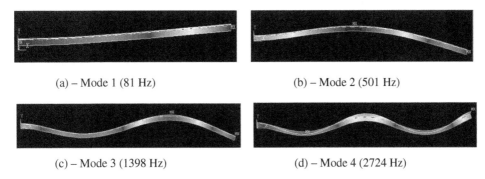

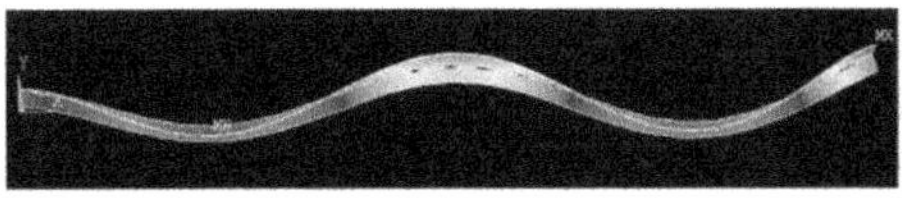

(d) – Mode 4 (2724 Hz)

Figure 8.12 Comparisons between the frequency response of aluminum beams/PSOL as obtained by different methods. (a) Mode 1 (81 Hz). (b) Mode 2 (501 Hz). (c) Mode 3 (1398 Hz). (d) Mode 4 (2724 Hz).

DTF, and ANSYS. Excellent agreement is evident between the different methods. Also displayed on the figure is the frequency response of the undamped beam/PSOL assembly when G_v = 1E5(1 + 0i) Pa for comparison purposes.

Figure 8.12 displays the mode shapes of the first four modes of vibration of the beam/slotted PSOL assembly. The displayed mode shapes are corresponding to natural frequencies of 81.0, 501, 1398, and 2724 Hz, respectively.

8.2.6 Summary

This section has presented the development of a finite element modeling of the PSOL damping treatments. The predictions of the model are validated against the predictions of a DTF model of the PSOL and the predictions of a model built using the commercial finite element software ANSYS.

The developed theoretical techniques present invaluable tools for designing effective PSOL treatments that can act as simple and effective means for enhancing the damping characteristics of conventional constrained layer damping.

8.3 Functionally Graded Damping Treatments

Conventionally, the viscoelastic cores of PCLD treatments are made of materials that have uniform shear modulus. Under such conditions, it is well-recognized that these treatments are only effective near their edges where the shear strains attain their highest values. In order to enhance the damping characteristics of the PCLD treatments, the cores can be manufactured from functionally graded viscoelastic materials (FGVEMs) that have an optimally selected gradient of the shear modulus over the length of the treatments. With such optimized distribution of the shear modulus, the shear strain can be enhanced, and the energy dissipation can be maximized.

The theory governing the vibration of beams treated with PCLD, that has functionally graded viscoelastic cores, is presented in this section using the FEM.

8.3.1 Background of Functionally Graded Constrained Layer Damping

PCLD is widely accepted in various vibration damping applications because of its simplicity, reliability, and effectiveness. Accordingly, extensive efforts have been devoted to the design of optimal PCLD treatments for various vibrating structures. These efforts aim primarily at optimizing the length (e.g., Plunkett and Lee 1970; Hajela and Lin 1991; Mantena et al. 1991; Demoret and Torvik 1995), shape (e.g., Lin and Scott 1987; Lumsdaine and Scott 1996), placement (e.g., Spalding and Mann 1995; Kruger et al. 1997; Ro and Baz 2002), material ingredients (e.g., Alberts and Xia 1995; Koratkara et al. 2003), and topology (e.g., Oh et al. 2000a; Lumsdaine 2002; Kim and Kim 2004; Pai 2004) of the PCLD.

In most of these studies, the focus was placed on constrained layer damping treatments with viscoelastic cores that have uniform shear modulus. For such treatments, it is well-known that the treatments are only effective near their edges where the shear strains attain their highest values. Therefore, several attempts have been made to overcome such deficiencies by introducing local discontinuities in the viscoelastic cores either by impregnating the cores with different types of fibers (e.g., Alberts and Xia 1995; Koratkara et al. 2003) or by changing the topology and/or the cellular structure of the cores themselves (e.g., Yi et al. 2000; Oh et al. 2000a; Lumsdaine 2002; Kim and Kim 2004; Pai 2004).

In this section, a radically different approach is adopted whereby the viscoelastic cores are manufactured from functionally graded materials (FGM) that have optimally selected gradient of the shear modulus along the treatments. This approach is different from that proposed by Venkataraman and Sankar (2001) where the viscoelastic treatments are functionally graded across the thickness that would be practically difficult for thin treatments. With the proposed longitudinal grading approach, the treatments can be very thin and when the shear modulus distribution is optimized, the shear strain can be enhanced, the energy dissipation can be maximized, and the size of the treatment can be dramatically reduced in both the thickness and length directions.

8.3.2 Concept of Constrained Layer Damping with Functionally Graded Viscoelastic Cores

The concept of the PCLD with FGVEM can best be understood by considering the viscoelastic core shown in Figure 8.13. In the figure, the shear modulus of the VEM is symmetrically graded along its length with the dark zones indicating high shear modulus and the lighter zones designating lower shear modulus. Other grading strategies that can also be useful are only limited by our imagination. With proper optimization of the shear modulus gradient, the total energy $W_{d_{FGVEM}}$ dissipated in the treatment, which is given by

$$W_{d_{FGVEM}} = \int AG'(x)\eta_v \gamma^2 dx, \tag{8.34}$$

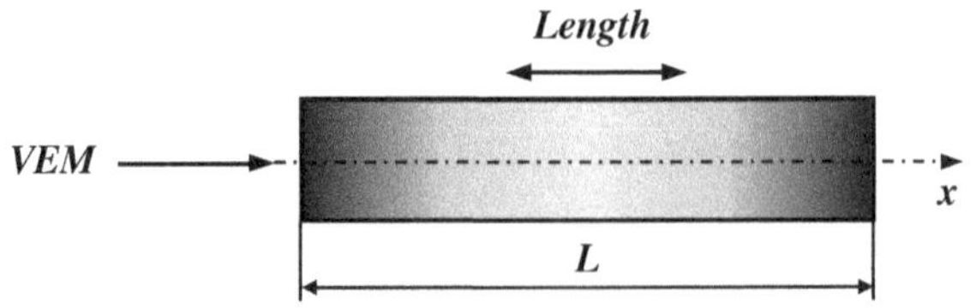

Figure 8.13 Functionally graded viscoelastic core.

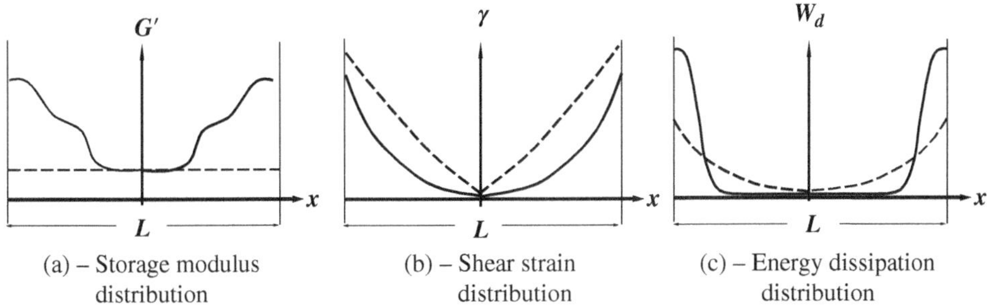

Figure 8.14 Characteristics of functionally graded damping treatments (—— Conventional PCLD, – – – PCLD/FGVEM). (a) storage modulus distribution, (b) shear strain distribution, and (c) energy dissipation distribution.

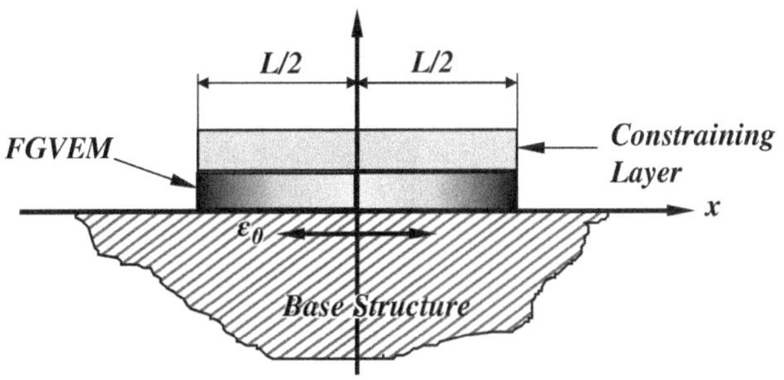

Figure 8.15 Structure/constrained FGVEM assembly.

can be enhanced because it is the product of the shear modulus and the square of the shear angle integrated over the length L of the treatment as shown in Figure 8.14. Note that A and η_v define the cross sectional area and loss factor of the VEM. In this study, both A and η_v are considered constant along the length of the treatment.

8.3.3 Finite Element Model

8.3.3.1 Quasi-Static Model of the Passive Constrained Damping Layer of Plunkett and Lee (1970)

In Section 6.2.1.1, the quasi-static model of a conventional PCLD treatment was presented. In this section, the model is extended to PCLD treatments with functionally graded viscoelastic cores using the theory of finite elements. A schematic drawing of the model is shown in Figure 8.15. Equations (6.2) through (6.5) are modified to account for the variation of the shear modulus along the VEM treatment to take the following forms:

Potential Energy:

$$PE = \frac{1}{2} E_2 h_2 b \int_{-L/2}^{L/2} u_{,x}^2 \, dx + \frac{1}{2} h_1 b \int_{-L/2}^{L/2} G'(x) \gamma^2 \, dx, \tag{8.35}$$

Kinetic Energy:

$$KE = \frac{1}{2} mb \int_{-L/2}^{L/2} \dot{u}^2 dx, \tag{8.36}$$

and

Dissipated Energy:

$$W_d = -h_1 b \int_{-L/2}^{L/2} \tau_d \gamma \, dx = -h_1 b \eta_g i \int_{-L/2}^{L/2} G'(x) \gamma^2 \, dx \tag{8.37}$$

where E_2, h_2, and b denote Young's modulus, thickness, and width of the constraining layer, respectively. Note that the storage shear modulus $G'(x)$ varies along the VEM length. Also, η_g denotes the loss factor, and $i = \sqrt{-1}$. In Eq. (8.36), m is the mass/unit width and unit length of the constraining layer, and τ_d is the dissipative shear stress developed by the viscoelastic core. Furthermore, the shear strain γ is given by: $\gamma = (u - u_o)/h_1 = (u - \varepsilon_o x)/h_1$.

Using the conventional interpolation equation for rods, as given by Eq. (4.25), where $u = [N]\{\Delta_e\}$, and applying the Lagrangian equations, then Eqs. (8.35) through (8.37) yield the following equation of motion of the eth element of the PCLD/FGM:

$$[M_e]\{\ddot{\Delta}_e\} + [[K_{ce}] + [K_{ve}]]\{\Delta_e\} = [C_e]\varepsilon_0 \tag{8.38}$$

where

$$[M_e] = \frac{mh_1}{E_2 h_2} \int_0^{L_e} N^T N \, dx \, , [K_{ce}] = \int_0^{L_e} N_{,x}^T N_{,x} \, dx,$$

$$[K_{ve}] = \int_0^{L_e} B^{*-2}(x) N^T N \, dx, [C_e] = \int_0^{L_e} B^{*-2}(x) N x [1 + L_e(e - N_{elements}/2 - 1)] dx. \tag{8.39}$$

where x is the local coordinate associated with the eth element and $N_{elements}$ is the number of elements.

Under quasi-static conditions, Eq. (8.38) reduces to:

$$[[K_{ce}] + [K_{ve}]]\{\Delta_e\} = [C_e]\varepsilon_0 \tag{8.40}$$

For a given input excitation ε_0, the deflection $\{\Delta_e\}$ can be determined from:

$$\{\Delta_e\} = [[K_{ce}] + [K_{ve}]]^{-1}[C_e]\varepsilon_0 \tag{8.41}$$

Hence, the shear strain at any location can be determined using Eq. (6.1) as follows:

$$\{\gamma_e\} = (\{\Delta_e\} - \varepsilon_0 x)/h_1 \tag{8.42}$$

Accordingly, the energy dissipated in shearing the VEM can be computed from:

$$W_{d_{FGVEM}} = \int AG'(x) \eta_v \gamma^2 dx \tag{8.43}$$

Example 8.3 Using the FEM, described in Section 8.3.3.1, determine the normalized shear strain $(\gamma h_1/\varepsilon_0 L)$ distribution and normalized dissipated energy $W_{d_{FGVEM}}/(ALG_0\varepsilon_0^2)$ over the length of the *VEM* for:

a) PCLD with a viscoelastic core that has a uniform shear modulus of G_0.
b) PCLD with a functionally graded viscoelastic core such that:

$L/B_o = 1$ where $B_0 = \sqrt{h_1 h_2 E_2/G}$ and $G = G_0\left[1 + a\left(\frac{x}{L/2}\right)^2\right]$ with $a = 10$ and $G_0 = 1$.

Compare the obtained FEM results with those obtained by using the boundary value problem (BVP) solution method described in Example 6.1.

Solution

Figure 8.16 shows a comparison between the characteristics of conventional PCLD and PCLD with FGM viscoelastic core. Figure 8.16a displays the comparison between the shear strain distributions in the two cases. The figure indicates that the conventional PCLD experiences higher shear strains than the PCLD with FGM viscoelastic core. However, the important characteristics are shown in Figure 8.16b where the dissipated energy by the PCLD with FGM is seen to be considerably higher than that of the conventional PCLD. Such characteristics are obtained in spite of the lower shear strains of the PCLD/

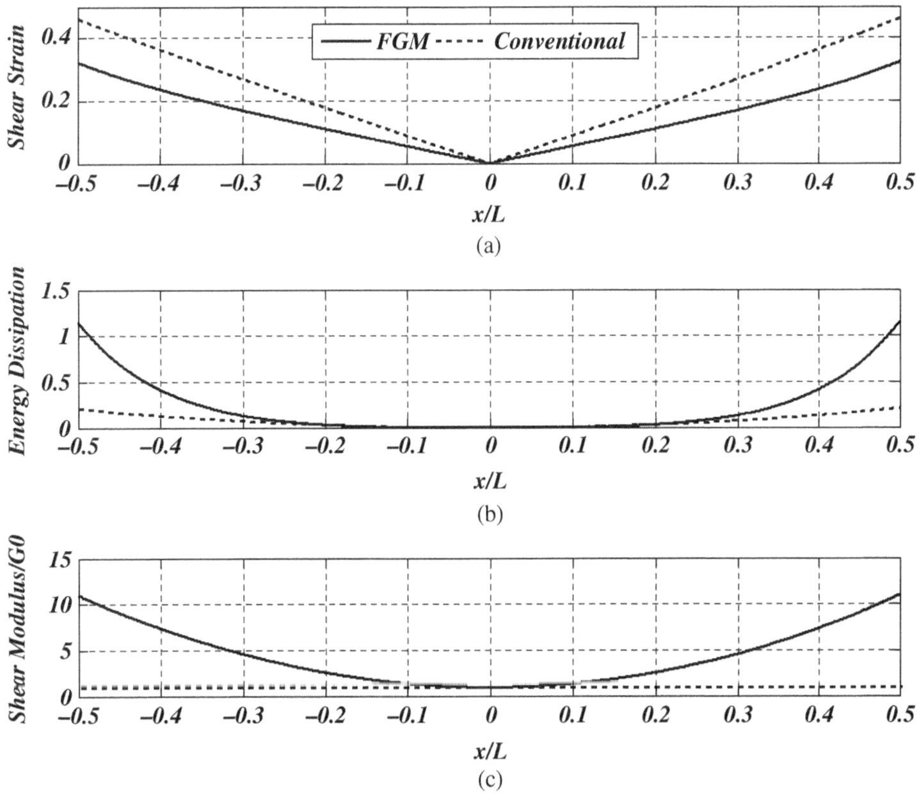

Figure 8.16 Comparison between the shear strain, energy dissipated, and shear modulus characteristics of conventional PCLD and PCLD with FGM viscoelastic core.

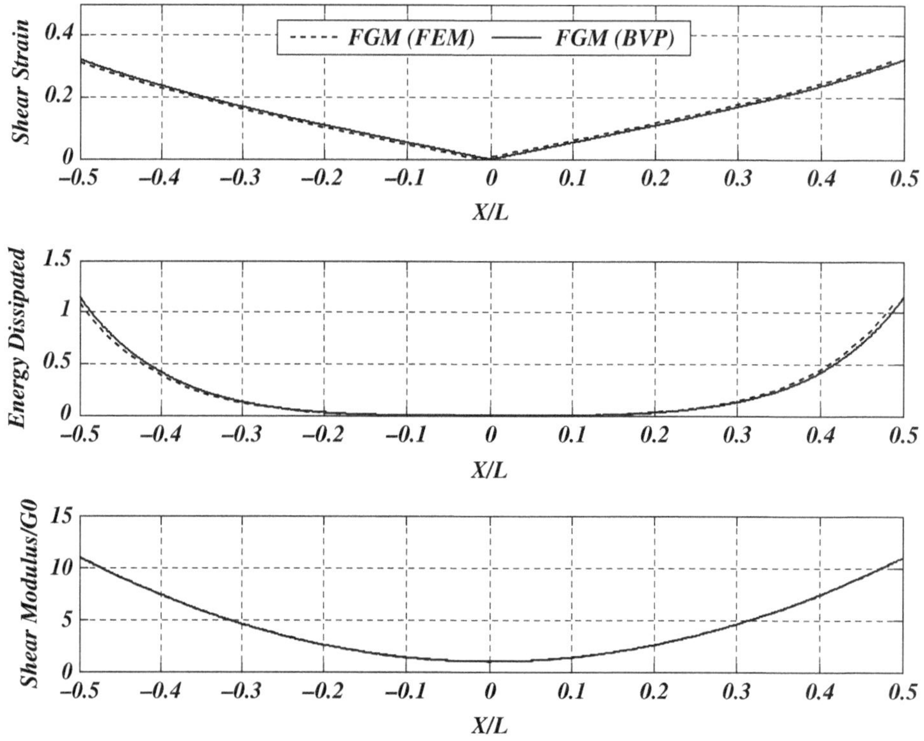

Figure 8.17 Comparison between the characteristics of conventional PCLD and PCLD with FGM viscoelastic core as obtained by the FEM and the boundary value problem (BVP) solution methods.

FGM treatment. This is attributed to the fact that graded shear modulus is higher for the PCLD/FGM treatment than the conventional PCLD treatment. Hence, the important parameter is neither the shear strain nor the shear modulus alone, but rather the product of the square of the shear strain and the shear modulus as the resulting product, when integrated over the VEM length, quantifies its ability to dissipate energy. Accordingly, grading and gradually increasing the shear modulus along the VEM compensates significantly for the drop in the developed shear strain.

Figure 8.17 displays a comparison between the characteristics of the conventional PCLD and PCLD with FGM viscoelastic core as obtained by the FEM and the BVP solution methods. Excellent agreement is evident between the two methods.

The FEM, however, provides a much more powerful tool for predicting the characteristics of more complex FGM configurations.

Example 8.4 Compare the characteristics of two treatments of PCLD with FGM viscoelastic cores where in the first treatment the shear modulus of the viscoelastic core is graded in a continuous manner such that:

$$L/B_0 = 1 \text{ where } B_0 = \sqrt{h_1 h_2 E_2/G} \text{ and } G = G_0\left[1 + a\left(\frac{x}{L/2}\right)^2\right] \text{ with } a = 10 \text{ and } G_0 = 1\,\text{N}\,\text{m}^{-2}.$$

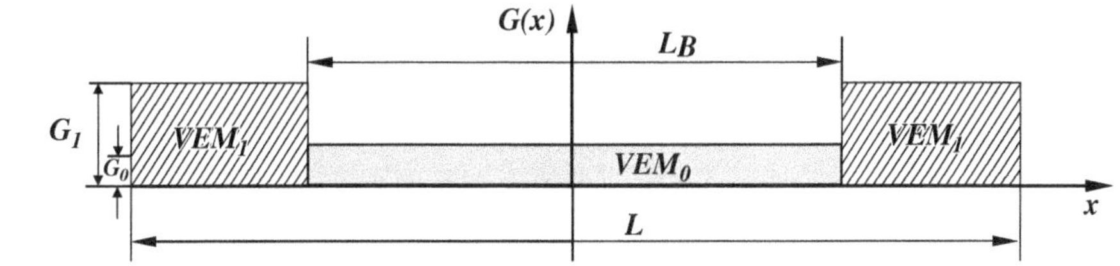

Figure 8.18 Shear modulus distribution for the discrete PCLD/FGM treatment.

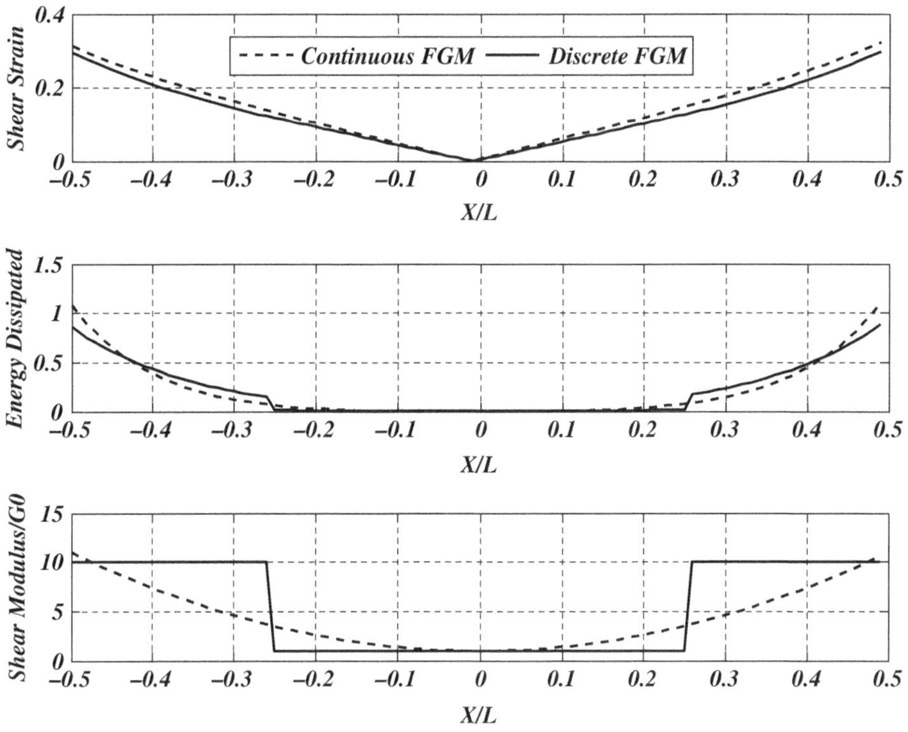

Figure 8.19 Comparison between the characteristics of conventional PCLD and PCLD with FGM viscoelastic core as obtained by the FEM method.

In the second treatment, the shear treatment is manufactured from two discrete materials that have the shear modulus distributions shown in Figure 8.18.

In the discrete PCLD/FGM, it is assumed that $G_1 = 10\,\mathrm{N\,m^{-2}}$ and $G_2 = 1\,\mathrm{N\,m^{-2}}$.

Solution

Figure 8.19 shows a comparison between the characteristics of the continuous and discrete PCLD with FGM viscoelastic cores. The figure indicates clearly that the two types of treatments behave very similarly and exhibit similar energy dissipation characteristics.

However, it is important to note that the use of the discrete functional grading of the VEM is favored because of its practical implications by virtue of the ease of its manufacturing and placement.

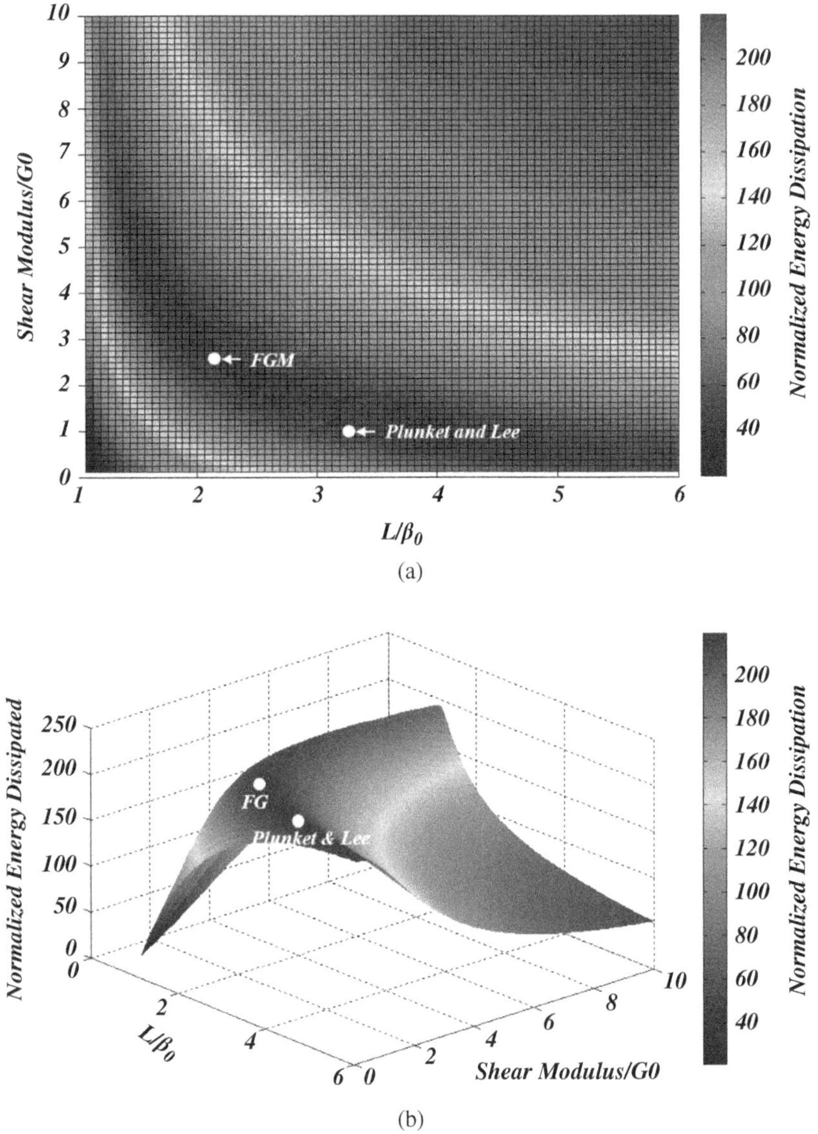

Figure 8.20 Contours of the normalized energy dissipation characteristics of FGVEM as a function of the treatment length L/β_0 and the shear modulus/G_0.

Figure 8.20a,b display the contours of the normalized energy dissipation in the plane of L/β_0 – shear modulus/G_0. The figure indicates that a FGVEM can produce the same energy dissipation as that produced by an optimal VEM with a uniform shear modulus but with a much smaller length. In particular, the optimal VEM, according to Plunkett and Lee (1970), produces an energy dissipation of 200 when $L/\beta_0 = 3.26$ while an optimal FGVEM produces the same energy dissipation with $L/\beta_0 = 2.12$ when the shear modulus ration $G_1/G_0 = 10$ and $L/L_B = 2$. Such a result suggests that the use of FGVEM can be an effective means for maximizing the energy dissipation/unit volume of the VEM when compared to a conventional VEM. For the considered example, the same energy dissipation is obtained by a FGVEM treatment that is about 35% shorter in length than a conventional VEM treatment with a uniform shear modulus.

8.3.3.2 Dispersion Characteristics of Passive Constrained Damping Layer with Uniform and Functionally Graded Cores

The dispersion characteristics relate the frequency and the wave number, which quantifies the wave propagation behavior of the damping treatment. These characteristics can be used to enable the predictions of disturbance propagation and energy transport as well as to generate the phase velocity, group velocity, attenuation, real wave number, and angle of incidence (Manconi and Mace 2008; Pavlakovic et al. 1997).

8.3.3.2.1 Passive Constrained Damping Layer with Uniform Cores

The dynamics of the unexcited (i.e., $u_0 = 0$) PCLD system, as given by Eq. (6.8), is simplified to the following expression:

$$mh_1/G^*\ddot{u} = B^{*^2} u_{,xx} - u \tag{8.44}$$

where m is the mass/unit width and unit length of the constraining layer, h_1 is the thickness of the viscoelastic layer, $B^* = \sqrt{h_1 h_2 E_2/G^*}$ is a characteristics complex length of the passive treatment, h_2 is the thickness the constraining layer, E_2 is its Young's modulus, G^* is the complex modulus of the VEM, and u is the longitudinal deflections of the constraining layer.

A time harmonic disturbance at frequency ω can propagate through the PCLD structure as follows:

$$u = Ue^{i(kx-\omega t)} \tag{8.45}$$

where k is the wave number and U is the wave amplitude. Substituting Eq. (8.45) into (8.44) gives:

$$(mh_1/G^*)\,\omega^2 = B^{*^2} k^2 + 1 \tag{8.46}$$

Note that G^* and B^* can be written, as given by Eqs. (6.A.1) and (6.A.2), as follows:

$$G^* = Ge^{i\theta} \text{ and } B^* = B_0 e^{-i\frac{\theta}{2}} \tag{8.47}$$

where $\theta = \tan^{-1}(\eta_g)$, $B_0 = \sqrt{\dfrac{h_1 h_2 E_2}{G}}$, and η_g = loss factor of the *VEM*.

Substituting Eq. (8.47) into Eq. (8.46) gives:

$$\bar{k}e^{-i\theta} = \sqrt{e^{-i\theta}\Omega^2 - 1} \tag{8.48}$$

where $\bar{k} = B_0 k$ and $\Omega^2 = (mh_1/G)\,\omega^2$. Note that Eq. (8.48) defines the dispersion characteristics of the PCLD with uniform VEM core.

The relationship exhibits nonlinear behavior, hence the treatment is dispersive in nature. This is unlike the characteristics of non-dispersive materials, where the wave propagation velocity that is equal to $c = \omega/k$ (or $\Omega/\bar{k}$) is independent of both ω (or Ω) and k (or $\bar{k}$) Hence, the wave propagation velocity c is constant.

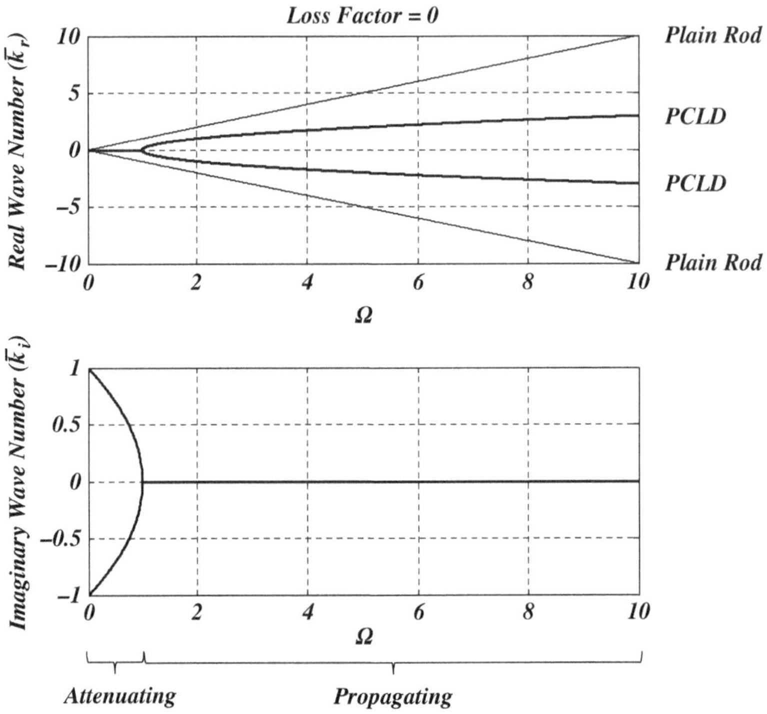

Figure 8.21 The dispersion characteristics of the PCLD with uniform VEM core for $\eta_g = 0$.

Example 8.5 Determine the dispersion characteristics of a PCLD with uniform VEM core as function of the loss factor η_g of the VEM.

Solution

The dispersion Eq. (8.48) is plotted, in the plane $\Omega - \bar{k}$, for different values of the loss factor η_g of the VEM, that is, for different values of $\theta = \tan^{-1}(\eta_g)$.

Figure 8.21 displays the real k_r and imaginary k_i components of the wave number as function of the frequency Ω for $\eta_g = 0$. Hence, the wave number k is given by:

$$k = k_r + i k_i \tag{8.49}$$

Hence, Eq. (8.45) can be rewritten as:

$$u = Ue^{i(k_r x - \omega t)}\, e^{-k_i x} \tag{8.50}$$

This indicates that the real component of the wave number k_r enables waves to spatially propagate without attenuation whereas the imaginary component of the wave number k_i contributes to the spatial attenuation of waves.

Figure 8.21 indicates that for $\Omega > 1$, the wave number is a pure real number representing wave propagation in the positive and negative x directions. However, for $\Omega < 1$, the wave number becomes a pure imaginary number suggesting that the waves are attenuated in both directions of the x-axis.

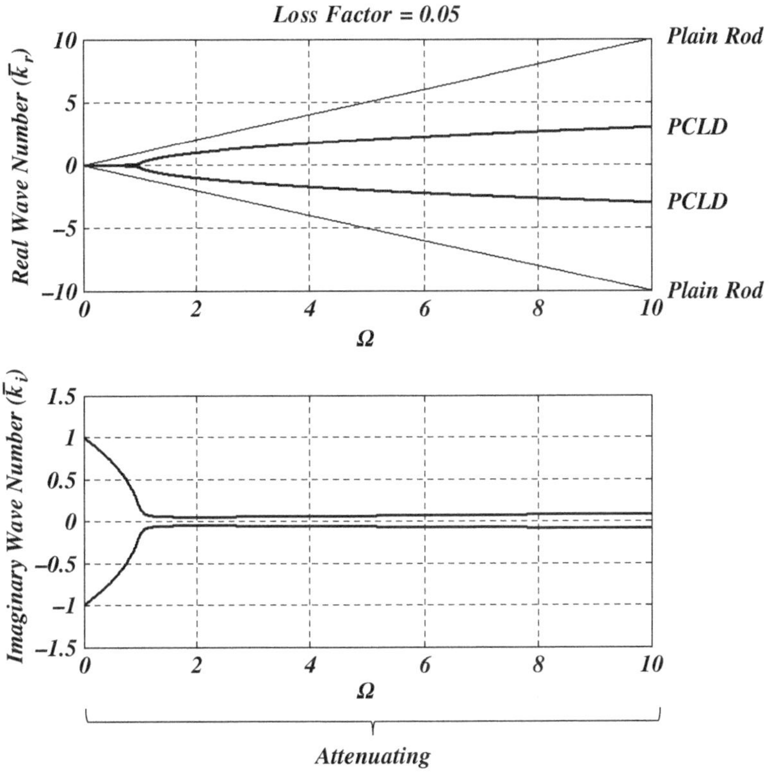

Figure 8.22 The dispersion characteristics of the PCLD with uniform VEM core for $\eta_g = 0.05$.

Note that Eq. (8.44) is very similar to the equations describing strings under and rods on elastic foundation as outlined by Mace and Manconi (2012).

The figure displays also the dispersion characteristics of a plain rod for comparison purposes. Such characteristics indicate that the dispersion curves are linear and that the wave number is purely real. Hence, the plain rod is a non-dispersive system.

Figures 8.22 through 8.24 show the corresponding dispersion characteristics for PCLD with uniform VEM when η_g varies from 0.05 to 1. Note that in all these three cases, the dispersive characteristics have wave numbers with high imaginary components. This results in attenuation throughout the entire frequency range (Figure 8.23).

8.3.3.2.2 Passive Constrained Damping Layer with Functionally Graded Cores

For the unexcited (i.e., $\varepsilon_0 = 0$) passive constrained damping layer with functionally graded cores *PCLD/FGM*, the equation of motion as described by Eq. (8.38) is simplified to assume the following form:

$$[M_e]\{\ddot{\Delta}_e\} + [[K_{ce}] + [K_{ve}]]\{\Delta_e\} = [0] \tag{8.51}$$

where $[M_e]$ is the mass matrix, $[K_{ce}]$ is the elastic stiffness matrix, $[K_{ve}]$ is the stiffness matrix of the VEM accounting for the functionally graded properties, and $\{\Delta_e\}$ is the nodal deflection vector. These matrices are defined in Eq. (8.39).

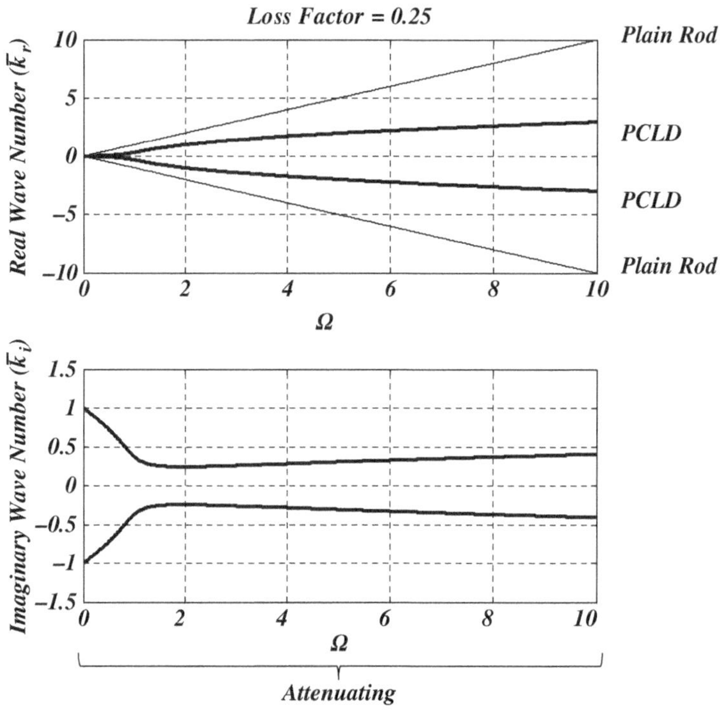

Figure 8.23 The dispersion characteristics of the PCLD with uniform VEM core for η_g = 0.25.

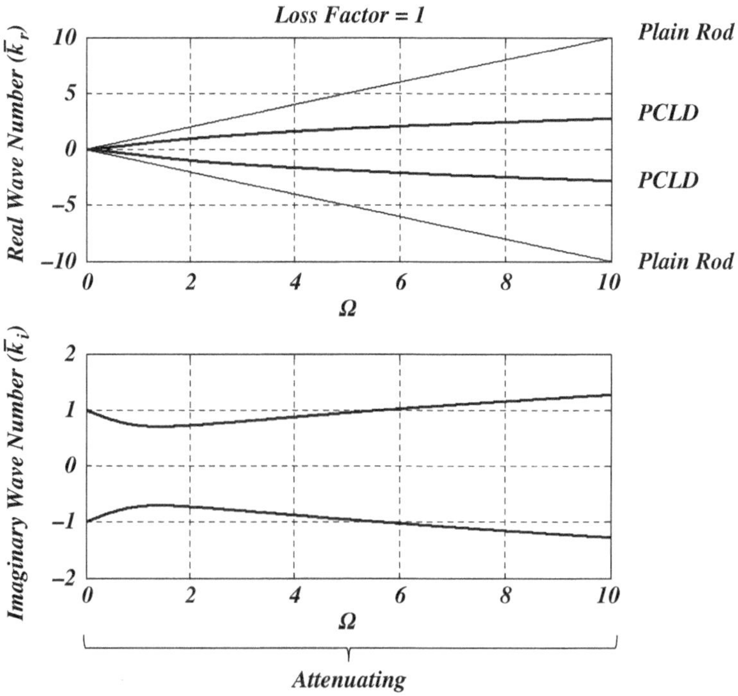

Figure 8.24 The dispersion characteristics of the PCLD with uniform VEM core for η_g = 1.0.

The dispersion characteristics of the PCLD/FGM can be extracted using the semi-analytical finite element (SAFE) method adopted, for example, by Shorter (2004) and Bartoli et al. (2006).

In this approach, it is assumed that the longitudinal deflection u of the constraining layer as follows:

$$u = U e^{i(kx-\omega t)} = [N]\{\Delta^e\}e^{i(kx-\omega t)} \tag{8.52}$$

where k and ω are the wave number and the frequency, respectively. Also, $[N]$ and $\{\Delta^e\}$ denote the interpolating function and the nodal deflection vector, respectively.

Hence, the kinetic and potential energies of the PCLD/FEM, as given by Eqs. (8.35) through (8.37), can be rewritten as follows

$$T = \frac{1}{2}\{\Delta^e\}^T e^{-i(kx-\omega t)} mb\omega^2 \int_{-L/2}^{L/2} [N]^T[N]\,dx\,\{\Delta^e\}e^{i(kx-\omega t)},$$

and

$$U = \frac{1}{2}\{\Delta^e\}^T e^{-i(kx-\omega t)} \left[E_2 h_2 bk^2 \int_{-L/2}^{L/2} [N_{,x}]^T[N_{,x}]\,dx\right]\{\Delta^e\}e^{i(kx-\omega t)}$$
$$+\frac{1}{2}\{\Delta^e\}^T e^{-i(kx-\omega t)} \left[\frac{b}{h_1}\int_{-L/2}^{L/2} G^*(x)\,[N]^T[N]\,dx\right]\{\Delta^e\}e^{i(kx-\omega t)} \tag{8.53}$$

Then, the Lagrangian formulation of the equation of motion yields:

$$\left[\omega^2[M]-[K_{ve}]\right]\{\Delta^e\} = k^2[K_{ce}]\{\Delta^e\} \tag{8.54}$$

where $[M]$, $[K_{ve}]$, and $[K_{ce}]$ are the mass matrix, the stiffness matrix of the VEM, and the elastic stiffness matrix, respectively. These matrices are given by:

$$[M] = \left[\frac{m}{E_2 h_2}\int_{-L/2}^{L/2}[N]^T[N]dx\right],\quad [K_{ve}] = \int_{-L/2}^{L/2} B^{*^2}(x)\,[N]^T[N]\,dx,$$
$$[K_{ce}] = \int_{-L/2}^{L/2}[N_{,x}]^T[N_{,x}]\,dx. \tag{8.55}$$

For a given frequency ω, Eq. (8.54) can be cast in the following eigenvalue problem as follows:

$$\left[\omega^2[M]-[K_{ve}]\right]^{-1}[K_{ce}]\{\Delta^e\} = \frac{1}{k^2}\{\Delta^e\},$$

or

$$A\{\Delta^e\} = \lambda^2\{\Delta^e\} \tag{8.56}$$

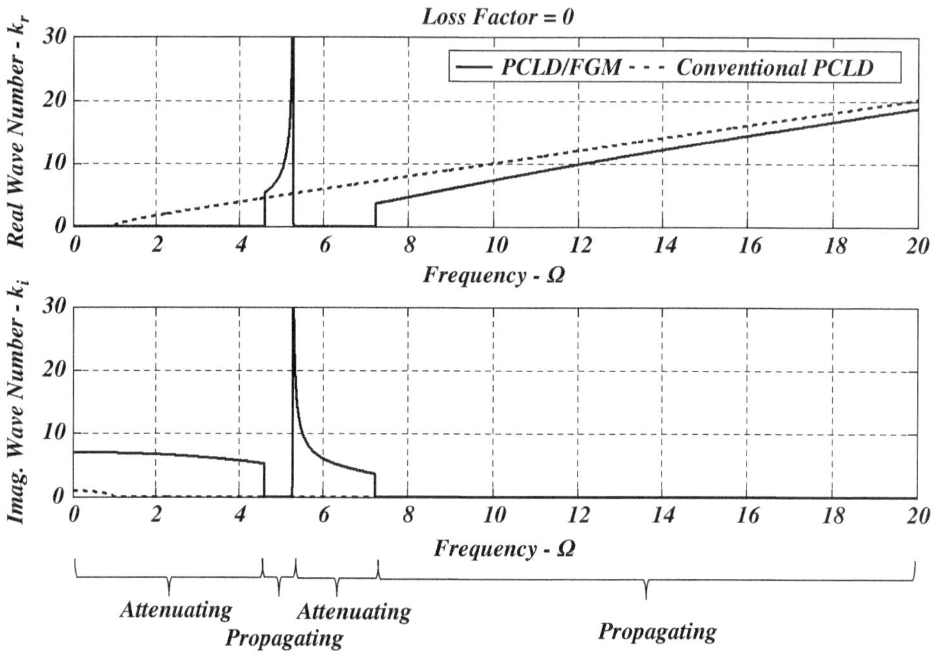

Figure 8.25 The dispersion characteristics of the PCLD with FGM VEM core for $\eta_g = 0.0$.

where $A = [\omega^2[M] - [K_{ve}]]^{-1}[K_{ce}]$ and $\lambda^2 = \frac{1}{k^2}$.

Hence, for a given frequency ω, the eigenvalue problem of Eq. (8.56) can be solved for the eigenvalue λ or the wave number k. Accordingly, the dispersion characteristics $\omega - k$ can be plotted.

Example 8.6 Determine the dispersion characteristics of a PCLD with functionally graded VEM core as function of the loss factor η_g of the VEM.

Compare the characteristics of the PCLD/FGM with those of PCLD with uniform VEM cores. For the PCLD/FGM, assume that $L/B_o = 1$ where $B_0 = \sqrt{h_1 h_2 E_2/G}$ and $G = G_0\left[1 + a\left(\frac{x}{L/2}\right)^2\right]$ with $a = 10$ and $G_o = 1\ \text{N m}^{-2}$.

Solution

Figures 8.25 through 8.28 show the corresponding dispersion characteristics for PCLD with uniform VEM when η_g varies from 0.0 to 1.0. Note that in all these figures, the dispersive characteristics of the PCLD/FGM have wave numbers with imaginary components higher than the corresponding cases of the PCLD with uniform VEM. This suggests that the PCLD/FGM has high spatial energy dissipation characteristics over a wide range of excitation frequencies as compared to the conventional PCLD (Figures 8.26 and 8.27).

In this manner, the PCLD/FGM demonstrates its effectiveness in spatial attenuation of the wave propagation along structural systems.

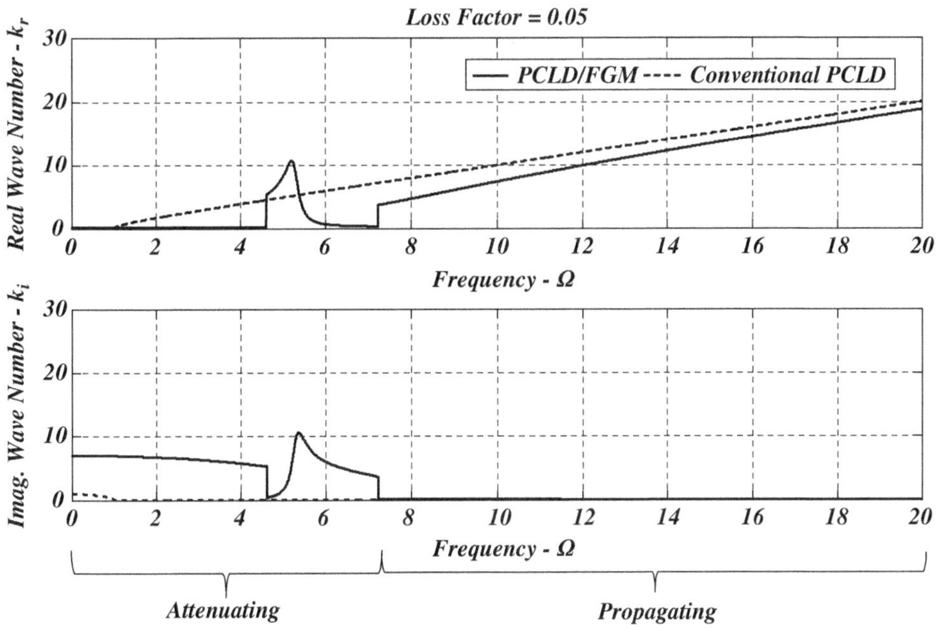

Figure 8.26 The dispersion characteristics of the PCLD with FGM VEM core for $\eta_g = 0.05$.

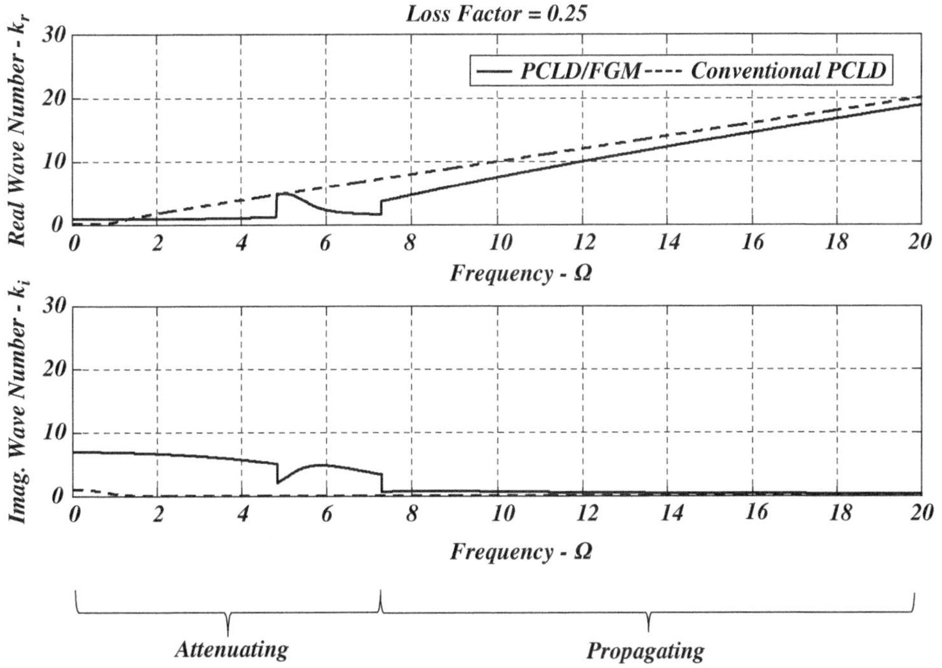

Figure 8.27 The dispersion characteristics of the PCLD with FGM VEM core for $\eta_g = 0.25$.

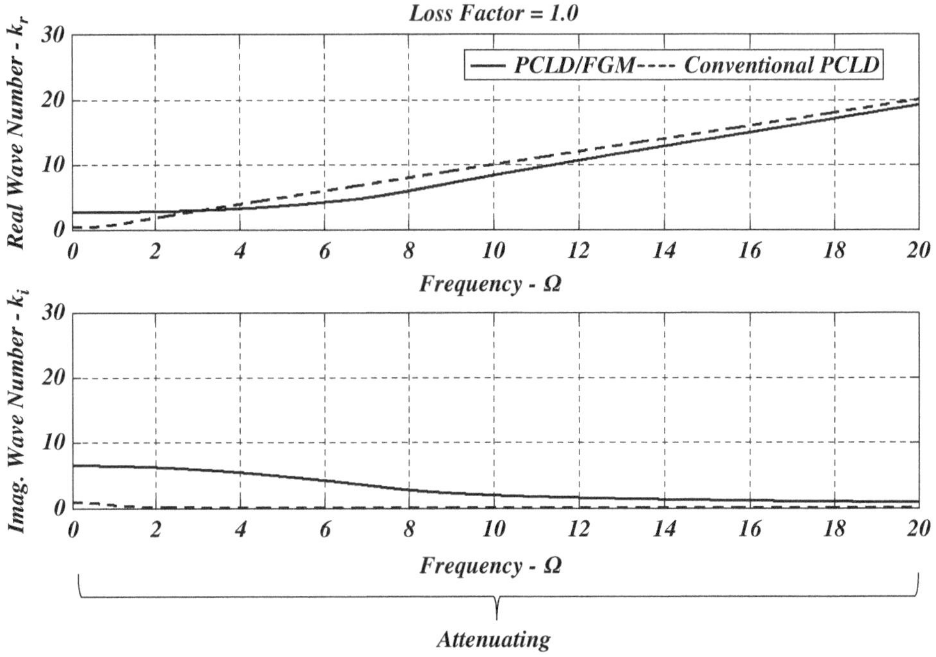

Figure 8.28 The dispersion characteristics of the PCLD with FGM VEM core for $\eta_g = 1.0$.

8.3.4 Summary

In this section, the idea of constrained viscoelastic layer with longitudinally graded properties has been introduced. The damping layer is made of a viscoelastic material whose storage shear modulus is variable along its length. The viscoelastic core is sandwiched between the base metal and the constraining layer. Grading can be utilized to maximize the shear energy dissipated in the viscoelastic layer. Continuous and discrete theoretical models are developed to model the characteristics of rod/PCLD/FGVEM. The models are exercised to determine the optimal energy dissipation and the dispersion characteristics of rods/PCLD/FGVEM. Furthermore, the obtained results demonstrate the effectiveness of the class of PCLD/FGVEM in producing similar energy dissipation as conventional PCLD with uniform VEM while using shorter treatment lengths. Also, the PCLD/FGVEM treatments are shown to exhibit higher imaginary wave numbers leading to higher spatial energy dissipation characteristics than conventional PCLD with uniform VEM.

8.4 Passive and Active Damping Composite Treatments

8.4.1 Passive Composite Damping Treatments

As typical viscoelastic materials exhibit high damping characteristics at usually low stiffness, hence these are not of interest in any structural applications. Conversely, structural

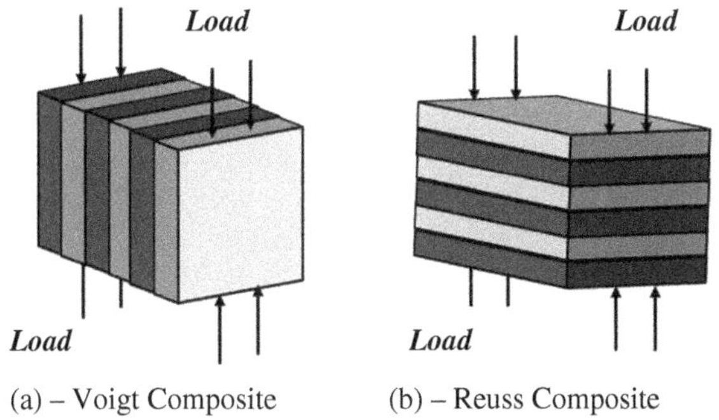

Figure 8.29 Configurations of damped structural composites (Phase 1 Phase 2). (a) Voigt composite. (b) Reuss composite.

materials, alloys, and ceramics have high stiffness but relatively low loss factors, hence incapable of damping out structural vibrations. Such contradicting characteristics are displayed in the loss factor-storage modulus map of Figure 3.5.

In this section, stiff material with high loss are developed by combining the attractive attributes of structural and viscoelastic materials to form composites that can simultaneously support structural loads and provide high vibration damping capabilities.

Figure 8.29 displays two possible configurations of these composites. The first configuration is called "Voigt composites" where the load is shared between the different layers of the composite as shown in Figure 8.29a. The second configuration is called "Reuss composites" where each layer experiences the same load as shown in Figure 8.29b.

i) ***Voigt composites***, the elastic modulus of the composite E_c can be determined in terms of the elastic moduli E_1 and E_2 of the constituents using the "rule of mixtures" as follows:

$$E_c = E_1 V_1 + E_2 V_2 \tag{8.57}$$

where V_1 and V_2 denote the volume fractions of phase 1 and 2, respectively, with $V_1 + V_2 = 1$.

Using the correspondence principle (Christensen 1982), the elastic relationship given by Eq. (8.57) can be converted to a steady-state harmonic viscoelastic relationship by replacing Young's moduli E_i ($i = c,1,2$) by the complex moduli $E_i^*(i\omega)$ with ω denoting the angular frequency of the harmonic loading. This conversion process gives

$$E_c^* = E_1^* V_1 + E_2^* V_2 \tag{8.58}$$

with $E_i^* = E_i' + iE_i' = E_i'(1 + i\eta_i)$ with $i = c,1,2$. Note that E_i' = storage modulus, E_i' = loss modulus, and η_i = loss factor = E_i'/E_i'.

Equation (8.58) can be expanded and rewritten as follows:

$$E_c'(1 + i\eta_c) = E_1'(1 + i\eta_1)V_1 + E_2'(1 + i\eta_2)V_2 \tag{8.59}$$

Equating the real and imaginary parts on both sides of Eq. (8.59) gives:

$$E_c' = E_1' V_1 + E_2' V_2 \tag{8.60}$$

and

$$E_c'\eta_c = E_1'\eta_1 V_1 + E_2'\eta_2 V_2 \tag{8.61}$$

Combining Eqs. (8.60) and (8.61) gives:

$$\eta_c = E_1'\eta_1 V_1 + \frac{E_2'\eta_2 V_2]}{E_1' V_1 + E_2' V_2} \tag{8.62}$$

In dimensionless form, Eq. (8.61) reduces to:

$$\frac{\eta_c}{\eta_1} = \frac{1 + (E_2'/E_1')(\eta_2/\eta_1)(V_2/V_1)}{1 + (E_2'/E_1')(V_2/V_1)} \tag{8.63}$$

ii) Reuss composites, the elastic modulus of the composite E_c can be determined in terms of the elastic moduli E_1 and E_2 of the constituents using the "rule of mixtures" as follows:

$$\frac{1}{E_c^*} = \frac{V_1}{E_1^*} + \frac{V_2}{E_2^*}$$

or

$$\frac{1}{E_c'(1 + i\eta_c)} = \frac{V_1}{E_1'(1 + i\eta_1)} + \frac{V_2}{E_2'(1 + i\eta_2)} \tag{8.64}$$

Equating the real and imaginary parts on both sides of Eq. (8.64) gives:

$$E_c' = E_1'\bar{E}_2'\left[\frac{(1-\eta_1\eta_2)(\bar{E}_2' V_1 + V_2) + (\eta_1 + \eta_2)(\bar{E}_2' V_1\eta_2 + V_2\eta_1)}{(\bar{E}_2' V_1 + V_2)^2 + (\bar{E}_2' V_1\eta_2 + V_2\eta_1)^2}\right] \tag{8.65}$$

and

$$\eta_c = \left[\frac{-(1-\eta_1\eta_2)(\bar{E}_2' V_1\eta_2 + V_2\eta_1) + (\eta_1 + \eta_2)(\bar{E}_2' V_1 + V_2)}{(1-\eta_1\eta_2)(\bar{E}_2' V_1 + V_2) + (\eta_1 + \eta_2)(\bar{E}_2' V_1\eta_2 + V_2\eta_1)}\right] \tag{8.66}$$

where $\bar{E}_2' = \dfrac{E_2'}{E_1'}$.

Example 8.7 Predict the properties of Voigt and Reuss composites containing Steel as phase 1, with $E_1' = 200$ GPa, $\eta_1 = 0.001$ and a viscoelastic elastomer as phase 2, with $E_2' = 0.020$ GPa, $\eta_2 = 1.0$.

Solution

The properties of the Voigt composites are determined using Eqs. (8.60) and (8.62). Also, Eqs. (8.65) and (8.66) are used to compute the properties of the Reuss composites.

Figure 8.30 shows that the magnitude of the complex modulus-loss factor map of the composite has an enclosed region which is bounded from above by the Reuss characteristics and from below by the Voigt characteristics. It is interesting to note that Roscoe (1969) has mathematically established bounds for the storage and loss moduli E' and E' of the complex modulus of composites and has shown these bounds are equivalent to the Voigt and Reuss relations.

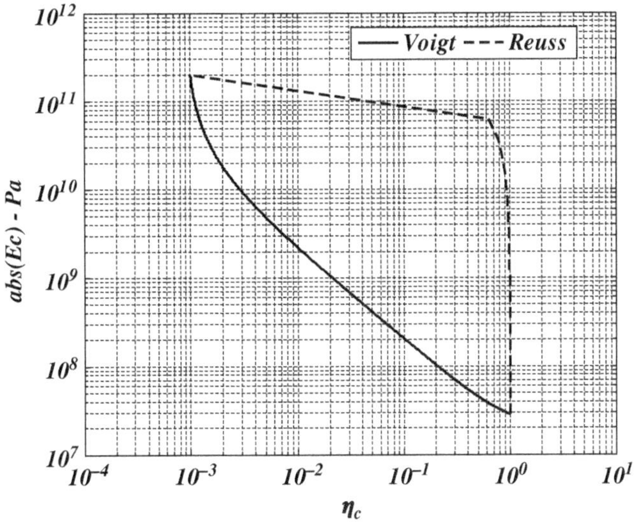

Figure 8.30 The absolute of the complex modulus-loss factor chart for Voigt and Reuss composites.

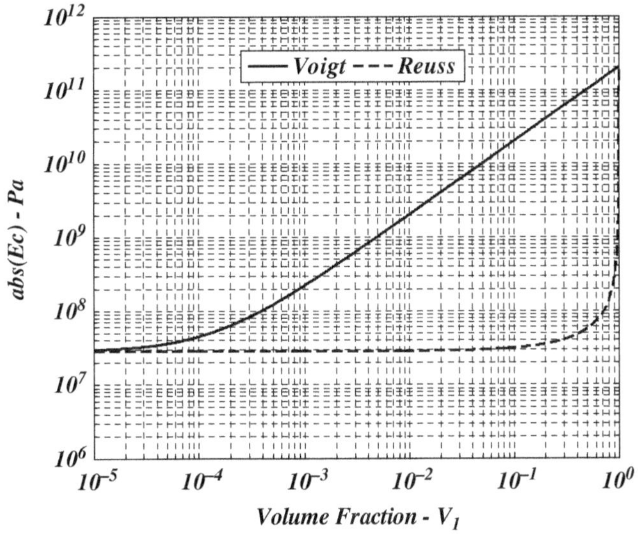

Figure 8.31 The effect of the volume fraction V_1 on the absolute of the complex modulus of Voigt and Reuss composites.

In Figure 8.31 displays the effect of the volume fraction V_1 on the absolute of the complex modulus of Voigt and Reuss composites. The figure indicates that the Reuss structure permits higher complex modulus than the Voigt configuration.

Figure 8.32 suggests that the Reuss structure permits higher loss factors than the Voigt structure whereas the Reuss structure is not as strong as the Voigt configuration.

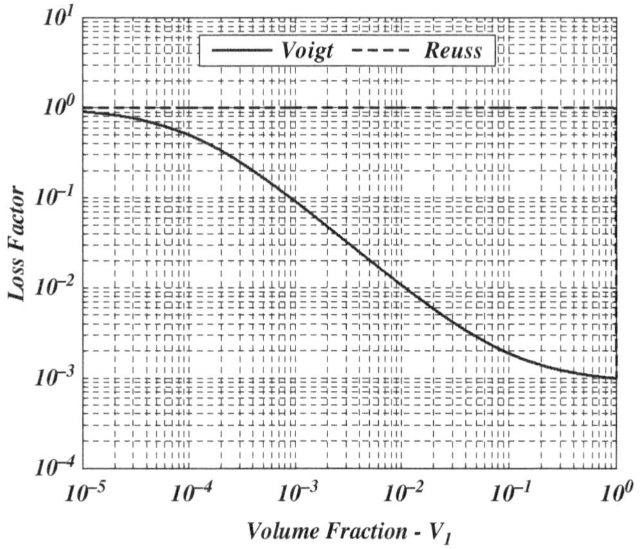

Figure 8.32 The effect of the volume fraction V_1 on the loss factor of Voigt and Reuss composites.

8.4.2 Active Composite Damping Treatments

In this section, a class of APDCs is introduced. This treatment consists of an array of piezoelectric rods obliquely embedded, at an angle θ, across the thickness of a VEM as shown in Figure 8.33. The piezoelectric rods are polarized along their length as defined by the poling direction that coincides with the local coordinate axis 3. It is important to note that the APDC has been investigated by Smith and Auld (1991), Chan and Unsworth (1989), Hayward and Hossack (1990), Shields (1997), Shields et al. (1998), Baz and Tempia (2004), and Arafa and Baz (2000).

The local coordinate system (1, 2, 3) is inclined at an angle θ to the global coordinate system (x, y, z). In the considered APDC treatment, the rods are activated by applying a control voltage along the z-direction and across the electrodes that are deposited parallel to the x–y plane as shown in Figure 8.33.

In this way, the piezoelectric rods induce both shear and compressional strains in the VEM as displayed in Figure 8.34. Hence, the overall damping characteristics of the composite can be enhanced. The APDC treatment may be either bonded to the surface of a structure or embedded within a laminated composite to control its vibration. The damping characteristics of the composite are controlled by the piezoelectric rods, whose longitudinal strains are adjusted in response to the structural vibrations so as to enhance energy dissipation and improve the dynamic behavior of the system.

The main merits of the APDC, with inclined piezo-rods, can best be quantified by considering the combined electromechanical coupling factors due to compression k_{33}^2 and shear k_{15}^2, which are given by:

$$k_{33}^2 = 1 - \frac{s_{33}^D}{s_{33}^E} = \frac{\textit{Compression Energy}}{\textit{Electrical Energy}} \tag{8.67}$$

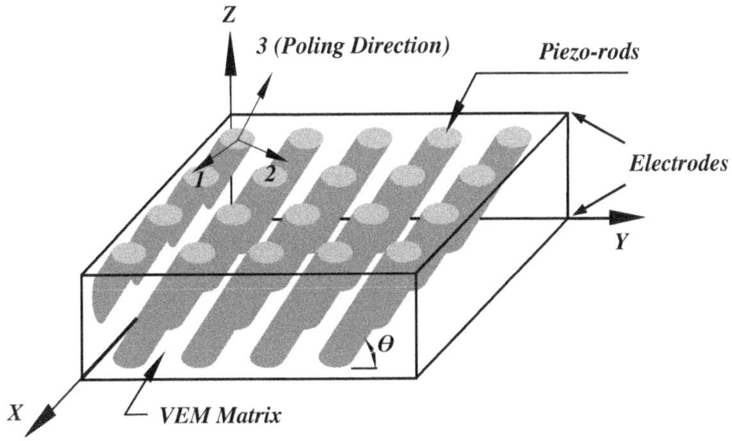

Figure 8.33 The active piezoelectric damping composites (APDC).

and,

$$k_{15}^2 = 1 - \frac{s_{44}^D}{s_{44}^E} = \frac{Shear\ Energy}{Electrical\ Energy} \tag{8.68}$$

where s_{ii}^E and s_{ii}^D are the compliance coefficients under short and open-circuit conditions, respectively. Note that when ii = 33 it denotes compression loading and when ii = 44 it defines shear loading.

Hence, the overall coupling factor k^2 is given by:

$$k^2 = k_{33}^2 + k_{15}^2 = \frac{Compression + shear\ Enegries}{Electrical\ Energy} \tag{8.69}$$

The proof of Eq. (8.67) is given in the Appendix. A similar approach can be adopted to proof Eq. (8.68).

Figure 8.35 shows the effect of the inclination angle of the PZT rods on the compressive, shear, and overall coupling factors for PZT-5H rods embedded inside a soft polyurethane matrix with a volume fraction of 15%. It is evident that the coupling factor due to shear attains a maximum when the inclination angle is zero and vanishes when the angle becomes 90°. The reverse occurs when we consider the coupling factor due to

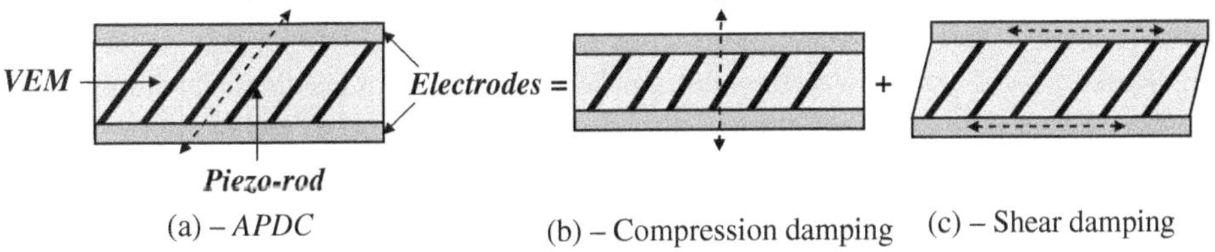

Figure 8.34 The compression and shear components of the Active piezoelectric damping composites (APDCs). (a) APDC. (b) Compression damping. (c) Shear damping.

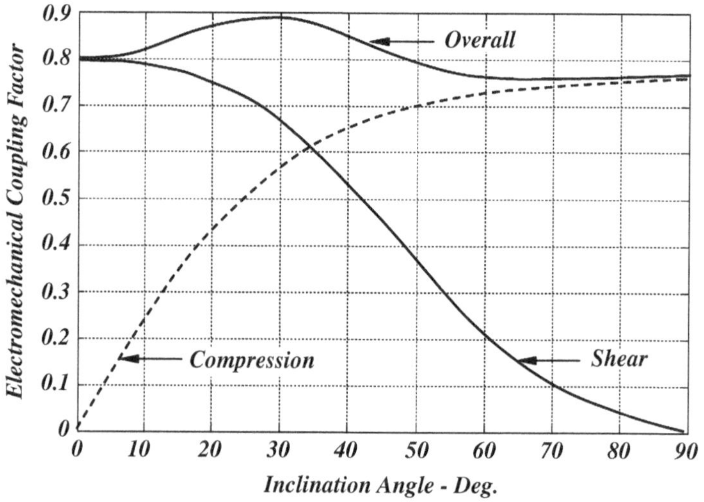

Figure 8.35 Effect of inclination angle on electromechanical coupling factors of APDC with 15% PZT-5H with soft polyurethane matrix.

compression. Hence, the combined effects result in an overall coupling factor which attains a maximum at an angle of nearly 28°. At such an optimal angle, the overall coupling factor of the APDC is 16% higher than that of the compression APDC. One should emphasize here that the coupling factor of 88% for the APDC is almost 98% higher than that of solid ceramics (Reader and Sauter 1993). Hence, one can readily understand the primary importance of the APDC treatments. Note that from Table 8.1, the compression coupling factors k_{33}^2 for solid PZT-5 and PZT-5H are 0.306 and 0.443. Using these ceramics in compression APDC increases the coupling factors k_{33}^2 to 0.715 and 0.77, which conforms with the findings in the literature.

Emphasis is placed here on utilizing the attractive characteristics of the APDC, with inclined piezo-rods, to control the dynamic behavior of beams treated with discrete APDC patches. This effort aims to investigate the effect of the design parameters of the APDC treatments on system response.

8.4.3 Finite Element Modeling of Beam with APDC

8.4.3.1 Model and Main Assumptions

The FEM briefly outlined in this section describes the behavior of elastic beams treated with APDC patches. Figure 8.36 shows a beam treated with a single APDC patch. The beam length is L and it is divided into n finite elements. The APDC patch is bonded to the ith element and is sandwiched between the base beam and an elastic constraining layer. Figure 8.37 depicts the geometry and deflections of the beam/APDC element.

It is assumed, in here, that the transverse displacements of the constraining layer and base beam, at any cross section, are unequal. It is also assumed that the longitudinal stresses in the VEM of the APDC are negligible. The constraining layer, the base beam, and piezoelectric rods are assumed to be elastic and dissipate no energy, whereas the VEM is assumed to be linearly viscoelastic. The polymer matrix is assumed to be

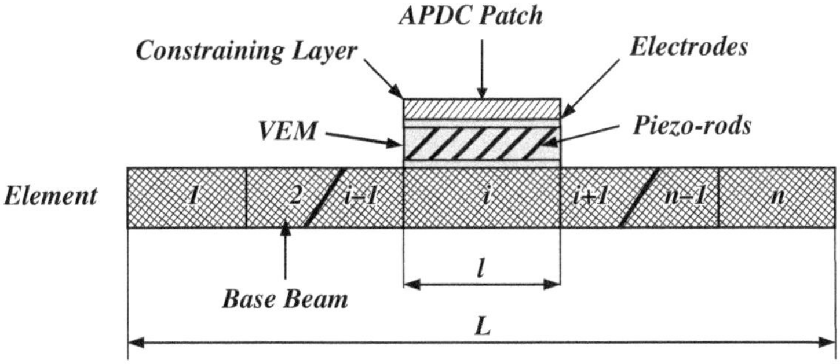

Figure 8.36 Finite element of a beam/APDC composite.

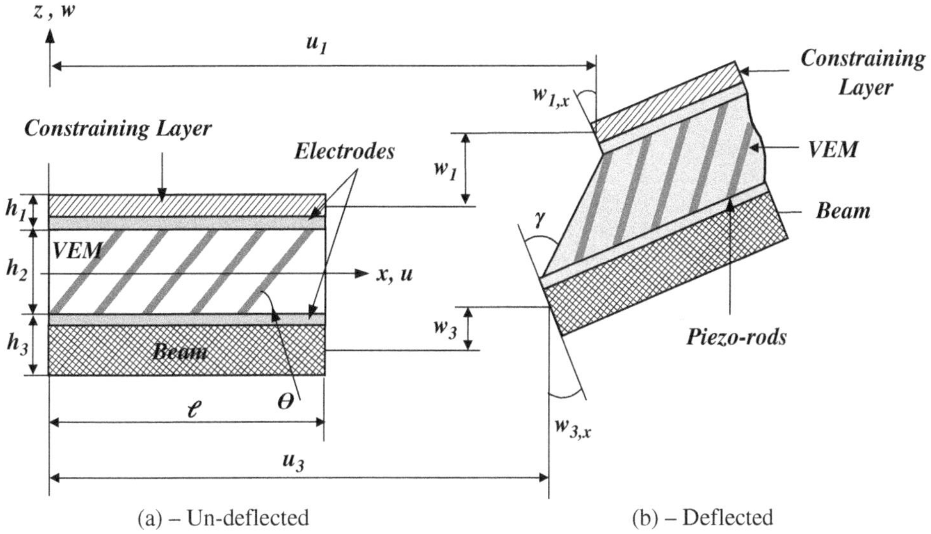

Figure 8.37 Geometry and deflections of beam/APDC. (a) Undeflected. (b) Deflected.

piezoelectrically inactive. All layers of the beam/APDC system are assumed to be perfectly bonded.

8.4.3.2 Kinematics

The basic kinematic variables of the beam/APDC system are the longitudinal, transverse, and angular displacements of the upper and lower beams. From the geometry depicted in Figure 8.36, the shear strain at the mid-plane of the viscoelastic core may be expressed in terms of the beam displacements as,

$$\gamma = \frac{1}{h_2}\left[(u_1 - u_3) + \left(\frac{h_1 + h_2}{2}\right)w_{1,x} + \left(\frac{h_2 + h_3}{2}\right)w_{3,x}\right] \tag{8.70}$$

where the u_i and h_i denote the longitudinal deflection and thickness of the ith layer ($i = 1$ for constraining layer, $i = 2$ for the *VEM* layer, and $i = 3$ for the beam). Also, the subscript

x denotes partial differentiation with respect to x. The transverse displacement w_2 of the viscoelastic core is assumed to have a linear variation across the thickness, or,

$$w_2 = \frac{w_1 + w_3}{2} + \frac{w_1 - w_3}{h_2} z \tag{8.71}$$

where w_i denotes the transverse deflection of the ith layer and z is the distance from the neutral axis.

8.4.3.3 Degrees of Freedom and Shape Functions

The beam/APDC elements considered in this study are one-dimensional elements which are bounded by two nodes. The element consists of an APDC patch, sandwiched between two elastic beam elements. Each node has six degrees of freedom to describe the longitudinal, transverse, and angular displacements of the upper and lower beams. The spatial distributions of the longitudinal displacements u_1, u_3, and transverse displacements w_1, w_3 over any element are assumed to take the form,

$$\begin{aligned} &u_1 = a_1 + a_2 x,\ u_3 = a_3 + a_4 x,\ w_1 = a_5 + a_6 x + a_7 x^2 + a_8 x^3, \\ &w_3 = a_9 + a_{10} x + a_{11} x^2 + a_{12} x^3. \end{aligned} \tag{8.72}$$

where the constants $\{a_1, a_2, \ldots, a_{12}\}$ are determined in terms of the 12 components of the nodal degrees of freedom vector $\{\delta_e\}$ of an element with nodes i and j, where,

$$\{\delta_e\} = \{u_{1i}, u_{3i}, w_{1i}, w_{1,xi}, w_{3i}, w_{3,xi}, u_{1j}, u_{3j}, w_{1j}, w_{1,xj}, w_{3j}, w_{3,xj}\}^T \tag{8.73}$$

Hence, the deflection vector $\{\delta\} = \{u_1, u_3, w_1, w_{1,x}, w_3, w_{3,x}\}^T$ at any point along an element may be expressed in terms of the nodal degrees of freedom vector as follows:

$$\{\delta\} = \{[A_1], [A_2], [A_3], [A_4], [A_5], [A_6]\}^T \{\delta_e\} \tag{8.74}$$

where $[A_1]$, $[A_2]$, $[A_3]$, $[A_4]$, $[A_5]$, and $[A_6]$ are the spatial interpolating vectors corresponding to u_1, u_3, w_1, $w_{1,x}$, w_3, and $w_{3,x}$, respectively.

Note that the normal strain in the APDC is expressed as:

$$\gamma = \frac{1}{h_2}\left[([A_1] - [A_3]) + \left(\frac{h_1 + h_2}{2}\right)[A_4] + \left(\frac{h_2 + h_3}{2}\right)[A_6]\right]\{\delta_e\} = \{B_2\}\{\delta_e\} \tag{8.75}$$

Equations (8.74) and (8.75) are combined to form the strain vector $\{S\}$ of the APDC as follows:

$$\{S\} = \begin{Bmatrix} \varepsilon_{zz} \\ \gamma \end{Bmatrix} = \begin{Bmatrix} B_1 \\ B_2 \end{Bmatrix}\{\delta_e\} = [B]\{\delta_e\} \tag{8.76}$$

For the untreated section of the beam, a one-dimensional, two-node beam element with two degrees of freedom per node is used.

8.4.3.4 System Energies

8.4.3.4.1 Strain Energy (PE)

The total strain energy of the beam/APDC system includes:

8.4.3.4.2 Energy Due to Extension of the Beam (U_{ext})

$$U_{ext} = \frac{1}{2}\sum_{i=1,3} E_i h_i b_i \int_0^l \left(\frac{\partial u_i}{\partial x}\right)^2 dx \tag{8.77}$$

where E_i and b_i denote Young's modulus and the width of the ith layer.

8.4.3.4.3 Energy Due to Bending of the Beam (U_{ben})

$$U_{ben} = \frac{1}{2}\sum_{i=1,3} E_i I_i \int_0^l \left(\frac{\partial^2 w_i}{\partial x^2}\right)^2 dx \tag{8.78}$$

where I_i denotes flexural rigidity of the ith layer.

8.4.3.4.4 Energy of the APDC (Up)

$$U_p = \frac{1}{2}\int_0^l \{T\}^T [Y]^{-1}\{T\} dV \tag{8.79}$$

where $\{T\}$, $[Y]$, and V denote the stress, elasticity tensor, and the volume of the APDC.
Details of the derivation of the constitutive equation of the APDC treatment are given in the Appendix (Baz and Tempia 2004) such that:

$$\{T\} = [Y]\{S\} - \{d\}E_z \tag{8.80}$$

where E_z defines the electric field across the APDC.

Hence, combining Eqs. (8.79) and (8.80) gives:

$$U_p = \frac{1}{2}\int_0^l ([Y]\{S\} - \{d\}E_z)^T [Y]^{-1}([Y]\{S\} - \{d\}E_z) dV \tag{8.81}$$

Equations (8.77), (8.78), and (8.81) yield the following expression of the total strain energy of the beam/APDC system:

$$\begin{aligned} PE &= U_{ext} + U_{ben} + U_p \\ &= \frac{1}{2}\{\delta_e\}^T [K_e]\{\delta_e\} + E_z^2 \int_0^L \{d\}^T [Y]^{-1}\{d\} dx - h_2 b E_2 \int_0^L \{d\}^T [B] dx \{\delta_e\} \end{aligned} \tag{8.82}$$

where $[K_e]$ is the extension and bending stiffness matrix of the element.

8.4.3.4.5 Kinetic Energy (KE)

The KE of the beam/APDC system is given by:

$$KE = \frac{1}{2}\sum_{i=1,3} m_i b_i \int_0^l \left[\left(\frac{\partial u_i}{\partial t}\right)^2 + \left(\frac{\partial w_i}{\partial t}\right)^2\right] dx + \frac{1}{2} m_2 b \int_0^l \left[\left(\frac{\partial w_2}{\partial t}\right)^2\right] dx$$
$$= \frac{1}{2}\{\dot{\delta}_e\}^T [M_e]\{\dot{\delta}_e\} \tag{8.83}$$

where m_i denotes the mass per unit length of the ith layer and $[M_e]$ is the mass matrix of the beam/APDC system.

8.4.3.5 Equations of Motion

Using the Lagrangian equations yields the equation of motion of the beam/APDC system as follows:

$$[M_o]\{\ddot{\delta}_o\} + [K_o]\{\delta_o\} = \{F_o\} \tag{8.84}$$

where $[M_o]$, $[K_o]$, and $\{\delta_o\}$ denote the overall mass matrix, the overall stiffness matrix, and the overall nodal deflection vector. Also,$\{F_o\}$ defines the vector of the control forces and moments generated by the APDC, given by:

$$\{F_o\} = h_2 b \int_0^L [B]^T \{d\} E_z dx \tag{8.85}$$

Equation (8.84) describes the dynamics/control of a single beam/APDC element. Assembly of the corresponding equations for the different elements and applying the proper boundary conditions yields the equations of motion for the entire beam/APDC system. The resulting equations are then utilized to predict the dynamic response of both the open-loop and closed-loop systems.

8.4.3.6 Control Law

A proportional plus derivative controller is considered whereby the electric field is generated such that:

$$\{F_o\} = -\{K_g\}\{\delta_o\} \tag{8.86}$$

where $\{K_g\}$ is the control gain given by:

$$\{K_g\} = (g_p + g_d s)\{\bar{C}\} \tag{8.87}$$

with g_p and g_d are the proportional and derivative control gains. Also, s defines the Laplace complex number and $\{\bar{C}\}$ is the vector denoting the degree(s) of freedom upon which the control effort is based.

Combining Eqs. (8.84) through (8.87) yields:

$$[M_o]\{\ddot{\delta}_o\} + [K_o]\{\delta_o\} = -h_2 b \int_0^L [B]^T \{d\}\left(g_p + g_d \frac{d}{dt}\right)\{\bar{C}\} dx \{\delta_o\}$$
$$= -[K_p]\{\delta_o\} - [C]\{\dot{\delta}_o\} \tag{8.88}$$

where $[K_P]$ and $[C]$ are the augmented effective stiffness and damping matrices that are generated by the feedback control actions. These matrices are given by:

$$[K_P] = h_2 b \int_0^L [B]^T \{d\}\, dx\, g_p \{\bar{C}\} \tag{8.89}$$

and

$$[C] = h_2 b \int_0^L [B]^T \{d\}\, dx\, g_d \{\bar{C}\} \tag{8.90}$$

Example 8.8 Consider the aluminum cantilever beam shown in Figure 8.38. The beam width (b) is 25.4 mm and thickness (h_3) is 2.0 mm. The beam is treated with one APDC patch that is bonded at its root. The volume fraction of the piezoelectric rods is taken to be 30%. Also, the thickness (h_2) of the polymer matrix of the APDC is 3.175 mm and the thickness (h_1) of the aluminum constraining layer is 2.0 mm. The length l of the APDC is 0.254 m. The physical properties of the piezoelectric rods of the APDC are listed in Tables 8.2 and 8.3. Compare the performance of beams treated with APDC with K_g = 0 and 4E8 V m^{-1}.

Solution

Figure 8.39 shows the modes and mode shape of the cantilevered beam/APDC composite.

Figure 8.40 shows the frequency response and the corresponding control voltage of the beam/APDC system for APDC patches with different piezo-rod orientation angles.

A summary of the performance characteristics of the beam/APDC composite is displayed in Figure 8.41. In Figure 8.41a, the effect of the piezo-rod orientation angle on the peak deflections of the beam tip is shown for the first and the second modes. The corresponding control voltages are displayed in Figure 8.41b.

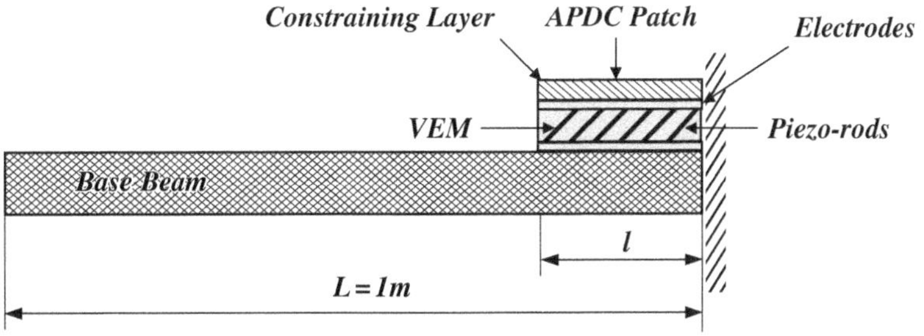

Figure 8.38 Schematic drawing of a cantilevered beam/APDC Composite.

Table 8.2 Properties of the APDC rods PZT-5H*.

Property	c_{11}^E [GN m^{-2}]	c_{12}^E [GN m^{-2}]	c_{13}^E [GN m^{-2}]	c_{33}^E [GN m^{-2}]	c_{23}^E [GN m^{-2}]	c_{22}^E [GN m^{-2}]	e_{33} [C m^{-2}]	e_{15} [C m^{-2}]	e_{13} [C m^{-2}]	$\frac{\varepsilon_{33}^S}{\varepsilon_0}$	$\frac{\varepsilon_{11}^S}{\varepsilon_0}$
Value	151	98	96	124	14	26.5	27	20	25.1	1500	1700

*Source: Smith and Auld (1991).

Table 8.3 Physical properties of the polymer matrix.

Property	Hard polyurethane	Soft polyurethane
C_{11} [GN m^{-2}]	3.0	0.016 67
C_{12} [GN m^{-2}]	2.9	0.016 64

Source: Smith and Auld (1991)

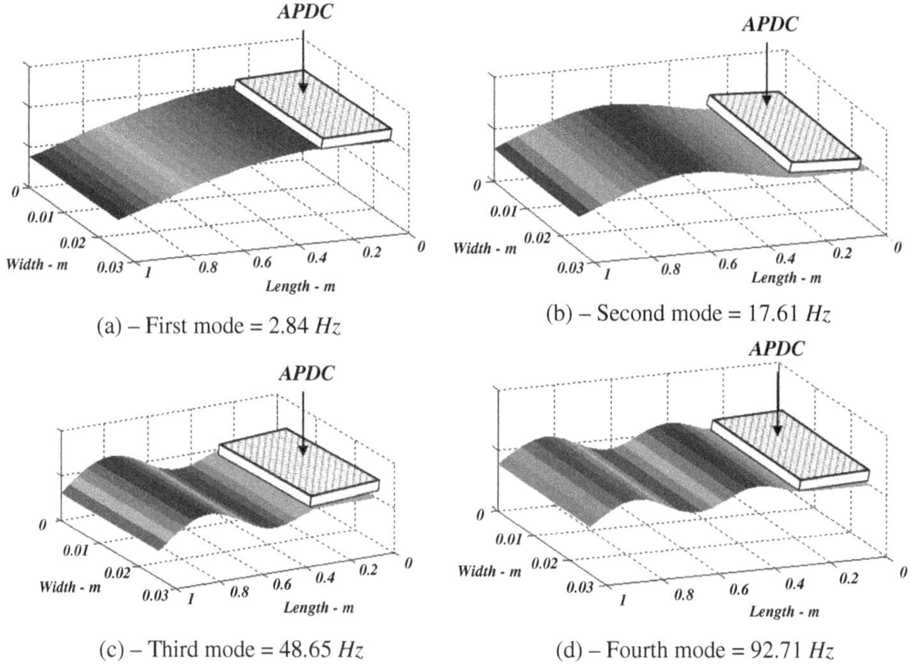

Figure 8.39 Modes and mode shapes of a cantilevered beam/APDC Composite. (a) First mode = 2.84 Hz, (b) second mode = 17.6a Hz, (c) third mode = 48.65 Hz, and (d) fourth mode = 92.71 Hz.

Figures 8.39 and 8.40 clearly indicate that the best control action, as quantified by the lowest amplitude of vibration and control voltage, is achieved when the APDC patch has piezo-rods inclined at an angle of 30°. This conclusion conforms with the characteristics displayed in Figure 8.35 where the overall electromechanical coupling factor of the APDC attains a maximum at a piezo-rod angle of about 28°. At such an orientation angle, the energy dissipation due to the combined compression and shear effects assumes a maximum value.

Example 8.9 Investigate the effect of the piezo-rod orientation angle on the energy dissipation characteristics of the APDC. Assume that the volume fraction of the piezoelectric rods is taken to be 30% and the thickness (h_2) of the polymer matrix of the APDC is 3.175 mm. Assume also that the length l of the APDC is 0.254 m and that the physical properties of the piezoelectric rods of the APDC are listed in Tables 8.2 and 8.3.

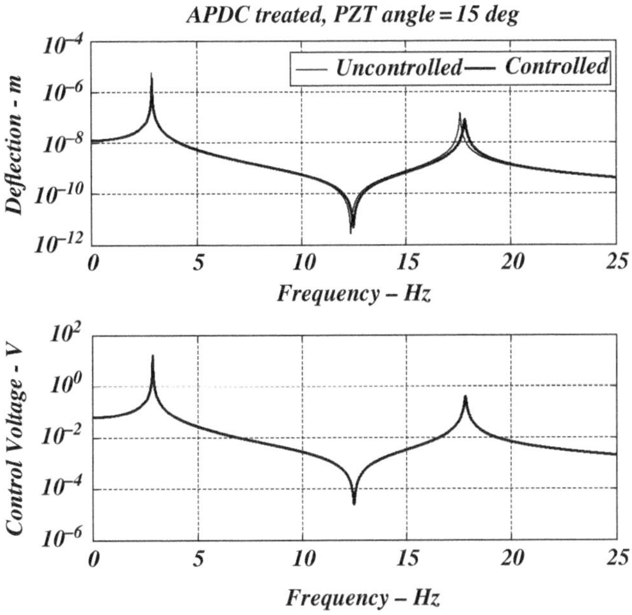

(a)

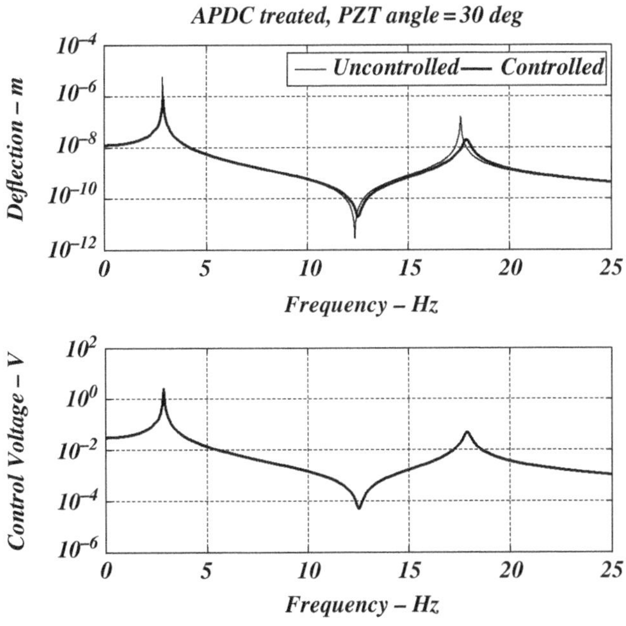

(b)

Figure 8.40 (a–f) Frequency response and control voltage of the beam/APDC composite for piezo-rods with different orientation angles.

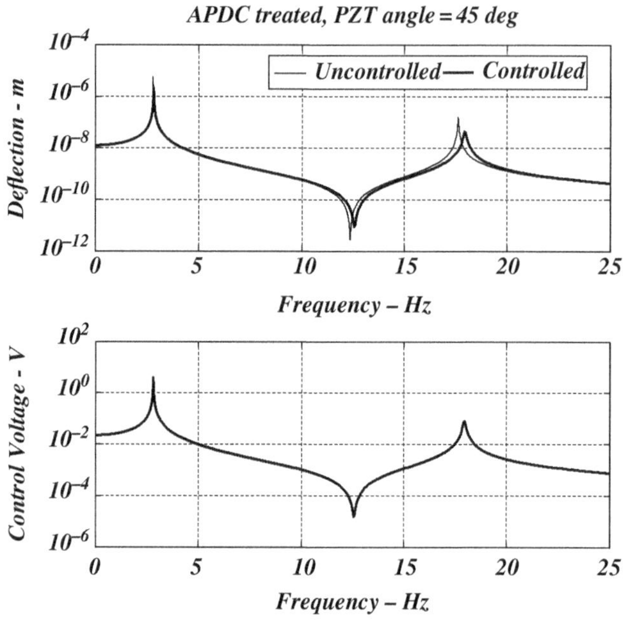

(c)

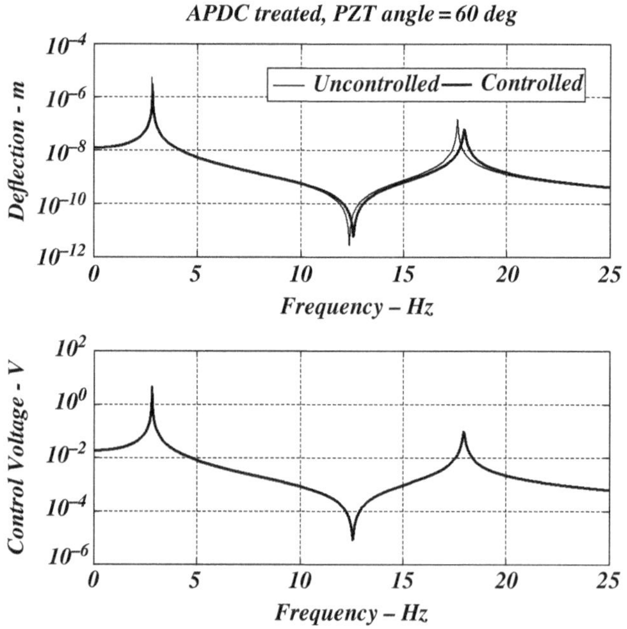

(d)

Figure 8.40 (Continued)

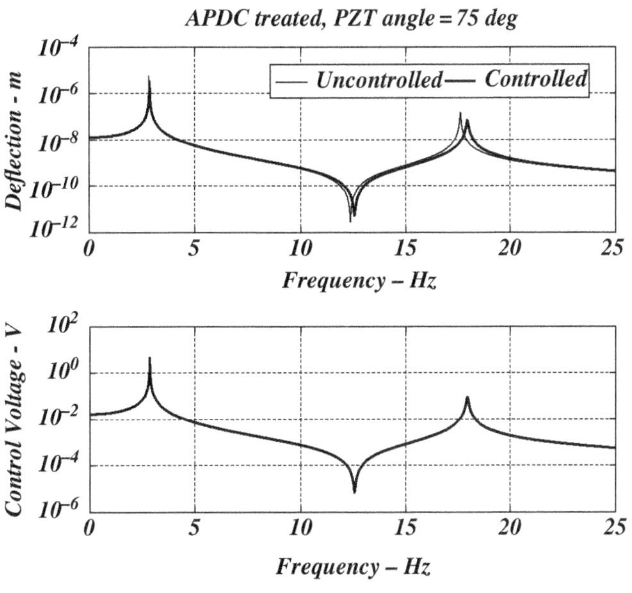

(e)

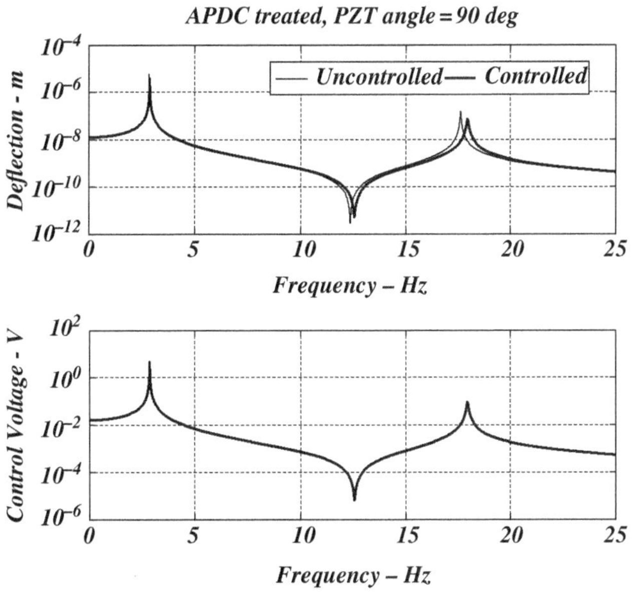

(f)

Figure 8.40 (Continued)

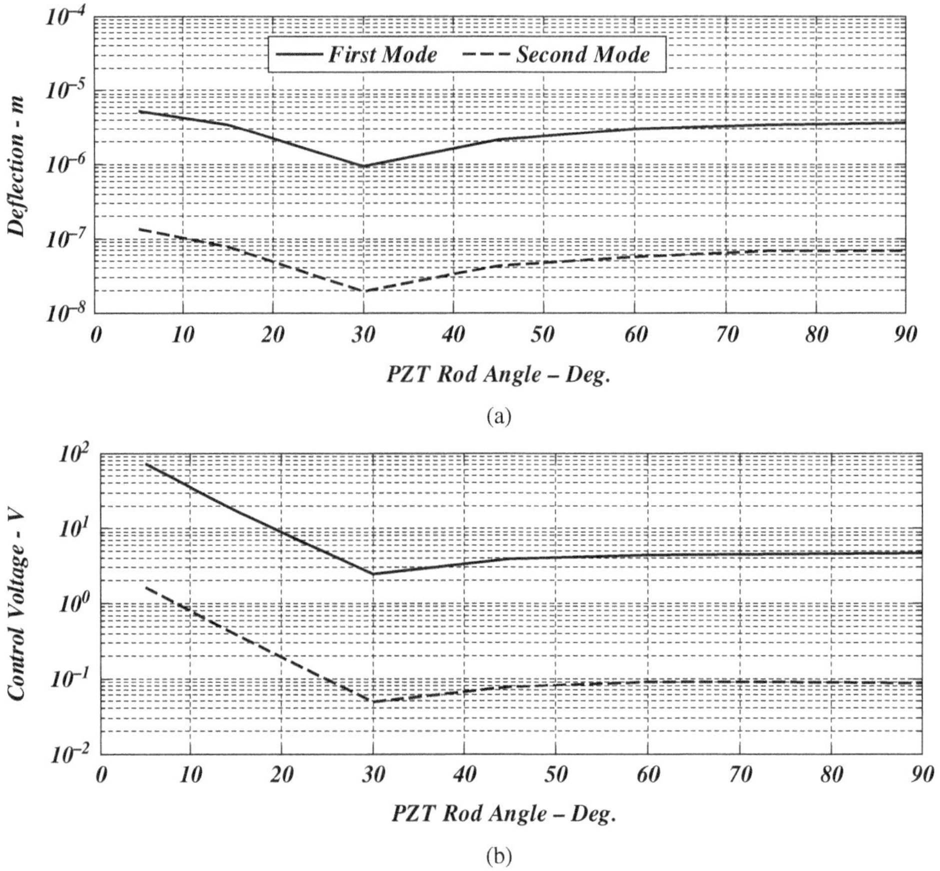

Figure 8.41 The effect of piezo-rod orientation angle on the amplitudes of vibration and control voltage at the (a) first and (b) second modes of vibration of the beam/APDC composite.

Solution

Figure 8.42 displays the finite element of the APDC patch. The patch is fixed at its bottom side and is excited across its thickness by a sinusoidal voltage at a frequency of 1000 Hz.

Figure 8.43 displays the time history of the APDC patch, during a full cycle, when it is excited by a sinusoidal excitation voltage. The figure indicates that the APDC patch undergoes a sequence of shear-compression and reversed shear-expansion during half the voltage cycle followed by a sequence of shear-expansion and reversed shear-compression for the remaining excitation cycle. Such a coordinated sequence of events is the reason behind the enhanced energy dissipation characteristics of the APDC treatment.

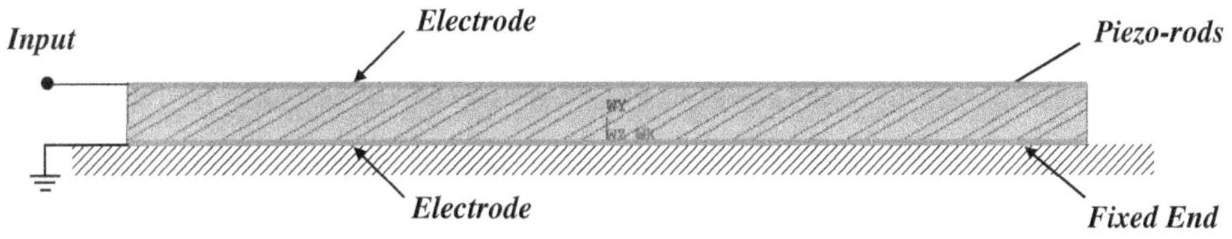

Figure 8.42 Finite element of APDC patch.

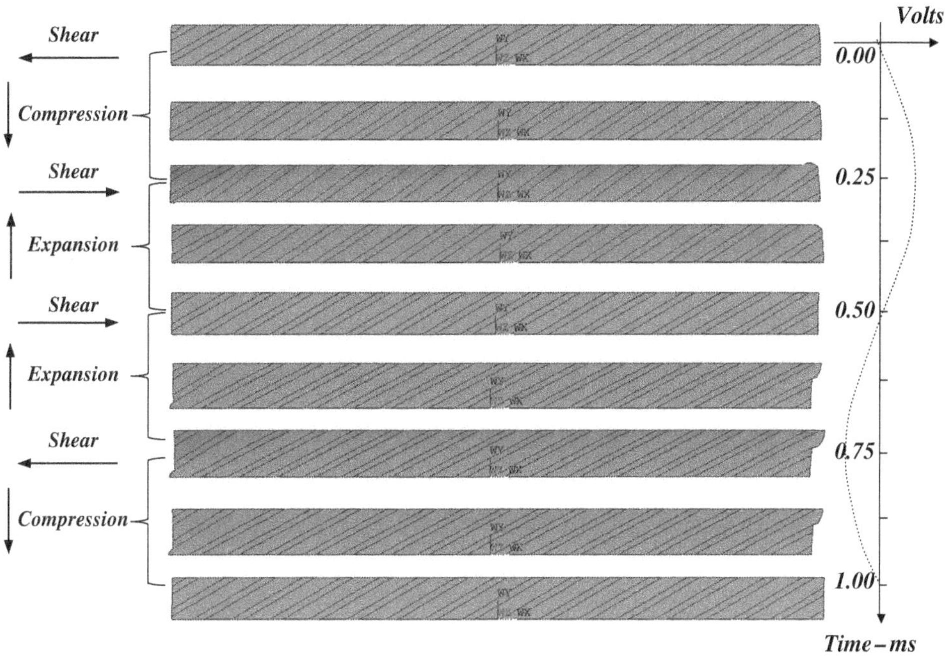

Figure 8.43 Time response of the APDC patch to a sinusoidal input voltage.

Figure 8.44 displays some of the mode shapes of the APDC patch indicating various shear modes and compression modes of the patch.

Figure 8.45 displays the effect of the piezo-rod inclination angle on the hysteresis characteristics of the APDC patches. The figure indicates that there is an optimal inclination angle of 30° where an optimal mix between the shear and compression damping exists. At such an inclination angle, the overall electromechanical coupling factor of the APDC patch attains its maximum value.

The energy dissipation characteristics of the APDC patches, as quantified by the area enclosed by the hysteresis loops are displayed in Figure 8.46. Such characteristics

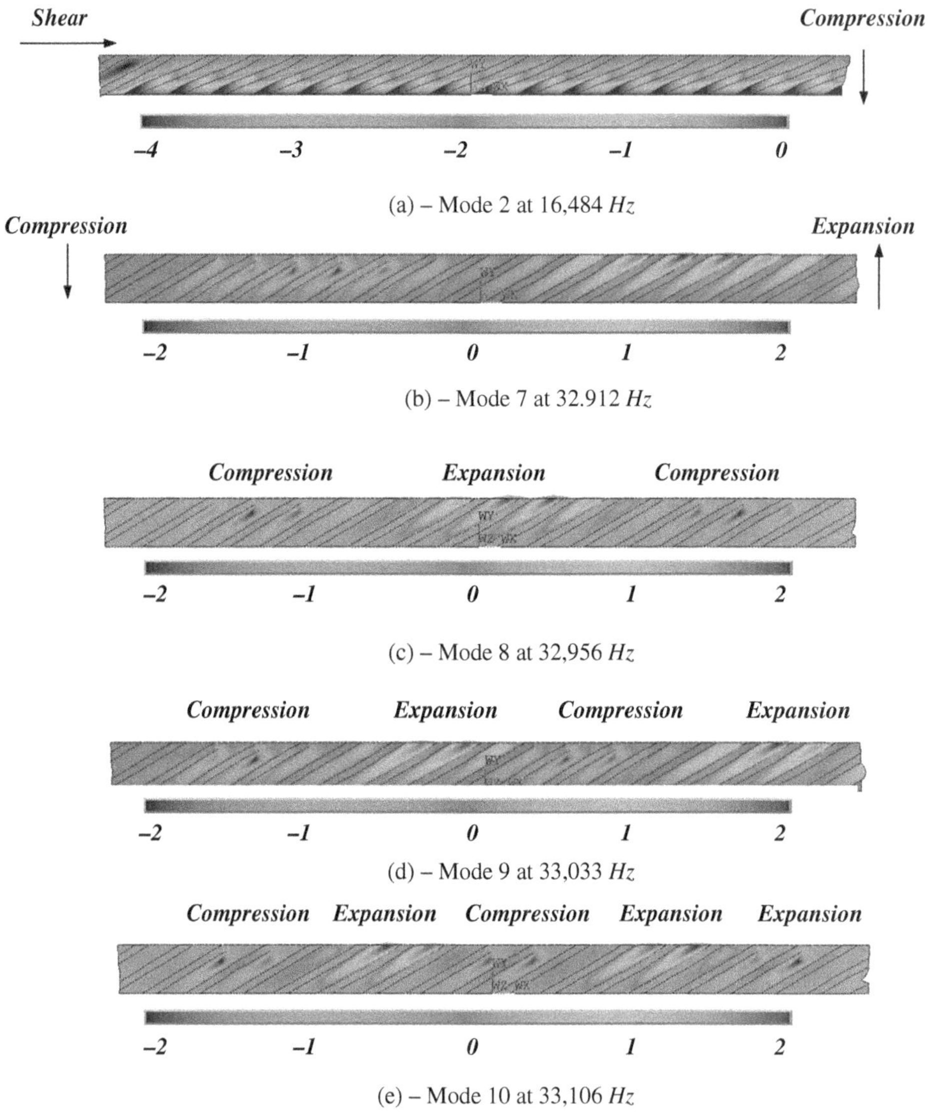

Figure 8.44 Some of the modes and mode shapes of the APDC patch. (a) Mode 2 at 16,484 Hz. (b) Mode 7 at 32.912 Hz. (c) Mode 8 at 32,956 Hz. (d) Mode 9 at 33,033 Hz. (e) Mode 10 at 33,106 Hz.

emphasize the findings that there is an optimal inclination angle of 30° where an optimal mix between the shear and compression damping exists.

8.4.4 Summary

Section 8.4 has presented a finite element analysis of the dynamics and control of elastic beams, partially treated with discrete patches of APDC. The ability of APDC to attenuate flexural vibrations of elastic beams was investigated for patches with different piezo-rod

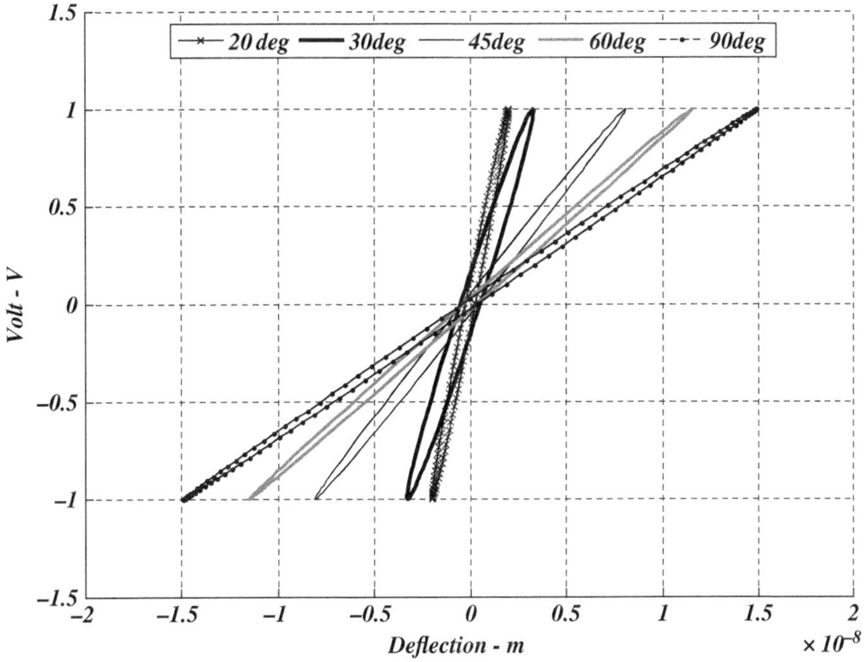

Figure 8.45 The effect of the piezo-rod inclination angle on the hysteresis characteristics the APDC patch.

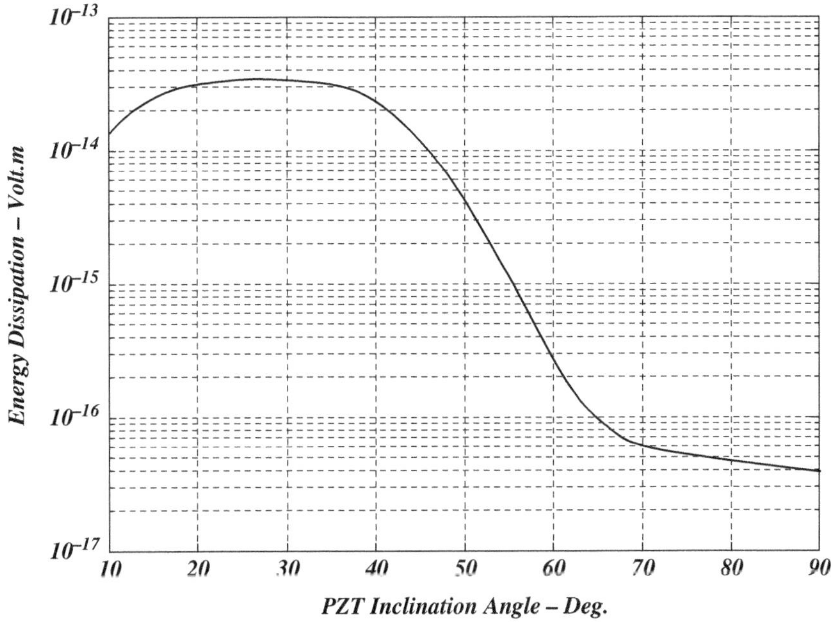

Figure 8.46 The effect of the piezo-rod inclination angle on the energy dissipation characteristics the APDC patch.

orientation angles. It is demonstrated that maximum amplitude attenuation and minimum control voltages are achieved when the piezo-rod inclination angles are about 30°. At such inclination angle, there is an optimal mix between the shear and compression damping effects that contributes to the maximum energy dissipation characteristics of the APDC as a result of optimal overall electromechanical coupling factor of the APDC patch.

8.5 Magnetic Damping Treatments

This section presents the fundamentals of controlling the structural vibration of beams treated with a new class of PMCs. The PMC belong the class of constrained damping treatments, where a viscoelastic core is sandwiched between two layers of stiffer material. In the PMC, the constraining layers are made of permanent magnets whose interaction forces aim to enhancing the damping characteristics of the viscoelastic.

8.5.1 Magnetic Constrained Layer Damping Treatments

The concept of the magnetic constrained layer damping (MCLD) can best be understood by considering first the multi-segment PCLD treatment shown in Figure 8.47. The undeflected configuration of the structure/PCLD system is shown in Figure 8.47a whereas Figure 8.47b shows the deflected configuration under the action of an external bending moment M_e. Due to such loading, shear strains of γ_T and γ_B are induced in the top and bottom viscoelastic layers, respectively. Increasing these shear strains is essential to enhancing the energy dissipation characteristics of the damping treatment. A preferred way for increasing the shear strains is to replace the conventional constraining layers by magnetic constraining layers, which are properly arranged and designed.

Figure 8.48 shows two possible arrangements of the magnetic constraining layers where the inter-layer interaction is either in repulsion as in Figure 8.48a or in attraction as in Figure 8.48b. Figure 8.48a shows that an MCLD, with layers in repulsion, has strains γ_{Tr} and γ_{Br} which are lower than the strains γ_T and γ_B of conventional PCLD treatments. Hence, it is not beneficial to arrange the magnetic layers in repulsion because of their low

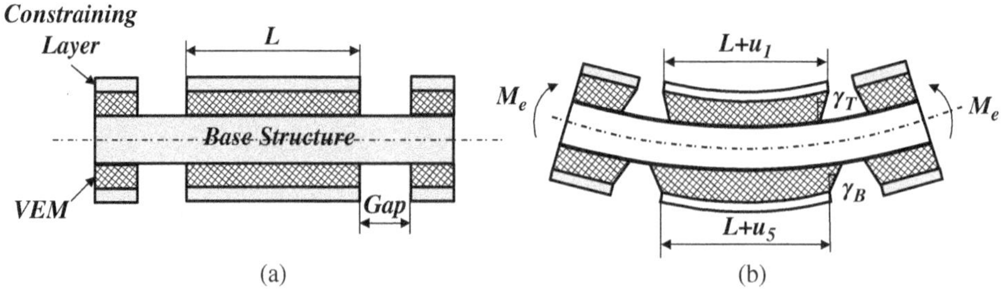

Figure 8.47 Conventional multi-segment passive constrained layer damping: (a) undeflected configuration and (b) deflected configuration.

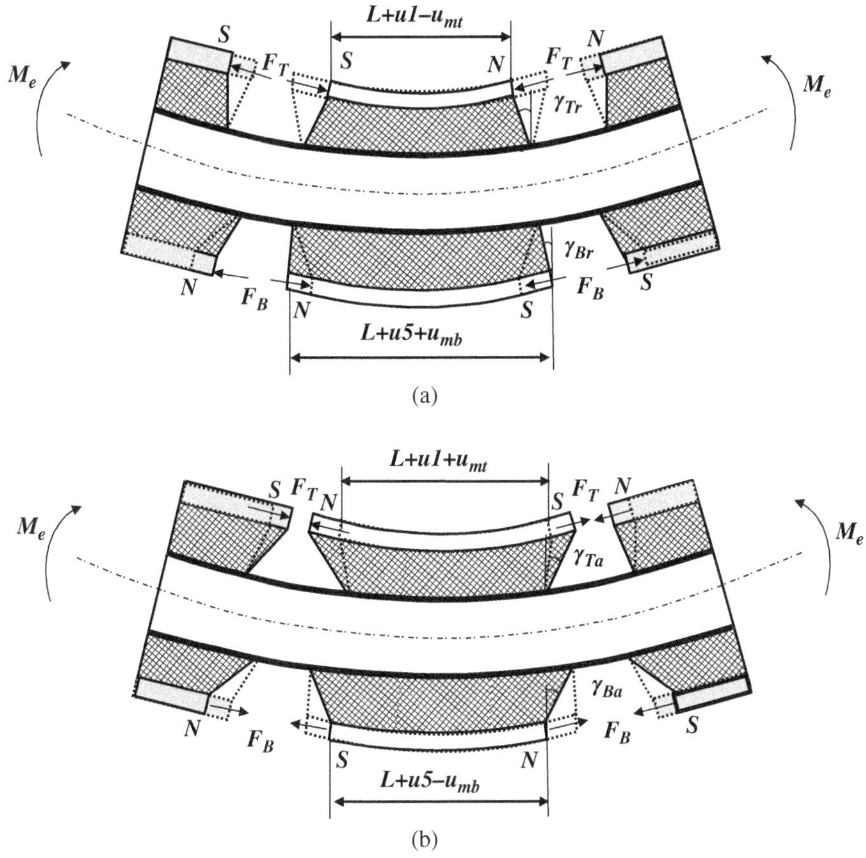

Figure 8.48 Possible arrangements of the magnetic constrained layer damping: (a) layers in repulsion and (b) layers in attraction. (PCLD – dashed lines and MCLD – solid lines).

energy dissipation characteristics. This is in spite of the fact that such an MCLD arrangement induces in-plane tensile loads in the base structure that tend to enhance its stiffness.

However, it is evident that the resulting shear strains γ_{Ta} and γ_{Ba} of the attraction arrangement are much higher than the strains γ_{Tr} and γ_{Br} of the repulsion arrangement. Note also that the strains γ_{Ta} and γ_{Ba} exceed the strains γ_T and γ_B of the PCLD treatment with conventional constraining layers. Therefore, significant improvement of the damping characteristics can be achieved by using MCLD treatments with magnetic layers in attraction. Note that the improved damping characteristics are attributed directly to the fact that the energy dissipation W_d in the viscoelastic core is given by:

$$W_d = -G'\eta A \int_0^L \gamma^2\, dx \tag{8.91}$$

where G' = storage shear modulus, η = loss factor, A = surface area, and L = length of the viscoelastic core. Hence, increasing the shear strain γ results in significant increase in the energy dissipated in the viscoelastic core. Such improved damping exists in the ACLD and APDC treatments but at the expense of the complexities associated with the use of piezo-sensors, piezo-actuators, control circuitry, and/or external energy sources. Hence, the use of the MCLD improves the damping characteristics of conventional PCLD treatments and achieves such improved characteristics in a much simpler and efficient way than the ACLD and APDC treatments.

8.5.2 Analysis of Magnetic Constrained Layer Damping Treatments

The concept and operation of the MCLD can best be understood by the analysis of the simplified configuration shown in Figure 8.49. This simplified configuration is an extension of the Plunkett and Lee configuration for the analysis of PCLD presented in Chapter 6.

The equations of motion and the associated boundary conditions are obtained as follows:

8.5.2.1 Equation of Motion

From the geometry of Figure 8.49, the shear strain γ in the viscoelastic core is given by:

$$\gamma = (u - u_o)/h_1 \tag{8.92}$$

where u and u_o are the longitudinal deflections of the constraining layer and base structure, respectively. Also, h_1 denotes the thickness of the viscoelastic layer.

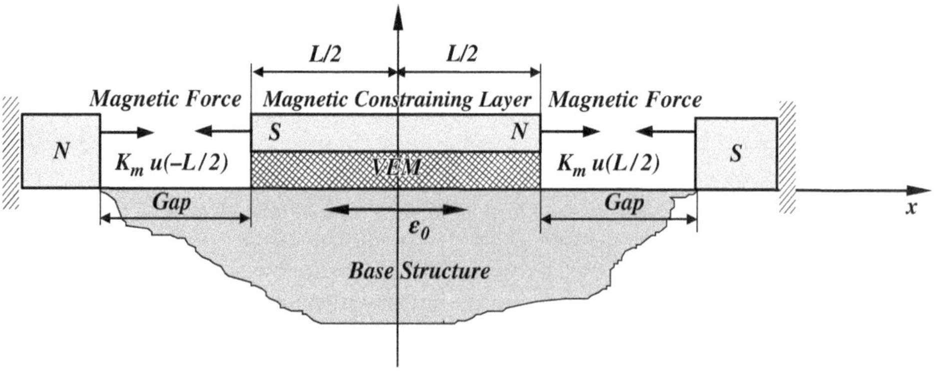

Figure 8.49 Structure with magnetic constrained layer damping.

The potential energy PE associated with the extension of the constraining layer, the shearing of the viscoelastic layer and the energy generated by the interaction between the magnetic constraining layers are:

$$PE = \frac{1}{2} E_2 h_2 b \int_{-L/2}^{L/2} u_{,x}^2 dx + \frac{1}{2} G' h_1 b \int_{-L/2}^{L/2} \gamma^2 dx + \frac{1}{2} K_m u(L/2)^2 + \frac{1}{2} K_m u(-L/2)^2 \tag{8.93}$$

where E_2, h_2, and b denote Young's modulus, thickness, and width of the constraining layer, respectively. Also, the subscript x denotes partial differentiation with respect to x. In Eq. (8.93), it is assumed that the behavior of the viscoelastic layer is linear and described in terms of the complex modulus $G^* = G'(1 + i\eta_g)$ with G', η_g and i denoting the storage shear modulus, loss factor, and $i = \sqrt{-1}$. Note that K_m is the equivalent magnetic stiffness which quantifies the interaction between the magnetic constraining layers. The magnetic energy appears only at the boundaries of the constraining layer.

The KE associated with the longitudinal deflection u is given by:

$$KE = \frac{1}{2} mb \int_{-L/2}^{L/2} \dot{u}^2 dx \tag{8.94}$$

where m is the mass/unit width and unit length of the constraining layer. In Eq. (8.94), the rotary inertia of the viscoelastic layer is neglected and also the inertia of the base structure is not considered.

The work W_d dissipated in the viscoelastic core is given by:

$$W_d = -h_1 b \int_{-L/2}^{L/2} \tau_d \gamma \, dx \tag{8.95}$$

where τ_d is the dissipative shear stress developed by the viscoelastic core. It is given by:

$$\tau_d = \left(G'\eta_g/\omega\right)\dot{\gamma} = \left(G'\eta_g\right)\gamma i \tag{8.96}$$

where ω denotes the excitation frequency of the base structure. In Eq. (8.96), $(G'\eta_g/\omega)$ quantifies the equivalent viscous damping of the VEM (Nashif et al. 1985).

The equations and boundary conditions governing the operation of the *MCLD* system are obtained by applying Hamilton's Principle (Meirovitch 1967):

$$\int_{t_1}^{t_2} \delta(KE - PE) dt + \int_{t_1}^{t_2} \delta(W_d) \, dt = 0 \tag{8.97}$$

where δ (.) denotes the first variation in the quantity inside the parentheses. Also, t denotes time with t_1 and t_2 defining the integration time limits.

From Eqs. (8.92) through (8.97), we have

$$\int_{t_1}^{t_2}\left[-mb\int_{-L/2}^{L/2}\ddot{u}\delta u\,dx - E_2h_2b[u_{,x}\delta u]_{-L/2}^{L/2} + E_2h_2b\int_{-L/2}^{L/2}u_{,xx}\delta u dx\right]dt$$
$$+\int_{t_1}^{t_2}[-K_m u(L/2)\delta u(L/2) - K_m u(-L/2)\delta u(-L/2)]dt \tag{8.98}$$
$$-\int_{t_1}^{t_2}\left[G'b/h_1\int_{-L/2}^{L/2}(u-u_0)\delta u\,dx + G'b/h_1\eta_g i\int_{-L/2}^{L/2}(u-u_0)\delta u dx\right]dt = 0$$

Hence, the resulting equation of the MCLD system is:

$$mh_1/G^*\ddot{u} = B^{*^2}u_{,xx} - (u-u_0) \tag{8.99}$$

with the following boundary conditions:

$$u_{,x} = \mp\bar{K}\,u(x) \quad at \quad x = \pm L/2 \tag{8.100}$$

where $\bar{K} = K_m/(E_2h_2b)$ denotes a dimensionless magnetic stiffness and $B^* = \sqrt{h_1h_2E_2/G^*}$ is a characteristic complex length of the passive treatment. It is important here to note that the second order partial differential Eq. (8.8) describing the PCLD system is the same as that obtained by Plunkett and Lee (1970) if the inertia of the constraining layer is set to zero.

Neglecting the inertia term in Eq. (8.99) gives the following quasi-static equilibrium equation:

$$B^{*2}u_{,xx} - u = -u_0 \quad \text{or} \quad B^{*^2}u_{,xx} - u = -\varepsilon_0 x \tag{8.101}$$

which is subjected to the boundary conditions given by Eq. (8.101).

8.5.2.2 Response of the MCLD Treatment

Equation (8.101) has the following general solution:

$$u = a_1e^{-x/B^*} + a_2e^{x/B^*} + \varepsilon_0 x \tag{8.102}$$

where a_1 and a_2 can be determined from the boundary conditions giving:

$$a_1 = -a_2 = \frac{1}{2}\varepsilon_0\left(\frac{\bar{K}L}{2}+1\right)\Big/\left[\bar{K}\sinh\left(\frac{L}{2B^*}\right) + \frac{1}{B^*}\cosh\left(\frac{L}{2B^*}\right)\right] \tag{8.103}$$

Hence, u and γ are given by

$$u = \varepsilon_0\left[x - \sinh\left(\frac{x}{B^*}\right)\left(\frac{\bar{K}L}{2}+1\right)\Big/\left(\bar{K}\sinh\left(\frac{L}{2B^*}\right) + \frac{1}{B^*}\cosh\left(\frac{L}{2B^*}\right)\right)\right] \tag{8.104}$$

and

$$\gamma = -\frac{\varepsilon_0 L}{h_1}\sinh\left(\frac{x/L}{B^*/L}\right)\left(\frac{\bar{K}L}{2}+1\right)\Big/\left(\bar{K}L\sinh\left(\frac{L}{2B^*}\right) + \frac{L}{B^*}\cosh\left(\frac{L}{2B^*}\right)\right) \tag{8.105}$$

Note that when $\bar{K} = 0$, the magnetic constraining layer reduces to a conventional constraining layer. Hence, Eq. (8.105) reduces to Eq. (6.11).

Example 8.10 Determine the equivalent loss factor of the MCLD as a function of the two parameters $\bar{K}$ and L/B_0.

Plot also the iso-contours of equivalent loss factor of the MCLD in the $\bar{K}-L/B_0$ plane.

Solution

Figure 8.50 displays the damping characteristics of the MCLD plotted on the Equivalent loss factor – L/B_0 plane.

In Figure 8.50a, the loss factor is plotted for values of $\bar{K}$ of – 4, 0, and 4. Note that when $\bar{K} = -4$, the MCLD is arranged with magnets in attraction. Also when $\bar{K} = 4$, the MCLD is provided with magnets in repulsion. At $\bar{K} = 0$, the MCLD acts as a conventional PCLD.

The figure indicates clearly that the MCLD with magnets in attraction has the highest equivalent loss factor among the other considered two arrangements (MCLD with magnets in repulsion and PCLD). In particular, MCLD in attraction generates a maximum loss factor of 9 at $L/B_0 = 3.8$ whereas the MCLD in repulsion produces a peak loss factor of at 2.5 at $L/B_0 = 5$. These loss factors are much higher than that produced by the optimal PCLD treatment which is 1.09 at $L/B_0 = 3.26$.

In effect, Figure 8.50a indicates that the MCLD whether in attraction or repulsion is more effective than the PCLD treatment. However, for peak loss factor, the MCLD requires longer treatment.

But, If the length of the treatment is maintained the same as that of the optimal PCLD treatment, the MCLD in attraction still outperforms the PCLD and generates loss factor of 7.32, that is, eight times that of the PCLD. Furthermore, the MCLD in repulsion with $L/B_0 = 3.26$ also produces 2.1 times that of the optimal PCLD.

Similar characteristics, but with more dramatic enhancement for the damping factor, are obtained when $\bar{K}$ of – 10, 0, and 10. Note that optimal loss factor becomes 48 when the MCLD has $\bar{K} = -10$ and 4 when the MCLD has $\bar{K} = 10$.

Figure 8.51 displays the iso-contours of the loss factor of the MCLD in the $L/B_0-\bar{K}$ plane and emphasizes the effectiveness of the MCLD in attraction as compared to the MCLD in repulsion or the PCLD treatment.

8.5.3 Passive Magnetic Composites

In this section, another class of PMC treatments is presented where damping characteristics are enhanced without the need for sensors, actuators, associated circuitry as well as any external energy sources.

The PMC consists in a constrained layer treatment type, where the constraining layers are made of strips whose magnetic properties enhance the compression and shear in the viscoelastic core. The resulting system acts similarly to a Den Hartog absorber where multi-modeled vibrations can be damped out. The effectiveness of the PMC treatments in enhancing the damping of the base structure is investigated and demonstrated via several numerical examples.

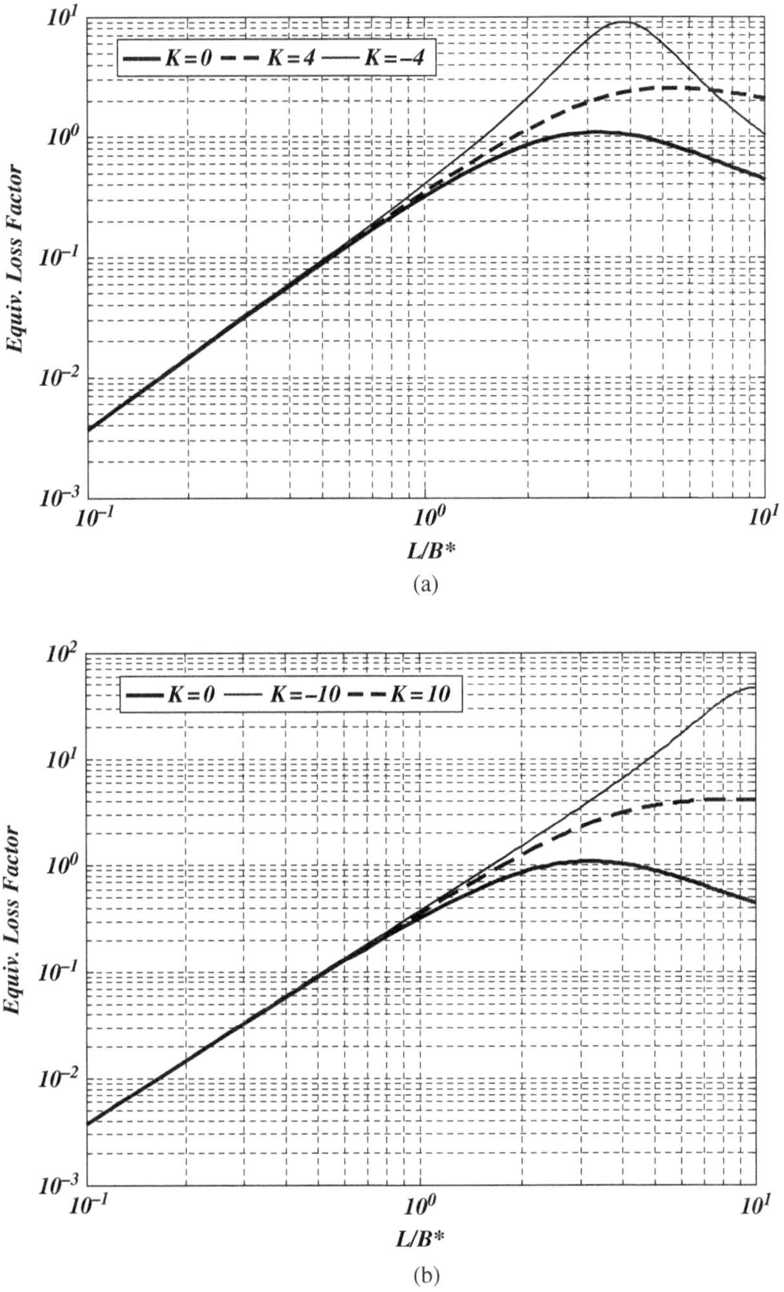

Figure 8.50 The equivalent loss factor of the magnetic constrained layer damping as a function of (a) L/B_0 (b) and $\bar{K}$.

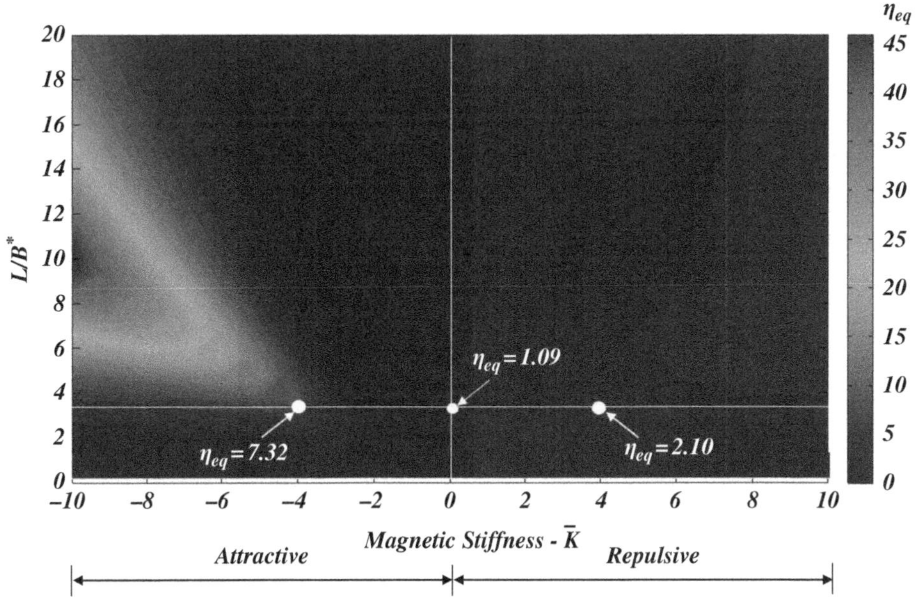

Figure 8.51 The iso-contours of the equivalent loss factor of the magnetic constrained layer damping in the $L/B_0 - \bar{K}$ plane.

8.5.3.1 Concept of Passive Magnetic Composite Treatment

The concept of PMC can be best understood by considering its simplified representation shown in Figure 8.52. In Figure 8.52a, the undeflected configuration of the beam/PMC system is presented. This structure can be viewed as a simple Den Hartog absorber, consisting ofa mass, spring, and viscous damper mounted on the base structure, as shown in Figure 8.52b. The PMC treatment is a sandwich damping system, where the damping is obtained from the combined compression and shear deformation of the viscoelastic layer. Increasing these deformations is essential to enhancing the energy dissipation characteristics of the damping treatment. Such an increase is achieved by replacing the conventional constraining layers by magnetic constraining layers arranged in attraction as shown in Figure 8.52a.

The PMC treatments can be employed to control the vibration of beams and plates as shown in Figure 8.51c and 8.52d.

8.5.3.2 Finite Element Modeling of Beams with PMC Treatment

8.5.3.2.1 Determination of the Magnetic Properties of the Region Surrounding the Constraining Layers

The magnetic properties of the region surrounding the constraining layers are obtained by the numerical computation of the magnetic vector potential (Moon 1984).

8.5.3.2.1.1 Mathematical Formulation In a static magnetic problem, the relations between the magnetic field **H**, the magnetic induction **B**, the current density **I**, and the magnetization **M** are given by Maxwell's equations (Brown 1966 and Griffith 1995):

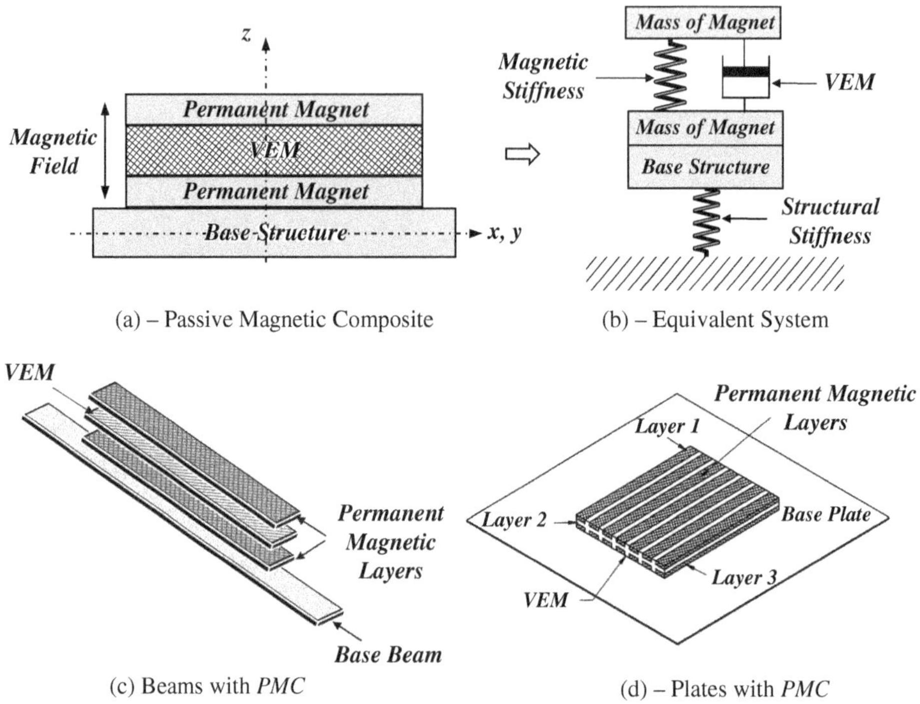

Figure 8.52 Concept of the passive magnetic composite (PMC) treatments of beams and plates. (a) Passive magnetic composite. (b) Equivalent system. (c) Beams with PMC. (d) Plates with PMC.

$$curl\,\mathbf{H} = \mathrm{I}, \tag{8.106}$$

$$div\,\mathbf{B} = 0, \tag{8.107}$$

and

$$\mathbf{B} = \mu_0(\mathbf{H} + \mathbf{M}). \tag{8.108}$$

where μ_0 is the permeability in vacuum. The relation between **M** and **H** in regions occupied by isotropic materials can be expressed as:

$$\mathbf{M} = (\mu_r - 1)\cdot\mathbf{H} \tag{8.109}$$

where μ_r is the relative permeability of the material. Permanent magnets are generally made of anisotropic material and therefore the magnetization **M** is given by:

$$\mathbf{M} = M(B_m)\cdot\mathbf{m} \tag{8.110}$$

m being the direction of the magnetization and B_m the induction in that direction.

The magnetic properties of a region can be determined by minimizing the magnetic energy (Kamminga 1975), defined as:

$$W = \int_V \left(\int_0^{\mathbf{B}} \mathbf{H}\cdot d\mathbf{B} \right) dV \tag{8.111}$$

with V denoting the volume of the region where the magnetic properties have to be evaluated. Equation (8.111) can be rewritten in terms of the magnetic vector potential **A**, which is defined as:

$$\mathbf{B} = curl\,\mathbf{A} \tag{8.112}$$

For modeling beams, lying in the x–y-plane, we will assume that the vector potential **A** is perpendicular to the plane of the region, that is:

$$\mathbf{A} = A(x,y)\cdot\mathbf{k} \tag{8.113}$$

where **k** is a unit vector in the z-direction. From Eqs. (8.112) and (8.113), the magnetic induction **B** becomes:

$$\begin{aligned}\mathbf{B} &= B_x(x,y)\cdot\mathbf{i} + B_y(x,y)\cdot\mathbf{j}\\ &= -\frac{\partial}{\partial y}(A(x,y))\cdot\mathbf{i} + \frac{\partial}{\partial y}(A(x,y))\cdot\mathbf{j}\end{aligned} \tag{8.114}$$

and the magnetization **M** is:

$$\mathbf{M}(B_m) = M\left(B_x\cos\alpha_m + B_y\sin\alpha_m\right)\cdot\mathbf{m} \tag{8.115}$$

where α_m is the angle between the direction of magnetization $\boldsymbol{m}$ and the x-axis.

As there are no current windings, then the current density **I** in Eq. (8.106) will be equal to zero, and the magnetic field is generated only by the presence of the permanent magnets. Hence, the total magnetic energy is equal to the sum of the energy W_{an} generated by the anisotropic material in the considered domain, that is, the permanent magnets and of the energy W_{is} isotropic of the non-magnetic material, that is, aluminum, viscoelastic material, and air. Expressions of W_{an} and W_{is} are given by:

$$\begin{aligned}W_{an} &= b\int_G\left(\frac{1}{\mu_0}\int_0^{\mathbf{B}}\mathbf{B}\cdot d\mathbf{B} - \int_0^{\mathbf{B}}\mathbf{M}\cdot d\mathbf{B}\right)dxdy\\ &= b\int_G\left(\frac{1}{2\mu_0}\left[\left(\frac{\partial A}{\partial x}\right)^2 + \left(\frac{\partial A}{\partial y}\right)^2\right] - \int_0^{\mathbf{B_m}}\mathbf{M}(\xi)\cdot d\xi\right)dxdy,\end{aligned} \tag{8.116}$$

and

$$W_{is} = b\int_G\left(\int_0^{\mathbf{B}}\frac{\xi}{\mu_0\mu_r(\xi)}\cdot d\xi\right)dxdy \tag{8.117}$$

where b is the off plane region width and G is the area of the considered region. The relative permeability μ_r is considered approximately constant and equal to 1 for all the non-magnetic materials in the region.

The total magnetic energy of the domain will obviously be:

$$W = W_{is} + W_{an} \tag{8.118}$$

8.5.3.2.1.2 Finite Element Formulation Equations (8.116) and (8.117) express the magnetic energy as a function of the magnetic potential **A** (or magnetic induction **B**). The magnetic properties of the region G will be determined by finding the values of **A** giving W a stationary value (Kamminga 1975). To this purpose, the region is divided into triangular finite elements where the behavior of the potential **A** is assumed to be linear, as follows:

$$\mathbf{A}^e = a^e + b^e \cdot u + c^e \cdot v \tag{8.119}$$

where $[u\ v]$ represent the local coordinates on the eth element. The global coordinates $[x, y]$ can be expressed in terms of local coordinates and as functions of the nodal coordinates $[x_i, y_i]$ as:

$$[x y] = \sum_{i=1}^{3} \beta_i(u,v) \cdot [x_i y_i] \tag{8.120}$$

where $\beta_i(u, v)$ are the functions defining the isoparametric transformation (Bathe 1996).

The isoparametric transformation in Eq. (8.120) can be also used to express the potential $\mathbf{A}^e$ in terms of the vertex potentials $\mathbf{A}_i^e$ ($i = 1, 2, 3$):

$$\mathbf{A}^e = \sum_{i=1}^{3} \beta_i(u,v) \cdot A_i = \{\beta_1\ \ \beta_2\ \ \beta_3\} \cdot \begin{Bmatrix} A_1^e \\ A_2^e \\ A_3^e \end{Bmatrix} = \boldsymbol{\beta} \cdot \{\mathbf{A}^e\} \tag{8.121}$$

From the definition of the vector potential **A** given in Eq. (8.121) and using the isoparametric transformation, the magnetic induction **B** in each element can be also expressed as a function of the potentials in the nodes (Silvester and Ferrari 1996):

$$\mathbf{B}^e = \sum_{i=1}^{3} grad\beta_i \cdot A_i = \mathbf{J}^{-1} \cdot \mathbf{D} \cdot \begin{bmatrix} A_1^e \\ A_2^e \\ A_3^e \end{bmatrix} = \mathbf{J}^{-1} \cdot \mathbf{D} \cdot \{\mathbf{A}^e\} \tag{8.122}$$

where **J** is the Jacobian matrix of the transformation, defined in two-dimensional Cartesian coordinates as:

$$\mathbf{J} = \begin{bmatrix} \dfrac{\partial x}{\partial u} & \dfrac{\partial y}{\partial u} \\ \dfrac{\partial x}{\partial v} & \dfrac{\partial y}{\partial v} \end{bmatrix} \tag{8.123}$$

and the matrix **D** is given by:

$$\mathbf{D} = \begin{bmatrix} \dfrac{\partial \beta_1}{\partial u} & \dfrac{\partial \beta_2}{\partial u} & \dfrac{\partial \beta_3}{\partial u} \\ \dfrac{\partial \beta_1}{\partial v} & \dfrac{\partial \beta_2}{\partial v} & \dfrac{\partial \beta_3}{\partial v} \end{bmatrix} \tag{8.124}$$

Equation (8.118) can be used to express the magnetic energy W as a function of the magnetic potentials in the nodes of the meshed region. In particular, the contribution to the magnetic energy from eth element of anisotropic material is found to be:

$$\Delta W_{an}^{e} = \frac{1}{2}\cdot\{\mathbf{A}^{e}\}^{T}\cdot\mathbf{S}_{an}^{e}\cdot\{\mathbf{A}^{e}\} - \{\mathbf{Mg}^{e}\}^{T}\cdot\{\mathbf{A}^{e}\} \tag{8.125}$$

where has the following expression:

$$\mathbf{S}_{an}^{e} = \frac{1}{\mu_0}\cdot b\cdot\Delta^{e}\cdot\mathbf{D}^{T}\cdot\mathbf{J}^{-1^{T}}\cdot\mathbf{J}^{-1}\cdot\mathbf{D} \tag{8.126}$$

By analogy with structural finite element analysis, this can be regarded as the magnetic stiffness matrix of the element.

The global magnetization vector $\{\mathbf{Mg^e}\}$ of the element is defined as follows:

$$\{\mathbf{Mg}^{e}\} = b\cdot\Delta^{e}\cdot M(B_m)\cdot[\cos\alpha_m \quad \sin\alpha_m]\cdot\mathbf{J}^{-1}\cdot\mathbf{D} \tag{8.127}$$

In Eqs. (8.126) and (8.127), Δ^{e} is the area of the eth element and B_m is:

$$B_m = [\cos\alpha_m \quad \sin\alpha_m]\cdot\mathbf{J}^{-1}\cdot\mathbf{D}\cdot\{\mathbf{A}^{e}\} \tag{8.128}$$

The contribution to the magnetic energy W from an element of the isotropic non-magnetic materials can be expressed in a similar way.

The approximate value of the magnetic energy over the region G will be given by the sum of all the contributions from the elements belonging to anisotropic and isotropic materials:

$$W \cong \sum_{k} \Delta W_{is}^{k} + \sum_{l} \Delta W_{an}^{l} \tag{8.129}$$

where k and l indicate the number of isotropic and anisotropic elements, respectively. The summation in Eq. (8.129) requires the assembly of all the elements magnetic stiffness matrices and of the nodal vector potentials, in a procedure typical of finite element analysis. The resulting expression of the energy in matrix form will be:

$$W = \frac{1}{2}\cdot\{\mathbf{A}\}^{T}\cdot(\mathbf{S}_{an} + \mathbf{S}_{is})\cdot\{\mathbf{A}\} - \{\mathbf{Mg}\}^{T}\cdot\{\mathbf{A}\} \tag{8.130}$$

where $\mathbf{S}_{an}$ and $\mathbf{S}_{is}$ represent, respectively, the global magnetic stiffness matrices for anisotropic and isotropic material. In particular $\{\mathbf{Mg}\}$ will have zero components in positions corresponding to elements belonging to non-magnetic materials. The values of the magnetic potential minimizing the magnetic energy can be found by partially differentiating Eq. (8.130) with respect to the vertex potentials:

$$\frac{\partial W}{\partial\{\mathbf{A}\}} = 0 \tag{8.131}$$

The resulting set of algebraic equations will therefore be:

$$(\mathbf{S}_{an} + \mathbf{S}_{is})\cdot\{\mathbf{A}\} - \{\mathbf{Mg}\} = 0 \tag{8.132}$$

Equation (8.132) is generally nonlinear in the potentials $\{\mathbf{A}\}$ because of the nonlinear magnetization characteristic of permanent magnets and of the relative permeability of isotropic materials (as given by Eqs. (8.126) and (8.127)). If appropriate constant values for the magnetization and the relative permeability are assumed, the set of equations becomes linear and an approximate solution can be easily found.

8.5.3.2.1.3 Determination of the Magnetic Force Once the potentials in the nodes of the region have been found, the value of the magnetic induction in each element can be obtained using Eq. (8.122).

The potentials in the nodes of the meshed region can be also used for the determination of the magnetic forces acting on the magnets of the analyzed region. In particular, the computation of the magnetic forces is based on the local application of the Virtual Work Principle. In that regard, the magnetic force in the s direction acting on a movable part of the considered region is equal to the derivative of the magnetic energy with respect to s (Coulomb and Meunir 1984):

$$F_s = -\frac{\partial W}{\partial s} = -\frac{\partial}{\partial s}\left(\int_V \left(\int_0^{\mathbf{B}} \mathbf{H}\cdot d\mathbf{B}\right) dV\right) \tag{8.133}$$

where s measures the virtual displacement of the movable part in the s direction. During this virtual translation the magnetic potential is assumed to remain constant. The evaluation of the force as a partial derivative of the magnetic energy is given in Appendix 8.B.

8.5.3.2.2 Finite Element Model of the PMC Treatment

The beam considered is composed of a base beam, two magnetized constraining layers and a sandwiched viscoelastic core is shown in Figure 8.52c. A FEM is developed to describe the interaction between the dynamics of the plane beam, viscoelastic layer, and constraining layers. The model in particular has been formulated in order to take into account the dissipation effect due to the compression and shear of the viscoelastic layer. Both full and partial treatments for the beam have been considered.

The structural finite element model presented hereafter is coupled with the magnetic finite element model of the region surrounding the constraining layers.

8.5.3.2.2.1 Geometry and Basic Kinematic Assumptions It is assumed that the shear strains in the constraining layers and in the beam as well as the longitudinal stresses in the viscoelastic cores are negligible. A linear viscoelastic behavior for the viscoelastic core is assumed, while the constraining layers and the beam are assumed to be elastic. The magnetic layer 3 and the base beam are considered to be perfectly bonded together forming a single equivalent layer. Hence, the original four-layer beam/*PMC* assembly can be then studied as a three-layer beam. The transverse displacement of the constraining magnetic layer 1 is assumed to be different than the transverse displacement of the base beam/magnet layer, so that the effect of the compression of the viscoelastic core may be considered. Finally, both the transverse and longitudinal displacements in the viscoelastic core are assumed to vary linearly along the thickness.

From the geometry of Figure 8.53 the shear strain in the viscoelastic layer can be expressed as:

$$\gamma = \frac{1}{h_2}\left[u_1 - u_3 + \frac{d}{2}\cdot(w_{,x1} + w_{,x3})\right] \tag{8.134}$$

where the subscripts denote the spatial derivative, u_1 and u_3 are the longitudinal displacements of the constraining layer and of the base beam and where the slope of the

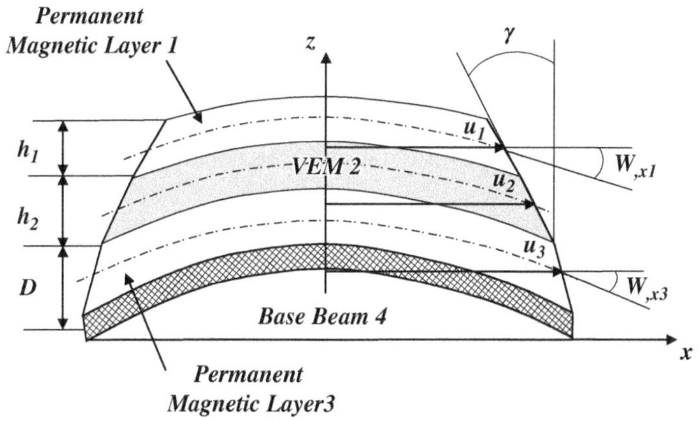

Figure 8.53 The passive magnetic composite (PMC) treatments of beams (u1. magnetic layer, u2. viscoelastic layer, u3. magnetic layer).

deflection line of the viscoelastic core has been expressed as the average of the slopes $w_{,x1}$ and $w_{,x3}$ of the constraining layers. The parameter d is given by:

$$d = h_2 + \frac{1}{2} h_1 + D \tag{8.135}$$

where D is the distance from the neutral axis of the base beam/magnet layer to the viscoelastic layer interface, h_1 denotes the thickness of the constraining layer and h_2 is the thickness of the viscoelastic layer.

The longitudinal deflection u_2 of the viscoelastic core can be also given in terms of the longitudinal displacements u_1 and u_3 and of the slopes of the deflection line of the constraining layers $w_{,x1}$ and $w_{,x3}$:

$$u_2 = \frac{1}{2}\left[u_1 + u_3 + \left(\frac{h_1}{2} - D\right)\cdot\left(\frac{w_{,x1} + w_{,x3}}{2}\right)\right] \tag{8.136}$$

8.5.3.2.2.2 Degrees of Freedom and Shape Functions The PMC elements considered are one-dimensional elements, bounded by two nodal points. Each node has six degrees of freedom to describe the longitudinal displacements u_1, the vertical deflection w_1 and the slope $w_{,x1}$ of the constraining magnetic layer 1, the longitudinal displacement u_3, the vertical deflection w_3, and the slope $w_{,x3}$ of the base beam/magnet layer. The degrees of freedom of each PMC element are then:

$$\{\delta^e\} = \{u_{1i}, w_{1i}, w_{,x1i}, u_{3i}, w_{3i}, w_{,x3i}, u_{1j}, w_{1j}, w_{,x1j}, u_{3j}, w_{3j}, w_{,x3j}\}^T \tag{8.137}$$

i and j denoting the left and right node, respectively.

The assumed shape functions giving the displacements over the element are assumed to be:

$$\begin{aligned} u_1 &= a_1 x + a_2,\ w_1 = a_3 x^3 + a_4 x^2 + a_5 x + a_6,\\ u_3 &= a_7 x + a_8,\ w_3 = a_9 x^3 + a_{10} x^2 + a_{11} x + a_{12}. \end{aligned} \tag{8.138}$$

Equation (8.138) can be expressed in the following compact form:

$$\{u_1, w_1, w_{,x1}, u_3, w_3, w_{,x3}\}^T = [\{\mathbf{N}_1\},\{\mathbf{N}_2\},\{\mathbf{N}_3\},\{\mathbf{N}_4\},\{\mathbf{N}_5\},\{\mathbf{N}_6\}]^T.\{a_1,\ldots,a_{12}\}^T \tag{8.139}$$

Each constant a_i can be expressed in terms of the nodal displacements:

$$\{\boldsymbol{\delta}^e\} = \mathbf{T}\cdot\{a_1, a_1,\ldots,a_{12}\}^T \tag{8.140}$$

where **T** is a transformation matrix obtained imposing the value of the shape functions at the boundaries of the element. The deflections at any location of the element can therefore be expressed as a function of the nodal displacements using **N** and **T** matrices:

$$\{\mathbf{u}\} = \mathbf{N}\cdot\mathbf{T}^{-1}\cdot\left\{\boldsymbol{\delta}^{(e)}\right\} \tag{8.141}$$

where

$$\mathbf{N} = [\{\mathbf{N}_1\},\{\mathbf{N}_2\},\{\mathbf{N}_3\},\{\mathbf{N}_4\},\{\mathbf{N}_5\},\{\mathbf{N}_6\}]^T \tag{8.142}$$

8.5.3.2.2.3 Effects of the Magnetic Forces on the Structure The magnetic forces given by Eq. (8.133) and Appendix 8.B are utilized to determine the interaction between the two permanent magnetic constrained layers. These forces have components along the axial direction x and the transverse direction z. The forces are included in the finite element model to act on the appropriate degrees of freedom in such a way that accounts for their effect on the dynamic behavior of the beam/PMC assembly.

For the component of the force along the x-direction, it contributes only to the stretch of the magnetic layers along the longitudinal direction. The effect of this force component is accounted for through the addition of the geometric stiffness matrix of the treated beam as outlined in Appendix 8.B.

As for the magnetic force component along the z-direction, the force is in general a nonlinear function of the distance between the interacting magnets (Brown 1966; Griffith 1995). Note that this distance, as measured along the z-direction, is equal to the initial viscoelastic core thickness plus the relative vertical displacement of the two magnetic layers. Since the encountered displacements are relatively small, the magnetic forces $\mathbf{F_m}$ can be accurately linearized around the initial thickness of the viscoelastic core $(w_1 - w_3)_0$ such that:

$$\mathbf{F_m} = \mathbf{F_{m0}} + \left.\frac{\partial \mathbf{F_m}}{\partial(w_1 - w_3)}\right|_0 \cdot (w_1 - w_3) \tag{8.143}$$

In Eq. (8.143), the magnetic force $\mathbf{F_m}$ is described as the sum of an initial static force $\mathbf{F_{m0}}$ and a linear dynamic force that is proportional to the relative displacement $(w_1 - w_3)$. The constant of proportionality $\left.\frac{\partial \mathbf{F_m}}{\partial(w_1 - w_3)}\right|_0$ is the linearized magnetic stiffness that quantifies the interaction between the magnetic layers.

Note that the static force $\mathbf{F_{m0}}$ is computed directly from the finite element analysis of the region surrounding the constraining layers and then used to determine the resulting displacements. The variation of the force with the relative displacement is utilized to determine the linearized magnetic stiffness and the associated dynamic magnetic forces.

8.5.3.2.2.4 Equation of Motion The equation of motion of the beam with PMC treatment can be obtained by applying Hamilton's Principle (Meirovitch 1967), expressed as follows:

$$\int_{t_1}^{t_2} \delta(T-U)\cdot dt + \int_{t_1}^{t_2} \delta W_m \cdot dt = 0 \tag{8.144}$$

where δ (.) denotes the first variation, t_1 and t_2 are the initial and final time, T and U are the total kinetic and strain energies of the element and W_m is the work done by the magnetic forces. The derivation and the expression of the total kinetic and potential energies are given in Appendix 8.B.

The work done by the magnetic forces can be expressed as:

$$W_m = \int_0^L \frac{F_{m_z}}{L}\cdot(w_1 - w_3)\cdot dx \tag{8.145}$$

where F_{mz} is the z component of the magnetic force expressed in Eq. (8.143). Substituting the expression given in Eq. (8.143) for the z component into Eq. (8.145) yields:

$$W_m = \int_0^L \frac{F_{m0_z}}{L}\cdot(w_1 - w_3)\cdot dx + \int_0^L \frac{1}{L}\cdot\frac{\partial F_{m_z}}{\partial(w_1 - w_3)}\cdot(w_1 - w_3)^2\cdot dx \tag{8.146}$$

Introducing the finite element notation into Eq. (8.144), and applying Hamilton's Principle leads to the following equation of motion for the PMC element:

$$\mathbf{M}^e\cdot\left\{\ddot{\boldsymbol{\delta}}^e\right\}^T + \left(\mathbf{K}^e + \mathbf{K}^e_{\mathbf{geo}} - \mathbf{K}^e_{\mathbf{m}}\right)\cdot\{\boldsymbol{\delta}^e\} = F^e_{m0_z}\{\mathbf{b}^e\} \tag{8.147}$$

where $\mathbf{M}^e$ and $\mathbf{K}^e$ denote the element mass and stiffness matrices as given in Appendix 8. B. Also, $\mathbf{K}_{geo}{}^e$ is the element geometric matrix counting for the effect of the axial component of the magnetic forces, $\mathbf{K}_m{}^e$ is an additional magnetic stiffness matrix, $F^e_{m0_z}$ is the static magnetic force and $\{\mathbf{b}^{(e)}\}$ is a vector defining on which degrees of freedom the force is applied.

The overall mass and stiffness matrices $\mathbf{M}_{ov}$ and $\mathbf{K}_{ov}$ can be obtained assembling all the element matrices. Applying the proper boundary conditions yields the overall equation for the entire beam. If harmonic base excitation is considered, the equation of motion of the beam will have the following expression:

$$\left(\mathbf{K} + \mathbf{K_{geo}} - \mathbf{K_m} - \omega^2\cdot\mathbf{M}\right)\cdot\{\boldsymbol{\delta}\} = -\begin{Bmatrix} \dots \\ \mathrm{K}_{ov}(j,1) - \omega^2\cdot\mathrm{M}_{ov}(j,1) \\ \dots \end{Bmatrix}\cdot w_0 \quad j = 2,..,n \tag{8.148}$$

where $\mathbf{M}$ and $\mathbf{K}$ denote the overall mass and stiffness matrices after applying the boundary conditions, $\{\boldsymbol{\delta}\}$ is the overall nodal displacements vector, w_0 denotes the amplitude of the vertical motion of the base, ω is the frequency of the motion and n is the number of unconstrained degrees of freedom.

Example 8.11 Determine the performance characteristics of the beam/PMC assembly shown in Figure 8.54. The beam is a cantilever aluminum beam that has a thickness = 5.08 mm, width = 25.4 mm, and length 254 mm. The viscoelastic material used in the core is 1.524 mm thick, it has a storage modulus $E' = 0.5$ MN m^{-2}, loss factor $\eta = 0.27$, and density $\rho = 150$ kg m^{-3}. The magnetic layers are made of 12.2 mm thick magnetic strips, having Young's modulus $E = 0.6$ GN m^{-2}, density $\rho = 3543$ kg m^{-3}, residual magnetic induction $B_r = 0.19$ T, and coercive force $H_c = 151{,}197.5$ Am^{-1}.

Note that the length of the VEM and the magnetic layers is 228.6 mm.

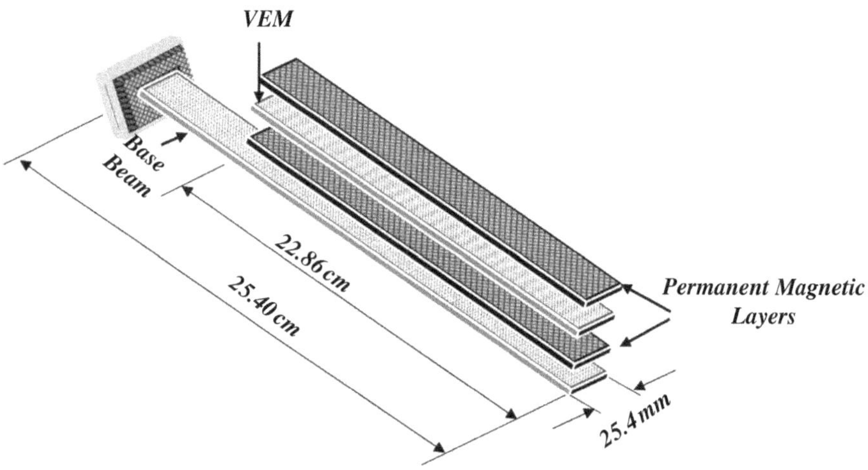

Figure 8.54 The passive magnetic composite (PMC) treatments of beams.

Solution

Figure 8.55 displays the first two modes and corresponding mode shapes of the beam/PMC assembly. The figure indicates that the first mode, 41.27 Hz, belongs to the shear of the VEM layer whereas the second mode, 45.2 Hz, is associated with the transverse bending of the base beam.

Figure 8.56 displays the frequency response of the beam/PMC assembly when the magnetic layers are in attraction. The figure displays also, for comparison purposes, the

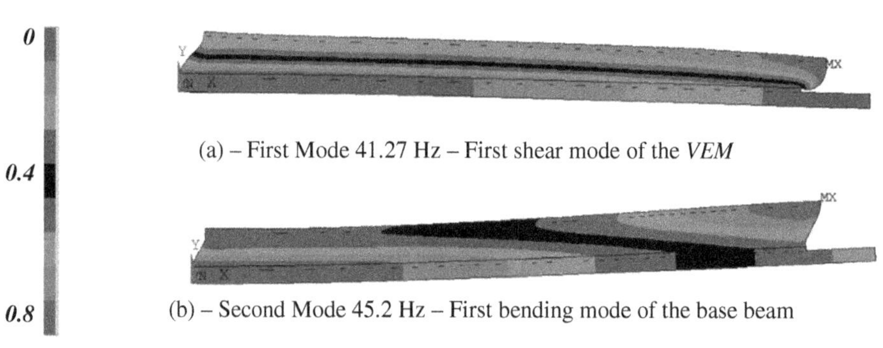

Figure 8.55 The first two modes and mode shapes of the beam/PMC assembly.

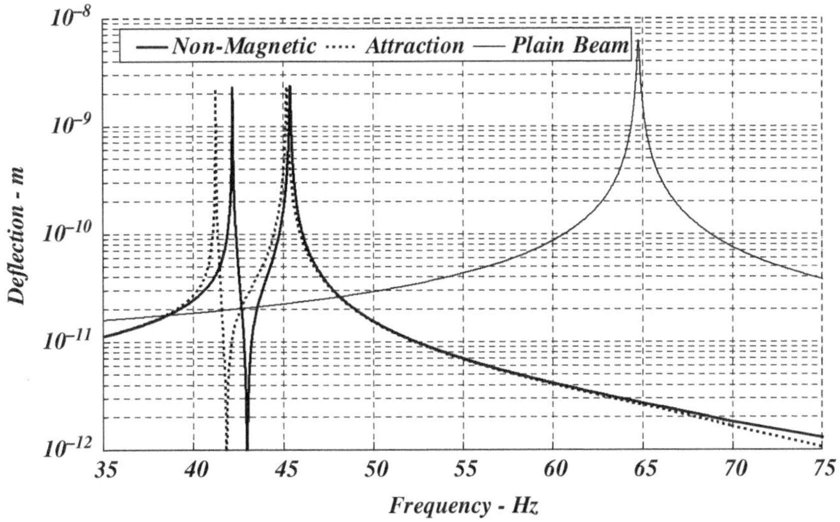

Figure 8.56 The frequency response of the beam/PMC assembly with the magnetic layers in attraction.

corresponding characteristics of a plain untreated beam and the characteristics of a beam with a non-magnetic PMC layers.

It can be seen that the use of the PMC treatment has resulted in splitting the single first mode of vibration of the beam into two modes of vibration. This pattern is similar to the use of classical dynamic absorbers in discrete vibrating systems. Note that the first of these two modes belongs to the shear of the VEM and the second is associated with the transverse bending of the beam. Also, it is observed that the use of the attraction arrangement has resulted in softening the assembly as compared with the non-magnetic PMC arrangement. Specifically, the first two modes of the PMC in attraction are reduced to 41.27 and 45.20 Hz as compared to the 42.22 and 45.41 Hz for the non-magnetic PMC arrangement.

Figure 8.57 displays a close-up view of the characteristics of the beam/PMC where the two modes of the VEM and beam are clearly identified for non-magnetic PMC and PMC where the magnetic layers are in attraction.

Analysis of the displayed results indicates that the use of PMC in attraction has resulted in increasing the damping ratio of the beam to 0.0664% as compared to the 0.0551% for the non-magnetic PMC. The influence on the damping ratio of the VEM mode is negligible.

Figure 8.58 displays the vectors of the magnetic field H around the PMC treatment.

Figure 8.59 displays the corresponding characteristics of the beam/PMC where the two modes of the VEM and beam are clearly identified for non-magnetic PMC and PMC where the magnetic layers are in repulsion. The displayed results indicate that the use of the repulsion arrangement has stiffened the assembly as compared with the non-magnetic PMC arrangement. Specifically, the first two modes of the PMC in repulsion are increased to 44.58 and 48.26 Hz as compared to the 42.22 and 45.41 Hz for the non-magnetic PMC arrangement.

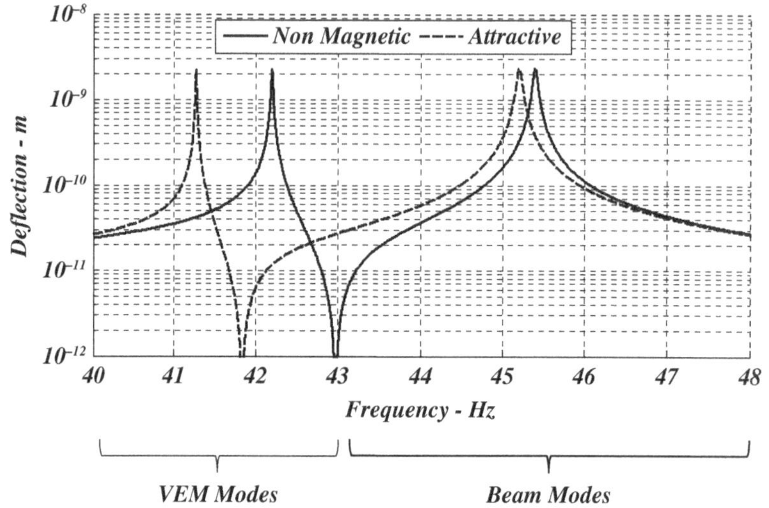

Figure 8.57 The frequency response of the beam/PMC assembly with the magnetic layers in attraction.

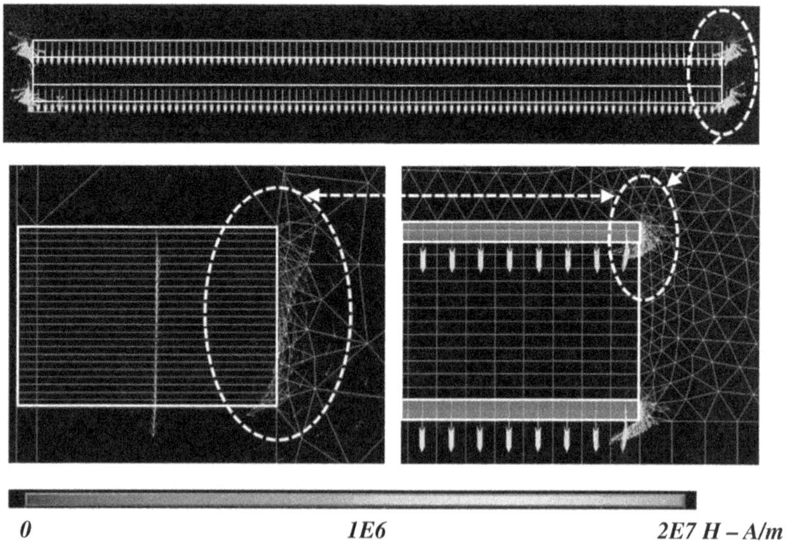

Figure 8.58 The vector of the magnetic field *H* around the beam/PMC assembly with the magnetic layers in attraction.

Also, analysis of the displayed results indicates that the use of PMC in repulsion has reduced the damping ratio of the beam to 0.0212% as compared to the 0.0664% for the PMC in attraction.

Figure 8.60 displays the corresponding vectors of the magnetic field *H* around the PMC treatment.

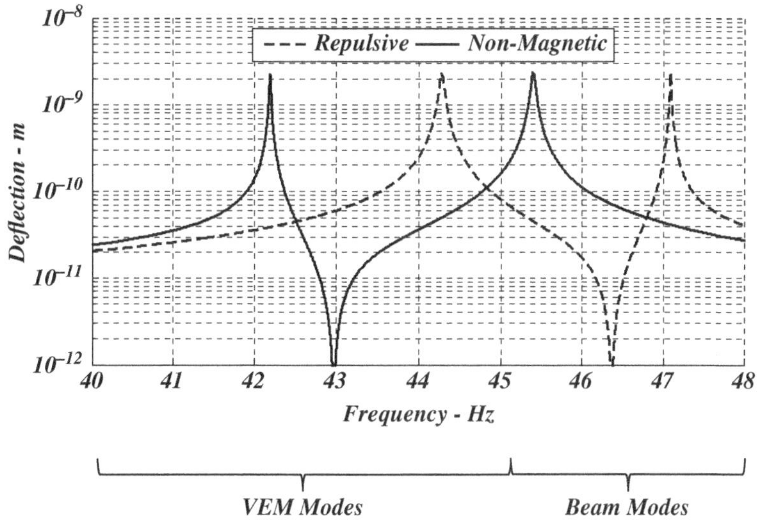

Figure 8.59 The frequency response of the beam/PMC assembly with the magnetic layers in repulsion.

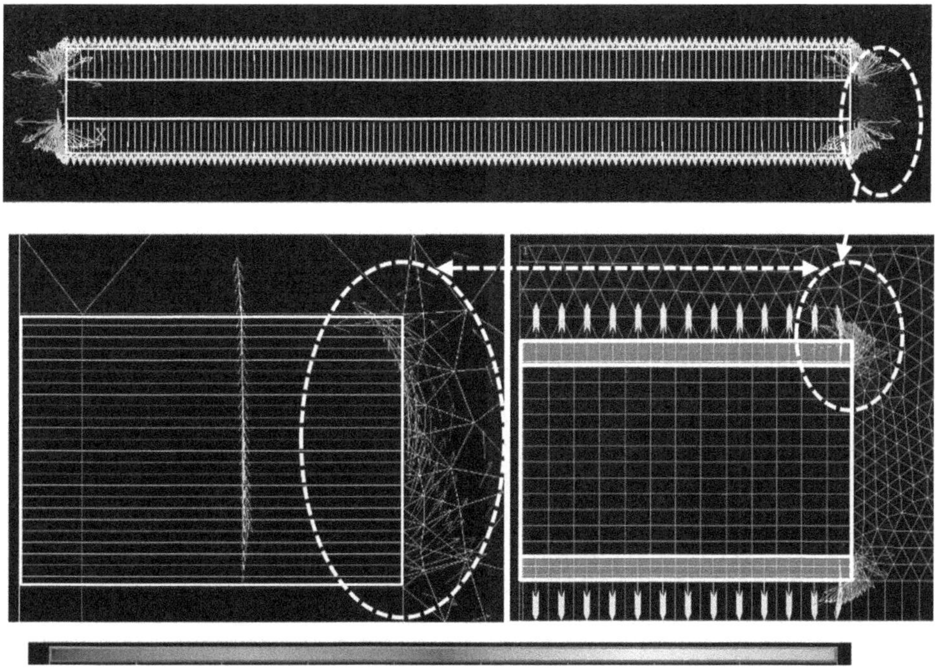

Figure 8.60 The vector of the magnetic field H around the beam/PMC assembly with the magnetic layers in repulsion.

Table 8.4 Characteristics of beam/PMC in attraction and repulsion as compared to non-magnetic PMC treatment.

Arrangement	Non-magnetic		Attractive		Repulsive	
Parameter	Frequency (Hz)	Damping ratio (ζ)	Frequency (Hz)	Damping ratio (ζ)	Frequency (Hz)	Damping ratio (ζ)
Mode 1	42.22	0.000 237	41.27	0.000 242	44.58	0.000 669
Mode 2	45.41	0.000 551	45.20	0.000 664	48.26	0.000 212

A summary of the modes of vibration and the modal damping ratios of the beam/PMC assembly with magnetic layers in attraction and repulsion is listed in Table 8.4 in comparison with the characteristics of the non-magnetic PMC layers.

8.5.4 Summary

In this section, the concept of vibration control using PMC treatments has been presented. The damping characteristic of the treatment have been evaluated both for magnets in attraction and repulsion. The PMC treatments allow for eliminating the need for the sensors, actuators, and associated circuitry as well as any external energy sources.

A FEM has been developed to describe the dynamic characteristics of magnetic composites. The performance of the PMC treatments for beams indicates that magnets in attraction configurations are much more effective in reducing vibration than the magnets in repulsion.

8.6 Negative Stiffness Composites

NSCs have attracted attention for the past decade as these composites have been recognized to potentially exhibit enhanced vibrational damping characteristics. Such improved damping characteristics are attributed to the fact that negative stiffness elements tend to assist rather than resist the deformation of the composite under load. This increased deformation significantly contributes to enhancing the energy stored internally in the NSC.

Accordingly, extensive efforts have been exerted to develop and to understand the unique behavior of such a class of composites. Examples of these include the pioneering work of Lakes (2001), Lakes et al. (2001), and Wang and Lakes (2004) where the concept of high damping composites with a negative stiffness phase has been introduced and investigated. Since then, negative stiffness mechanisms have been explored in numerous vibration isolation systems. Distinct among these efforts, is the work of Platus (1999) where *Minus K* technologies have implemented negative stiffness elements into isolation tables to attenuate low frequency vibrations. Also, the concept has been extended to periodic meta-material configurations as reported, for example, by Nadkarni et al. (2014) in an attempt to mitigate solitary wave propagation. In all these applications, the negative stiffness elements have been primarily made of unstable snap-through elements.

In this section, the emphasis is placed on using magnetic composites in attraction to generate the necessary negative stiffness inclusions. In this manner, the concepts developed in Section 8.5 represent the foundation necessary for understanding the behavior of the negative stiffness magnetic composites.

8.6.1 Motivation to Negative Stiffness Composites

The concept of NSC can best be understood by considering a simplified version of the composite that has a single degree of freedom as shown in Figure 8.61.

8.6.1.1 Sinusoidal Excitation

The equation of motion of the system is:

$$m\ddot{x} + c\dot{x} + (k - k_m)x = f_o \sin(\omega t) \tag{8.149}$$

Hence, the steady-state response is given by:

$$\frac{X_o}{f_o} = \frac{1}{\sqrt{([k-k_m] - m\omega^2)^2 + (c\omega)^2}} \tag{8.150}$$

Assuming that the parameters: mass m, damping coefficient c, and structural stiffness k are all constants. Only k_m is a variable negative stiffness, which can be tuned as desired.

Accordingly, the energy dissipated during a full oscillation cycle, as given by Table 2.14 for viscous damping, is as follows:

$$E_v = \pi c \omega X_o^2 \tag{8.151}$$

Substituting Eq. (8.150) into (8.151) gives:

$$E_v = \frac{\pi c \omega f_o^2}{([k-k_m] - m\omega^2)^2 + (c\omega)^2} = \frac{\pi \left(\frac{c\omega}{m}\right) f_o^2}{\left(\omega_n^2 - \omega^2\right)^2 + \left(\frac{c\omega}{m}\right)^2}. \tag{8.152}$$

where $\omega_n = \sqrt{\dfrac{[k-k_m]}{m}}$ = natural frequency of system with augmented negative stiffness.

Equation (8.152) can be rewritten assuming that:

$$\bar{C} = \frac{c}{m\,\omega_{\max}}, \bar{\omega} = \frac{\omega}{\omega_{\max}}, \text{ and } \bar{\omega}_n = \frac{\omega_n}{\omega_{\max}} \tag{8.153}$$

where $\omega_{\max}$ is a selected maximum frequency.

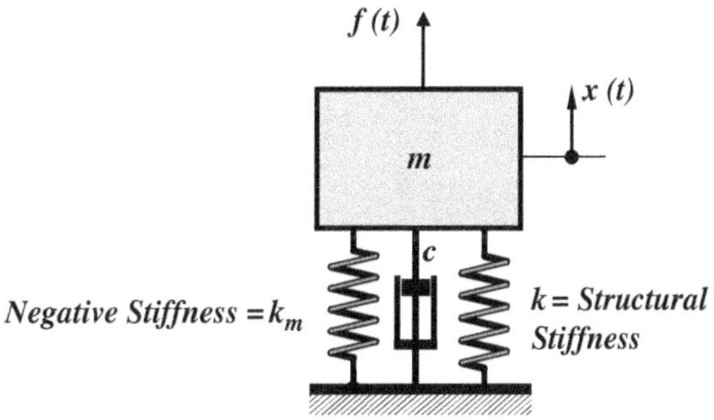

Figure 8.61 Single degree of freedom system augmented with a negative stiffness element.

Then, Eq. (8.152) reduces to:

$$E_v = \frac{\pi\left(\frac{c\,\omega}{m}\right)\frac{f_o^2}{m}}{\left(\omega_n^2-\omega^2\right)^2+\left(\frac{c\,\omega}{m}\right)^2} \tag{8.154}$$

This equation can be cast in the following dimensionless form:

$$\frac{E_v}{\left(\frac{\pi f_o^2}{m\,\omega_{\max}^2}\right)} = \frac{\left(\frac{c}{m\,\omega_{\max}}\frac{\omega}{\omega_{\max}}\right)}{\left[\left(\frac{\omega_n}{\omega_{\max}}\right)^2-\left(\frac{\omega}{\omega_{\max}}\right)^2\right]^2+\left[\left(\frac{c}{m\,\omega_{\max}}\right)\left(\frac{\omega}{\omega_{\max}}\right)\right]^2}. \tag{8.155}$$

Substituting from Eq. (8.153) and letting $\bar{E}_v = \frac{E_v}{\left(\frac{\pi f_o^2}{m\,\omega_{\max}^2}\right)}$, then Eq. (8.155) reduces to:

$$\bar{E}_v = \frac{\bar{C}\,\bar{\omega}}{\left[\bar{\omega}_n^2-\bar{\omega}^2\right]^2+\left[\bar{C}\,\bar{\omega}\right]^2} \tag{8.156}$$

Example 8.12 Determine the effect of the excitation frequency $\bar{\omega}$ on the energy dissipation characteristics $\bar{E}_v$ of *NSC* for different $0<\bar{\omega}_n\le 1$ and for $\bar{C}=1$.

Solution

Figure 8.62 displays the effect of varying the excitation frequency $\bar{\omega}$ on the energy dissipation characteristics $\bar{E}_v$ of *NSC* for values of the effective natural frequency $\bar{\omega}_n$ between 0.05 and 1. The displayed characteristics are drawn for $\bar{C}=1$.

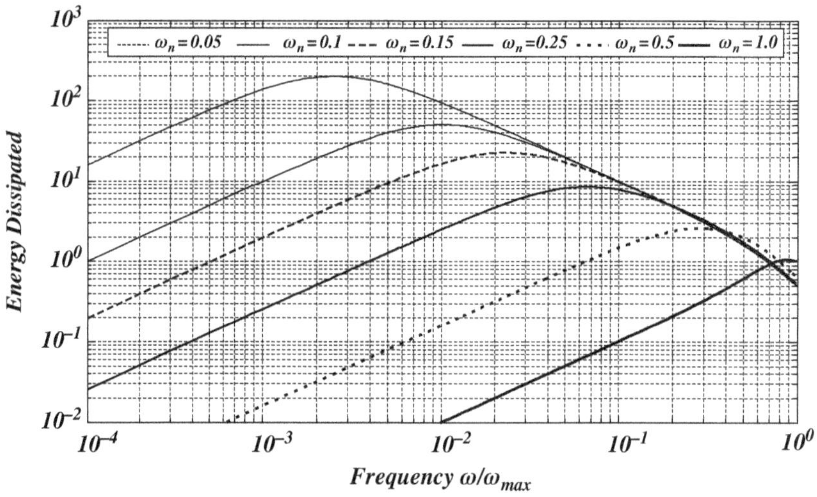

Figure 8.62 Energy dissipation characteristics of a single degree of freedom system augmented with a negative stiffness element for $\bar{C}=1$.

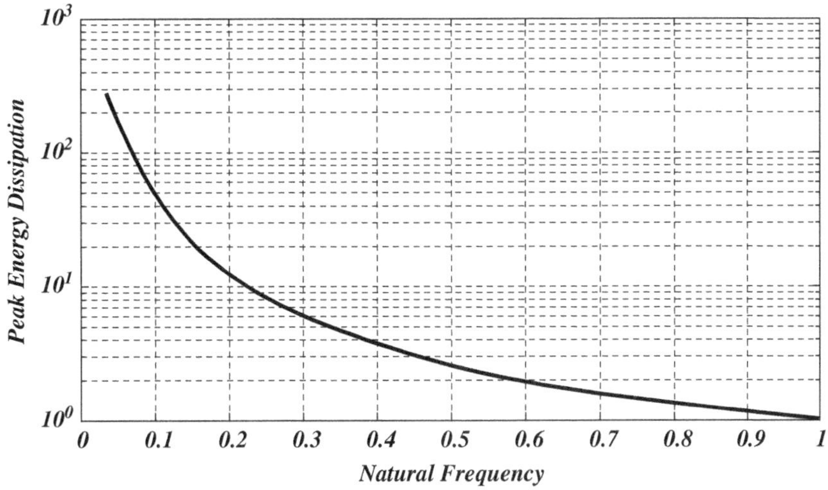

Figure 8.63 Effect of effective natural frequency $\bar{\omega}_n$ on the peak energy dissipation of the NSC system augmented with a negative stiffness element for $\bar{C} = 1$.

The figure indicates that decreasing the effective natural frequency of the NSC system results in enhancing the energy dissipation characteristics considerably. For example, reducing $\bar{\omega}_n$ from 1 to 0.05 has resulted in increasing the peak energy dissipation $\bar{E}_{v_{\max}}$ by two orders of magnitudes from 1.066 to 197.6.

Figure 8.63 summarizes the effect of the effective natural frequency $\bar{\omega}_n$ on the peak energy dissipation $\bar{E}_{v_{\max}}$.

Note that the energy dissipation reaches a peak when $d\bar{E}_v/d\bar{\omega} = 0$. This requires that:

$$\bar{\omega}^4 - \frac{1}{3}\left(-\bar{C}^2 + 2\bar{\omega}_n^2\right)\bar{\omega}^2 - \frac{1}{3}\bar{\omega}_n^4 = 0 \tag{8.157}$$

Equation (8.157) has a real positive solution $\bar{\omega}_*^2$ such that:

$$\bar{\omega}_*^2 = \frac{1}{6}\left(-\bar{C}^2 + 2\bar{\omega}_n^2\right) + \sqrt{\left[\frac{1}{6}\left(-\bar{C}^2 + 2\bar{\omega}_n^2\right)\right]^2 + \frac{1}{3}\bar{\omega}_n^4} \tag{8.158}$$

Hence, the peak energy dissipation $\bar{E}_{v_{\max}}$ is given by:

$$\bar{E}_{v_{\max}} = \frac{\bar{C}\bar{\omega}_*}{\left[\bar{\omega}_n^2 - \bar{\omega}_*^2\right]^2 + \left[\bar{C}\bar{\omega}_*\right]^2} \tag{8.159}$$

Example 8.13 Plot the pole-zero map of the NSC system for different values of the effective natural frequency $\bar{\omega}_n$ and for $\bar{C} = 1$.

Solution

As

$$m\ddot{x} + c\dot{x} + (k - k_m)x = f_o$$

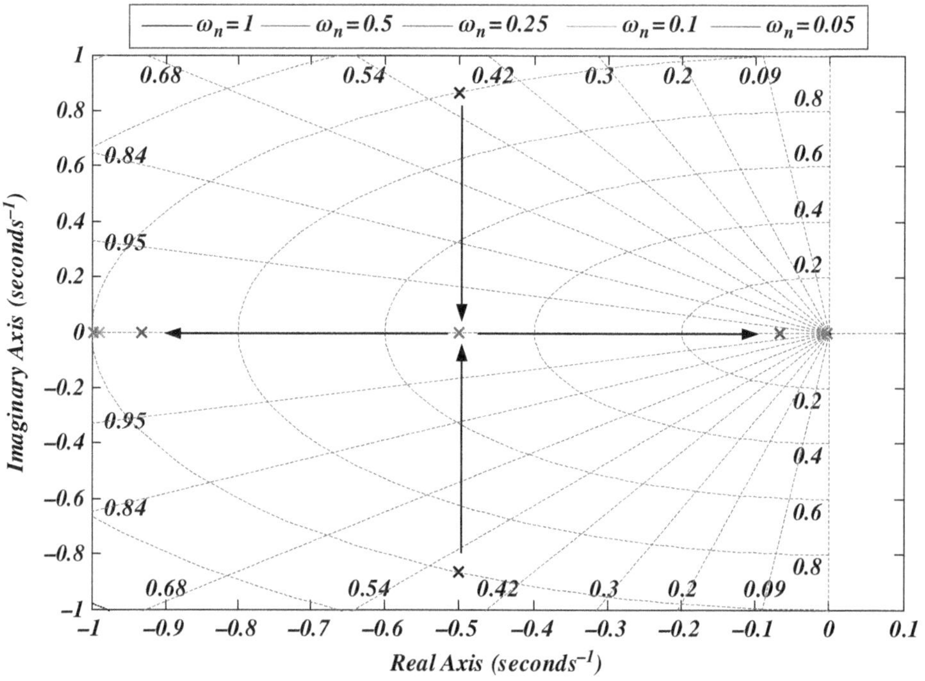

Figure 8.64 Effect of effective natural frequency $\bar{\omega}_n$ on the poles of the NSC system for $\bar{C} = 1$.

Using the Laplace transform and the dimensionless quantities of Eq. (8.153), this equation reduces into the following dimensionless form:

$$\frac{X_o}{f_o/\left(m\ \omega_{\max}^2\right)} = \frac{1}{\bar{s}^2 + \bar{C}\bar{s} + \bar{\omega}_n^2}$$

where $\bar{s} = s/\omega_{\max}$.

Figure 8.64 displays poles of the characteristics equation of the system response for different values of the effective natural frequency $\bar{\omega}_n$ and for $\bar{C} = 1$.

The figure indicates that the poles migrate from the imaginary axis as for $\bar{\omega}_n = 1$ with damping ratio of 0.5 to the real axis for values of $\bar{\omega}_n \leq 0.5$ leading to a critically damped system with a damping ratio of 1.0.

Example 8.14 Determine the hysteresis characteristics of the *NSC* system for different values of the effective natural frequency $\bar{\omega}_n$ for an excitation frequency $\bar{\omega}$ = 0.1 and for $\bar{C} = 1$.

Solution

Using the dimensionless parameters of Eq. (8.153), the equation of motion of the system (8.149) reduces to:

$$\ddot{\bar{x}} + \bar{C}\dot{\bar{x}} + \bar{\omega}_n^2\bar{x} = \sin\left(\bar{\omega}\tau\right) \tag{8.160}$$

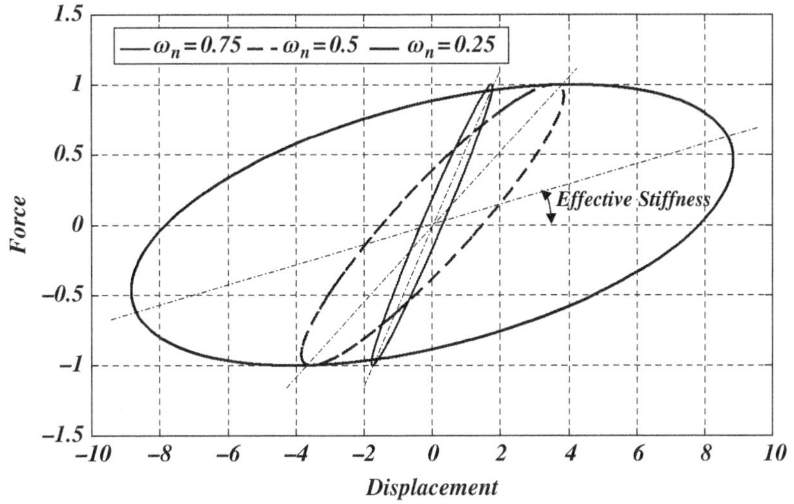

Figure 8.65 The hysteresis characteristics of the NSC system augmented with a negative stiffness element for $\bar{\omega} = 0.1$ and $\bar{C} = 1$.

where $\ddot{\bar{x}} = \dfrac{\ddot{x}}{x_0\ \omega_{\max}^2}, \dot{\bar{x}} = \dfrac{\dot{x}}{x_0\ \omega_{\max}}, \bar{x} = \dfrac{x}{x_0}, x_0 = \dfrac{f_o}{m\ \omega_{\max}^2},$ and $\tau = \bar{\omega} t$.

The system has a steady-state solution $\bar{x}_0$ given by:

$$\bar{x}_0 = \bar{X}_0 \sin(\bar{\omega}\tau - \varphi) \tag{8.161}$$

where $\bar{X}_0 = X_o/x_0$, and the phase angle φ is given by:

$$\varphi = \tan^{-1}\left(\frac{\bar{C}\bar{\omega}}{\bar{\omega}_n^2 - \bar{\omega}^2}\right) \tag{8.162}$$

Figure 8.65 displays the force $\bar{f} = \sin(\bar{\omega}\tau)$ versus $\bar{x}_0 = \bar{X}_0 \sin(\bar{\omega}\tau - \varphi)$ for different values of $\bar{\omega}_n$ when $\bar{\omega} = 0.1$ and for $\bar{C} = 1$.

The figure emphasizes the results obtained in Example 8.12 that showed decreasing the effective natural frequency of the NSC system results in enhancing the energy dissipation characteristics considerably as quantified by the area enclosed inside the hysteresis loops. Note that the enhancement of the energy dissipation is achieved at the expense of softening the NSC system. This is demonstrated by considering the stiffness for $\bar{\omega}_n = 0.25$, as indicated in Figure 8.65, which is clearly much lower than that for $\bar{\omega}_n = 0.75$.

A balance between the enhanced energy dissipation and the reduced stiffness characteristics can be achieved through the application of proper optimization techniques.

Example 8.15 Determine the optimal effective natural frequency $\bar{\omega}_n^*$ for the single degree of freedom system NSC by maximizing the following performance index, I:

$$I = \left(\int_0^1 \bar{E}_v\, d\bar{\omega}\right)\bar{\omega}_n \tag{8.163}$$

for different values of the parameter, $\bar{C}$.

Solution

Note that the performance index I is formed of two parts. The first part of I is $\int_0^1 \bar{E}_v \, d\bar{\omega}$, which quantifies the total energy dissipation over a frequency band $0 < \bar{\omega} = \omega/\omega_{max} < 1$. This part represents the area under the $\bar{E}_v - \bar{\omega}$ characteristics curve for any particular $\bar{\omega}_n$.

The second part of the performance index I is the effective natural frequency $\bar{\omega}_n$.

In this manner, maximizing the performance index I guarantees a simultaneous maximization of both the total energy dissipation and the stiffness of the NSC.

Figure 8.66 displays the optimal characteristics for the NSC for values of $\bar{C}$ = 0.1, 1, and 10.

Table 8.5 summarizes the optimal values of the effective natural frequency $\bar{\omega}_n^*$ for different values of the parameter $\bar{C}$. The table includes also the optimal performance index.

8.6.1.2 Impact Loading

The equation of motion of the system under unit impulse loading can be written by modifying Eq. (8.160) to assume the following form:

$$\ddot{\bar{x}} + \bar{C}\dot{\bar{x}} + \bar{\omega}_n^2 \bar{x} = \delta(\tau) \tag{8.164}$$

where $\delta(\tau)$ is a unit impulse loading.

The corresponding unit impulse response h of Eq. (8.164) is given by:

$$h = \frac{e^{-\frac{1}{2}\bar{C}\tau}}{\sqrt{\left(\frac{1}{2}\bar{C}\right)^2 - (\bar{\omega}_n)^2}} \sinh\left(\tau\sqrt{\left(\frac{1}{2}\bar{C}\right)^2 - (\bar{\omega}_n)^2}\right) \tag{8.165}$$

Example 8.16 Determine the unit impulse response for the single degree of freedom system NSC for different values of $\bar{\omega}_n$ and $\bar{C} = 1$.

Solution

Using Eq. (8.165), the unit impulse response of the system is computed and displayed in Figure 8.67 for different values of $\bar{\omega}_n$ and $\bar{C} = 1$.

The displayed results indicate that the oscillatory response of the system when $\bar{\omega}_n$ = 1 becomes critically damped when $\bar{\omega}_n \le 0.5$ as is supported by the results shown in Figure 8.64. However, the peak unit impulse response increases as the NSC becomes softer.

This pattern is similar to what has been observed when the NSC system is subjected to sinusoidal oscillation (Examples 8.12 and 8.13).

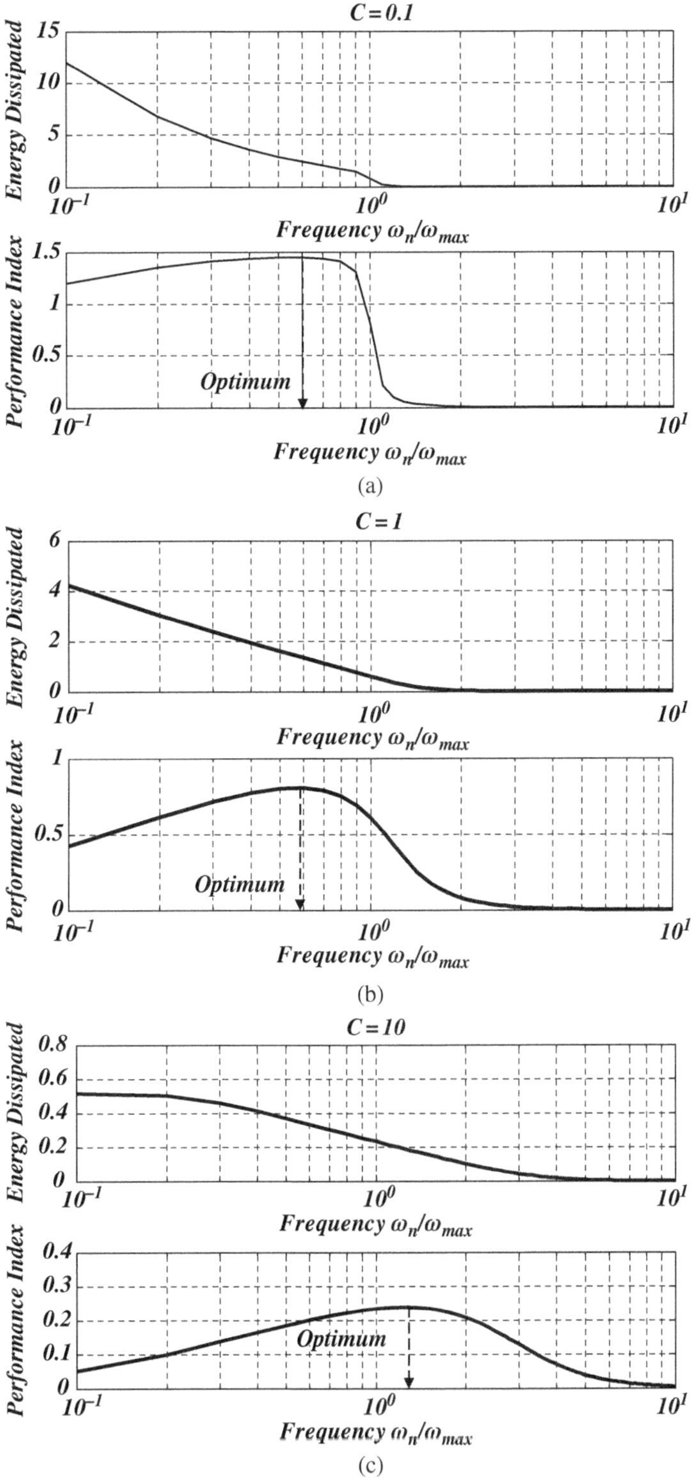

Figure 8.66 The optimal characteristics of the NSC system augmented with a negative stiffness element for different values of $\bar{C}$. (a) C = 0.1, (b) C = 1.0, and (c) C = 10.

Table 8.5 Optimal characteristics of the NSC system for different values of $\bar{C}$.

$\bar{C}$	0.1	1	2	5	10
$\bar{\omega}_n^*$	0.5819	0.5846	0.702	0.8983	1.2
Optimal performance index	1.447	0.805	0.5626	0.344	0.239

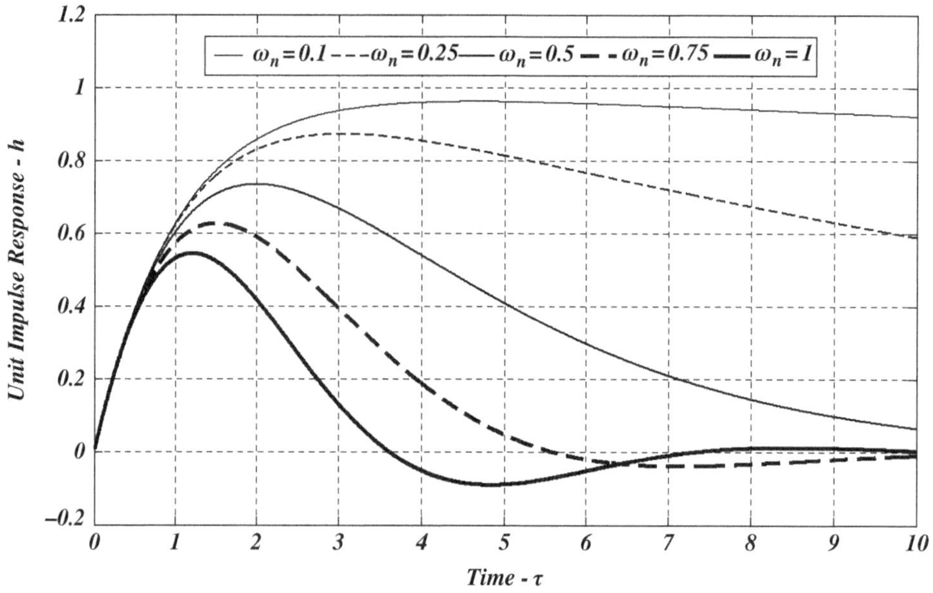

Figure 8.67 The unit impulse response of the NSC system augmented with a negative stiffness element for different values of $\bar{\omega}_n$ and $\bar{C} = 1$.

8.6.1.3 Magnetic Composite with Negative Stiffness Inclusions

The concept of NSCs can best be understood by considering a simplified version of the composite that has negative stiffness spring elements in the form of an array of permanent magnets arranged in attraction as shown in Figure 8.68.

8.6.1.3.1 Force Between Two Bar Magnets

The interaction forces between two bar magnets are predicted using Gilbert's model that assumes these forces are generated by magnetic charges near the poles. The model adequately predicts the forces even for closely interacting magnets where the magnetic field is rather complex.

Vokoun et al. (2009) developed an analytical expression for computing the force between two cylindrical magnets, of radius R, area A, and length L, assuming uniform magnetization. The force F between two identical and aligned bar magnets are given by:

$$\bar{F} = F \Big/ \left[\frac{B_o^2 A^2 (L^2 + R^2)}{\pi \mu_0} \right] = \left[\frac{1}{\bar{x}^2} + \frac{1}{(\bar{x}+2)^2} - \frac{2}{(\bar{x}+1)^2} \right] \tag{8.166}$$

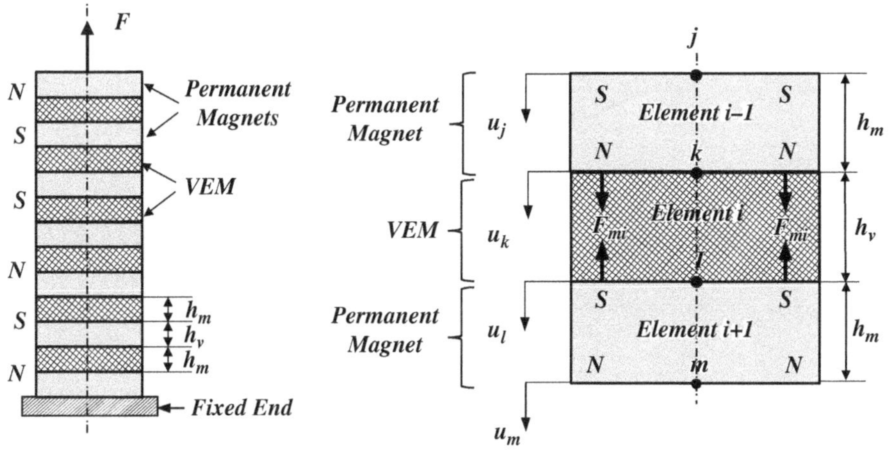

Figure 8.68 Magnetic composite with negative stiffness inclusions. (a) Entire composite. (b) Finite element of composite.

where x = spacing between the magnets, $\bar{x} = x/L$, B_0 = the magnetic flux density = $\frac{1}{2}\mu_0 M$ with M = the magnetization (A/m) and μ_0 = the permeability of space = $4\pi \times 10^{-7}\ T\,\text{m}\,\text{A}^{-1}$.

Figure 8.69 displays the relationship $\bar{F} - \bar{x}$ along with the asymptotes for small and large spacing. It is evident that for small spacing the magnetic force $\bar{F}$ is inversely proportional to $\bar{x}^2$ whereas for large spacing $\bar{F}$ is inversely proportional to $\bar{x}^4$.

Linearization of $\bar{F}$ around $\bar{x} = 1$ gives:

$$\bar{F} = \bar{F}_0 + \left.\frac{\partial \bar{F}}{\partial \bar{x}}\right|_{\bar{x}=1} (\bar{x} - 1) = \frac{1}{54}[33 - 85(\bar{x} - 1)] \tag{8.167}$$

Or,

$$F = \frac{87}{54}\left[\frac{B_o^2 A^2 (L^2 + R^2)}{\pi \mu_0}\right] - \frac{85}{54}\left[\frac{B_o^2 A^2 (L^2 + R^2)}{\pi \mu_0}\right]\bar{x} = F_s + K_m \bar{x} \tag{8.168}$$

Hence, the negative stiffness resulting from accounting for the magnetic forces K_m can be written as:

$$K_m = -\frac{85}{54}\left[\frac{B_o^2 A^2 (L^2 + R^2)}{\pi \mu_0}\right] \tag{8.169}$$

8.6.1.3.2 Finite Element Model

The finite element of the NSC composite can be expressed using, for example, Eq. (4.10) for a single GHM mini-oscillator, as follows:

$$\begin{bmatrix} M & 0 \\ 0 & \dfrac{K_N \alpha_n}{\omega_n{}^2} \end{bmatrix} \begin{Bmatrix} \ddot{x} \\ \ddot{z} \end{Bmatrix} + \begin{bmatrix} 0 & 0 \\ 0 & \dfrac{2\zeta_n K_N \alpha_n}{\omega_n} \end{bmatrix} \begin{Bmatrix} \dot{x} \\ \dot{z} \end{Bmatrix} + \begin{bmatrix} K_N + K_N\alpha_n & -K_N\alpha_n \\ -K_N\alpha_n & K_N\alpha_n \end{bmatrix} \begin{Bmatrix} x \\ z \end{Bmatrix} = \begin{Bmatrix} f \\ 0 \end{Bmatrix} \tag{8.170}$$

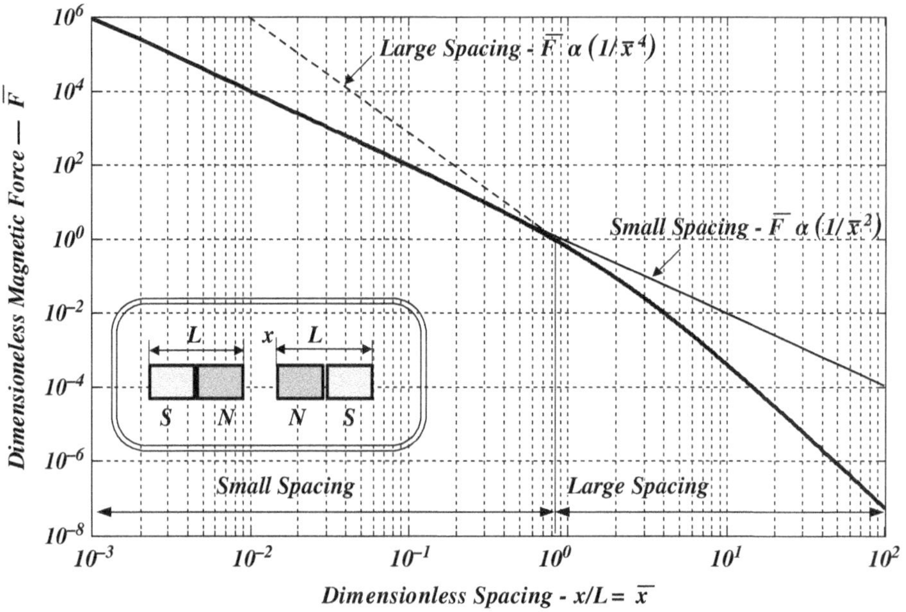

Figure 8.69 The dimensionless magnetic force between two magnetic layers as function of the dimensionless axial spacing between the layers.

where K_N = net stiffness matrix = $K - K_m$, where K denotes the structural stiffness.

Example 8.17 Determine the hysteresis characteristics for the NSC composites shown in Figure 8.68 when it is excited by a sinusoidal force $\boldsymbol{F}$ at a frequency of 1 Hz. The physical and geometrical characteristics of the NSC composite are listed in Table 8.6. The FEM is divided into N_T = 17 elements of equal length. Eight of VEM elements (N_V = 8) and N_M = 9 for the magnetic layers.

The VEM is described by the GHM modeling approach with one mini-oscillator with the following parameters: $E_0 = 9GPa$, $\alpha_1 = 10$, $\zeta_1 = 5$, $\omega_1 = 8,000 rad/s$.

Solution

Figure 8.70 displays the hysteresis characteristics for the NSC for different values of the magnetic flux density B_0, that is, different values of the stiffness ratio K_m/K.

The displayed results indicate that increasing the ratio K_m/K results in increasing the hysteresis characteristics of the NSC. This trend is in line with the observed characteristics for a single mass system as indicated in Figure 8.65.

Table 8.6 The physical and geometrical characteristics of the NSC composite.

Parameter	R (m)	Element length L (m)	B_o (T)	μ_o (T m A^{-1})	N_T	N_v	N_M
Value	0.006 25	0.006 25	3×10^6	$4\pi \times 10^{-7}$	17	8	9

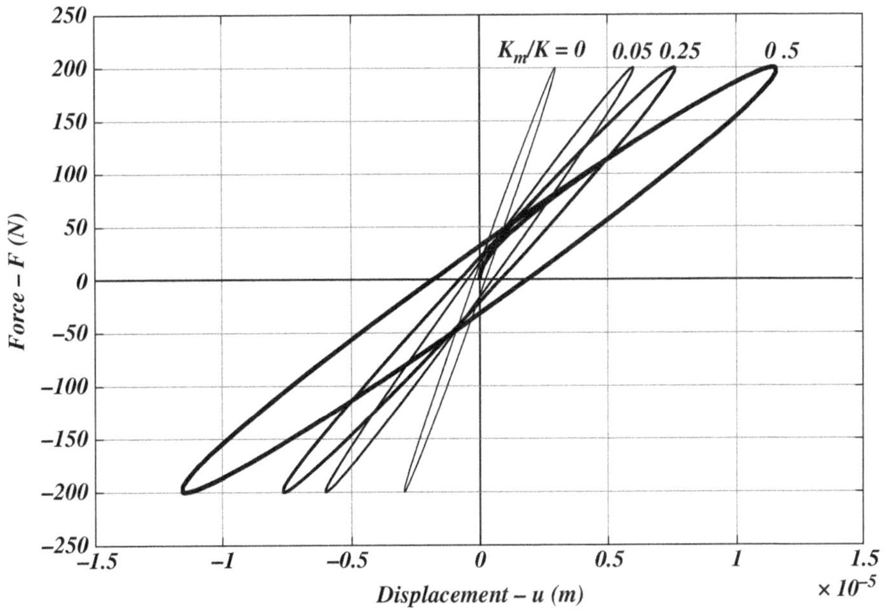

Figure 8.70 The hysteresis characteristics of the NSC system augmented with a negative stiffness element with different B_0.

Example 8.18 Compute the static deflections of the NSC composite shown in Figure 8.68 by using ANSYS. Determine these deflections when the magnets are arranged both in attraction and in repulsion. Note that the NSC composite is mounted in a cantilever arrangement with its bottom end fixed to a rigid foundation.

The physical and geometrical parameters of the composite are as listed in Example 8.17.

Solution

Figure 8.71a displays the finite element mesh of the NSC composite and the arrangements of the magnets and the VEM.

Figure 8.71b shows the distribution of the deflection along the composite when the magnetic layers are arranged in repulsion. The corresponding distribution when the magnetic layers are arranged in attraction is shown in Figure 8.71c.

Example 8.19 Compute the effect of the arrangement of the magnetic layers on the modes of vibration of the NSC composite shown in Figure 8.68 by using ANSYS. The NSC composite is mounted in a cantilever arrangement with its bottom end fixed to a rigid foundation. Determine these modes when the magnets are arranged both in attraction and in repulsion.

The physical and geometrical parameters of the composite are as listed in Example 8.17.

Determine also the frequency response of the NSC composite when it is excited at its free end by a force $F = 200\,N$.

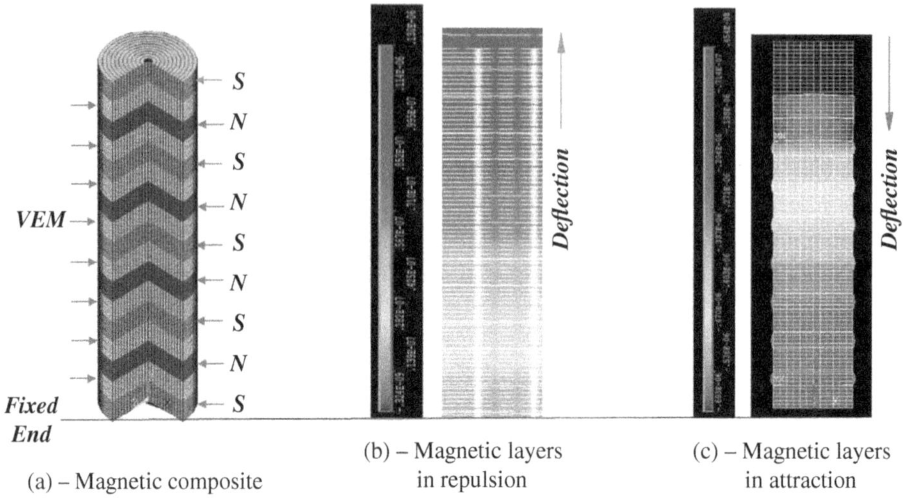

Figure 8.71 Finite element of magnetic composites with magnetic layers arranged in attraction and repulsion. (a) Magnetic composite, (b) magnetic layers in repulsion, and (c) magnetic layers in attraction.

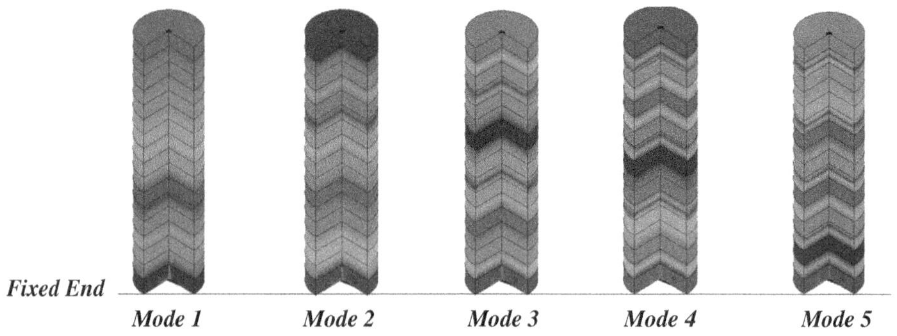

Figure 8.72 Mode shapes of magnetic composites.

Solution

Figure 8.72 displays the mode shapes of the first five natural frequencies of the NSC composite.

Table 8.7 lists the natural frequencies of the composite when the magnets are arranged either in repulsion and attraction as compared to the corresponding to the natural frequencies of the composite when all the layers are non-magnetic.

The displayed results indicate that arranging the magnetic layers in repulsion increases the natural frequencies of the composites and in attraction decreases the natural frequencies of the composites as compared to the non-magnetic composites.

Table 8.7 Natural frequencies of the composite with the magnetic layers in attraction and repulsion as compared to non-magnetic layers.

	Natural frequencies (Hz)		
Mode number	**Non-magnetic**	**Repulsion (M = 3E6 A m^{-1})**	**Attraction (M = 3E6 A m^{-1})**
1	566.56	574.48	369.99
2	1670.3	1693.9	1105.5
3	2690.1	2729.4	1774.0
4	3583.1	3638.0	2348.0
5	4321.8	4388.3	2805.0
6	4893.7	4965.9	3124.0
7	5297.1	5373.9	3393.0

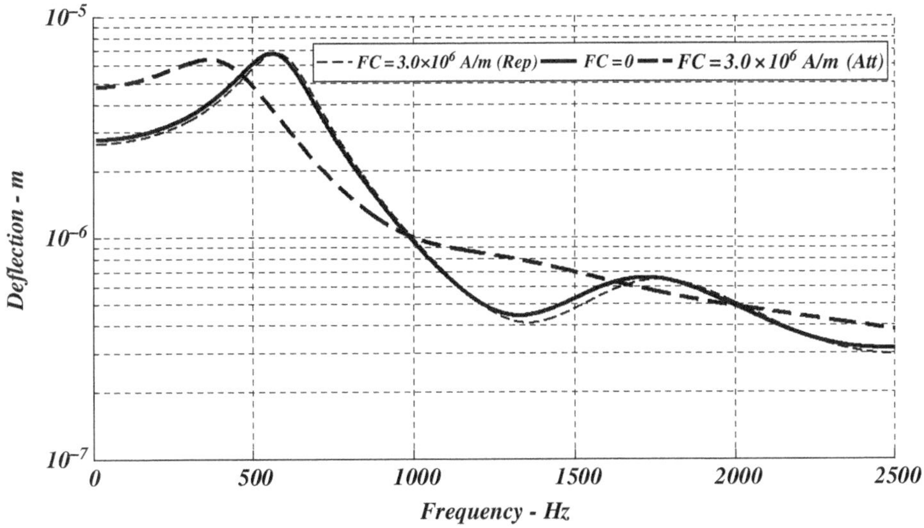

Figure 8.73 Frequency response of non-magnetic composite and magnetic composites with magnets in attraction and repulsion.

Figure 8.73 displays the frequency response of the magnetic composite when the magnets are arranged either in repulsion and attraction compared to the corresponding natural frequencies of the composite when all the layers are non-magnetic.

The effect of the arrangement of the magnetic layers in the composite on the modal damping ratio is listed in Table 8.8 for layers in attraction and repulsion as compared to non-magnetic layers. It is evident that the effective modal damping ratios are enhanced considerably when the magnetic layers are arranged in attraction, that is, when the effect of the negative stiffness becomes significant.

Figure 8.74 displays the magnetic field distributions inside the magnetic composites. Figure 8.74a shows the distribution when the magnetic layers are arranged in attraction

Table 8.8 Damping characteristics of magnetic composite with layers in attraction and repulsion as compared to non-magnetic layers.

Arrangement	Non-magnetic		Attractive		Repulsive	
Parameter	Frequency (Hz)	Damping ratio (ζ)	Frequency (Hz)	Damping ratio (ζ)	Frequency (Hz)	Damping ratio (ζ)
Mode 1	561	0.211	363	0.505	580	0.205
Mode 2	1712	0.185	1220	1.000	1773	0.166

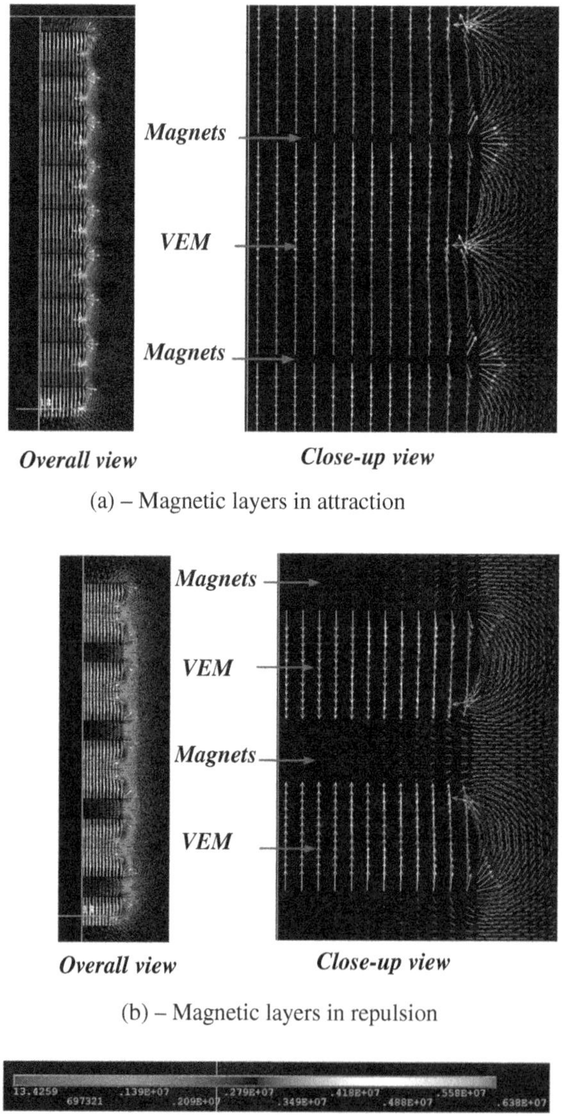

Figure 8.74 Magnetic fields in magnetic composites with magnetic layers arranged in attraction and repulsion. (a) Magnetic layers in attraction. (b) Magnetic layers in repulsion.

Table 8.9 Effect of k_r on the stable equilibrium points.

k_r	0	0.25	0.5
Stable point #1	0	0	0
Stable point #2	2	1.76	1.54

whereas Figure 8.74b displays the corresponding distribution when the magnetic layers are arranged in repulsion.

8.7 Summary

This chapter has presented five classes of advanced damping treatments. The considered treatments include: the stand-off damping treatment, functionally graded damping treatment, APDCs, PMCs, and NSCs. All five of these classes of damping treatments aim to provide novel and innovative means for improving damping behavior over that of conventional constrained damping layer treatments.

The theoretical models of these treatments are presented, in detail, and numerical examples are included to outline their important characteristics and provide means for optimizing their performance. The developed theoretical models are also validated by using the commercial finite element package ANSYS.

References

Alberts, T.E. and Xia, H.C. (1995). Design and analysis of fiber enhanced viscoelastic damping polymers. *Journal of Vibration and Acoustics* 117 (4): 398–404.

Arafa, M. and Baz, A. (2000). Dynamics of active piezoelectric damping composites. *Journal of Composites Eng.: Part B* 31: 255–264.

Bartoli, I., Marzania, A., di Scale, F.L., and Viola, E. (2006). Modeling wave propagation in damped waveguides of arbitrary cross-section. *Journal of Sound and Vibration* 295: 685–707.

Bathe, K.J. (1996). *Finite Element Procedures in Engineering Analysis*. Englewood Cliffs, New York: Prentice Hall.

Baz, A. and Poh, S. (2000). Performance characteristics of magnetic constrained layer damping. *Journal of Shock & Vibration* 7 (2): 18–90.

Baz, A. and Tempia, A. (2004). Active piezoelectric damping composites. *Journal of Sensors and Actuators: A. Physical* 112 (2–3): 340–350.

Brown, W.F. (1966). *Magnetoelastic Interactions*. New York: Springer-Verlag.

Chan, H.L.W. and Unsworth, J. (1989). Simple model for piezoelectric ceramic/polymer 1-3 composites used in ultrasonic transducer applications. *IEEE Transactions on Ultrasonics, Ferroelectrics, and Frequency Control* 36 (4): 434–441.

Christensen, R.M. (1982). *Theory of Viscoelasticity: An Introduction*, 2e. New York: Academic Press.

Coulomb, J.L. and Meunier, G. (1984). Finite element implementation of virtual work principle for magnetic or electric force and torque computation. *IEEE Transactions on Magnetics* 20 (5): 1894–1896.

Demoret K. and Torvik P., "Optimal length of constrained layers on a substrate with linearly varying strains", Proceedings of the ASME Design Engineering Technology Conference, Boston, MA, ASME DE, Vol. 84, No. 3, pp. 719–926, 1995.

Falugi M., "Analysis of five layer viscoelastic constrained layer beam", Proceedings of Damping '91, San Diego, CA, Vol. II, Paper No. CCB, 1991.

Falugi M., Moon Y., and Arnold R., "Investigation of a four layer viscoelastic constrained layer damping system," USAF/WL/FIBA/ASIAC, Report No. 189.1A, 1989.

Garrison, M.R., Miles, R.N., Sun, J., and Bao, W. (1994). Random response of a plate partially covered by a constrained layer damper. *Journal of Sound and Vibration* 172: 231–245.

Griffith, D.J. (1995). *Introduction to Electrodynamics*. New Delhi: Prentice-Hall of India.

Hajela, P. and Lin, C.-Y. (1991). Optimal design of viscoelastically damped beam structures. *Applied Mechanics Reviews* 44 (11S): S96–S106.

Hayward, G. and Hossack, J.A. (August 1990). Unidimensional Modeling of 1-3 composite transducers. *Journal of the Acoustical Society of America* 88 (2): 599–608.

IEEE STANDARD, ANSI/IEEE 176-1987 – IEEE Standard on Piezoelectricity, The Institute of Electrical and Electronics Engineers, Inc., New York, NY, March 1987.

Kamminga, W. (1975). Finite-element solution for devices with permanent magnets. *Journal of Applied Physics* 8: 841–855.

Kim T. W. and Kim J. H., "Eigenvalue sensitivity based topological optimization for passive constrained layer damping", Proceedings of the 45th AIAA/ASME/ASCE/AHS/ASC Structures, Structural Dynamics and Materials Conference, Palm Springs, CA, Paper No. AIAA-2004-1904, 19–22 April, 2004.

Koratkar, N., Wei, B., and Ajayan, P. (2003). Multifunctional structural reinforcement featuring carbon nanotube films. *Composites Science and Technology* 63 (11): 1525–1531.

Kruger, D.H., Mann, A.J. III, and Wiegandt, T. (1997). Placing constrained layer damping patches using reactive shearing structural intensity measurements. *Journal of the Acoustical Society of America* 101 (4): 2075–2082.

Lakes, R.S. (2001). Extreme damping in compliant composites with a negative-stiffness phase. *Philosophical Magazine Letters* 81 (2): 95–100.

Lakes, R.S., Lee, T., Bersie, A., and Wang, Y.C. (2001). Extreme damping in composite materials with negative-stiffness inclusions. *Nature* 410 (6828): 565–567.

Lin, T.-C. and Scott, R.A. (1987). Shape optimization of damping layers. *Proceedings of the 58th Shock and Vibration Symposium, Huntsville, AL* 1: 395–409.

Lumsdaine A., "Topology optimization of constrained damping layer treatments", Proceedings of the ASME International Mechanical Engineering Congress & Exposition Conference, New Orleans, LA, Paper No.: IMECE2002-39021, pp. 149-156, 2002.

Lumsdaine A. and Scott R. A., "Optimal design of constrained plate damping layers using continuum finite elements", Proceedings of the ASME Noise Control and Acoustics Division, 1996 ASME International Mechanical Engineering Congress and Exposition Conference, Atlanta, GA, pp. 159–168, 1996.

Mace, B.R. and Manconi, E. (2012). Wave motion and dispersion phenomena: veering, locking and strong coupling effects. *Journal of the Acoustical Society of America* 131 (2): 1015–1028.

Manconi E. and Mace B. R., "Wave propagation in viscoelastic laminated composite plates using a wave/finite element method", Paper E139, Proceedings of the 7th European Conference on Structural Dynamics, EURODYN 2008, Southampton. UK, 7–8 July 2008.
Mantena, P.R., Gibson, R., and Hwang, S. (1991). Optimal constrained viscoelastic tape lengths for maximizing damping in laminated composites. *Journal of AIAA* 29 (10): 1678–1685.
Mead, D.J. (1998). *Passive Vibration Control*. London: Wiley.
Meirovitch, L. (1967). *Analytical Methods in Vibrations*. New York, NY: Macmillan Publishing Co., Inc.
Meirovitch, L. (2010). *Fundamentals of Vibration*. Long Grove, IL: Waveland.
Moon, F. (1984). *Magneto-Solid Mechanics*. New York, NY: Wiley.
Nadkarni, N., Daraio, C., and Kochmann, D.M. (2014). Dynamics of periodic mechanical structures containing Bistable elastic elements: from elastic to solitary wave propagation. *Physical Review E* 90: 023204.
Nashif, A.D., Jones, D.I.G., and Henderson, J.P. (1985). *Vibration Damping*. Wiley.
Oh, J., Ruzzene, M., and Baz, A. (1999). Control of the dynamic characteristics of passive magnetic composites. *Composites Engineering, Part B* 30 (7): 739–751.
Oh, J., Poh, S., Ruzzene, M., and Baz, A. (2000a). Vibration control of beams using electromagnetic compressional damping treatment. *ASME Journal of Vibration & Acoustics* 122 (3): 235–243.
Oh, J., Ruzzene, M., and Baz, A. (2000b). An analysis of passive magnetic composites for suppressing the vibration of beams. *International Journal of Applied Electromagnetics and Mechanics* 11: 95–116.
Omer, A. and Baz, A. (2000). Vibration control of plates using electromagnetic damping treatment. *Journal of Intelligent Material Systems and Structures* 11 (10): 791–797.
Pai R., Lumsdaine A., and Parsons M. J., "Design and fabrication of optimal constrained layer damping topologies", Proceedings of the SPIE Vol. 5386, SPIE, Bellingham, WA, Smart Structures and Materials 2004: Damping and Isolation (eds K.-W. Wang and W. W. Clark), 2004.
Parin, M., Rogers, L., Moon, Y.I., and Falugi, M. (1989). Practical stand-off damping treatment for sheet metal. *Proceeding of Damping '89* II, Paper No. IBA.
Pavlakovic, B., Lowe, M., Alleyne, O., and Cawley, P. (1997). DISPERSE: a general purpose program for creating dispersion curves. In: *Review of Progress in Quantitative and Nondestructive Evaluation* (ed. D.O. Thompson and D.E. Chimenti). New York: Plenum Press.
Platus D. L., "Negative-stiffness-mechanism vibration isolation systems", Proceedings of SPIE, Vol. 3786, p. 98, 1999.
Plunkett, R. and Lee, C.T. (1970). Length optimization for constrained viscoelastic layer damping. *The Journal of the Acoustical Society of America* 48 (1): 150–161.
Reader, W.T. and Sauter, D.F. (1993). Piezoelectric composites of the 1-3 type used as underwater sound sources. *The Journal of the Acoustical Society of America* 93 (4): 2305. and Proceedings of DAMPING '93 Conference, San Francisco, CA, pp. 1–18, 1993.
Ro, J. and Baz, A. (2002). Vibration control of plates using self-sensing active constrained layer damping networks. *Journal of Vibration and Control* 8 (8): 833–845.

Rogers, L. and Parin, M. (1995). Experimental results for stand-off passive vibration damping treatment. *Passive Damping and Isolation, Proceedings SPIE Smart Structures and Materials* 2445: 374–383.

Roscoe, R. (1969). Bounds for the real and imaginary parts of the dynamic moduli of composite viscoelastic systems. *Journal of the Mechanics and Physics of Solids* 17: 17–22.

Ruzzene, M., Oh, J., and Baz, A. (2000). Finite element Modeling of magnetic constrained layer damping. *Journal of Sound and Vibration* 236 (4): 657–682.

Shields W. H., "Active control of plates using compressional constrained layer damping", Ph. D. Thesis, The Catholic University of America, March 1997.

Shields, W., Ro, J., and Baz, A. (1998). Control of sound radiation from a plate into an acoustic cavity using active piezoelectric-damping composites. *Smart Materials and Structures* 7: 1–11.

Shorter, P.J. (2004). Wave propagation and damping in linear viscoelastic laminates. *Journal of the Acoustical Society of America* 115 (5): 1917–1925.

Silvester, P.P. and Ferrari, R.L. (1996). *Finite Elements for Electrical Engineers*. Cambridge University Press.

Smith, W.A. and Auld, B.A. (1991). Modeling 1-3 composite piezoelectrics: thickness-mode oscillations. *IEEE Transactions on Ultrasonics, Ferroelectrics, and Frequency Control* 38 (1): 40–47.

Spalding, A.B. and Mann, J.A. III (1995). Placing small constrained layer damping patches on a plate to attain global or local velocity changes. *Journal of the Acoustical Society of America* 97: 3617–3624.

Tao, Y., Morris, D.G., Spann, F., and Haugse, E. (1999). Low frequency noise reducing structures using passive and active damping methods. *Passive Damping and Isolation, Proceedings of SPIE Smart Structures and Materials* 3672.

Venkataraman S. and Sankar B., "Analysis of sandwich beams with functionally graded core", Proceedings of the 42nd AIAA/ASME/ASCE/AHS/ASC Structures, Structural Dynamics, and Materials Conference and Exhibition, Seattle, WA, April 16–19 2001, Paper No: AIAA-2001-1281, Anaheim, CA, 2001.

Vokoun, D., Beleggia, M., Heller, L., and Sittner, P. (2009). Magnetostatic interactions and forces between cylindrical permanent magnets. *Journal of Magnetism and Magnetic Materials* 321: 3758–3763.

Wang, Y.C. and Lakes, R.S. (2004). Extreme stiffness systems due to negative stiffness elements. *American Journal of Physics* 72: 40.

Whittier J. S., "The effect of configurational additions using viscoelastic interfaces on the damping of a cantilever beam", Wright Air Development Center, WADC Technical. Report 58–568, 1959.

Yang, B. and Tan, C.A. (1992). Transfer functions of one-dimensional distributed parameter systems. *Journal of Applied Mechanics* 59: 1009–1014.

Yellin, J.M., Shen, I.Y., Reinhall, P.G., and Huang, P. (2000). An analytical and experimental analysis for a one-dimensional passive stand-off layer damping treatment. *ASME Journal of Vibration and Acoustics* 122: 440–447.

Yi, Y., Park, S., and Youn, S. (2000). Design of microstructures of viscoelastic composites for optimal damping characteristics. *International Journal of Solids & Structures* 37: 4791–4810.

8.A Matrices of the Models of a Passive Stand-Off Layer

8.A.1 Distributed Transfer Function Model

$$\mathbf{F}(\mathbf{s}) = \begin{bmatrix} 0 & 1 & 0 & 0 & 0 & 0 & 0 & 0 \\ 0 & 0 & 1 & 0 & 0 & 0 & 0 & 0 \\ 0 & 0 & 0 & 1 & 0 & 0 & 0 & 0 \\ -s^2\left(\frac{a_3}{a_3 - a^2{}_1}\right) & 0 & \frac{a_3}{(a_3 - a_1^2)}\left[\varepsilon - \frac{\beta\varepsilon a_1}{a_3}\right] & 0 & 0 & \frac{a_3}{(a_3 - a_1^2)}\left[\varepsilon\beta - \frac{\varepsilon\beta^2 a_1}{a_3}\right] & 0 & \frac{a_3}{(a_3 - a_1^2)}\left[\frac{c_2^2 s^2 a_1}{a_3} + \frac{\beta^2\varepsilon a_1}{a_3} - \beta\varepsilon\right] \\ 0 & 0 & 0 & 0 & 0 & 1 & 0 & 0 \\ 0 & \frac{+\varepsilon\beta}{a_2} & 0 & 0 & \frac{c_1^2 s^2 + \varepsilon\beta^2}{a_2} & 0 & \frac{-\varepsilon\beta^2}{a_2} & 0 \\ 0 & 0 & 0 & 0 & 0 & 0 & 0 & 1 \\ 0 & \frac{-\beta\varepsilon}{a_3} & 0 & \frac{a_1}{a_3} & \frac{-\varepsilon\beta^2}{a_3} & 0 & \frac{c_2^2 s^2 + \varepsilon\beta^2}{a_3} & 0 \end{bmatrix}$$

$$M(s) = \begin{bmatrix} 1 & 0 & 0 & 0 & 0 & 0 & 0 & 0 \\ 0 & 1 & 0 & 0 & 0 & 0 & 0 & 0 \\ 0 & 0 & 0 & 0 & 0 & 1 & 0 & 0 \\ 0 & 0 & 0 & 0 & 0 & 0 & 1 & 0 \\ 0 & 0 & 0 & 0 & 0 & 0 & 0 & 0 \\ 0 & 0 & 0 & 0 & 0 & 0 & 0 & 0 \\ 0 & 0 & 0 & 0 & 0 & 0 & 0 & 0 \\ 0 & 0 & 0 & 0 & 0 & 0 & 0 & 0 \end{bmatrix},$$

$$N(s) = \begin{bmatrix} 0 & 0 & 0 & 0 & 0 & 0 & 0 & 0 \\ 0 & 0 & 0 & 0 & 0 & 0 & 0 & 0 \\ 0 & 0 & 0 & 0 & 0 & 0 & 0 & 0 \\ 0 & 0 & 0 & 0 & 0 & 0 & 0 & 0 \\ 0 & 0 & 1 & 0 & 0 & 0 & 0 & 0 \\ 0 & 0 & 0 & 0 & 1 & 0 & 0 & 0 \\ 0 & 0 & 0 & 0 & 0 & 0 & 0 & 1 \\ 0 & \varepsilon\left(1 - \frac{\beta a_1}{a_3}\right) & 0 & -\frac{1}{\mathrm{K}} & \varepsilon\left(\beta - \frac{\beta^2 a_1}{a_3}\right) & 0 & \varepsilon\left(\frac{\beta^2 a_1}{a_3} - \beta\right) & 0 \end{bmatrix}$$

and

$$\gamma(s) = \{P(s) \ 0 \ 0 \ 0 \ 0 \ 0 \ 0 \ 0\}^T$$

8.A.2 Finite Element Model

The stiffness matrix [$\boldsymbol{K}$] is given by $K = [K_1\, K_2\, K_3\, K_4\, K_5\, K_6\, K_7\, K_8]$ where $\boldsymbol{K_i}$ values are as follows

$$K_1 = \begin{bmatrix} K_1\dfrac{d^2v_1}{dx^2}\dfrac{d^2v_1}{dx^2} + K_2\dfrac{dv_2}{dx}\dfrac{dv_1}{dx} + s^2v_1v_1 \\ K_1\dfrac{d^2v_1}{dx^2}\dfrac{d^2v_2}{dx^2} + K_2\dfrac{dv_2}{dx}\dfrac{dv_2}{dx} + s^2v_1v_2 \\ -K_8\dfrac{dv_1}{dx}\theta_1 \\ a_1\dfrac{d^3v_1}{dx^3}\phi_1 + K_8\dfrac{dv_1}{dx}\phi_1 \\ K_1\dfrac{d^2v_1}{dx^2}\dfrac{d^2v_3}{dx^2} + K_2\dfrac{dv_2}{dx}\dfrac{dv_3}{dx} + s^2v_1v_3 \\ K_1\dfrac{d^2v_1}{dx^2}\dfrac{d^2v_4}{dx^2} + K_2\dfrac{dv_2}{dx}\dfrac{dv_4}{dx} + s^2v_1v_4 \\ -K_8\dfrac{dv_1}{dx}\theta_2 \\ a_1\dfrac{d^3v_1}{dx^3}\phi_2 + K_8\dfrac{dv_1}{dx\phi_2} \end{bmatrix}, \quad K_2 = \begin{bmatrix} K_1\dfrac{d^2v_2}{dx^2}\dfrac{d^2v_1}{dx^2} + K_2\dfrac{dv_2}{dx}\dfrac{dv_1}{dx} + s^2v_2v_1 \\ K_1\dfrac{d^2v_2}{dx^2}\dfrac{d^2v_2}{dx^2} + K_2\dfrac{dv_2}{dx}\dfrac{dv_2}{dx} + s^2v_2v_2 \\ -K_8\dfrac{dv_2}{dx}\theta_1 \\ a_1\dfrac{d^3v_2}{dx^3}\phi_1 + K_8\dfrac{dv_2}{dx}\phi_1 \\ K_1\dfrac{d^2v_2}{dx^2}\dfrac{d^2v_3}{dx^2} + K_2\dfrac{dv_2}{dx}\dfrac{dv_3}{dx} + s^2v_2v_3 \\ K_1\dfrac{d^2v_2}{dx^2}\dfrac{d^2v_4}{dx^2} + K_2\dfrac{dv_2}{dx}\dfrac{dv_4}{dx} + s^2v_2v_4 \\ -K_8\dfrac{dv_2}{dx}\theta_2 \\ a_1\dfrac{d^3v_2}{dx^3}\phi_2 + K_8\dfrac{dv_2}{dx}\phi_2 \end{bmatrix},$$

$$K_3 = \begin{bmatrix} K_3\dfrac{d\theta_1}{dx}v_1 \\ K_3\dfrac{d\theta_1}{dx}v_2 \\ a_2\dfrac{d\theta_1}{dx}\dfrac{d\theta_1}{dx} + K_5\theta_1\theta_1 \\ K_6\theta_1\phi_1 \\ K_3\dfrac{d\theta_1}{dx}v_3 \\ K_3\dfrac{d\theta_1}{dx}v_4 \\ a_2\dfrac{d\theta_1}{dx}\dfrac{d\theta_2}{dx} + K_5\theta_1\theta_2 \\ K_6\theta_1\phi_2 \end{bmatrix}, \quad K_4 = \begin{bmatrix} K_4\dfrac{d\phi_1}{dx}v_1 \\ K_4\dfrac{d\phi_1}{dx}v_2 \\ K_6\phi_1\theta_1 \\ a_3\dfrac{d\phi_1}{dx}\dfrac{d\phi_1}{dx} + K_7\phi_1\phi_1 \\ K_4\dfrac{d\phi_1}{dx}v_3 \\ K_4\dfrac{d\phi_1}{dx}v_4 \\ K_6\phi_1\theta_2 \\ a_3\dfrac{d\phi_1}{dx}\dfrac{d\phi_2}{dx} + K_7\phi_2\phi_1 \end{bmatrix},$$

$$K_5 = \begin{bmatrix} K_1 \frac{d^2 v_3}{dx^2}\frac{d^2 v_1}{dx^2} + K_2 \frac{dv_4}{dx}\frac{dv_1}{dx} + s^2 v_3 v_1 \\ K_1 \frac{d^2 v_3}{dx^2}\frac{d^2 v_2}{dx^2} + K_2 \frac{dv_4}{dx}\frac{dv_2}{dx} + s^2 v_3 v_2 \\ -K_8 \frac{dv_3}{dx}\theta_1 \\ a^3{}_1 \frac{d^3 v_3}{dx^3}\phi_1 + K_8 \frac{dv_3}{dx}\phi_1 \\ K_1 \frac{d^2 v_3}{dx^2}\frac{d^2 v_3}{dx^2} + K_2 \frac{dv_4}{dx}\frac{dv_3}{dx} + s^2 v_3 v_3 \\ K_1 \frac{d^2 v_3}{dx^2}\frac{d^2 v_4}{dx^2} + K_2 \frac{dv_4}{dx}\frac{dv_4}{dx} + s^2 v_3 v_4 \\ -K_8 \frac{dv_3}{dx}\theta_2 a^3{}_1 \frac{d^3 v_3}{dx^3}\phi_2 + K_8 \frac{dv_3}{dx}\phi_2 \end{bmatrix}, \quad K_6 = \begin{bmatrix} K_1 \frac{d^2 v_4}{dx^2}\frac{d^2 v_1}{dx^2} + K_2 \frac{dv_4}{dx}\frac{dv_1}{dx} + s^2 v_4 v_1 \\ K_1 \frac{d^2 v_4}{dx^2}\frac{d^2 v_2}{dx^2} + K_2 \frac{dv_4}{dx}\frac{dv_2}{dx} + s^2 v_4 v_2 \\ -K_8 \frac{dv_4}{dx}\theta_1 \\ a^3{}_1 \frac{d^3 v_4}{dx^3}\phi_1 + K_8 \frac{dv_4}{dx}\phi_1 \\ K_1 \frac{d^2 v_4}{dx^2}\frac{d^2 v_3}{dx^2} + K_2 \frac{dv_4}{dx}\frac{dv_3}{dx} + s^2 v_4 v_3 \\ K_1 \frac{d^2 v_4}{dx^2}\frac{d^2 v_4}{dx^2} + K_2 \frac{dv_4}{dx}\frac{dv_4}{dx} + s^2 v_4 v_4 \\ -K_8 \frac{dv_4}{dx}\theta_2 \\ a^3{}_1 \frac{d^3 v_4}{dx^3}\phi_2 + K_8 \frac{dv_4}{dx}\phi_2 \end{bmatrix},$$

$$K_7 = \begin{bmatrix} K_3 \frac{d\theta_2}{dx} v_1 \\ K_3 \frac{d\theta_2}{dx} v_2 \\ a_2 \frac{d\theta_2}{dx}\frac{d\theta_1}{dx} + K_5 \theta_2 \theta_1 \\ K_6 \theta_2 \phi_1 \\ K_3 \frac{d\theta_2}{dx} v_3 \\ K_3 \frac{d\theta_2}{dx} v_4 \\ a_2 \frac{d\theta_2}{dx}\frac{d\theta_2}{dx} + K_5 \theta_2 \theta_2 \\ K_6 \theta_2 \phi_2 \end{bmatrix}, \quad K_8 = \begin{bmatrix} K_4 \frac{d\phi_2}{dx} v_1 \\ K_4 \frac{d\phi_2}{dx} v_2 \\ K_6 \phi_2 \theta_1 \\ a_3 \frac{d\phi_2}{dx}\frac{d\phi_1}{dx} + K_7 \phi_2 \phi_1 \\ K_4 \frac{d\phi_2}{dx} v_3 \\ K_4 \frac{d\phi_2}{dx} v_4 \\ K_6 \phi_2 \theta_2 \\ a_3 \frac{d\phi_2}{dx}\frac{d\phi_2}{dx + K_7 \phi_2 \phi_2} \end{bmatrix}$$

8.B The Electromechanical Coupling Factor of One Piezoelectric Rod

In the one-dimensional case, these equations, as given in the IEEE Standard on Piezoelectricity (1987), are:

$$S_3 = s_{33}^E T_3 + d_{33} E_3, \tag{8.B.1}$$

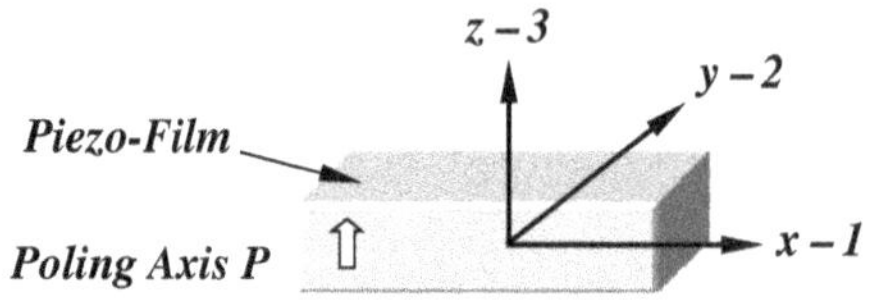

Figure 8.B.1 Coordinate system of piezoelectric film.

and

$$D_3 = d_{33}\, T_3 + \varepsilon_{33}^T E_3 \tag{8.B.2}$$

Equation (8.B.1) describes the mechanical strain S_3, in direction 3, due to a mechanical stress T_3, in direction 3, and an electrical field E_3, in direction 3. In Eq. (8.B.2), the electric displacement D_3, in direction 3, is generated by the mechanical stress T_3 and the electric field E_3. The directions assigned in the constitutive equations are defined in Figure 8.B.1 in relation to the poling axis P of the piezo-film.

From Eq. (8.B.2), the electric field intensity, E_3, can be written as:

$$E_3 = \frac{1}{\varepsilon_{33}^T} D_3 - \frac{d_{33}}{\varepsilon_{33}^T T_3} \tag{8.B.3}$$

Eliminating E_3 from Eq. (8.B.1) gives:

$$S_3 = s_{33}^E\left(1 - k_{33}^2\right) T_3 + g_{33} D_3,$$

or

$$S_3 = s_{33}^D T_3 + g_{33} D_3. \tag{8.B.4}$$

where $s_{33}^D = s_{33}^E\left(1 - k_{33}^2\right) =$ open-circuit compliance, $k_{33}^2 = d_{33}^2 / \left(s_{33}^E \varepsilon_{33}^T\right)$ electromechanical coupling factor (EMCF) and $g_{33} = d_{33}/\varepsilon_{33}^T =$ piezoelectric constant.

Note that the EMCF can be written as:

$$k_{33}^2 = 1 - s_{33}^D / s_{33}^E = d_{33}^2 / \left(s_{33}^E \varepsilon_{33}^T\right) \tag{8.B.5}$$

Under stress-free conditions, $T_3 = 0$. Hence, Eqs. (8.B.1) and (8.B.2) reduce to:

$$S_3 = d_{33} E_3, \quad \text{and} \quad D_3 = \varepsilon_{33}^T E_3 \tag{8.B.6}$$

Then, the compression energy $U_{compression}$ is given by:

$$U_{compression} = \frac{1}{2} bt \int_0^L \frac{1}{s_{33}^E} S_3^2 dx = \frac{btd_{33}^2 E_3^2 L}{2\, s_{33}^E} \tag{8.B.7}$$

Also, the electrical energy $U_{electrical}$ is given by:

$$\begin{aligned} U_{electrical} &= \frac{1}{2} bt \int_0^L D_3 E_3 dx \\ &= \frac{1}{2} bt \int_0^L \varepsilon_{33}^T E_3^2 dx = \frac{1}{2} btL \varepsilon_{33}^T E_3^2 \end{aligned} \tag{8.B.8}$$

Therefore, from Eqs. (8.B.5), (8.B.7), and (8.B.8) we have:

$$k_{31}^2 = \frac{d_{31}^2}{s_{11}^E \, \varepsilon_{33}^T} = \frac{Compression \quad Energy}{Electrical \quad Energy} \tag{8.B.9}$$

8.C Constitutive Equations of APDC

The constitutive equations of the APDC are derived according to Baz and Tempia (1998). The local coordinates of the piezoelectric rods are denoted by 1, 2, and 3, whereas the global system coordinates are denoted by x, y, and z. The constitutive equations for the component phases give the stress and electric displacement as functions of the strain and electric field. For points within the polymer phase, which is assumed to be isotropic and homogeneous, we have,

$$\left\{T_g^m\right\} = \left[c_g^m\right]\left\{S_g^m\right\} \tag{8.C.1}$$

where the superscript m denotes polymeric matrix, subscript g denotes global coordinates and $[c]$is the elastic stiffness matrix. For the piezoelectric rods, the constitutive equations in local coordinates may be written as,

$$\left\{T_l^p\right\} = \left[c_l^p\right]\left\{S_l^p\right\} - \left\{e_l^p\right\}\left\{E_l\right\} \tag{8.C.2}$$

where the superscript p denotes piezoelectric phase and subscript l denotes local coordinates. Using appropriate rotation matrices, the local, and global coordinate systems may be related, and we obtain,

$$\left\{T_g^p\right\} = [c]\left\{S_g^p\right\} - \left\{e_g^p\right\}\left\{E_g\right\} \tag{8.C.3}$$

Adopting the assumptions of Baz and Tempia (1998), we then obtain the lateral strains in the polymer phase as

$$S_x^m = AS_z + BS_{yz} - CE_z \tag{8.C.4}$$

$$S_y^m = LS_z + MS_{yz} - OE_z \tag{8.C.5}$$

where A, B, C, L, M, and O are constant coefficients given by,

$$A = \frac{\left[c_{13}^p - c_{13}^m\right]\left[c_{22}^m + f\cdot c_{22}^p\right] - \left[c_{12}^m + f\cdot c_{12}^p\right]\left[c_{23}^p - c_{23}^m\right]}{\left[c_{22}^m + f\cdot c_{22}^p\right]\left[c_{11}^m + f\cdot c_{11}^p\right] - \left[c_{12}^m + f\cdot c_{12}^p\right]^2}, \tag{8.C.6}$$

$$B = c_{14}^p\left[c_{22}^m + f\cdot c_{22}^p\right] - \left[c_{12}^m + f\cdot c_{12}^p\right]\frac{c_{24}^p}{\left[c_{22}^m + f\cdot c_{22}^p\right]\left[c_{11}^m + f\cdot c_{11}^p\right] - \left[c_{12}^m + f\cdot c_{12}^p\right]^2}, \tag{8.C.7}$$

$$C = \frac{c_{19}^p\left[c_{22}^m + f\cdot c_{22}^p\right] - \left[c_{12}^m + f\cdot c_{12}^p\right]c_{29}^p}{\left[c_{22}^m + f\cdot c_{22}^p\right]\left[c_{11}^m + f\cdot c_{11}^p\right] - \left[c_{12}^m + f\cdot c_{12}^p\right]^2}, \tag{8.C.8}$$

$$L = \frac{\left[c_{23}^{p} - c_{23}^{m}\right]\left[c_{11}^{m} + f\cdot c_{11}^{p}\right] - \left[c_{12}^{m} + f\cdot c_{12}^{p}\right]\left[c_{13}^{p} - c_{13}^{m}\right]}{\left[c_{22}^{m} + f\cdot c_{22}^{p}\right]\left[c_{11}^{m} + f\cdot c_{11}^{p}\right] - \left[c_{12}^{m} + f\cdot c_{12}^{p}\right]^{2}}, \tag{8.C.9}$$

$$M = \frac{c_{24}^{p}\left[c_{11}^{m} + f\cdot c_{11}^{p}\right] - \left[c_{12}^{m} + f\cdot c_{12}^{p}\right]c_{14}^{p}}{\left[c_{22}^{m} + f\cdot c_{22}^{p}\right]\left[c_{11}^{m} + f\cdot c_{11}^{p}\right] - \left[c_{12}^{m} + f\cdot c_{12}^{p}\right]^{2}}, \tag{8.C.10}$$

and

$$O = \frac{c_{29}^{p}\left[c_{11}^{m} + f\cdot c_{11}^{p}\right] - \left[c_{12}^{m} + f\cdot c_{12}^{p}\right]c_{19}^{p}}{\left[c_{22}^{m} + f\cdot c_{22}^{p}\right]\left[c_{11}^{m} + f\cdot c_{11}^{p}\right] - \left[c_{12}^{m} + f\cdot c_{12}^{p}\right]^{2}} \tag{8.C.11}$$

where the constants c_{ij} are the coefficients of the stiffness matrix $[c]$ of Eq. (8.C.3). These coefficients are given in terms of the elements of the stiffness and rotation matrices.

Once the strains in one phase are determined, the strains in the other phase are obtained by,

$$S_{x,y}^{p} = -f\, S_{x,y}^{m} \tag{8.C.12}$$

where

$$f = \frac{(1-\mu)}{\mu} \tag{8.C.13}$$

and μ is the volume fraction of the piezoelectric rods. In this way, the stress in the z-direction, shear strain in the x–z plane and electric field in the z-direction may be obtained in terms of the corresponding strains and applied electric field, and this serves as the overall constitutive law of the APDC.

8.D Magnetic Forces in the Passive Magnetic Composite

The magnetic energy expression is a sum of integrals over all the finite elements (Eq. (8.129)). Each of the integrals depends on the coordinates of the nodes. The evaluation of the derivative of the coordinates of the nodes allows the evaluation of the derivative of each element integral. In order to compute the x component of the force, the derivative with respect to x of the nodal coordinates $[x_i\ y_i]$ such that:

$$\frac{\partial x_i}{\partial x} = p, \quad \frac{\partial y_i}{\partial x} = 0 \tag{8.D.1}$$

with $p = 1$ for the nodes belonging to the movable part and $p = 0$ for the nodes of all the other elements. The contribution to the magnetic energy given by eth element is:

$$\Delta W^{e} = V^{e}\cdot \int_{0}^{\mathbf{B}} \frac{1}{\mu_0}\cdot \mathbf{B}\cdot d\mathbf{B} \tag{8.D.2}$$

with V^e denoting the volume of the element, which, in the two-dimensional case, can be expressed as the product of the off plane width b of the meshed region, times its area Δ^e. Integrating Eq. (8.D.2), and introducing the finite element notation, we then have:

$$\Delta W^e = \frac{1}{2\cdot\mu_0}\cdot b\cdot\Delta^e\cdot\mathbf{B}^{eT}\cdot\mathbf{B}^e \tag{8.D.3}$$

Expressing the magnetic induction as a function of the magnetic potential through Eq. (8.122), Eq. (8.D.3) becomes:

$$\Delta W^e = \frac{b\cdot\Delta^e}{2\cdot\mu_0}\cdot\{\mathbf{A}^e\}^T\cdot\left[\mathbf{D}^T\cdot\mathbf{J}^{-1T}\cdot\mathbf{J}^{-1}\cdot\mathbf{D}\right]\cdot\{\mathbf{A}^e\} \tag{8.D.4}$$

The contribution of the element e to the global force can then be obtained from Eqs. (8.133) and (8.D.4) as follows:

$$F^e = -\frac{\partial\Delta W^e}{\partial s} = -\frac{b}{2\cdot\mu_0}\cdot\{\mathbf{A}^e\}^T\cdot\frac{\partial}{\partial s}\left[\mathbf{D}^T\cdot\mathbf{J}^{-1T}\cdot\mathbf{J}^{-1}\cdot\mathbf{D}\cdot\Delta^e\right]\cdot\{\mathbf{A}^e\} \tag{8.D.5}$$

This equation reduces to:

$$F^e = -\frac{b}{2\cdot\mu_0}\cdot\{\mathbf{A}^e\}^T\cdot\mathbf{D}^T\cdot\mathbf{J}^{-1T}\cdot\left[\left(\mathbf{J}^{-1}\cdot\frac{\partial\mathbf{J}}{\partial s}+\frac{\partial\mathbf{J}^T}{\partial s}\mathbf{J}^{-1}\right)\cdot\Delta^e+\frac{\partial\Delta^e}{\partial s}\right]\cdot\mathbf{J}^{-1}\cdot\mathbf{D}\cdot\{\mathbf{A}^e\} \tag{8.D.6}$$

where it can be shown that (Coulomb and Meunier 1984):

$$\frac{\partial\mathbf{J}}{\partial s} = \mathbf{D}\cdot\begin{bmatrix}\frac{\partial x_1}{\partial s} & \frac{\partial y_1}{\partial s}\\ \frac{\partial x_2}{\partial s} & \frac{\partial y_2}{\partial s}\\ \frac{\partial x_3}{\partial s} & \frac{\partial y_3}{\partial s}\end{bmatrix}, \tag{8.D.7}$$

and:

$$\frac{\partial\Delta^{(e)}}{\partial s} = -|\mathbf{J}|^{-1}\cdot\frac{\partial|\mathbf{J}|}{\partial \mathbf{s}}\cdot\Delta^{(e)}. \tag{8.D.8}$$

The final expression for the contribution of the eth element to the global forces acting on the movable part of the region is given by:

$$F^e = -\frac{b}{2\cdot\mu_0}\cdot\{\mathbf{A}^e\}^T\cdot\mathbf{D}^T\cdot\mathbf{J}^{-1T}\cdot\left[\left(\mathbf{J}^{-1}\cdot\frac{\partial\mathbf{J}}{\partial s}+\frac{\partial\mathbf{J}^T}{\partial s}\mathbf{J}^{-1}\right)-|\mathbf{J}|^{-1}\cdot\frac{\partial|\mathbf{J}|}{\partial s}\right]\cdot\mathbf{J}^{-1}\cdot\mathbf{D}\cdot\{\mathbf{A}^e\}\cdot\Delta^e \tag{8.D.9}$$

8.E Stiffness and Mass Matrices Passive Magnetic Composite (PMC)

The mass and stiffness matrices of a beam-PMC element can be obtained expressing the total strain and kinetic energies of the treatment. In particular the kinematic expressions are used, while Eq. (8.140) can be considered to express the energies in matrix form.

8.E.1 Strain Energies

The strain energies associated with the various layers of the PMC treatment have the following expressions:

Extension of constrained layer 1

$$U_1 = \frac{1}{2} \cdot E_1 A_1 \cdot \{\boldsymbol{\delta}^e\}^T \cdot \mathbf{T}^{-1T} \cdot \left(\int_0^L \{\mathbf{N}_1'\}^T \cdot \{\mathbf{N}_1'\} \cdot dx \right) \cdot \mathbf{T}^{-1} \cdot \{\boldsymbol{\delta}^e\} \tag{8.E.1}$$

where E_1 and A_1 are the Young's modulus and the area of cross section of the constraining layer, respectively, and L is the length of the element.

Extension of the base beam/magnet layer

$$U_2 = \frac{1}{2} \cdot (E_3 A_3 + E_1 A_1) \cdot \{\boldsymbol{\delta}^e\}^T \cdot \mathbf{T}^{-1T} \cdot \int_0^L \{\mathbf{N}_4'\}^T \cdot \{\mathbf{N}_4'\} \cdot dx \cdot \mathbf{T}^{-1} \cdot \{\boldsymbol{\delta}^e\} \tag{8.E.2}$$

where E_3 and A_3 are the Young's modulus and the area of cross section of the base beam. It is assumed that the constraining layer applied on the base beam is identical to the constraining layer 1.

Extension of the viscoelastic layer

$$U_3 = \frac{1}{2} \cdot E_2 A_2 \cdot \{\boldsymbol{\delta}^e\}^T \cdot \mathbf{T}^{-1T} \cdot \int_0^L \{\mathbf{N}_7'\}^T \cdot \{\mathbf{N}_7'\} \cdot dx \cdot \mathbf{T}^{-1} \cdot \{\boldsymbol{\delta}^e\} \tag{8.E.3}$$

where E_2 and A_2 are the Young's modulus and the area of cross section of viscoelastic layer 2, and $\{\mathbf{N}_7\}$ is an interpolating vector obtained considering Eq. (8.134).

Shear of the viscoelastic layer

$$U_4 = \frac{1}{2} \cdot G_2 A_2 \cdot \{\boldsymbol{\delta}^e\}^T \cdot \mathbf{T}^{-1T} \cdot \int_0^L \{\mathbf{N}_8'\}^T \cdot \{\mathbf{N}_8'\} \cdot dx \cdot \mathbf{T}^{-1} \cdot \{\boldsymbol{\delta}^e\} \tag{8.E.4}$$

where G_2 is the shear modulus of viscoelastic layer and $\{\mathbf{N}_8\}$ is an interpolating vector obtained considering Eqs. (8.134).

Compression of the viscoelastic layer

$$U_5 = \frac{1}{2} \cdot \frac{E_2 b}{h_2} \cdot \{\boldsymbol{\delta}^e\}^T \cdot \mathbf{T}^{-1T} \cdot \int_0^L \{\mathbf{N}_9\}^T \cdot \{\mathbf{N}_9\} \cdot dx \cdot \mathbf{T}^{-1} \cdot \{\boldsymbol{\delta}^e\} \tag{8.E.5}$$

where $\{\mathbf{N}_9\}$ is a vector obtained considering the difference of the transverse displacements of the constraining layer 1 and of the base beam/magnet layer, that is:

$$\{\mathbf{N}_9\} = \{\mathbf{N}_5\} - \{\mathbf{N}_2\} \tag{8.E.6}$$

Bending of the constraining layer

$$U_6 = \frac{1}{2} \cdot E_1 I_1 \cdot \{\boldsymbol{\delta}^e\}^T \cdot \mathbf{T}^{-1T} \cdot \int_0^L \left\{\mathbf{N}_2''\right\}^T \cdot \left\{\mathbf{N}_2''\right\} \cdot dx \cdot \mathbf{T}^{-1} \cdot \{\boldsymbol{\delta}^e\} \tag{8.E.7}$$

where $E_1 I_1$ is the flexural rigidity of layer 1.

Bending of the base beam/magnet layer

$$U_7 = \frac{1}{2} \cdot (E_3 I_3 + E_1 I_1) \cdot \{\boldsymbol{\delta}^e\}^T \cdot \mathbf{T}^{-1T} \cdot \int_0^L \left\{\mathbf{N}_5''\right\}^T \cdot \left\{\mathbf{N}_5''\right\} \cdot dx \cdot \mathbf{T}^{-1} \cdot \{\boldsymbol{\delta}^e\} \tag{8.E.8}$$

where $E_3 I_3$ is the flexural rigidity of the base beam.

Bending of the viscoelastic layer

$$U_8 = \frac{1}{2} \cdot E_2 I_2 \cdot \{\boldsymbol{\delta}^e\}^T \cdot \mathbf{T}^{-1T} \cdot \int_0^L \left\{\mathbf{N}_{10}''\right\}^T \cdot \left\{\mathbf{N}_{10}''\right\} \cdot dx \cdot \mathbf{T}^{-1} \cdot \{\boldsymbol{\delta}^e\} \tag{8.E.9}$$

where $E_2 I_2$ is the flexural rigidity of the viscoelastic core and $\left\{\mathbf{N}_{10}''\right\}$ is an interpolating vector expressed as:

$$\left\{\mathbf{N}_{10}''\right\} = \frac{1}{h_2}\left[\{\mathbf{N}_1'\} - \{\mathbf{N}_2'\} + \left(\frac{h_1}{2} + D\right) \cdot \left(\frac{\{\mathbf{N}_2''\} + \{\mathbf{N}_5''\}}{2}\right)\right] \tag{8.E.10}$$

The total strain energy will be given by the sum of all the contributions listed previously:

$$U = \sum_{i=1}^{8} U_i = \frac{1}{2} \cdot \{\boldsymbol{\delta}^e\}^T \cdot \mathbf{K}^e \cdot \{\boldsymbol{\delta}^e\} \tag{8.E.11}$$

where $\mathbf{K}^e$ is the stiffness matrix of the element.

8.E.2 Kinetic Energies

The total KE of the element is given by the sum of the following contributions:

Constraining layer 1

$$T_1 = \frac{1}{2} \cdot \rho_1 A_1 \cdot \left\{\dot{\boldsymbol{\delta}}^e\right\}^T \cdot \mathbf{T}^{-1T} \cdot \int_0^L \left(\{\mathbf{N}_1\}^T \cdot \{\mathbf{N}_1\} + \{\mathbf{N}_2\}^T \cdot \{\mathbf{N}_2\}\right) \cdot dx \cdot \mathbf{T}^{-1} \cdot \left\{\dot{\boldsymbol{\delta}}^e\right\} \tag{8.E.12}$$

where ρ_1 is the density of the constraining layer 1.

Base beam/magnet

$$T_3 = \frac{1}{2}\cdot(\rho_3 A_3 + \rho_1 A_1)\cdot\left\{\dot{\boldsymbol{\delta}}^e\right\}^T\cdot\mathbf{T}^{-1T}\cdot\int_0^L\left(\{\mathbf{N}_4\}^T\cdot\{\mathbf{N}_4\} + \{\mathbf{N}_5\}^T\cdot\{\mathbf{N}_5\}\right)\cdot dx\cdot\mathbf{T}^{-1}\cdot\left\{\dot{\boldsymbol{\delta}}^e\right\} \tag{8.E.13}$$

where ρ_3 is the density of the base beam.

Viscoelastic layer 2

$$T_4 = \frac{1}{2}\cdot\rho_2 A_2\cdot\left\{\dot{\boldsymbol{\delta}}^e\right\}^T\cdot\mathbf{T}^{-1T}\cdot\int_0^L\left(\{\mathbf{N}_{11}\}^T\cdot\{\mathbf{N}_{11}\} + \{\mathbf{N}_{12}\}^T\cdot\{\mathbf{N}_{12}\}^T\right)\cdot dx\cdot\mathbf{T}^{-1}\cdot\left\{\dot{\boldsymbol{\delta}}^e\right\} \tag{8.E.14}$$

where ρ_2 is the density of the viscoelastic layer 2 and $\left\{\mathbf{N}_{11}''\right\}$ and $\left\{\mathbf{N}_{12}''\right\}$ are interpolating vectors expressed as:

$$\{\mathbf{N}_{11}\} = \frac{1}{2}\left[\{\mathbf{N}_1\} + \{\mathbf{N}_4\} + \left(\frac{h_1}{2} - D\right)\cdot\left(\frac{\{\mathbf{N}_2'\} + \{\mathbf{N}_5'\}}{2}\right)\right] \tag{8.E.15}$$

$$\{\mathbf{N}_{12}\} = \frac{\{\mathbf{N}_2\} + \{\mathbf{N}_5\}}{2} \tag{8.E.16}$$

The total KE will be given by the sum of the terms expressed here:

$$T = \sum_{i=1}^{5} T_i = \frac{1}{2}\cdot\left\{\dot{\boldsymbol{\delta}}^e\right\}^T\cdot\mathbf{M}^e\cdot\left\{\dot{\boldsymbol{\delta}}^e\right\} \tag{8.E.17}$$

where $\mathbf{M}^e$ is the mass matrix of the element.

8.E.3 Work Done by the Magnetic Forces

From Eq. (8.146), and introducing the shape function can be expressed:

$$\begin{aligned} W_m = {} & \frac{F^e_{m_{0_z}}}{L}\cdot\int_0^L\{\mathbf{N}_9\}^T\cdot dx\cdot\mathbf{T}^{-1}\cdot\{\boldsymbol{\delta}^e\} \\ & + \frac{1}{L}\cdot\partial\frac{F^e_{m_z}}{\partial(w_1 - w_3)}\cdot\{\boldsymbol{\delta}^e\}^T\cdot\mathbf{T}^{-1T}\int_0^L\cdot\{\mathbf{N}_9\}^T\cdot\{\mathbf{N}_9\}\cdot dx\cdot\mathbf{T}^{-1}\cdot\{\boldsymbol{\delta}^e\} \end{aligned} \tag{8.E.18}$$

which can be written in compact form as follows:

$$W_m = \frac{F^e_{m_{0z}}}{L} \cdot \int_0^L \{\mathbf{N}_9\}^T \cdot dx \cdot \mathbf{T}^{-1} \cdot \{\boldsymbol{\delta}^e\}$$
$$+ \frac{1}{L} \cdot \partial \frac{F^e_{m_z}}{\partial(w_1 - w_3)} \cdot \{\boldsymbol{\delta}^e\}^T \cdot \mathbf{T}^{-1T} \cdot \mathbf{K_m} \cdot \mathbf{T}^{-1} \cdot \{\boldsymbol{\delta}^e\} \tag{8.E.19}$$

8.E.4 Geometric Stiffness Matrix

When a beam element is subjected to an axial force in addition to flexural loading, its stiffness matrix has to be modified to consider the effect of the axial load on the bending behavior. The strain energy associated to this effect is given by:

$$U_a = \frac{1}{2} \cdot P \cdot \int_0^L \left(\frac{\partial w}{\partial x}\right)^2 dx \tag{8.E.20}$$

where P is the axial load on the element. Introducing the finite element formulation, then:

$$U_a = \frac{1}{2} \cdot P \cdot \{\boldsymbol{\delta}^e\}^T \cdot \mathbf{T}^{-1T} \cdot \int_0^L \{\mathbf{N'}_4\}^T \cdot \{\mathbf{N'}_4\} \cdot dx \cdot \mathbf{T}^{-1} \cdot \{\boldsymbol{\delta}^e\}$$
$$= \frac{1}{2} \cdot \{\boldsymbol{\delta}^e\}^T \cdot \mathbf{K_{geo}} \cdot \{\boldsymbol{\delta}^e\} \tag{8.E.21}$$

where $\mathbf{K_{geo}}$ is the geometric stiffness matrix of the element.

Problems

8.1 Consider the beam/PSOL assembly shown in Figure 8.6 (in Example 8.1). The main geometrical and physical parameters of the beam/PSOL assembly are listed in Table 8.1. The assembly is mounted in a simply-supported manner at its ends A and B. The beam is excited at its mid-span by a sinusoidal excitation in the transverse direction with amplitude of 1 mm.

Determine:

a) the natural frequencies and mode shapes of the beam/PSOL assembly.
b) the frequency response of the mid-span point of the assembly using the DTF method.

8.2 Compare the results obtained for the beam/PSOL assembly of Problem 8.1 by developing an ANSYS finite element program for the free-free beam as shown in Figure 8.6.

a) Compare the natural frequencies and mode shapes of the beam/PCLD as predicted by ANSYS with the theoretical predictions as obtained in Problem 1 by the DTF method.

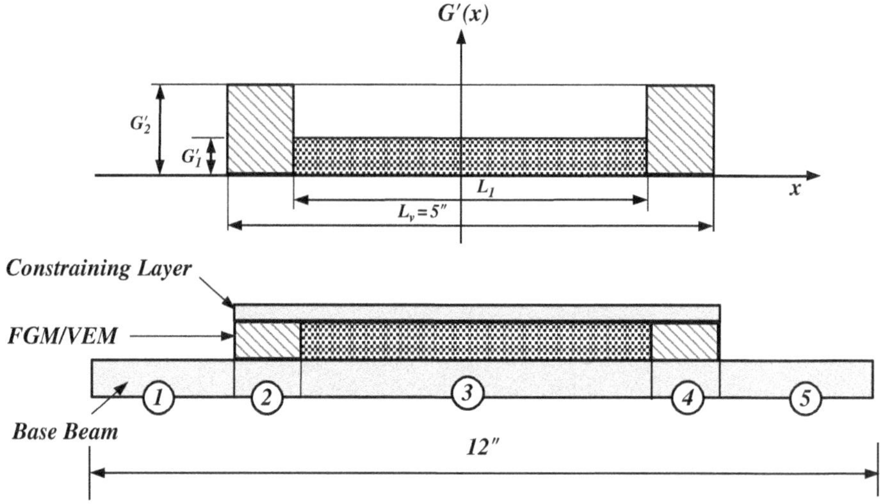

Figure P8.1 Functionally graded constrained layer damping treatment of a free-free beam.

b) Also, establish comparisons between the predictions of the frequency response of the beam/PSOL of ANSYS with that of Problem 1 by the DTF method.

8.3 Consider the functionally graded constrained layer damping treatment of a free-free beam as shown in Figure P8.1. The figure shows that the beam and the constraining layer are sandwiching two different viscoelastic materials. The VEM treatment partially covered the beam and is placed symmetrically around the beam's center. The beam has five sections; 1 and 5 are untreated, Section 3 is treated with VEM_1 while Sections 2 and 4 are treated with VEM_2.

The base beam is manufactured from aluminum that is 12″ long, 1″ wide, and 0.040″ thick. The constraining layer is also made of aluminum that is 0.020″ thick. All the VEM sections have a total length of 5″ and thickness of 0.020″ with the VEM_1 has complex shear modulus given by $G_1 = 6(1 + i)$ MN m^{-2} and $G_2 = G_2'$ $(1 + i)$ MN m^{-2}.

Develop a FEM for the beam/PCLD with a FGM VEM using the concepts developed in Chapter 4, Section 4.3.

Assume that the beam is excited at one end with a unit sine sweep from 0 to 500 Hz, determine and plot the energy dissipated D contours, in the entire viscoelastic core, as a function of the dimensionless ratios L_1/L_v and G_2'/G_1'. Find the optimal values of these ratios that maximize the energy dissipated. The energy dissipated D is determined from:

$$D = \eta \int_0^L A_2 G_2' \gamma^2 dx$$

8.4 Confirm the results obtained in Problem 8.3 by developing an ANSYS finite element program for the functionally graded constrained layer damping treatment of the free-free beam as shown in Figure P8.1.

Compare the modes and mode shapes of the beam/PCLD as predicted by ANSYS with the theoretical predictions as obtained in Problem 8.3. Also, establish comparisons between the predictions of the contours of the energy dissipation of ANSYS with that of Problem 8.3.

8.5 Predict the properties of the composite, shown in Figure P8.2a for different orientation angle θ. The composite contains of Steel as phase A, with $E'_A = 200$ GPa, $\eta_A = 0.001$ and a viscoelastic elastomer as phase B, with $E'_B = 0.020$ GPa, $\eta_B = 1.0$.

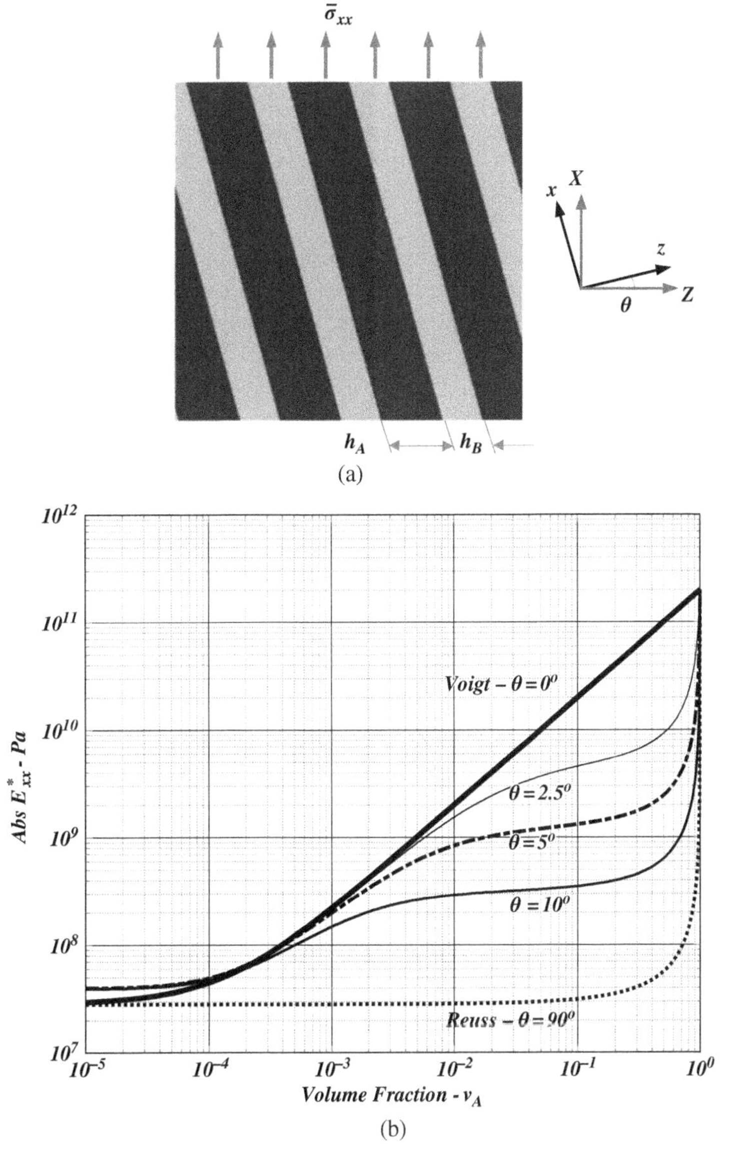

Figure P8.2 Effect of volume fraction on the effective modulus of a steel and VEM composite.

Show that:

$$E_{xx}^* = \frac{1}{s_{11}^*} = \left[\frac{\cos^4\theta}{E_{Voigt}^*} + \frac{\sin^4\theta}{E_{Reuss^*}} + \cos^2\theta\sin^2\theta\left(\frac{1}{G_{xz}^*} + 2s_{13}^*\right)\right]^{-1},$$

where $G_{xz}^* = 1/\left(\frac{\nu_A}{G_A^*} + \frac{1-\nu_A}{G_B^*}\right)$ and $s_{13}^* = \frac{\nu_A \upsilon_A + \nu_B \upsilon_B - \upsilon_A \upsilon_B}{\nu_A \upsilon_B E_A^* + \nu_B \upsilon_A E_B^* - \nu_A E_A^* - \nu_B E_B^*}.$

Note that $E_i^* = E_i'(1 + i\eta_i)$ and $G_i^* = G_i'(1 + i\eta_i)$ are the complex and shear moduli respectively of layer $i = A, B$. Also, ν_A denotes the volume fraction of the Ath layer, and υ_i is Poisson's ratio of the ith layer.

Plot and impose the obtained characteristics on the Voigt and Reuss composite limits as drawn in the abs(E_{xx}^*) – volume fraction ν_A as indicated in Figure P8.2b. (Hint, see: Harris B., *Engineering Composite Materials,* Appendix 1, pp. 184–187, The Institute of Materials, London, 1999, and Liu, B., Feng, X., and Zhang, S.M., "The effective Young's modulus of composites beyond the Voigt estimation due to the Poisson effect," *Composites Science and Technology,* Vol. 69, 2198–2204, 2009).

8.6 For the composite, shown in Figure P8.2a, determine and plot the effect of the orientation angle θ on the characteristics on the loss factor (η) – volume fraction ν_A as indicated in Figure P8.3. Impose on the displayed characteristics the Voigt and Reuss composite limits as drawn in Figure 8.32.

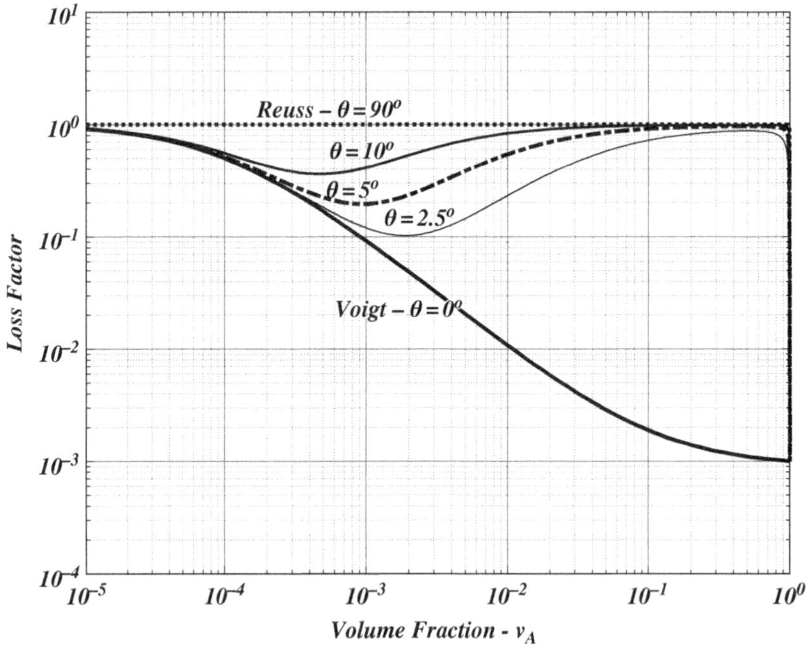

Figure P8.3 Effect of volume fraction on the effective loss factor of a steel and VEM composite.

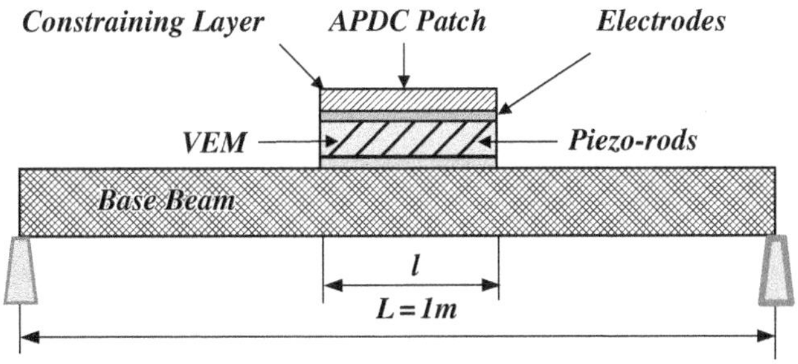

Figure P8.4 A simply supported beam with partial APDC treatment.

8.7 Consider the simply supported aluminum beam shown in Figure P8.4. The beam width (b) is 25.4 mm and thickness (h_3) is 2.0 mm. The beam is treated with one APDC patch that is bonded at its root. The volume fraction of the piezoelectric rods is taken to be 30%. Also, the thickness (h_2) of the polymer matrix of the APDC is 3.175 mm and the thickness (h_1) of the aluminum constraining layer is 2.0 mm. The length l of the APDC is 0.254 m. The physical properties of the piezoelectric rods of the APDC are listed in Tables 8.2 and 8.3. Compare the performance of beams treated with APDC with K_g = 0 and 4E8 V m^{-1}.

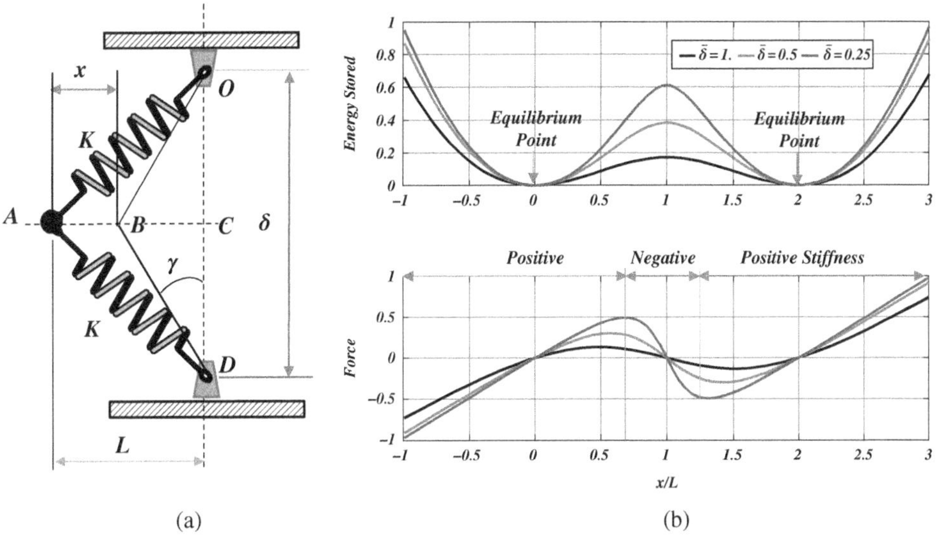

Figure P8.5 A two-spring assembly with positive and negative equivalent stiffness.

8.8 For the two-spring assembly displayed in Figure P8.5a, which is intended to generate either positive or negative effective stiffness depending on the deflection x as indicated in Figure P8.5b.

Show that the deflections d experienced by the springs due to a motion of the attachment point A at distance x is given by:

$$\bar{d} = \frac{d}{L} = \frac{(OA - OB)}{L} = \sqrt{1 + \bar{\delta}^2} - \sqrt{(1 - \bar{x})^2 + \bar{\delta}^2}$$

with $\bar{x} = x/L$, and $\bar{\delta} = \frac{1}{2}(\delta/L)$ where δ and L are as indicated on the figure by the distances OD and AC, respectively. Assume that the springs have an initial free length of $L\sqrt{\left(1+\bar{\delta}^2\right)} = OA$.

Determine and plot, as shown in Figure P8.5b, the energy stored $\left(E/KL^2 = \frac{1}{2}\bar{d}^2\right)$ and the force $(F/KL = \bar{d}\sin\gamma)$ generated by each spring, along the *x-direction*, as a function of the dimensionless motion $\bar{x}$ for different values of the separation distance $\bar{\delta}$. Note that K is the stiffness of the springs.

Indicate the effect of $\bar{\delta}$ on the width of the zones of negative stiffness of the two-spring assembly.

Show that assembly has stable equilibrium points occurring at $\bar{x} = x/L = 0, 2$ independent of the value of $\bar{\delta}$.

8.9 For the three-spring assembly displayed in Figure P8.6a, which is intended to enhance the positive effective stiffness component as compared to the two-spring assembly of Figure 8.5a in order to enable its use in vibration isolators and to enhance its ability to support a load.

Determine and plot, as shown in Figure P8.6b, the energy stored $\left(E/KL^2 = \bar{d}^2 + \frac{1}{2}k_r\bar{x}^2\right)$ and the force $(F/KL = \bar{d}\sin\gamma + k_r\bar{x})$ generated by the three springs, along the x-direction, as a function of the dimensionless motion $\bar{x}$ for a separation distance $\bar{\delta} = 0.25$. Note that K is the stiffness of the inclined springs, K_s is the stiffness of the base spring, and $k_r = K_s/K$.

Show that the effect of k_r on the location of the stable equilibrium points of the three-spring assembly, is as shown in Table 8.9, as compared with that of the two-spring case for $\bar{\delta} = 0.25$.

Show also that the equivalent stiffness K_{eq} of the three-spring assembly is given by:

$$K_{eq}/K = 1 + k_r - \frac{\bar{\delta}^2\sqrt{1+\bar{\delta}^2}}{\left[(1-\bar{x})^2 + \bar{\delta}^2\right]^{3/2}}$$

8.10 The three-spring assembly of Figure P8.6 is augmented with a damper C_s to form the main building block of the vibration isolation system shown in Figure P8.7. The system is intended to reduce the vibration of the mass M resulting from base excitation $\bar{y} = A\sin(\omega t - \varphi)$ where $\bar{y} = y/L$.

Show that the second order Taylor series expansion of the equivalent stiffness $K_{eq_{Approx}}$, around $\bar{x} = 1$, is given by:

$$K_{eq_{Approx}}/K = \left[1 + k_r - \sqrt{1+\bar{\delta}^2}\,\bar{\delta}\right] + \frac{3}{2}\frac{\sqrt{1+\bar{\delta}^2}}{\bar{\delta}^3}(\bar{x}-1)^2$$

Show that a comparison between $\left(K_{eq_{Approx}}/K\right)$ and (K_{eq}/K) for $k_r = 0$ and $\bar{\delta} = 3$ is as shown in Figure P8.8.

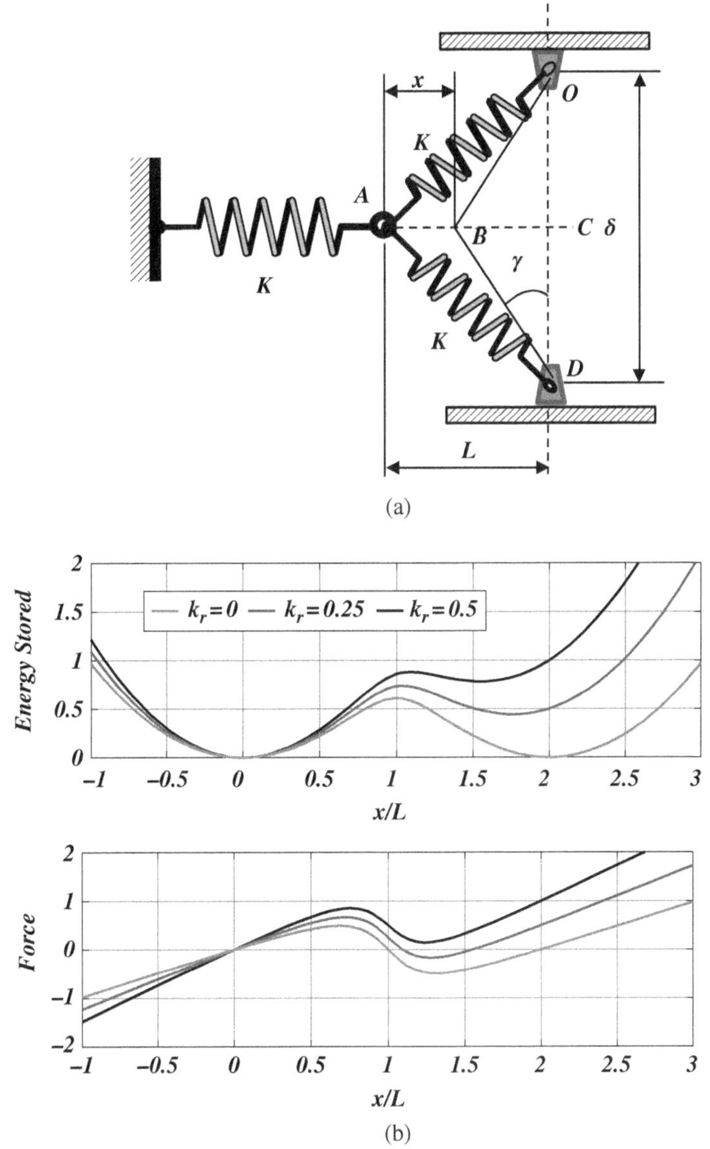

Figure P8.6 A three-spring assembly with positive and negative equivalent stiffness.

Show that the equation of motion of the mass M is given by:

$$M\ddot{z} + C_s\dot{z} + \alpha z + \beta z^3 = -A\omega^2 \sin(\omega t - \varphi)$$

where $z = \bar{x} - 1 - \bar{y} z = x - 1 - y, \alpha = K\left[1 + k_r - \sqrt{1 + \bar{\delta}^2}/\bar{\delta}\right], \beta = (3K/2)\sqrt{1 + \bar{\delta}^2}/\bar{\delta}^3.$

If the solution z is assumed to be: $z = Z\sin(\omega t)$, then show that the equation of motion reduces to:

$$-\omega^2 MZ\sin(\omega t) + \omega C_s Z\cos(\omega t) + \alpha Z\sin(\omega t) + \beta Z^3 \sin^3(\omega t) = -A\omega^2 M\sin(\omega t - \varphi)$$

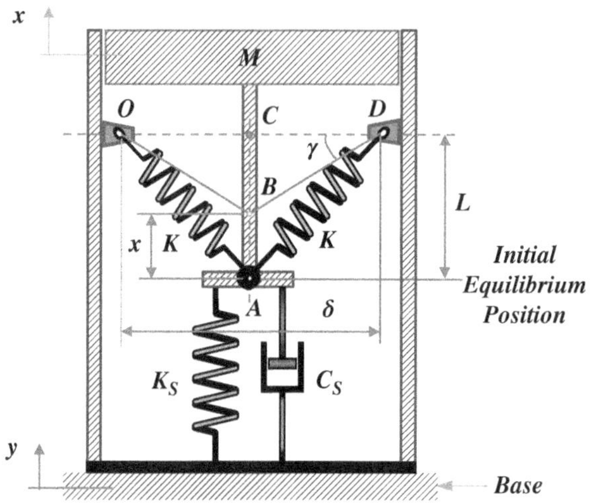

Figure P8.7 Vibration isolation system consisting of a three-spring/damper assembly.

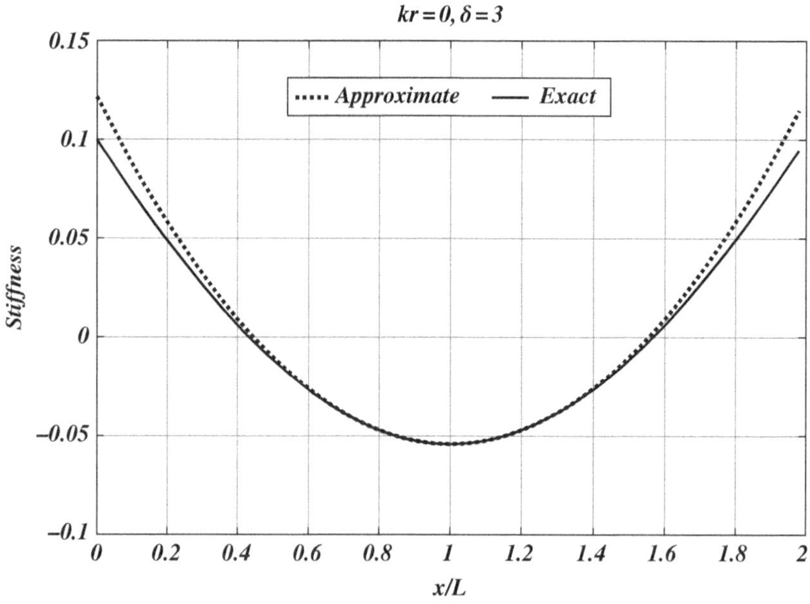

Figure P8.8 Comparison between exact and approximate equivalent stiffness of the vibration isolation system.

Let $\sin^3(\omega t) \simeq \frac{3}{4}\sin(\omega t)$ and expand $\sin(\omega t - \varphi) = \sin(\omega t)\cos(\varphi) - \cos(\omega t)\sin(\varphi)$. By equating the coefficients of the $\sin(\omega t)$ and $\cos(\omega t)$ on both sides along with eliminating φ between the resulting two equations, show that:

$$\left[\left(\alpha - \omega^2 M\right)Z + \frac{3}{4}\beta Z^3\right]^2 + \left[\omega C_s Z\right]^2 = \left[A\omega^2 M\right]^2$$

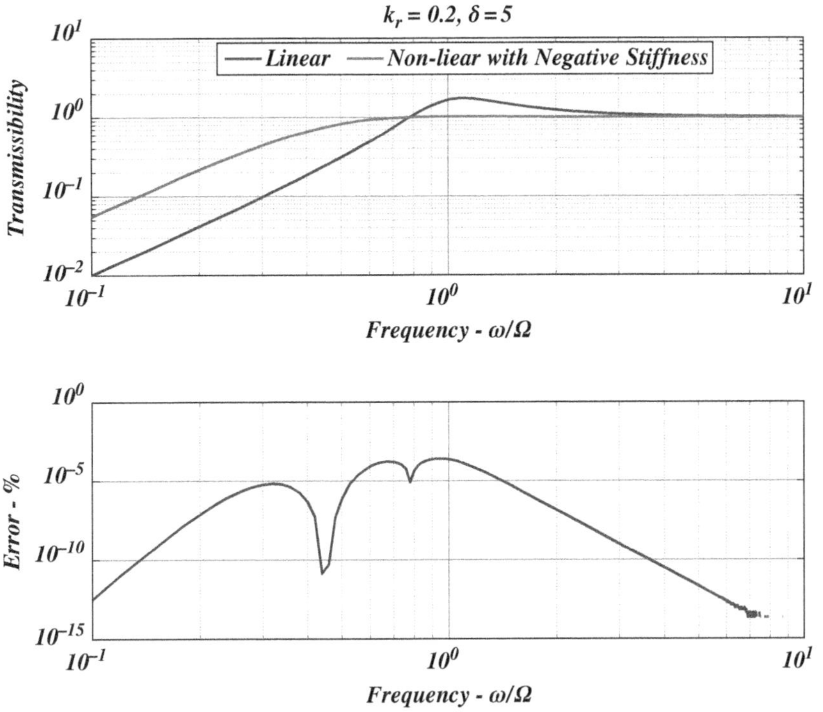

Figure P8.9 Transmissibility T_{NS} of the isolator when $k_r = 0.2$ and $\bar{\delta} = 5$.

and show that the transmissibility T_{NS} is given, in dimensionless form, by:

$$T_{NS} = \frac{Z}{A} = \frac{\Omega^2}{\sqrt{\left[(\bar{\alpha} - \Omega^2) + \frac{3}{4}\bar{\beta}Z^2\right]^2 + [2\zeta\Omega]^2}}$$

where $\bar{\alpha} = \left[1 + k_r - \sqrt{1 + \bar{\delta}^2}/\bar{\delta}\right], \bar{\beta} = (3/2)\sqrt{1 + \bar{\delta}^2}/\bar{\delta}^3, 2\zeta\Omega = C_s/M,$

and $\Omega = \omega/\sqrt{K/M}$.

Check that the transmissibility T_{NS} of the isolator when $k_r = 0.2$ and $\bar{\delta} = 5$ is as displayed in Figure P8.9. Note that, because of the dependence of the transmissibility on Z, then it is best to use an iterative algorithm to accurately determine it. Compute the prediction error of the transmissibility and compare the prediction against the performance of linear isolator with transmissibility T_L given by:

$$T_L = \frac{\Omega^2}{\sqrt{\left[(1 - \Omega^2)\right]^2 + [2\zeta\Omega]^2}}$$

9

Vibration Damping with Shunted Piezoelectric Networks

9.1 Introduction

Shunted piezoelectric networks have been used extensively as effective means for damping structural vibration and associated noise radiation. Their effectiveness stems from their behavior characteristics that resemble that of conventional viscoelastic materials. Furthermore, these networks are light in weight, easy to use, and can have tunable characteristics. In this chapter, the theory governing the operation of these shunted piezoelectric networks is presented. Applications to the control of a single as well of as multimodes of vibration are discussed using the theory of finite elements.

The theory of operation these shunted piezoelectric networks has been developed by numerous investigators such as: Hagood et al. (1990), Hagood and von Flotow (1991), Edberg et al. (1991), Law et al. (1995), Lesieutre and Davis (1997), Park and Inman (1999), Tsai and Wang (1999), and Moheimani et al. (2001). Comprehensive reviews of the field of shunted piezoelectric networks are presented by Gripp and Rade (2018) as well as by Yan et al. (2017).

9.2 Shunted Piezoelectric Patches

Piezoelectric patches bonded to the vibrating structure are utilized to convert the vibrational energy of the structure into electrical energy. The generated electrical energy is then dissipated in shunted electric networks, as shown in Figure 9.1, which are tuned in order to maximize their energy dissipation characteristics (Lesieutre 1998). These electric networks are usually resistive, inductive, and/or capacitive.

9.2.1 Basics of Piezoelectricity

In Appendix 6.B of Chapter 6, the constitutive equations of one-dimensional piezoelectric patches are given by

$$\begin{Bmatrix} S_1 \\ D_3 \end{Bmatrix} = \begin{bmatrix} s_{11}{}^E & d_{31} \\ d_{31} & \varepsilon_{33}{}^T \end{bmatrix} \begin{Bmatrix} T_1 \\ E_3 \end{Bmatrix} \tag{9.1}$$

Active and Passive Vibration Damping, First Edition. Amr M. Baz.

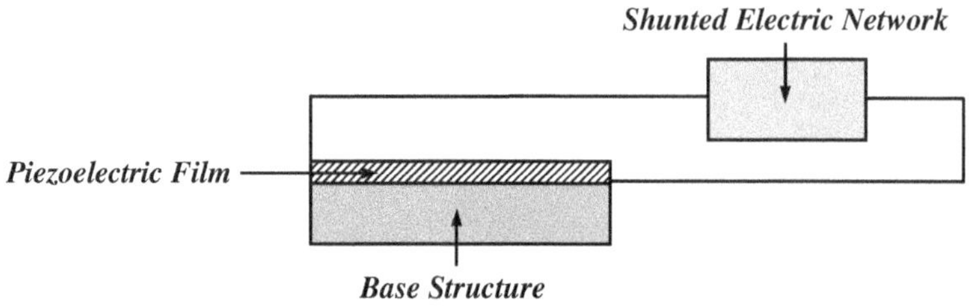

Figure 9.1 Shunted piezoelectric treatments.

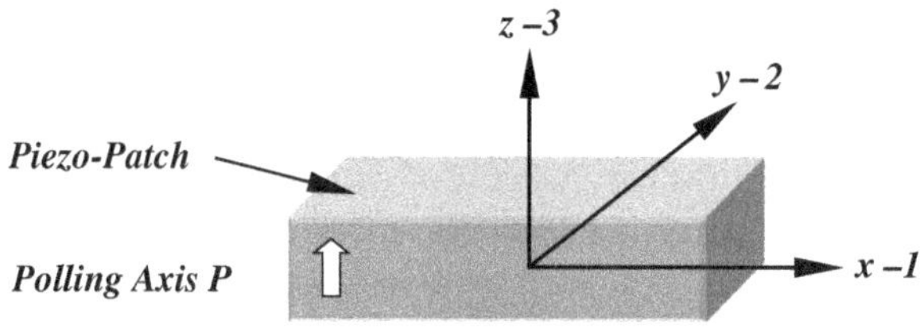

Figure 9.2 Coordinate system of piezoelectric patches.

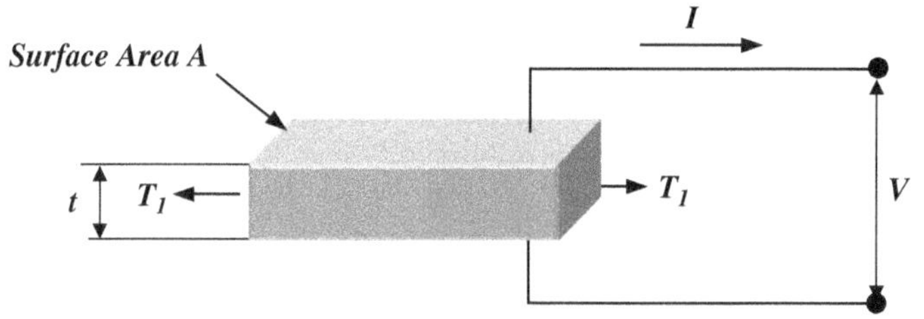

Figure 9.3 Piezoelectric patch.

which describes the mechanical strain S_1, in direction 1, due to a mechanical stress T_1, in direction 1, and an electrical field E_3, in direction 3. In Eq. (9.1), the electric displacement D_3, in direction 3, is generated by the mechanical stress T_1, and the electric field E_3. The directions assigned in the constitutive equations are defined in Figure 9.2 in relation to the poling axis P of the film according to the IEEE STD 176 (1987).

Consider the piezo-patch shown in Figure 9.3.

Then,

$$D_3 = \frac{1}{A}Q = \frac{1}{A}\int Ids \tag{9.2}$$

where Q, A, I, and s are the charge, surface area of piezo-patch, current, and Laplace complex number, respectively.

In the Laplace domain, Eq. (9.2) becomes

$$D_3 = \frac{I}{As} \tag{9.3}$$

Also, as

$$E_3 = \frac{V}{t} \equiv \text{Electric Field} \tag{9.4}$$

with t and V denoting the thickness of the piezo-patch and the voltage across the patch.

Substituting Eqs. (9.3) and (9.4) into Eq. (9.1), it reduces to

$$\begin{Bmatrix} S_1 \\ I \end{Bmatrix} = \begin{bmatrix} s_{11}^E & \dfrac{d_{31}}{t} \\ Ad_{31}s & \dfrac{\varepsilon_{33}^T As}{t} \end{bmatrix} \begin{Bmatrix} T_1 \\ V \end{Bmatrix} = \begin{bmatrix} s_{11}^E & \dfrac{d_{31}}{t} \\ Ad_{31}s & Y^D \end{bmatrix} \begin{Bmatrix} T_1 \\ V \end{Bmatrix} \tag{9.5}$$

where $Y^D = \dfrac{A\,\varepsilon_{33}^T}{t}s$ denotes the admittance, which is the reciprocal of the impedance Z^D.

Consider the two following boundary conditions:

9.2.1.1 Effect of Electrical Boundary Conditions

*i) **Short-circuit piezo-patch***

This implies that $V = 0$, reducing the first row of Eq. (9.5) to

$$S_1 = s_{11}^E\,T_1$$

that is, the piezo-patch has a short-circuit compliance of s_{11}^E.

*ii) **Open-circuit piezo-patch***

This implies that $I = 0$, reducing the second row of Eq. (9.5) to

$$-d_{31}T_1 = \frac{\varepsilon_{33}^T}{t}V \tag{9.6}$$

Eliminating V between Eq. (9.6) and the first row of Eq. (9.5) gives

$$S_1 = s_{11}^E\left(1 - \frac{d_{31}^2}{s_{11}^E\,\varepsilon_{33}^T}\right)T_1 = s_{11}^E\left(1 - k_{31}^2\right)T_1 = s_{11}^D\,T_1 \tag{9.7}$$

that is, the piezo-patch has open-circuit compliance of s_{11}^D. In Eq. (9.7), k_{31}^2 denotes the electromechanical coupling factor of the piezo-patch. k_{31}^2 defines the efficiency of converting electrical energy into mechanical energy as given in the appendix. It is always <1 and ranges usually between 0.3 and 0.5 for typical ceramic piezoelectric materials.

Accordingly, open-circuit compliance s_{11}^D is much smaller than the short-circuit compliance s_{11}^E. This implies that the open-circuit Young's modulus is much higher than that under short-circuit conditions. Hence, changing the electrical boundary conditions on the piezo-patch results in significant changes in the mechanical properties of the patch as summarized in Figure 9.4.

9.2.1.2 Effect of Mechanical Boundary Conditions

*i) **Stress-free piezo-patch***

This implies that $T_1 = 0$, reducing the second row of Eq. (9.5) to

$$I = \frac{A\,\varepsilon_{33}^T}{t}s\,V = C^T s\,V = Y^D\,V \tag{9.8}$$

Short-Circuit Conditions	Open-Circuit Conditions
Piezo-patch → ▭ Short	Piezo-patch → ▭ Open
Compliance $= s_{11}^E$	Compliance $= s_{11}^D = s_{11}^E(1-k_{31}^2) < s_{11}^E$

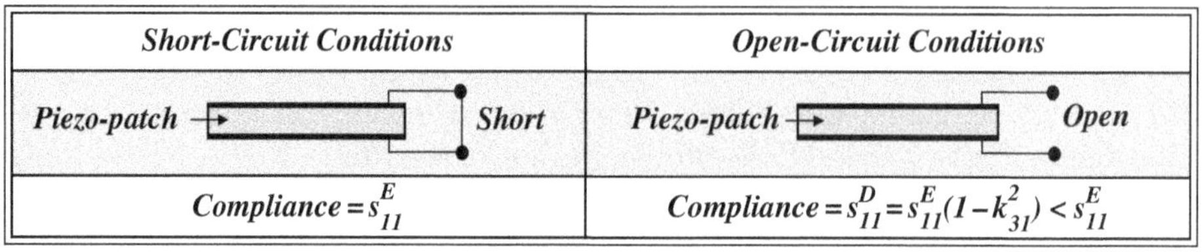

Figure 9.4 Effect of electrical boundary conditions on mechanical properties of a piezoelectric patch.

Stress-Free Conditions	Strain-Free Conditions
Free-Free piezo-patch	Fixed-Fixed piezo-patch
Capacitance $= C^T$	Capacitance $= C^S = C^T(1-k_{31}^2) < C^T$

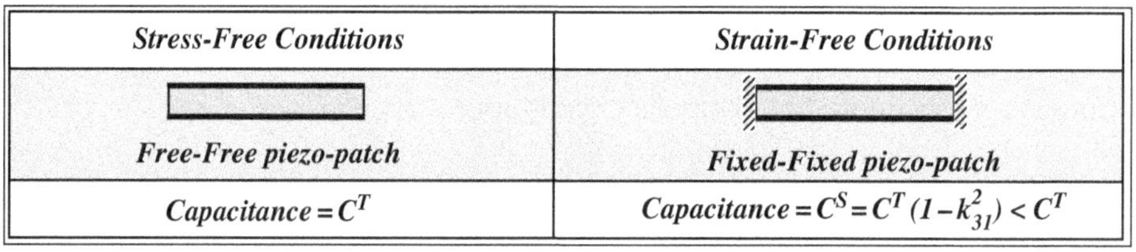

Figure 9.5 Effect of mechanical boundary conditions on electrical properties of a piezoelectric patch.

that is, the piezo-patch has stress-free capacitance $C^T = \dfrac{A\,\varepsilon_{33}^T}{t}$. In Eq. (9.8), Y^D denotes the admittance that is the reciprocal of the impedance Z^D.

ii) Strain-free piezo-patch

This implies that $S_1 = 0$, reducing the first row of Eq. (9.5) to

$$-s_{11}^E T_1 = \frac{d_{31}}{t} V \tag{9.9}$$

Eliminating T_1 between Eq. (9.9) and the second row of Eq. (9.5) gives

$$I = C^T\left(1-k_{31}^2\right)V = C^S V \tag{9.10}$$

that is, the piezo-patch has strain-free capacitance C^S, which is much smaller than its stress-free capacitance C^T. Hence, changing the mechanical boundary conditions on the piezo-patch results in significant changes in the electrical properties of the patch as summarized in Figure 9.5.

9.2.2 Basics of Shunted Piezo-Networks

Figure 9.6 shows a piezo-patch with a shunted circuit that has an admittance of Y^{SH}.

Then, the admittance of the parallel combination of a piezo-patch and a shunted circuit is given by

$$Y^{EL} = Y^D + Y^{SH} \tag{9.11}$$

Accordingly, Eq. (9.5) can be modified to describe the interaction between the piezo-patch and shunted circuit as follows

$$\begin{Bmatrix} S_1 \\ I \end{Bmatrix} = \begin{bmatrix} s_{11}^E & \dfrac{d_{31}}{t} \\ Ad_{31}s & Y^{EL} \end{bmatrix} \begin{Bmatrix} T_1 \\ V \end{Bmatrix} \tag{9.12}$$

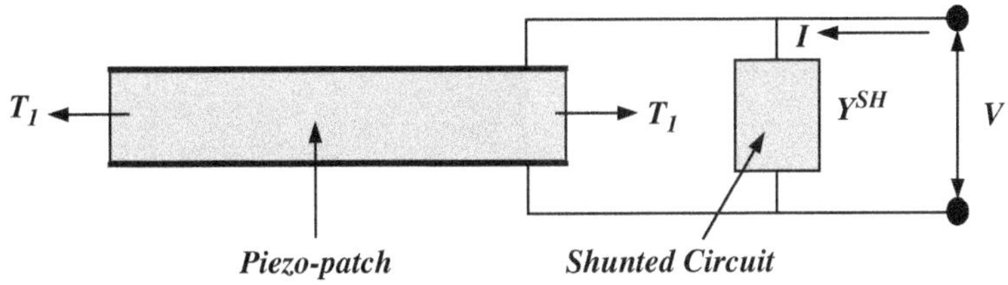

Figure 9.6 Piezoelectric patch with a shunted circuit.

The effect of the shunted network on the constitutive relations of the piezo-network can be determined from Eq. (9.12) by considering $I = 0$ (open circuit), then the second row of Eq. (9.12) reduces to

$$Ad_{31}sT_1 = -Y^{EL}V \tag{9.13}$$

Then, eliminating V between Eq. (9.13) and the first row of Eq. (9.12) gives

$$S_1 = s_{11}^E T_1 - \frac{d_{31}Ad_{31}s}{t}\frac{Y^D}{Y^D}\frac{}{Y^{EL}}T_1 = s_{11}^E\left[1 - \frac{d_{31}^2}{\varepsilon_{33}^T s_{11}^E}\frac{Y^D}{Y^{EL}}\right]T_1 = s^{SH}T_1$$

where s^{SH} denotes the compliance of the shunted network that is given by

$$s^{SH} = s_{11}^E\left[1 - k_{31}^2\bar{Z}^{EL}\right] \tag{9.14}$$

where $\bar{Z}^{EL} = \frac{Y^D}{Y^{EL}}$ and $k_{31}^2 = \frac{d_{31}^2}{\varepsilon_{33}^T s_{11}^E}$.

Comparing Eqs. (9.7) and (9.14) indicates that the compliance of the piezo-patch alone is real whereas the compliance of the shunted network can be complex depending on the configuration of the shunted circuit. Accordingly, the piezo-patch alone has an elastic modulus of $1/s_{11}^D$ whereas the piezo-shunted network has a complex modulus of $1/s^{SH}$. Hence, the piezo-shunted network has a storage modulus as well as a loss factor and acts as a conventional viscoelastic damping material.

A better quantification of the effect of shunting a piezo-patch on its mechanical characteristics can be achieved by considering the concept of "mechanical impedance" Z^{ME}, which is defined as

$$Z^{ME} = F\dot{x} = \frac{Force}{Velocity} \quad or \quad Z^{ME}(s) = \frac{F}{sx} = \frac{Stiffness}{s} \tag{9.15}$$

For a one-dimensional piezo-patch, one can determine the patch stiffness from

$$Stiffness = \frac{A_{cs}}{s^p L}$$

where A_{cs}, L, and s^p are the cross sectional area, length, and compliance of the patch, respectively. The mechanical impedance of the patch becomes

$$Z^{ME} = \frac{A_{cs}}{s^p Ls} \tag{9.16}$$

Let $\bar{Z}^{ME} = (Z^{ME})^{SH}/(Z^{ME})^{D}$ = ratio of the mechanical impedance of the shunted network $(Z^{ME})^{SH}$, to that of the open-circuit piezo-patch $(Z^{ME})^{D}$.

Then, from Eqs. (9.7), (9.14), and (9.16), $\bar{Z}^{ME}$ can be written as

$$\bar{Z}^{ME} = \frac{(Z^{ME})^{SH}}{(Z^{ME})^{D}} = \frac{\frac{A_{cs}}{s^{SH}Ls}}{\frac{A_{cs}}{s^{D}Ls}} = \frac{s^{D}}{s^{SH}} = \frac{s_{11}^{E}\left(1-k_{31}^{2}\right)}{s_{11}^{E}\left(1-k_{31}^{2}\bar{Z}^{EL}\right)} \tag{9.17}$$

The mechanical impedance and, in turn, the complex modulus characteristics will be determined here for three types of shunted circuits. These circuits include the following.

9.2.2.1 Resistive-Shunted Circuit

Figure 9.7 shows a piezo-patch coupled with a shunted resistance R. The admittance $Y^{SH} = 1/R$ and the admittance of the parallel combination of a piezo-patch and a shunted circuit is given by

$$Y^{EL} = C^{T}s + \frac{1}{R},$$

or

$$Z^{EL} = \frac{R}{RC^{T}s+1} \tag{9.18}$$

Let $\bar{Z}^{EL} = \frac{Z^{EL}}{Z^{D}}$, to denote the impedance ratio of the shunted network to that of the piezo-patch alone, then

$$\bar{Z}^{EL} = Z^{EL}Y^{D} = \frac{R}{RC^{T}s+1}Y^{D} = \frac{RC^{T}s}{RC^{T}s+1} \tag{9.19}$$

Substituting Eq. (9.19) into Eq. (9.17) and using Eq. (9.10) gives

$$\bar{Z}^{ME} = \frac{1-k_{31}^{2}}{1-k_{31}^{2}\dfrac{RC^{T}s}{RC^{T}s+1}} = \frac{\left(1-k_{31}^{2}\right)+\left(1-k_{31}^{2}\right)RC^{T}s}{1+\left(1-k_{31}^{2}\right)RC^{T}s} = \frac{\left(1-k_{31}^{2}\right)+RC^{S}s}{1+RC^{S}s} \tag{9.20}$$

$\bar{Z}^{ME}$ can be determined in the frequency domain, by replacing s by $i\omega$ in Eq. (9.20) to give

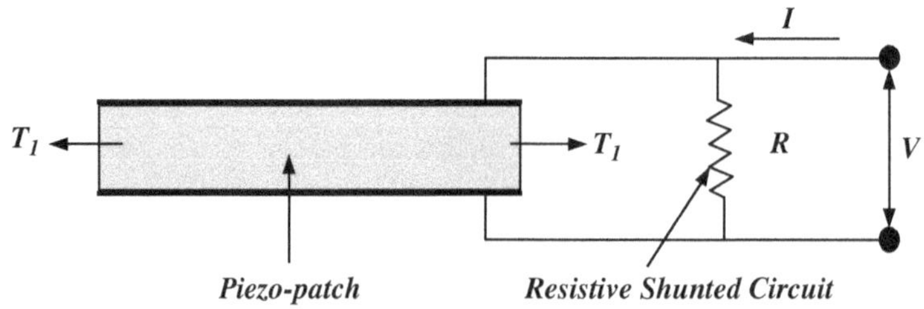

Figure 9.7 Piezoelectric patch with a resistive-shunted circuit.

$$\bar{Z}^{ME} = \frac{(1-k_{31}^2) + RC^S\omega i}{1 + RC^S\omega i}$$

Let $\Omega = RC^S\omega$ be a dimensionless frequency, then this equation reduces to

$$\bar{Z}^{ME} = \frac{(1-k_{31}^2) + \Omega i}{1 + \Omega i} = \left[1 - \frac{k_{31}^2}{1+\Omega^2}\right]\left[1 + \frac{k_{31}^2\Omega}{1-k_{31}^2+\Omega^2} i\right] = E'[1+\eta i] \tag{9.21}$$

where $E' = \left[1 - \frac{k_{31}^2}{1+\Omega^2}\right]$ = dimensionless storage modulus, and $\eta = \frac{k_{31}^2\Omega}{1-k_{31}^2+\Omega^2}$ = loss factor.

The loss factor attains a maximum when: $\frac{\partial\eta}{\partial\Omega} = 0$. This occurs at a frequency Ω^* given by

$$\Omega^* = \sqrt{1-k_{31}^2}$$

and the corresponding maximum value of the loss factor is $\eta_{max} = \frac{k_{31}^2}{2\sqrt{1-k_{31}^2}}$

It is therefore essential to use a piezo-patch with high electromechanical coupling factor to achieve the highest possible loss factor.

9.2.2.2 Resistive and Inductive Shunted Circuit

Figure 9.8 shows a piezo-patch coupled with a shunted resistance R and inductance L. The admittance $Y^{SH} = 1/(R + Ls)$ and the admittance of the parallel combination of a piezo-patch and a shunted circuit is given by

$$Y^{EL} = C^T s + \frac{1}{(R+Ls)}, \quad \text{or} \quad \bar{Z}^{EL} = \frac{C^T s(R+Ls)}{C^T s(R+Ls)+1} \tag{9.22}$$

Substituting Eq. (9.22) into Eq. (9.17) and replacing s with $i\omega$ gives

$$\bar{Z}^{ME} = \left[1 - \frac{k_{31}^2(1-\bar{L}\Omega^2)}{(1-\bar{L}\Omega^2)^2+\Omega^2}\right]\left[1 + \left(\frac{k_{31}^2\Omega}{(1-\bar{L}\Omega^2)^2+\Omega^2-k_{31}^2(1-\bar{L}\Omega^2)}\right)i\right] = E'(1+\eta i) \tag{9.23}$$

where $\bar{L} = \frac{L}{R^2C^s}$, $E' = \left[1 - \frac{k_{31}^2(1-\bar{L}\Omega^2)}{(1-\bar{L}\Omega^2)^2+\Omega^2}\right]$, and $\eta = \left(\frac{k_{31}^2\Omega}{(1-\bar{L}\Omega^2)^2+\Omega^2-k_{31}^2(1-\bar{L}\Omega^2)}\right)$.

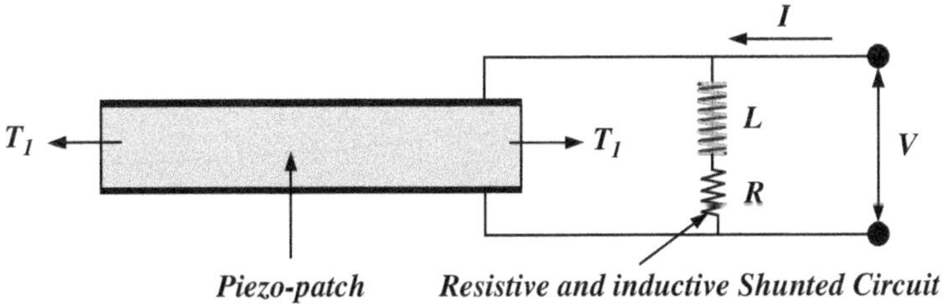

Figure 9.8 Piezoelectric patch with a resistive and inductive shunted circuit.

Resonant conditions occurs when $\left(1-\bar{L}\Omega^2\right)=0$, or $L=\dfrac{1}{C^s\omega^2}$ with $\eta_{\max}=\dfrac{k_{31}^2}{\Omega}=k_{31}^2\sqrt{\bar{L}}$.

Example 9.1 Determine the storage modulus E' and loss factor η for resistive as well as resistive and inductive shunted networks as a function of the frequency Ω. Assume that $k_{31}^2 = 0.5$ and $\bar{L} = 1$.

Solution

Figure 9.9 shows plots of the storage modulus E' and loss factor η for resistive as well as resistive and inductive shunted networks as obtained from Eqs. (9.21) and (9.23), respectively.

It is clear that storage modulus and loss factor characteristics shown in Figure 9.9 resemble that of conventional viscoelastic materials as shown in Chapters 2–4. The figure indicates also that the use of inductance in the shunted circuit enhances the storage modulus at high frequencies and improves the loss factor at low frequencies.

Example 9.2 Determine the size of the inductance L that is necessary to achieve resonant conditions in a R–L shunted network for different frequencies. Assume that the piezo-patch has $\varepsilon_{33}^T = 1.3054E^{-8}$ farad m^{-1}, $A = 0.001\,25$ m^2, $t = 0.000\,187\,5$ m, and $k_{31}^2 = 0.38$.

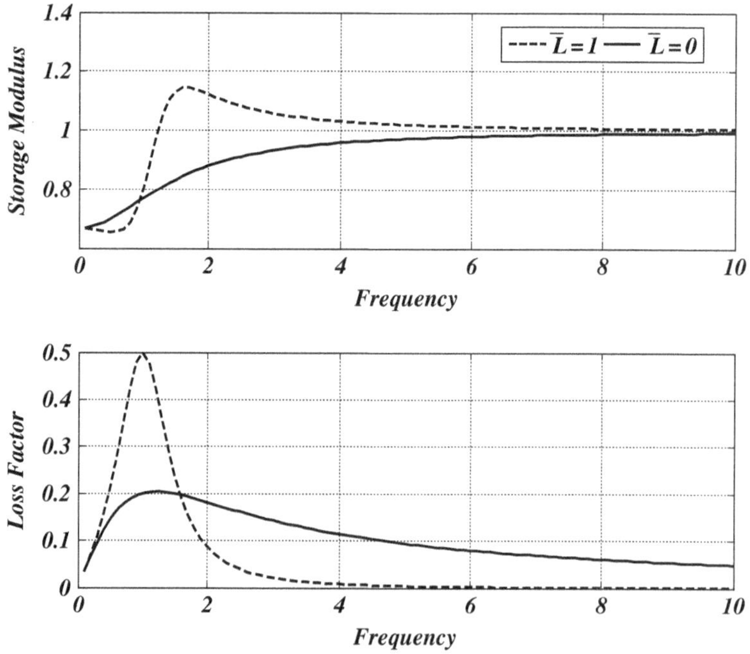

Figure 9.9 Storage modulus and loss factor of resistive as well as resistive and inductive shunted circuits.

Solution

As $C^s = \frac{\varepsilon_{33}^T A}{t}\left(1-k_{31}^2\right) = 5.3956E\text{-}8$ Farad and $L = \frac{1}{C^s(2\pi f)^2}$ where f is the frequency, then,
$L = \frac{1}{C^s(2\pi f)^2} = \frac{4.6947\ \text{E5}}{f^2}$ Henry.

Table 9.1 lists the inductance L for different frequencies f.

It is clear from the table that if the frequency f is small then L becomes very large to be practical and it is necessary to electronically synthesize the inductor.

9.2.2.3 Resistive, Capacitive, and Inductive Shunted Circuit

Figure 9.10 shows a piezo-patch coupled with a shunted resistance R, capacitance C_e, and inductance L. The admittance $Y^{SH} = \frac{C_e s}{C_e s(Ls+R)+1}$ and the admittance of the parallel combination of a piezo-patch and a shunted circuit is given by

$$Y^{EL} = \frac{C^T s[C_e s(Ls+R)+1] + C_e s}{[C_e s(Ls+R)+1]},$$

and

$$\bar{Z}^{EL} = \frac{C^T s[C_e s(Ls+R)+1]}{C^T s\left[C_e s(Ls+R)+1\right] + + C_e s} \tag{9.24}$$

Substituting Eq. (9.24) into Eq. (9.17) and replacing s with $i\omega$ gives

$$\bar{Z}^{ME} = E'(1+\eta i) \tag{9.25}$$

Table 9.1 Effect of resonant frequency f on the inductance L.

f (Hz)	1000	100	10	1
L (Henry)	0.469	46.95	4694	469,470

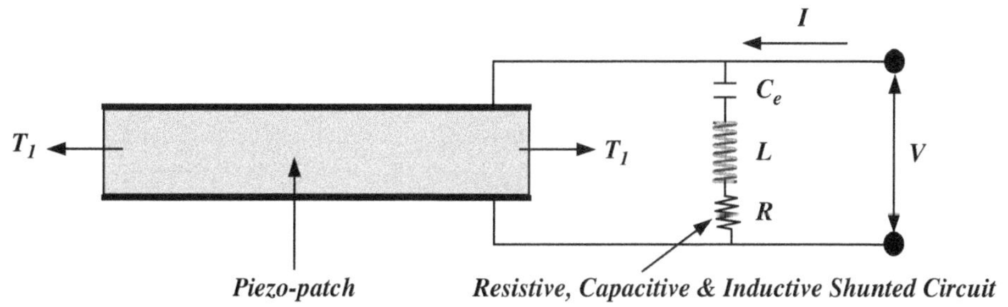

Figure 9.10 Piezoelectric patch with a resistive and inductive shunted circuit.

where $\bar{L} = \dfrac{L}{R^2 C^s}$, $C_r = \dfrac{C^T}{C_e}$, $E' = \left[1 - \dfrac{k_{31}^2\left([C_r+1] - k_{31}^2 C_r - \bar{L}\Omega^2\right)}{\left([C_r+1] - k_{31}^2 C_r - \bar{L}\Omega^2\right)^2 + \Omega^2}\right]$, and

$$\eta = \left(\frac{k_{31}^2\Omega}{\left([C_r+1] - k_{31}^2 C_r - \bar{L}\Omega^2\right)^2 + \Omega^2 - k_{31}^2\left([C_r+1] - k_{31}^2 C_r - \bar{L}\Omega^2\right)}\right)$$

Note that when $C_r = 0$, these equations reduce to those of the R–L shunted network and with $\bar{L} = 0$, the equations reduce to those of the R shunted network.

Example 9.3 Determine the effect of using R–C shunted networks on the storage modulus and loss factor of a piezo-patch. Consider both positive as well as negative values of the shunted capacitance. Assume that $k_{31}^2 = 0.5$.

Solution

Figure 9.11 shows the effect of using positive shunted capacitance C_e on the storage modulus and loss factor of a piezo-patch as obtained from Eqs. (9.25). Note that C_r is taken = 1 and $\bar{L} = 0$.

The figure indicates that adding a positive shunted capacitance does not affect the storage modulus at high frequencies but enhances it at low frequencies. However, the use of a positive shunted capacitance results in significant deterioration of the loss factor.

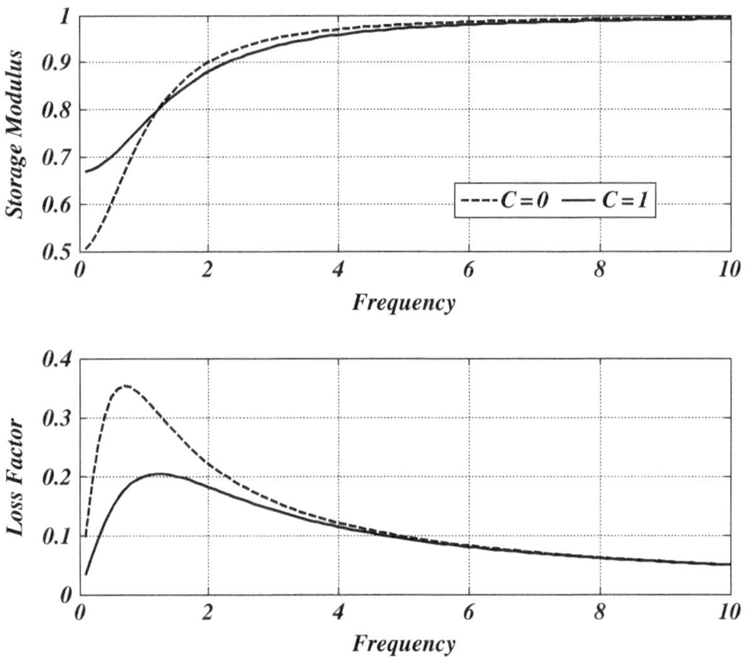

Figure 9.11 Storage modulus and loss factor of resistive as well as resistive and capacitive shunted circuits ($C_r = 1$).

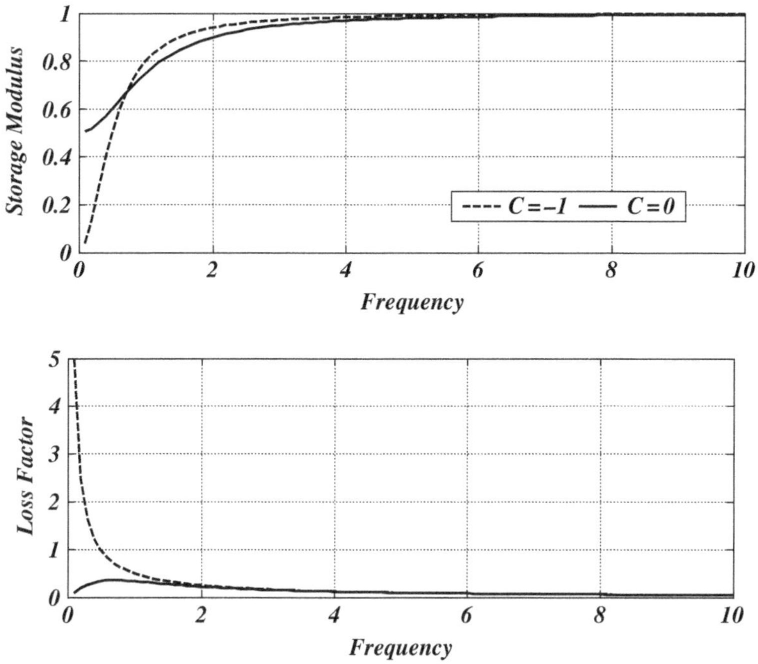

Figure 9.12 Storage modulus and loss factor of resistive as well as resistive and capacitive shunted circuits ($C_r = -1$).

Figure 9.12 shows the effect of using negative shunted capacitance C_e on the storage modulus and loss factor of a piezo-patch as obtained from Eqs. (9.25). Note that C_r is taken = −1 and $\bar{L} = 0$.

The figure indicates that a positive shunted capacitance does not affect the storage modulus at high frequencies but degrades it at low frequencies. However, the use of a negative shunted capacitance results in significant improvement of the loss factor, particularly at low frequencies. Such a result is extremely important for vibration control applications where most of the dominant modes of vibration occur usually at low frequencies.

It is, however, necessary to electronically synthesize such negative capacitances.

9.2.3 Electronic Synthesis of Inductances and Negative Capacitances

9.2.3.1 Synthesis of Inductors

As noted in Example 9.2, it is essential to electronically synthesize inductors in order to obtain compact shunted networks that can attenuate low frequency vibrations. Figure 9.13 shows one way of synthesizing the inductors (Deliyannis et al. 1999).

From the circuit diagram shown in Figure 9.13

$$\textbf{Loop 1}: \frac{V_i - V_o}{R} = i_i,$$

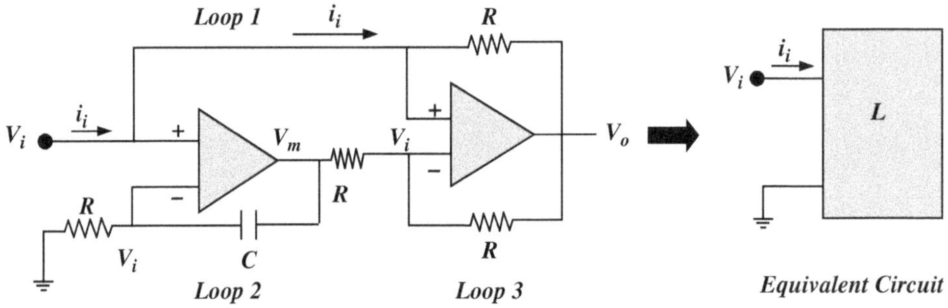

Figure 9.13 Electronically synthesized inductor.

$$\textbf{\textit{Loop 2}}: (V_m - V_i)Cs = \frac{V_i}{R},$$

$$\text{and } \textbf{\textit{Loop 3}}: \frac{(V_m - V_i)}{R} + \frac{(V_o - V_i)}{R} = 0$$

Eliminate V_o and V_m from the above equations gives

$$V_i = \left(R^2 Cs\right) i_i \quad \text{or} \quad V_i = Lsi_i, \text{that is}, \quad L = R^2\, C \tag{9.26}$$

Hence, the circuit of Figure 9.13 can be used to synthesize an inductor L using a combination of resistors R, a capacitance C, and two operational amplifiers. The value of the synthesized inductor L depends on the values of R and C as given by Eq. (9.26).

9.2.3.2 Synthesis of Negative Capacitances

As noted in Example 9.3, it is essential to electronically synthesize negative capacitance to enhance the damping of low frequency vibrations. Figure 9.14 shows one way of synthesizing these negative capacitances (Deliyannis et al. 1999).

The circuit diagram shown in Figure 9.14 yields:

$$\textbf{\textit{Loop 1}}: \frac{V_o - V_i}{R_f} = i_c = \frac{V_i}{R_c},$$

$$\text{and } \textbf{\textit{Loop 2}}: i_i = i_r + i_c = \frac{V_i}{R} + (V_i - V_o)Cs,$$

Eliminating V_o from these equations gives

$$i_i = -1/\left[\left(R_f/R_s\right)Cs\right] V_i \quad \text{or} \quad C_s = -\left(R_f/R_s\right)C \tag{9.27}$$

Hence, the circuit of Figure 9.14 can be used to synthesize a negative capacitance C_s using a combination of resistors R_f and R_s, a capacitance C, and one operational amplifier. The value of the synthesized capacitance C_s depends on the values of R_f, R_s, and C, as given by Eq. (9.27).

9.2.4 Why Negative Capacitance Is Effective?

Consider the equivalent circuit of a piezo-patch with a shunted circuit as shown in Figure 9.15.

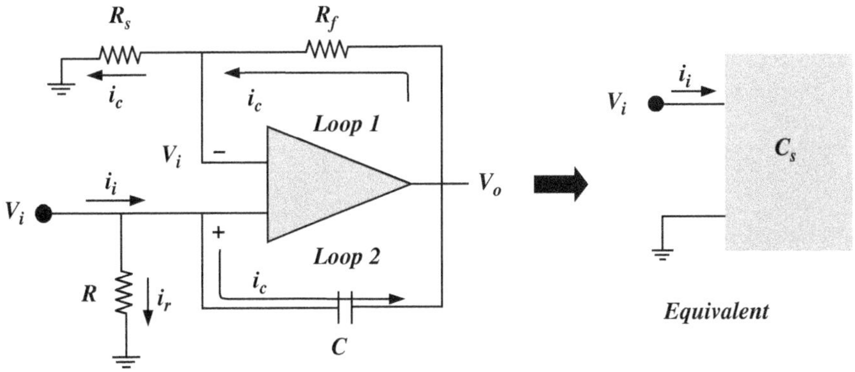

Figure 9.14 Electronically synthesized negative capacitance.

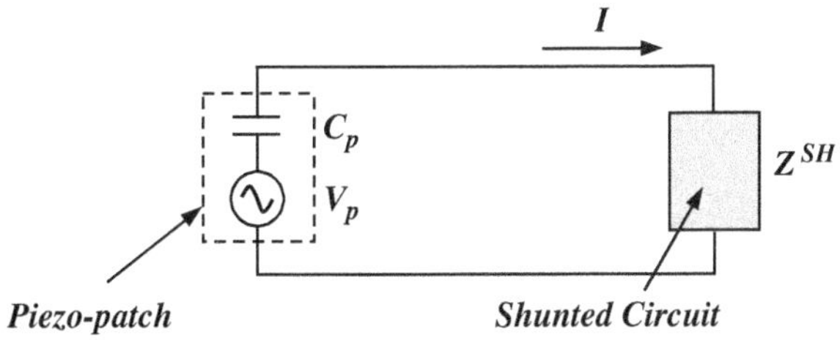

Figure 9.15 Piezo-patch with a shunted network.

Then, the electrical impedances of the piezo-patch Z_p and the shunted circuit Z_{SH} can generally be written as

$$Z_p = R_p + iX_p \quad and \quad Z^{SH} = R_{SH} + iX_{SH} \tag{9.28}$$

where R and X denote the resistance and reactance, respectively. Also, subscripts p and SH denote piezo and shunted, respectively.

Then, $Z_{total} = (R_p + R_{SH}) + i(X_p + X_{SH})$ and $I = \dfrac{V_p}{Z_{total}}$

Accordingly, the power dissipated P can be determined from

$$P = R_{SH} I^2 = R_{SH} \frac{V_p^2}{\left[(R_p + R_{SH})^2 + (X_p + X_{SH})^2\right]} \tag{9.29}$$

Then, P is a maximum when:

$$\frac{\partial P}{\partial X_{SH}} = 0 \quad \text{or} \quad \frac{R_{SH}\, V_p^2 2(X_p + X_{SH})}{\left[(R_p + R_{SH})^2 + (X_p + X_{SH})^2\right]^2} = 0, \; i.e. \; X_p = -X_{SH}$$

This results in a power dissipation of

$$P^o = R_{SH} \frac{V_p^2}{\left[(R_p + R_{SH})^2\right]} \tag{9.30}$$

and $\frac{\partial P}{\partial R_{SH}} = 0$ or $2R_{SH}\left[R_p + R_{SH}\right] - \left[R_p + R_{SH}\right]^2 = 0, i.e.\ R_p = R_{SH}$

This makes the maximum power dissipated to be given by

$$P^{oo} = \frac{V_p^2}{4R_{SH}} \tag{9.31}$$

If the piezo-patch has an impedance given by

$$Z_p = R_p - iX_p = R_p - i\frac{1}{C_p\omega}$$

Then, for maximum energy dissipation in the shunted network, the circuit must have an impedance Z_{SH} given by

$$Z^{SH} = R_{SH} + iX_{SH} = R_p + i\frac{1}{C_p\omega}$$

that is, $Z^{SH} = Z_p^*$, which implies that the shunted circuit must have a capacitance equals to the negative of the piezo-patch capacitance. This condition for maximum energy dissipation is called also the condition for "impedance matching" between the piezo-patch and the shunted circuit.

Note also that the maximum power dissipated P^{oo} in the shunted circuit does not require large shunted resistance. On the contrary, the smaller the resistance, the higher will be the maximum power dissipation as suggested by Eq. (9.31). This concept has been utilized by several investigators to achieve effective structural damping (Forward 1979; Browning and Wynn 1996; Wu 1998 and 2000).

9.2.5 Effectiveness of the Negative Capacitance from a Control System Perspective

Figure 9.16 includes three configurations of single degree of freedom (SDOF) systems supported on piezoelectric patches. In Figure 9.16a, the patch is resistively shunted using

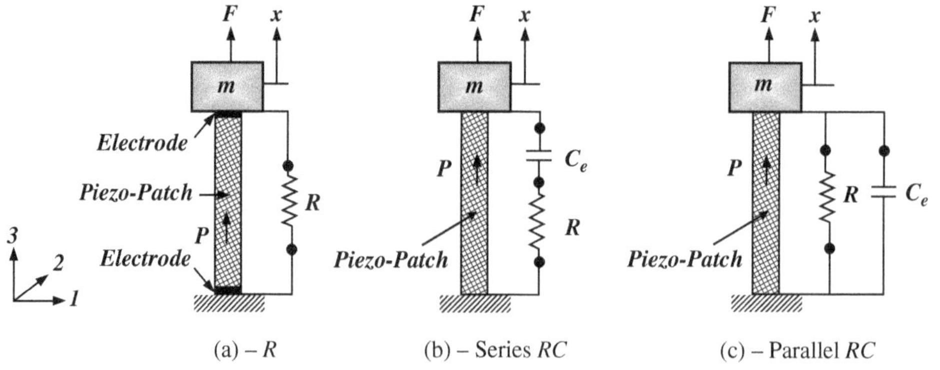

Figure 9.16 A *SDOF* system with three shunting configurations. (a) *R*. (b) Series *RC*. (c) Parallel *RC*.

a resistor R. In Figure 9.16b,c, the patches are shunted using capacitive elements C_e that are placed either in series or in parallel with the piezo-patches, respectively.

In the considered three configurations, it is assumed that the piezo-patch is fixed at one end and connected to a mass m at the other end. The piezo-patch is assumed also to operate in the 33 mode according to the coordinate axis shown in the figure and that the poling axis P is lined up with the three-direction.

From the material presented in Sections 9.2.1 through 9.2.3, the equation of motion of the SDOF systems shown in Figure 9.16 can be written in the Laplace domain as:

$$ms^2X + Z^{ME}sX = F \tag{9.32}$$

where X = Laplace transform of x, s = Laplace complex number, and $Z^{ME}s = \frac{A}{s^{SH}L}$ with and s^{SH} can be determined by modifying Eq. (9.14) to be as follows:

$$s^{SH} = s_{33}^{E}\left[1 - k_{33}^{2}\bar{Z}^{EL}\right] \tag{9.33}$$

Note that A and L are the cross sectional area and length of the piezo-patch, respectively. Also, $\bar{Z}^{EL} = \frac{Z^{EL}}{Z^{D}}$ = the impedance ratio of the shunted network to that of the piezo-patch alone and k_{33}^{2} = electromechanical coupling factor (EMCF). Let K^{SC} = stiffness of short-circuited piezo-patch = $\frac{A}{s_{33}^{E}L}$, then combining Eqs. (9.32) and (9.33) gives:

$$X = \frac{F}{ms^2 + \frac{K^{SC}}{1 - k_{33}^{2}\bar{Z}^{EL}}} \tag{9.34}$$

Equation (9.34) can be put in a dimensionless transfer function form as follows:

$$\text{or } \frac{X}{X_{ST}} = \frac{\left(1 - k_{33}^{2}\bar{Z}^{EL}\right)}{\bar{s}^2\left(1 - k_{33}^{2}\bar{Z}^{EL}\right) + 1} \tag{9.35}$$

where $X_{ST} = F/K^{SC}$ = static deflection and $\bar{s} = s/\omega_n$ with $\omega_n = \sqrt{K^{SC}/m}$ = short-circuit natural frequency of piezo = patch/mass system.

Table 9.2 lists $\bar{Z}^{EL}$ for the three shunting configurations of Figure 9.16, the corresponding transfer functions X/X_{ST}, and their characteristic equations. The table lists also the characteristics equations after putting them in a form suitable for plotting the root locus diagrams of the three configurations.

Example 9.4 Determine the effect of using R, R/C series, and R/C parallel shunted networks on the root locus plots of the characteristic equations of the three configurations for values of $0 < r < \infty$ and $\alpha = -1$, that is, when the shunted capacitance C_e is negative and matches that of the piezo-patch or $C_e = -C^T$. Assume that $k_{33}^{2} = 0.5$. Determine also the maximum attainable damping ratio for each configuration.

Solution

In order to plot the root locus diagram, the characteristic equation of each configuration can be written in the following form:

Table 9.2 Dynamic characteristics of the shunting configurations.

Shunting configuration	R	Series R/C	Parallel R/C
$\bar{Z}^{EL}$	$\frac{\Omega\bar{s}}{\Omega\bar{s}+1}$	$\frac{\alpha\Omega\bar{s}}{\alpha\Omega\bar{s}+(1+\alpha)}$	$\frac{\Omega\bar{s}}{\Omega(1+\alpha)\bar{s}+1}$
X/X_{ST}	$\frac{(1-k_{33}^2)\Omega\bar{s}+1}{(1-k_{33}^2)\Omega\bar{s}^3+\bar{s}^2+\Omega\bar{s}+1}$	$\frac{(1-k_{33}^2)(\alpha\Omega\bar{s}+1)+\alpha}{\bar{s}^2[(1-k_{33}^2)(\alpha\Omega\bar{s}+1)+\alpha]+\alpha\Omega\bar{s}+(1+\alpha)}$	$\frac{(1+\alpha-k_{33}^2)\Omega\bar{s}+1}{(1+\alpha-k_{33}^2)\Omega\bar{s}^3+\bar{s}^2+\Omega(1+\alpha)\bar{s}+1}$
Characteristic equation	$(1-k_{33}^2)\Omega\bar{s}^3+\bar{s}^2+\Omega\bar{s}+1=0$	$\bar{s}^2\left[(1-k_{33}^2)(\alpha\Omega\bar{s}+1)+\alpha\right]+\alpha\Omega\bar{s}+(1+\alpha)=0$	$(1+\alpha-k_{33}^2)\Omega\bar{s}^3+\bar{s}^2+\Omega(1+\alpha)\bar{s}+1=0$
Characteristic equation for root locus plots	$1+\frac{1}{\Omega}\frac{\bar{s}^2+1}{\bar{s}[(1-k_{33}^2)\bar{s}^2+1]}=0$	$1+\frac{1}{\Omega}\frac{\left[(1-k_{33}^2)+\alpha\right]\bar{s}^2+(1+\alpha)}{\alpha\bar{s}[(1-k_{33}^2)\bar{s}^2+1]}=0$	$1+\frac{1}{\Omega}\frac{\bar{s}^2+1}{\bar{s}[(1+\alpha-k_{33}^2)\bar{s}^2+(1+\alpha)]}=0$

$\Omega = RC^T\omega_n$, $\bar{s}=s/\omega_n$, $\alpha = C_e/C^T$.

$$\text{Characteristic equation} = 1+K\frac{N(s)}{D(s)}=0$$

where $K = 1/r$. The characteristic equation has roots equal to:

i) roots of D(s) when $K = 0$, and
ii) roots of N(s) when $K = \infty$.

The roots of $D(s)$ are called poles and those of $N(s)$ are called zeros. Table 9.3 lists the poles and zeros of the three configurations for $\alpha = -1$ and $k_{33}^2 = 0.5$.

Figure 9.17 displays the root locus plots of the three configurations using the MATLAB command "rlocus (N,D)", where N and D denote the vectors of the coefficients of the numerator and denominator polynomials arranged in descending order of s. For example, for the R configuration, the MATLAB commands are:

```
>>N= [1 0 1];
>>D= [0.5 0 1 0];
>>rlocus (N,D)
```

It can be seen that the maximum attainable damping ratios for the configurations are as listed in Table 9.4.

Table 9.3 Dynamic characteristics of the shunting configurations.

Shunting configuration	R	Series R/C	Parallel R/C
Poles	$\bar{s}=0, \pm 1/\sqrt{1-k_{33}^2}j$ $=0, \pm 1.404j$	$\bar{s}=0, \pm 1/\sqrt{1-k_{33}^2}j$ $=0, \pm 1.404j$	$\bar{s}=0,0,0$
Zeros	$\bar{s}=\pm j$	$\bar{s}=0,0$	$\bar{s}=\pm j$

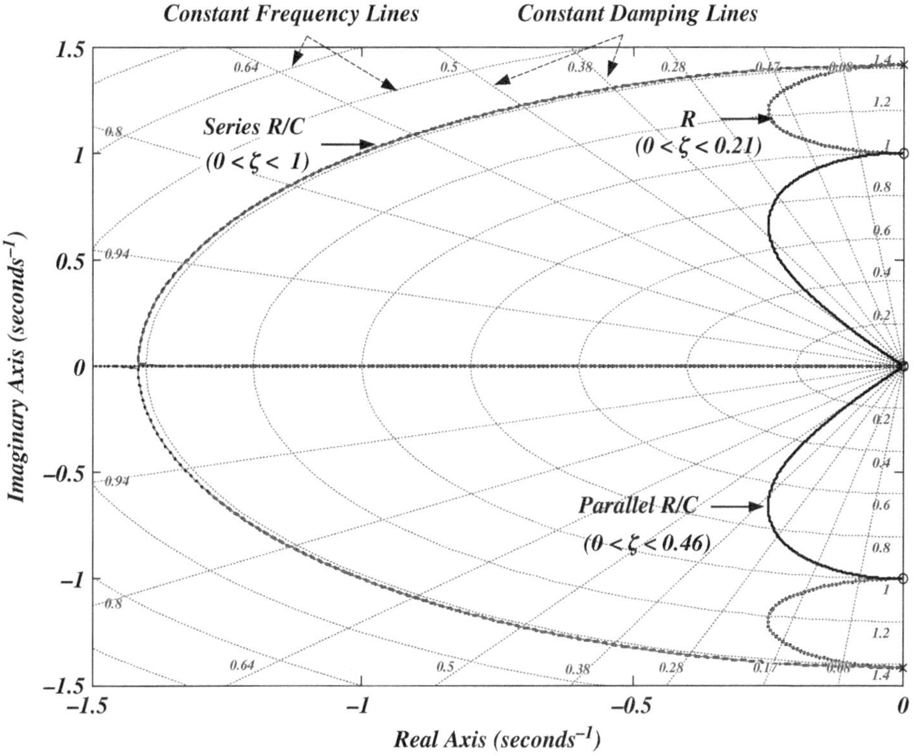

Figure 9.17 Root locus plots of the three shunting configurations (········ *R*, ▬ ▬ ▪ series *R/C*, ——— parallel *R/C*).

Table 9.4 Maximum attainable damping ratios of the shunting configurations.

Shunting configuration	*R*	Series R/C	Parallel R/C
Maximum ζ	0.21	1	0.46

9.2.6 Electrical Analogy of Shunted Piezoelectric Networks

Analysis of shunted piezoelectric networks coupled to vibrating structures using the electrical analogy approach presents a viable means that enables easily the modeling of such a multi-field system where mechanical and electrical fields are coupled. Figure 9.18 shows the electrical analog of the coupled rod/shunted piezo-network designating the mechanical and electrical domains.

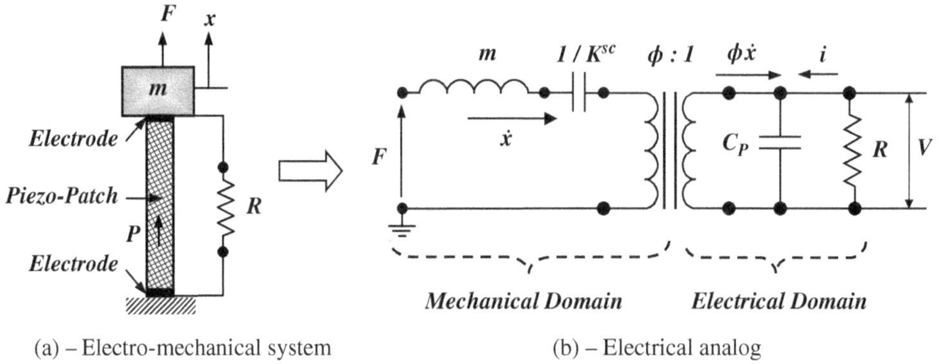

Figure 9.18 Electrical analog of a rod/shunted piezo-network. (a) Electromechanical system. (b) Electrical analog.

Extracting the equivalent electrical analog requires modifying Eq. (9.5) to take the following form:

$$\begin{Bmatrix} x \\ I \end{Bmatrix} = \begin{bmatrix} \dfrac{s_{11}^E L}{A} & d_{31} \\ d_{31}s & \dfrac{\varepsilon_{33}^T As}{L} \end{bmatrix} \begin{Bmatrix} F \\ V \end{Bmatrix} \tag{9.36}$$

where S_1 and T_1 are replaced by x/L and F/A with L and A denoting the length and cross sectional area of the piezo-element, respectively.

Expanding Eq. (9.36) gives:

$$x = C^{sc}F + d_{31}V, \tag{9.37}$$

and

$$I = d_{31}sF + C_p^T sV. \tag{9.38}$$

$C^{sc} = \dfrac{s_{11}^E L}{A} = 1/K^{sc}$ = compliance of the piezo-element and $C_p^T = \dfrac{\varepsilon_{33}^T A}{L}$ = capacitance of the piezo-element.

Under free strain conditions, $x = 0$ and Eq. (9.37) gives:

$$F = -(d_{31}/C^{sc})V = \phi V \tag{9.39}$$

where ϕ denotes the transformer turning ratio that transforms voltage into force. It is given by:

$$\phi = -\left(\frac{d_{31}A}{s_{11}^E L}\right) \tag{9.40}$$

Differentiating Eq. (9.37) with respect to the time, and using Eqs. (9.38) and (9.39) gives:

$$i + \dot{x}\phi = C_p^s \dot{v} \tag{9.41}$$

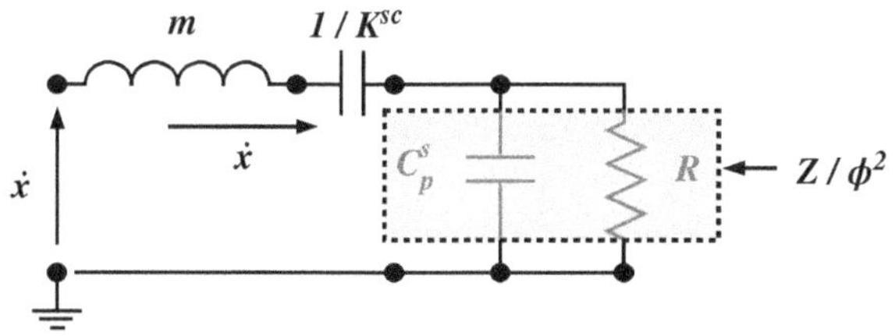

Figure 9.19 Simplified electrical analog of a rod/shunted piezo-network.

where $C_p^s = \left(1 - k_{31}^2\right) C_p^T$. Equation (9.41) can be used to simplify the equivalent electrical analog circuit to reduce it to the form shown in Figure 9.19.

Analysis of the simplified circuit yields the following dynamic equation:

$$\left[ms^2 + K^{sc} + \phi^2 \frac{Rs}{RC_p^s + 1} \right] X = F \tag{9.42}$$

Dividing by K^{sc} and putting in a dimensionless form, Eq. (9.42) reduces to:

$$\frac{X}{X_{ST}} = \frac{\left(1 - k_{33}^2\right) r\bar{s} + 1}{\left(1 - k_{33}^2\right) r\bar{s}^3 + \bar{s}^2 + r\bar{s} + 1} \tag{9.43}$$

where $r = RC^T \omega_n$, $\bar{s} = s/\omega_n$, and $\omega_n = \sqrt{K^{SC}/m}$. Note that Eq. (9.43) is the same as that listed in the second column of Table 9.2.

9.3 Finite Element Modeling of Structures Treated with Shunted Piezo-Networks

Two finite element approaches will be presented in this section for describing the interaction between shunted piezo-networks and base structures.

9.3.1 Equivalent Complex Modulus Approach of Shunted Piezo-Networks

In this approach, the dynamics of the shunted piezo-networks will be accounted for using the equivalent complex modulus derived in Section 9.2.2 for different types of shunted circuits.

In this regard, only the structural degrees of freedom will be considered as the electrical degrees of freedom are condensed when generating the equivalent complex modulus of the shunted piezo-networks.

Consider the ith finite element of a rod treated with a shunted piezo-network as shown in Figure 9.20.

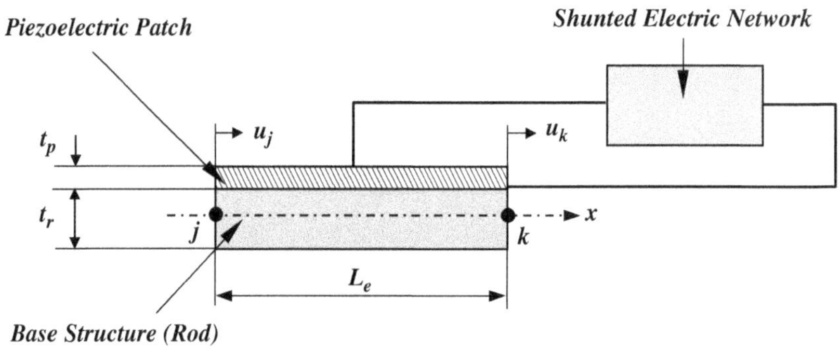

Figure 9.20 Finite element of a rod/shunted piezo-network.

The rod/shunted piezo-network system has only two structural degrees of freedom $\{\Delta_i\} = \{u_j, u_k\}^T$. The longitudinal displacement $u(x)$ of the system, at any position x, is given by the following shape function

$$u(x) = \{N\}\{\Delta_i\} \tag{9.44}$$

where $\{N\} = \{(1\text{-}x/L_e)\ x/L_e\}$ where L_e is the element length.

The potential (PE) of the element is given by

$$PE = \frac{1}{2}(E_r t_r + E_p t_p)b\int_0^{L_e} u_x^2 dx = \frac{1}{2}\{\Delta_i\}^T (E_r t_r + E_p t_p)b\int_0^{L_e} \{N\}_x^T\{N\}_x dx\{\Delta_i\}$$
$$= \frac{1}{2}\{\Delta_i\}^T[[K_r] + [K_p]]\{\Delta_i\} \tag{9.45}$$

where $[K_r] = (E_r t_r)b\int_0^{L_e} \{N\}_x^T\{N\}_x dx$ and $[K_p] = (E_p t_p)b\int_0^{L_e} \{N\}_x^T\{N\}_x dx.$

Note that $E_p = E'(1 + i\eta)$ is the equivalent complex modulus of the shunted piezo-network with E' and η as given in Section 9.1.2 for different shunting networks. Also, t_r and t_p denote the thickness of the rod and the piezo-patch, respectively.

Similarly, and kinetic energy (KE) of the element is given by

$$KE = \frac{1}{2}\left(\rho_r t_r + \rho_p t_p\right)b\int_0^{L_e} \dot{u}^2 dx = \frac{1}{2}\{\dot{\Delta}_i\}^T\left(\rho_r t_r + \rho_p t_p\right)b\int_0^{L_e} \{N\}^T\{N\}dx\{\dot{\Delta}_i\}$$
$$= \frac{1}{2}\{\dot{\Delta}_i\}^T[[M_r] + [M_p]]\{\dot{\Delta}_i\} \tag{9.46}$$

where $[M_r] = (\rho_r t_r)b\int_0^{L_e} \{N\}^T\{N\}dx$ and $[M_p] = \left(\rho_p t_p\right)b\int_0^{L_e} \{N\}^T\{N\}dx$ with ρ_r and ρ_p denoting the density of the rod and piezo-patch, respectively.

Hence, the equation of motion of the ith element becomes

$$[[M_r] + [M_p]]\{\ddot{\Delta}_i\} + [[K_r] + [K_p]]\{\Delta_i\} = \{F_i\} \tag{9.47}$$

where $\{F_i\}$ is the force vector acting on the ith element.

The equation of motion of the entire rod/shunted piezo-network system can then be determined by assembling the mass and stiffness matrices of the individual elements to yield:

$$[M_o]s^2\{\Delta\} + [K_{r_o} + K_{p_o}]\{\Delta\} = \{F_o\} \tag{9.48}$$

where $[M_o], [K_{r_o}]$, and $[K_{p_o}]$ are the overall mass matrix, overall structural stiffness matrix, and overall piezo-network stiffness matrix. Also, $\{\Delta\}$ and $\{F_o\}$ denote the deflection and load vectors of the entire rod/ shunted piezo-network assembly.

The boundary conditions are imposed on the structure to eliminate the rigid body modes and then the remaining system can be solved using the exact complex eigenvalue problem solvers, the Golla–Hughes–McTavish (GHM) method (Chapter 4), or the Modal Strain Energy (MSE) Method (Chapter 5).

A similar approach can be adopted when the shunted piezo-networks are used to control the vibration for beams, plates, or shells.

Example 9.5 Determine the longitudinal response of the fixed-free rod shown in Figure 9.21 when it is subjected to a unit sinusoidal loading acting at its free end. The rod is made of aluminum with length of 1 m, width of 0.025 m, and thickness of 0.025 m. A lead zirconate titanate (PZT)-5H piezo-patch is bonded to the rod such that its length = 1 m, width = 0.025 m, and thickness = 0.0025 m. The physical parameters of the piezo-patch are given in the following table:

Material	s_{11}^E (m^2 N^{-1})	d_{31} (m V^{-1})	ε_{33}^T (Farad m^{-1})	Density (kg m^{-3})
PZT-5H	0.165E-10	−274 E-12	2.49E-8	7300

PZT, lead zirconate titanate (ceramic).

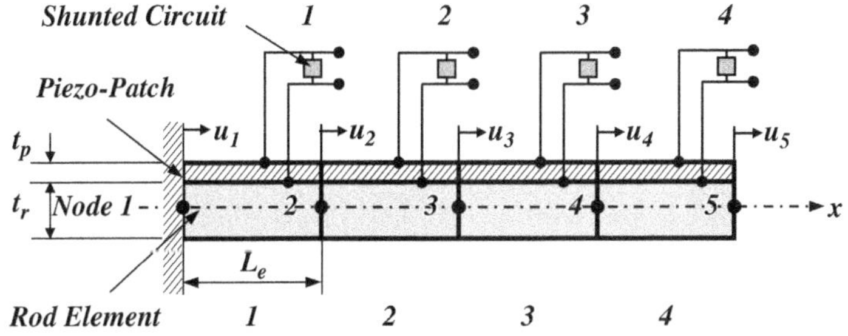

Figure 9.21 A four-element rod/piezo-patch system.

Assume that the rod is divided into four finite elements with $L_e = 0.25$ m. Determine the response of its free end (Node 5) when the piezo-patch is:

a) short-circuited.
b) shunted using a resistance $R = 744\ \Omega$.
c) shunted using R–L circuit with $R = 744\ \Omega$ and $\bar{L} = 1.224$.
d) shunted using R–C circuit with $R = 744\ \Omega$ and $C_e = -\ C^T$.

Solution

Figure 9.22 shows the response of the free end of the rod for the different shunting conditions.

Note that R is tuned to be 744 Ω to target the second mode at 3800 Hz such that $R = \sqrt{(1-k_{31}^2)}/[C^S(2\pi\,3{,}800)] = 744\Omega$ with $C^S = 0.05087$ μFarad and $k_{31}^2 = 0.1827$.
Also, $\bar{L}$ is tuned to be 1.224 to target the second mode such that

$$L = \frac{1}{C^S(2\pi\,3{,}800)^2} = 0.0345 \text{ Henrys and } \bar{L} = \frac{L}{R^2C^S} = \frac{0.0345}{744^2\,5.087E-8} = 1.224.$$

The figure indicates that the R and the R–L shunt circuits have been effective in damping out the high frequency modes of vibration. However, the R–C shunt circuit with an impedance-matched negative capacitance ($C_e = -C^T$) has been more effective in attenuating the vibration over a wider frequency band particularly at low frequencies.

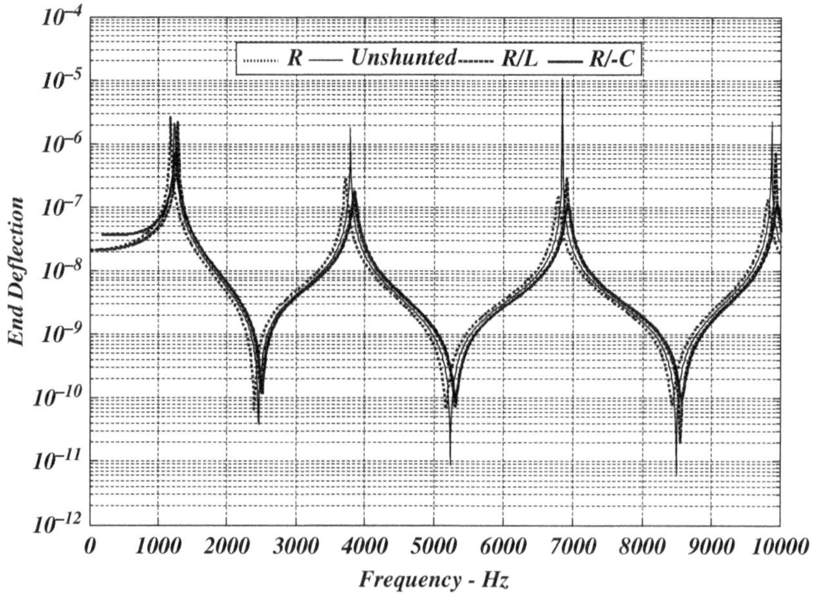

Figure 9.22 Response of the rod/piezo-patch system for different shunting conditions.

9.3.2 Coupled Electromechanical Field Approach of Shunted Piezo-Networks

In this approach, both the structural as well as the electrical degrees of freedom will be simultaneously considered to describe the interaction between the dynamics of the shunted piezo-networks and the vibrating structure. The equations governing such an interaction are based on rewriting the constitutive Eq. (9.1) for the piezo-patch as follows

$$\begin{Bmatrix} T_1 \\ E_3 \end{Bmatrix} = \begin{bmatrix} \dfrac{1}{s_{11}^E(1-k_{31}^2)} & -\dfrac{d_{31}}{s_{11}^E(1-k_{31}^2)\,\varepsilon_{33}^T} \\ -\dfrac{d_{31}}{s_{11}^E(1-k_{31}^2)\,\varepsilon_{33}^T} & \dfrac{1}{\varepsilon_{33}^T(1-k_{31}^2)} \end{bmatrix} \begin{Bmatrix} S_1 \\ D_3 \end{Bmatrix} = \begin{bmatrix} c_{11}^D & -h_{31} \\ -h_{31} & \dfrac{1}{\varepsilon_{33}^S} \end{bmatrix} \begin{Bmatrix} S_1 \\ D_3 \end{Bmatrix} \tag{9.49}$$

where $k_{31}^2 = \dfrac{d_{31}^2}{s_{11}^E\,\varepsilon_{33}^T}$, $h_{31} = \dfrac{d_{31}}{s_{11}^E(1-k_{31}^2)\,\varepsilon_{33}^T}$, $\varepsilon_{33}^S = \varepsilon_{33}^T(1-k_{31}^2)$, and $c_{11}^D = 1/s_{11}^D = 1/s_{11}^E(1-k_{31}^2)$.

This form of constitutive relationship will be used to generate the potential and kinetic energies of an element of the rod/shunted piezo-network in terms of the structural degrees of freedom $\{\Delta\} = \{u_j\ u_k\}^T$ and the electrical degrees of freedom Q_i (the charge of the ith element). Note that the longitudinal displacement $u(x)$ of the system, at any position x, is given by the following shape function

$$u(x) = \{N\}\{\Delta_i\} \tag{9.50}$$

where $\{N\} = \{(1-x/L_e)\ x/L_e\}$ where L_e is the element length.

The KE of the element can now be determined from

$$\begin{aligned} KE &= \frac{1}{2}b\left(t_r\rho_r + t_p\rho_p\right)\int_0^l \dot{u}^2 dx \\ &= \frac{1}{2}\{\dot{\Delta}_i\}^T[[M_r] + [M_p]]\{\dot{\Delta}_i\} = \frac{1}{2}\{\dot{\Delta}_i\}^T[M]\{\dot{\Delta}_i\} \end{aligned} \tag{9.51}$$

where $[M_r] = (\rho_r t_r)b\int_0^{L_e}\{N\}^T\{N\}dx$ = mass matrix of rod,

$[M_p] = (\rho_p t_p)b\int_0^{L_e}\{N\}^T\{N\}dx$ = mass matrix of piezo-patch and $[M] = [M_r] + [M_p]$ with $\rho_r$and ρ_p denoting the density of the rod and piezo-patch, respectively.

The potential energy PE of the element is given by

$$PE = \frac{1}{2}\int_V S_1\left(T_{1_p} + T_{1_r}\right)dv + \frac{1}{2}\int_V D_3 E_3 dv \tag{9.52}$$

where $T_{1_p} = \frac{1}{s_{11}^D}S_1 - h_{31}D_3$ = stress on piezo-patch and $T_{1_r} = E_r S_1$ = stress on rod with E_r denoting Young's modulus of the rod.

Substituting from Eq. (9.49) into Eq. (9.52) gives

$$\begin{aligned} PE = &\frac{1}{2}(bt_r)\int_0^{L_e} S_1 E_r S_1 dx + \frac{1}{2}(bt_p)\int_0^{L_e} S_1\left[c_{11}^D S_1 - h_{31}D_3\right]dx \\ &+ \frac{1}{2}(bt_p)\int_0^{L_e} D_3\left[-h_{31}S_1 + \frac{1}{\varepsilon_{33}^s}D_3\right]dx \end{aligned} \tag{9.53}$$

Note that $S_1 = u_x = \{N_x\}\{\Delta_i\}$ and $D_3 = Q_i/(bL_e)$ = constant along the element, then Eq. (9.53) reduces to

$$PE = \frac{1}{2}\{\Delta_i\}^T[K_s]\{\Delta_i\} - \frac{t_p}{L_e}h_{31}Q_i\{1\ \ -1\}\{\Delta_i\} + \frac{1}{2}\frac{Q_i^{\,2}t_p}{bL_e\,\varepsilon_{33}^s} \tag{9.54}$$

where $[K_s] = b\left(t_r E_r + t_p c_{11}^D\right)\int_0^{L_e}\{N_x\}^T\{N_x\}dx$ = stiffness matrix of rod/piezo-patch system.

The virtual work δW associated with the shunted network and the external loads $\{F_i\}$ is given by

$$\delta W = -\left(L\dot{I} + RI + \frac{1}{C_e}\int Idt\right)\delta Q_i = -\left(L\ddot{Q}_i + R\dot{Q}_i + \frac{1}{C_e}Q_i\right)\delta Q_i + \{F_i\}\delta\{\Delta_i\} \tag{9.55}$$

Now, the equations governing the dynamics of the rod/piezo-patch system can be extracted using the Lagrangian dynamics to give

$$\frac{d}{dt}\frac{\partial KE}{\partial\{\dot{\Delta}_i\}} + \frac{\partial PE}{\partial\{\Delta_i\}} = \{F_i\} \quad \text{and} \quad \frac{d}{dt}\frac{\partial KE}{\partial\{\dot{Q}_i\}} + \frac{\partial PE}{\partial\{Q_i\}} = -\left\{L\ddot{Q}_i + R\dot{Q}_i + \frac{1}{C_e}Q_i\right\}$$

These equations can be written in the following matrix form

$$\begin{bmatrix}[M] & 0\\ 0 & L\end{bmatrix}\begin{Bmatrix}\{\ddot{\Delta}_i\}\\ \ddot{Q}_i\end{Bmatrix} + \begin{bmatrix}0 & 0\\ 0 & R\end{bmatrix}\begin{Bmatrix}\{\dot{\Delta}_i\}\\ \dot{Q}_i\end{Bmatrix} + \begin{bmatrix}[K_s] & -\frac{t_p}{L_e}h_{31}\begin{Bmatrix}1\\ -1\end{Bmatrix}\\ -\frac{t_p}{L_e}h_{31}\begin{Bmatrix}1\\ -1\end{Bmatrix}^T & \left(\frac{1}{C_e} + \frac{1}{C^s}\right)\end{bmatrix}\begin{Bmatrix}\{\Delta_i\}\\ Q_i\end{Bmatrix} = \begin{Bmatrix}\{F_i\}\\ 0\end{Bmatrix} \tag{9.56}$$

or

$$[M_i]\{\ddot{X}_i\} + [C_i]\{\dot{X}_i\} + [K_i]\{X_i\} = \{f_i\} \tag{9.57}$$

where $[M_i] = \begin{bmatrix} [M] & 0 \\ 0 & L \end{bmatrix}$, $[C_i] = \begin{bmatrix} 0 & 0 \\ 0 & R \end{bmatrix}$, $[K_i] = \begin{bmatrix} [K_s] & -\frac{t_p}{L_e} h_{31} \begin{Bmatrix} 1 \\ -1 \end{Bmatrix} \\ -\frac{t_p}{L_e} h_{31} \begin{Bmatrix} 1 \\ -1 \end{Bmatrix}^T & \left(\frac{1}{C_e} + \frac{1}{C^S}\right) \end{bmatrix}$,

$\{X_i\} = \begin{Bmatrix} \{\Delta_i\} \\ Q_i \end{Bmatrix}$, and $\{f_i\} = \begin{Bmatrix} \{F_i\} \\ 0 \end{Bmatrix}$.

The equation of motion of the entire rod/shunted piezo-network system can then be determined by assembling the mass and stiffness matrices of the individual elements to yield:

$$[M_o]\{\ddot{X}\} + [C_o]\{\dot{X}\} + [K_o]\{X\} = \{F_o\} \tag{9.58}$$

where $[M_o]$, $[C_o]$, and $[K_o]$ are the overall mass, damping, and stiffness matrices of the entire rod/ shunted piezo-network assembly. Also, $\{X\}$ and $\{F_o\}$ denote the structural and electrical degrees of freedom as well as load vectors of the entire assembly.

Note that bonding the shunted piezo-patch to the rod has resulted in adding damping to the rod as quantified by the damping matrix $[C_o]$. It has also modified the stiffness and mass matrices.

Once the boundary conditions are imposed on the structure to eliminate the rigid body modes and then, the remaining system can be directly solved using the real eigenvalue problem solvers because all the mass, damping, and mass matrices of the assembly are real and symmetric matrices.

Efficient solution of Eq. (9.58) may require condensation of the electrical degrees of freedom and retaining only the structural degrees of freedom.

Example 9.6 Determine the longitudinal response of the fixed-free rod shown in Figure 9.21 using the coupled electromechanical field approach and compare the results with those obtained using the complex modulus approach. Consider only the case when the piezo-patch is shunted with an *R–L* circuit such that $R = 744\ \Omega$ and $\bar{L} = 1.224$.

Solution

Figure 9.23 shows the response of the free end of the rod as calculated by using the complex modulus and coupled field approaches. It can be seen that the two approaches yield similar results.

Example 9.7 Consider the cantilevered plate shown in Figure 9.24. Two sets of resistively and inductively shunted piezo-patches are placed symmetrically near the fixed end to control the first mode of vibration of the plate. The dimensions of the aluminum plate and the *PZT*-4 piezoelectric patches are given in Figure 9.24. The shunted resistances R_n of the piezo-patch sets 1 and 2 are tuned to the first mode such that

$$R_n = \sqrt{(1 - k_{31}^2/(C^S\omega_n)} \quad n = 1 \tag{9.59}$$

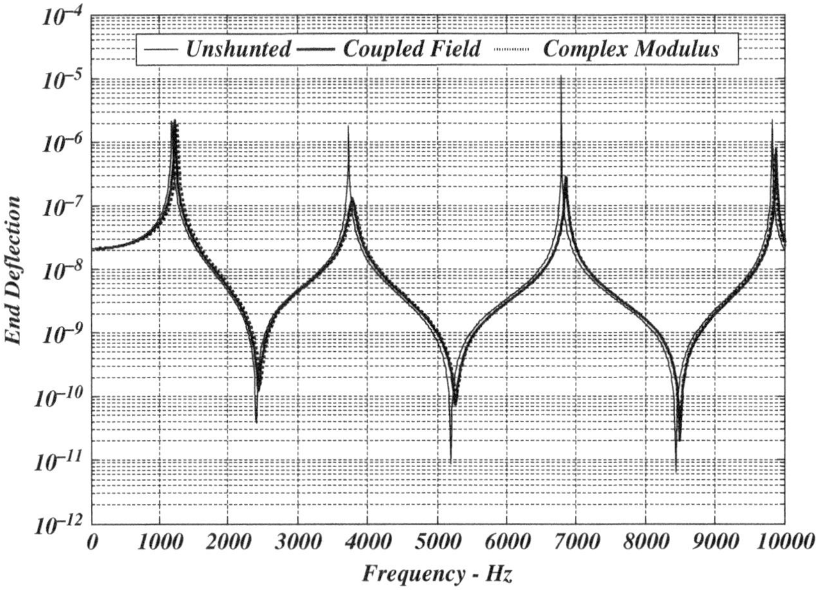

Figure 9.23 Response of the rod/piezo-patch system using complex modulus and coupled field approaches.

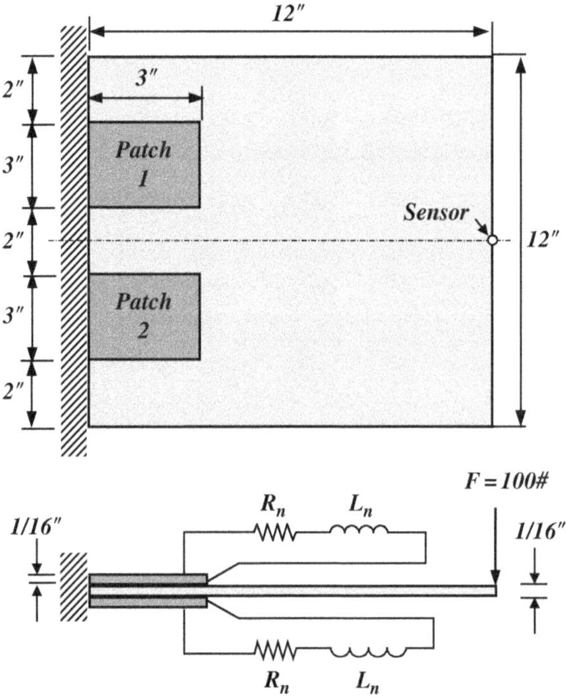

Figure 9.24 Plate with two sets of resistively and inductively shunted piezo-patches to control the first mode of vibration.

and

$$L_n = 1/\left(C^S \omega_n^2\right) \quad n = 1 \tag{9.60}$$

Using the ANSYS finite element software package, determine:

i) Frequency response for plate with unshunted piezo-patches (i.e., $R_n = 0$, $L_n = 0$).
ii) Frequency response for plate with tuned shunted piezo-patches (i.e., $R_n \neq 0$ as given by Eq. (9.59) and $L_n = 0$).
iii) Frequency response for plate with tuned shunted piezo-patches (i.e., R_n as given by Eq. (9.59) and L_n as given by Eq. (9.60)).

Solution

The plate/piezo-patches assembly is modeled using ANSYS. The first four natural frequencies and the corresponding mode shapes are determined as shown in Figure 9.25. The first four natural frequencies are found to be: 24.14, 47.49, 117.64, and 142.53 Hz, respectively. The displayed mode shapes indicate that the first natural frequencies correspond to first bending, first torsion, two-axes bending, and second bending modes, respectively.

From such an identification process, it is expected that the piezo-patches should be effective in damping out the vibration at the first and fourth modes and become less effective in controlling the second and third modes.

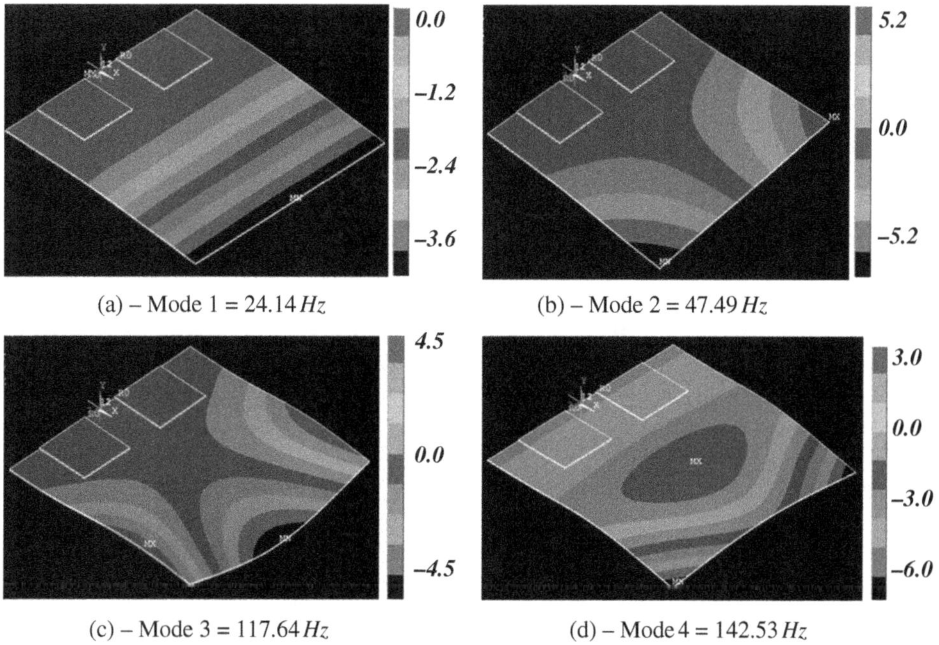

Figure 9.25 Mode shapes of the plate with two sets of resistively and inductively shunted piezo-patches. (a) Mode 1 = 24.14 Hz. (b) Mode 2 = 47.49 Hz. (c) Mode 3 = 117.64 Hz. (d) Mode 4 = 142.53 Hz.

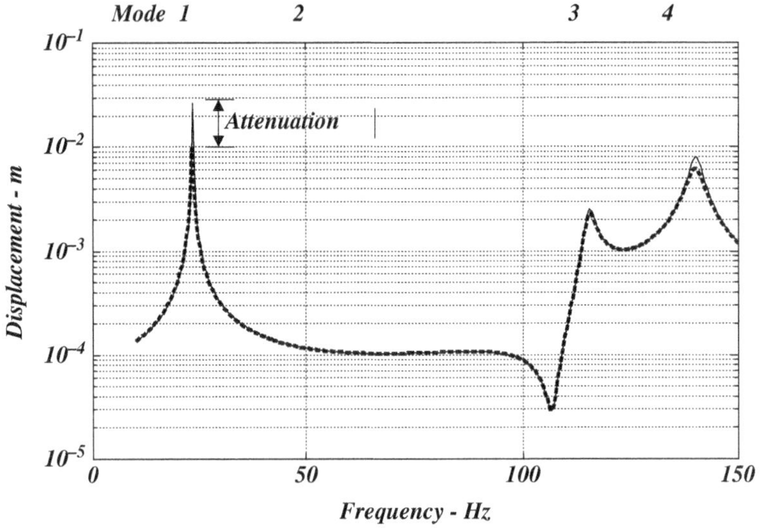

Figure 9.26 Frequency response of the plate with two sets of resistively shunted piezo-patches (——— uncontrolled, ········ controlled).

Figure 9.26 shows a comparison between the frequency response of the uncontrolled plate and the plate when the all the four piezo-patches are shunted with resistors tuned to the first mode. It can be seen that attenuation of about 3.5 times is achieved for the first mode.

Note that the second mode does not appear in the displayed response as the sensor is placed at the nodal line as shown in Figures 9.24 and 9.25b.

Figure 9.27 displays a comparison between the frequency response of the uncontrolled plate and the plate when the all the four piezo-patches are shunted with resistors and inductors that are tuned to the first mode. It can be seen that attenuation of about five times is achieved for the first mode.

9.4 Active Shunted Piezoelectric Networks

9.4.1 Basic Configurations

In this section, three basic shunted circuits are considered. In the first configuration, shown in Figure 9.28a, the shunting circuit is switched between the open and short-circuit states and it is denoted by the "OC-SC" configuration. Performance of this configuration serves as the datum against which the performance of the "RS" and "OC-RS" configurations is evaluated. Figure 9.28b shows the resistive shunting (RS) configuration that was presented in Sections 9.2.2.1, 9.2.5, and 9.2.6.

The third configuration, shown in Figure 9.28c, is provided with actively controlled switching capabilities in order to enhance their damping characteristics (Clark 2000;

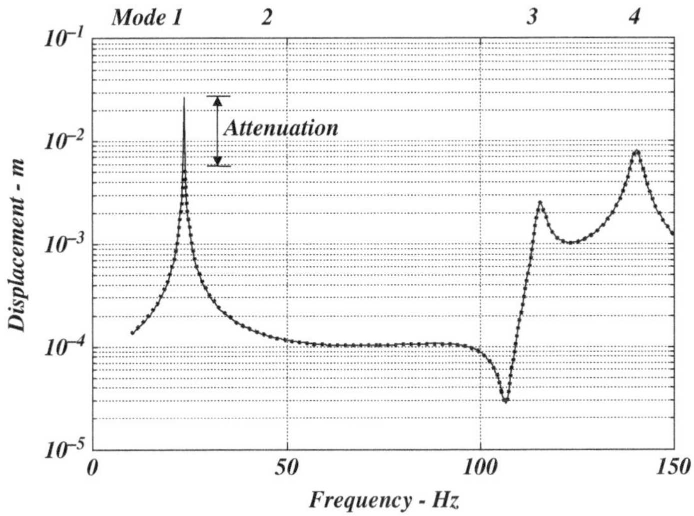

Figure 9.27 Frequency response of the plate with two sets of resistively and inductively shunted piezo-patches (uncontrolled, controlled).

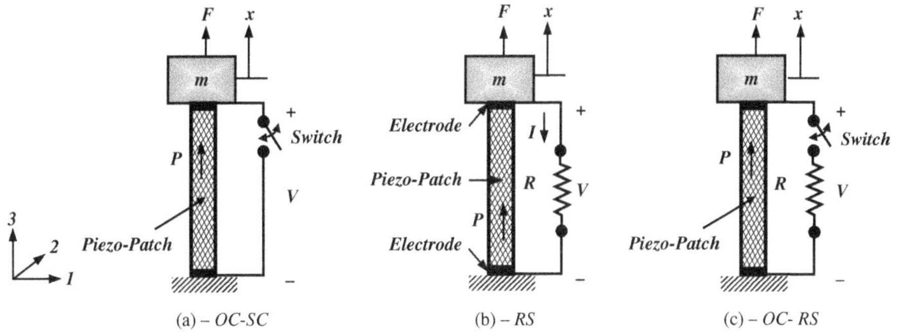

Figure 9.28 Three shunting configurations. (a) OC-SC. (b) RS. (c) OC-RS.

Corr and Clark 2002; Itoh et al. 2011; Mokrani et al. 2012; Neubauer et al. 2013). In such a configuration, as the switching takes place between the open circuit and resistive shunt states, it is called the "OC-RS" configuration.

In the considered three configurations, it is assumed that the piezo-patch is fixed at one end and connected to a mass m at the other end. The piezo-patch is assumed also to operate in the *33* mode according to the coordinate axis shown in the figure and that the poling axis P is lined up with the three-direction.

The dynamic equations of the different configurations can be derived using the relationships listed in Table 9.5 that are obtained from Figure 9.4, Eq. (9.7), and Eq. (9.20). The ultimate goal is to utilize these equations to establish a comparison between the energy dissipation characteristics of three configurations.

Table 9.5 Compliance and stiffness ratio of different shunting configurations.

Circuit	Short circuit (SC)	Open circuit (OC)	Resistive shunt (RS)
Compliance (s)	$s = s_{33}^{E}$	$s = s_{33}^{E}\left(1-k_{33}^{2}\right)$	$s = s_{33}^{E}\left[1-\dfrac{k_{33}^{2}\,RC_{p}^{T}s}{\left(RC_{p}^{T}s+1\right)}\right]$
K/K^{SC}	1	$1/\left(1-k_{33}^{2}\right)$	$1\Big/\left[1-\dfrac{k_{33}^{2}\,RC_{p}^{T}s}{\left(RC_{p}^{T}s+1\right)}\right]$

9.4.2 Dynamic Equations

9.4.2.1 Short-Circuit Configuration

The equation of motion of the mass m can be written as:

$$\left(ms^{2}+K^{sc}\right)X = F \tag{9.61}$$

Dividing by K^{SC}, this equation can be put in the following dimensionless form:

$$\left(\bar{s}^{2}+1\right)X = X_{ST} \tag{9.62}$$

where, s = Laplace complex number, $\bar{s} = s/\omega_n$, $\omega_n = \sqrt{K^{SC}/m}$, and $X_{ST} = F/K^{SC}$.

9.4.2.2 Open-Circuit Configuration

Using Eq. (9.7), the equation of motion of the OC configuration can be easily written in the following dimensionless form:

$$\left(\bar{s}^{2}+\frac{1}{1-k_{33}^{2}}\right)X = X_{ST} \tag{9.63}$$

9.4.2.3 Resistive-Shunted Configuration

Using Eq. (9.20), the corresponding equation of motion of the RS configuration can be written as follows:

$$\left[\bar{s}^{2}+1+k_{33}^{2}\frac{r\bar{s}}{(1-k_{33}^{2})\Omega\bar{s}+1}\right]X = X_{ST} \tag{9.64}$$

where $\Omega = RC^{T}\omega_n$.

9.4.3 More on the Resistive Shunting Configuration

Equation (9.64) represents a second order dynamical system that has a dimensionless mass = 1, dimensionless stiffness = 1, and a damping coefficient $C(\bar{s})$ such that:

$$C(\bar{s}) = k_{33}^{2}\frac{\Omega}{(1-k_{33}^{2})\,\Omega\bar{s}+1} \tag{9.65}$$

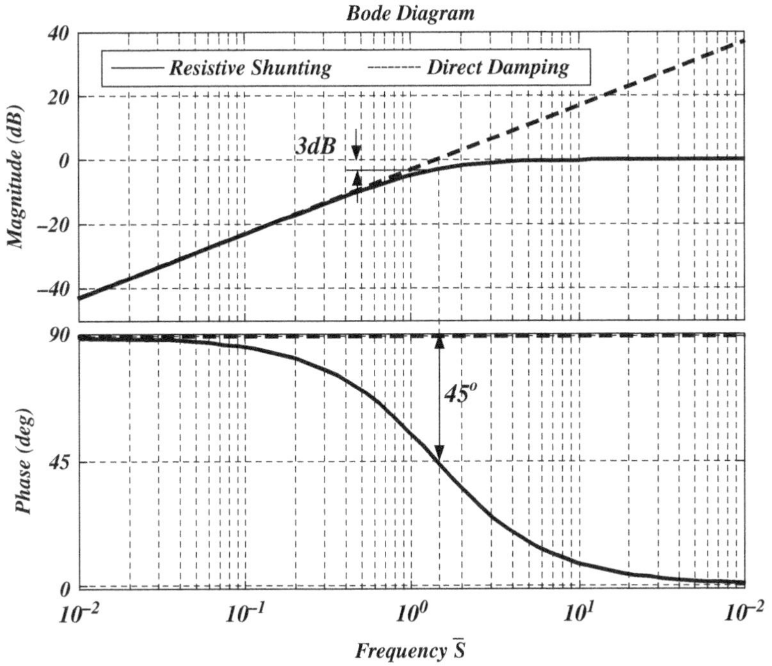

Figure 9.29 Damping effect of resistive shunting as compared to direct damping.

Hence, Eq. (9.64) reduces to:

$$\left[\bar{s}^2 + C(\bar{s})\bar{s} + 1\right]X = X_{ST} \tag{9.66}$$

Figure 9.29 displays a comparison between the damping effects $C(\bar{s})\bar{s}$ for the RS and a direct damping effect of $k_{33}^2\,\Omega\,\bar{s}$ when Ω assumes its optimal value of 1.414 and $k_{33}^2 = 0.5$. It is evident that the RS acts as a delayed damping effect that produces a magnitude shift of 3 dB and phase angle of 45° as compared to the direct damping effect.

Note that the magnitude and phase angle shifts come about because $C(\bar{s})\bar{s}$ is not just an imaginary number but rather is a complex number given by:

$$C(\bar{s})\bar{s}\big|_{\bar{s}=i\bar{\omega}} = K'(1 + \eta i) \tag{9.67}$$

where $\bar{\omega} = \omega/\omega_n$, $K' = \dfrac{k_{33}^2(1 - k_{33}^2)(\Omega\bar{\omega})^2}{1 + \left(1 - k_{33}^2\right)^2(\Omega\bar{\omega})^2}$, and $\eta = \dfrac{1}{(1 - k_{33}^2)\,\Omega\bar{\omega}}$ with ω = excitation frequency.

Hence, Eq. (9.66) can be rewritten as:

$$\left[\bar{s}^2 + \bar{C}\,\bar{s} + (1 + K')\right]X = X_{ST} \tag{9.68}$$

with $\bar{C} = \dfrac{k_{33}^2\,\Omega}{1 + \left(1 - k_{33}^2\right)^2(\Omega\bar{\omega})^2}$.

Equation (9.68) indicates that the RS adds damping and increases the structural stiffness compared to the stiffness under short-circuit conditions.

Example 9.8 Show that the stiffness at open circuit condition $K^{OC} = \frac{1}{1-k_{33}^2}$ is always larger than the stiffness with RS $K^{RS} = 1 + K'$ for any values of k_{33}^2 and $r\bar{\omega}$.

Solution

The stiffness at open circuit condition K^{OC} can be written as:

$$K^{OC} = \frac{1}{1-k_{33}^2} = 1 + \frac{k_{33}^2}{1-k_{33}^2} \tag{a}$$

Also, the stiffness with RS $K^{RS} = 1 + K'$ can be written as:

$$K^{RS} = 1 + K' = 1 + \frac{k_{33}^2(1-k_{33}^2)(\Omega\bar{\omega})^2}{1+(1-k_{33}^2)^2(\Omega\bar{\omega})^2}$$

$$= 1 + \frac{k_{33}^2}{(1-k_{33}^2) + \frac{1}{(1-k_{33}^2)(\Omega\bar{\omega})^2}} \tag{b}$$

Hence, $K^{RS} = 1 + \frac{k_{33}^2}{(1-k_{33}^2) + \frac{1}{(1-k_{33}^2)(\Omega\bar{\omega})^2}} < K^{OC} = 1 + \frac{k_{33}^2}{1-k_{33}^2}$.

9.4.4 Open-Circuit to Resistive Shunting (OC-RS) Configuration

The characteristics of the OC-RS configuration can now be easily considered as it is simply a combination of the characteristics of the open-circuit and RS configurations given by Eqs. (9.63) and (9.68), respectively.

9.4.4.1 Dynamic Equations

The time domain equations representing the dynamics of the OC-RS configuration are as follows:

$$\textit{When operating in OC mode}: \ddot{x} + K^{OC}x = x_{ST} \tag{9.69}$$

$$\text{where } K^{OC} = \frac{1}{1-k_{33}^2}$$

and

$$\textit{When operating in RS mode}: \ddot{x} + \bar{C}\dot{x} + K^{RS}x = x_{ST} \tag{9.70}$$

where $K^{RS} = 1 + \frac{k_{33}^2}{(1-k_{33}^2) + \frac{1}{(1-k_{33}^2)(\Omega\bar{\omega})^2}}$

9.4.4.2 Switching Between OC and RS Modes

The switching between the OC and RS modes of operation is carried out in such a manner to ensure monotonic decrease of system's total energy, that is, guarantee system stability.

To develop the globally stable switching strategy, a Lyapunov function V is defined such that:

$$V = \frac{1}{2}\{x\,\dot{x}\}\begin{bmatrix} K^{RS} & 0 \\ 0 & 1 \end{bmatrix}\begin{Bmatrix} x \\ \dot{x} \end{Bmatrix} = \frac{1}{2}K^{RS}x^2 + \frac{1}{2}\dot{x}^2 \tag{9.71}$$

The defined Lyapunov function is equal to the sum of the potential and kinetic energies of the system under the RS mode of operation. Then, the rate of change of the Lyapunov function can be determined by differentiating Eq. (9.71) with respect to time and substituting from Eq. (9.69) for the OC mode and Eq. (9.70) for the RS mode. Without loss of generality, if $x_{st} = 0$, then the time derivative of V reduces to:

$$\textit{When operating in OC mode: } \dot{V} = \left(K^{RS}x + \ddot{x}\right)\dot{x} = \left(K^{RS} - K^{OC}\right)x\dot{x} \tag{9.72}$$

As $(K^{RS} - K^{OC}) < 0$, then $\dot{V} < 0$ for $x\dot{x} > 0$.

$$\textit{When operating in RS mode: } \dot{V} = \left(K^{RS}x + \ddot{x}\right)\dot{x} = \left(K^{RS} - K^{RS}\right)x\dot{x} - \bar{C}\dot{x}^2 = -\bar{C}\dot{x}^2 \tag{9.73}$$

Then, for $x\dot{x} < 0$, the operation is switched to *RS* mode as $\dot{V} < 0$. In summary, when the system is attempting to move away from equilibrium, that is, $x\dot{x} > 0$, the shunted network is switched to the high-stiffness state (open circuit). However, when the system starts moving toward the equilibrium, that is, $x\dot{x} < 0$, the network is switched to the low-stiffness and/or dissipative state (RS). In this manner and during a full vibration cycle, the switching occurs four times, once after each quarter of a cycle.

Example 9.9 Compare the phase-plane plots $(x - \dot{x})$ for a mass supported on a piezo-element, which is shunted using:
RS and *b*. switched resistive shunting (OC-RS).
For values of $r\bar{\omega} = 0.5$ and 1.414. Assume that $k_{33}^2 = 0.5$ and the mass is subjected to initial conditions $x_0 = 1$, and $\dot{x}_0 = 1$.

Solution

The characteristics with the RS shunting is obtained by integrating Eq. (9.68) while those with the OC-RS shunting is obtained by integrating Eqs. (9.63) and (9.68) using the globally stable switching strategy outlined in Section 9.4.4.2.

Figure 9.30a,b shows the phase-plane plots at $r\bar{\omega} = 0.5$ and 1.414, respectively. The two figures show comparisons between the performance with RS and OC-RS shunting.

It is evident that the characteristics under RS shunting is smooth whereas that with the OC-RS has obvious discontinuities at the switching points between the OC and RS modes of operations.

Furthermore, in Figure 9.30a where $\Omega\bar{\omega} = 0.5$, the use of OC-RS shunting has resulted in faster damping of the vibration than the RS shunting. However, the reverse is true when $\Omega\bar{\omega} = 1.414$. The reasons behind such distinct differences between the performance of the two shunting configurations will be best understood by considering their energy dissipation characteristics in Section 9.4.5.

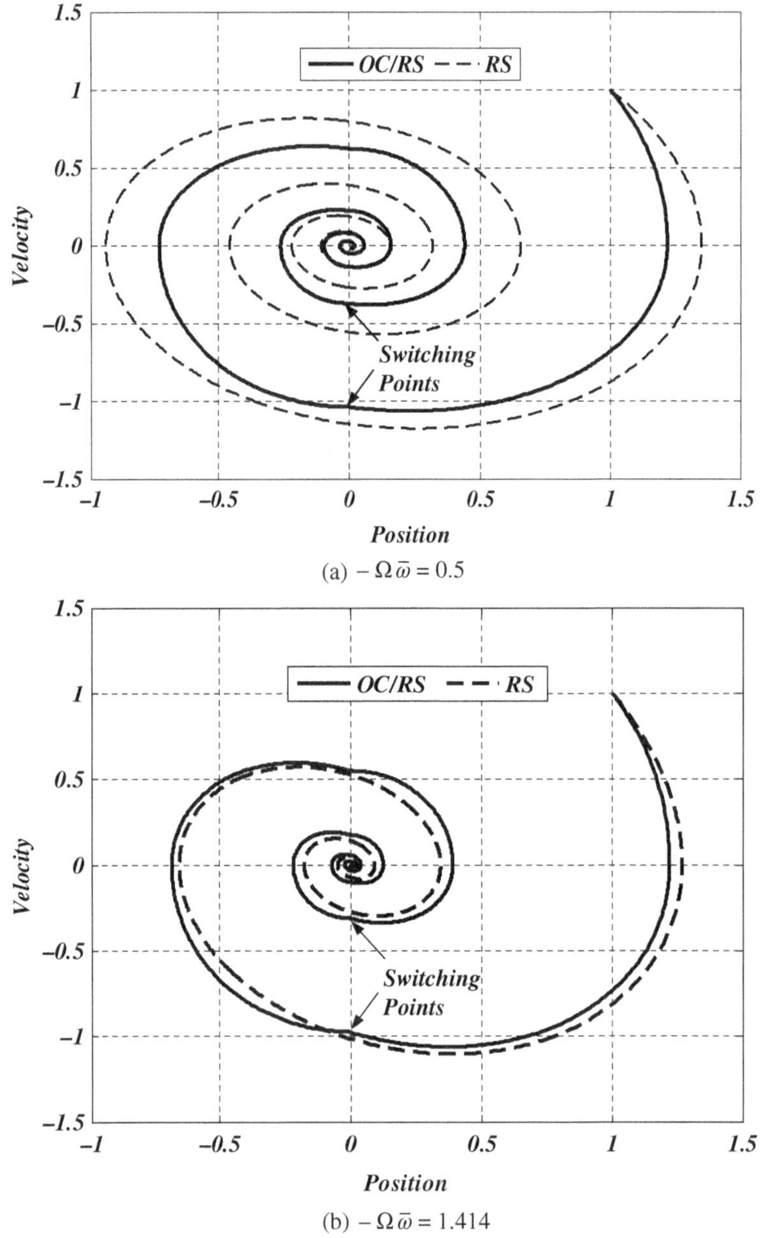

Figure 9.30 Comparisons between the phase-plane plots for resistive shunting and switched resistive shunting for different values of dimensionless frequency $r\bar{\omega}$. (a) $\Omega\bar{\omega} = 0.5$. (b) $\Omega\bar{\omega} = 1.414$.

9.4.5 Energy Dissipation of Different Shunting Configurations

Computing the energy dissipation characteristics of the different shunting configurations is essential to determining their effective and optimal regions of operation.

9.4.5.1 Energy Dissipation with Resistive Shunting

The energy dissipated W_D^{RS} per cycle of sinusoidal oscillation with RS can be determined using Eq. (9.70) that represents a second order system with damping coefficient $\bar{C}$ and stiffness $K^{RS} = (1 + K')$. From Chapter 2, Table 2.14, the energy dissipated is given by:

$$W_D^{RD} = \pi \bar{C} \bar{\omega} X_0^2 \tag{9.74}$$

where X_0 denotes the magnitude of oscillation.

Hence, substituting from Eq. (9.70) into (9.74) gives:

$$W_D^{RS} = \frac{\pi\, k_{33}^2 \Omega \bar{\omega}}{1 + (1 - k_{33}^2)^2 (\Omega \bar{\omega})^2} X_0^2 \tag{9.75}$$

An equivalent loss factor η^{RS} can be determined for the *RS* configuration by normalizing the dissipated energy W_D^{RS} with respect to the elastic energy W_e^{RS} given by:

$$W_e^{RS} = 2K^{RS} X_0^2 \tag{9.76}$$

From Eqs. (9.62), (9.75), and (9.76), the loss factor η^{RS} can be determined as follows:

$$\eta^{RS} = \frac{\pi\, k_{33}^2 \Omega \bar{\omega}}{2K^{RS}\left[1 + \left(1 - k_{33}^2\right)^2 (\Omega \bar{\omega})^2\right]} \tag{9.77}$$

9.4.5.2 Energy Dissipation with OC-RS Switched Shunting

The energy dissipated $W_D^{OC/RS}$ per cycle of sinusoidal oscillation with OC-RS can be determined using Eqs. (9.69) and (9.70) along with the globally stable switching strategy that is schematically represented in Figure 9.31.

$$W_D^{OC/RS} = \int_0^{\tau/4} K^{OS} x\dot{x}dt + \int_{\tau/4}^{\tau/2} \left(\bar{C}\dot{x}^2 + K^{RS} x\dot{x}\right)dt + \int_{\tau/2}^{3\tau/4} K^{OS} x\dot{x}dt + \int_{3\tau/4}^{\tau} \left(\bar{C}\dot{x}^2 + K^{RS} x\dot{x}\right)dt \tag{9.78}$$

Let $x = X_0 \sin \bar{\omega} t$, then Eq. (9.78) reduces to:

$$W_D^{OC/RS} = \left(K^{OC} - K^{RS}\right) X_0^2 + \frac{\pi}{2}\left(\frac{k_{33}^2 \Omega \bar{\omega}}{1 + (1 - k_{33}^2)^2 (\Omega \bar{\omega})^2}\right) X_0^2 \tag{9.79}$$

The equivalent loss factor $\eta^{OC/RS}$ can be determined for the OC-RS configuration by normalizing the dissipated energy $W_D^{OC/RS}$ with respect to the nominal elastic energy $W_e^{OC/RS}$ given by

$$W_e^{OC/RS} = K^{OC} X_0^2 + K^{RS} X_0^2 \tag{9.80a}$$

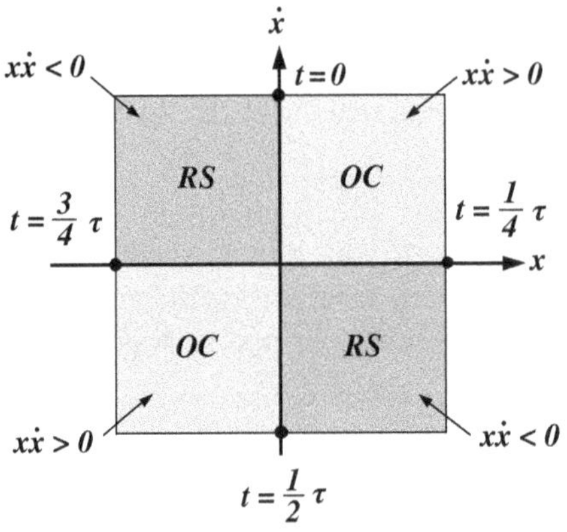

Figure 9.31 Switching strategy for the OC-RS configuration.

This yields:

$$\eta^{OC/RS} = \frac{1}{(K^{OC} + K^{RS})}\left[\left(\frac{1}{1-k_{33}^2} - K^{RS}\right) + \frac{\pi}{2}\left(\frac{k_{33}^2\,\Omega\bar{\omega}}{1 + (1-k_{33}^2)^2(\Omega\bar{\omega})^2}\right)\right] \tag{9.80b}$$

Example 9.10 Compare the loss factors for RS and switched resistive shunting (OC-RS) as a function of $\Omega\bar{\omega}$ when $k_{33}^2 = 0.5$.

Solution

The loss factors for the RS and switched resistive shunting (OC-RS) configurations as predicted by Eqs. (9.77) and (9.80b) are plotted as a function of $\Omega\bar{\omega}$ when $k_{33}^2 = 0.5$.

Figure 9.32 displays a comparison between the loss factors of the two configurations. It can be seen that the switched resistive shunting (OC-RS) configuration is superior to the RS configuration at low dimensionless frequencies $\Omega\bar{\omega}$ up to 0.707. However, for higher frequencies, the RS configuration performs much better than the switched resistive shunting (OC-RS) configuration in damping structural vibrations.

9.5 Multi-Mode Vibration Control with Shunted Piezoelectric Networks

9.5.1 Multi-Mode Shunting Approaches

In this section, shunted piezoelectric networks are utilized to provide multi-mode control of the vibration of flexible structures. Extensive work has been carried out in this regards including the work by Hollkamp (1994), whereby a single piezoelectric element is coupled with an inductive–resistive–capacitive (R–L–C) shunting network to damp out multiple vibration modes as shown in Figure 9.33a. The main concept behind

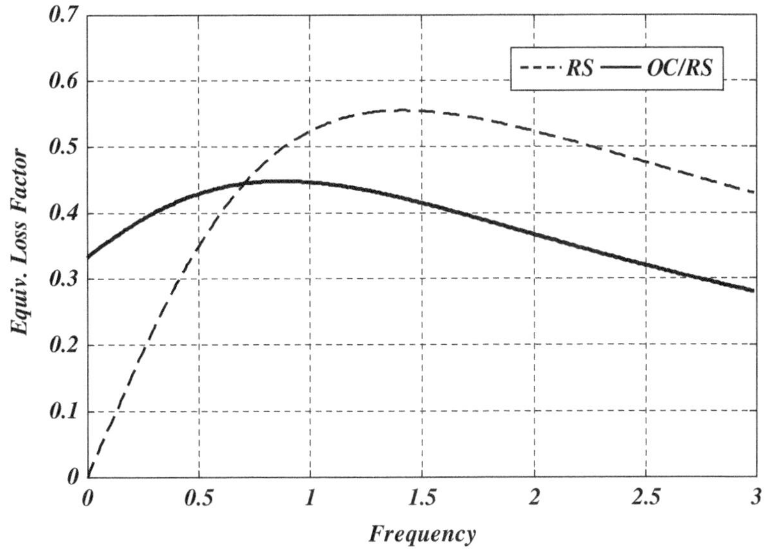

Figure 9.32 Equivalent loss factors for the RS and OC-RS configurations.

Hollkamp's configuration is to couple the piezo-element with an electric network that has as many resonances as the structural modes that need to be suppressed. However, the tuning of the R–L–C network poses serious challenges as tuning one electrical resonance to suppress a particular mode tends to detune the rest of the network. Therefore, the design parameters of the network were determined by numerical optimization aiming at minimizing the weighted vibration energy. In 1998 and 1999, Wu introduced an innovative method for implementing the multi-mode damping using a single shunted piezoelectric element as shown in Figure 9.33b. In his approach, "current blocking" networks consisting of parallel L–C branches are placed in series with each shunted parallel R–L branch network, which is intended to control a particular mode. In this manner, the coupling between the individual parallel shunt networks was prevented and it was possible to derive closed-form analytical expressions to determine the design parameters of the shunted. Although this approach was effective in enabling the tuning of each mode separately, it required the use of a large number of components, particularly as the number of modes to be controlled increased. This is clearly manifested by the large number of "current blocking" branches shown in Figure 9.33b.

In 2003, Behrens et al. presented another practical approach to the design of a multi-mode piezoelectric shunting network. This approach was called the "current flowing" method, in contrast with the "current blocking" method presented by Wu (1998, 2000). In this method, additional series L–C branches are added in series with each L–R shunt branch as displayed in Figure 9.33c. The "current flowing" shunt requires a smaller number of components and requires no floating inductors when compared with the "current blocking" method.

The idea behind the "current flowing" shunt is to allow a single frequency content ω_i of the current generated by the piezo-element to flow in each branch. This is in contrast to blocking N-1 components as in Wu's configurations where N denotes the number of dominant modes to be suppressed. Such a simple filtering approach is achieved by using

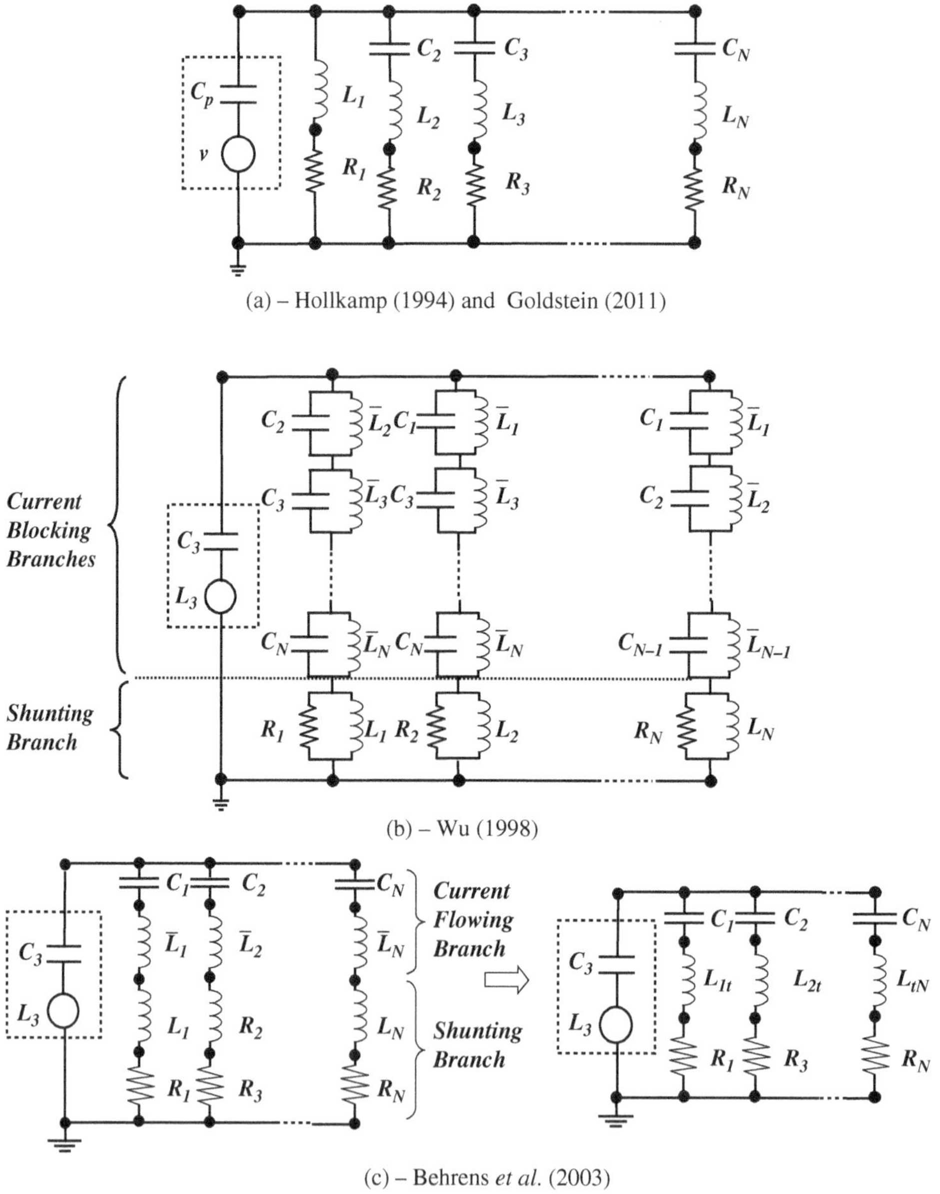

Figure 9.33 Typical shunted networks for multi-mode vibration control. (a) Hollkamp (1994) and Goldstein (2011). (b) Wu (1998). (c) Behrens et al. (2003).

a series capacitor–inductor circuit $C_i - \bar{L}_i$ as shown in Figure 9.33c. The series $C_i - \bar{L}_i$ is tuned to pass only the structural resonant frequency ω_i and block all the other frequencies. Damping the structural vibration at the resonant frequency ω_i is carried out by tuning the shunting branch $C_p - L_i$ and $L_i - R_i$ to ω_i as outlined in Section 9.2.

It is important to note that if the filtering inductance $\bar{L}_i$ and the shunting inductance L_i are combined such that: $L_{ti} = \bar{L}_i + L_i$, then Behrens et al.'s configuration reduces to nearly the Hollkamp's configuration. However, now the physical meaning of each component in the Behrens et al.'s configuration is associated with a particular mode that can be tuned separately. This is unlike the Hollkamp's configuration where the components are highly coupled between all the considered modes. Furthermore, one additional distinction is that the shunt circuit proposed by Hollkamp includes only one resistor–inductor circuit for the first mode, while in Behrens et al. approach a resistor–inductor–capacitor circuit is used to shunt each mode.

9.5.2 Parameters of Behrens et al.'s Multi-Mode Shunting Network

Because of the simplicity of the Behrens et al. network, the emphasis is placed in this section on introducing the expressions necessary for selecting the design parameters of its different components.

9.5.2.1 Components of the Current Flowing Branches

The inductances $\bar{L}_i$ and capacitances C_i are tuned to allow the current flow at the ith natural frequency ω_i of the flexible structure/piezo-element system such that:

$$\bar{L}_i = 1/\left(\omega_i^2 C_i\right) \text{ for } i = 1,\ldots,N \tag{9.81}$$

For simplicity, the values of the capacitances C_i are specified and the inductances $\bar{L}_i$ are computed to satisfy Eq. (9.81).

9.5.2.2 Components of the Shunting Branches

The inductances L_i are tuned into C_P according to the equations developed in Section 9.2.2.2, or:

$$L_i = 1/\left(\omega_i^2 C_p^S\right) \text{ for } i = 1,\ldots,N \tag{9.82}$$

Hence,

$$L_{ti} = \left(C_i + C_p^S\right)/\left(\omega_i^2 C_i C_p^S\right) \text{ for } i = 1,\ldots,N \tag{9.83}$$

The resistances R_i are selected according to:

a) Optimize the loss factor of RS:

This requires that: $\Omega^* = \sqrt{1-k_{31}^2} = R_i\left(1-k_{31}^2\right)C^T\omega_i$

$$\text{that is}, R_i = \sqrt{1-k_{31}^2}\,C^T\omega_i \tag{9.84}$$

b) Optimize the H_2 norm of the Controlled System:

Behrens et al. (2003) proposed the use of an optimization approach where the H_2 norm of the controlled system is minimized.

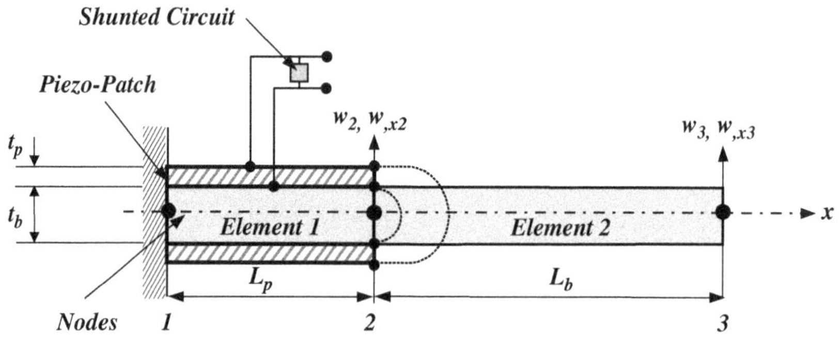

Figure 9.34 A beam with a single shunted piezo-element.

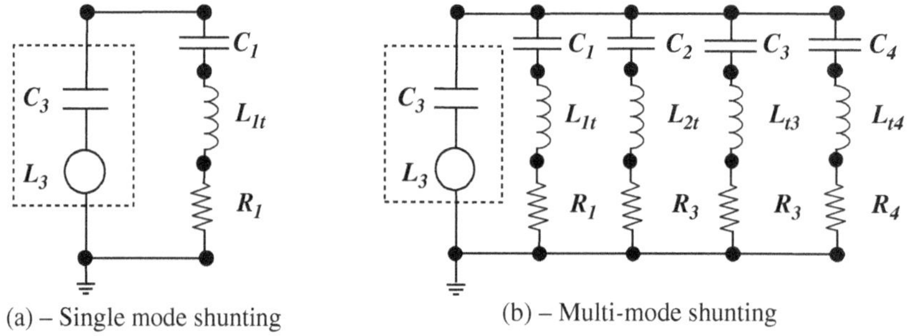

Figure 9.35 Single and multi-mode shunted networks. (a) Single mode shunting. (b) Multi-mode shunting.

Example 9.11 Consider the cantilevered beam shown in Figure 9.34 that is controlled by two symmetrically placed piezoelectric patches using the "current flowing" shunt network of Behrens et al. displayed in Figure 9.35.

The basic geometrical and physical properties of the beam and piezo-patch are listed next.

Beam system					
Parameter	Length (cm)	Thickness (cm)	Width (cm)	Density (kg m^{-3})	Young's modulus (GPa)
Beam	30.0	1.00	2.50	7800	210.00
Piezoelectric patch	12.0	0.25	2.50	1800	118.34

Piezoelectric patch				
Parameter	d_{31} (m V^{-1})	k_{31}^2	ε_{33}^T (Coulomb mV^{-1})	C^T (nf)
Value	274E-12	0.3041	2.92E-8	70.12

If the beam is divided into two elements as shown in Figure 9.34, design the shunting networks shown in Figure 9.35 to control the vibration of the first mode and all the four modes of vibration of the beam/piezo-patch system. Plot the corresponding frequency response for the two cases in comparison with the response of the uncontrolled beam.

Solution

The system is modeled using the theory of finite elements resulting in the following four natural frequencies:

ω_i= 132, 657, 2293, and 4877.3 Hz

The shunting capacitances C_i (i = 1,...,4) are selected to be 50 *nanofarad.* Then, Eqs. (9.83) and (9.84) are used to compute the shunting total inductances L_{ti} and resistances R_i. These values are listed here:

MODE	1	2	3	4
C_i (nf)	50	50	50	50
L_{ti} (Henry)	58.86	2.38	0.195	0.0431
R_i (Ohms)	20 611	4411	1186.3	557.8

Figure 9.36 displays the frequency response of the beam/piezo-patch system when the first mode of vibration is controlled using the shunted network shown in Figure 9.35a. It is evident that the considered shunting network has proven to be very effective in

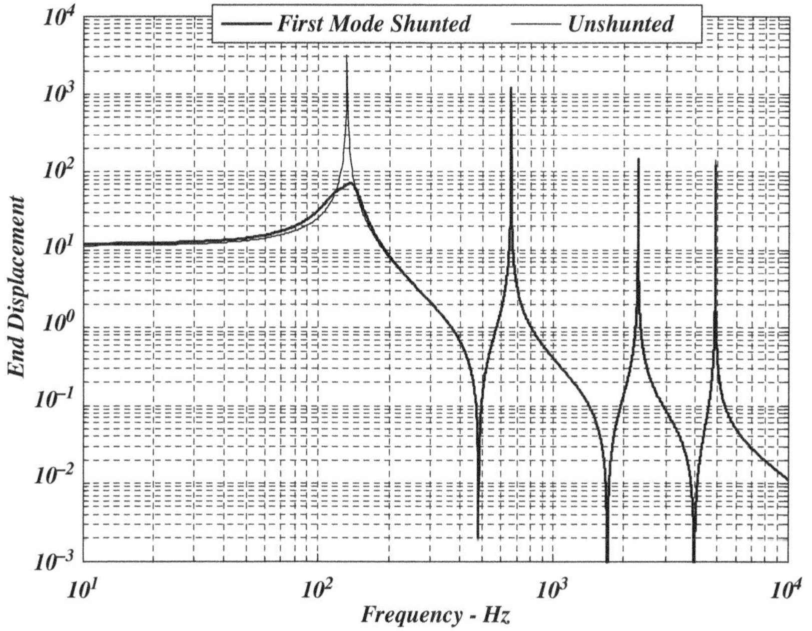

Figure 9.36 Frequency response of a system with a single mode shunted network.

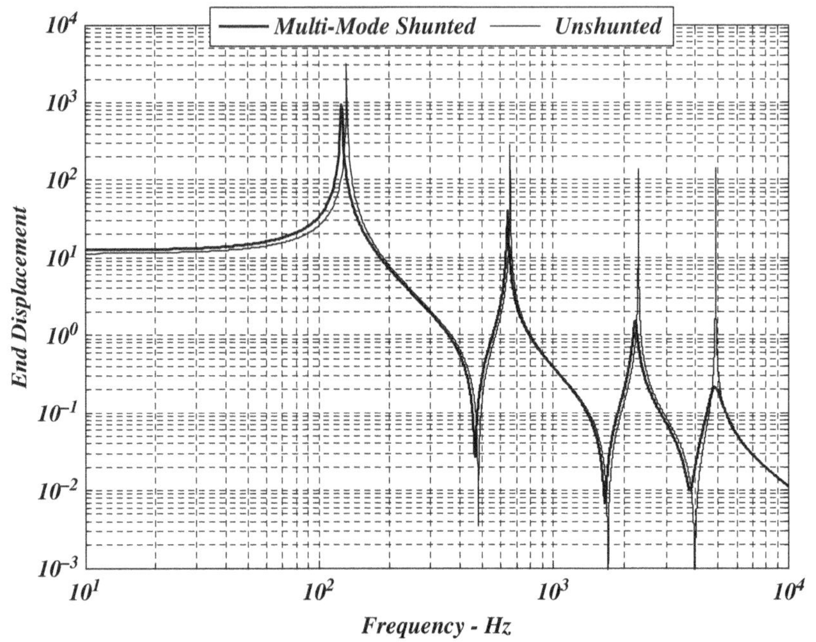

Figure 9.37 Frequency response of a system with a four mode shunted network.

attenuating the vibration of the first mode. However, it has not affected the remaining three modes at all.

Figure 9.37 shows the corresponding frequency response of the beam/piezo-patch system when all the four modes of vibration are controlled using the shunted network shown in Figure 9.35b. The considered shunting network has effectively attenuated the vibration of all the four modes. However, the attenuation at the first mode has not been as good as when the first mode is controlled alone.

9.6 Summary

This chapter has presented the basics of using piezoelectric patches as effective means for controlling structural vibrations. The equations governing the performance of shunted piezo-patches are presented for different types of shunting circuits. It is shown that shunted piezo-patches behave like conventional viscoelastic materials. Expressions for the complex modulus of shunted piezo-patches are derived for three types of shunting circuits. The use of shunting circuits with matched capacitive elements is shown to generate the maximum energy dissipation particularly at low vibration frequencies. Practical issues related to the implementation of the negative shunting capacitance as well as inductances are presented.

The complex modulus approach is utilized to integrate the dynamics of the shunted piezo-patches with the dynamics of the structures to which they are bonded to. The combined dynamics are modeled using the theory of finite elements. Another finite element approach is presented that relies on using the coupled electromechanical field formulation. Comparison between the two approaches is presented for rods experiencing longitudinal vibrations.

The presented concepts can be easily generalized for more complex structures such as beams, plates, and shells.

References

Baz, A. (2009). The structure of an active acoustic metamaterial with tunable effective density. *New Journal of Physics* 11, 123010.

Baz, A. (2010). An active acoustic metamaterial with tunable effective density. *ASME Journal of Vibration & Acoustics* 132 (4, 041011 1-9).

Behrens, S., Moheimani, S.O.R., and Fleming, A.J. (2003). Multiple mode current flowing passive piezoelectric shunt controller. *Journal of Sound and Vibration* 266 (5): 929–942.

Browning D. and Wynn W., Vibration damping system using active negative capacitive shunt circuit with piezoelectric reaction mass actuator", United States Patent, No. 5,558,477, filed December 4 1994 and issued September 24 1996.

Clark, W.W. (2000). Vibration control with state-switched piezoelectric materials. *Journal of Intelligent Material Systems and Structures* 11 (4): 263–271.

Corr, L.R. and Clark, W.W. (2002). Comparison of low-frequency piezoceramic shunt techniques for structural damping. *Smart Materials and Structures* 11 (3): 370–376.

Deliyannis, T., Sun, Y., and Fidler, J.K. (1999). *Continuous Time Active Filter Design*. Boca Raton, FL: CRC Press.

Edberg D. L., Bicos A. S., and Fechter J. S., "On piezoelectric energy conversion for electronic passive damping enhancement", Proceedings of Damping '91, San Diego, CA, WL-TR-91-3078, Vol. 2, pp. GBA1-10, 1991.

Forward R. L., "Electromechanical transducer-coupled mechanical structure with negative capacitance compensation circuit, United States Patent, No. 4,158,787, filed May 1 1998, and issued June 19 1979.

Goldstein, A. (2011). Self-tuning multimodal piezoelectric shunt damping. *Journal of the Brazilian Society of Mechanical Sciences and Engineering* 33 (4): 428–436.

Gripp, J.A.B. and Rade, D.A. (2018). Vibration and noise control using shunted piezoelectric transducers: A review. *Mechanical Systems and Signal Processing* 112: 359–383.

Hagood, N.W. and von Flotow, A. (1991). Damping of structural vibrations with piezoelectric materials and passive electrical networks. *Journal of Sound and Vibration* 146 (2): 243–268.

Hagood, N.W., Chung, W.H., and von Flotow, A. (1990). Modeling of piezoelectric actuator dynamics for active structural control. *Journal of Intelligent Material Systems and Structures* 1: 327–354.

Hollkamp, J.J. (1994). Multimodal passive vibration suppression with piezoelectric materials and resonant shunts. *Journal of Intelligent Material Systems and Structures* 5: 49–57.

IEEE STD 176 (1987). *IEEE Standard on Piezoelectricity*. New York, NY: The Institute of Electrical and Electronics Engineers.

Itoh, T., Shimomura, T., and Okubo, H. (2011). Semi-active vibration control of smart structures with sliding mode control. *Journal of System Design and Dynamics* 5 (5): 716–726.

Law, H., Rossiter, P., Koss, L., and Simon, G. (1995). Mechanisms in damping of mechanical vibration by piezoelectric ceramic-polymer composite materials. *Journal of Materials Science* 30: 2648–2655.
Lesieutre, G.A. (1998). Vibration damping and control using shunted piezoelectric materials. *The Shock and Vibration Digest* 30 (3): 187–195.
Lesieutre G. and Davis C., "Can a coupling coefficient of a piezoelectric device be higher than those of its active material?", in Proceedings of SPIE, Smart Structures and Integrated Systems (ed. M. Regelburgge), Vol. 30417, pp. 281–292, 1997.
Moheimani, S.O.R., Fleming, A.J., and Behrens, S. (2001). Highly resonant controller for multimode piezoelectric shunt damping. *Electronics Letters* 37 (25): 1505–1506.
Mokrani, B., Rodrigues, G., Ioan, B., Bastaits, R., and Preumont, A. (2012). Synchronized switch damping on inductor and negative capacitance. *Journal of Intelligent Material Systems and Structures* 23 (18): 2065–2075.
Neubauer, M., Han, X., and Wallaschek, J. (2013). On the maximum damping performance of piezoelectric switching techniques. *Journal of Intelligent Material Systems and Structures* 24 (6): 717–728.
Park C. H. and Inman D. J., "A uniform model for series R-L and Parallel R-L shunt circuits and power consumption", Proceedings of SPIE, Smart Structures and Integrated Systems (ed. N. Wereley), Vol. 3668, pp. 797–804, 1999.
Tsai, M.S. and Wang, K.W. (1999). On the structural damping characteristics of active piezoelectric actuators with passive shunt. *Journal of Sound and Vibration* 221 (1): 1–22.
Velazquez, C.A. (1995). *Electromechanical surface damping combining constrained layer and shunted piezoelectric materials with passive electrical networks of second order*. New York: Masters Thesis, Department of Mechanical Engineering, Rochester Institute of Technology, Rochester.
Wu, S. (1998). Method for multiple mode piezoelectric shunting with single PZT transducer for vibration control. *Journal of Intelligent Material Systems and Structures* 9(12): 991–998.
Wu S., "Broadband piezoelectric shunts for structural vibration control," United States Patent, No. 6,075,309, filed July 18 and issued April 11 2000.
Yan, B., Wang, K., Hu, Z. et al. (2017). Shunt damping vibration control technology: a review. *Applied Sciences* 7: 494. doi: 10.3390/app7050494.

9.A Electromechanical Coupling Factor

The coupling between the input electrical energy and the output mechanical energy is quantified by the EMCF.

The efficiency *Eff* of conversion of electrical energy into mechanical energy can be determined from:

$$Eff = \frac{U_{elastic}}{U_{electrical}} \tag{9.A.1}$$

where the elastic energy $U_{elastic}$ is given by:

$$U_{elastic} = \frac{1}{2}\int stress^{*} starin \, \mathrm{dv} \tag{9.A.2}$$

From first row of Eq. (9.1), with $T_1 = 0$, the strain $= d_{31}E_3$ and the corresponding stress $= \frac{d_{31}E_3}{s_{11}^E}$. Note also that dv defines an infinitesimal volume, then Eq. (9.A.2) reduces to

$$U_{elastic} = \frac{1}{2}\left(\frac{d_{31}E_3}{s_{11}^E}\right)(d_{31}E_3)At = \frac{1}{2}\frac{d_{31}^2}{s_{11}^E}AtE_3^2 \qquad (9.A.3)$$

Also, the electrical energy $U_{electrical}$ is given by:

$$U_{electrical} = \frac{1}{2}QV = \frac{1}{2}\int D_3 dA\, E_3 t \qquad (9.A.4)$$

where Q, V, and D_3 are the charge, voltage, and electrical displacement, respectively. But, for stress-free piezo-patch, that is, $T_1 = 0$, then the second row of Eq. (9.1) gives

$$D_3 = \varepsilon_{33}^T E_3 \qquad (9.A.5)$$

Substituting Eq. (9.A.5) into Eq. (9.A.4) gives

$$U_{electrical} = \frac{1}{2}\varepsilon_{33}^T E_3^2 At \qquad (9.A.6)$$

Then, Eqs. (9.A.2) and (9.A.6) give

$$Eff = \frac{\frac{1}{2}\frac{d_{31}^2}{s_{11}^E}E_3^2 At}{\frac{1}{2}\varepsilon_{33}^T E_3^2 At} = \frac{d_{31}^E}{s_{11}^E\,\varepsilon_{33}^T = k_{31}^2} \qquad (9.A.7)$$

Accordingly, the EMCF k_{31}^2 is equal to the efficiency of converting the electrical input energy into an elastic energy.

The physical properties of different piezoelectric materials are given in Table 9.A.1. The corresponding values of the EMCF are listed in Table 9.A.2.

Table 9.A.1 Physical parameters of typical piezoelectric materials.

Material	s_{11}^E (m^2 N^{-1})	d_{31} (m V^{-1})	ε_{33}^T (Farad m^{-1})	Density (kg m^{-3})
PZT-4	0.159E-10	−180 E-12	1.50E-8	7600
PZT-5H	0.165E-10	−274 E-12	2.49E-8	7300
PVDF	0.500E-9	23 E-12	110E-12	1780

PZT, lead zirconate titanate (ceramic); PVDF, polyvinylidene fluoride (polymer).

Table 9.A.2 Typical values of the EMCF (k_{31}^2).

PZT-4	*PZT-5H*	*PVDF*
0.137	0.185	0.0096

PZT, lead zirconate titanate (ceramic); PVDF, polyvinylidene fluoride (polymer).

Problems

9.1 Determine the dimensionless mechanical impedance $\bar{\bar{Z}}^{ME}$ of the two shunted piezoelectric networks shown in Figure P9.1. Note that $\bar{\bar{Z}}^{ME}$ is defined as:

$$\bar{\bar{Z}}^{ME} = Z^{ME\ Shunted} / Z^{ME\ Open-Circuit}.$$

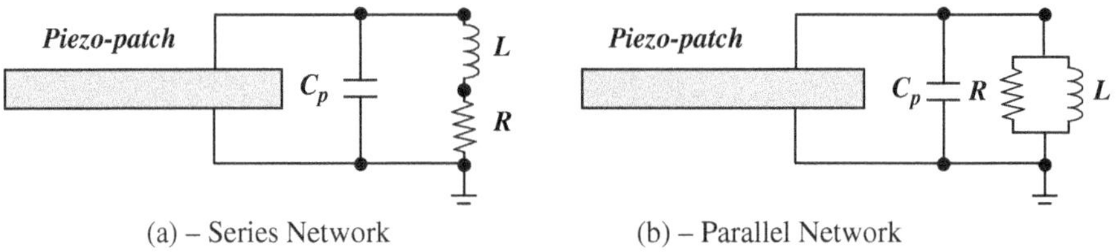

Figure P9.1 Series and parallel R–L shunted networks. (a) Series network. (b) Parallel network.

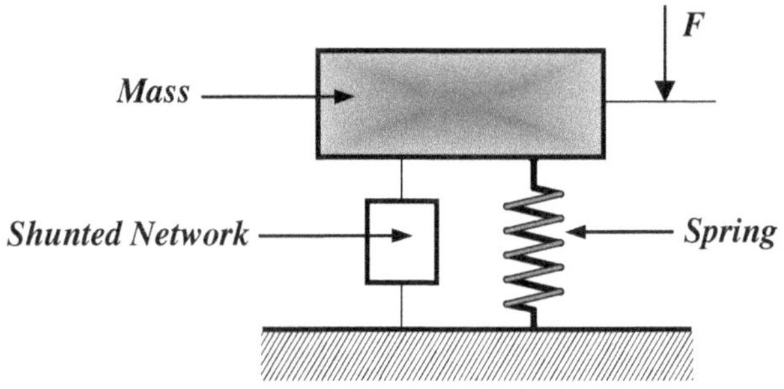

Figure P9.2 A spring-mass system damped by a shunted piezoelectric network.

Derive expressions for the equivalent storage modulus and loss factor for the two networks. Determine the tuning frequencies ω^* at which the loss factor is a maximum. Show that these frequencies are given by:

(a) $\omega^*_{series} = 1/\sqrt{LC^s}$ and (b) $\omega^*_{series} = \sqrt{R^2/(R^2LC^s - L^2)}$ where $C^s = C^T(1 - k^2)$ with k = EMCF.

9.2 Assume that the vibration of the spring-mass system shown in Figure P9.2 is damped out using the concept of shunted piezoelectric networks.

Determine the dimensionless deflection X/X_{st} of the mass M when the shunted network is: (a) a series L–R circuit, and (b) a parallel L–R circuit. Note that $X_{st} = F/(K_s + K_{sn})$ such that F = applied force, K_s = spring stiffness, and K_{sn} = equivalent stiffness of shunted network.

Show that if $K^2_{coupling} = \dfrac{K_{sn}}{K_s + K_{sn}}\left(\dfrac{k^2}{1-k^2}\right)$ with k = EMCF, then:

$$(a)\ \left(\frac{X}{X_{st}}\right)_{series} = \frac{\left(\gamma^2+\delta^2\Omega\gamma+\delta^2\right)}{\left(1+\gamma^2\right)\left(\gamma^2+\delta^2\Omega\gamma+\delta^2\right)+K^2_{coupling}\left(\gamma^2+\delta^2\Omega\gamma\right)},$$

$$\text{and } (b)\ \left(\frac{X}{X_{st}}\right)_{parallel} = \frac{\left(\Omega\gamma^2+\gamma+\delta^2\Omega\right)}{\left(1+\gamma^2\right)\left(\gamma+\Omega\gamma^2+\delta^2\Omega\right)+K^2_{coupling}\left(\Omega\gamma^2\right)}.$$

where $\Omega = RC^s\omega_n$, $C^s = C^T(1-k^2)$, $\omega_n = \sqrt{\dfrac{(K_s+K_{sn})}{M}}$, $\gamma = s/\omega_n$, and $\delta = \omega_e/\omega_n$.

Also, $\omega_e = 1/\sqrt{LC^s}$ and s is the Laplace complex number.

9.3 Optimal tuning of the shunted network of Problem 9.2 occurs when the poles of the transfer function X/X_{st} coincide. Let the poles be $\gamma_{1,2} = a + bi$ and $\gamma_{3,4} = a - bi$, and show for the series L–R network that:

$$\delta^2\Omega = -4a,$$

$$\left(1+\delta^2\right)+K^2_{coupling} = 6a^2+2b^2,$$

$$\delta^2\Omega\left(1+K^2_{coupling}\right) = -4a\left(a^2+b^2\right),$$

$$\delta = a^2+b^2.$$

Eliminate a and b from these four equations to give:

$$\delta^* = 1+K^2_{coupling} \text{ and } \Omega^* = 2\sqrt{K^2_{coupling}/\left(1+K^2_{coupling}\right)^3}$$

Derive similar expressions for the parallel L–R Network.

9.4 Show that the active circuit of Figure P9.3 has an input impedance Z given by:
$Z = V_i/i_i = -1/Cs$

Figure P9.3 Active circuit to synthesize a negative capacitance.

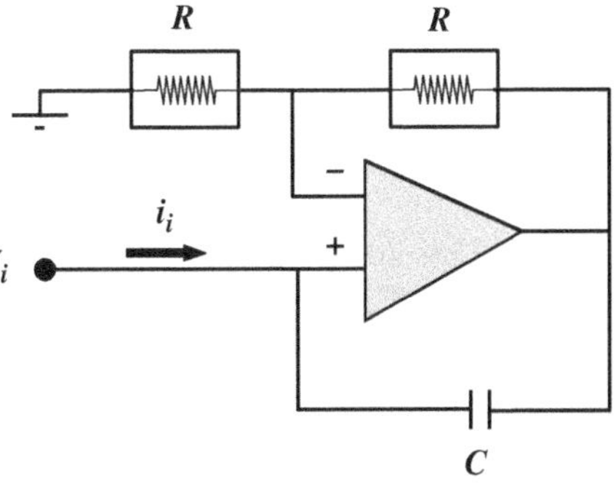

9.5 Consider a simple model of a composite damping treatment consisting of a conducting polymer layer and piezo-ceramic block that are parallel to the direction of the applied load as shown in Figure P9.4. Assume that the conducting polymer layer acts as a shunting electric resistance R when combined with the piezo-ceramic block. The polymer layer has complex modulus $E_1^* = E_1(1+i\eta_1)$ with E_1 and η_1 are the storage modulus and loss factor ($\eta_1 = 2\,\zeta_1$). The complex modulus of the piezo-

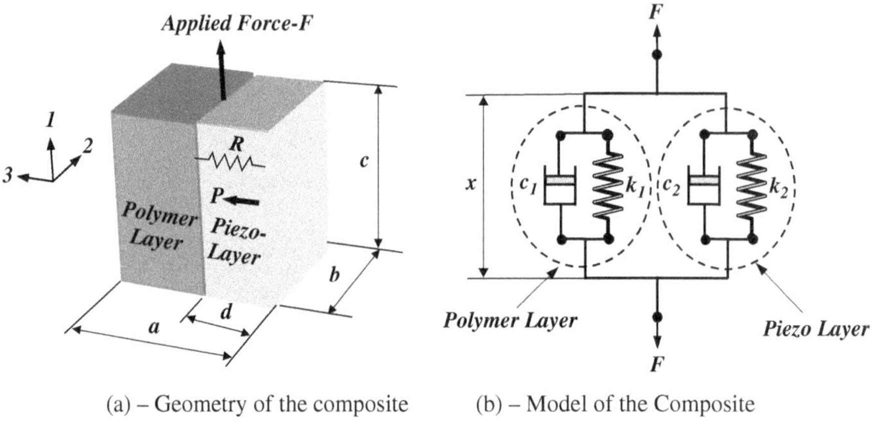

Figure P9.4 A composite damping treatment consisting of a conducting polymer layer and piezo-ceramic block that are parallel to the direction of the applied load. (a) Geometry of the composite. (b) Model of the composite.

layer is given by $E_2^* = E_2(1 + i\eta_2)$ where the storage modulus and loss factor E_2 and η_2 ($\eta_2 = 2\,\zeta_2$) are determined from Eq. (9.21) as follows:

$$E_2 s_{11}^D = \left[1 - \frac{k_{31}^2}{1+\Omega^2}\right], \; \eta_2 = \frac{k_{31}^2 \Omega}{1 - k_{31}^2 + \Omega^2} \text{ with } \Omega = RC^T\omega$$

with s_{11}^D, and k_{31}^2 denoting the compliance and the EMCF. Also, C^T $(= bc\,\varepsilon_{33}^T/d)$ denotes the stress-free capacitance of the piezo-layer with ε_{33}^T is the permittivity and ω is the excitation frequency. The subscripts 1 and 3 are associated with the coordinate system displayed in Figure P9.4 relative to the poling direction P. Note that for the considered composite:

The displacement of the polymer and piezo-layers: $x_t = x$,
The total stiffness of the composite: $k_t = k_1 + k_2$,
The total energy dissipated by the composite: $U_t = U_1 + U_2 = 2\pi\zeta_t k_t x^2$,
The energy dissipated by the polymer layer: $U_1 = 2\pi\zeta_1 k_1 x^2$,
The energy dissipated by the piezo-layer: $U_2 = 2\pi\zeta_2 k_2 x^2$.

Also, the stiffnesses k_1 and k_2 of the polymer and piezo-layer are given by:

$$k_1 = \frac{E_1 A_1}{c} \text{ and } k_2 = \frac{E_2 A_2}{c} \text{ with } A_1 = b(a-d),\; A_2 = bd$$

Show that the total damping ratio of the composite ζ_t can be determined from:

$$\zeta_t = \frac{1}{1+q}(q\zeta_1 + \zeta_2)$$

where $q = k_1/k_2 = p\left(\frac{1}{\nu} - 1\right)$ with $p = E_1/E_2$ and ν = the piezo-layer volume fraction $= \frac{V_2}{V_1 + V_2} = \frac{d}{a}$. Note that V_1 and V_2 are the volumes of the polymer and the piezo-layers.

Determine and plot the total damping ratio of the composite ζ_t as a function of Ω for different values of the resistance R when the volume fraction $\nu = 0.5$.

In the plots consider the geometrical and physical parameters of the composite layers listed in Table P9.1.

Table P9.1 The geometrical and physical parameters of the parallel composite layers.

s^E_{11} ($m^2\,N^{-1}$)	k^2_{31}	ε^T_{33} (Farad m^{-1})	E_1 ($N\,m^{-2}$)	ζ_1	a (m)	b (m)	c (m)
0.12E-10	0.12	1.50E-8	20E6	0.5	0.005	0.025	0.05

9.6 Consider a simple model of a composite damping treatment consisting of a conducting polymer layer and piezo-ceramic block that are perpendicular to the direction of the applied load as shown in Figure P9.5. Assume that the conducting polymer layer acts as a shunting electric resistance R when combined with the piezo-ceramic block. The polymer layer has complex modulus $E_1^* = E_1(1 + i\eta_1)$ with E_1 and η_2 are the storage modulus and loss factor ($\eta_2 = 2\,\zeta_2$). The complex modulus of the piezo-layer is given by $E_2^* = E_2(1 + i\eta_2)$ where the storage modulus and loss factor E_2 and ζ_2 are determined from Eq. (9.21) as follows:

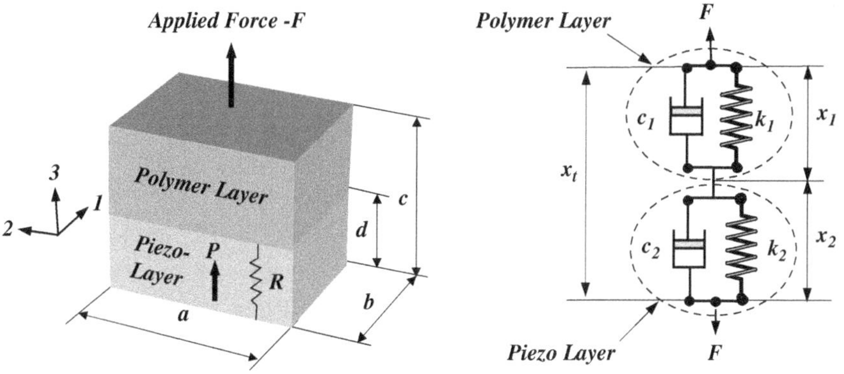

(a) – Geometry of the composite (b) – Model of the Composite

Figure P9.5 A composite damping treatment consisting of a conducting polymer layer and piezo-ceramic block that are perpendicular to the direction of the applied load (a) Geometry of the composite. (b) Model of the composite.

$$E_2 s^D_{33} = \left[1 - \frac{k^2_{33}}{1 + \Omega^2}\right], \; \zeta_2 = \frac{1}{2}\frac{k^2_{33}\Omega}{1 - k^2_{33} + \Omega^2} \text{ with } \Omega = RC^T\omega$$

with s^D_{33}, and k^2_{33} denoting the compliance and the EMCF. Also, $C^T\left(= ab\,\varepsilon^T_{33}/d\right)$ denotes the stress-free capacitance of the piezo-layer and ω is the excitation frequency.

Note that for the considered composite:

The displacement of the polymer and piezo-layers: $x_t = x_1 + x_2$,
The total stiffness of the composite: $1/k_t = 1/k_1 + 1/k_2$,
The total energy dissipated by the composite: $U_t = U_1 + U_2 = 2\pi\zeta_t F^2/k_t$,
The energy dissipated by the polymer layer: $U_1 = 2\pi\zeta_1 F^2/k_1$,
The energy dissipated by the piezo-layer: $U_2 = 2\pi\zeta_2 F^2/k_2$.

Also, the stiffnesses k_1 and k_2 of the polymer and piezo-layer are given by: $k_1 = \dfrac{E_1 A_1}{c-d}$ and $k_2 = \dfrac{E_2 A_2}{d}$ with $A_1 = ab = A_2$

Show that the total damping ratio of the composite ζ_t can be determined from:

$$\zeta_t = \frac{1}{1+q}(\zeta_1 + q\zeta_2)$$

where $q = k_1/k_2 = p\left(\dfrac{v}{1-v}\right)$ with $p = E_1/E_2$ and v = the piezo-layer volume fraction $= \dfrac{V_2}{V_1+V_2} = \dfrac{d}{c}$. Note that V_1 and V_2 are the volumes of the polymer and the piezo-layers.

Determine and plot the total damping ratio of the composite U_t as a function of Ω for different values of the resistance R when the volume fraction $v = 0.5$. In the plots consider the geometrical and physical parameters of the composite layers listed in Table P9.2.

Table P9.2 The geometrical and physical parameters of the perpendicular composite layers.

s_{11}^E (m^2 N^{-1})	k_{33}^2	ε_{33}^T (Farad m^{-1})	E_1 (N m^{-2})	ζ_1	a (m)	b (m)	c (m)
0.16E-10	0.49	1.50E-8	20E6	0.5	0.05	0.025	0.005

9.7 Consider a structural system consisting of a spring-damper (E, c) as shown in Figure P9.6a. Determine the equivalent mechanical impedance of the system Z^{ME} as given by:

$$Z^{ME} = \frac{\sigma}{\dot{\varepsilon}}$$

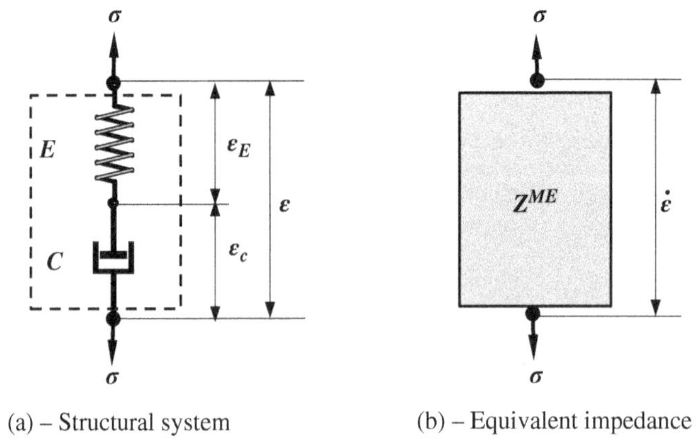

Figure P9.6 Spring damper in series and their equivalent mechanical impedance (a) Structural system. (b) Equivalent impedance.

The system is connected to an impedance-matched shunting system as shown in Figure 9.7. Show that this system would consist of a spring-damper combination ($-E$, c), that is, will have a spring with negative stiffness in the same manner as having a negative capacitance shunting of piezo-patches to ensure maximum energy

Figure P9.7 Spring-damper system connected to an impedance-matched shunting system.

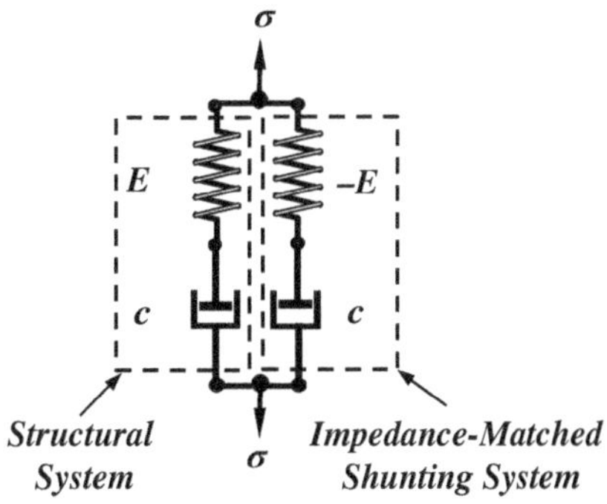

dissipation (see Section 9.2.4). In this regard, the force, and velocity are considered as the mechanical analogies of voltage and current. This leads to the spring and damper becoming analogous to the electrical capacitance and resistance.

Show also that the complex modulus of the assembly is given by:

$$E^* = \frac{2E^2Cs}{E^2 - C^2s^2}$$

Determine and plot the equivalent storage modulus and loss factor of the assembly as function of the frequency ω. Discuss the obtained results.

9.8 Consider the equivalent electrical analog of an acoustic cavity with bonded piezoelectric patch. The model of the cavity is represented by the capacitance C_S and inductance L_S, whereas the structural model of the piezo-patch is represented by the capacitance C_D and inductance L_D as shown in Figure P9.8 (Baz 2009, 2010).

The electric impedance Z_P consists of the electric capacitance of the piezo-patch C_P and an R shunting network as shown in Figure P9.9.

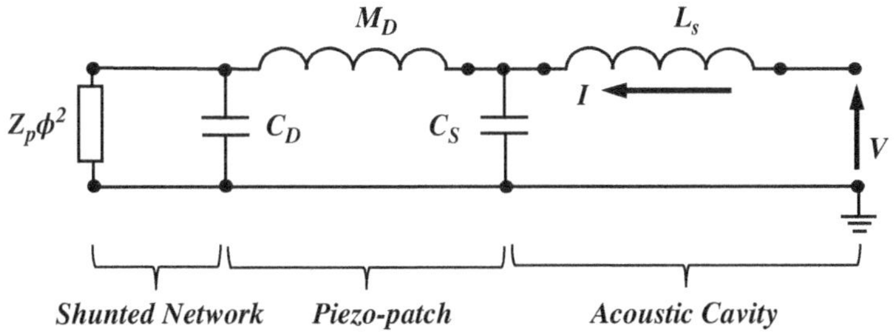

Figure P9.8 The equivalent electrical analog of an acoustic cavity with bonded piezoelectric patch.

Figure P9.9 The electric impedance of a piezo-patch with resistance shunting.

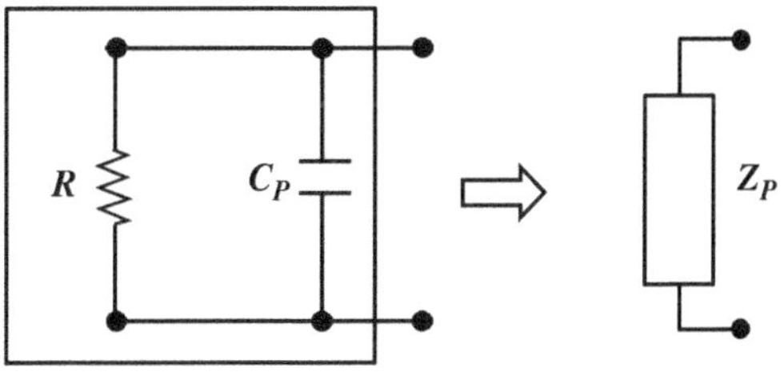

Table P9.3 The parameters of structure and piezo-patch assembly.

Parameter	C_P	C_C	L_C	A
Value	18.24 nF	1.85E-15 $m^4s^2\ kg^{-1}$	24,069 $kg\ m^{-4}$	1 m^2
Parameter	φ	C_D	M_D	l
Value	138.3 $Pa\ V^{-1}$	1.52E-13 $m^4s^2\ kg^{-1}$	13,456 $kg\ m^{-4}$	1 m

The parameters of structure and piezo-patch assembly are listed in Table P9.3. Determine the response of the structure and piezo-patch assembly when subjected to sinusoidal input voltage excitation V. The response is measured by the magnitude of the transfer function $T = \frac{V/l}{sl}$ = effective density of the acoustic cavity where s is the Laplace complex number.

Show that the *real* $(T) -\omega$ characteristics for shunting resistors R = *0, 1E4,* and *1E6Ω* is as shown in Figure P9.10. Note that, within the frequency band where the

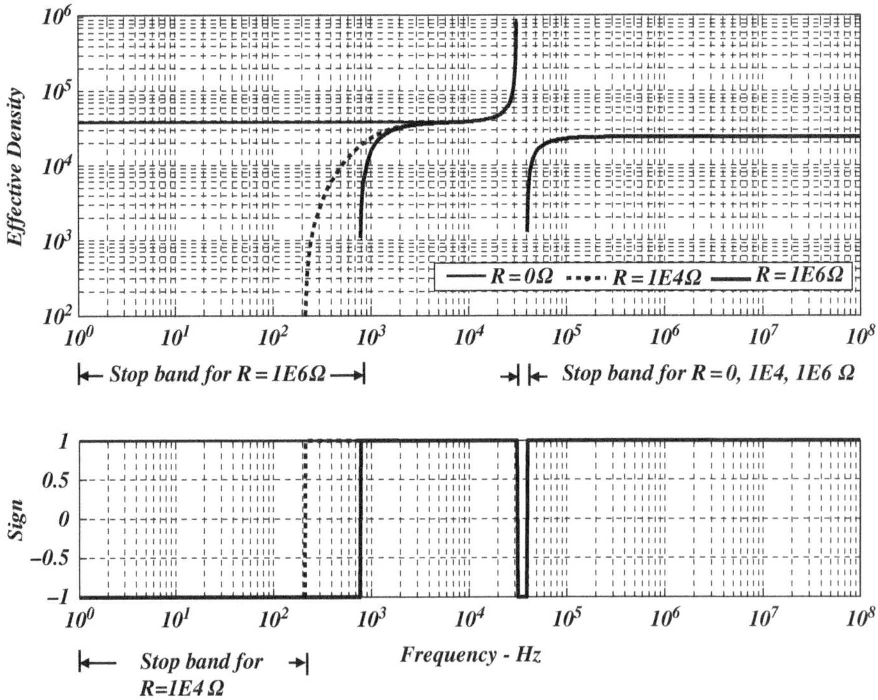

Figure P9.10 The effective density of the acoustic cavity as function of frequency when using resistive shunting.

sign of the *real* (T) is negative, the waves will cease to propagate along the cavity. This band is called the "Stop Band."

9.9 For the dynamical system shown in Figure P9.8 and described in Problem 9.9, let the electric impedance Z_P consists of the electric capacitance of the piezo-patch C_P and R–C shunting network as shown in Figure P9.11.

Determine the response of the structure and piezo-patch assembly when subjected to sinusoidal input voltage excitation V. The response is measured by the magnitude of the transfer function $T = (V/I)/(sI)$ = effective density of the acoustic cavity where s is the Laplace complex number.

Determine the *real* $(T) - \omega$ characteristics for the following shunting combination of R–C:

a) $R = 0\,\Omega$ and $C = -20$ nf.
b) $R = 0\,\Omega$ and $C = 20$ nf.
c) $R = 1E6\Omega$ and $C = 20$ nf.

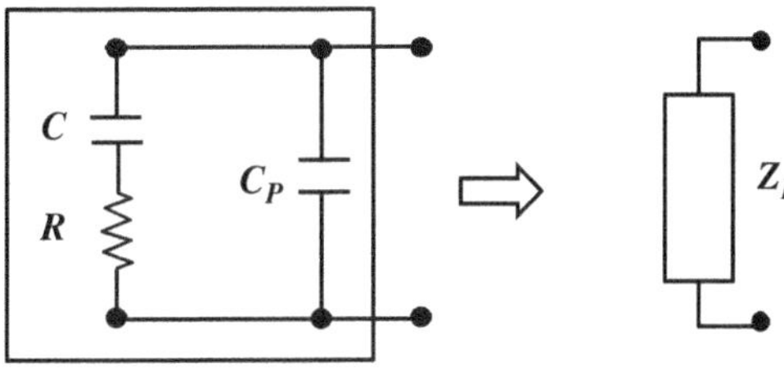

Figure P9.11 The electric impedance of a piezo-patch with resistance and capacitive shunting.

Show that the *real* $(T) - \omega$ characteristics are as shown in Figure P9.12.

9.10 For the two-mode shunting network shown in Figure P9.13, show that the dimensionless electrical impedance $\bar{Z}^{EL}$ is given by (Velazquez 1995):

$$\bar{Z}^{EL} = \frac{Ns}{\Delta}$$

where $N = s^3 + \left(\frac{R_2}{L_2} + \frac{R_1}{L_1}\right)s^2 + \left(\frac{R_1 R_2}{L_1 L_2} + \frac{1}{CL_2} + \frac{1}{CL_1}\right)s + \left(\frac{R_1 + R_2}{CL_1 L_2}\right)$

$$\Delta = s^4 + \left(\frac{R_2}{L_2} + \frac{R_1}{L_1}\right)s^3 + \left(\frac{R_1 R_2}{L_1 L_2} + \frac{1}{CL_2} + \frac{1}{CL_1 + \frac{1}{C_p L_1}}\right)s^2 + \left(\frac{R_1 + R_2}{CL_1 L_2} + \frac{R_2}{C_p L_1 L_2}\right)s + \frac{1}{CC_p L_1 L_2}$$

Show then that the non-dimensional mechanical impedance $\bar{Z}^{ME}$ for the multi-mode shunting is given by:

$$\bar{Z}^{ME} = 1 + \frac{k_{ij}^2 (Ns - \Delta)}{\Delta - k_{ij}^2 Ns}$$

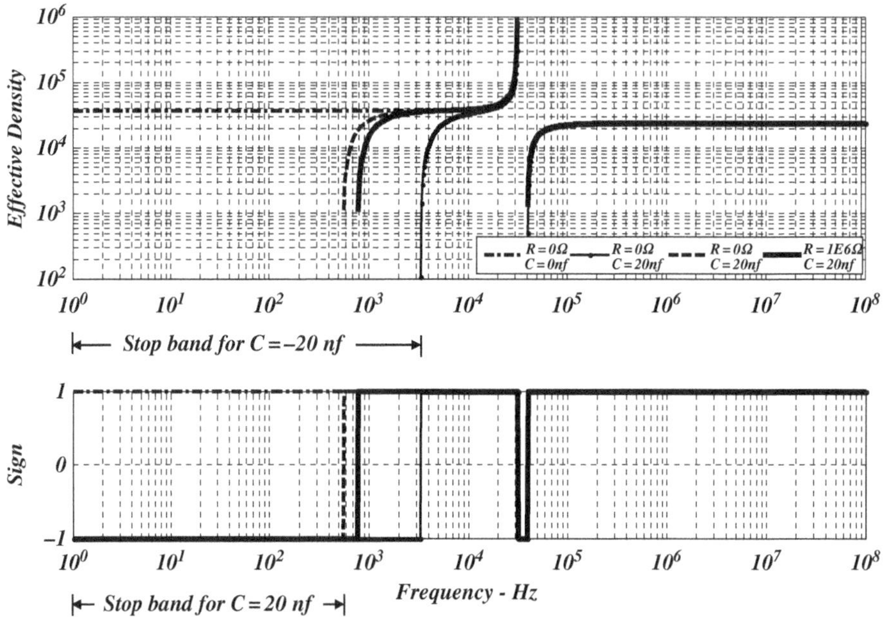

Figure P9.12 The effective density of the acoustic cavity as function of frequency when using a resistive and capacitive shunting.

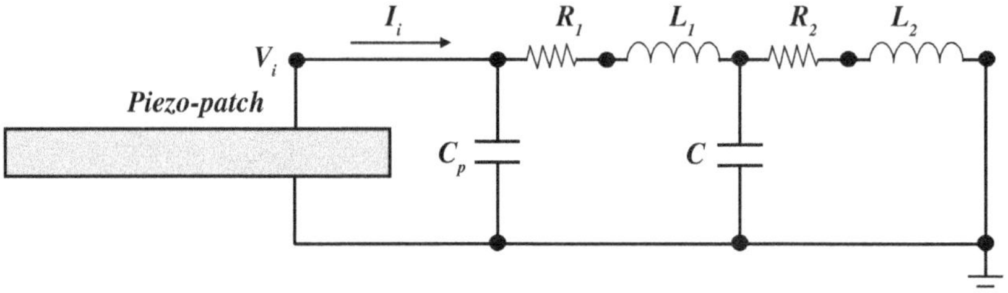

Figure P9.13 A two-mode shunting network.

Derive an expression of the non-dimensional effective complex modulus E^* of the piezoelectric patch and show that it is given by:

$$E^* = \frac{k_{31}^2\delta_1^2(\gamma^2 + \gamma\delta_2^2\Omega_2 + \delta_2^2)}{\gamma^4 + \gamma^3(\delta_2^2\Omega_2 + \delta_1^2\Omega_1) + \gamma^2(\delta_1^2\Omega_1\delta_2^2\Omega_2 + \delta_1^2[1+\mu] + \delta_2^2) + \gamma\delta_1^2\delta_2^2(\Omega_1 + \Omega_2[1+\mu]) + \delta_1^2\delta_2^2}$$

where $\gamma = \beta i, \beta = \omega/\omega_o, \bar{\omega}_i^2 = 1/(\bar{C}_i L_i), \delta_i^2 = (\omega_i/\omega_o)^2$,

$$\bar{C}_1 = \left(1 - k_{31}^2\right)C_p^T, \bar{C}_2 = C, \Omega_i = \bar{C}_i R^i \omega_o, \mu = \bar{C}_1/C_2 = \frac{L_2}{L_1}\left(\frac{\delta_2}{\delta_1}\right)^2.$$

10

Vibration Control with Periodic Structures

10.1 Introduction

Periodic structures, whether passive or active, are structures that consist of identical substructures, or cells, connected in an identical manner as shown in Figures 10.1 and 10.2. Because of their periodic nature, these structures exhibit unique dynamic characteristics that make them act as mechanical filters for wave propagation. As a result, waves can propagate along the periodic structures only within specific frequency bands called the "pass bands" and wave propagation is completely blocked within other frequency bands called the "stop bands" as shown in Figure 10.3a. The spectral width and location of these bands are fixed for passive periodic structures, and are tunable in response to the structural vibration for active periodic structures as indicated in Figure 10.3b (Baz 2001).

The theory of periodic structures was originally developed for solid state applications (Brillouin 1946) and extended, in the early 1970s, to the design of mechanical structures (Mead 1970; Cremer et al. 1973). Since then, the theory has been extensively applied to a wide variety of structures such as spring-mass systems (Faulkner and Hong 1985); periodic rods (Ruzzene and Baz 2000); periodic beams (Mead 1970; Mead and Markus 1983; Roy and Plunkett 1986; Faulkner and Hong 1985); stiffened plates (Sen Gupta 1970; Mead 1971, 1986; Mead and Yaman 1991); ribbed shells (Mead and Bardell 1987; Ruzzene and Baz 2001); and space structures.

Apart from their unique filtering characteristics, the ability of periodic structures to transmit waves from one location to another within the pass bands can be greatly reduced when the ideal periodicity is disrupted or disordered (Hodges 1982; Hodges and Woodhouse 1983). This results in the well-known phenomenon of "localization," whereby the effects of an external disturbance are localized at (or confined to) the structural zone surrounding it.

In the case of passive structures, the aperiodicity (or the disorder) can result from *unintentional* material, geometric, and manufacturing variability (Cai and Lin, 1991). However, in the case of active periodic structures the aperiodicity can be *intentionally* introduced by proper tuning of the controllers of the individual substructure or cell (Baz 2001; Chen et al. 2000).

With such unique filtering/localization characteristics of the periodic/aperiodic structures, as summarized in Figure 10.4, it would be possible to passively or actively control the wave propagation both in the spectral/spatial domains in an attempt to stop/confine the propagation of undesirable disturbances.

Active and Passive Vibration Damping, First Edition. Amr M. Baz.

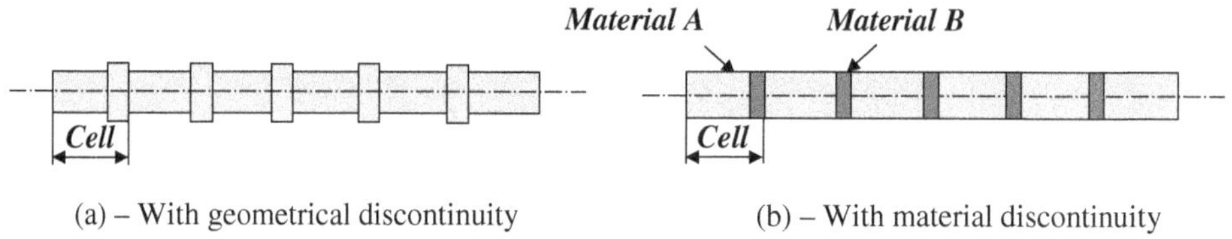

Figure 10.1 Typical examples of passive periodic structures: (a) with geometrical discontinuity and (b) with material discontinuity.

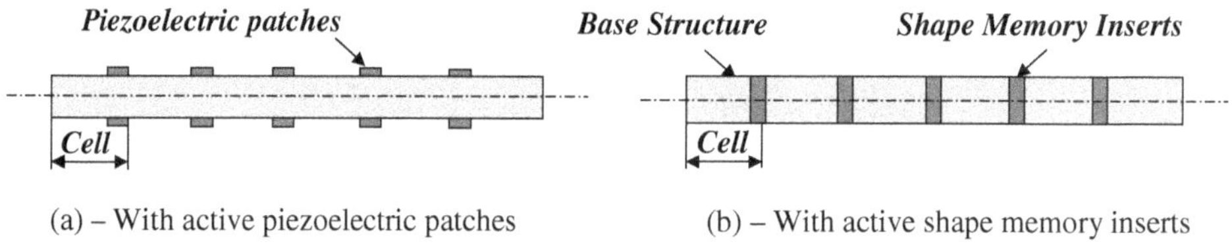

Figure 10.2 Typical examples of active periodic structures: (a) with active piezoelectric patches and (b) with active shape memory inserts.

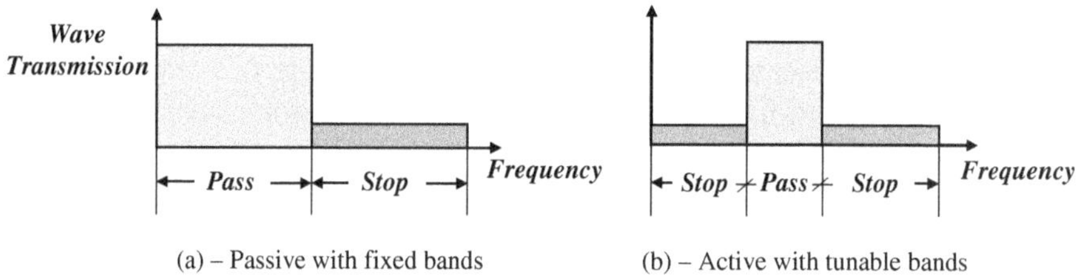

Figure 10.3 Pass and stop bands of passive and active periodic structures. (a) Passive with fixed bands and (b) active with tunable bands.

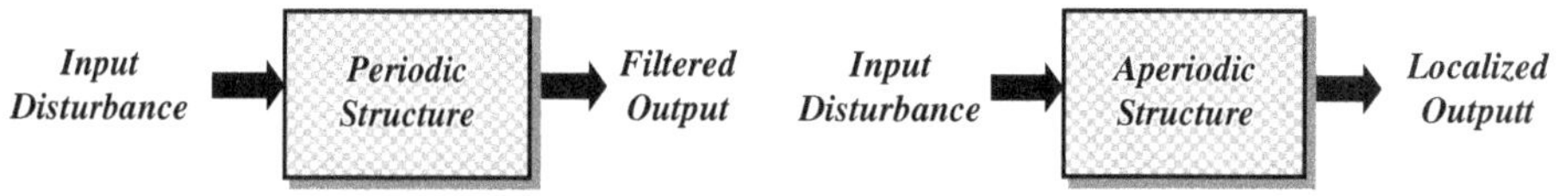

Figure 10.4 Basic characteristics of (a) periodic and (b) aperiodic structures.

Here, the emphasis is placed on studying the dynamics of one-dimensional periodic/aperiodic structures in their passive and active modes of operation in order to demonstrate their unique filtering/localization capabilities.

10.2 Basics of Periodic Structures

10.2.1 Overview

The dynamics of one-dimensional periodic structures are determined using the transfer matrix method. The basic characteristics of the transfer matrices of periodic structures are presented and related to physics of wave propagation along these structures. The

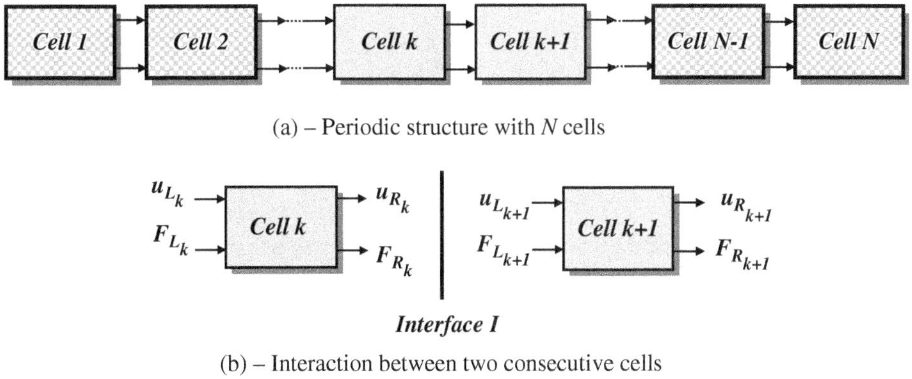

Figure 10.5 One-dimensional periodic structure. (a) Periodic structure with N cells and (b) interaction between two consecutive cells.

methodologies for determining the pass and stop bands, the natural frequencies, the mode shapes, and frequency response of periodic structures are introduced. Illustrated examples are given to demonstrate the application of these methodologies to various periodic structures including spring-mass systems, rods, and beams.

10.2.2 Transfer Matrix Method

10.2.2.1 The Transfer Matrix

Consider the generic one-dimensional periodic structure shown in Figure 10.5.

The undamped dynamics of the kth cell are determined from the following finite element expression:

$$\begin{bmatrix} M_{LL} & M_{LR} \\ M_{RL} & M_{RR} \end{bmatrix} \begin{Bmatrix} \ddot{u}_{L_k} \\ \ddot{u}_{R_k} \end{Bmatrix} + \begin{bmatrix} K_{LL} & K_{LR} \\ K_{RL} & K_{RR} \end{bmatrix} \begin{Bmatrix} u_{L_k} \\ u_{R_k} \end{Bmatrix} = \begin{Bmatrix} F_{L_k} \\ F_{R_k} \end{Bmatrix} \tag{10.1}$$

where M_{ij} and K_{ij} are appropriately partitioned matrices of the mass and stiffness matrices. Also, u and F define the deflection and force vectors with subscripts L_i and R_i denoting the left and right sides of the kth cell.

For a sinusoidal excitation at a frequency ω, Eq. (10.1) reduces to:

$$\begin{bmatrix} K_{LL} - M_{LL}\omega^2 & K_{LR} - M_{LR}\omega^2 \\ K_{RL} - M_{RL}\omega^2 & K_{RR} - M_{RR}\omega^2 \end{bmatrix} \begin{Bmatrix} u_{L_k} \\ u_{R_k} \end{Bmatrix} = \begin{Bmatrix} F_{L_k} \\ F_{R_k} \end{Bmatrix},$$

or

$$\begin{bmatrix} K_{d_{LL}} & K_{d_{LR}} \\ K_{d_{RL}} & K_{d_{RR}} \end{bmatrix} \begin{Bmatrix} u_{L_k} \\ u_{R_k} \end{Bmatrix} = \begin{Bmatrix} F_{L_k} \\ F_{R_k} \end{Bmatrix} \tag{10.2}$$

where K_d is the dynamic stiffness matrix of the kth cell.

Eq. (10.2) is rearranged to take the following form:

$$\begin{Bmatrix} u_{R_k} \\ F_{R_k} \end{Bmatrix} = \begin{bmatrix} -K_{d_{LR}}^{-1}K_{d_{LL}} & K_{d_{LR}}^{-1} \\ -K_{d_{RR}}K_{d_{LR}}^{-1}K_{d_{LL}} + K_{d_{RL}} & K_{d_{RR}}K_{d_{LR}}^{-1} \end{bmatrix} \begin{Bmatrix} u_{L_k} \\ F_{L_k} \end{Bmatrix} \tag{10.3}$$

Considering now the compatibility and equilibrium conditions at the interface between the kth and the $k + 1$th cells, yields the following expressions:

$$u_{R_k} = u_{L_{k+1}} \; and \; F_{R_k} = -F_{L_{k+1}} \tag{10.4}$$

Substituting these conditions into Eq. (10.3), it reduces to:

$$\begin{Bmatrix} u_{L_{k+1}} \\ F_{L_{k+1}} \end{Bmatrix} = \begin{bmatrix} -K_{d_{LR}}^{-1}K_{d_{LL}} & K_{d_{LR}}^{-1} \\ K_{d_{RR}}K_{d_{LR}}^{-1}K_{d_{LL}} - K_{d_{RL}} & -K_{d_{RR}}K_{d_{LR}}^{-1} \end{bmatrix} \begin{Bmatrix} u_{L_k} \\ F_{L_k} \end{Bmatrix} \tag{10.5}$$

In a more compact form, Eq. (10.5) can be rewritten as:

$$\begin{Bmatrix} u_L \\ F_L \end{Bmatrix}_{k+1} = \begin{bmatrix} t_{11} & t_{12} \\ t_{21} & t_{22} \end{bmatrix} \begin{Bmatrix} u_L \\ F_L \end{Bmatrix}_k \; or \; Y_{k+1} = [T_k]\, Y_k \tag{10.6}$$

where Y and $[T_k]$ denotes the state vector = $\{u_L \; F_L\}^T$ and the transfer matrix of the kth cell. Note that the transfer matrix relates the state vector at the left end of $k + 1$th cell to that at the left end of the kth cell. For exactly periodic structures, $[T_k]$ is the same for all cells and Eq. (10.6) reduces to:

$$Y_{k+1} = [T]\, Y_k \tag{10.7}$$

However, for aperiodic structures, $[T_k]$ varies from cell to cell to appropriately account for the nature of aperiodicity.

10.2.2.2 Basic Properties of the Transfer Matrix

The transfer matrix $[T]$ has very interesting characteristics and is called a "simplectic matrix." Among these characteristics are:

i) Determinant of [T] = 1:

Proof From Eq. (10.5) and the symmetry of the dynamic stiffness matrix, we get:

$$\begin{aligned} \det[T] &= \det\left(\begin{bmatrix} -K_{d_{LR}}^{-1}K_{d_{LL}} & K_{d_{LR}}^{-1} \\ K_{d_{RR}}K_{d_{LR}}^{-1}K_{d_{LL}} - K_{d_{RL}} & -K_{d_{RR}}K_{d_{LR}}^{-1} \end{bmatrix}\right) \\ &= \det\left(\left[K_{d_{LR}}^{-1}K_{d_{LL}}\left(K_{d_{RR}}K_{d_{LR}}^{-1}\right)^T - K_{d_{LR}}^{-1}\left(K_{d_{RR}}K_{d_{LR}}^{-1}K_{d_{LL}}\right)^T + K_{d_{LR}}^{-1}\left(K_{d_{RL}}\right)^T\right]\right) \\ &= \det\left(\left[K_{d_{LR}}^{-1}K_{d_{LL}}\left(K_{d_{RR}}K_{d_{LR}}^{-1}\right)^T - K_{d_{LR}}^{-1}\left(K_{d_{RR}}K_{d_{LR}}^{-1}K_{d_{LL}}\right)^T + K_{d_{LR}}^{-1}\left(K_{d_{RL}}\right)^T\right]\right) \\ &= \det(I) = 1 \quad \textbf{QED} \end{aligned} \tag{10.8}$$

ii) $[T]^{-T} = J\,[T]\,J^{-1}$

$$\text{where } J = \begin{bmatrix} 0 & I \\ -I & 0 \end{bmatrix}$$

Proof From Eq. (10.6);

$$\text{As } [T] = \begin{bmatrix} t_{11} & t_{12} \\ t_{21} & t_{22} \end{bmatrix}, \text{ then } [T]^{-1} = \begin{bmatrix} t_{22} & -t_{12} \\ -t_{21} & t_{11} \end{bmatrix} \qquad \text{(a)}$$

But

$$J[T]J^{-1} = [0\ I; -I\ 0] \begin{bmatrix} t_{11} & t_{12} \\ t_{21} & t_{22} \end{bmatrix} [0\ -I; I\ 0] \qquad \text{(b)}$$

From (a) and (b):

$$[T]^{-T} = J\,[T]\,J^{-1}\ \textbf{\textit{QED}}. \qquad (10.9)$$

iii) Eigenvalues of [T] are λ and λ^{-1}

Proof The eigenvalue problem of $[T]$ can be written as:

$$[T]Y_k = \lambda Y_k \qquad (10.10)$$

Equation (10.10) can be rewritten as:

$$J[T]J^{-1}JY_k = \lambda J\,Y_k \qquad (10.11)$$

where $J = \begin{bmatrix} 0 & I \\ -I & 0 \end{bmatrix}$. Using Eq. (10.9) and let $X_k = J\,Y_k$, then Eq. (10.11) reduces to:

$$[T]^{-T}X_k = \lambda X_k,$$

or

$$[T]^{T}X_k = \lambda^{-1} X_k. \qquad (10.12)$$

Equation (10.12) indicates that λ^{-1} is an eigenvalue of the matrix $[T]^T$ that means that it is also an eigenvalue of the matrix $[T]$. The equation also indicates that X_k is the corresponding eigenvector. Accordingly, if λ and Y_k are an eigenvalue and an eigenvector for $[T]$ then λ^{-1} and X_k are companion eigenvalue and an eigenvector for $[T]$.

iv) Mathematical Meaning of eigenvalue λ of [T]

Combining Eqs. (10.7) and (10.10) gives:

$$Y_{k+1} = \lambda Y_k \qquad (10.13)$$

indicating that the eigenvalue λ of the matrix $[T]$ is the ratio between the state vectors at two consecutive cells.

Hence, one can reach the following conclusions:

a) If $|\lambda| = 1$, then $Y_{k+1} = Y_k$ and the state vector propagates along the structure as is. This condition defines a "pass band" condition.

and

b) If $|\lambda| < 1$, then $Y_{k+1} < Y_k$ and the state vector is attenuated as it propagates along the structure. This condition defines a "stop band" condition.
A further explanation of the physical meaning of the eigenvalue λ can be extracted by rewriting it as:

$$\lambda = e^{\mu} = e^{\alpha + i\beta} \tag{10.14}$$

where μ is defined as the "propagation constant," which is a complex number whose real part (α) represents the logarithmic decay of the state vector and its imaginary part (β) defines the phase difference between the adjacent cells.

For example, rewriting Eq. (10.13) as:

$$\{u_L \; F_L\}^T{}_{k+1} = e^{\alpha + i\beta} \{u_L \; F_L\}^T{}_k \tag{10.15}$$

and considering only the jth components u_{L_j} of the deflection vector u_L, at cells k and $k+1$, which can be written as:

$$u_{L_{jk+1}} = U_{L_{jk+1}} e^{i\varphi_{jk+1}} \text{ and } u_{L_{jk}} = U_{L_{jk}} e^{i\varphi_{jk}} \tag{10.16}$$

where $U_{L_{jn}}$ *and* φ_{j_n} denote the amplitude and phase shift of the jth component u_{L_j} at the nth cell.

From Eqs. (10.15) and (10.16), we get:

$$\ell n\left(u_{L_{jk+1}} / u_{L_{jk}}\right) = \ell n\left(U_{L_{jk+1}} / U_{L_{jk}}\right) + i\left(\varphi_{jk+1} - \varphi_{jk}\right) = \alpha + i\beta \tag{10.17}$$

Eq. (10.17) indicates that:

$$\alpha = \ell n\left(U_{L_{jk+1}} / U_{L_{jk}}\right) = \text{logarithmic decay of amplitude},$$

and

$$\beta = \left(\varphi_{jk+1} - \varphi_{jk}\right) = \text{phase difference between the adjacent cells}$$

Therefore, the equivalent conditions for the pass and the stop bands can be written in terms of the propagation constant parameters (α and β) as follows:

a) If $\alpha = 0$ (i.e., μ is imaginary), then we have the "*Pass Band*" as there is no amplitude attenuation.

b) If $\alpha \neq 0$ (i.e., μ is real or complex), then we have the "*Stop Band*" as there is amplitude attenuation defined by the value of α.

In this case, a purely elastic periodic structure acts as if it is a damped structure that stops rather than attenuates the structural vibration. For such unique behavior, the periodic structures can be very effective in impeding the propagation of undesirable waves.

v) *Physical Meaning of the eigenvalues λ and λ⁻¹ of [T]*

A better insight into the physical meaning of the eigenvalues λ *and* λ^{-1} can be gained by considering the following transformation of the cell dynamics into the wave mode component domain:

$$Y_k = \begin{Bmatrix} u_L \\ F_L \end{Bmatrix}_k = \Phi\, W_k = \Phi \begin{Bmatrix} w_L^r \\ w_L^L \end{Bmatrix}_k \text{ and } Y_{k+1} = \begin{Bmatrix} u_L \\ F_L \end{Bmatrix}_{k+1} = \Phi\, W_{k+1} = \Phi \begin{Bmatrix} w_L^r \\ w_L^L \end{Bmatrix}_{k+1} \tag{10.18}$$

where Φ is the eigenvector matrix of the transfer matrix $[T]$. Also, W_k is the wave mode component vector that has the right-going wave component w^r and left-going wave component w^L.

Substituting Eq. (10.18) into Eq. (10.7), it reduces to:

$$Y_{k+1} = \Phi \begin{Bmatrix} w_L^r \\ w_L^L \end{Bmatrix}_{k+1} = [T]\, Y_k = [T]\, \Phi \begin{Bmatrix} w_L^r \\ w_L^L \end{Bmatrix}_k,$$

or

$$\begin{Bmatrix} w_L^r \\ w_L^L \end{Bmatrix}_{k+1} = \Phi^{-1}[T]\,\Phi \begin{Bmatrix} w_L^r \\ w_L^L \end{Bmatrix}_k \tag{10.19}$$

Note that the matrix $\Phi^{-1}[T]\,\Phi$ reduces to the matrix of eigenvalues of the matrix $[T]$, that is:

$$\begin{Bmatrix} w_L^r \\ w_L^L \end{Bmatrix}_{k+1} = diag\,(\lambda_1, \lambda_2, \ldots, \lambda_{2n-1}, \lambda_{2n}) \begin{Bmatrix} w_L^r \\ w_L^L \end{Bmatrix}_k$$

But, because of the particular nature of the eigenvalues of $[T]$ as they appear in pairs (λ, λ^{-1}), then this equation reduces to:

$$\begin{Bmatrix} w_L^r \\ w_L^L \end{Bmatrix}_{k+1} = \begin{bmatrix} \Lambda & 0 \\ 0 & \Lambda^{-1} \end{bmatrix} \begin{Bmatrix} w_L^r \\ w_L^L \end{Bmatrix}_k \tag{10.20}$$

where

$$\Lambda = diag\,[\lambda_1, \lambda_2, ..] \; and \; \Lambda^{-1} = diag\,[\lambda_1^{-1}, \lambda_2^{-1}, ..].$$

Equation (10.20) can be expanded to give:

$$w_{L_{k+1}}^r = \Lambda\, w_{L_k}^r \; and \; w_{L_{k+1}}^L = \Lambda^{-1}\, w_{L_k}^L \tag{10.21}$$

For the jth component of w, we have:

$$w_{L_{j_{k+1}}}^r = \lambda_j w_{L_{j_k}}^r \; and \; w_{L_{j_{k+1}}}^L = \lambda_j^{-1} w_{L_{j_k}}^L \tag{10.22}$$

that is, the eigenvalue λ_j is the ratio between the amplitude of the right-going waves, whereas λ_j^{-1} defines the ratio between the amplitude of the left-going waves as shown in Figure 10.6.

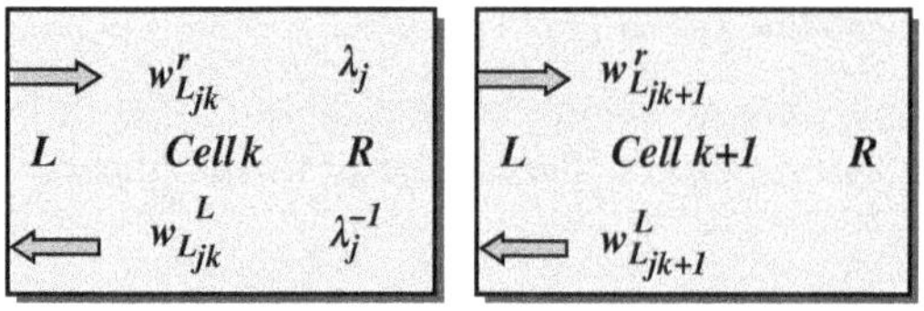

Figure 10.6 The propagation of right- and left-going waves.

Therefore, if $|\lambda_j| < 1$, then the pair $(\lambda_j, \lambda_j^{-1})$ define attenuation (or stop bands) along the wave propagation direction from cell k to cell $k+1$. Also, if $|\lambda_j| = 1$, then the pair $(\lambda_j, \lambda_j^{-1})$ define propagation without attenuation (i.e., pass bands).

One question remains here to be answered and that is: why the eigenvector matrix Φ transforms the state vector $Y_k = \{u_L\ F_L\}^T$ to the wave mode component vector $W_k = \{w_L^r\ w_L^L\}^T$ as governed by Eq. (10.18). An attempt to answer this question is illustrated in Example 10.1 for the case of rods undergoing longitudinal vibrations.

Example 10.1 Consider the rod of Figure 10.7 to show that the eigenvector matrix Φ, of the transfer matrix $[T]$, transforms the state vector $Y_k = \{u_L\ F_L\}^T$ to the wave mode component vector $W_k = \{w_L^r\ w_L^L\}^T$ as governed by Eq. (10.18).

Solution

As the equation of motion of the rod is given by:

$$u_{xx} = (\rho/E)\, u_{tt}$$

where u is the longitudinal deflection, ρ is the density and E is Young's modulus.

Then, assuming a solution: $u = U(x)\, e^{i\omega t}$, reduces the equation of motion to:

$$U_{xx} + (\rho/E)\, \omega^2\, U = 0,$$

or

$$U_{xx} + k^2\, U = 0 \tag{10.23}$$

where k = wave number = $\sqrt{\rho/E}\,\omega$.

Equation (10.23) can be written in the following state-space form:

$$\frac{d}{dx}\begin{Bmatrix} U \\ U_x \end{Bmatrix} = \begin{bmatrix} 0 & 1 \\ -k^2 & 0 \end{bmatrix}\begin{Bmatrix} U \\ U_x \end{Bmatrix} = A\begin{Bmatrix} U \\ U_x \end{Bmatrix}$$

that has a solution:

$$\begin{Bmatrix} U \\ U_x \end{Bmatrix}_x = e^{Ax}\begin{Bmatrix} U \\ U_x \end{Bmatrix}_0.$$

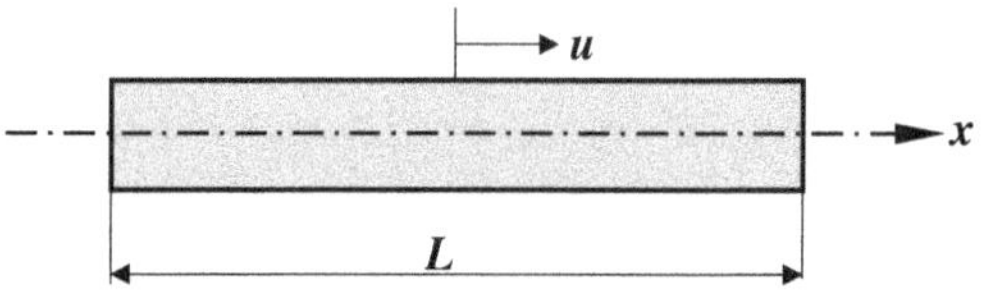

Figure 10.7 Rod undergoing longitudinal vibrations.

This solution can be put in a transfer matrix form by setting $U_x = F/EA$ and extracting e^{Ax} using symbolic manipulation software (such as Mathematica) to give:

$$\begin{Bmatrix} U \\ F/EA \end{Bmatrix}_x = \begin{bmatrix} \dfrac{\left(1+e^{-2ikx}\right)}{2e^{-ikx}} & \dfrac{\left(1-e^{-2ikx}\right)}{2ike^{-ikx}} \\ \dfrac{ik\left(1-e^{-2ikx}\right)}{2e^{-ikx}} & \dfrac{\left(1+e^{-2ikx}\right)}{2e^{-ikx}} \end{bmatrix} \begin{Bmatrix} U \\ F/EA \end{Bmatrix}_0,$$

or,

$$\begin{Bmatrix} U \\ F \end{Bmatrix}_x = \begin{bmatrix} \dfrac{\left(1+e^{-2ikx}\right)}{2e^{-ikx}} & \dfrac{1}{EA}\dfrac{\left(1-e^{-2ikx}\right)}{2ike^{-ikx}} \\ EA\dfrac{ik\left(1-e^{-2ikx}\right)}{2e^{-ikx}} & \dfrac{\left(1+e^{-2ikx}\right)}{2e^{-ikx}} \end{bmatrix} \begin{Bmatrix} U \\ F \end{Bmatrix}_0 = [T]\begin{Bmatrix} U \\ F \end{Bmatrix}_0 \quad (10.24)$$

Accordingly, the transfer matrix $[T]$ is given by in a more compact form as:

$$[T] = \begin{bmatrix} \cos(kx) & \dfrac{1}{Z\omega}\sin(kx) \\ -Z\omega\sin(kx) & \cos(kx) \end{bmatrix} \quad (10.25)$$

where $Z = A\sqrt{E\rho}$, which is called impedance of the rod.

Using symbolic manipulation software (such as Mathematica) indicates that $[T]$ has the following two eigenvalues:

$$\lambda_1 = e^{-ikx} \; and \; \lambda_2 = e^{ikx} = \lambda^{-1}$$

Note that as $\lambda_{1,2} = e^{\mp ikx} = \cos(kx) \mp i\sin(kx)$ *then* $|\lambda_{1,2}| = 1$, that is, defining pass band. This means that the uniform rod will transmit any incident waves without any attenuation.

Note also that λ_2 is the reciprocal of λ_1 to confirm property ***c*** of the transfer matrices as given in Section 10.2.2.2.

Symbolic manipulation software indicates also that $[T]$ has the following two eigenvectors, which correspond to $\lambda_{1,2}$, respectively:

$v_1 = \{1 \; -EAki\}^T$ and $v_2 = \{1 \; EAki\}^T$ with eigenvector matrix $\Phi = \begin{bmatrix} 1 & 1 \\ -EAki & EAki \end{bmatrix}$.

Now, the eigenvector matrix Φ is used to define the following transformations:

$$\begin{Bmatrix} u \\ F \end{Bmatrix}_x = \begin{bmatrix} 1 & 1 \\ -EAki & EAki \end{bmatrix} \begin{Bmatrix} w_1 \\ w_2 \end{Bmatrix}_x \quad \text{and} \quad \begin{Bmatrix} u \\ F \end{Bmatrix}_0 = \begin{bmatrix} 1 & 1 \\ -EAki & EAki \end{bmatrix} \begin{Bmatrix} w_1 \\ w_2 \end{Bmatrix}_0 \quad (10.26)$$

The physical meaning of the vectors $\{w_1 \; w_2\}_x^T$ and $\{w_1 \; w_2\}_o^T$ will become apparent shortly.

The first part of Eq. (10.26) gives:

$$u(x) = w_1(x) + w_2(x) \quad (10.27)$$

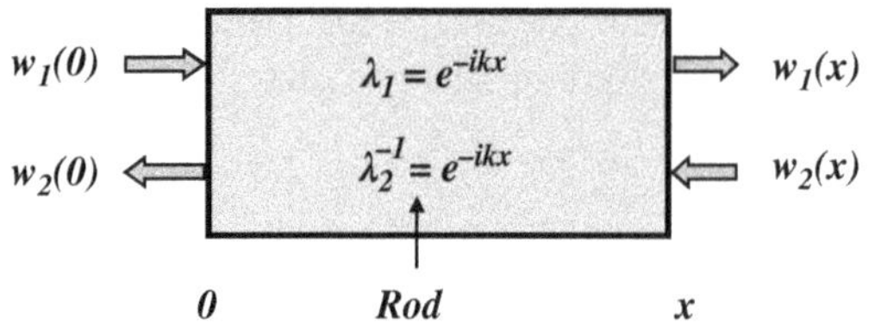

Figure 10.8 Right and left-going waves in rods.

Combining Eqs. (10.24) with (10.26) gives, after some manipulation, the following expression:

$$\begin{Bmatrix} w_1 \\ w_2 \end{Bmatrix}_x = \begin{bmatrix} e^{-ikx} & 0 \\ 0 & e^{ikx} \end{bmatrix} \begin{Bmatrix} w_1 \\ w_2 \end{Bmatrix}_0 \tag{10.28}$$

Accordingly,

$$w_1(x) = e^{-ikx} w_1(0) \text{ and } w_2(x) = e^{ikx} w_2(0) \tag{10.29}$$

Hence, substituting for $w_1(x)$ and $w_2(x)$ from Eq. (10.29) into Eq. (10.27), it reduces to:

$$u(x) = e^{-ikx} w_1(0) + e^{ikx} w_2(0) \tag{10.30}$$

Equation (10.30) indicates that the longitudinal displacement $u(x)$ consists of two wave components: one propagating to the right = $w_1(x) = e^{-ikx}\, w_1(0)$ and one propagating to the left = $w_2(x) = e^{ikx}\, w_2(0)$. Therefore, the vector $\{w_1\ w_2\}_x^T$ physically denotes a vector = $\{w^r\ w^L\}_x^T$ with $w^r(x)$ and $w^L(x)$ defining the right and left-going waves at location x.

Note that the vector $\{w^r\ w^L\}_x^T = \{w_1\ w_2\}_x^T$ is generated from the vector $\{\mathrm{w}_1\ \mathrm{w}_2\}_0^{\mathrm{T}}$ using the propagation Eq. (10.29) with $\{w_1\ w_2\}_0^T$ defining a vector = $\{w^r\ w^L\}_0^T$ where $w^r(0)$ and $w^L(0)$ denote right- and left-going waves at location 0.

A graphical representation of Eq. (10.29) is displayed in Figure 10.8.

In summary, the present example has proved that the eigenvector matrix Φ, of the transfer matrix $[T]$, transforms the state vector $Y = \{u\ F\}^T$ to the wave mode component vector $W = \{w^r\ \ w^L\}^T$ as governed by Eq. (10.26) that is equivalent to Eq. (10.18).

Example 10.2 Consider the following transfer matrix $[T]$ of a one-dimensional periodic structure such that $[T] = \begin{bmatrix} t_{11} & t_{12} \\ t_{21} & t_{22} \end{bmatrix}$

Show that the sum of the eigenvalues of $[T]$ is given by:

$\lambda_1 + \lambda_2 = \lambda + \lambda^{-1} = t_{11} + t_{22} = 2\cosh(\mu)$

Solution

The eigenvalues of $\{T]$ can be determined from

$$\det(\lambda[I] - [T]) = 0$$

or

$$\lambda^2 - (t_{11} + t_{22})\lambda - t_{12}t_{21} + t_{11}t_{22} = 0$$

but as the $det([T])$ = 1, from Section 10.2.2.2.i, then this equation reduces to

$$\lambda^2 - (t_{11} + t_{22})\lambda + 1 = 0$$

Hence, from the theory of quadratic equations, the roots λ_1 and λ_2 of these equations must be such that:

$$\lambda_1 + \lambda_2 = t_{11} + t_{22} \text{ and } \lambda_1\lambda_2 = 1$$

But as $\lambda_1 = 1/\lambda_2 = \lambda$ (Section 10.2.2.2.iii), and $\lambda = e^{\mu}$(as given by Eq. (10.14)), then

$$\lambda_1 + \lambda_2 = t_{11} + t_{22} = e^{\mu} + e^{-\mu} = 2\cosh(\mu) \tag{10.31}$$

Example 10.3 Consider the following transfer matrix $[T]$ of a one-dimensional periodic structure such that

$$[T] = \begin{bmatrix} t_{11} & t_{12} \\ t_{21} & t_{22} \end{bmatrix}$$

Show that $[T]^N$ is given by:

$$[T]^N = c_1\left[[T] + [T]^{-1}\right] + c_2\left[[T] - [T]^{-1}\right]$$

where $c_1 = \dfrac{\cosh(N\mu)}{2\cosh(\mu)}$ and $c_2 = \dfrac{\sinh(N\mu)}{2\sinh(\mu)}$

Solution

As

$$[T] + [T]^{-1} = (e^{\mu} + e^{-\mu})[I] = 2\cosh(\mu)[I]$$

and

$$[T] - [T]^{-1} = (e^{\mu} - e^{-\mu})[I] = 2\sinh(\mu)[I]$$

then,

$$c_1\left[[T] + [T]^{-1}\right] = \cosh(N\mu)[I] \text{ and } c_2\left[[T] - [T]^{-1}\right] = \sinh(N\mu)[I]$$

or

$$\begin{aligned} c_1\left[[T] + [T]^{-1}\right] + c_2\left[[T] - [T]^{-1}\right] &= \left[\cosh(N\mu) + \sinh(N\mu)\right][I] \\ &= \frac{1}{2}\left[\left(e^{N\mu} + e^{-N\mu}\right) + \left(e^{N\mu} - e^{-N\mu}\right)\right][I] = e^{N\mu}[I] = [T]^N \end{aligned} \tag{10.32}$$

10.3 Filtering Characteristics of Passive Periodic Structures

10.3.1 Overview

The methodologies for determining the pass and stop bands of passive periodic structures are introduced here. Illustrated examples are given to demonstrate the application of these methodologies to rods as examples of one-dimensional periodic structures.

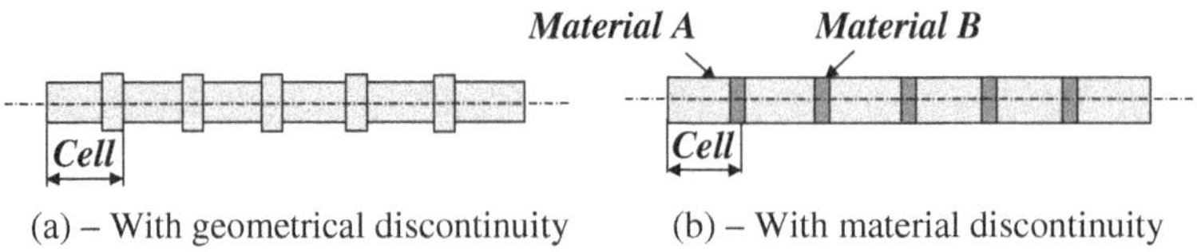

Figure 10.9 Passive periodic rods. (a) With geometrical discontinuity and (b) with material discontinuity.

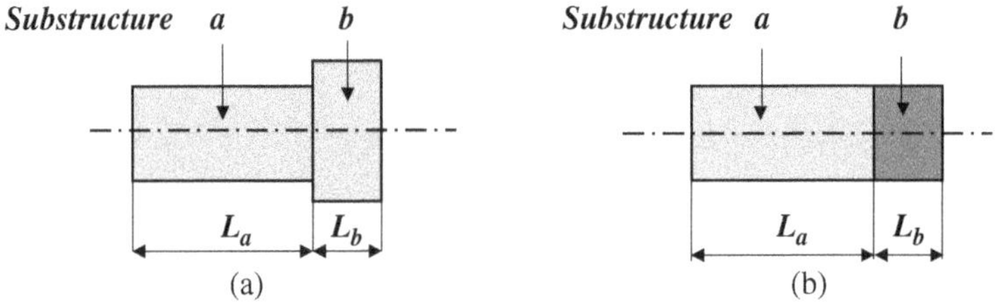

Figure 10.10 Unit cells of passive periodic rods.

10.3.2 Periodic Rods in Longitudinal Vibrations

Consider the longitudinal vibrations of the periodic rods shown in Figure 10.9.

These rods are made of assemblies of the periodic cells shown in Figure 10.10a,b. Each of these unit cells consists of two substructures a and b, which can be either of the same material with different cross sections (Figure 10.10a), or made of different materials with the same cross sections (Figure 10.10b).

The dynamic characteristics of the individual substructure (a or b) can be described by its transfer matrix $[T_s]$, as defined by Eq. (10.25), as follows:

$$[T_s] = \begin{bmatrix} \cos(k_s L_s) & \dfrac{1}{Z_s \omega}\sin(k_s L_s) \\ -Z_s \omega \sin(k_s L_s) & \cos(k_s L_s) \end{bmatrix} \text{ with } s = a, b \tag{10.33}$$

Combining the transfer matrices of the substructures a and b, yields the transfer matrix $[T]$ for the unit cell as follows:

$$[T] = [T_b][T_a] \tag{10.34}$$

The pass and stop bands characteristics of the periodic rod can be determined by investigating the eigenvalues of the transfer matrix $[T]$ for different combinations of the longitudinal rigidities $E_s A_s$ and dimensionless frequencies $k_s L_s$.

Example 10.4 Determine the pass and stop band characteristics for the periodic rod shown in Figure 10.11, which has the following geometrical and physical properties.

Material	E (GN m^{-2})	ρ (kg m^{-3})	A (m^2)	Case 1–L (m)	Case 2–L (m)
a	210	7800	0.000625	0.025	0.04
b	0.025	1200	0.000625	0.025	0.01

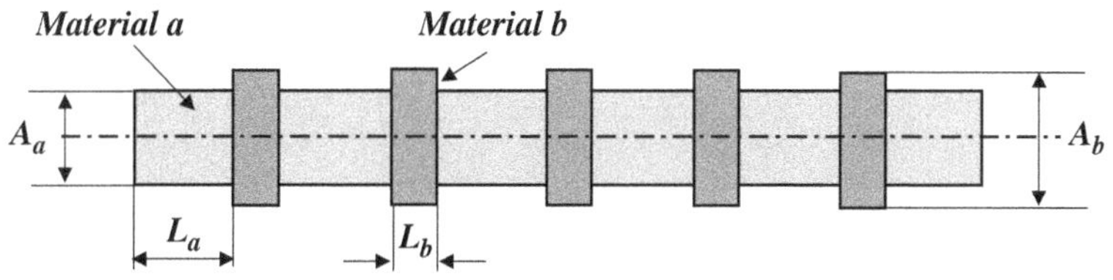

Figure 10.11 Periodic rod with material and geometrical discontinuities.

Solution

Figure 10.12a,b shows the magnitude of the eigenvalues of the transfer matrix $[T] = [T_b]\,[T_a]$ as function of the frequency. The figures identify also the pass and stop bands for the two cases.

Note that the width and location of the stop and pass bands depends primarily on the physical and geometrical properties of the periodic rod.

Example 10.5 Determine the pass and stop band characteristics for the periodic rod shown in Figure 10.11 that has the following geometrical and physical properties.

Material	E (GN m^{-2})	ρ (kg m^{-3})	A (m^2)	L (m)
a (elastic)	210	7800	0.000625	0.025
b (viscoelastic)	$E'(1+\eta i)$	1200	0.000625	0.025

where $E'(1+\eta i) = 0.025\text{E}9\left(1+\alpha_1\frac{-\Omega^2+2i\Omega}{(1-\Omega^2)+2i\Omega}\right)$ with $\Omega = (\omega/\omega_1)$, $\alpha_1 = 39$, and $\omega_1 = 19{,}058$ rad/s.

Solution

Figure 10.13 shows the magnitude of the eigenvalues of the transfer matrix $[T] = [T_b]\,[T_a]$ as function of the frequency. The figure indicates that presence of damping in the periodic rod has resulted in eliminating the pass bands completely and extended the stop band over the considered frequency band.

In this example, the unit cell of the periodic structure is assumed to generically consist of an elastic substructure bonded to a viscoelastic substructure. Specifically, the latter substructure can be made of either a conventional viscoelastic material or a shunted piezoelectric network (Thorp et al. 2001).

10.4 Natural Frequencies, Mode Shapes, and Response of Periodic Structures

10.4.1 Natural Frequencies and Response

The modal parameters and the response of periodic structures can be determined using the transfer matrix approach presented in Section 10.3. The state vectors at the beginning and the end of the structure are related by the system transfer matrix $[T_N]$:

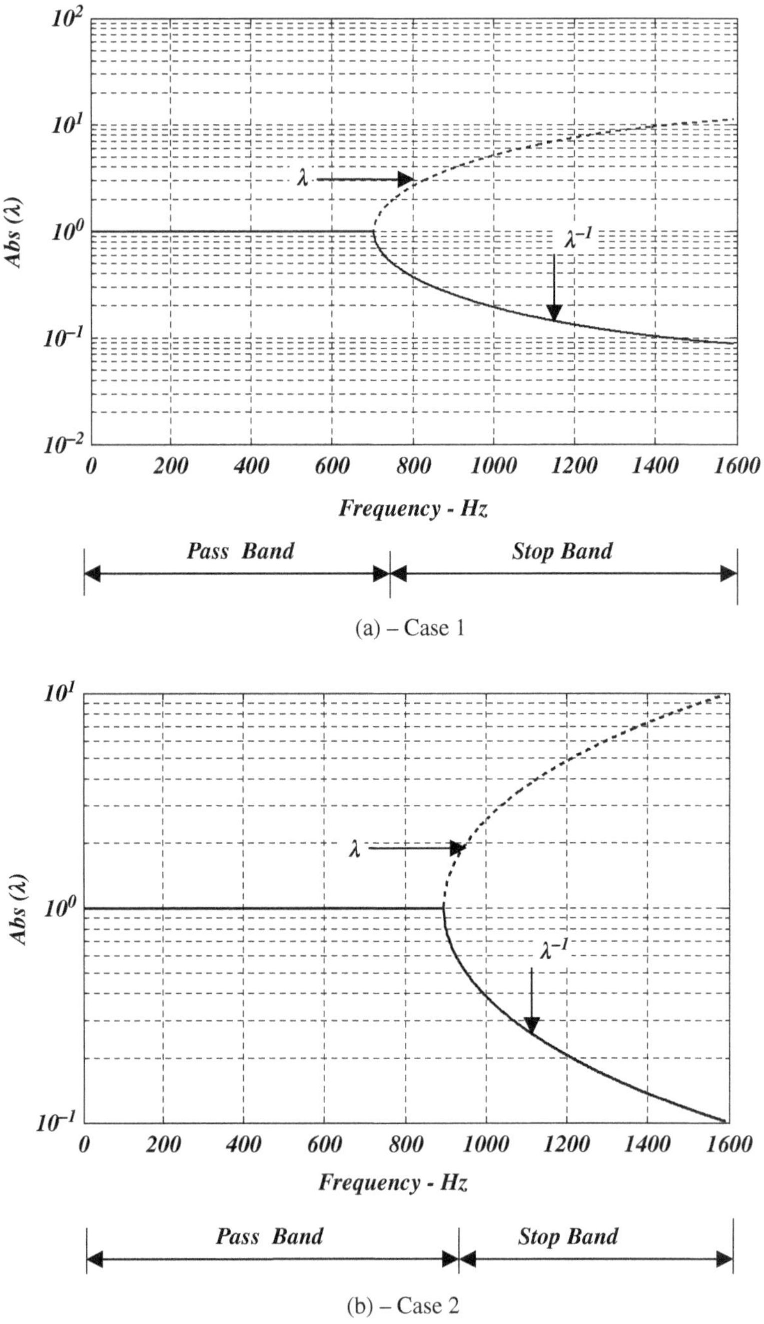

Figure 10.12 Pass and stop bands of an elastic periodic rod. (a) Case 1 and (b) Case 2.

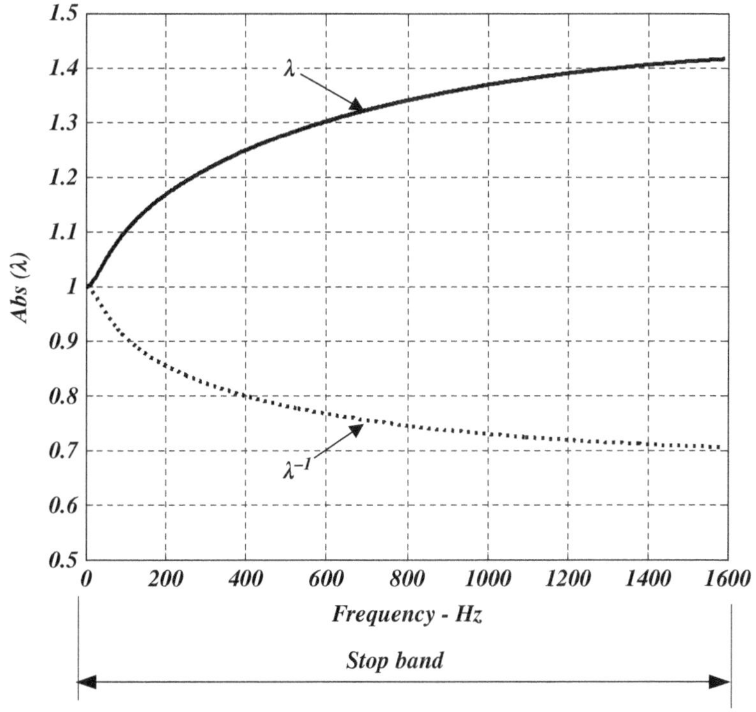

Figure 10.13 Stop band of an elastic-viscoelastic periodic rod.

$$Y_N = [T_N]\cdot Y_0 \text{ or} \begin{Bmatrix} x_N \\ F_N \end{Bmatrix} = \begin{bmatrix} T_{N11} & T_{N12} \\ T_{N21} & T_{N22} \end{bmatrix} \cdot \begin{Bmatrix} x_0 \\ F_0 \end{Bmatrix} \tag{10.35}$$

where subscripts "*o*" and "*N*" denote, respectively, the beginning and the end of the periodic structure.

Equation (10.35) can be expanded as follows:

$$x_N = T_{N11}\cdot x_0 + T_{N12}\cdot F_0, \ \text{and} F_N = T_{N21}\cdot x_0 + T_{N22}\cdot F_0 \tag{10.36}$$

where $\{x_0, x_N\}$ and $\{F_0, F_N\}$ are the amplitudes of the displacements and forces at the beginning and the end of the structure.

Equation (10.36) can be used to extract the natural frequencies of the periodic structure after imposing the boundary conditions as listed in Table 10.1.

For example, for a structure fixed at end "*N*" and excited at its free end "*0*" by a harmonic force (f_0), the displacement x_0 at the free end and the force at the fixed end F_N can be obtained from Eq. (10.36) as follows:

$$x_0 = -T_{N11}^{-1}\, T_{N12} f_0 \ \text{ and } \ F_N - \left(-T_{N21}\, T_{N11}^{-1}\, T_{N12} + T_{N22}\right) f_0 \tag{10.37}$$

Equation (10.37) is used to determine the unknown displacement and force at the two ends of the structure, for the considered set of boundary conditions. The state vectors at any intermediate point can be now determined using the transfer matrices of the individual cells (Eq. (10.13)).

Table 10.1 Equations for computing the natural frequencies of periodic structures at different boundary conditions.

No.	End 0	End *N*	Conditions	Equations
1	Free	Free	$F_0 = 0$ and $F_N = 0$	$T_{N21} = 0$
2	Free	Fixed	$F_0 = 0$ and $x_N = 0$	$T_{N11} = 0$
3	Fixed	Free	$x_0 = 0$ and $F_N = 0$	$T_{N22} = 0$
4	Fixed	Fixed	$x_0 = 0$ and $x_N = 0$	$T_{N12} = 0$

Example 10.6 Determine the natural frequencies for a fixed-free periodic rod that consists of 10 cells, each of which is as shown in Figure 10.11 and has the following geometrical and physical properties.

Material	*E* (GN m^{-2})	ρ (kg m^{-3})	*A* (m^2)	*L* (m)
A	210	7800	0.000625	0.025
B	0.025	1200	0.000625	0.025

Solution

The transfer matrix of the substructures *a* and *b* are given, according to Eq. (10.33), by

$$[T_a] = \begin{bmatrix} \cos(a\omega) & \dfrac{1}{Z_a\omega}\sin(a\omega) \\ -Z_z a\sin(a\omega) & \cos(a\omega) \end{bmatrix} = \begin{bmatrix} t_{11_a} & t_{12_a} \\ t_{21_a} & t_{22_a} \end{bmatrix},$$

and

$$[T_b] = \begin{bmatrix} \cos(b\omega) & \dfrac{1}{Z_b\omega}\sin(b\omega) \\ -Z_b b\sin(b\omega) & \cos(b\omega) \end{bmatrix} = \begin{bmatrix} t_{11_b} & t_{12_b} \\ t_{21_b} & t_{22_b} \end{bmatrix},$$

where a = 4.818E-6 and b = 1.732E-4.

Accordingly, the transfer matrix $[T]$ of the unit cell is given by

$$[T] = [T_b][T_a] = \begin{bmatrix} t_{11} & t_{12} \\ t_{21} & t_{22} \end{bmatrix} = \begin{bmatrix} (t_{11_b}t_{11_a} + t_{12_b}t_{21_a}) & t_{12} \\ t_{21} & (t_{21_b}t_{12_a} + t_{22_b}t_{22_a}) \end{bmatrix}$$

that is,

$$t_{11} = \cos(a\omega)\cos(b\omega) - \frac{Z_a}{Z_b}\sin(a\omega)\sin(b\omega), \tag{10.38}$$

and

$$t_{22} = \cos(a\omega)\cos(b\omega) - \frac{Z_b}{Z_a}\sin(a\omega)\sin(b\omega). \tag{10.39}$$

Then, using Eqs. (10.9) and (10.32), one can easily show that the transfer matrix $[T_N]$ of the entire rod is given by

$$[T_N] = ([T_b][T_a])^N \begin{bmatrix} t_{N11} & t_{N12} \\ t_{N21} & t_{N22} \end{bmatrix}$$

with

$$t_{N22} = \cosh(N\mu) + \frac{\sinh(N\mu)}{2\sinh(\mu)}(t_{22} - t_{11}).$$

The natural frequencies of the periodic rod can be determined, according to Table 10.1, from

$$t_{N22} = \cosh(N\mu) + \frac{\sinh(N\mu)}{2\sinh(\mu)}(t_{22} - t_{11}) = 0 \tag{10.40}$$

with t_{11} and t_{22} are as given by Eqs. (10.38) and (10.39).

Note that Eq. (10.40) is an equation in the frequency ω. Solution of Eq. (10.40) for the natural frequencies of the periodic rod requires a search for the frequencies that satisfy Eq. (10.40) with an additional constraint that these frequencies must lie within the pass bands of the periodic rod. Figure 10.14 shows a flow chart of the search procedure for the natural frequencies.

Figure 10.15a shows the real part α of the propagation parameter μ as a function of the frequency and defines accordingly the pass band of the periodic rod. Within this band, the boundary condition given by Eq. (10.40) is displayed also as a function of the frequency in Figure 10.15b. The natural frequencies that satisfy Eq. (10.40) are marked on Figure 10.15b.

Table 10.2 shows a comparison between the natural frequencies obtained using the theory of periodic structures and the theory of finite elements.

It can be seen that there is a close agreement between the predictions of the theory of periodic structures and the theory of finite elements.

For the sake of completion, Figure 10.16 shows the real part α of the propagation parameter μ and the boundary condition given by Eq. (10.40) for a uniform rod made of material a only. It is clear that the rod exhibits only a pass band as indicated in Example 10.1. The figure displays also the natural frequencies that satisfy Eq. (10.40). Within the considered frequency range, the first two modes are 2594.25 and 7783.2 Hz, as predicted by the theory of periodic structures, compared to 2595.04 and 7801.13 Hz, as predicted by the theory of finite elements.

10.4.2 Mode Shapes

Computation of the mode shapes corresponding to the natural frequencies obtained in Section 10.4.1 can be carried out using Eq. (10.36). For the case of a fixed-free periodic rod, let $x_0 = 0$ and $F_0 = 1$ to give

$$x_N = T_{N12}(\omega_n) \cdot F_0,$$

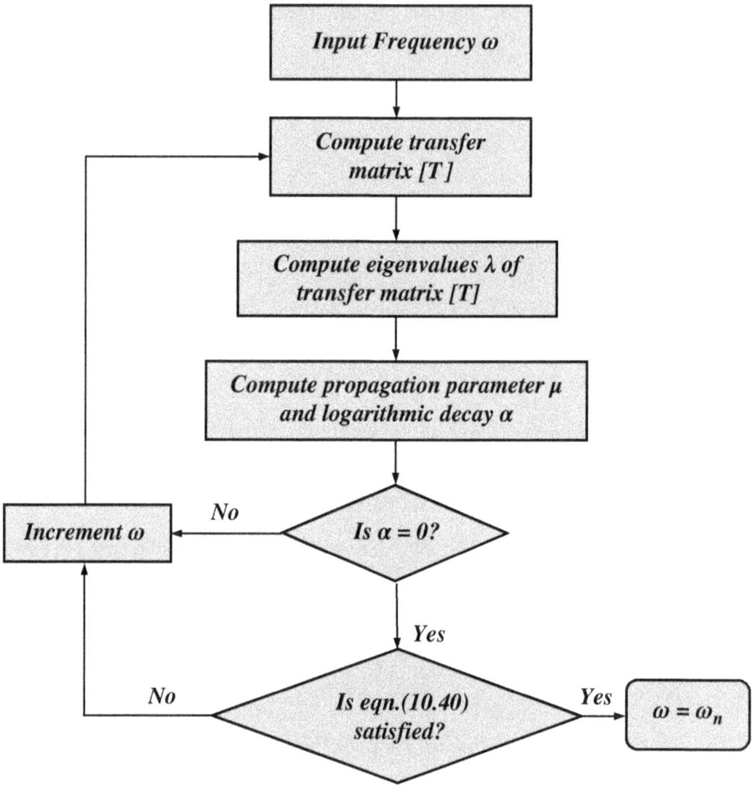

Figure 10.14 Search for the natural frequencies of the periodic rod.

with $T_{N12}(\omega_n)$ is the partitioned transfer matrix T_{N12} evaluated at the natural frequency ω_n.

The state vectors at any intermediate point can now be determined using the transfer matrices of the individual cells (Eq. (10.32)) as follows

$$\begin{Bmatrix} x_L \\ F_L \end{Bmatrix}_k = [T(\omega_n)]^{-1} \begin{Bmatrix} x_L \\ F_L \end{Bmatrix}_{k+1}$$

with $k = N\text{-}1, \ldots, 1$

The vector $\{x_0, x_1, .., x_N\}$is the mode shape corresponding to the natural frequency ω_n.

Example 10.7 Determine the mode shapes corresponding to the first four natural frequencies of the fixed-free periodic rod of Example 10.6. Use the theory of periodic structures presented in Section 10.4.2. Plot these modes shapes and compare the result with the predictions of a finite element model of the rod.

Solution

Figure 10.17 shows the mode shapes of the first four natural frequencies of the periodic rod as obtained by the theory of periodic structures and the theory of finite elements.

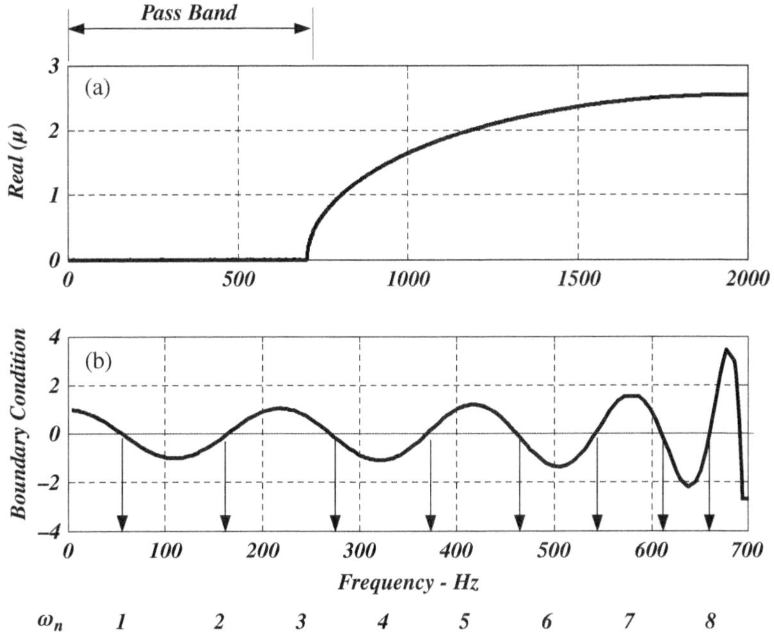

Figure 10.15 The logarithmic decay (a) and the natural frequencies (b) of the periodic rod.

Table 10.2 The natural frequencies of the periodic rod (Hz).

Mode	1	2	3	4	5	6	7	8
Theory of Periodic Structures	55	164	269.5	370	462	542.5	609.6	657
Finite Element Theory	55.04	164.07	269.85	370.04	462.08	543.16	610.39	692.3

Figure 10.17 indicates excellent agreement between the predictions of the mode shapes by the theory of periodic structures and the theory of finite elements. Note that the piecewise linear behavior displayed in the figure is attributed to the fact that the deflections are computed only at the boundaries of the cells.

10.5 Active Periodic Structures

In this section, the theory governing the operation of active periodic structures is introduced and numerical examples are presented to illustrate their tunable filtering and localization characteristics (Baz, 2001). The examples considered include periodic/aperiodic spring-mass systems controlled by piezoelectric actuators.

The presented examples emphasize the unique potential of the active periodic structures in controlling the wave propagation both in the spectral and spatial domains in an attempt to stop/confine the propagation of undesirable disturbances.

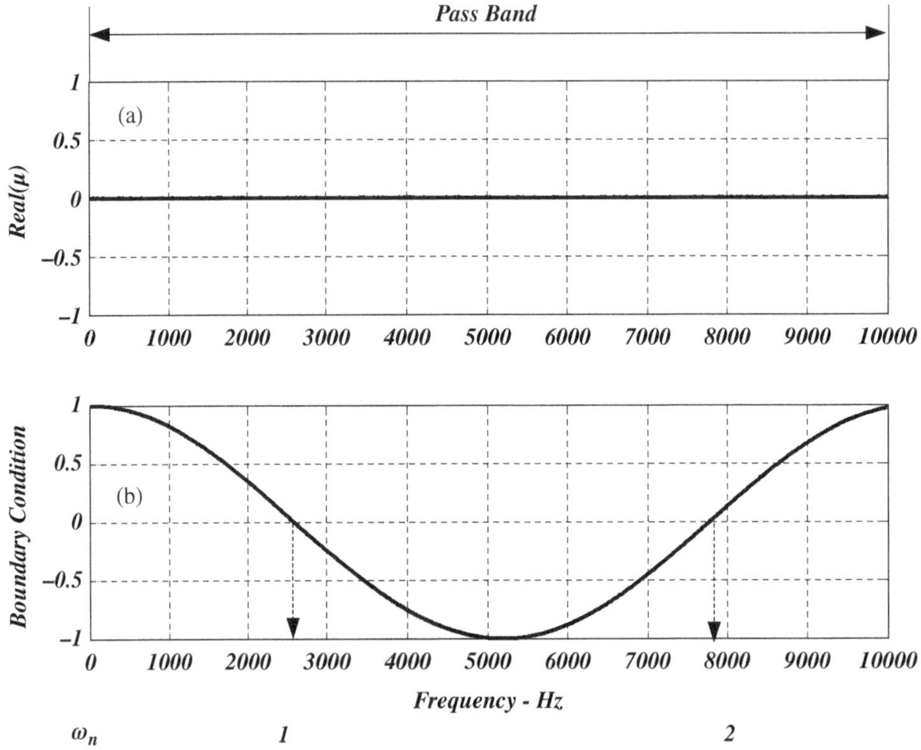

Figure 10.16 The logarithmic decay (a) and the natural frequencies of a uniform rod (b).

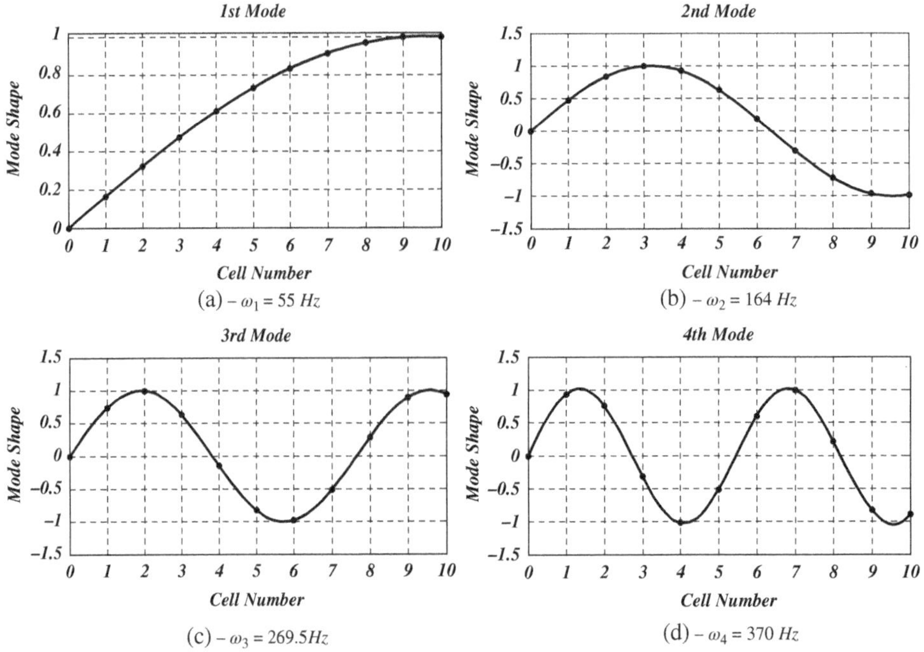

Figure 10.17 Mode shapes of the periodic rod.

10.5.1 Modeling of Active Periodic Structures

Figure 10.18a shows a periodic rod consisting of a stepped base structure which is augmented with active piezoelectric inserts placed periodically along the rod length. The idealized equivalent spring-mass system, shown in Figure 10.18b, is made of identical cells each of which consists of an assembly of passive and active sub-cells. Each sub-cell consists of three masses connected and separated by a passive spring and an active piezoelectric spring as indicated in Figure 10.18c,d.

The dynamics of such one-dimensional periodic systems are determined using the classical transfer matrix method presented in Sections 10.2–10.4. The developed transfer matrices are utilized to determine the pass and stop bands, natural frequencies, mode shapes and frequency response of the system for different control gains and disorder levels generated by randomizing the control gains of the individual cells.

10.5.2 Dynamics of One Cell

Considering the free-body diagram of one cell of the periodic spring-mass system shown in Figure 10.18c; one can describe the dynamics of the passive and active sub-cells as follows:

10.5.2.1 Dynamics of the Passive Sub-Cell

The equation of motion of the passive sub-cell is given by:

$$\begin{bmatrix} m & 0 \\ 0 & m \end{bmatrix} \begin{Bmatrix} \ddot{x}_L \\ \ddot{x}_I \end{Bmatrix} + \begin{bmatrix} k_s & -k_s \\ -k_s & k_s \end{bmatrix} \begin{Bmatrix} x_L \\ x_I \end{Bmatrix} = \begin{Bmatrix} F_L \\ F_I \end{Bmatrix}$$

where m and k_s are half the mass of the step and stiffness of the passive spring. These parameters are given by $m = \frac{1}{2} t_m b L_m$ and $k_s = t_s b E_s / L_s$ where t, b, and L denote the thickness, width, and length with subscripts m and s denote the mass and base structure, respectively. Also, x and F define the deflection and force with subscripts L and I_s denoting the left and interface sides of the passive sub-cell.

For a sinusoidal motion at a frequency ω, this equation of motion reduces to:

$$\begin{bmatrix} k_s - m\omega^2 & -k_s \\ -k_s & k_s - m\omega^2 \end{bmatrix} \begin{Bmatrix} x_L \\ x_I \end{Bmatrix} = \begin{Bmatrix} F_L \\ F_{Is} \end{Bmatrix} \tag{10.41}$$

10.5.2.2 Dynamics of the Active Sub-Cell

The constitutive equations of the active piezoelectric spring are given by (Agnes 1999):

$$\begin{Bmatrix} E_p \\ T_p \end{Bmatrix} = \begin{bmatrix} 1/\varepsilon^s & -h_p \\ -h_p & C^D \end{bmatrix} \begin{Bmatrix} D_p \\ S_p \end{Bmatrix} \tag{10.42}$$

where E_p, D_p, T_p, and S_p are the electrical field intensity, electrical displacement, stress, and strain of the piezo-spring, respectively. Also, ε^s, h_p, and C^D define the electrical permittivity, piezo-coupling constant, and elastic modulus.

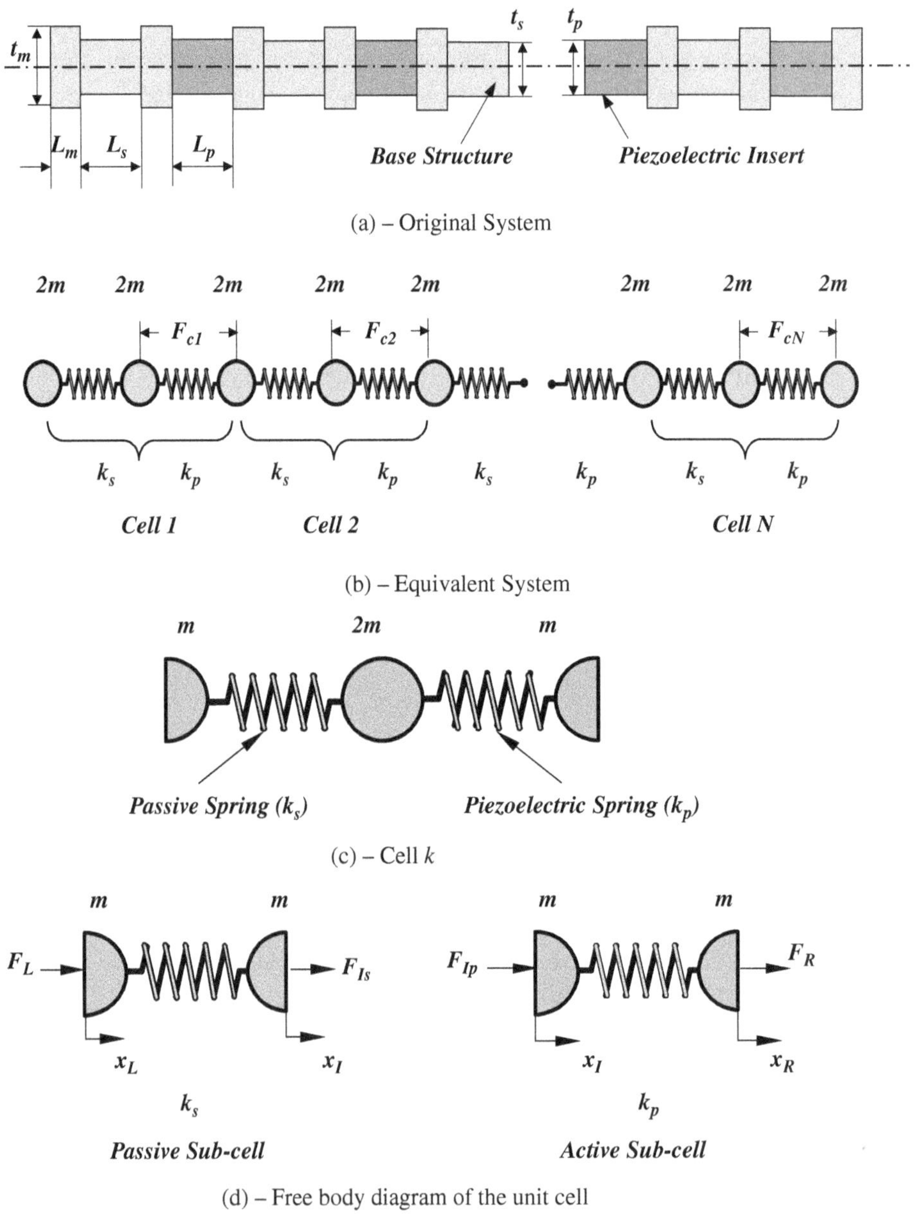

Figure 10.18 An active periodic spring-mass system. (a) Original system, (b) equivalent system; (c) cell *k*; and (d) free-body diagram of the unit cell.

Equation (10.42) can be rewritten in terms of applied voltage V, piezo-force F_p, electrical charge Q_p, and net deflection $(x_R - x_I)$ as follows:

$$\begin{Bmatrix} V/t_p \\ F_p/bt_p \end{Bmatrix} = \begin{bmatrix} 1/\varepsilon^s & -h_p \\ -h_p & C^D \end{bmatrix} \begin{Bmatrix} Q_p/bL_p \\ (x_R - x_I)/L_p \end{Bmatrix} \tag{10.43}$$

Eliminating the charge Q_p from Eq. (10.43) gives:

$$F_p = -h_p \varepsilon^s b V_p + \left[bt_p \left(C^D - h_p^2 \varepsilon^s \right) / L_p \right] (x_R - x_I) \tag{10.44}$$

Let the piezo-voltage V_p be generated according to the following control law:

$$V_p = -K_g (x_R - x_L) \tag{10.45}$$

Then, Eq. (10.44) reduces to:

$$\begin{aligned} F_p &= \left\{ h_p \varepsilon^s b K_g + \left[bt_p \left(C^D - h_p^2 \varepsilon^s \right) / L_p \right] \right\} (x_R - x_I) \\ &= (k_{pc} + k_{ps}) \{-1 \quad 1\} \begin{Bmatrix} x_I \\ x_R \end{Bmatrix} \end{aligned} \tag{10.46}$$

where $k_{pc} = h_p \varepsilon^s b K_g$ and $k_{ps} = \left[bt_p \left(C^D - h_p^2 \varepsilon^s \right) / L_p \right]$ with k_{pc} and k_{ps} denoting the active piezo-stiffness due to the control gain, K_g, and the structural piezo-stiffness.

Equation (10.46) can be used to generate the force vector $\{F_{Ip} \quad F_R\}^T$ acting on the piezo-spring rewritten as:

$$\begin{Bmatrix} F_{Ip} \\ F_R \end{Bmatrix} = (k_{pc} + k_{ps}) \begin{Bmatrix} -1 \\ 1 \end{Bmatrix} \{-1 \quad 1\} \begin{Bmatrix} x_I \\ x_R \end{Bmatrix} = \begin{bmatrix} k_p & -k_p \\ -k_p & k_p \end{bmatrix} \begin{Bmatrix} x_I \\ x_R \end{Bmatrix} \tag{10.47}$$

with $k_p = (k_{pc} + k_{ps})$ is the total stiffness of the piezo-spring.

Hence, the dynamic equations of sinusoidal motions of the active sub-cell is given by:

$$\begin{bmatrix} k_p - m\omega^2 & -k_p \\ -k_p & k_p - m\omega^2 \end{bmatrix} \begin{Bmatrix} x_I \\ x_R \end{Bmatrix} = \begin{Bmatrix} F_{Ip} \\ F_R \end{Bmatrix} \tag{10.48}$$

10.5.2.3 Dynamics of the Entire Cell

The dynamics of the entire cell can be determined by the assembly of the dynamic equations of the passive and active sub-cells that are given by Eqs. (10.41) and (10.48), respectively. This yields the following dynamic equations:

$$\begin{Bmatrix} F_L \\ F_I \\ F_R \end{Bmatrix} = \begin{bmatrix} k_s - m\omega^2 & -k_s & 0 \\ -k_s & k_s + k_p - 2m\omega^2 & -k_p \\ 0 & -k_p & k_p - m\omega^2 \end{bmatrix} \begin{Bmatrix} x_L \\ x_I \\ x_R \end{Bmatrix} \tag{10.49}$$

where $F_I = F_{Is} + F_{Ip}$ denotes the total interface force.

Using Guyan's reduction to eliminate the interface degree of freedom x_I yields the following condensed dynamic equations:

$$\begin{bmatrix} K_{d_{LL}} & K_{d_{LR}} \\ K_{d_{RL}} & K_{d_{RR}} \end{bmatrix} \begin{Bmatrix} x_{L_k} \\ x_{R_k} \end{Bmatrix} = \begin{Bmatrix} F_{L_k} \\ F_{R_k} \end{Bmatrix} \tag{10.50}$$

where

$$K_{d_{LL}} = k_s[-1/(1 + r_k - 2R^2) + (1 - R^2)],$$
$$K_{d_{LR}} = -k_s r_k/(1 + r_k - 2R^2) = K_{d_{RL}},$$
$$K_{d_{RR}} = k_s\left[-r_k^2/(1 + r_k - 2R^2) + (r_k - R^2)\right].$$

with $r_k = k_p/k_s$ = stiffness ratio and $R = \omega/\sqrt{k_s/m}$ = frequency ratio. Note that the stiffness ratio r_k can be written as:

$$r_k = k_{ps}/k_s + k_{pc}/k_s = r_{ks} + r_{kc} \tag{10.51}$$

where r_{ks} and r_{kc} denote the structural stiffness ratio and the control stiffness ratio.

In Eq. (10.51), an additional subscript k is added to denote the dynamics of the kth cell.

10.5.2.4 Dynamics of the Entire Periodic Structure

The dynamics of the entire periodic structure can be determined by rearranging Eq. (11) to take the following form:

$$\begin{Bmatrix} x_{R_k} \\ F_{R_k} \end{Bmatrix} = \begin{bmatrix} -K_{d_{LR}}^{-1} K_{d_{LL}} & K_{d_{LR}}^{-1} \\ -K_{d_{RR}} K_{d_{LR}}^{-1} K_{d_{LL}} + K_{d_{LR}} & K_{d_{RR}} K_{d_{LR}}{}^{-1} \end{bmatrix} \begin{Bmatrix} x_{L_k} \\ F_{L_k} \end{Bmatrix} \tag{10.52}$$

Consider now the compatibility and equilibrium conditions at the interface between the kth and the $k + 1$th cells, which yields the following expressions:

$$x_{R_k} = x_{L_{k+1}} \text{ and } F_{R_k} = -F_{L_{k+1}} \tag{10.53}$$

Substituting these conditions into Eq. (10.52), it reduces to:

$$\begin{Bmatrix} x_{L_{k+1}} \\ F_{L_{k+1}} \end{Bmatrix} = \begin{bmatrix} -K_{d_{LR}}^{-1} K_{d_{LL}} & K_{d_{LR}}^{-1} \\ K_{d_{RR}} K_{d_{LR}}^{-1} K_{d_{LL}} - K_{d_{LR}} & -K_{d_{RR}} K_{d_{LR}}^{-1} \end{bmatrix} \begin{Bmatrix} x_{L_k} \\ F_{L_k} \end{Bmatrix} \tag{10.54}$$

In a more compact form, Eq. (10.54) can be rewritten as:

$$\begin{Bmatrix} x_L \\ F_L \end{Bmatrix}_{k+1} = \begin{bmatrix} t_{11} & t_{12} \\ t_{21} & t_{22} \end{bmatrix} \begin{Bmatrix} x_L \\ F_L \end{Bmatrix}_k \text{ or } Y_{k+1} = [T_k]\, Y_k \tag{10.55}$$

where Y and $[T_k]$ denote the state vector = $\{x_L\ F_L\}^T$ and the transfer matrix of the kth cell. Note that the transfer matrix relates the state vector at the left end of $k + 1$th cell to that at the left end of the kth cell. For exactly periodic structures, $[T_k]$ is the same for all cells and Eq. (10.55) reduces to:

$$Y_{k+1} = [T]\, Y_k \tag{10.56}$$

However, for aperiodic structures, $[T_k]$ varies from cell to cell to appropriately account for the nature of aperiodicity. In this section, the aperiodicity can be easily introduced by varying the control gain K_g. Such flexibility is made possible by providing the periodic structure with active control capabilities.

The wave propagation as well as the pass and stop band characteristics of the active periodic structure can be analyzed by studying the eigenvalues of the transfer matrix $[T]$ using the approaches outlined in Sections 10.2–10.4.

Example 10.8 Determine the filtering characteristics of the passive periodic spring-mass system, shown in Figure 10.18, when $k_s = 1$, $r_{ks} = 1$ and $r_{kc} = 0$ (i.e., $K_g = 0$).

Solution

The system Eqs. (10.50), (10.54–10.56) are used to compute the transfer matrix $[T]$. The eigenvalues λ of the matrix $[T]$ are computed and plotted against the dimensionless frequency parameter, R.

Equation (10.14) is then used to compute the attenuation parameter α and the phase shift parameter β. These propagation parameters are plotted also against R.

The resulting characteristics are shown in Figure 10.19. Figure 10.19a displays the absolute of the eigenvalues ($|\lambda|$) of the transfer matrix $[T]$ as a function of the dimensionless frequency R. The figure indicates that $|\lambda| = 1$ for $R < 1.4$ and $|\lambda| \neq 1$

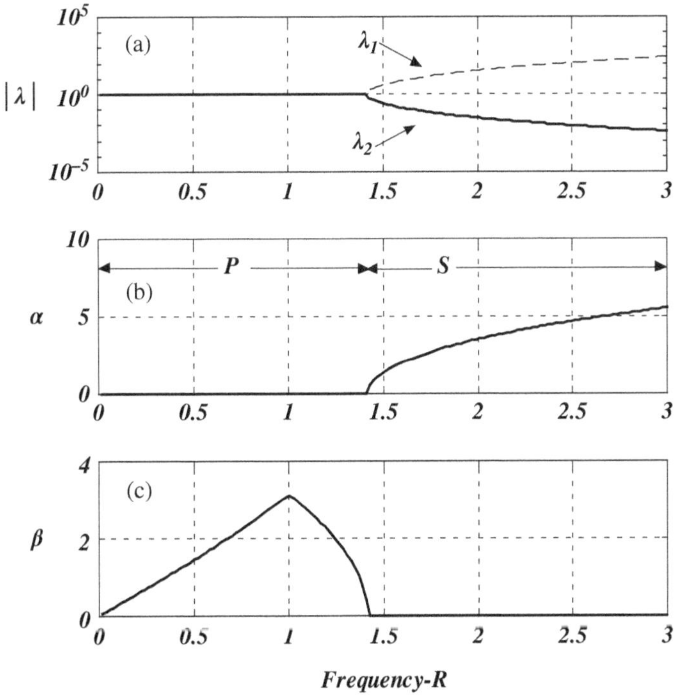

Figure 10.19 Filtering characteristics of the passive periodic spring-mass system with $r_{ks} = 1$ and $r_{kc} = 0$. (P = Pass band, S = Stop band).

for $R > 1.4$. Accordingly, the passive system has a *pass band* for $R < 1.4$ and a *stop band* for $R > 1.4$. In other words, the system acts as a "low pass" filter with a cut-off frequency of 1.4.

For the sake of completion, Figure 10.19b and c are included to display the real and imaginary parts (α and β) of the propagation constant μ. It is very clear that for $R <$ 1.4, the attenuation parameter $\alpha = 0$, that is, the system has no apparent damping. This results in complete propagation of the waves or a "pass band."

But for $R > 1.4$, the attenuation parameter $\alpha \neq 0$, that is, the system has an apparent-damping that results in attenuating the propagation of waves in a manner equivalent to the presence of actual damping. This occurs over a broad frequency range that is called a "stop band." Under these conditions, the system behaves as a "low pass filter."

Example 10.9 Determine the filtering characteristics of the active periodic spring-mass system, shown in Figure 10.18, when $k_s = 1$, $r_{ks} = 1$ and for the following values of r_{kc}:

$$\text{a.} r_{kc} = 2, \text{b.} r_{kc} = 5, \text{c.} r_{kc} = -0.75, \text{ and } r_{kc} = iR$$

Solution

Figure 10.20–d displays the corresponding filtering characteristics of the active periodic structure with the control gain K_g selected such that $r_{kc} = 2$, 5, -0.75, and iR, respectively.

In Figure 10.20a,b, the filtering characteristics exhibit two pass bands and two stop bands. For example, when $r_{kc} = 5$, the pass bands occur for $0 < R < 1$ and $2.4 < R < 2.65$, and the stop bands occur for $1 < R < 2.4$ and $2.65 < R < \infty$.

It is important to note that increasing K_g increases the width of the notch filtering and shifts the highest cut-off frequency to a higher frequency R.

Figure 10.20c shows the corresponding characteristics when K_g is selected such that $r_{kc} = -0.75$ indicating a positive proportional feedback control action. This action results in reducing the width of the notch filtering and shifts the highest cut-off frequency to a lower frequency R.

With such simple adjustment of the control gain K_g, it would be possible to tune the filtering characteristics of the structure in response to external excitations in order to stop the propagation of these excitations along the structure.

It is worth noting here that the passive periodic structures have usually high cut-off frequencies. Hence, the effectiveness of the passive periodic structures is limited only to the blocking of the propagation of high frequency excitations. But, once provided with active control capabilities, these cut-off frequencies can be lowered considerably and the active structures become effective in stopping the propagation of low-frequency excitations as well.

Figure 10.20d displays the filtering characteristics of the active periodic structure when $r_{kc} = iR$ indicating a negative derivative feedback. The figure indicates that structure exhibits stop band characteristics over the entire frequency range.

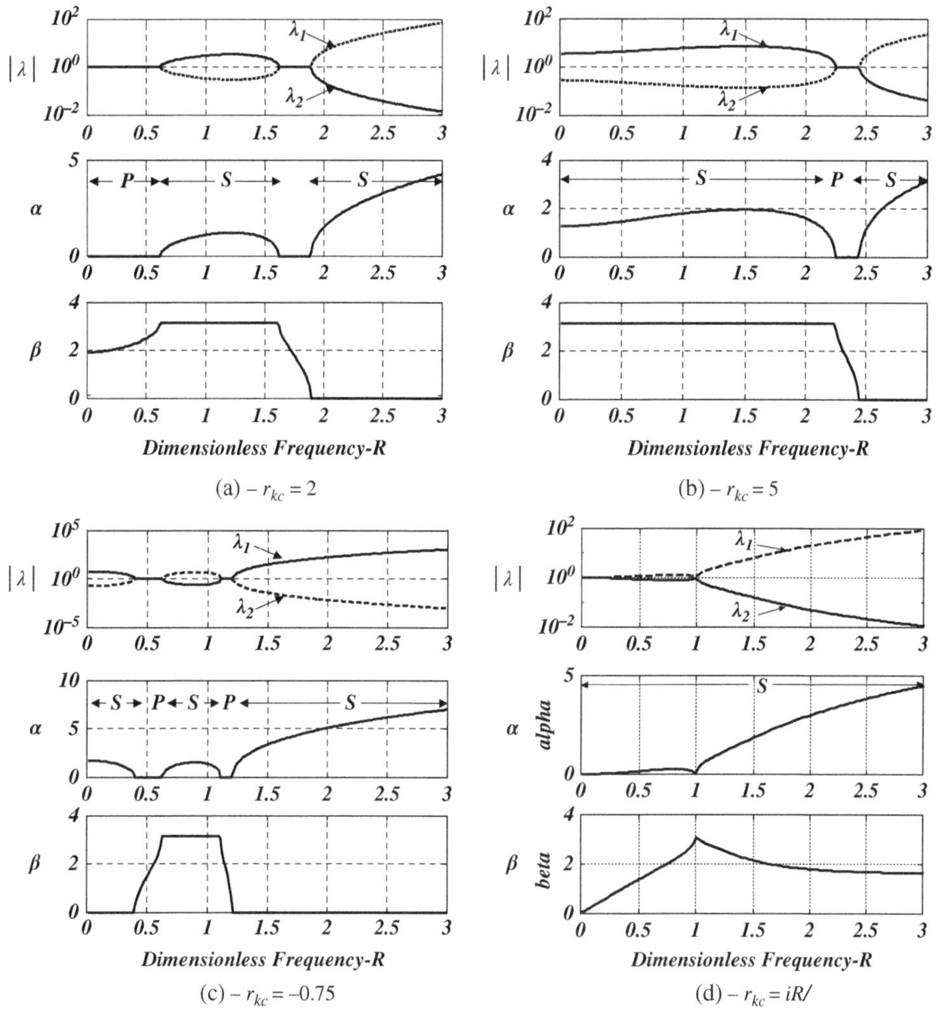

Figure 10.20 Filtering characteristics of the active periodic spring-mass system with $r_{ks} = 1$ and for different values of r_{kc} (P = Pass band, S = Stop band).

10.6 Localization Characteristics of Passive and Active Aperiodic Structures

10.6.1 Overview

In this section, the localization effects arising from the presence of unintentional irregularity (or disorder) in the geometrical parameters of the periodic structure are investigated. Particular emphasis is placed on studying the effect of the disorder levels and excitation frequencies on the response of the structure as well as on the extent of the achieved localization.

In this regard, the work of Mead and Lee (1984), Pierre (1988), Luongo (1992), Langley (1994), Mester and Benaroya (1995), and Mead (1996) present the necessary background and pioneering work in that area.

10.6.2 Localization Factor

The performance of the disordered periodic structure can be quantified using the "localization factor – γ." The origin of the *localization factor* can best be understood by considering Figure 10.6 and rewriting Eq. (10.20) as follows

$$\begin{Bmatrix} w_{L_k}^L \\ w_{L_{k+1}}^r \end{Bmatrix} = \begin{bmatrix} 0 & \Lambda \\ \Lambda & 0 \end{bmatrix} \begin{Bmatrix} w_{L_k}^r \\ w_{L_{k+1}}^L \end{Bmatrix} = \begin{bmatrix} 0 & t^L \\ t^r & 0 \end{bmatrix} \begin{Bmatrix} w_{L_k}^r \\ w_{L_{k+1}}^L \end{Bmatrix} = [S] \begin{Bmatrix} w_{L_k}^r \\ w_{L_{k+1}}^L \end{Bmatrix} \tag{10.57}$$

or,

$$w_{L_k}^L = t^L w_{L_{k+1}}^L \text{ and } w_{L_{k+1}}^r = t^r w_{L_k}^r \tag{10.58}$$

Accordingly, the coefficients t^L and t^r have the physical meaning of the transmission coefficients associated with wave propagation from the left and right as well as the eigenvalue of the transfer matrix [T]. The matrix [S] is known as the "scattering matrix."

For a periodic structure made of N cells, Eqs. (10.57) and (10.58) become

$$\begin{Bmatrix} w_{L_1}^L \\ w_{L_N}^r \end{Bmatrix} = \begin{bmatrix} 0 & t^{L^N} \\ t^{r^N} & 0 \end{bmatrix} \begin{Bmatrix} w_{L_1}^r \\ w_{L_N}^L \end{Bmatrix} \tag{10.59}$$

and

$$w_{L_1}^L = t^{L^N} w_{L_N}^L = \lambda^N w_{L_N}^L \text{ and } w_{L_N}^r = t^{r^N} w_{L_1}^r = \lambda^N w_{L_1}^r \tag{10.60}$$

Equation (10.60) can be rewritten as

$$\left(w_{L_N}^r / w_{L_1}^r\right) = \lambda^N,$$

or

$$ln\left(\left|w_{L_N}^r / w_{L_1}^r\right|\right) = N \cdot ln(|\lambda|) = N \cdot \alpha \cdot N$$

which reduces to

$$\alpha = \frac{1}{N}\sum^{N} \ln(|\lambda|) = \text{logarithmic decay} \tag{10.61}$$

Equation (10.61) indicates that the "average logarithmic decay" in a periodic structure with N cells is the same as the logarithmic decay in the individual cell.

Note that the diagonal terms in the scattering matrix are zero for the case of a purely periodic structure. In the case of aperiodic structures, these terms become non-zero, because the transfer matrices of the individual cells are no longer the same due to the irregularities. Hence, a matrix Φ that diagonalizes the nominal transfer matrix will not necessarily diagonalize the disordered transfer matrix and Eq. (10.59) takes the following form

$$\begin{Bmatrix} w_{L_1}^L \\ w_{L_N}^r \end{Bmatrix} = \begin{bmatrix} r_N^r & t_N^L \\ t_N^r & r_N^L \end{bmatrix} \begin{Bmatrix} w_{L_1}^r \\ w_{L_N}^L \end{Bmatrix} \tag{10.62}$$

with the coefficients r_N^L and r_N^r physically denoting the reflection coefficients associated with the left and right propagating waves.

Due to the reflections at the interfaces of dissimilar cells, wave propagation in aperiodic chain is attenuated even if the excitation frequency is within the pass band and the system remains undamped. This is in contrast to periodic structures where the scattering matrix includes transmission matrices only.

The phenomenon of attenuation of wave propagation in aperiodic structures was reported first by Anderson (1958) in atomic lattices and then by Hodges (1982) in engineering dynamical systems.

The performance of the actively disordered periodic structure is quantified, for example, by using the localization factor – γ that is defined according to Cai and Lin (1991) as follows:

$$\gamma = \frac{1}{N} \cdot \sum_{k=1}^{N} \log(|\lambda_k|) \tag{10.63}$$

where λ_k is the eigenvalue of the kth cell. Also, according to Eq. (10.63), the localization factor γ defines the average exponential decay for an assembly of N cells. For a perfect periodic structure $|\lambda_k|$ is constant for every cell. In a pass band, $|\lambda_k|$ is equal to one and the localization factor is correspondingly equal to zero.

Equation (10.36) can be used to determine the unknown displacement and force at the two ends of the structure, for any considered set of boundary conditions. The state vectors at any intermediate point can be now determined using the transfer matrices of the individual cells (Eq. (10.33)).

The irregularity is introduced here by adding a disorder term to one or all of the geometrical parameters of the structure as, for example, the length of one of the sub-cells such that:

$$L_{ka_t} = L_{ka_o} + d_k \tag{10.64}$$

where L_{ka_t} and L_{ka_o} denote the total and the nominal length of the kth sub-cell. In Eq. (10.64), d_k defines a random disorder that is normally distributed with zero mean and variance of σ.

Alternatively, the irregularity can be introduced by adding a disorder term to one or all of the control gain K_g of the individual sub-cells such that:

$$K_{gk} = K_{gk_o} + \delta K_{gk} \tag{10.65}$$

where K_{gk} and K_{gk_o} denote the total and the nominal control gain of the kth sub-cell. In Eq. (10.65), δK_{gk} defines a random disorder which is normally distributed with zero mean and variance of σ.

Example 10.10 Determine the localization factor γ for the idealized periodic rod model shown in Figure 10.21 when various disorder levels are introduced in the control gains of the sub-cells. The disorder is assumed to have a Gaussian distribution with a mean control gain $r_{kc_0} = 2$ and variance $\sigma = 0.25$, 1, and 3.

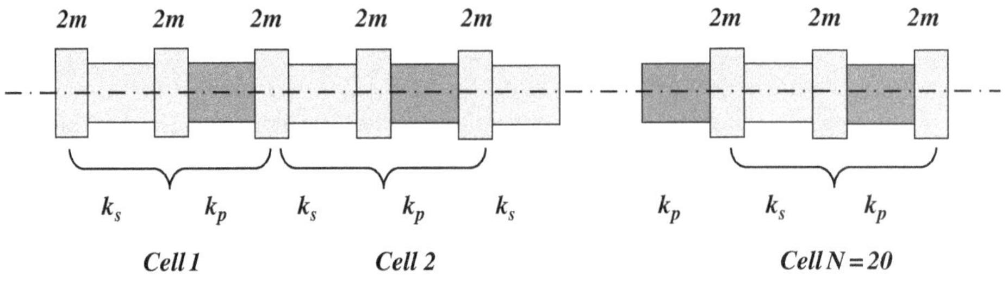

Figure 10.21 An active periodic rod.

Solution

The localization factor – γ is calculated for the aperiodic structure using Eq. (10.63) as follows:

$$\gamma = \frac{1}{N} \cdot \sum_{k=1}^{N} \log(|\lambda_k|)$$

where λ_k is the eigenvalue of the kth cell.

Figure 10.22 shows that the localization factor becomes different from zero within the confined boundaries of the stop bands.

Note that the localization factors concentrate at the boundaries of the stop bands, indicating that the disorder extends the range of frequency where longitudinal waves are attenuated. The width of the attenuating bands and the value of the localization factors increase with the level of disorder. In particular, Figure 10.22c shows that for high levels of irregularity, with $\sigma = 3$, the first and the second pass bands have been completely eliminated.

Example 10.11 Determine the response of the fixed-free idealized periodic rod model shown in Figure 10.23 when subjected to a unit sinusoidal force acting longitudinally at its free end. Investigate the effect of introducing various disorder levels, in the control gains of the sub-cells, on the response of the rod as well as extent of the achieved localization. Assume that the nominal control gain is $r_{kc_0} = 2$, $k_s = 1$, and $\omega_n = 1$. Also, assume that the sinusoidal excitation occurs at excitation frequencies of $R = 0.9$, 1.75, and 1.95. Note that these excitation frequencies are selected to be near the boundaries of the stop bands as shown in Figure 10.20a. At these frequencies, waves can propagate along the structure when no disorder is considered.

Solution

From Eq. (10.36) with $x_O = 0$,

$$x_N = T_{N12} \cdot F_O, \text{ and } F_N = T_{N22} \cdot F_O$$

Eliminating F_O between these equations gives

$$x_N = T_{N12} \cdot T_{N22}^{-1} F_N$$

Figure 10.22 Effect of disorder level (sigma = σ) on the localization factor (P = Pass band, S = Stop band). (a) σ = 0.25, (b) σ = 1 and (c) σ = 3.

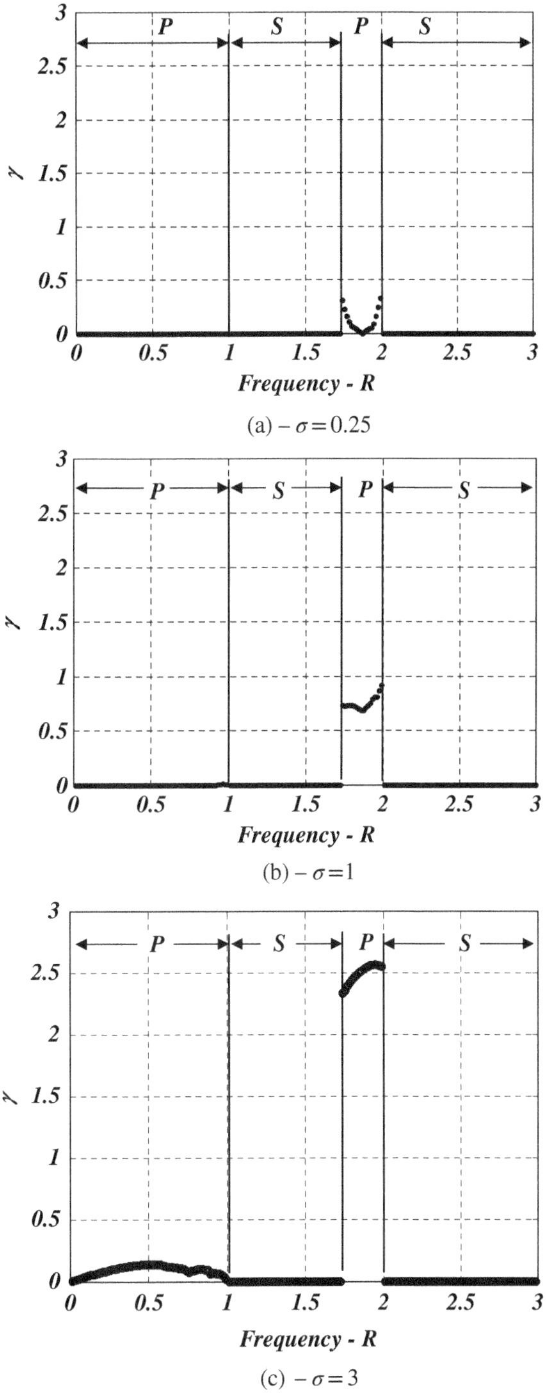

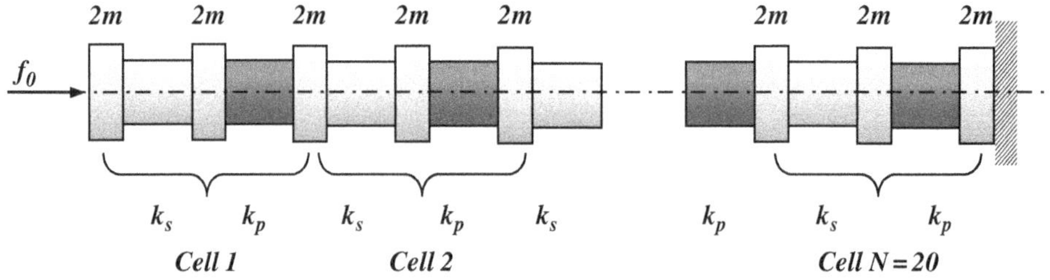

Figure 10.23 An active free-fixed periodic rod with sinusoidal end loading.

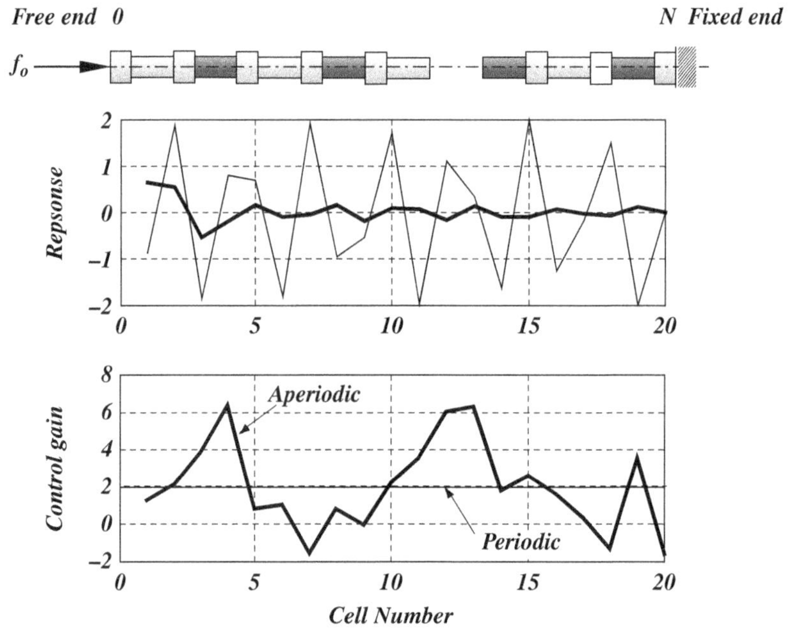

Figure 10.24 Vibration localization at frequency R = 0.9 with disorder level σ=2.5 (periodic ——, aperiodic ▬▬).

This equation gives the displacement x_N at the free end of the rod in terms of the magnitude of the input excitation force F_N. The state vectors at any intermediate point can be now determined using the transfer matrices of the individual cells (Eq. (10.33)).

Figure 10.24 emphasizes that the vibration energy is spatially spread over the entire periodic structure and is confined to the zone near the excitation end for the aperiodic structure. This localization of the vibration energy is achieved easily by randomly distributing the control gain, r_{kc}. Similar performance characteristics can, however, be obtained in passive periodic structures by randomly varying the geometric parameters of the different cells. But, this task is rather cumbersome to be practical.

The phenomenon of vibration localization is demonstrated also in Figure 10.25a,b for the excitation frequencies of R = 1.75 and 1.95. However, in these cases only small

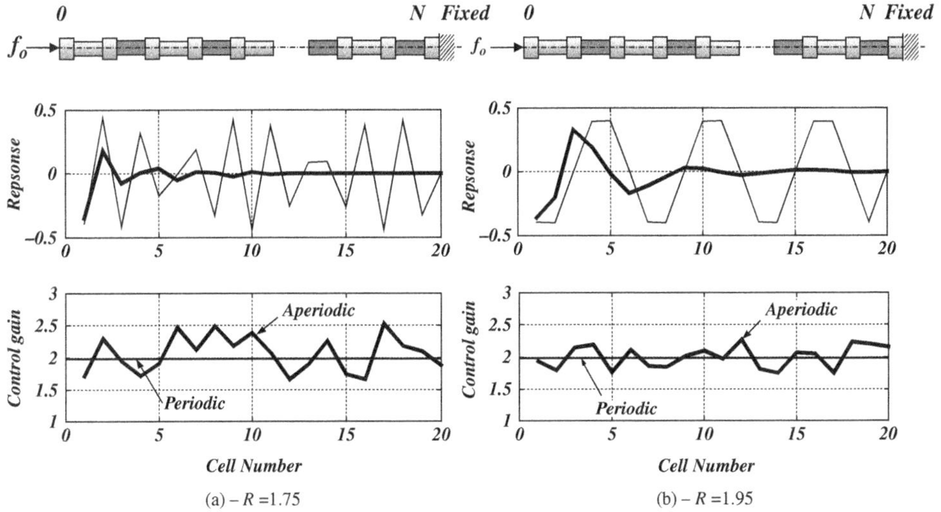

Figure 10.25 Vibration localization with disorder level σ = 0.25 (periodic ——, aperiodic ▬▬). (a) R = 1.75 and (b) R = 1.95.

disorder levels (σ= 0.25) were necessary to stop the propagation of the waves along the structure. This again emphasizes the importance of providing structures with active control capabilities in order to tailor the disorder level to match the nature of the external excitation.

Example 10.12 Determine the time and frequency response of the fixed-free idealized periodic rod model shown in Figure 10.26 when subjected to a unit impulse force acting longitudinally at its free end. Using Morlet wavelet analysis as described in Appendix 10.A, illustrate the stop band and pass band characteristics of the periodic rod for different values of the control gain such that r_{kc} = 0 and 5. Assume that k_s = 1 and ω_n = 1. Comment on the results.

Solution

For general loading, Eq. (10.49) can be rewritten as:

$$\begin{bmatrix} m & 0 & 0 \\ 0 & 2m & 0 \\ 0 & 0 & m \end{bmatrix} \begin{Bmatrix} \ddot{x}_L \\ \ddot{x}_I \\ \ddot{x}_R \end{Bmatrix} + \begin{bmatrix} k_s & -k_s & 0 \\ -k_s & k_s + k_p & -k_p \\ 0 & -k_p & k_p \end{bmatrix} \begin{Bmatrix} x_L \\ x_I \\ x_R \end{Bmatrix} = \begin{Bmatrix} F_L \\ F_I \\ F_R \end{Bmatrix} \tag{a}$$

Dividing by k_s gives:

$$\frac{1}{\omega_n^2} \begin{bmatrix} 1 & 0 & 0 \\ 0 & 2 & 0 \\ 0 & 0 & 1 \end{bmatrix} \begin{Bmatrix} \ddot{x}_L \\ \ddot{x}_I \\ \ddot{x}_R \end{Bmatrix} + k_s \begin{bmatrix} 1 & -1 & 0 \\ -1 & 1 + r_k & -r_k \\ 0 & -r_k & r_k \end{bmatrix} \begin{Bmatrix} x_L \\ x_I \\ x_R \end{Bmatrix} = \frac{1}{k_s} \begin{Bmatrix} F_L \\ F_I \\ F_R \end{Bmatrix} \tag{b}$$

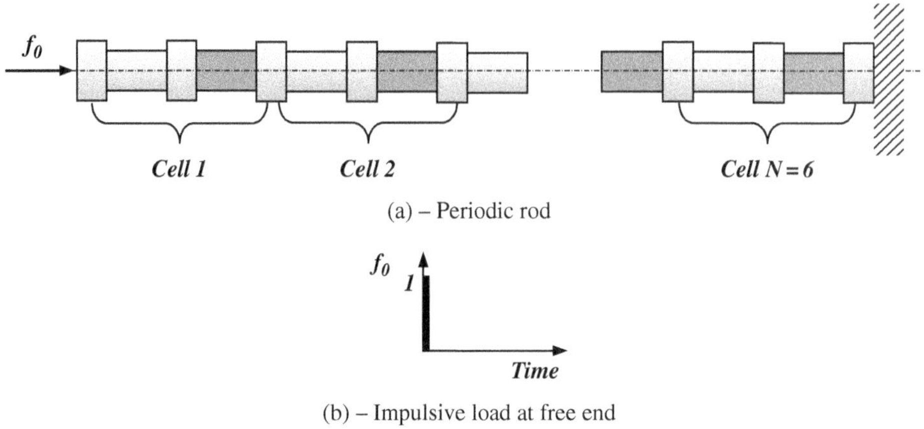

Figure 10.26 An active free-fixed periodic rod subject to an impulsive loading. (a) Periodic rod and (b) impulsive load at the free end.

or

$$[M]\begin{Bmatrix} \ddot{x}_L \\ \ddot{x}_I \\ \ddot{x}_R \end{Bmatrix} + [K]\begin{Bmatrix} x_L \\ x_I \\ x_R \end{Bmatrix} = \frac{1}{k_s}\begin{Bmatrix} F_L \\ F_I \\ F_R \end{Bmatrix}$$

where

$$[M] = \frac{1}{\omega_n^2}\begin{bmatrix} 1 & 0 & 0 \\ 0 & 2 & 0 \\ 0 & 0 & 1 \end{bmatrix} \text{ and } [K] = k_s\begin{bmatrix} 1 & -1 & 0 \\ -1 & 1+r_k & -r_k \\ 0 & -r_k & r_k \end{bmatrix}$$

Now, the theory of finite elements can be utilized to assemble the overall mass and stiffness matrices $[M_o]$ and $[K_o]$ to formulate the overall equations of motion of the periodic rod. After imposing the boundary conditions, these equations can be written as follows:

$$[M_o]\{\ddot{X}\} + [K_o]\{X\} = \frac{1}{k_s}\{F\} \tag{c}$$

where

$$\{X\} = \{x_{L_1}\ x_{L_2} \ldots.\ x_{LN}\}^T \text{ and } \{F\} = \{F_{L_1}\ F_{L_2} \ldots.\ F_{LN}\}^T.$$

Figures 10.27 and 10.28 show the characteristics of the passive and active periodic rods, respectively.

In Figure 10.27a, the time response of the free end of the rod is shown while Figure 10.27b displays the corresponding Fast Fourier Transform (FFT) indicating a sharp drop in the response for frequencies R higher than 1.4. This indicates the presence of a stop band starting at $R > 1.4$. This result matches the predictions of the eigenvalues of the transfer matrix that are displayed in Figure 10.27c.

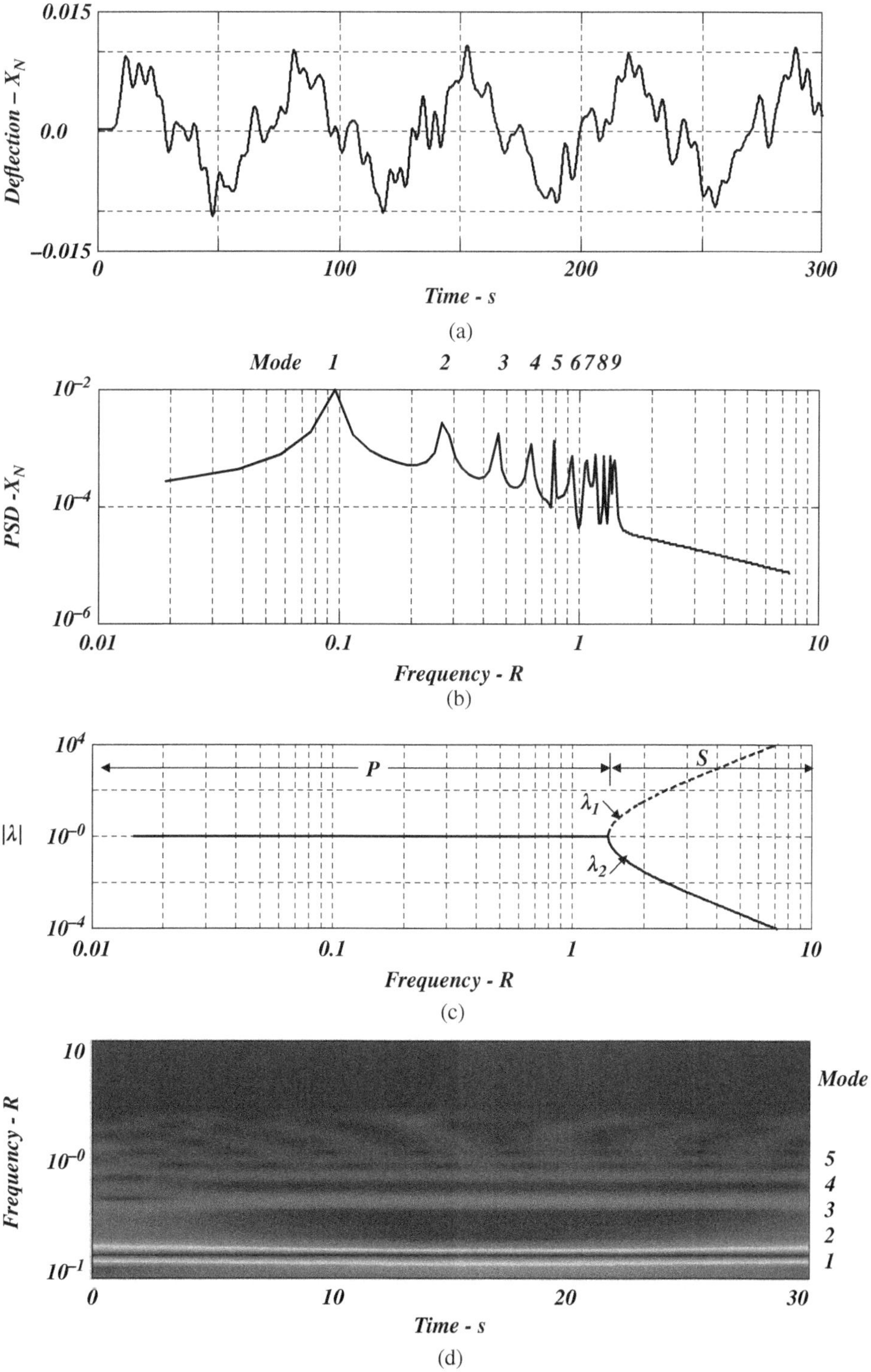

Figure 10.27 Characteristics of passive periodic rod ($r_{kc} = 0$). (a) Deflection vs time, (b) PSD vs frequency, (c) wavelength vs frequency, and (d) frequency vs time.

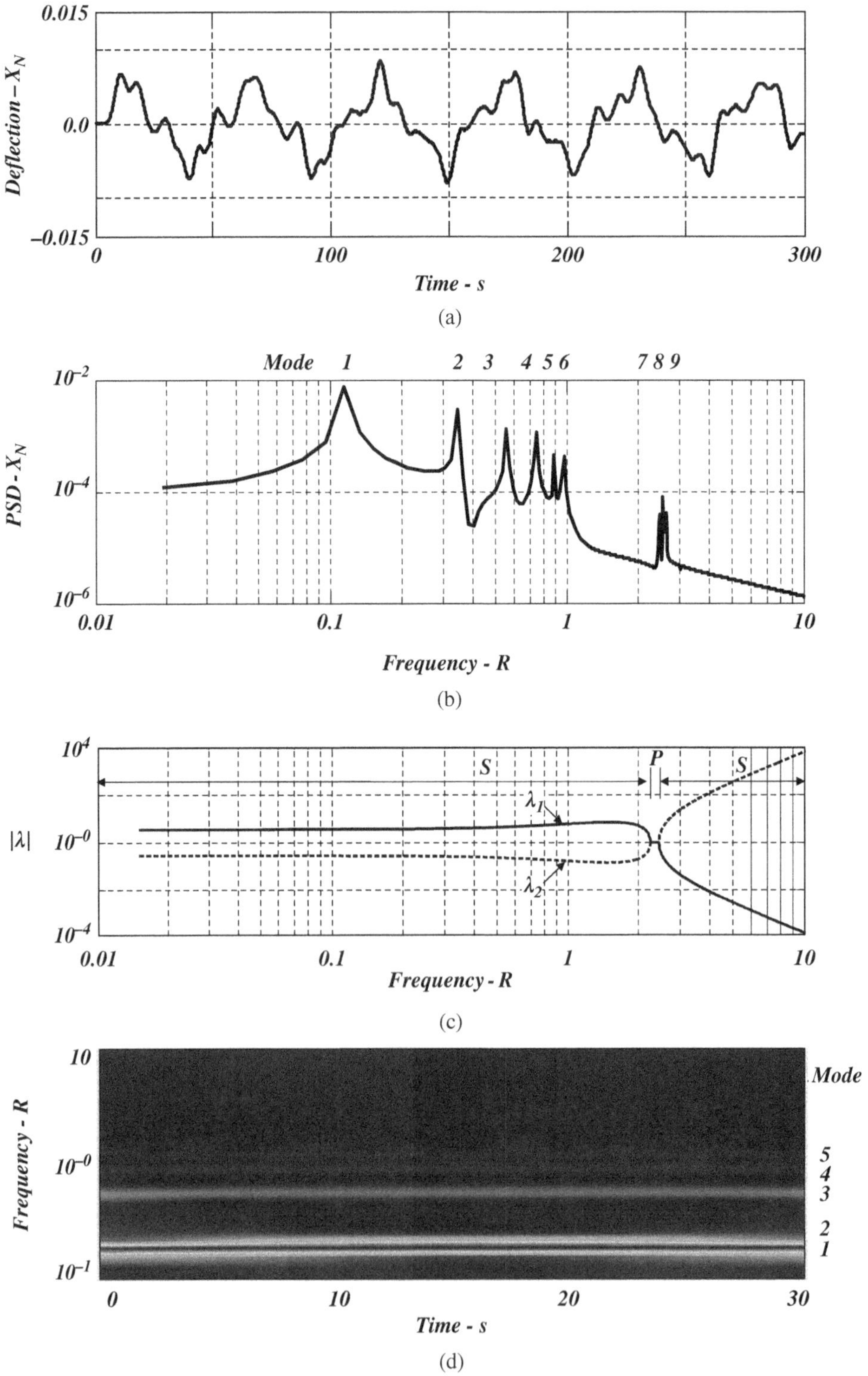

Figure 10.28 Characteristics of active periodic rod ($r_{kc} = 5$).(a) Deflection vs time, (b) PSD vs frequency, (c) wavelength vs frequency, and (d) frequency vs time.

Figure 10.27d shows a spectrogram obtained by applying the continuous Morlet wavelet transform (WT), described in Appendix 10.A, in order to display the frequency content of the response of the free end of the rod as a function of time. It is evident that the spectrogram identifies clearly the modes of vibration of the rod and these modes coincide with those identified by the FFT as shown in Figure 10.27c.

Introducing the control action, with $r_{kc} = 5$, has resulted in the characteristics shown in Figure 10.28. Figure 10.28a,b indicate a decrease in the amplitude of response of the rod both in the time and frequency domains. Such a decrease is evident for frequencies $R < 1$ due to the apparent damping resulting from operation within the stop band predicted by the eigenvalues of the transfer matrix in Figure 10.28c. Such apparent damping is quantified by the "attenuation parameter" α shown in Figure 10.20b. For values of $1 < R < 2$, the "attenuation parameter" α reaches a peak plateau resulting in a significant attenuation of the wave propagation.

More importantly, Figure 10.28b shows also that high frequency modes appear around $R = 2.4$ where the active periodic rod has a pass band as can be identified clearly by the eigenvalues of the transfer matrix displayed in Figure 10.28c.

The spectrogram of Figure 10.28d identifies also the modes of vibration of the rod in a manner that conforms to those identified by the FFT as shown in Figure 10.28c.

10.7 Periodic Rod with Periodic Shunted Piezoelectric Patches

Shunted piezoelectric patches are periodically placed along rods, as shown in Figure 10.29, to control the longitudinal wave propagation in these rods. The resulting periodic structure is capable of filtering the propagation of waves over specified stop bands. The location and width of the stop bands can be tuned, using the shunting capabilities of the piezoelectric materials, in response to external excitations and to compensate for any structural uncertainty.

10.7.1 Transfer Matrix of a Plain Rod Element

The transfer matrix of the rod element of the ith cell is given by Eq. (10.25) as follows:

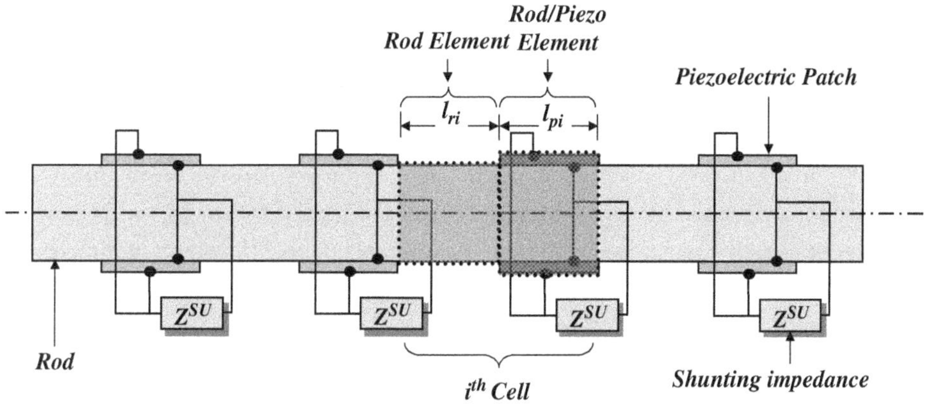

Figure 10.29 A rod with periodic shunted piezoelectric patches.

$$[T_{ri}] = \begin{bmatrix} \cos(k_{ri}l_{ri}) & \frac{1}{Z_{ri}\omega}\sin(k_{ri}l_{ri}) \\ -Z_{ri}\omega\sin(k_{ri}l_{ri}) & \cos(k_{ri}l_{ri}) \end{bmatrix} \tag{10.66}$$

where $Z_{ri} = A_{ri}\sqrt{E_{ri}\rho_{ri}}$= impedance of the rod element.

It is evident from Eq. (10.66) that changing the material properties (E_{ri},ρ_{ri}) or the geometry (A_{ri}) along the rod length generates a discontinuity in the rod impedance. Controlling the impedance of periodic structures can modify the wave propagation characteristics along the rod.

10.7.2 Transfer Matrix of a Rod/Piezo-Patch Element

The transfer matrix of the rod/piezo-patch element of the ith cell can be obtained in a manner similar to the plain rod by using appropriate expressions for the wave number k_{ci} and impedance z_{ci} of the composite element such that Eq. (10.66) is modified to assume the following form:

$$[T_{ci}] = \begin{bmatrix} \cos(k_{ci}l_{ci}) & \frac{1}{Z_{ci}\omega}\sin(k_{ci}l_{ci}) \\ -Z_{ci}\omega\sin(k_{ci}l_{ci}) & \cos(k_{ci}l_{ci}) \end{bmatrix} \tag{10.67}$$

where $Z_{ci} = A_{ci}\sqrt{E_{ci}\rho_{ci}}$= impedance of the rod/piezo-patch element.

The wave number k_{ci} and impedance Z_{ci} of the composite element are determined using the "rule of mixtures" of composites such that:

$$Z_{ci} = \sqrt{\left(\rho_{ri}A_{ri} + \rho_{pi}A_{pi}\right)\left(E_{ri}A_{ri} + E_{pi}^{SU}A_{pi}\right)} \tag{10.68}$$

and

$$k_{ci} = \omega\sqrt{\left(\rho_{ri}A_{ri} + \rho_{pi}A_{pi}\right)/\left(E_{ri}A_{ri} + E_{pi}^{SU}A_{pi}\right)} \tag{10.69}$$

where A_{ri} and A_{pi} denote, respectively, the cross sections of the rod and of the piezo-patch, ρ_{ri} and ρ_{pi} are the densities of the rod and of the piezo, and E_{ri} is the Young's modulus of the rod material. Also, E_{pi}^{SU} denotes the complex modulus of the shunted piezo-patch. It is determined by combining Eqs. (9.7) and (9.14) to yield:

$$E_{pi}^{SU} = \bar{E}_{pi}^{SU}(1 + \eta j) = E_{pi}^{D}\left[1 - \frac{k_{ii}^2}{1 + sC_{pi}^{S}Z_i^{SU}}\right] \tag{10.70}$$

where η = loss factor, $\bar{E}_{pi}^{SU}$= storage modulus, E_{pi}^{D} = Young's modulus of the unshunted piezo-patch at open circuit, k_{ii}^2 = electromechanical coupling factor, $C_{pi}^{S} = C_{pi}^{T}\left(1 - k_{ii}^2\right)$ = capacitance at zero strain, s = Laplace complex number, and Z_i^{SU} = shunted impedance. For inductive-resistive shunting Z_i^{SU} is given by:

$$Z_i^{SU} = Ls + R \tag{10.71}$$

The values of shunting components L and R are determined to ensure optimal damping characteristics as outlined in Table 10.3 and as described in Sections 9.2.2.1 and 9.2.2.2.

Table 10.3 Optimal shunting parameters.

Shunting Circuit	Resistive	Resistive/inductive
Tuning Frequency	$\omega_{tuning} = \sqrt{1-k_{ii}^2}/\left(R \cdot C_{pi}^S\right)$	$\omega_{tuning} = 1/\sqrt{L \cdot C_{pi}^S}$

10.7.3 Transfer Matrix of a Unit Cell

The transfer matrix of a unit cell of a rod that is treated with periodically distributed piezo-patches can be determined by combining Eqs. (10.66) and (10.67) to give:

$$[T_i] = [T_{ci}][T_{ri}] \tag{10.72}$$

Extracting the eigenvalues of the transfer matrix $[T_i]$ will provide a means for identifying the wave propagation characteristics along the periodic rod with piezo-patches.

Example 10.13 Consider a 57.9 cm long cantilevered rod is considered. The rod is made of molded plastic and has a rectangular cross section that is 7.24 cm wide and 1.2 cm thick. The rod is provided with four pairs of piezoelectric patches placed symmetrically and periodically along its length as shown in Figure 10.29. The piezoelectric patches are 7.24 cm long, 7.24 cm wide, and 0.0267 cm thick. Their electromechanical coupling factor is k_{31} = 0.44 and their open circuit capacitance is C^S_{pi} = 390 nf. The material properties and the geometry of the rod and the piezoelectric patches are summarized, respectively, in Tables 10.4 and 10.5.

Determine the wave propagation characteristics of the rod with and without shunting the piezo-patches using inductive-resistive shunting circuits.

Solution

Equations (10.66–10.72) are used to compute the transfer matrix of the unit cell of the periodic rod with piezo-patches for different shunting strategies. The corresponding eigenvalues of the transfer matrix are determined and the attenuation and phase shift parameters (α and β) are extracted using Eq. (10.14).

Figure 10.30 displays the effect of the shunting strategies on the attenuation parameterα. In Figure 10.30a, the effect of the shunting resistance R on α is shown when the shunting inductance L = 0. It can be seen that increasing R beyond 25 Ω results in insignificant improvement in the attenuation parameter α. The effect of the shunting inductance L on the attenuation parameter α is displayed in Figure 10.30b when the shunting resistance R is maintained at 25Ω. A peak attenuation parameter value is obtained when L = 0.4 mH.

Figure 10.31 displays the attenuation and phase shift parameters (α and β) when the piezo-patches are shunted using R = 25 Ω and L = 0 H along with the corresponding parameters of the unshunted piezo-patches for comparison.

Table 10.4 Geometrical and physical properties of a rod.

Young's modulus (N/m^2)	Density (kg/m^3)	Length (cm)	Width (cm)	Thickness (cm)
3.6E9	1726	57.9	7.24	1.2

Table 10.5 Geometrical and physical properties of piezoelectric patches.

Young's modulus (N/m^2)	Density (kg/m^3)	Length (cm)	Width (cm)	Thickness (cm)	Coupling factor k_{31}	Capacitance C^S_{pi} (nF)
6.2E10	7800	7.24	7.24	0.0267	0.44	390

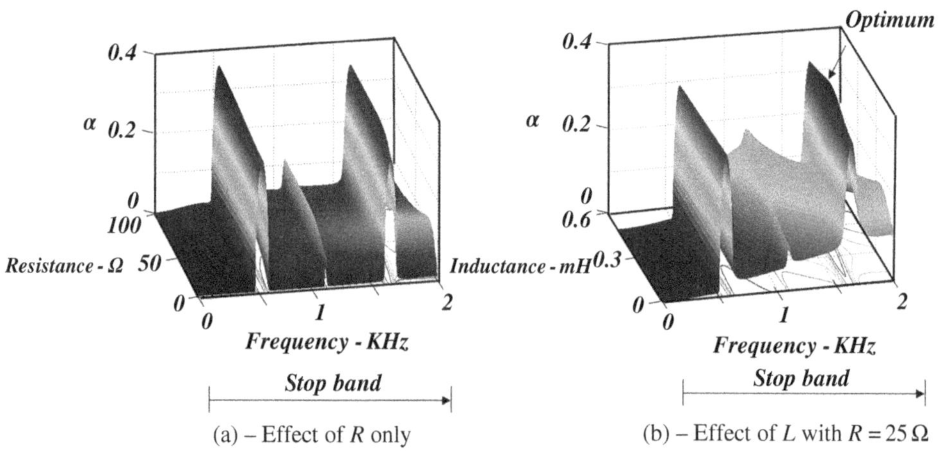

Figure 10.30 A rod with periodic shunted piezoelectric patches. (a) Effect of R only and (b) effect of L with R = 25Ω.

It can be seen that the periodic rod with unshunted piezo-patches has three stop bands as marked in Figure 10.31. However, providing the piezo-patches with resistive shunting has resulted in stopping wave propagation over the considered frequency band as the attenuation parameter $\alpha > 0$ over the entire range.

Using an inductive and resistive shunting circuit with R = 25 Ω and L = 0.4 mH has improved the attenuation parameter over the considered frequency band as shown in Figure 10.31.

10.8 Two-Dimensional Active Periodic Structure

10.8.1 Dynamics of Unit Cell

Figures 10.33 and 10.34 show an idealized 2D active periodic structure. The idealized structure is a spring-mass system consisting of a two-dimensional array of masses that are coupled by a set of active and passive springs. The active springs are placed in a periodic manner along the x-axis. Such a placement strategy is considered to illustrate the characteristics of the 2D periodic structure. Similar characteristics can be obtained with other periodic placements of the active springs.

In the considered spring-mass system, a mass located at the ith row and jth column, has a single degree of freedom w_{ij} in the out-of-the x–y plane direction. The spectral equation of motion of that mass can be written, at a frequency ω, as

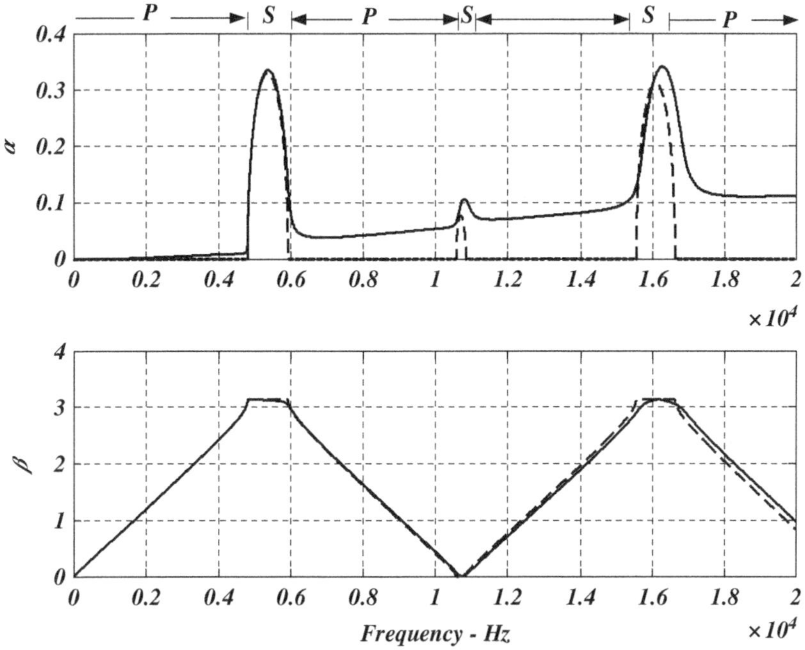

Figure 10.31 Propagation constants for resistive shunting (····· $R = 0\Omega$, —— $R = 25\Omega$).

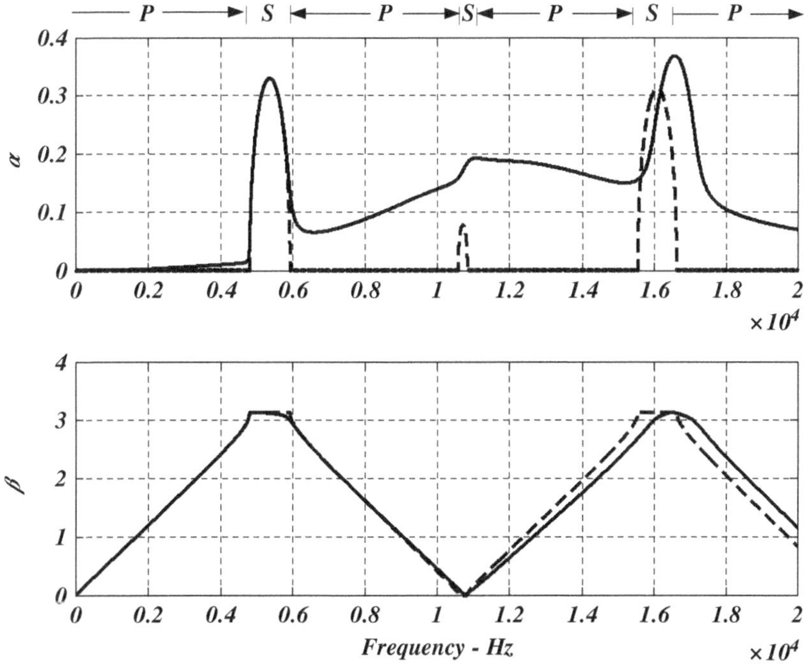

Figure 10.32 Propagation constants for resistive/inductive shunting····· ($R = 0\Omega$ and $L = 0\ mH$, —— $R = 25\Omega$ $L = 0.4\ mH$).

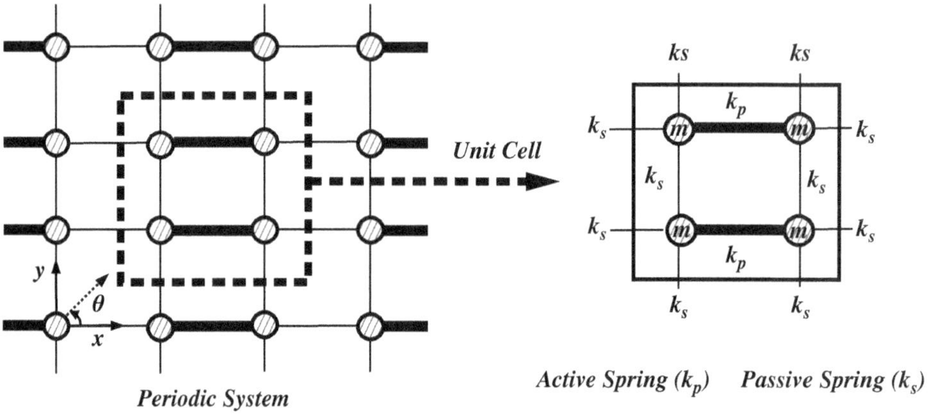

Figure 10.33 A 2D periodic spring-mass system.

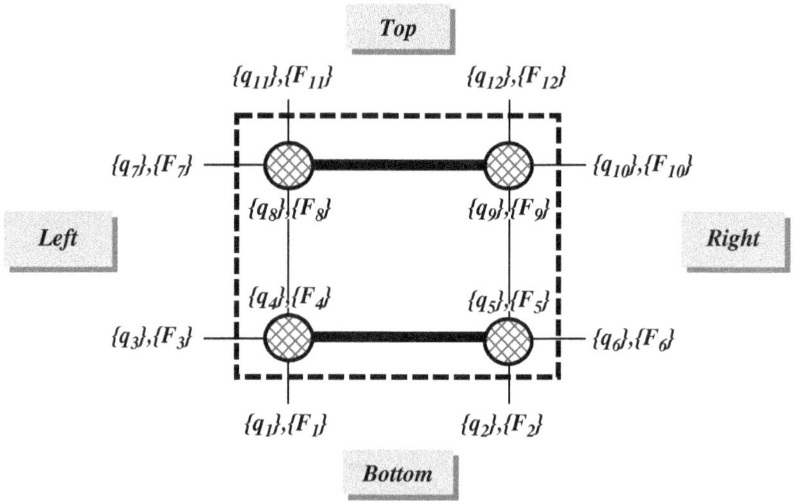

({F_i} = generalized force at i^{th} port, {q_i} = generalized deflection at i^{th} port)

Figure 10.34 Deflections and forces acting on a unit cell.

$$-\omega^2 m w_{ij} + (3k_s + k_p)w_{ij} - k_s(w_{i-1,j} + w_{i+1,j} + w_{i,j-1}) - k_p w_{i,j+1} = F_{ij} \tag{10.73}$$

Similar equations of motion can be written for the other four nodes in unit cell can be obtained.

In Eq. (10.73), k_s and k_p denote the stiffness of the passive and active springs, respectively. Note that k_p is defined as the sum of the stiffness components of the active spring due to its structural rigidity k_{ps} and control gain k_{pc}, respectively, that is,

$$k_p = (k_{ps} + k_{pc}) \tag{10.74}$$

This definition implies that the active spring generates a force $F_{p_{ij}}$ by feeding back the spring deflection as follows:

$$F_{p_{ij}} = -k_{pc}(w_{ij} - w_{ij+1}) \tag{10.75}$$

Hence, k_{pc} defines in effect the gain of the controller.
Let r_k be the ratio of the active to the passive stiffness, then

$$r_k = k_p/k_s = k_{ps}/k_s + k_{pc}/k_s = r_{ks} + r_{kc} \tag{10.76}$$

where r_{ks} and r_{kc} denote the structural stiffness ratio and the control stiffness ratio respectively.

Equations (10.73–10.76) can be used to assemble the equation of motion of the entire cell to yield the following matrix equation

$$\left(-\omega^2[M] + [K]\right)\{q\} = \{F\} \tag{10.77}$$

where

$$[K] = \begin{bmatrix} k_s & 0 & 0 & -k_s & 0 & 0 & 0 & 0 & 0 & 0 & 0 & 0 \\ 0 & k_s & 0 & 0 & -k_s & 0 & 0 & 0 & 0 & 0 & 0 & 0 \\ 0 & 0 & k_s & -k_s & 0 & 0 & 0 & 0 & 0 & 0 & 0 & 0 \\ -k_s & 0 & -k_s & 3k_s+k_p & -k_p & 0 & 0 & -k_s & 0 & 0 & 0 & 0 \\ 0 & -k_s & 0 & -k_p & 3k_s+k_p & -k_s & 0 & 0 & -k_s & 0 & 0 & 0 \\ 0 & 0 & 0 & 0 & -k_s & k_s & 0 & 0 & 0 & 0 & 0 & 0 \\ 0 & 0 & 0 & 0 & 0 & 0 & k_s & -k_s & 0 & 0 & 0 & 0 \\ 0 & 0 & 0 & -k_s & 0 & 0 & -k_s & 3k_s+k_p & -k_p & 0 & -k_s & 0 \\ 0 & 0 & 0 & 0 & -k_s & 0 & 0 & -k_p & 3k_s+k_p & -k_s & 0 & -k_s \\ 0 & 0 & 0 & 0 & 0 & 0 & 0 & 0 & -k_s & k_s & 0 & 0 \\ 0 & 0 & 0 & 0 & 0 & 0 & 0 & -k_s & 0 & 0 & k_s & 0 \\ 0 & 0 & 0 & 0 & 0 & 0 & 0 & 0 & -k_s & 0 & 0 & k_s \end{bmatrix}$$

and

$$[M] = \begin{bmatrix} 0 & 0 & 0 & 0 & 0 & 0 & 0 & 0 & 0 & 0 & 0 & 0 \\ 0 & 0 & 0 & 0 & 0 & 0 & 0 & 0 & 0 & 0 & 0 & 0 \\ 0 & 0 & 0 & 0 & 0 & 0 & 0 & 0 & 0 & 0 & 0 & 0 \\ 0 & 0 & 0 & m & 0 & 0 & 0 & 0 & 0 & 0 & 0 & 0 \\ 0 & 0 & 0 & 0 & m & 0 & 0 & 0 & 0 & 0 & 0 & 0 \\ 0 & 0 & 0 & 0 & 0 & 0 & 0 & 0 & 0 & 0 & 0 & 0 \\ 0 & 0 & 0 & 0 & 0 & 0 & 0 & 0 & 0 & 0 & 0 & 0 \\ 0 & 0 & 0 & 0 & 0 & 0 & 0 & m & 0 & 0 & 0 & 0 \\ 0 & 0 & 0 & 0 & 0 & 0 & 0 & 0 & m & 0 & 0 & 0 \\ 0 & 0 & 0 & 0 & 0 & 0 & 0 & 0 & 0 & 0 & 0 & 0 \\ 0 & 0 & 0 & 0 & 0 & 0 & 0 & 0 & 0 & 0 & 0 & 0 \\ 0 & 0 & 0 & 0 & 0 & 0 & 0 & 0 & 0 & 0 & 0 & 0 \end{bmatrix} \tag{10.78}$$

with the vectors $\{q\}$ and $\{F\}$ denote the deflection and force vectors as defined in Figure 10.34 and are given by

$$\{q\} = \{q_1, \ldots, q_{12}\}^T$$
$$= \{w_{i+2,j}, w_{i+2,j+1}, w_{i+1,j-1}, w_{i+1,j}, w_{i+1,j+1}, w_{i+1,j+2}, w_{i,j-1}, w_{i,j}, w_{i,j+1}$$
$$w_{i,j+2}, w_{i-1,j}, w_{i-1,j+1}\}^T$$

and

$$\{F\} = \{F_1, \ldots, F_{12}\}^T$$
$$= \{F_{i+2,j}, F_{i+2,j+1}, F_{i+1,j-1}, F_{i+1,j}, F_{i+1,j+1}, F_{i+1,j+2}, F_{i,j-1}, F_{i,j}, F_{i,j+1}, F_{i,j+2}, F_{i-1,j}, F_{i-1,j+1}\}^T \tag{10.79}$$

10.8.2 Formulation of Phase Constant Surfaces

The propagation of plane wave motion through two-dimensional periodic structures is described according to *Bloch's Theorem* (Hussein 2009; Hussein et al. 2014) as follows

$$w(n,x) = W(x)e^{(\mu_x n_x + \mu_y n_y)}, \tag{10.80}$$

where μ_x and μ_y are propagation constants. Also, n_x and n_y denote integers that represent the position of the unit cell within the structure as shown in Figure 10.35.

The propagation constant, μ, can be written in the form $\mu_j = \alpha_j + j\beta_j$, where α_j and β_j are attenuation and phase constants, respectively. Unlike in the case of one-dimensional periodic structure, it is difficult to present graphically the complete range of possible values of μ_x and μ_y as functions of frequency. For this reason, results are normally presented only for the pass bands of the structure, which are described in terms of "phase constant surfaces" on which μ_x and μ_y are purely imaginary (i.e., $\alpha_x = \alpha_y = 0$). The usual analysis approach is to specify the phase constants β_x and β_y and then solve for the wave propagation frequency, ω. There are multiple solutions for ω at each combination of β_x and β_y, and therefore solutions produce a series of "phase constant surfaces."

In two-dimensional periodic structures, stop bands are identified by the existence of frequency gaps between two consecutive phase constant surfaces. The evaluation and analysis of the phase constant surfaces therefore provides the tools required to fully analyze wave propagation in two-dimensional periodic structures, and to explore their

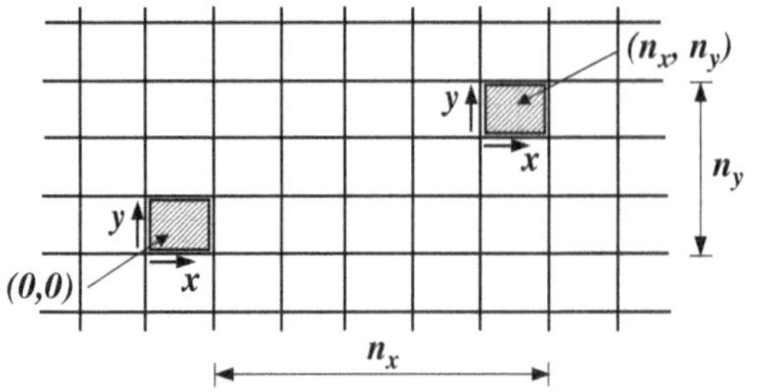

Figure 10.35 Schematic drawing of a two-dimensional periodic.

application as mechanical filters capable of impeding or confining the propagation of waves in limited regions of the structures.

To generate the multiple phase constant surfaces, the wave motion is characterized by the Bloch's relationships that relate the deflections and forces at the edges of a unit cell to the corresponding deflections and forces at the edges of the adjacent cells as follows:

$$\{q_6\} = e^{\mu_x}\{q_3\},\ \{q_{10}\} = e^{\mu_x}\{q_7\},\ \{q_{11}\} = e^{\mu_x}\{q_1\},\ \text{and}\ \{q_{12}\} = e^{\mu_x}\{q_2\}$$

and

$$\{F_6\} = e^{\mu_x}\{F_3\},\ \{F_{10}\} = e^{\mu_x}\{F_7\},\ \{F_{11}\} = e^{\mu_x}\{F_1\},\ \text{and}\{F_{12}\} = e^{\mu_x}\{F_2\}. \quad (10.81)$$

Equation (10.81) is used to condense the original deflection and force vectors, $\{q\}$ and $\{F\}$, to the reduced-order $\{q_r\}$ and $\{F_r\}$, as follows:

$$\{q\} = [A]\{q_r\} and \{F\} = [A]\{F_r\}, \quad (10.82)$$

where $\{q_r\} = \{q_1\, q_2\, q_3\, q_4\, q_5\, q_7\, q_8\, q_9\}^T$ and $\{F_r\} = \{F_1\, F_2\, F_3\, F_4\, F_5\, F_7\, F_8\, F_9\}^T$ with the matrix $[\boldsymbol{A}]$ defined as follows

$$[A] = \begin{bmatrix} 1 & 0 & 0 & 0 & 0 & 0 & 0 & 0 \\ 0 & 1 & 0 & 0 & 0 & 0 & 0 & 0 \\ 0 & 0 & 1 & 0 & 0 & 0 & 0 & 0 \\ 0 & 0 & 0 & 1 & 0 & 0 & 0 & 0 \\ 0 & 0 & 0 & 0 & 1 & 0 & 0 & 0 \\ 0 & 0 & e^{\mu_x} & 0 & 0 & 0 & 0 & 0 \\ 0 & 0 & 0 & 0 & 0 & 1 & 0 & 0 \\ 0 & 0 & 0 & 0 & 0 & 0 & 1 & 0 \\ 0 & 0 & 0 & 0 & 0 & 0 & 0 & 1 \\ 0 & 0 & 0 & 0 & 0 & e^{\mu_x} & 0 & 0 \\ e^{\mu_y} & 0 & 0 & 0 & 0 & 0 & 0 & 0 \\ 0 & e^{\mu_y} & 0 & 0 & 0 & 0 & 0 & 0 \end{bmatrix} \quad (10.83)$$

Then, the equation of motion (10.77) reduces to

$$\left(-\omega^2\left[M_r\left(\mu_x,\mu_y\right)\right] + \left[K_r\left(\mu_x,\mu_y\right)\right]\right)\{q_r\} = [A]^{*^T}[A]\{F_r\}, \quad (10.84)$$

where the reduced mass and stiffness matrices of the cell are defined in terms of the propagation constant pair μ_x and μ_y as $[\boldsymbol{M}]_r = [\boldsymbol{A}]^{*T}\,[\boldsymbol{M}][\boldsymbol{A}]$ and $[\boldsymbol{K}_r] = [\boldsymbol{A}]^{*T}[\boldsymbol{K}][\boldsymbol{A}]$. The values of the frequency corresponding to an assigned pair of propagation constants can be obtained by solving the following eigenvalue problem

$$\left|-\omega^2\left[M_r\left(\mu_x,\mu_y\right)\right] + \left[K_r\left(\mu_x,\mu_y\right)\right]\right| = 0 \quad (10.85)$$

Multiple solutions of the eigenvalue problem are presented in terms of "phase constant surfaces."

10.8.3 Filtering Characteristics

The filtering characteristics of the two-dimensional periodic structure are obtained by solving the eigenvalue problem, given by Eq. (10.85), for different values of the purely imaginary propagation constants ranging between 0 and π. The obtained solutions are used to generate the "phase constant surfaces" that define the frequencies corresponding to free wave propagation as well as the pass and stop bands of the two-dimensional periodic structure.

Example 10.14 Determine the filtering characteristics of the 2D passive periodic spring-mass system shown in Figure 10.33 when $r_{ks} = 1$ and $r_{kc} = 0$.

Solution

The resulting characteristics are presented in Figure 10.36 as quantified by the phase constant and directivity plots. Figure 10.36a shows the propagation surfaces with the gaps-between them indicating the stop bands. In Figure 10.36b, the dotted areas indicate pass bands while the white gaps between two consecutive pass bands indicate stop bands.

It can be seen that the locations of stop and pass bands are dependent on the direction of wave propagation, $\theta = \tan^{-1}(\beta_x/\beta_y)$, which demonstrates that the two-dimensional periodic structures can be utilized as directional mechanical filters.

Example 10.15 Determine the filtering characteristics of the actively controlled 2D periodic spring-mass system shown in Figure 10.33 when $r_{kc} = 2$ and 5.

Solution

Figures 10.37 and 10.38 display the corresponding filtering characteristics of the resulting active periodic structure with $r_{kc} = 2$ and 5 respectively indicating negative proportional feedback control action.

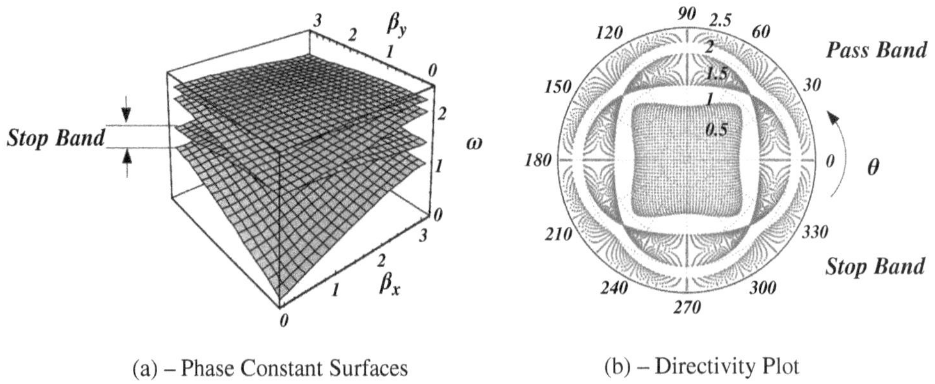

Figure 10.36 Filtering characteristics of the passive periodic structure with $r_{ks} = 1$ and $r_{kc} = 0$. (a) Phase constant surfaces and (b) directivity plot.

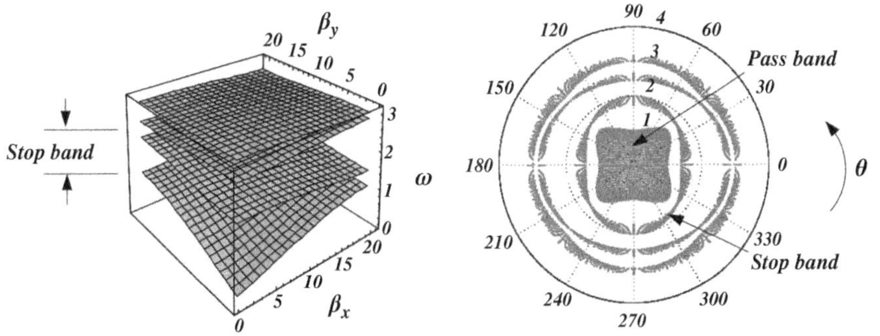

(a) – Phase Constant Surfaces (b) – Directivity Plot

Figure 10.37 Filtering characteristics of the active periodic structure with $r_{ks} = 1$ and $r_{kc} = 2$. (a) Phase constant surfaces and (b) directivity plot.

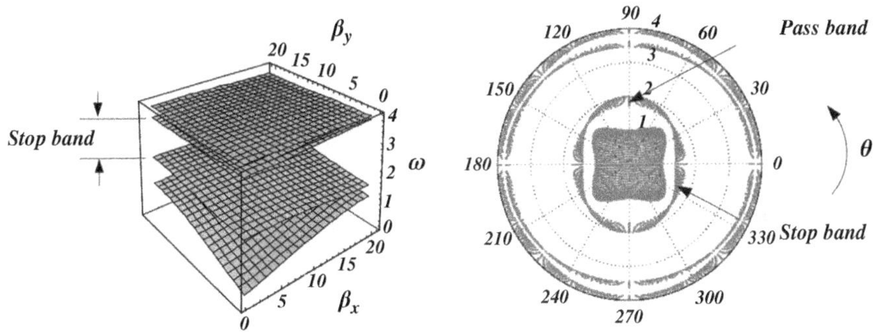

(a) – Phase Constant Surfaces (b) – Directivity Plot

Figure 10.38 Filtering characteristics of the active periodic structure with $r_{ks} = 1$ and $r_{kc} = 5$. (a) Phase constant surfaces and (b) directivity plot.

In these cases, the filtering characteristics that are quantified by the spectral width and directivity of the stop bands can be controlled by proper selection of the control gain r_{kc}.

It is evident that increasing r_{kc} increases the width of the stop bands significantly in all the wave propagation directions.

10.9 Periodic Structures with Internal Resonances

In this section, a special class of periodic structures is introduced because of their unusual response to elastic wave propagation as has been recently reported, for example, by Liu et al. (2005); Milton and Willis (2007); Huang and Sun (2011); Zhou and Hu (2013); and Hussein and Frazier (2013). This class of structures consists of rigid base structures containing cavities which house resonating masses connected to the cavity wall by springs. The macroscopic dynamical properties of the resulting periodic structures

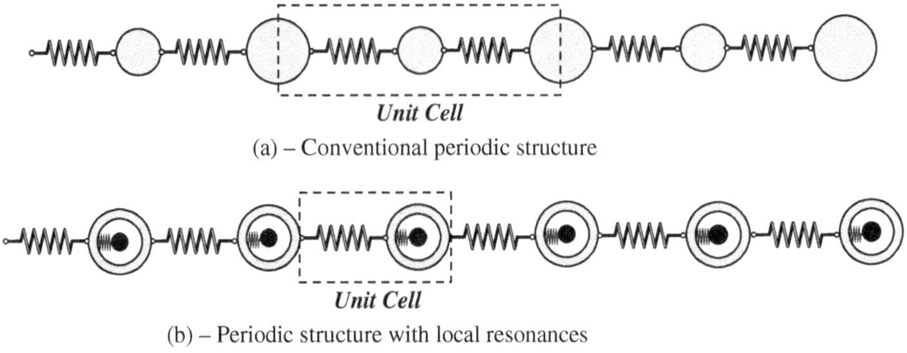

Figure 10.39 Periodic structures with and without local resonances: (a) conventional periodic structure and (b) periodic structure with local resonances.

depend on the resonant properties of substructures that contribute to the rise of interesting effects such as broad stop band characteristics.

Figure 10.39 shows typical schematic drawings of a 1D conventional periodic structure and a 1D periodic structure with local resonances.

The effect of introduction of the local resonances on the dynamic characteristics of these periodic structures can be understood by considering the unit cells shown in Figure 10.40 and extracting the eigenvalues of the corresponding transfer matrices using the approaches outlined in Sections 10.2–10.4.

10.9.1 Dynamics of Conventional Periodic Structure

The equations of motion of the conventional periodic structure are given by:

$$\begin{bmatrix} m_1 & 0 & 0 \\ 0 & \frac{1}{2}m_2 & 0 \\ 0 & 0 & \frac{1}{2}m_2 \end{bmatrix} \begin{Bmatrix} \ddot{x}_i \\ \ddot{x}_L \\ \ddot{x}_R \end{Bmatrix} + \begin{bmatrix} (k_1+k_2) & -k_1 & -k_2 \\ -k_1 & k_1 & 0 \\ -k_2 & 0 & k_2 \end{bmatrix} \begin{Bmatrix} x_i \\ x_L \\ x_R \end{Bmatrix} = \begin{Bmatrix} 0 \\ F_L \\ F_R \end{Bmatrix}$$

or

$$[M]\{\ddot{X}\} + [K]\{X\} = \{F\} \tag{10.86}$$

where m_1 and m_2 denote the periodic masses as shown in Figure 10.40a. Also, k_1 and k_2 denote the stiffness of the periodic springs connecting these masses. The independent degrees of freedom of the system are $X = \{x_L, x_i, x_R\}^T$ as displayed in Figure 10.40b. Note that $[M]$, $[K]$, and $\{F\}$ define the overall mass matrix, stiffness matrix, and load vector of the unit cell.

Using the static condensation method to condense the internal degree of freedom x_i gives:

$$x_i = R\begin{Bmatrix} x_L \\ x_R \end{Bmatrix} \text{ with } R = \frac{1}{(k_1+k_2)}[k_1 \;\; k_2] \tag{10.87}$$

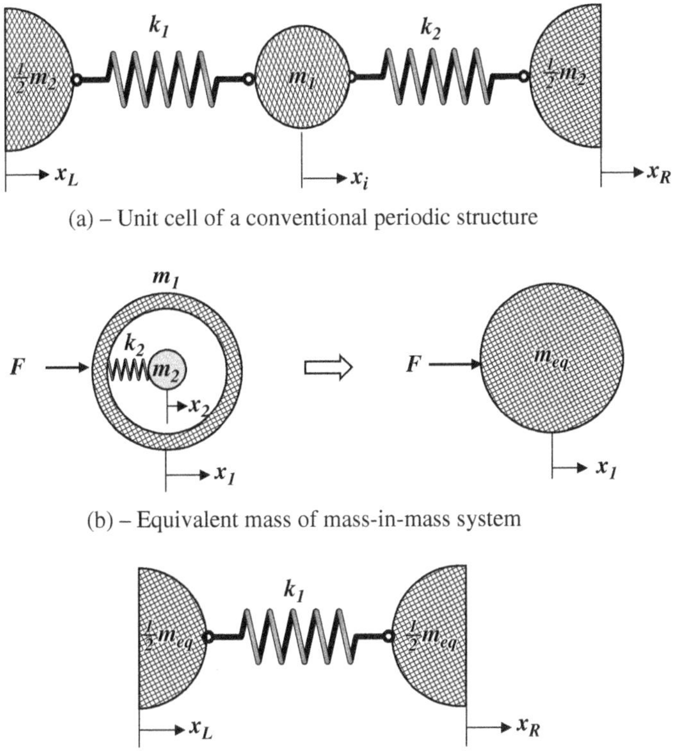

(a) – Unit cell of a conventional periodic structure

(b) – Equivalent mass of mass-in-mass system

(c) – Unit cell of a periodic structure with local resonator

Figure 10.40 Unit cells of periodic structures with and without local resonance. (a) Unit cell of a conventional periodic structure, (b) equivalent mass of mass-in-mass system, and (c) unit cell of a periodic structure with local resonator.

Hence, the entire vector of independent degrees of freedom of the system can be constructed from the boundary degrees of freedom $\{x_L \quad x_R\}^T$ as follows:

$$\begin{Bmatrix} x_i \\ x_L \\ x_R \end{Bmatrix} = \begin{bmatrix} R \\ I \end{bmatrix} \begin{Bmatrix} x_L \\ x_R \end{Bmatrix} = T_R \begin{Bmatrix} x_L \\ x_R \end{Bmatrix} \tag{10.88}$$

Using the transformation matrix T_R, as defined in Eq. (10.88), the dynamics of the conventional periodic structure are condensed, for sinusoidal excitation at a frequency ω, to the following form:

$$\begin{bmatrix} K_{d_{LL_c}} & K_{d_{LR_c}} \\ K_{d_{RL_c}} & K_{d_{RR_c}} \end{bmatrix} \begin{Bmatrix} x_L \\ x_R \end{Bmatrix} = \begin{Bmatrix} F_L \\ F_R \end{Bmatrix}$$

or

$$[[K_R] - \omega^2 [M_R]] \begin{Bmatrix} x_L \\ x_R \end{Bmatrix} = \begin{Bmatrix} F_L \\ F_R \end{Bmatrix} \tag{10.89}$$

with $[K_R] = T_R^T [K] T_R$ and $[M_R] = T_R^T [M] T_R$.

Then, the transfer matrix T_c of the conventional periodic structure is given by:

$$T_c = \begin{bmatrix} -K_{d_{LR_c}}^{-1} K_{d_{LL_c}} & K_{d_{LR_c}}^{-1} \\ K_{d_{RR_c}} K_{d_{LR_c}}^{-1} K_{d_{LL_c}} - K_{d_{LR_c}} & -K_{d_{RR_c}} K_{d_{LR_c}}^{-1} \end{bmatrix} \tag{10.90}$$

10.9.2 Dynamics of Periodic Structure with Internal Resonances

10.9.2.1 Equivalent Mass. Of the Mass-In-Mass Arrangement

In order to derive the equations of motion of the periodic structure with internal resonances, it is essential to extract the equivalent mass of the mass-in-mass arrangement as shown in Figure 10.40b.

The equations of motion of the mass-in-mass arrangement are:

$$m_1 \ddot{x}_1 + k_2(x_1 - x_2) = F$$

and

$$m_2 \ddot{x}_2 + k_2(x_2 - x_1) = 0 \tag{10.91}$$

Assuming sinusoidal motion with $F = F_0 \sin \omega t$, $x_1 = X_1 \sin \omega t$, and $x_2 = X_2 \sin \omega t$. Then, eliminating the internal degree of freedom x_2, this results in:

$$\left[m_1 + \frac{k_2}{k_2/m_2 - \omega^2} \right] \ddot{x}_1 = F$$

This means that the two-degree-of-freedom mass-in-mass arrangement behaves as a single-degree-of-freedom system with an equivalent mass m_{eq} given by:

$$m_{eq} = \left[m_1 + \frac{k_2}{k_2/m_2 - \omega^2} \right] \tag{10.92}$$

10.9.2.2 Transfer Matrix of the Mass-In-Mass Arrangement

The equations of motion of the unit cell shown in Figure 10.40c, using the equivalent mass, are given by:

$$\begin{bmatrix} \frac{1}{2} m_{eq} & 0 \\ 0 & \frac{1}{2} m_{eq} \end{bmatrix} \begin{Bmatrix} \ddot{x}_L \\ \ddot{x}_R \end{Bmatrix} + \begin{bmatrix} k_1 & -k_1 \\ -k_1 & k_1 \end{bmatrix} \begin{Bmatrix} x_L \\ x_R \end{Bmatrix} = \begin{Bmatrix} F_L \\ F_R \end{Bmatrix} \tag{10.93}$$

For sinusoidal excitation at a frequencyω, Eq. (10.93) reduces to:

$$\begin{bmatrix} k_1 - \frac{1}{2} m_{eq} \omega^2 & -k_1 \\ -k_1 & k_1 - \frac{1}{2} m_{eq} \omega^2 \end{bmatrix} \begin{Bmatrix} x_L \\ x_R \end{Bmatrix} = \begin{Bmatrix} F_L \\ F_R \end{Bmatrix} \tag{10.94}$$

In a more compact form, Eq. (10.94) can be written as:

$$\begin{bmatrix} K_{d_{LL_i}} & K_{d_{LR_i}} \\ K_{d_{RL_i}} & K_{d_{RR_i}} \end{bmatrix} \begin{Bmatrix} x_L \\ x_R \end{Bmatrix} = \begin{Bmatrix} F_L \\ F_R \end{Bmatrix} \tag{10.95}$$

Then, the transfer matrix T_i of the periodic structure with internal resonances is given by:

$$T_i = \begin{bmatrix} -K_{d_{LR_i}}^{-1} K_{d_{LL_i}} & K_{d_{LR_i}}^{-1} \\ K_{d_{RR_i}} K_{d_{LR_i}}^{-1} K_{d_{LL_i}} - K_{d_{LR_i}} & -K_{d_{RR_i}} K_{d_{LR_i}}^{-1} \end{bmatrix} \tag{10.96}$$

Example 10.16 Determine the filtering characteristics of the conventional periodic structure shown in Figure 10.39a when k_1 = 1 N/m, k_2 = 1 N/m, and m_1 = 1 *kg* for values of m_2 = 0.025 and 0.1 kg. Compare these characteristics with those of the periodic structure with internal resonances shown in Figure 10.39b.

Solution

Using the transfer matrices T_c and T_i given by Eqs. (10.90) and (10.96), the eigenvalues are computed for different excitation frequencies. Figures 10.41a and (10.41b) display these eigenvalues and identify the pass and stop bands of both the conventional periodic structure and the periodic structure with internal resonances.

The displayed results indicate that using periodic structure with internal resonances results in enhancing the stop band characteristics compared to those obtained with conventional periodic structures.

Example 10.17 Consider the 2D metamaterial plate-like configuration displayed in Figure 10.42. The plate is manufactured of aluminum and consists of assemblies of periodic cells with built-in local resonances. Each cell is made of a base plate-like structure that is provided with cavities filled by a viscoelastic membrane that supports a small mass to form a source of local resonance (Nouh et al. 2015). Table 10.6 lists the main geometrical properties of the aluminum plate and the local resonant masses.

Determine the propagation surfaces and directivity plots of plain aluminum plate and for the periodic plate with internal resonance sources shown in Figure 10.42.

Assume the VEM membranes that support the resonant masses have storage modulus of 10 MPa and loss factor of 0.4.

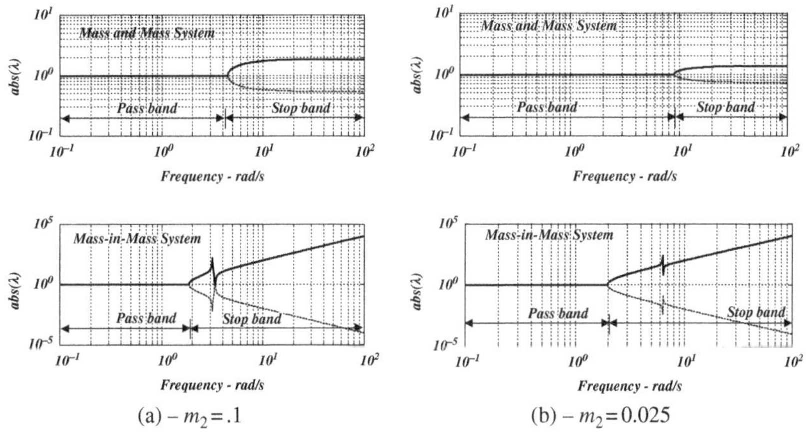

Figure 10.41 Filtering characteristics of periodic structures with and without local resonance. (a) m_2 = 0.1 and (b) m_2 = 0.025.

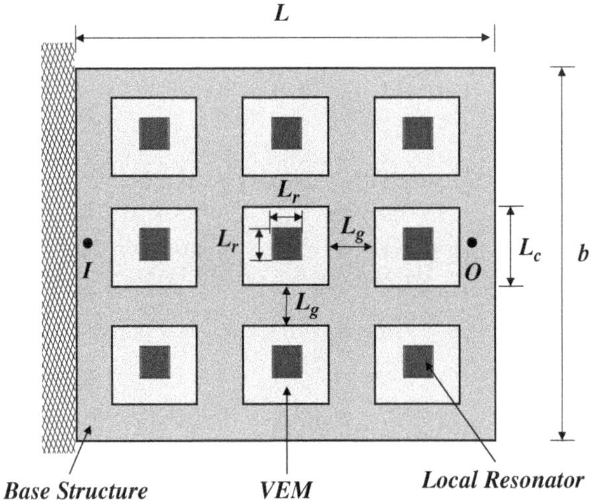

Figure 10.42 Metamaterial plate with periodic local resonances.

Table 10.6 Main geometrical parameters of the periodic metamaterial plate.

Length	b	L_c	L_r	L_g	L	Thickness (t) of plate and masses
Value (cm)	15.2	5.1	1.3	1.3	20.3	0.1524

Solution

Figure 10.43 shows a finite element model of a unit cell of the metameterial plate with periodic local resonances. The model consists primarily of quadrilateral square plate elements of dimensions $L_e \times L_e \times t$. The theory of modeling this class of plates has been presented in Section 4.7 for the general case when it is treated with constrained VEM. In this section, the plate is considered to a plain plate which is totally untreated.

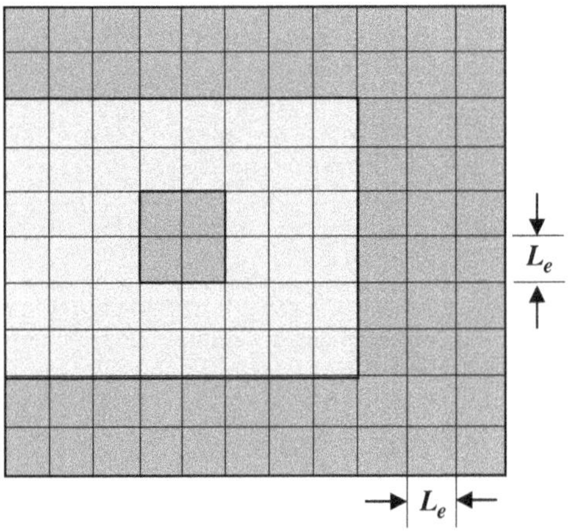

Figure 10.43 Finite element model of a unit cell of the metamaterial plate with periodic local resonances.

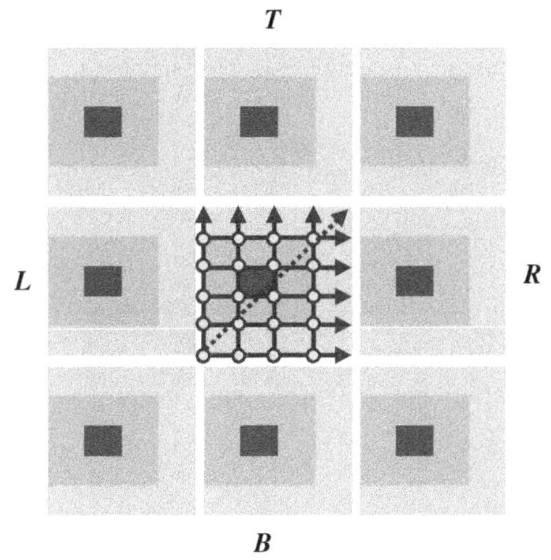

Figure 10.44 Periodic cells and mapping relations of the metamaterial plate (L = left, R = right, T = top, B = bottom).

Figure 10.44 displays unit cell along with its eight neighboring cells in order to enable the computation of the propagation surfaces and directivity plots of plain aluminum plate and for the periodic plate with internal resonance sources as outlined in Section 10.8.2.

Figure 10.45a shows the propagation surfaces computed for the unit cell displayed in Figure 10.43, consisting of an aluminum base with a viscoelastic membrane that houses a local mass, for a frequency range from 0 to 8 kHz. A planar view of these surfaces is shown in Figure 10.45b. A large stop band is observed between 1.5 and 5.9 kHz.

Figure 10.46 shows a close-up of the low-frequency range of the propagation surfaces. The gaps between these surfaces in the 0–1 kHz range can be noticed between 200 and 215, 390 and 450, between 610 and 720 Hz, and once again between 790 and 900 Hz.

Figure 10.47 shows the propagation surfaces mathematically computed for a structure built from a plain cell, which acts as a datum for comparison purposes. For a conventional metal plate, waves should be able to propagate freely in both directions across the plate at any excitation frequency. This is shown clearly by the overlapping

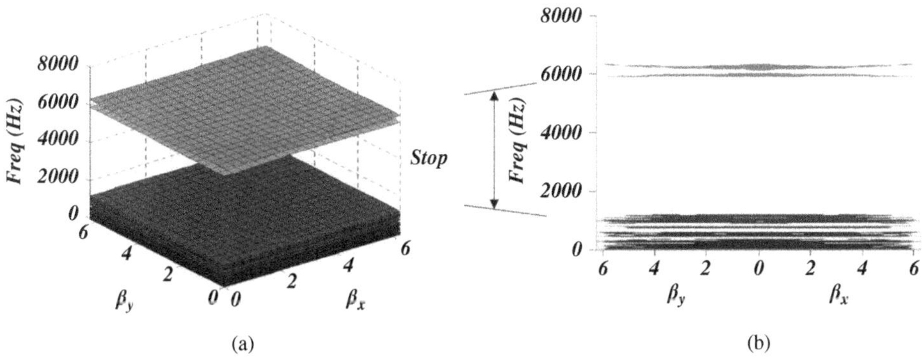

Figure 10.45 (a) 3D and (b) 2D plots of the propagation surfaces for a metamaterial plate with periodic local resonances for frequencies up to 8 kHz.

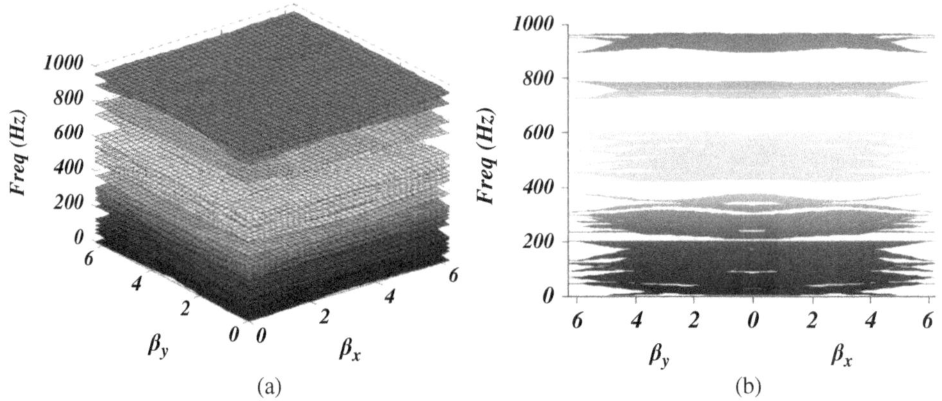

Figure 10.46 (a) 3D and (b) 2D plots of the propagation surfaces for a metamaterial plate with periodic local resonances for frequencies up to 1 kHz.

propagation surfaces throughout the entire considered frequency range allowing for no gaps between the surfaces to extend along the 0–2π range of β_x or β_y.

Figure 10.48 shows the vibration pattern of the metamaterial plate with the periodic local resonances when subject to an external excitation over the frequency range between 395 Hz until 460 Hz in the transverse z-direction. This range corresponds to the first low-frequency stop band range highlighted in Figure 10.45. It can be seen that this corresponds to the first natural mode of many of the local resonators. Hence, a large fraction of the vibration energy propagates directly into the locally resonant masses, which act as local absorbers of vibration, and thereby relieving the base structure completely from the vibration.

The frequency-dependent directional behavior of the wave propagation in this class of structures can be visualized by using the polar plots such as those displayed in

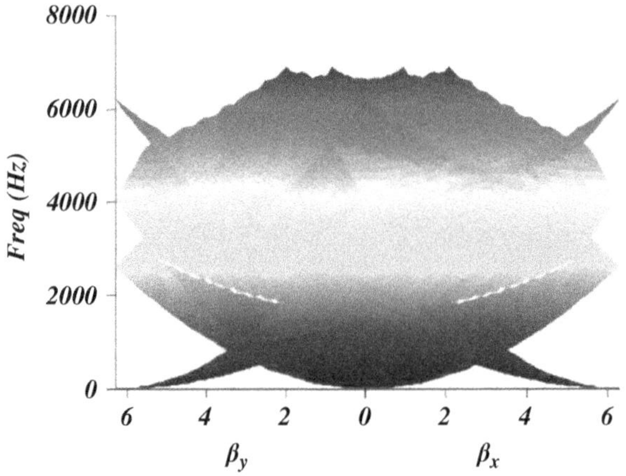

Figure 10.47 Propagation surfaces for a plain aluminum plate for vibration frequencies up to 8 kHz showing complete absence of stop bands.

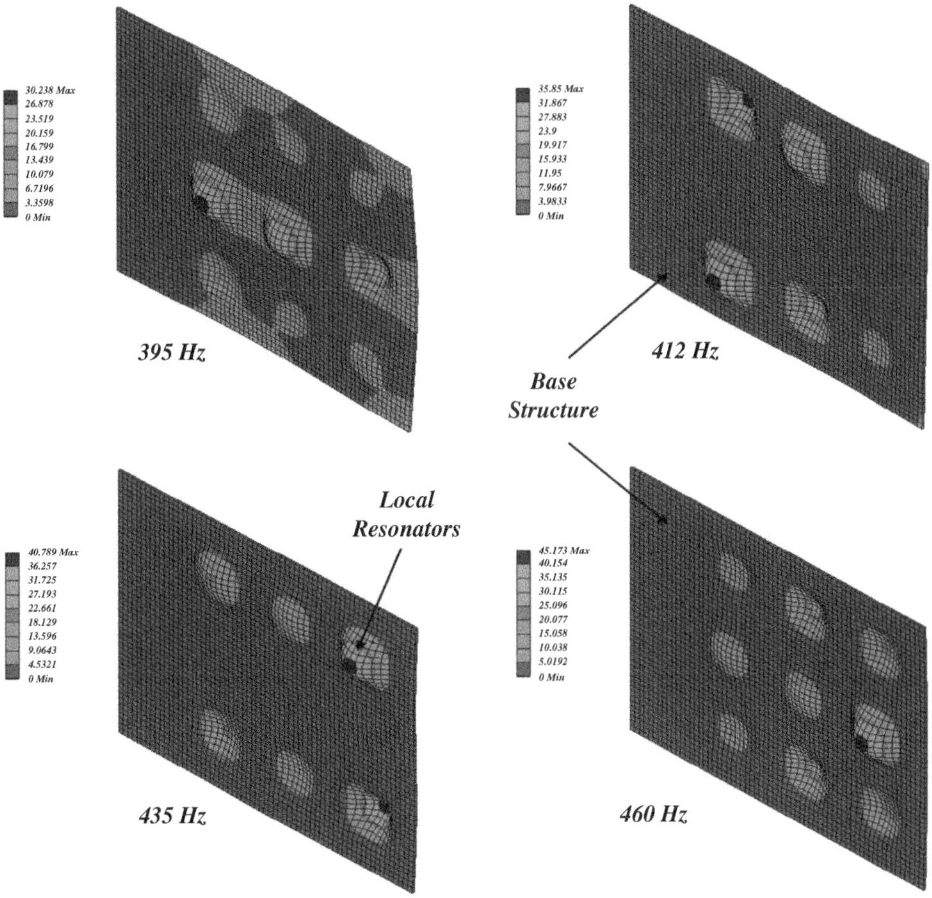

Figure 10.48 Vibration of the metamaterial plate with periodic local resonances during the first low-frequency stop band.

Figures 10.49a and 10.49b. In these diagrams, referred to as the directivity plots, the vibration frequency is plotted as a function of θ that is a measure of the direction of wave propagation given by$\tan^{-1}(\beta_x/\beta_y)$. Frequencies are quantified by directivity plots as constant radii circles, and hence a stop band frequency is indicated by a circle that does not contain a solution for θ along its entire circumference, as can be seen in Figure 10.49a. Finally, Figure 10.49b shows the directivity of wave propagation in a plain aluminum cell for comparison. As expected, it can be seen that for any chosen frequency, that is, any constant radius circle, there has to exist at least one solution for θ along the circle's perimeter.

Note that if the physical properties of the individual unit cells can be tuned, this class of metamaterial structures can be used to confine/block wave propagation at the developed stop bands, as shown here, as well as steer the propagating waves within the pass bands through this directional filtering mechanism.

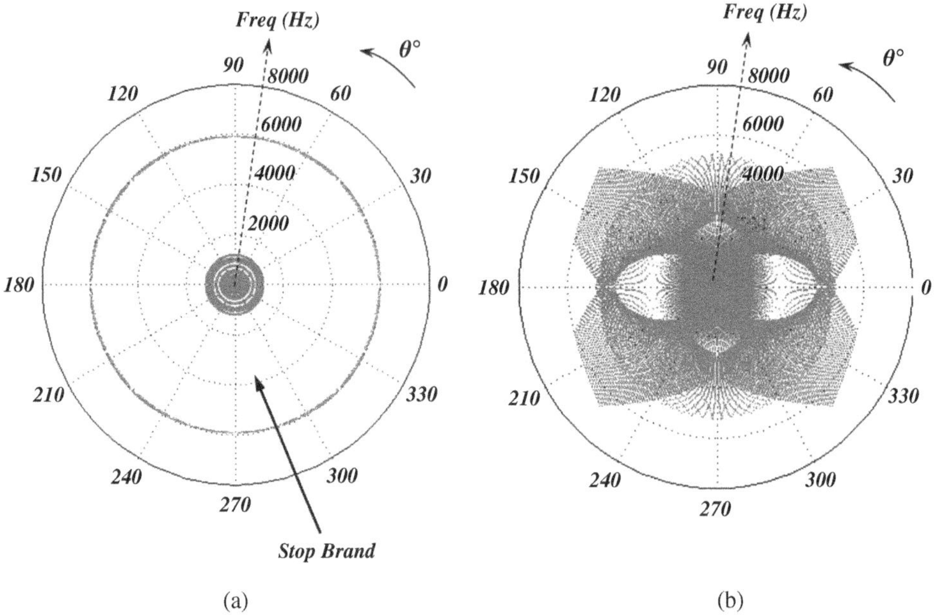

Figure 10.49 Directivity plots for the plate with (a) periodic local sources of resonance and (b) the plain aluminum plate for frequencies up to 8 kHz.

10.10 Summary

The dynamic characteristics of passive and active periodic structures are presented using the transfer matrix method. The theory governing the dynamics of these structures is introduced in order to demonstrate their unique mechanical filtering capabilities for wave propagation. As a result of these characteristics, waves are shown to propagate along the periodic structures only within specific frequency bands called the pass bands and wave propagation is completely blocked within other frequency bands called the stop bands. The spectral width and location of these bands are shown to be fixed for passive periodic structures, and tunable in response to the structural vibration for active periodic structures.

References

Agnes G., "Piezoelectric coupling of bladed-disk assemblies", Proceedings of the Smart Structures and Materials Conference on Passive Damping (ed. T. Tupper Hyde), Newport Beach, CA, SPIE-Vol. 3672, pp. 94–103, 1999.

Anderson, P.W. (1958). Absence of diffusion in certain random lattices. *Physical Review* 109: 1492–1505.

Baz, A. (2001). Active control of periodic structures. *ASME Journal of Vibration and Acoustics* 123: 472–479.

Brillouin, L. (1946). *Wave Propagation in Periodic Structures*, 2e. Dover.

Cai, G. and Lin, Y. (1991). Localization of wave propagation in disordered periodic structures. *AIAA Journal* 29 (3): 450–456.

Chen, T., Ruzzene, M., and Baz, A. (2000). Control of wave propagation in composite rods using shape memory inserts: theory and experiments. *Journal of Vibration and Control* 6 (7): 1065–1081.

Chui, C.K. (1992). An Introduction to Wavelets. In: *Wavelets Analysis and Applications*, vol. 1. Academic Press Inc.

Cremer, L., Heckel, M., and Ungar, E. (1973). *Structure-Borne Sound*. New York: Springer-Verlag.

Faulkner, M. and Hong, D. (1985). Free vibrations of a mono-coupled periodic system. *Journal of Sound and Vibration* 99 (1): 29–42.

Gopalakrishnan, S. and Mitra, M. (2010). *Wavelet Methods for Dynamical Problems: With Application to Metallic, Composite, and Nano-Composite Structures*. CRC Press.

Hodges, C.H. (1982). Confinement of vibration by structural irregularity. *Journal of Sound and Vibration* 82 (3): 441–444.

Hodges, C.H. and Woodhouse, J. (1983). Vibration isolation from irregularity in a nearly periodic structure: theory and measurements. *Journal of Acoustical Society of America* 74 (3): 894–905.

Huang, H.H. and Sun, C.T. (2011). A study of band-gap phenomena of two locally resonant acoustic metamaterials", Proceedings of IMechE Part N. *Journal Nanoengineering and Nanosystems* 224: doi: 10.1177/1740349911409981.

Hussein, M.I. (2009). Theory of damped Bloch waves in elastic media. *Physical Review B* 80: 212301.

Hussein, M.I. and Frazier, M.J. (2013). Metadamping: an emergent phenomenon in dissipative Metamaterials. *Journal of Sound and Vibration* 332: 4767–4774.

Hussein, M.I., Leamy, M.J., and Ruzzene, M. (2014). Dynamics of Phononic materials and structures: historical origins, recent progress, and future outlook. *Applied Mechanics Reviews* 66 (4): 040802–(1–38).

Langley, R.S. (1994). On the forced response of one-dimensional periodic structures: vibration localization by damping. *Journal of Sound and Vibration* 178 (3): 411–428.

Liu, Z., Chan, C.T., and Sheng, P. (2005). Analytic model of phononic crystals with local resonances. *Physical Review B* 71: 014103.

Luongo, A. (1992). Mode localization by structural imperfections in one-dimensional continuous systems. *Journal of Sound and Vibration* 155 (2): 249–271.

Mead, D.J. (1970). Free wave propagation in periodically supported, infinite beams. *Journal of Sound and Vibration* 11 (2): 181–197.

Mead, D.J. (1971). Vibration response and wave propagation in periodic structures. *ASME Journal of Engineering for Industry* 93: 783–792.

Mead, D.J. (1986). A new method of analyzing wave propagation in periodic structures; applications to periodic Timoshenko beams and stiffened plates. *Journal of Sound and Vibration* 114 (1): 9–27.

Mead, D.J. (1996). Wave propagation in continuous periodic structures: research contributions from Southampton. *Journal of Sound and Vibration* 190 (3): 495–524.

Mead, D.J. and Bardell, N.S. (1987). Free vibration of a thin cylindrical shell with periodic circumferential stiffeners. *Journal of Sound and Vibration* 115 (3): 499–521.

Mead, D.J. and Lee, S.M. (1984). Receptance methods and the dynamics of disordered one-dimensional lattices. *Journal of Sound and Vibration* 92 (3): 427–445.
Mead, D.J. and Markus, S. (1983). Coupled flexural-longitudinal wave motion in a periodic beam. *Journal of Sound and Vibration* 90 (1): 1–24.
Mead, D.J. and Yaman, Y. (1991). The harmonic response of rectangular sandwich plates with multiple stiffening: a flexural wave analysis. *Journal of Sound and Vibration* 145 (3): 409–428.
Mester, S. and Benaroya, H. (1995). Periodic and near periodic structures: review. *Shock and Vibration* 2 (1): 69–95.
Milton, G.W. and Willis, J.R. (2007). On modifications of Newton's second law and linear continuum elastodynamics. *Proceedings of the Royal Society of London Series A* 463: 855–880.
Nouh, M., Aldraihem, O., and Baz, A. (2015). Wave propagation in metamaterial plates with periodic local resonances. *Journal of Sound and Vibration* 341: 53–73.
Pierre, C. (1988). Mode localization and eigenvalue loci veering phenomena in disordered structures. *Journal of Sound and Vibration* 126 (3): 485–502.
Roy, A. and Plunkett, R. (1986). Wave attenuation in periodic structures. *Journal of Sound and Vibration* 114 (3): 395–411.
Ruzzene, M. and Baz, A. (2000). Control of wave propagation in periodic composite rods using shape memory inserts. *ASME Journal of Vibration and Acoustics* 122: 151–159.
Ruzzene, M. and Baz, A. (2001). Active control of wave propagation in periodic fluid-loaded shells. *Smart Materials and Structures* 10 (5): 893–906.
Sen Gupta, G. (1970). Natural flexural waves and the normal modes of periodically-supported beams and plates. *Journal of Sound and Vibration* 13: 89–111.
Thorp, O., Ruzzene, M., and Baz, A. (2001). Attenuation and localization of wave propagation in rods with periodic shunted piezoelectric patches. *Journal of Smart Materials & Structures* 10: 979–989.
Zhou, X. and Hu, G. (2013). Dynamic effective models of two-dimensional acoustic metamaterials with cylindrical inclusions. *Acta Mechanica* 224: 1233–1241.

10.A The Wavelet Transform

The *WT* of a signal $x(t)$ is an example of a time-scale decomposition obtained by dilating and translating along the time axis a chosen analyzing function (wavelet) (Chui 1992; Gopalakrishnan and Mitra 2010). The integral or continuous *WT* relative to some basic wavelet $\psi(t)$ is defined as:

$$W_\psi(a,b) = \frac{1}{\sqrt{a}} \int_{-\infty}^{\infty} x(t)\,\psi^*\left(\frac{t-b}{a}\right) dt \tag{10.A.1}$$

where b is a translation parameter used for positioning the function $\psi(t)$ over the time domain, and $a > 0$ is a scaling parameter dilating or contracting the function $\psi(t)$. The *WT* provides a flexible time-frequency window, which automatically narrows when observing high frequency phenomena and widens when studying low-frequency components (Chui 1992). The wavelet function used in this work is the *Morlet wavelet,* defined in the time domain as:

$$\psi(t) = e^{-\frac{t^2}{2}} e^{i\,\omega_w t} \tag{10.A.2}$$

The Morlet wavelet is a sinusoidal function, oscillating at the frequencyω_w, modulated by a Gaussian envelope of unit variance. Being composed of a modulated sinusoidal function, the Morlet wavelet is well suited for reproducing and analyzing signals in many applications.

As signal decomposition, the *WT* cannot be directly compared to a time-frequency representation. However, it can be shown that b represents a time parameter and that the dilation parameter a is strictly related to frequency. In the frequency domain, the Morlet wavelet becomes:

$$\psi(\omega) = \sqrt{2\pi}\, e^{-\frac{1}{2}(\omega - \omega_w)^2} \tag{10.A.3}$$

Equation (10.A.3) shows that the frequency domain formulation of the Morlet wavelet is a Gaussian function centered at $\omega = \omega_w$. Its dilated version is expressed as:

$$\psi(a\omega) = \sqrt{2\pi}\, e^{-\frac{1}{2}(a\,\omega - \omega_w)^2} \tag{10.A.4}$$

whose maximum is located at $a\,\omega = \omega_w$. Since $\omega_w = 1.875\,\pi$ is a fixed parameter defining the wavelet function, the center of the Gaussian curve and therefore the frequency of the analysis can be located by changing the dilation parameter as follows:

$$\omega = \frac{\omega_w}{a} \tag{10.A.5}$$

The scale parameter can be hence considered as the inverse of a frequency parameter and thus the *WT* can be classified as a time-frequency transform.

An alternative formulation of the continuous *WT* can be obtained transforming both the signal $x(t)$ and the wavelet function $\psi(t)$in the frequency domain:

$$W_g(a,b) = \sqrt{a} \int_{-\infty}^{\infty} X(\omega)\, \psi^*(a\,\omega)\, e^{ib\omega}\, d\omega \tag{10.A.6}$$

with $X(\omega)$ and $\psi^*(a\,\omega)\, e^{i\,b\,\omega_0}$ being the Fourier transforms of $x(t)$ and $\psi^*\left(\frac{t-b}{a}\right)$, respectively.

This formulation of the *WT* can be expressed in a discrete form as:

$$W\left(m, bn = \sqrt{m\,\Delta\,a} \sum_n X(f_n)\, \psi^*(m\,\Delta a f_n)\, e^{i\,\Delta b\, 2\pi\, n f_n}\right. \tag{10.A.7}$$

wheref_n is the discrete frequency and $\Delta\,a$ and $\Delta\,b$ are discrete increments of dilation and translation parameters. Equation (10.A.7) allows an easy implementation of the *WT*. The frequency domain formulation of the *WT* is particularly convenient when the signal to be analyzed is expressed in the frequency domain.

Problems

10.1 Consider the periodic spring-mass system shown in Figure P10.1a, determine the first three natural frequencies of the system using the symmetric and asymmetric

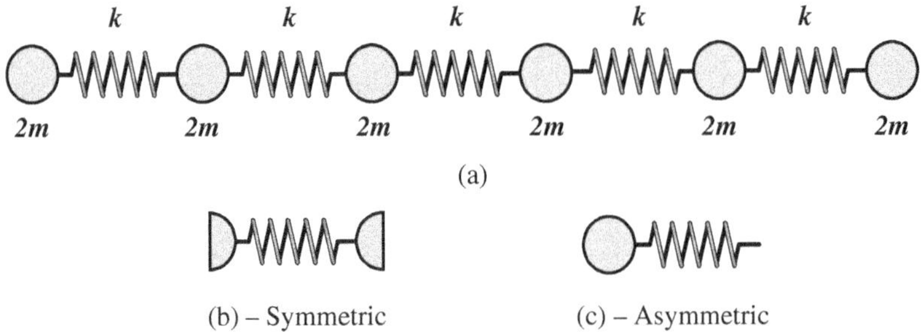

Figure P10.1 (a–c) A periodic spring-mass system.

configurations shown in Figures P10.1b,c. Consider the system to be *a*. fixed-free and *b*. free-free. Compare the results of the considered configurations.

Determine also the mode shapes corresponding to the different natural frequencies assuming the system to consist of 50 cells (N = 50). Comment on the shape of the system transfer matrix for both the symmetric and asymmetric configurations.

10.2 For the spring-mass system of Problem P10.1, determine and plot the distribution of the longitudinal displacement along the system. Assume that the system is fixed-free and that it is excited by a unit sinusoidal force applied at its free end. Consider three excitation frequencies coinciding with the first three natural frequencies of the system.

10.3 Consider the periodic spring-mass system shown in Figure P10.2, show that if the equation of motion of the nth mass is given by:

$$m\ddot{u}_n = K(u_{n+1} + u_{n-1} - 2u_n)$$

and the deflection u_n is given by: $u_n = u_0\, e^{i(nka - \omega t)}$, where k = wave number, a = spacing between masses, ω = frequency, and t = time, then:

$$\omega(k) = \sqrt{\frac{4K}{m}}\left|\sin\left(\frac{ka}{2}\right)\right|$$

describes the dispersion relationship of the periodic spring-mass system as shown in Figure P10.3.

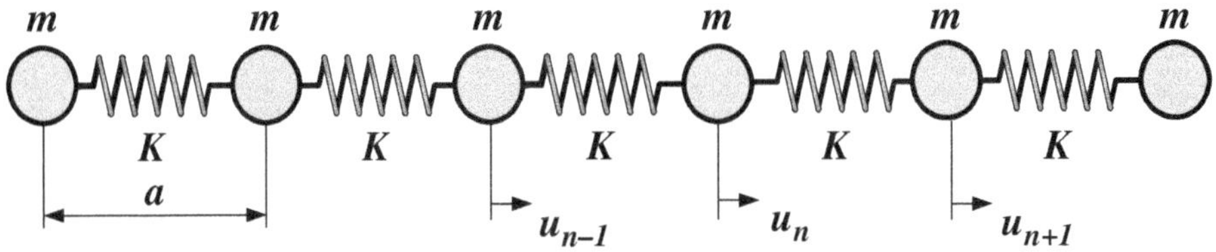

Figure P10.2 Mono-atomic periodic spring-mass system.

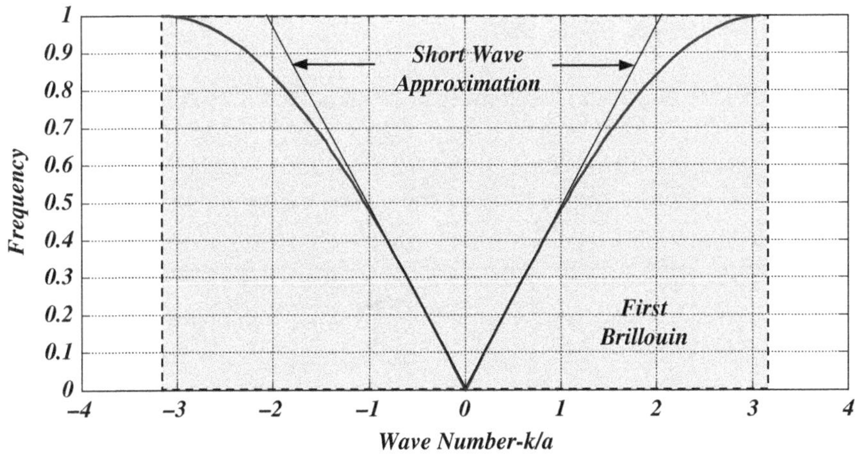

Figure P10.3 The dispersion relationship of the mono-atomic periodic spring-mass system.

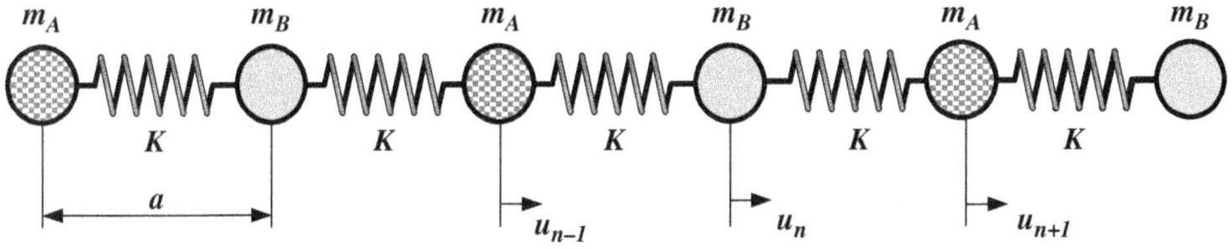

Figure P10.4 A diatomic periodic spring-mass system.

Show also that for short wave number k, then the wave propagation is non-dispersive with the following linear relationship between ω–k as shown in Figure P10.3:

$$\omega(k) = \sqrt{\frac{Ka^2}{m}}\,k$$

10.4 Consider the diatomic periodic spring-mass system shown in Figure P10.4, show that if the equations of motion of the masses m_A and m_B are given by:

$$m_A \ddot{u}_{2n} = K(u_{2n+1} + u_{2n-1} - 2u_{2n}),$$

and

$$m_B \ddot{u}_{2n+1} = K(u_{2n+2} + u_{2n} - 2u_{2n+1}).$$

and the deflections u_{2n} and u_{2n+1} are given by:

$$u_{2n} = u_1\, e^{i(2nka - \omega t)} \text{ and } u_{2n+1} = u_1\, e^{i([2n+1]ka - \omega t)},$$

where k = wave number, a = spacing between masses, ω = frequency, and t = time, then:

$$\omega^2(k) = \frac{K}{\mu}\left[1 \pm \sqrt{1 - 4\mu\sin^2(ka)}\right]$$

describes the dispersion relationship of the periodic spring-mass system as shown in Figure P10.5 with roots $\omega_{1,2}$. Note that: $\mu = \frac{m_A m_B}{m_A + m_B}$ is denoted by the effective mass. Also, the high frequency root ω_1 is denoted by the "optical mode" and the low-frequency root ω_2 is called the "acoustical mode."

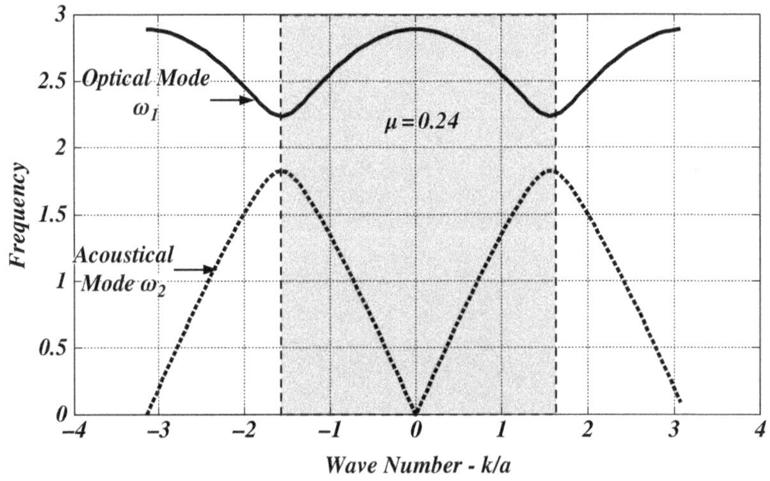

Figure P10.5 The dispersion relationship of the bi-atomic periodic spring-mass system.

10.5 Consider the periodic rod shown in Figure P10.6a, determine the natural frequencies of the entire system assuming the rod to be fixed-free and consists of 50 cells (N = 50). Use the asymmetric configurations shown in Figure P10.6b.

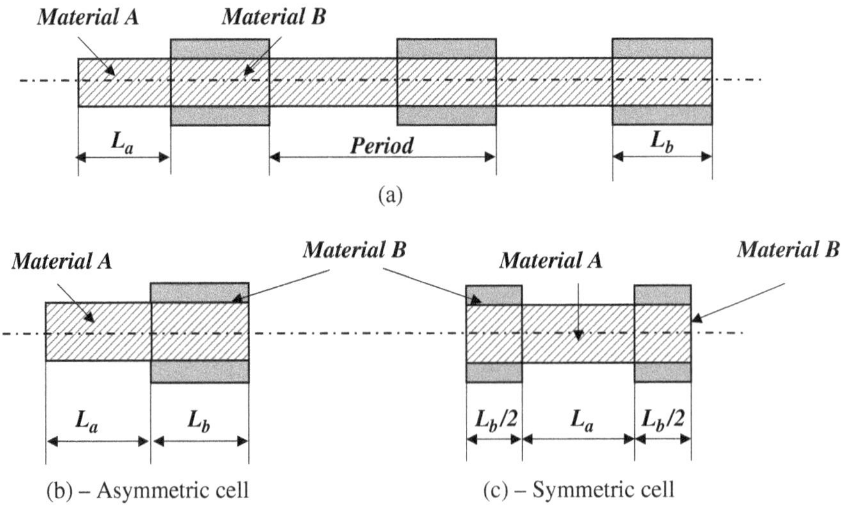

Figure P10.6 A periodic rod with material discontinuities.

Determine and plot the distribution of the longitudinal displacement along the system. Assume the system to be excited by a unit sinusoidal force applied at its free end. Consider three excitation frequencies coinciding with the first three natural frequencies of the system.

10.6 For the periodic rod of Figure P10.6, assume that the impedance ratio z_a/z_b varies along the rod in a normal random manner with zero mean and standard deviation σ. Determine and plot the distribution of the longitudinal displacement along the system. Assume the system to be excited by a unit sinusoidal force applied at its free end. Consider three excitation frequencies coinciding with the first three natural frequencies of the system. Assume different values of σ and determine the decay parameter γ as a function of σ.

10.7 Figure P10.7a shows a schematic drawing of the shear mode periodic mount that is made of identical cells in the longitudinal direction. Each cell can be divided into four elements (parts). These elements are numbered 1,2,3,4 from the right to the left as shown in Figure P10.7b. When the mount is subjected to longitudinal loading F, it deflected and assumes the configuration shown in Figure (P10.7c). The shear strain γ in the viscoelastic material is given by: where u_1 and u_3 are the longitudinal deflections of the aluminum core and outer aluminum layer, respectively. Also, h_2 defines the thickness of viscoelastic layer between the aluminum core and the outer layer.

Using the theory of finite elements, derive the equations of motion for the elements and determine the transfer matrix of the unit cell.

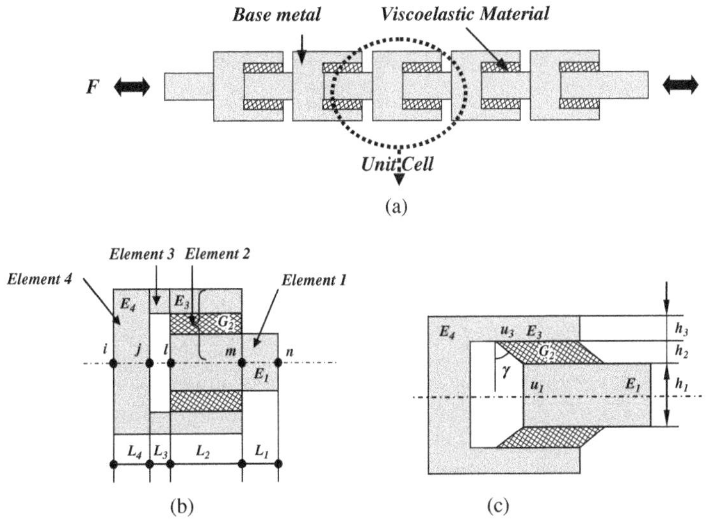

Figure P10.7 A schematic drawing of the shear mode periodic mount.

10.8 Consider the periodic shear mount shown in Figure (P10.8) is used to isolate the vibration of a payload mounted on the top platform from being transmitted to the foundation.

Table P10.1 Geometric properties.

Length (mm)		Thickness (mm)	
L_1	4.76	h_1	3.17
L_2	17.46	h_2	3.18
L_3	4.76	h_3	3.18
L_4	3.18	b	25.4

Table P10.2 Physical properties.

Material	Density (kg/m^3)	Modulus (MPa)
Aluminum	2700	70,000[a]
Viscoelastic layer	1201.4	$9.193 + 4.596i$[b]

a Young's modulus.
b Complex shear modulus ($G(1 + \eta i)$, $\eta = 0.5$).

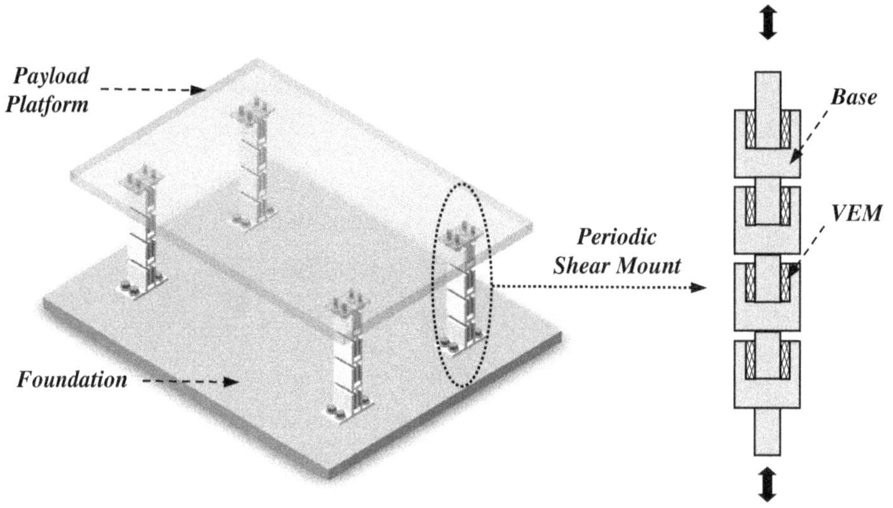

Figure P10.8 A periodic shear mount as vibration isolator.

The periodic mount is manufactured from an aluminum base material and a viscoelastic core (VEM) that have the geometrical and physical properties listed in Tables P10.1 and P10.2.

Using the transfer matrix developed in problem P10.7, determine the attenuation parameter α as function of the excitation frequency and extract the pass and stop band characteristics of the mount.

10.9 The periodic beam cell shown in Figure P10.9 consists of two sections such that each section is modeled by a single finite element. The physical and geometrical parameters of the two sections are: $m_a = 1$ kg/m, $EI_a = 1$ Nm^2, $L_a = 0.025$ m, $m_b = 1$ kg/m, $EI_b = 2$ Nm^2, and $L_b = 0.025$ m.

If the dynamic stiffness matrices K_{Da} and K_{Db} of the two sections are given by:

$$K_{Da} = \left[K_a - \omega^2 M_a\right] \quad \text{and} \quad K_{Db} = \left[K_b - \omega^2 M_b\right]$$

where

$$[K_i] = \frac{EI_i}{L_i^3}\begin{bmatrix} 12 & 6L_i & -12 & 6L_i \\ 6L_i & 4L_i^2 & -6L_i & 2L_i^2 \\ -12 & -6L_i & 12 & -6L_i \\ 6L_i & 2L_i^2 & -6L_i & 4L_i^2 \end{bmatrix}, [M_i] = \frac{m_i L_i}{420}\begin{bmatrix} 156 & 22L_i & 54 & -13L_i \\ 22L_i & 4L_i^2 & 13L_i & -3L_i^2 \\ 54 & 13L_i & 156 & -22L_i \\ -13L_i & -3L_i^2 & -22L_i & 4L_i^2 \end{bmatrix}$$

with $i = a, b$

Determine the transfer matrices of each section ($[T_a]$, $[T_b]$) and of the unit cell ($[T] = [T_b]\,[T_a]$). Show that the plot of the absolute value of the eigenvalues of $[T]$ as function of the frequency ω is as shown in Figure P10.10.

10.10 The periodic beam cell shown in Figure P10.9 consists of two sections (a, b) such that each section is modeled by the following distributed-parameter model:

$$W_{,xxxx} - \beta_i^4 W = 0 \quad \text{with } i = a, b$$

where $\beta_i^4 = m_i\omega^2/EI_i$. Put this distributed-parameter model into the following state-space model:

$$\frac{d}{dx}\begin{Bmatrix} W \\ W_{,x} \\ W_{,xx} \\ W_{,xxx} \end{Bmatrix} = \begin{bmatrix} 0 & 1 & 0 & 0 \\ 0 & 0 & 1 & 0 \\ 0 & 0 & 0 & 1 \\ \beta_i^4 & 0 & 0 & 0 \end{bmatrix}\begin{Bmatrix} W \\ W_{,x} \\ W_{,xx} \\ W_{,xxx} \end{Bmatrix}$$

Show that the state-space model can be reduced to the following transfer matrix formulation:

$$Y_x = [T_i]\, Y_0$$

where $[T_i] = e^{A_i x}$ with the matrix A_i given by:

$$A_i = \begin{bmatrix} 0 & 1 & 0 & 0 \\ 0 & 0 & \frac{1}{EI_i} & 0 \\ 0 & 0 & 0 & 1 \\ EI_i\beta_i^4 & 0 & 0 & 0 \end{bmatrix}$$

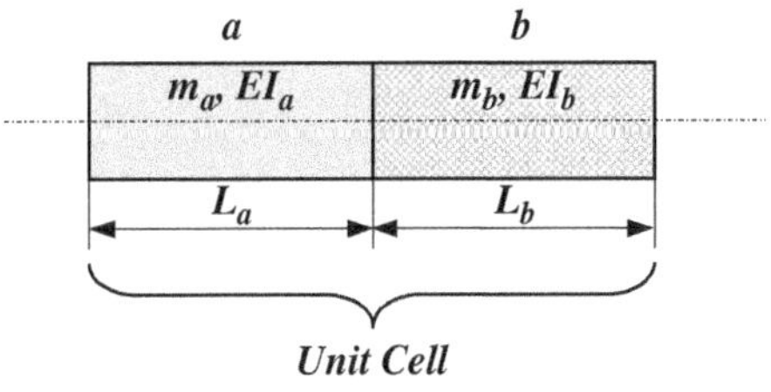

Figure P10.9 A periodic beam cell.

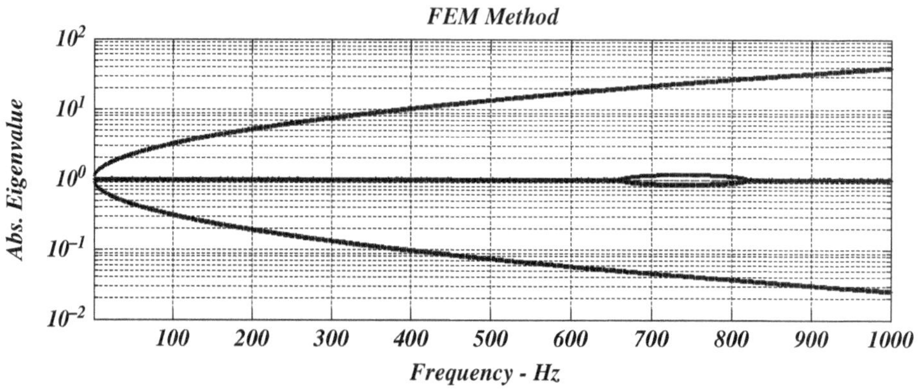

Figure P10.10 A plot of the absolute value of the eigenvalues of transfer matrix of a beam unit cell.

Also, Y_x and Y_0 are the state vectors given by:

$$Y_x = \{W \;\; W_{,x} \;\; M \;\; V\}_x^T, Y_0 = \{W \;\; W_{,x} \;\; M \;\; V\}_0^T$$

with M and V denoting the bending moment and shear force given by:

$$M = EI\, W_{,xx} \text{ and } V = EI\, W_{,xxx}$$

Determine the transfer matrices of each section ($[T_a]$, $[T_b]$) and of the unit cell ($[T] = [T_b]\,[T_a]$). Show that the plot of the absolute value of the eigenvalues of $[T]$ as function of the frequency ω is as shown in Figure P10.10.

11

Nanoparticle Damping Composites

11.1 Introduction

Modeling of the viscoelastic material (VEM) properties of composites consisting of a polymer matrix filled with nanoparticles is essential for predicting the behavior of these composites for different volumetric fractions and physical properties of the constituents. Examples of the nanomaterials include nanotubes and nanoparticles, such as those shown in Figure 11.1. Among the important nanomaterials that can significantly impact the damping characteristics of polymer composites are carbon nanotubes, carbon black (CB) particles, and piezoelectric particles (Aldraihem 2011; Aldraihem et al. 2007).

The modeling techniques of the viscoelastic properties of the polymeric composite materials aim primarily at reducing the heterogeneous composite medium to an equivalent homogenous, anisotropic continuum as illustrated in Figure 11.2. The development approaches of the equivalent properties of the homogenous medium from the geometrical and physical properties of the constituents of the microstructure belong to the well-established fields of "micromechanics" and "homogenization."

The most common approaches of *micromechanics* and *homogenization* can generally be classified as analytical or numerical methods as is briefly outlined in Figure 11.3. Among the adopted analytical methods are the Halpin–Tsai method, the self-consistent method, the Mori–Tanaka model, and the double-inclusion method. Also, the widely used numerical methods include the finite element-based methods, the method of macroscopic degrees of freedom (DOF), generalized method of cells (GMC), high-fidelity generalized method of cells (HFGMC), and the variational asymptotic method for unit cell homogenization (VAMUCH) (Sejnoha and Zeman 2013; Nemat-Nasser and Hori 1999).

Excellent reviews of these analytical and numerical methods are presented by Tucker and Liang (1999), Nemat-Nasser and Hori (1999), and Torquato (2001).

Besides these homogenization methods, bounding techniques for the stiffness of the composite have been developed by Voigt (1887), Reuss (1929), and Hashin and Shtrikman (1962).

In this chapter, particular emphasis is placed on the Mori–Tanaka method (*MTM*) (1973), which is based on Eshelby's equivalent inclusion technique (1957). In the MTM, it is assumed that the average strain in the individual inclusion is proportional to the average strain in the matrix by the concentration Eshelby's strain tensor that relates also the strain in the inclusion to the applied strain.

Active and Passive Vibration Damping, First Edition. Amr M. Baz.

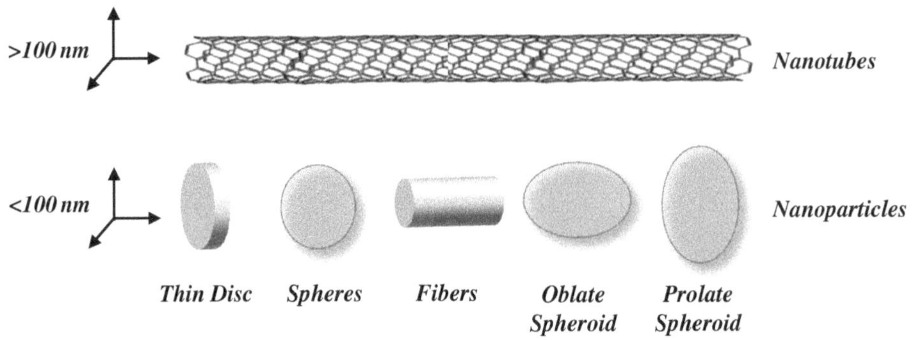

Figure 11.1 Different types of nanoscale inclusions.

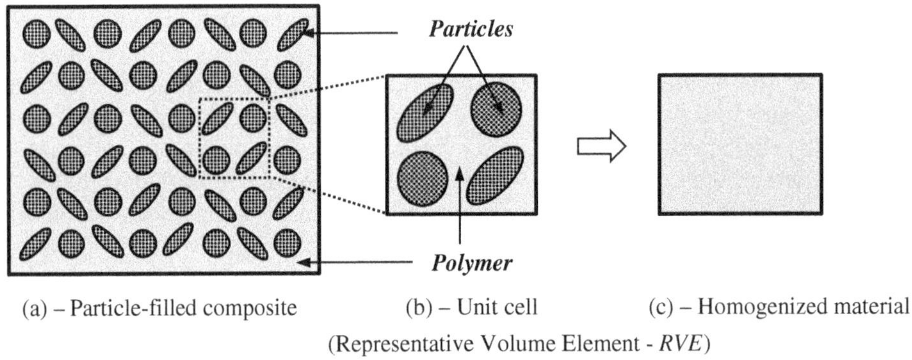

Figure 11.2 Equivalent homogenous, anisotropic continuum for a heterogeneous particle-filled composite. (a) particle-filled composite, (b) unit cell, and (c) homogenized material (Representative volume element – RVE).

It is important to note here that as the particle-filled composites are multi-scale in nature, as the scale of the individual constituents is of a much lower order of magnitude than that of the entire composite structure.

Hence, all these previously mentioned "homogenization" methods, by developing the effective homogeneous medium, simplify computational effort as the focus shifts to analysis of the macroscopic scale of the entire composite rather than on the microscopic scale of the individual constituents, which is computationally exhaustive.

11.2 Nanoparticle-Filled Polymer Composites

The particle-filled polymer composite considered in this section, shown in Figure 11.4, consists of a polymer matrix with embedded N nanoparticle phases. The constituents of the composite are assumed to be perfectly bonded. Furthermore, the nanoparticle

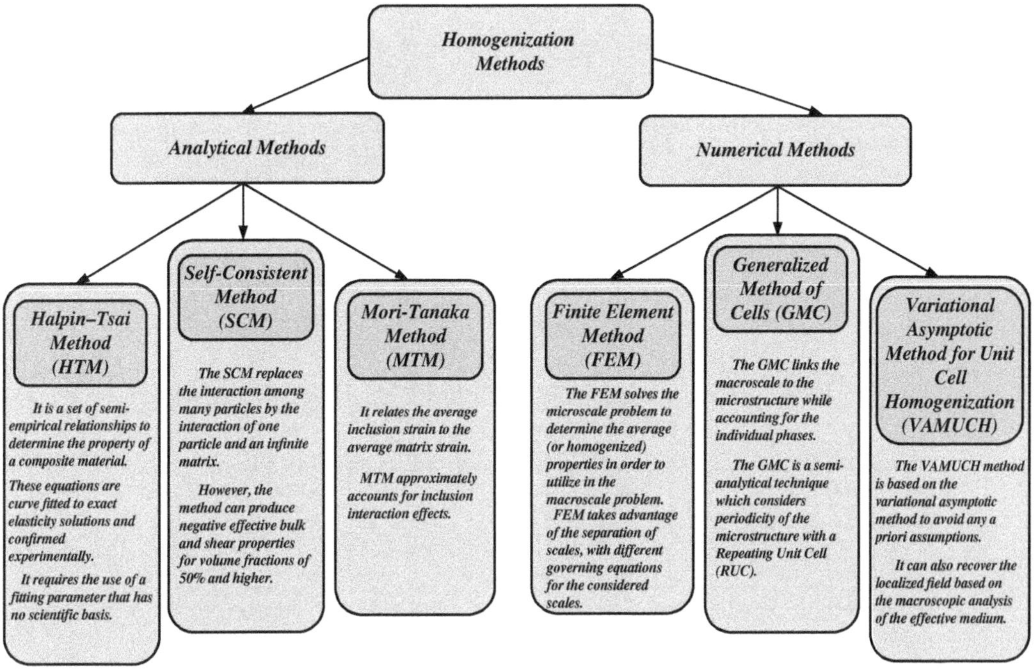

Figure 11.3 Main features of the common homogenization methods.

inclusions can be spheroidal in shape and assume any arbitrary orientation within the matrix.

In this section, all the nanoparticles are considered to be completely passive. But, in Section 11.3, some of these particles are imparted active capabilities such as in piezoelectric nanoparticles and others are utilized to make the polymer conductive. With such a combination, the matrix provides electric current paths and resistive loading in order to transmit and dissipate the electrical energy generated by the piezoelectric inclusions. In other words, the nanoparticle-filled composite acts as flexible shunted piezoelectric networks similar to those discussed in Chapter 9.

11.2.1 Composites with Unidirectional Inclusions

In this section, the theory governing the prediction of the constitutive characteristics of nanoparticle-filled composites is outlined. The composite is assumed to obey the conventional assumptions which are often used in the micromechanics analysis. Also, the composite is assumed to be homogeneous on the macro-scale.

The effective viscoelastic properties are derived using the approach introduced by Dvorak and Benveniste (1992) along with the correspondence principle. For a macro-scale homogeneous composite, a representative volume element (RVE) that contains

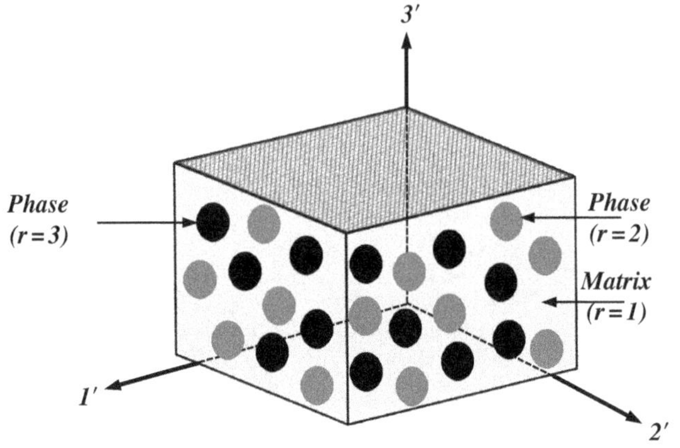

Figure 11.4 Constituents and global coordinate system of particle-filled polymer composite with $N = 3$ constituents.

N perfectly bonded phases can be selected. The volume averaged fields of the hybrid composite are given by:

$$[\bar{\varepsilon}] = \sum_{r=1}^{N} f_r [\bar{\varepsilon}_r], \tag{11.1}$$

and

$$[\bar{\sigma}] = \sum_{r=1}^{N} f_r [\bar{\sigma}_r]. \tag{11.2}$$

where $[\bar{\sigma}]$ and $[\bar{\varepsilon}]$ denote the overall average stress and strain fields, respectively. Also, f_r denotes the volume fraction of phase r. Furthermore, $[\bar{\sigma}_r]$ and $[\bar{\varepsilon}_r]$ denote the average stress and strain fields in phase r, respectively.

The correspondence constitutive equation of the overall composite is given by

$$[\bar{\sigma}] = [c^*]\,[\bar{\varepsilon}] \tag{11.3}$$

where $[c^*]$ is the overall complex stiffness matrix in the principle material (*1*-*2*-3) coordinate system.

Similarly, the correspondence constitutive equation in each phase can be written as

$$[\bar{\sigma}_r] = [c_r^*]\,[\bar{\varepsilon}_r] \tag{11.4}$$

where $[c_r^*]$ denotes the viscoelastic complex stiffness of phase r in the principle material coordinate system.

It has been shown that the overall average stress of the hybrid composite is given by $[\bar{\sigma}] = [\sigma^0]$ (Dvorak and Benveniste 1992). Similarly, the overall average strain of the hybrid composite is given by $[\bar{\varepsilon}] = [\varepsilon^0]$. Thus, under uniform far field $[\varepsilon^0]$ or $[\sigma^0]$ the volume-averages fields in phase r can be expressed as

$$[\bar{\varepsilon}_r] = [A_r^*]\,[\bar{\varepsilon}], \tag{11.5}$$

and

$$[\bar{\sigma}_r] = [B_r^*]\,[\bar{\sigma}]. \tag{11.6}$$

where $[A_r^*]$ and $[B_r^*]$ are the correspondence concentration factors of phase r.

Substituting Eq. (11.1) into Eq. (11.5) and Eq. (11.2) into Eq. (11.6), yields:

$$[\bar{\varepsilon}] = \sum_{r=1}^{N} f_r [A_r^*]\,[\bar{\varepsilon}] \;\rightarrow\; I = \sum_{r=1}^{N} f_r [A_r^*], \tag{11.7}$$

and

$$[\bar{\sigma}] = \sum_{r=1}^{N} f_r\,[B_r^*]\,[\bar{\sigma}] \;\rightarrow\; I = \sum_{r=1}^{N} f_r\,[B_r^*]. \tag{11.8}$$

Hence, Eqs. (11.7) and (11.8) yield the following expression:

$$\sum_{r=1}^{N} f_r [A_r^*] = \sum_{r=1}^{N} f_r [B_r^*] = [I] \tag{11.9}$$

where $[I]$ denotes the identity matrix.

Combining Eqs. (11.4) and (11.5), gives:

$$[\bar{\sigma}_r] = [c_r^*]\,[A_r^*]\,[\bar{\varepsilon}] \tag{11.10}$$

Substituting Eq. (11.10) into Eq. (11.2), yields:

$$[\bar{\sigma}] = \left[\sum_{r=1}^{N} f_r [c_r^*]\,[A_r^*]\right][\bar{\varepsilon}] \tag{11.11}$$

Comparing Eqs. (11.3) and (11.11), yields:

$$[c^*] = \sum_{r=1}^{N} f_r [c_r^*]\,[A_r^*],$$

or

$$[c^*] = f_1\,[c_1^*]\,[A_1^*] + \sum_{r=2}^{N} f_r [c_r^*]\,[A_r^*] \tag{11.12}$$

Using Eq. (11.9) yields:

$$\sum_{r=1}^{N} f_r\left[A_r^*\right] = f_1\left[A_1^*\right] + \sum_{r=2}^{N} f_r\left[A_r^*\right] = [I],$$

or

$$f_1\left[A_1^*\right] = [I] - \sum_{r=2}^{N} f_r\left[A_r^*\right]$$

i.e.

$$f_1\left[c_1^*\right]\left[A_1^*\right] = \left[c_1^*\right] - \sum_{r=2}^{N} f_r\left[c_1^*\right]\left[A_r^*\right] \tag{11.13}$$

Substituting Eq. (11.13) into Eq. (11.13), gives the following expression for the overall viscoelastic stiffness and compliance in the principle material coordinate system:

$$\left[c^*\right] = \left[c_1^*\right] + \sum_{r=2}^{N} f_r\left(\left[c_r^*\right] - \left[c_1^*\right]\right)\left[A_r^*\right] \tag{11.14}$$

In a similar manner, the corresponding expression for the overall viscoelastic compliance $[s^*]$ in the principle material coordinate system is given by:

$$\left[s^*\right] = \left[s_1^*\right] + \sum_{r=2}^{N} f_r\left(\left[s_r^*\right] - \left[s_1^*\right]\right)\left[B_r^*\right], \tag{11.15}$$

where $\left[c_1^*\right]$ and $\left[s_1^*\right]$ denote the complex stiffness and compliance of the matrix, respectively. Note that Eqs. (11.14) and (11.15) satisfy the internal consistency relationships; that is, the effective stiffness $[c^*]$ and compliance $[s^*]$ are the inverse of each other.

The correspondence concentration factors, $\left[A_r^*\right]$ and $\left[B_r^*\right]$, can be estimated by the various micromechanics tools such as the Eshelby's method, the self-consistent method, and the Mori–Tanaka approach, and so on. Among all those, the Mori–Tanaka approach is known to be one of the most elegant and accurate methods that can be used to estimate the overall elastic properties of aligned as well as randomly oriented short fiber composite (Tucker III and Liang 1999; Tandon and Weng 1986). Moreover, the Mori–Tanaka approach yields identical results when either a uniform elastic strain $[\varepsilon^0]$ or a uniform stress $[\sigma^0]$ field is prescribed at the boundary of the composite.

Theorem 11.1 Adopting the Mori–Tanaka approach (Dvorak and Benveniste 1992; Tucker III and Liang 1999), the correspondence concentration factors can be estimated as follows:

$$\left[A_s^*\right] = \left[T_s^*\right]\left[\sum_{r=1}^{N} f_r\left[T_r^*\right]\right]^{-1} \tag{11.16}$$

and

$$\left[B_s^*\right] = \left[W_s^*\right]\left[\sum_{r=1}^{N} f_r\left[W_r^*\right]\right]^{-1} \tag{11.17}$$

(a) – Stress-free state

(b) – Inclusion with stress-free subject to strain ε^T

(c) – Strained inclusion is reinserted inside the matrix

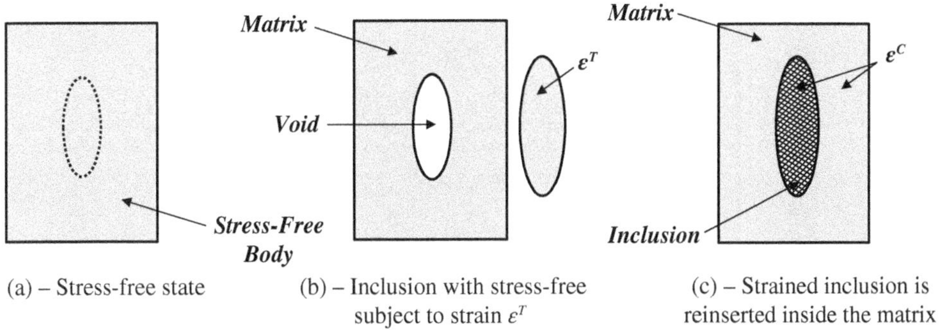

Figure 11.5 Eshelby inclusion problem. (a) Stress-free state, (b) inclusion with stress-free subject to strain ε^T, and (c) strained inclusion is reinserted inside the matrix.

with the correspondence dilute concentration matrices $\left[T_r^*\right]$ and $\left[W_r^*\right]$ given by

$$\left[T_r^*\right] = \left[\left[I\right] + \left[S^*\right]\left[c_1^*\right]^{-1}\left(\left[c_r^*\right] - \left[c_1^*\right]\right)\right]^{-1}, \tag{11.18}$$

and

$$\left[W_r^*\right] = \left[c_r^*\right]\left[T_r^*\right]\left[s_1^*\right]. \tag{11.19}$$

where $[S^*]$ is the "Eshelby tensor" which is function of the Poisson's ratio of the matrix and the shape and geometry of the inclusions.

Proof

Consider the stress-free body has stiffness c_1^* as shown in Figure 11.5a. A small region of the body, inside the dotted contour, is isolated as a separate body and is subjected to a uniform transformation strain ε^T as indicated in Figure 11.5b. This strained region is reinserted inside the body inducing a constrained strain field ε^C. The remaining part of the body is now called a "matrix" and the isolated strained zone is called an "inclusion." Analysis of this sequence of events is called the "Eshelby inclusion problem" (Eshelby 1957).

Inside the matrix, the stress $[\sigma_1]$ is given by the following expression:

$$\left[\sigma_1\right] = \left[c_1^*\right]\left[\varepsilon^C\right] \tag{11.20}$$

However, within the inclusion the stress $[\sigma_r]$ is given by the following expression:

$$\left[\sigma_r\right] = \left[c_1^*\right]\left(\left[\varepsilon^C\right] - \left[\varepsilon^T\right]\right) \tag{11.21}$$

According to Eshelby's approach, the constrained strain $[\varepsilon^C]$ is uniquely related to the transformation strain $[\varepsilon^T]$ such that:

$$\left[\varepsilon^C\right] = \left[S^*\right]\left[\varepsilon^T\right] \tag{11.22}$$

where $[S^*]$ is the "Eshelby tensor," which is a function of the Poisson's ratio of the matrix and the shape and geometry of the inclusions.

Furthermore, Eshelby developed an equivalence theorem between the problems of homogeneous inclusion and an inhomogeneous inclusion of the same shape. Such an equivalence can best be understood by considering the bodies shown in Figure 11.6a,

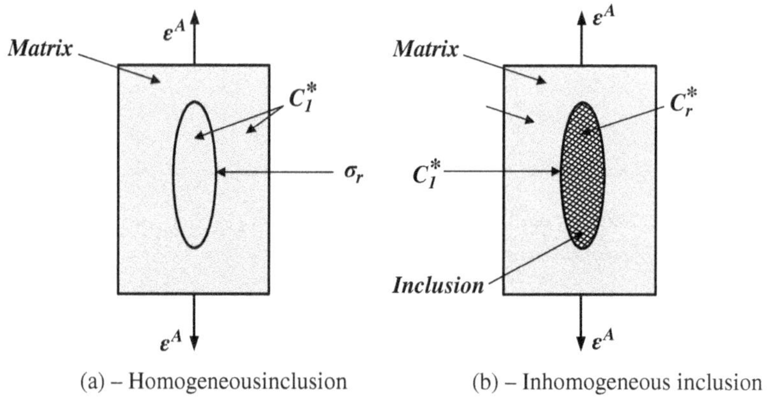

Figure 11.6 Eshelby equivalent inclusion problem. (a) homogeneous inclusion and (b) inhomogeneous inclusion.

b for the homogeneous and inhomogeneous inclusions, respectively. These inclusions have stiffnesses $\left[c_1^*\right]$ and $\left[c_r^*\right]$, respectively.

In Figure 11.6a, the homogeneous inclusion is subjected to a transformation strain $[\varepsilon^T]$ whereas the inhomogeneous inclusion of Figure 11.6b has no transformation strain. The two bodies of Figure 11.6 are subject to a uniform applied strain $[\varepsilon^A]$. Eshelby developed an expression for the transformation strain $[\varepsilon^T]$ that results in the same stress and strain distributions in both problems. Hence, the constitutive Eq. (11.21) for the homogeneous inclusion can be modified to account for the applied strain $[\varepsilon^A]$ to yield:

$$\textbf{\textit{For homogeneous inclusion}}: [\sigma_r] = \left[c_1^*\right]\left(\left[\varepsilon^A\right] + \left[\varepsilon^C\right] - \left[\varepsilon^T\right]\right) \tag{11.23}$$

Accordingly, the corresponding constitutive equation for the inhomogeneous inclusion can be written as follows:

$$\textbf{\textit{For inhomogeneous inclusion}}: [\sigma_r] = \left[c_r^*\right]\left(\left[\varepsilon^A\right] + \left[\varepsilon^C\right]\right) \tag{11.24}$$

The equivalence of the stresses between the two problems yields:

$$\left[c_1^*\right]\left(\left[\varepsilon^A\right] + \left[\varepsilon^C\right] - \left[\varepsilon^T\right]\right) = \left[c_r^*\right]\left(\left[\varepsilon^A\right] + \left[\varepsilon^C\right]\right) \tag{11.25}$$

Substituting Eq. (11.22) into Eq. (11.25) gives the following expression:

$$-\left[\left[c_1^*\right] + \left(\left[c_r^*\right] - \left[c_1^*\right]\right)[S^*]\right]\left[\varepsilon^T\right] = \left(\left[c_r^*\right] - \left[c_1^*\right]\right)\left[\varepsilon^A\right] \tag{11.26}$$

Equation (11.26) gives the expression of the transformation strain $[\varepsilon^T]$ which is necessary for the equivalence in terms of the applied strain transformation strain $[\varepsilon^A]$.

Note that, in the far field, the average strains in the matrix and the rth inclusion can be written as:

$$[\bar{\varepsilon}_1] = \left[\varepsilon^A\right] \tag{11.27}$$

and

$$[\bar{\varepsilon}_r] = \left[\varepsilon^A\right] + \left[\varepsilon^C\right] \tag{11.28}$$

Then, substituting Eq. (11.28) into Eq. (11.25), it reduces to:

$$[c_1^*]\left([\bar{\varepsilon}_r]-[\varepsilon^T]\right)=[c_r^*][\bar{\varepsilon}_r]$$

or

$$-[\varepsilon^T]=[c_1^*]^{-1}\left([c_r^*]-[c_1^*]\right)[\bar{\varepsilon}_r]$$

that is,

$$-[\varepsilon^T]=[c_1^*]^{-1}\left([c_r^*]-[c_1^*]\right)[\bar{\varepsilon}_r]$$

Hence,

$$-[S^*][\varepsilon^T]=[S^*][c_1^*]^{-1}\left([c_r^*]-[c_1^*]\right)[\bar{\varepsilon}_r] \qquad (11.29)$$

Combining Eqs. (11.22) and (11.29) yields:

$$-[\varepsilon^C]=[S^*][c_1^*]^{-1}\left([c_r^*]-[c_1^*]\right)[\bar{\varepsilon}_r] \qquad (11.30)$$

Substituting Eqs. (11.27) and (11.28) into Eq. (11.30), it reduces to:

$$[\bar{\varepsilon}_1]=\left(I+[S^*][c_1^*]^{-1}\left([c_r^*]-[c_1^*]\right)\right)[\bar{\varepsilon}_r]$$

or

$$[\bar{\varepsilon}_r]=\left(I+[S^*][c_1^*]^{-1}\left([c_r^*]-[c_1^*]\right)\right)^{-1}[\bar{\varepsilon}_1] \qquad (11.31)$$

In the Mori–Tanaka assumption, it is assumed that when many identical particles (inclusions) are embedded into the composite, the average particle strain is given by:

$$[\bar{\varepsilon}_r]=[T_r^*][\bar{\varepsilon}_1] \qquad (11.32)$$

that is, each particle experiences a far-field strain that is equal to the average strain of the matrix. In other words, a direct comparison between Eqs. (11.31) and (11.32) gives the following expression of the strain concentration factor:

$$[T_r^*]=\left[[I]+[S^*][c_1^*]^{-1}\left([c_r^*]-[c_1^*]\right)\right]^{-1} \qquad (11.33)$$

This completes the proof of Eq. (11.18).

Now for proving Eq. (11.16), two approaches can be adopted:

First Approach:

This approach is simple and aiming at showing that the two sides of the equation are equal. It begins by multiplying both sides by f_s and summing up over the number of phases $s = 1,\ldots N$ inside the composite to yield the following:

$$\begin{aligned}\sum_{s=1}^{N} f_s[A_s^*] &= \sum_{s=1}^{N} f_s[T_s^*]\left[\sum_{r=1}^{N} f_r[T_r^*]\right]^{-1} \\ &= \left[\sum_{s=1}^{N} f_s[T_s^*]\right]\left[\sum_{r=1}^{N} f_r[T_r^*]\right]^{-1} = [I]\end{aligned} \qquad (11.34)$$

that is, the right hand side of Eq. (11.34) is equal to an identity [*I*].

But, from Eq. (11.9), it is also seen that the left-hand side of Eq. (11.34) is also equal to an identity *I*. Hence, Eq. (11.16) is valid.

Second Approach:

From Eq. (11.1), as:

$$[\bar{\varepsilon}] = \sum_{r=1}^{N} f_r [\bar{\varepsilon}_r],$$

Then, substituting Eq. (11.32) in this equation, yields:

$$[\bar{\varepsilon}] = \left[\sum_{r=1}^{N} f_r [T_r^*]\right] [\bar{\varepsilon}_1] \quad \text{or,} \quad [\bar{\varepsilon}_1] = \left[\sum_{r=1}^{N} f_r [T_r^*]\right]^{-1} [\bar{\varepsilon}] \tag{11.35}$$

Pre-multiplying both sides by $[T_s^*]$ and using Eq. (11.32) gives the following expression:

$$[T_s^*][\bar{\varepsilon}_1] = [T_s^*]\left[\sum_{r=1}^{N} f_r [T_r^*]\right]^{-1} [\bar{\varepsilon}] \tag{11.36}$$

From Eqs. (11.32) and (11.5), Eq. (11.36) reduces to:

$$[T_s^*][\bar{\varepsilon}_1] = [\bar{\varepsilon}_s] = [T_s^*]\left[\sum_{r=1}^{N} f_r [T_r^*]\right]^{-1} [\bar{\varepsilon}]$$

$$\text{i.e.} \quad [\bar{\varepsilon}_s] = [T_s^*]\left[\sum_{r=1}^{N} f_r [T_r^*]\right]^{-1} [\bar{\varepsilon}] = [A_s^*][\bar{\varepsilon}] \quad or \quad [A_s^*] = [T_s^*]\left[\sum_{r=1}^{N} f_r [T_r^*]\right]^{-1} \quad \textbf{Q.E.D.}$$

Note that the Eshelby tensors for spheroidal inclusions, in an isotopic matrix, can be expressed as follows (Nemat-Nasser and Hori 1999)

$$[S] = \begin{bmatrix} S_{1111} & S_{1122} & S_{1133} & 0 & 0 & 0 \\ S_{1122} & S_{1111} & S_{1133} & 0 & 0 & 0 \\ S_{3311} & S_{3311} & S_{3333} & 0 & 0 & 0 \\ 0 & 0 & 0 & S_{1313} & 0 & 0 \\ 0 & 0 & 0 & 0 & S_{1313} & 0 \\ 0 & 0 & 0 & 0 & 0 & S_{1212} \end{bmatrix} \tag{11.37}$$

The elements of the Eshelby tensor are a function of the inclusion aspect ratio $\alpha = a_3/a_1$ and the Poisson's ratio ν_m of the matrix as tabulated in Table 11.1. where g_1 and g_2 appearing in the table are defined as follows:

$$g_1 = \frac{\alpha}{(1-\alpha^2)^{3/2}}\left(\cos^{-1}(\alpha) - \alpha\left(1-\alpha^2\right)^{1/2}\right) \tag{11.38}$$

and

$$g_2 = \frac{\alpha}{(\alpha^2-1)^{3/2}}\left(-\cosh^{-1}(\alpha) + \alpha\left(\alpha^2-1\right)^{1/2}\right) \tag{11.39}$$

Table 11.1 Elements of the Eshelby tensor for common inclusion shapes (Aldraihem 2011). Reproduced with permission of Elsevier.

Shape	Thin disc ($\alpha = 0$)	Sphere ($\alpha = 1$)	Fiber ($\alpha \rightarrow$)	Oblate spheroid ($0 < \alpha < 1$, $g = g_1$) and Prolate spheroid ($1 < \alpha <$, $g = g_2$)
S_{1111}	0	$\frac{7-5v_m}{15(1-v_m)}$	$\frac{5-4v_m}{8(1-v_m)}$	$\frac{3\alpha^2}{8(1-v_m)(\alpha^2-1)}+\frac{g}{4(1-v_m)}\left(1-2v_m-\frac{9}{4(\alpha^2-1)}\right)$
S_{1122}	0	$\frac{5v_m^{-1}}{15(1-v_m)}$	$\frac{4v_m^{-1}}{8(1-v_m)}$	$\frac{\alpha^2}{8(1-v_m)(\alpha^2-1)}-\frac{g}{4(1-v_m)}\left(1-2v_m+\frac{3}{4(\alpha^2-1)}\right)$
S_{1133}	0	$\frac{5v_m^{-1}}{15(1-v_m)}$	$\frac{v_m}{2(1-v_m)}$	$\frac{-\alpha^2}{2(1-v_m)(\alpha^2-1)}+\frac{g}{4(1-v_m)}\left(-1+2v_m+\frac{3\alpha^2}{(\alpha^2-1)}\right)$
S_{3311}	$\frac{v_m}{1-v_m}$	$\frac{5v_m^{-1}}{15(1-v_m)}$	0	$2\frac{v_m^{-1}}{2(1-v_m)}-\frac{1}{2(1-v_m)(\alpha^2-1)}+\frac{g}{2(1-v_m)}\left(1-2v_m+\frac{3}{2(\alpha^2-1)}\right)$
S_{3333}	1	$\frac{7-5v_m}{15(1-v_m)}$	0	$\frac{1-2v_m}{2(1-v_m)}+\frac{3\alpha^2-1}{2(1-v_m)(\alpha^2-1)}-\frac{g}{2(1-v_m)}\left(1-2v_m+\frac{3\alpha^2}{(\alpha^2-1)}\right)$
S_{1313}	$\frac{1}{2}$	$\frac{4-5v_m}{15(1-v_m)}$	$\frac{1}{4}$	$\frac{1-2v_m}{4(1-v_m)}-\frac{(\alpha^2+1)}{4(1-v_m)(\alpha^2-1)}-\frac{g}{8(1-v_m)}\left(1-2v_m-\frac{3(\alpha^2+1)}{(\alpha^2-1)}\right)$
S_{1212}	0	$\frac{4-5v_m}{15(1-v_m)}$	$\frac{3-4v_m}{8(1-v_m)}$	$\frac{\alpha^2}{8(1-v_m)(\alpha^2-1)}+\frac{g}{4(1-v_m)}\left(1-2v_m-\frac{3}{4(\alpha^2-1)}\right)$

11.2.2 Arbitrarily Oriented Inclusion Composites

Two processes are needed to find the effective properties of a hybrid composite with any desired inclusion orientation distribution. The first process is the evaluation of the effective properties for the hybrid composite with unidirectional aligned inclusions. This step is carried out in the previous section. The second process involves averaging the properties obtained from the first process in terms of the orientation distribution. Advani and Tucker III (1987) suggested a straightforward scheme for orientation averaging in short fiber composites. Furthermore, their averaging scheme can deliver reliable and accurate predictions for thermoelastic properties (Gusev et al. 2002). According to the approach of Advani and Tucker III (1987), the orientation average of $\Gamma(\theta, \phi)$ is denoted by $\langle\Gamma\rangle$ and is defined as follows:

$$\langle\Gamma\rangle = \int_{\theta_i}^{\theta_f}\int_{\varphi_i}^{\varphi_f} \Gamma(\theta,\phi)\,\Psi(\theta,\phi)\,\sin\phi\,d\varphi\,d\theta \tag{11.40}$$

where ϕ and θ denote the two Euler angles that are usually used to describe the spatial orientation of an inclusion as shown in Figure 11.7. The angles (θ_i, ϕ_i) and (θ_f, ϕ_f) denote the lower and upper limits, respectively.

The inclusion distribution region is defined by $(-\theta_0 = \theta_i) \leq \theta \leq (\theta_f = \theta_0)$ and $(\pi/2 - \phi_0 = \phi_i) \leq \phi \leq (\phi_f = \pi/2 + \phi_0)$, where θ_0 and ϕ_0 are prescribed angles. The symbol, $\Psi(\theta, \phi)$, denotes the probability distribution function. This function must satisfy specific conditions leading to:

$$\Psi(\theta, \phi) = \frac{1}{4\,\theta_0\,\sin\phi_0} \tag{11.41}$$

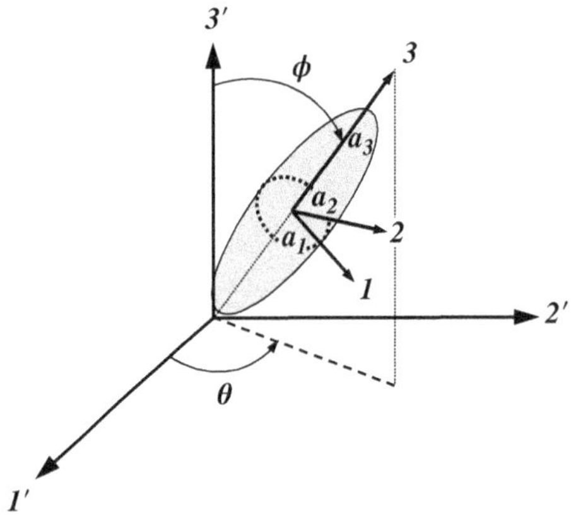

Figure 11.7 Geometry and coordinate systems of a spheroidal inclusion.

Using Eq. (11.41) into Eq. (11.40) yields the final expression of the orientation average:

$$\langle \Gamma \rangle = \frac{1}{4\,\theta_0\ \sin\phi_0} \int_{-\theta_0}^{\theta_0} \int_{\pi/2-\phi_0}^{\pi/2+\phi_0} \Gamma(\theta,\phi)\sin\phi\, d\phi\, d\theta \tag{11.42}$$

In this equation, the function $\Gamma(\theta, \phi)$ can be used to represent the transformed viscoelastic stiffness, compliance, or any other property tensor in the global (*1′-2′-3′*) coordinate system. An expression for the transformed viscoelastic stiffness, $[\bar{c}^*]$, is determined by employing the transformation (Wetherhold 1988):

$$[\bar{c}^*] = [R_1][c^*][R_1]^T \tag{11.43}$$

Similarly, the transformed viscoelastic compliance obtained by

$$[\bar{s}^*] = [R_2][s^*][R_2]^T \tag{11.44}$$

where the transformation matrices $[R_1]$ and $[R_2]$ are defined in Appendix 11.A, and superscript T denotes the transpose.

Substituting Eqs. (11.43) and (11.44) into Eq. (11.42), one obtains the orientation average of the viscoelastic stiffness:

$$\langle \bar{c}^* \rangle = \frac{1}{4\,\theta_0\ \sin\phi_0} \int_{-\theta_0}^{\theta_0} \int_{\pi/2-\phi_0}^{\pi/2+\phi_0} [R_1][c^*][R_1]^T \sin\phi\, d\phi\, d\theta \tag{11.45}$$

and the viscoelastic compliance:

$$\langle \bar{s}^* \rangle = \frac{1}{4\,\theta_0\ \sin\phi_0} \int_{-\theta_0}^{\theta_0} \int_{\pi/2-\phi_0}^{\pi/2+\phi_0} [R_2][s^*][R_2]^T \sin\phi\, d\phi\, d\theta \tag{11.46}$$

Expressions (11.45) and (11.46) do not guarantee satisfying the internal consistency relationships. In other words, the orientation averaged stiffness (11.45) and compliance (11.46) are not necessarily each other's inverse for all orientation distribution.

For predictions of the elastic properties, Hine et al. (2002, 2004) have shown that the orientation averaged stiffness and compliance equations, which are similar to (11.45) and

(11.46), provide significantly different results for the thermoelastic properties of short fiber composites. Furthermore, Hine and coworkers have demonstrated that the stiffness equation provides accurate predictions of the orientation averaged properties when compared with the results from their numerical simulation.

According to this discussion, Eq. (11.45) will be used to explicitly determine the orientation averaged viscoelastic properties of the hybrid composites. Furthermore, the viscoelastic properties will be determined by evaluating the limits of Eq. (11.45) as the angles θ_0 and ϕ_0 approach certain values. Three general orientations will be considered. In aligned orientations, all the inclusions are uni-directionally aligned along the $1'$-axis. To achieve this case, the values of θ_0 and ϕ_0 are set to approach zero. In two-dimensional random orientations, the inclusions are randomly oriented in the $1'$-$2'$ plane. Here, the values of θ_0 and ϕ_0 are set to approach π and zero, respectively. In three-dimensional random orientations, the inclusions are randomly oriented in all directions. In this orientation, the values of θ_0 and ϕ_0 are set to approach π and $\pi/2$, respectively. For these three orientations, the viscoelastic stiffness tensor can be written as:

$$\langle \bar{c}^* \rangle = \begin{bmatrix} c^*_{1'1'} & c^*_{1'2'} & c^*_{1'3'} & 0 & 0 & 0 \\ c^*_{1'2'} & c^*_{2'2'} & c^*_{2'3'} & 0 & 0 & 0 \\ c^*_{1'3'} & c^*_{2'3'} & c^*_{3'3'} & 0 & 0 & 0 \\ 0 & 0 & 0 & c^*_{4'4'} & 0 & 0 \\ 0 & 0 & 0 & 0 & c^*_{5'5'} & 0 \\ 0 & 0 & 0 & 0 & 0 & c^*_{6'6'} \end{bmatrix} \tag{11.47}$$

where the components of $\langle \bar{c}^* \rangle$ are listed in Table 11.2.

Table 11.2 Components of complex moduli $\langle \bar{c}^* \rangle$ for different orientations (Aldraihem 2011). Reproduced with permission of Elsevier.

$c^*_{i'j'}$	**Aligned orientation** $(\theta_0 \to 0, \phi_0 \to 0)$	**2D random orientation** $(\theta_0 \to \pi,\ \phi_0 \to 0)$	**3D random orientation** $(\theta_0 \to \pi, \phi_0 \to \pi/2)$
$c^*_{1'1'} =$	c^*_{33}	$\frac{1}{8}(3c^*_{11} + 3c^*_{33} + 4c^*_{44} + 2c^*_{13})$	$\frac{1}{15}(8c^*_{11} + 3c^*_{33} + 8c^*_{44} + 4c^*_{13})$
$c^*_{1'2'} =$	c^*_{13}	$\frac{1}{8}(c^*_{11} + c^*_{33} - 4c^*_{44} + 6c^*_{13})$	$\frac{1}{15}(c^*_{11} + c^*_{33} - 4c^*_{44} + 5c^*_{12} + 8c^*_{13})$
$c^*_{2'2'} =$	c^*_{11}	$\frac{1}{8}(3c^*_{11} + 3c^*_{33} + 4c^*_{44} + 2c^*_{13})$	$\frac{1}{15}(8c^*_{11} + 3c^*_{33} + 8c^*_{44} + 4c^*_{13})$
$c^*_{1'3'} =$	c^*_{13}	$\frac{1}{2}(c^*_{12} + c^*_{13})$	$\frac{1}{15}(c^*_{11} + c^*_{33} - 4c^*_{44} + 5c^*_{12} + 8c^*_{13})$
$c^*_{2'3'} =$	c^*_{12}	$\frac{1}{2}(c^*_{12} + c^*_{13})$	$\frac{1}{15}(c^*_{11} + c^*_{33} - 4c^*_{44} + 5c^*_{12} + 8c^*_{13})$
$c^*_{3'3'} =$	c^*_{11}	c^*_{11}	$\frac{1}{15}(8c^*_{11} + 3c^*_{33} + 8c^*_{44} + 4c^*_{13})$
$c^*_{4'4'} =$	c^*_{66}	$\frac{1}{4}(c^*_{11} - c^*_{12} + 2c^*_{44})$	$\frac{1}{30}(7c^*_{11} + 2c^*_{33} + 12c^*_{44} - 5c^*_{12} - 4c^*_{13})$
$c^*_{5'5'} =$	c^*_{44}	$\frac{1}{4}(c^*_{11} - c^*_{12} + 2c^*_{44})$	$\frac{1}{30}(7c^*_{11} + 2c^*_{33} + 12c^*_{44} - 5c^*_{12} - 4c^*_{13})$
$c^*_{6'6'} =$	c^*_{44}	$\frac{1}{8}(c^*_{11} + c^*_{33} + 4c^*_{44} - 2c^*_{13})$	$\frac{1}{30}(7c^*_{11} + 2c^*_{33} + 12c^*_{44} - 5c^*_{12} - 4c^*_{13})$

Several properties can be extracted from the three-dimensional moduli of the hybrid composite (11.47). For example, the complex modulus in the global coordinate system can be expressed as

$$\bar{c}_q^* = \bar{c}'_q + i\bar{c}''_q; q = 1', \ldots, 6' \tag{11.48}$$

and the loss factor is accordingly defined as

$$\eta_q = \frac{\bar{c}''_q}{\bar{c}'_q} \tag{11.49}$$

with modulus related to the compliance via the following expression:

$$c_q^* = 1/\bar{s}_q^*, \tag{11.50}$$

and

$$[\bar{s}^*] = \langle \bar{c}^* \rangle^{-1}. \tag{11.51}$$

where c'_q denotes the storage modulus, c''_q denotes the loss modulus, and subscript q denotes the contracted index.

Based on the theory presented in Section 11.2, the computation of the viscoelastic properties of particle-filled polymer composites is carried out according to the flow chart displayed in Figure 11.8.

Table 11.3 lists the material properties of typical nanoparticles which are commonly used in manufacturing nanoparticle polymer composites.

Example 11.1 Consider a carbon black/high-density polyethylene (CB/HDPE) composite. The HDPE polymer has the storage modulus and loss factor characteristics shown in Figure 11.9 (Zhang and Yi 2002) and CB nanoparticles that have the properties

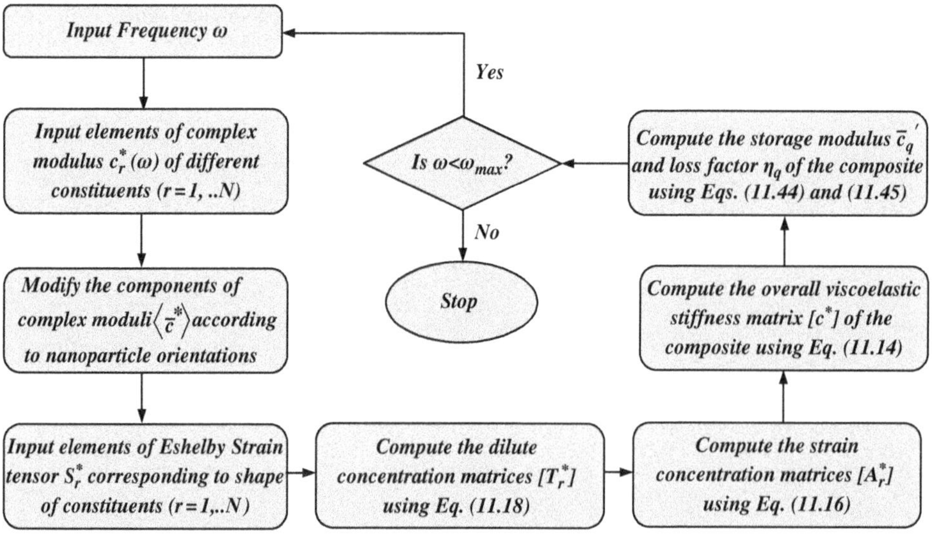

Figure 11.8 Flow chart for computing the viscoelastic properties of particle-filled polymer composites.

Table 11.3 Material properties of the ingredients of the hybrid composite (Aldraihem 2011). Reproduced with permission of Elsevier.

Property	Particle					
	PZT-5H[a]	PCM51[b]	MWCNT[c]	CB[d]	DGEBA-DDM[e]	Hercules 3501-6[f]
$c_{11}^E = c_{22}^E$ (GPa)	127.2	129.22	999.66	9.2885	2.4409	7.4879
c_{12}^E (GPa)	80.2	86.4	369.74	3.9808	1.6962	4.5894
$c_{13}^E = c_{23}^E$ (GPa)	84.67	83.06	369.74	3.9808	1.6962	4.5894
c_{33}^E (GPa)	117.44	116.9	999.66	9.2885	2.4409	7.4879
$c_{44}^E = c_{55}^E$ (GPa)	23	28.83	314.96	2.6538	0.3723	1.4493
c_{66}^E (GPa)	23.5	21.41	314.96	2.6538	0.3723	1.4493
k_{31}	0.388	0.37				
k_{33}	0.752	0.72				
k_{15}	0.675	0.72				
η_m	–	–			0.047	0.03

[a] Electro Ceramic Division, Morgan Matroc Inc., Bedford, OH.
[b] Noliac, http://www.noliac.com (PCM51 is also called NCE51).
[c] Tian and Wang (2008).
[d] Aldraihem et al. (2007).
[e] Pascault et al. (2002).
[f] Lesieutre et al. (1993).

listed in Table 11.3. Determine the effect of the volume fractions f_r of the CB nanoparticles on the viscoelastic characteristics of the composite. Assume $0 < f_r < 0.2$.

Solution

The isotropic stiffness matrix for the HDPE polymer is given by:

$$\bar{c}_1^* = \begin{bmatrix} c_{11}^* & c_{12}^* & c_{13}^* & 0 & 0 & 0 \\ c_{12}^* & c_{22}^* & c_{23}^* & 0 & 0 & 0 \\ c_{13}^* & c_{23}^* & c_{33}^* & 0 & 0 & 0 \\ 0 & 0 & 0 & c_{44}^* & 0 & 0 \\ 0 & 0 & 0 & 0 & c_{55}^* & 0 \\ 0 & 0 & 0 & 0 & 0 & c_{66}^* \end{bmatrix}$$

where $c_{11}^* = \frac{E^*(1-\nu)}{[(1+\nu)(1-2\nu)]}$, $c_{12}^* = \frac{\nu E^*}{[(1+\nu)(1-2\nu)]}$, $c_{22}^* = c_{11}^*$, $c_{33}^* = c_{11}^*$, $c_{13}^* = c_{12}^*$, $c_{44}^* = \frac{E^*}{2(1+\nu)}$, $c_{55}^* = c_{44}^*$, and $c_{66}^* = c_{44}^*$.

Note that $E^* = E_1'(1 + i\eta_1)$ where E^*= complex modulus of HDPE, E_1' = storage modulus, and η_1= loss factor. Values of E_1' and η_1 are displayed in Figure 11.5 as a function of frequency ω.

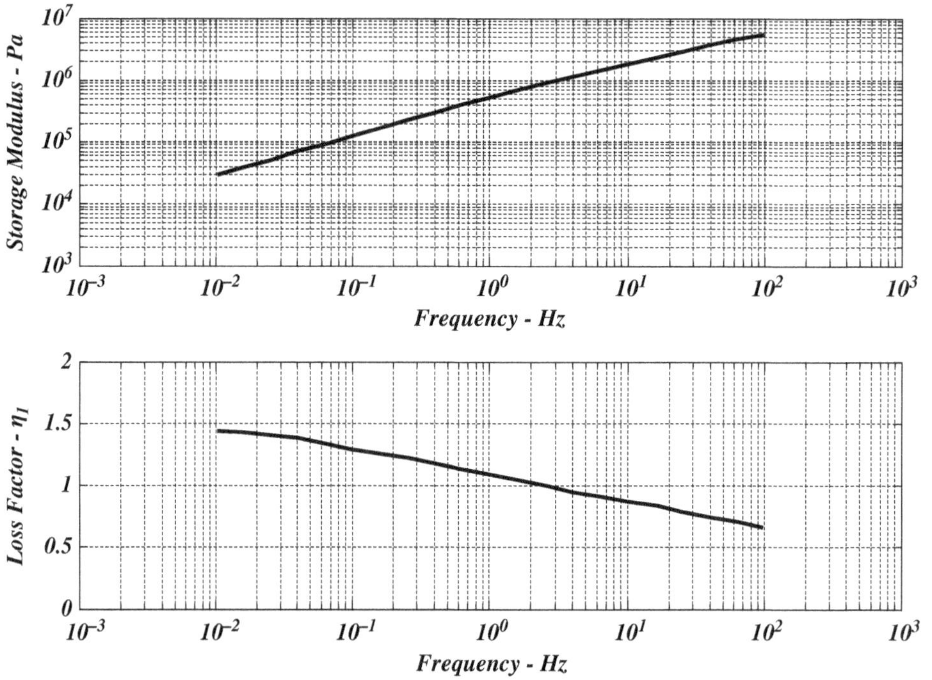

Figure 11.9 Complex modulus of high-density polyethylene (HDPE) polymer.

Similarly, for the CB nanoparticles, the stiffness matrix $\bar{c}_2^*$ of the second phase of the composite is given by:

$$\bar{c}_2^* = \begin{bmatrix} 9.2885 & 3.9808 & 3.9808 & 0 & 0 & 0 \\ 3.9808 & 9.2885 & 3.9808 & 0 & 0 & 0 \\ 3.9808 & 3.9808 & 9.2885 & 0 & 0 & 0 \\ 0 & 0 & 0 & 2.6538 & 0 & 0 \\ 0 & 0 & 0 & 0 & 2.6538 & 0 \\ 0 & 0 & 0 & 0 & 0 & 2.6538 \end{bmatrix} GPa$$

Note that $\bar{c}_2^*$ is a matrix with real elements.

Figure 11.10 displays the values of the complex modulus of the CB-filled HDPE polymer composite as predicted by the micromechanics approach outlined in Section 11.2 for CB volume fractions varying between 0 and 20%. These predictions are in excellent agreement with the corresponding experimental results shown in Figure 11.11.

It is important to note that plotting the storage modulus against the loss modulus results in a unified master line on which the characteristics of the CB/HDPE composite collapse for all the considered CB volume fractions. The equation of this line is:

$$E'' \cong E' \; MPa$$

The equation is valid for both the experimental results and micromechanics predictions. This leads to a loss factor $\eta \cong 1$.

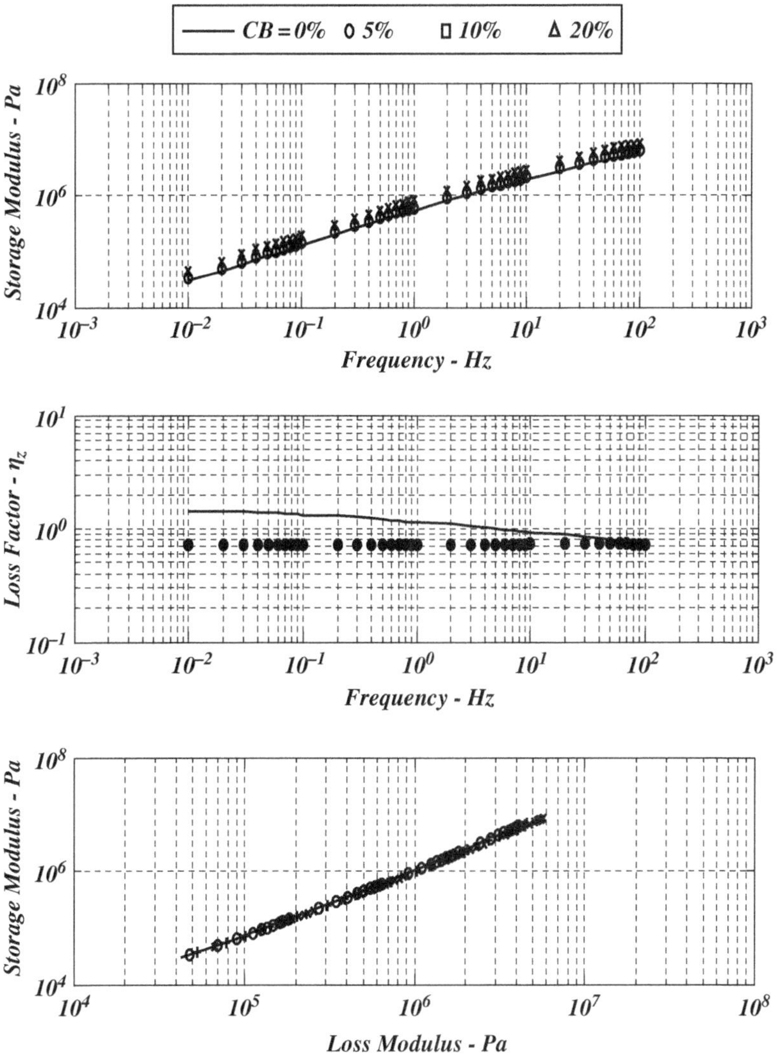

Figure 11.10 The predicted complex modulus of carbon black-filled high-density polyethylene polymer composite.

Furthermore, the displayed results indicate that increasing the CB concentration makes the composite stiffer, but at the expense of reducing its damping behavior as quantified by its loss factor.

Example 11.2 Investigate the relationship between the viscoelastic characteristics of the CB particle-filled high-density polyethylene (CB/HDPE) polymer composite and the corresponding limiting characteristics of the Voigt and Reuss models described in Chapter 8 (Section 8.4.1). Use the plots of the absolute modulus versus

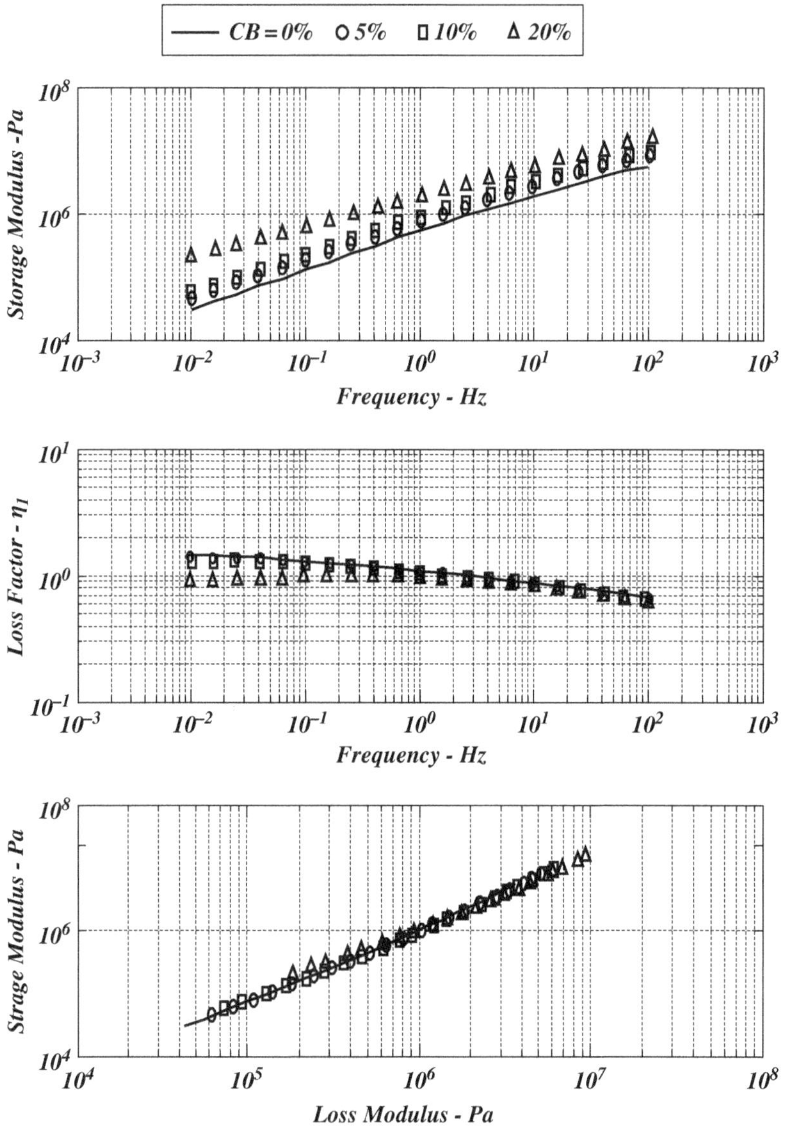

Figure 11.11 The experimental complex modulus of carbon black-filled high-density polyethylene polymer composite.

the CB volume fraction and loss factor versus the CB volume fraction to illustrate the relative characteristics for frequencies of 0.01, 1, and 100 Hz.

Solution

Consider the Voigt and Reuss composites shown in Figure 11.12. Let c_1^* and c_2^* denote the stiffness tensors of the HDPE and CB, respectively. These tensors are as described in Example 11.1.

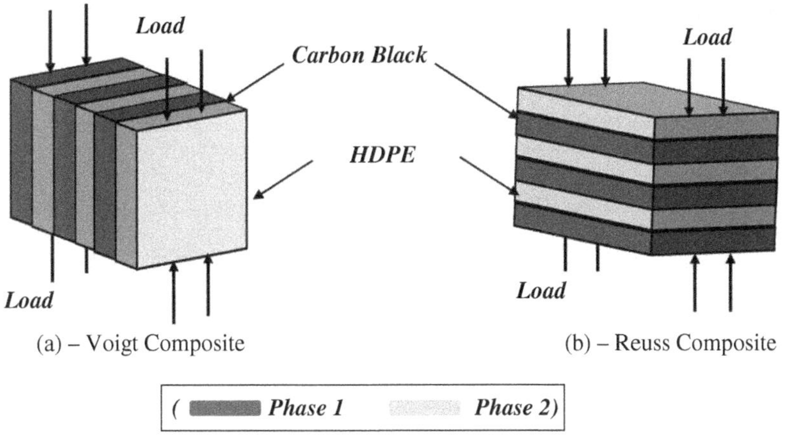

Figure 11.12 Configurations of damped structural composites (Phase 1 Phase 2). (a) Voigt composite and (b) Reuss composite.

Then, the effective stiffness tensor c_V^* of the Voigt model, which assumes the uniformity of the strain is given by:

$$c_V^* = \left[f_1 c_1^* + f_2 c_2^*\right] \tag{11.52}$$

where f_1 and f_2 are the volume fractions of HDPE and CB such that $f_1 + f_2 = 1$.

Similarly, the effective compliance tensor s_R^* of the Reuss model, which assumes the uniformity of the stress is given by:

$$s_R^* = \left[f_1 s_1^* + f_2 s_2^*\right] \tag{11.53}$$

where s_1^* and s_2^* are the compliances of HDPE and CB.

Hence, the effective stiffness tensor c_R^* of the Reuss model is given by:

$$c_R^* = \left[s_R^*\right]^{-1} \tag{11.54}$$

Using Eqs. (11.52) through (11.54) yields the characteristics displayed in Figure 11.13 for frequencies of 0.01, 1, and 100 Hz. It can be seen that the viscoelastic characteristics of the CB/HDPE composite lie inside the Voigt and Reuss bounds.

11.3 Comparisons with Classical Filler Reinforcement Methods

In this section, the predictions of the complex modulus of CB/polymer composites are determined using the classical filler reinforcement methods, which are either analytical in nature or based on finite element modeling of the RVE. These predictions are compared with those generated by the Eshelby's strain tensor approach outlined in Section 11.2.

The analytical approach presented here is based on the filler reinforcement theory developed by Smallwood (1944) and Guth (1945). Such a theory is limited to

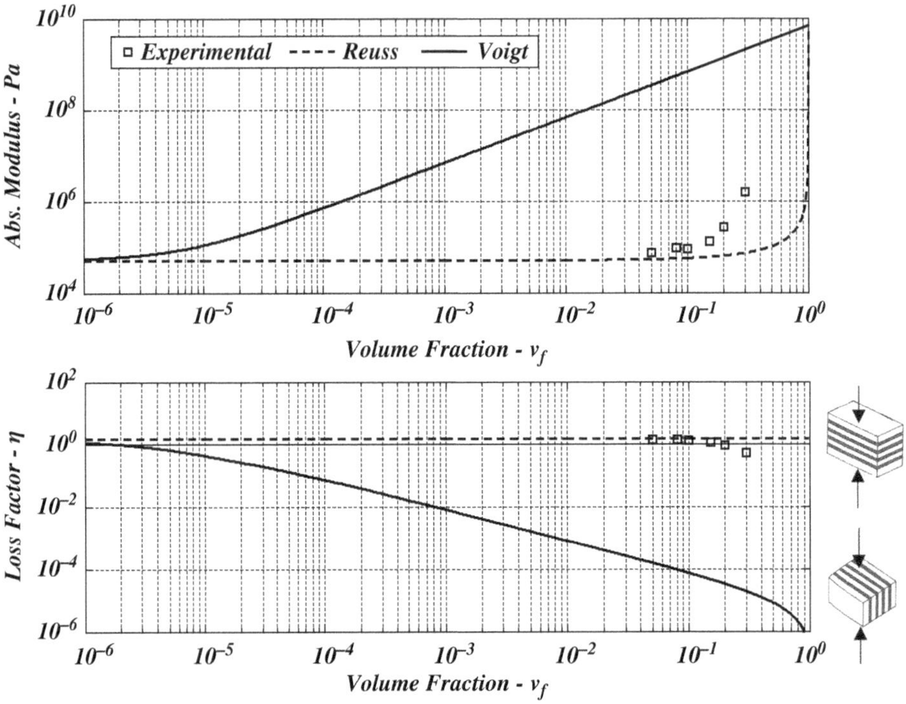

(a) – Frequency 0.01 *Hz*

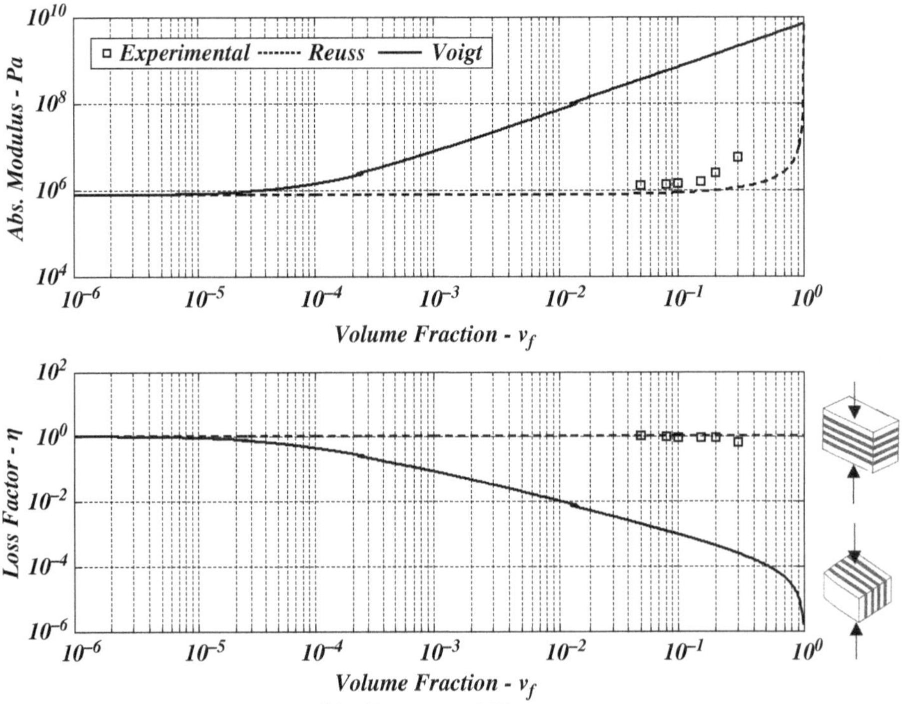

(b) – Frequency 1 *Hz*

Figure 11.13 Characteristics of CB-filled HDPE composite relative to the Voigt and Reuss bounds. Frequencies of (a) 0.01 Hz, (b) 1 Hz, and (c) 100 Hz.

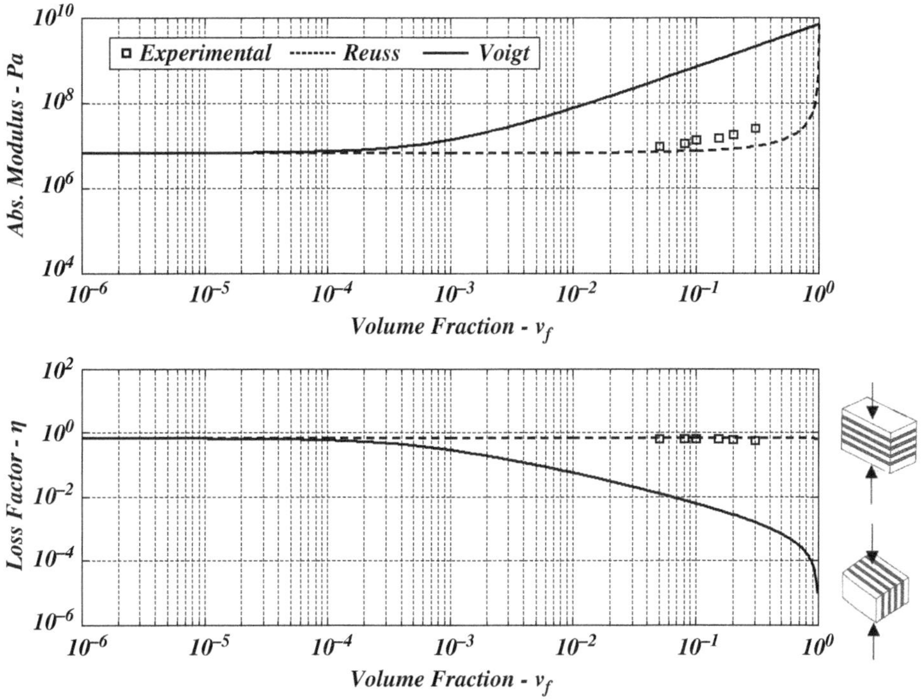

(c) – Frequency 100 *Hz*

Figure 11.13 (Continued)

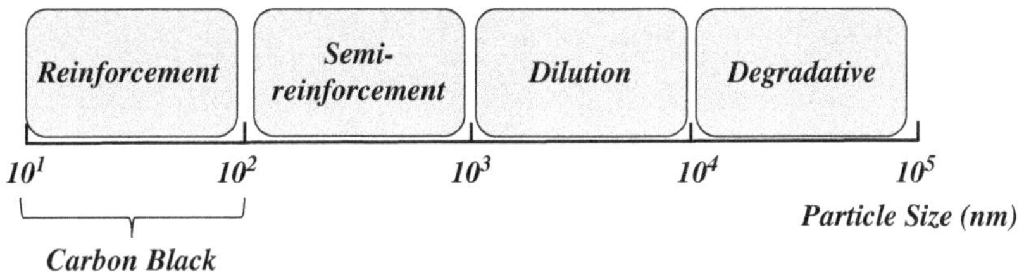

Figure 11.14 Nature of interactions between fillers and polymers (Leblanc 2002). Reproduced with permission of Elsevier.

nanoparticle fillers of diameters less than 100 nm, such as CB particles that usually have diameters of about 40 nm, as outlined in Figure 11.14.

The interaction between the CB particles and the host polymer is analytically quantified by the Smallwood–Guth model, which for solid spherical particles is given as follows:

$$E_\nu/E_0 = \left(1 + 2.5\nu + 14.1\nu^2\right) \tag{11.55}$$

where E_0 is the modulus of the polymer matrix, E_ν is the modulus of the CB/polymer composite, and ν is the volume fraction of the filler in the composite. Equation (11.55) applies to both the storage and loss moduli of the polymer.

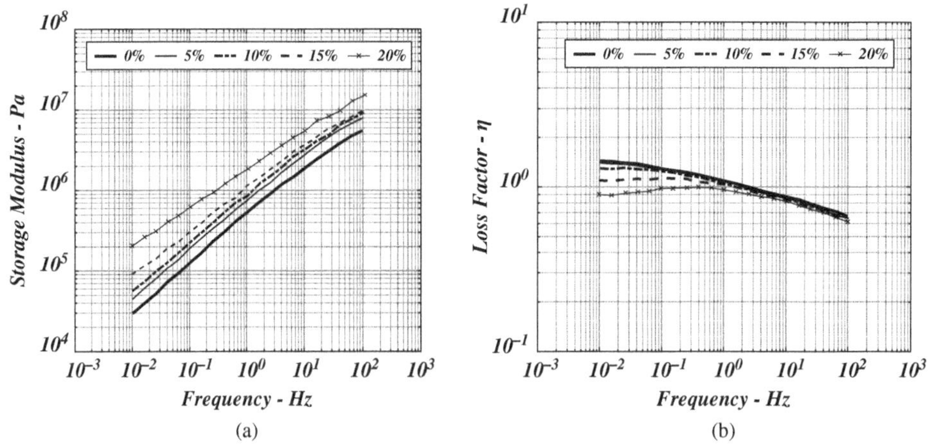

Figure 11.15 The complex modulus of CB/polymer composites (Zhang and Yi 2002). (a) Storage modulus. (b) Loss factor. Reproduced with permission of John Wiley & Sons.

Details of the derivation of Eq. (11.55) are given in Appendix 11.B.

Example 11.3 The storage modulus E' and the loss factor η of the HDPE CB/composite are shown in Figure 11.15 parts a and b, respectively. These characteristics are displayed for volume fractions v of the CB filler ranging between 0 and 20%. Show that using the Smallwood and Guth Eq. (11.55) makes the $E' - v$ characteristics collapses to form nearly a single line.

Also plot the Cole–Cole diagram of the storage modulus E' versus the loss modulus $E'' = \eta E'$ and discuss the obtained results.

Solution

Figures 11.16 and 11.17 show the $E' - v$ characteristics and the Cole–Cole plot of E' versus E'' following the use of the Smallwood–Guth Eq. (11.55). It can be seen that the equation results in collapsing the $E' - v$ and $E' - E''$ characteristics to form nearly single lines especially for CB volume fractions that are less than or equal to 15%.

These figures also display comparisons with the predictions obtained by using Eshelby's strain tensor approach. Close agreement is obvious between the predictions of the two approaches particularly for volume fractions less than 15%.

Example 11.4 Determine the reinforcement effect resulting from embedding CB particles, of different volume fractions, in HDPE. Use the finite element modeling of a RVE (representative volume element) of the CB/polymer composite as shown in Figure 11.18 for a square array arrangement. Compare these predictions with those generated by the Eshelby's strain tensor approach outlined in Section 11.2 and the Smallwood–Gush Equations given in Appendix 11.B.

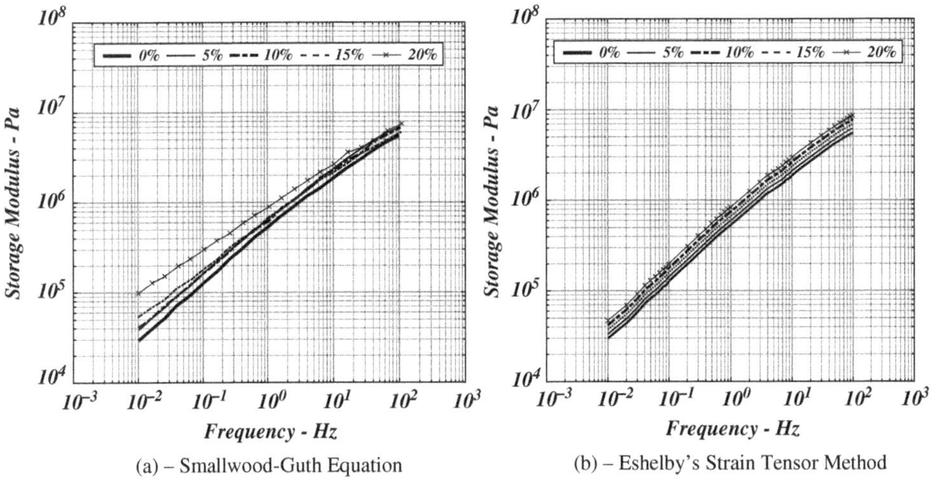

Figure 11.16 The reduced complex modulus of CB/polymer composites. (a) Smallwood–Guth equation. (b) Eshelby's strain tensor method.

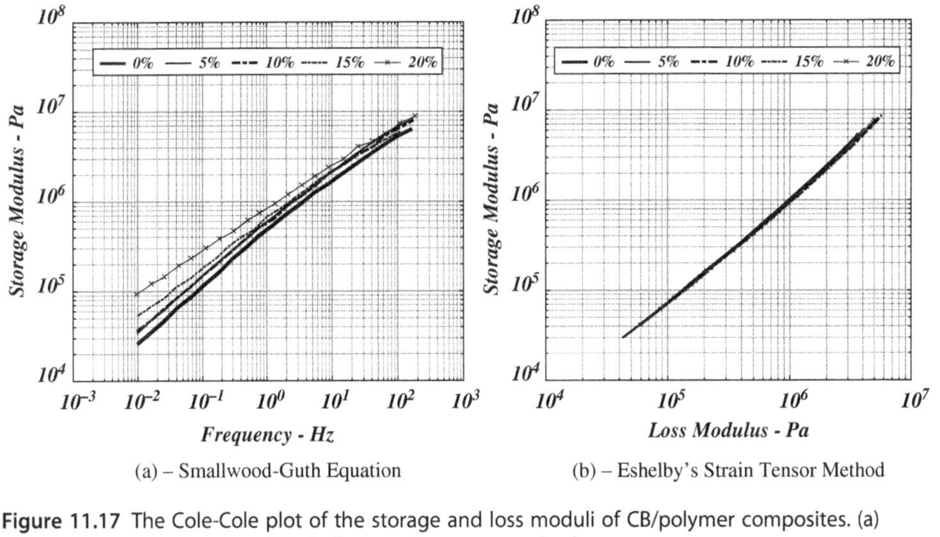

Figure 11.17 The Cole-Cole plot of the storage and loss moduli of CB/polymer composites. (a) Smallwood–Guth equation. (b) Eshelby's strain tensor method.

Solution

Figure 11.19a shows a finite element model of the RVE that is developed in ANSYS using SOLID186 elements. Figure 11.19b shows a unit cell with the associated boundary conditions that is used in analysis in order to simplify the computational effort (Brinson and Lin 1998).

Figure 11.20 displays a comparison between the deflections w in the z direction as predicted by Smallwood (1944) reinforcement theory and the ANSYS finite element of the

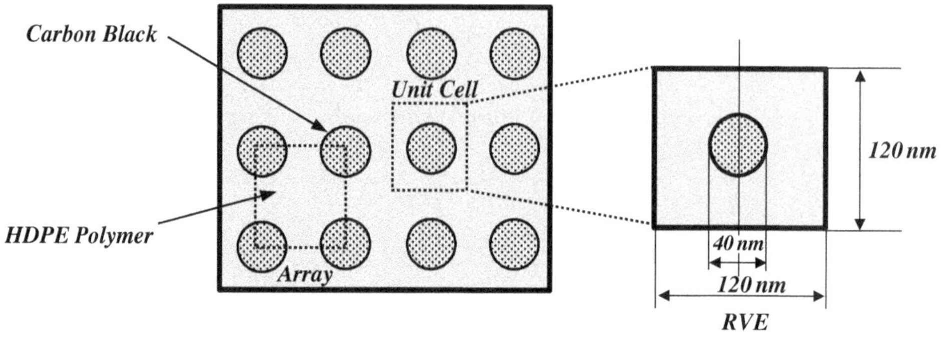

Figure 11.18 A representative volume element (RVE) of the CB/polymer composite.

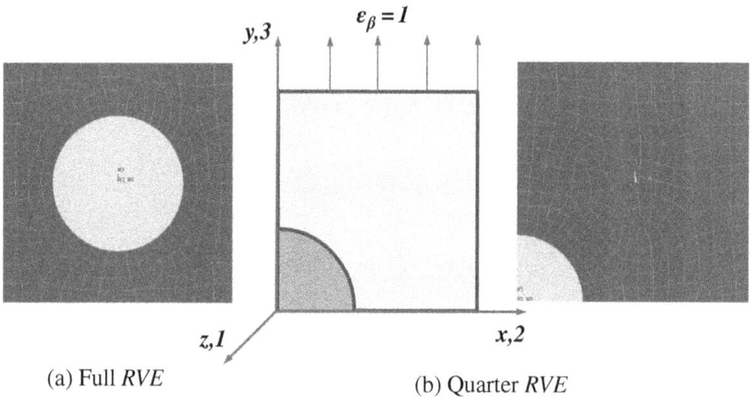

Figure 11.19 Finite element mesh of a representative volume element (RVE) of the CB/polymer composite. (a) Full RVE and (b) quarter RVE.

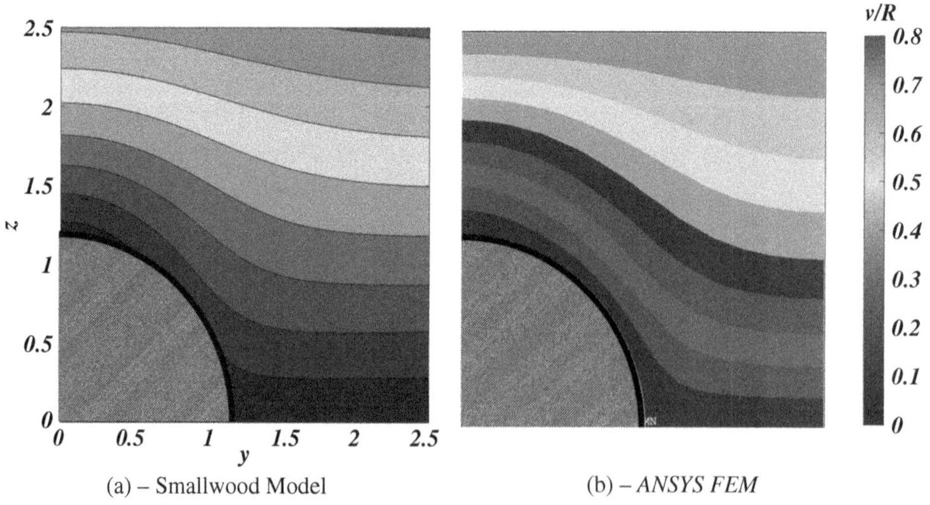

Figure 11.20 Comparison between the deflections in the *y* direction as predicted by Smallwood (1944) and ANSYS finite element of a representative volume element (RVE) of the CB/polymer composite. (a) Smallwood model, (b) ANSYS FEM.

RVE for a CB/polymer composite with a CB volume fraction of 20%. The Smallwood prediction is based on Eq. (11.B.9) given in the Appendix 11.B. The figures indicate that the contours of the deflection v as predicted by Smallwood model are in close agreement with those predicted by the ANSYS model.

Figure 11.21 displays the contours of the strain and stress, along the y direction, as predicted by the ANSYS finite element of an RVE of the CB/polymer composite with a CB volume fraction of 20%.

Figure 11.22 displays the contours of the deflection, along the z direction, as predicted by the ANSYS finite element of a RVE of the CB/polymer composite with a CB volume fraction ranging from 0 to 16.6%.

The effect of the volume fraction of the filler on the mechanical properties of a CB/polymer composite is determined using the approach outlined by Barbero (2013). In that approach, the components of the stiffness matrix $c_{\alpha\beta}$ are determined by using the average stress $\bar{\sigma}_\alpha$ and average strain $\bar{\varepsilon}_\beta$ such that:

$$\bar{\sigma}_\alpha = c_{\alpha\beta}\,\bar{\varepsilon}_\beta \tag{11.56}$$

where α and β = 1,…,6.

In Eq. (11.56), the average stresses $\bar{\sigma}_\alpha$ are evaluated by computing the stress field over the entire RVE when a unit strain ε_β is applied along the β direction such that:

$$c_{\alpha\beta} = \bar{\sigma}_\alpha = \frac{1}{V}\int_V \sigma_\alpha(x,y,z)\,dV \quad with\, \varepsilon_\beta = 1 \tag{11.57}$$

where V is the volume of the VRE.

The modulus of elasticity E_T and Poisson's ratio ν_T in the transverse directions (y and z) are extracted from (Barbero 2013) as follows:

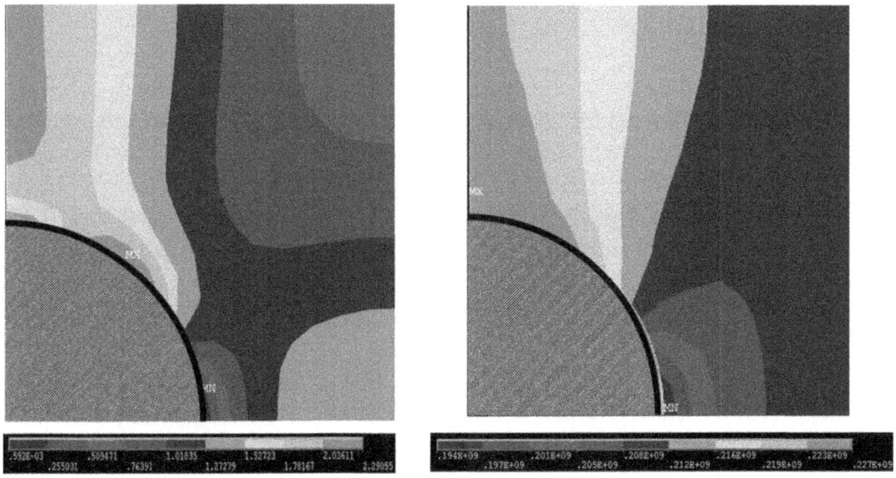

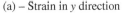

(a) – Strain in y direction (b) – Stress in the y direction

Figure 11.21 Contours of the strain and stress, in the y direction, as predicted by ANSYS finite element of a representative volume element (RVE) of the CB/polymer composite. (a) strain in the y direction, (b) stress in the y direction.

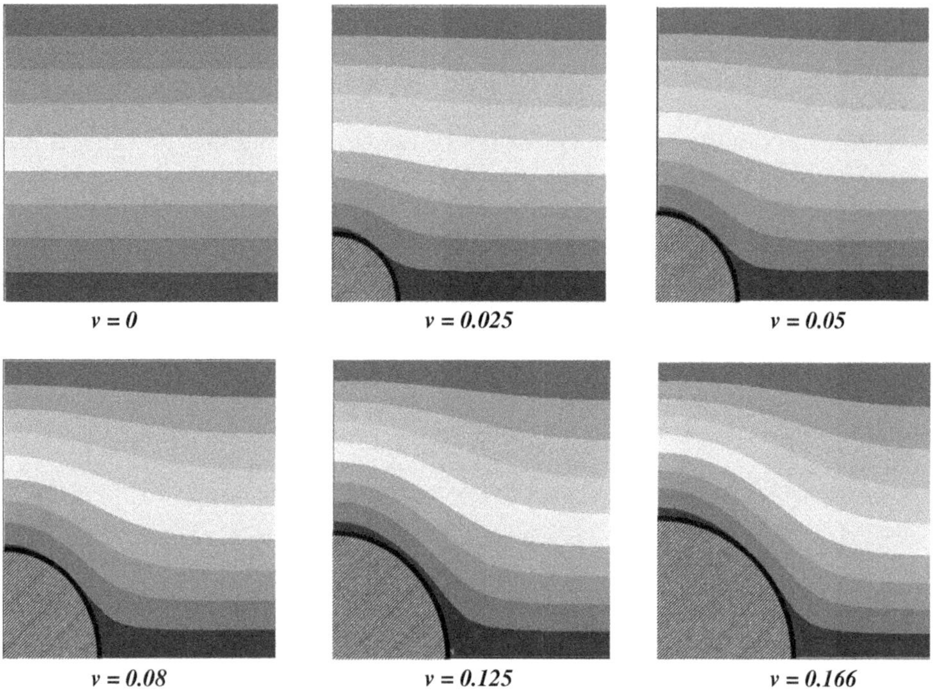

Figure 11.22 Contours of the deflection, in the *y* direction, as predicted by ANSYS finite element of a representative volume element (RVE) of the CB/polymer composite.

$$E_T = [c_{11}(c_{22} + c_{23}) - 2c_{12}c_{12}] - [(c_{22} - c_3)/(c_{11}c_{22} - c_{12}c_{21})] \tag{11.58}$$

and

$$\nu_T = [(c_{11}c_{23} - c_{12}c_{21})/(c_{11}c_{22} - c_{12}c_{21})] \tag{11.59}$$

Figure 11.23 displays comparisons between the effect of the filler volume fraction on the reinforcement, in the z direction, as predicted by the ANSYS Finite Element, Smallwood–Guth model, and Eshelby's method. The reinforcement effect is quantified by the dimensionless ratio of the elastic modulus E_z of the CB/polymer composite, in the z direction, relative to the storage modulus of the pristine polymer E_O; that is, E_z/E_O.

The figure suggests that the Smallwood–Guth model adequately predicts the reinforcement effect over a practical range of filler volume fraction. Such a prediction is also computationally simple and efficient when compared to the finite element method (FEM) and Eshelby's method.

11.4 Applications of Carbon Black/Polymer Composites

11.4.1 Basic Physical Characteristics

CB/polymer composites have attracted the attention for several decades primarily because of for their important use in manufacturing automotive tires. But, in this section, the emphasis is placed on their novel utilization in vibration damping applications. These

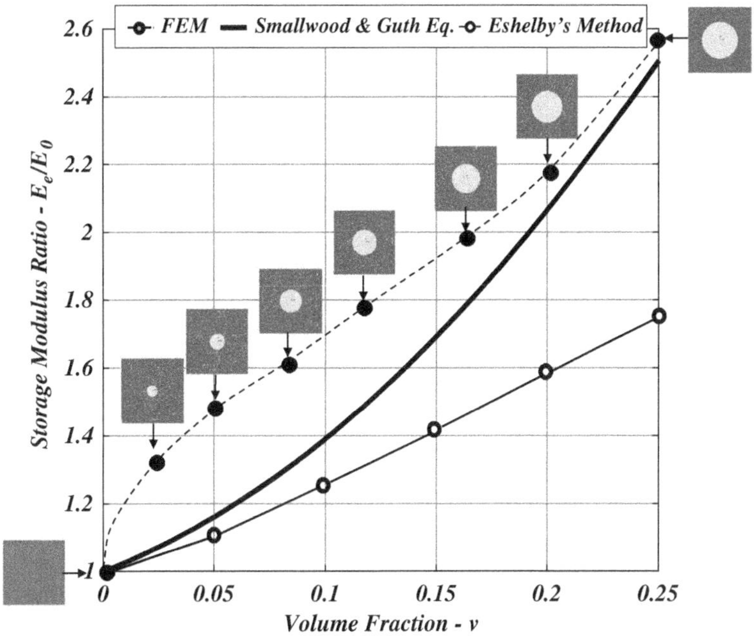

Figure 11.23 Comparison between the reinforcement of a filler, in the *y* direction, as predicted by the ANSYS finite element, Smallwood–Guth model, and Eshelby's method.

applications stem mainly from of the fact that embedding CB inside the polymer matrix renders it to be electrically conducting. The extent of the electrical conduction depends on the volume fraction of the CB filler into the composite.

An important metric that quantifies the conductivity of the polymer is the "Percolation Threshold." This threshold defines the concentration of the fillers that makes the polymer conductive. Generally speaking, as the concentration of the CB increases, the conductivity of the composite increases or the resistivity decreases as shown in Figure 11.24.

The figure indicates that the rate of increase of the conductivity, or decrease of resistivity, is slow at low concentrations as shown in region *A* of Figure 11.24. The resistivity drops rapidly with further increase in the CB as it goes into region *B* where the rate of change increases by more than 10 orders of magnitude. Further increase of the CB content fails to improve the resistivity as seen during region *C*.

It is important to physically understand the underlying phenomena that are behind such changes in the conductivity of the polymer matrix. At low CB content, the gap between the CB particles, where the electrons are transmitted, is very large and the resistivity of the composite is approximately that of the polymer matrix. As the concentration increases, the percolation threshold is reached where the resistivity starts to decrease abruptly as a function of the CB loading. In this region, region *B*, the gap between the CB particles is close but not touching. As a result, the electron must overcome the potential barrier and cross the gap between the CB particles.

In region *C*, where the CB loading is higher than the percolation threshold, the CB particles form chains that act as pure resistive conduits for conducting electricity as

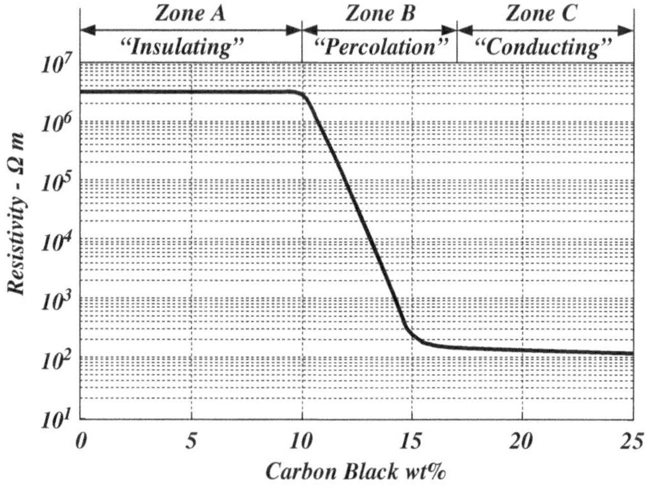

Figure 11.24 Effect of carbon black concentration on the resistivity of polymer/CB composites.

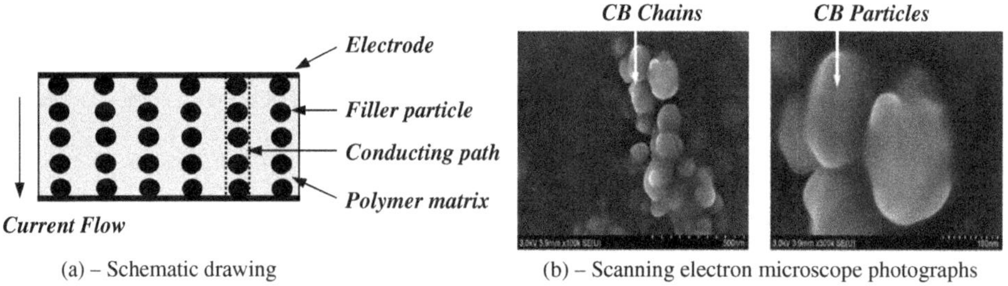

Figure 11.25 The microstructure of a CB/polymer composite. (a) schematic drawing, (b) scanning electron microscope photographs.

indicated in Figure 11.25. Figure 11.25a displays a schematic drawing of the microstructure of a CB/polymer composite whereas Figure 11.25b shows a scanning electron microscope photograph of the microstructure emphasizing the existence of CB chains.

Ding et al. (2013) and Wang et al. (2005) suggested that the equivalent electrical circuits of the CB/polymer composite depend, in general, on the CB volume fraction and, in particular, on the percolation characteristics as shown in Figure 11.26. In the figure, R_a, R_c, C_c, and L denote the CB aggregate resistance, the contact resistance, the contact capacitance, and inductance, respectively.

The full practical potential of the CB/polymer composites is usually utilized when operating in the first part of zone C where the transmission of electricity is governed primarily by the ohmic mechanism (R_a). Further increase of the CB content makes the mixing and manufacturing of a uniform CB composite rather difficult. More importantly, the complexity of the resulting equivalent electrical circuit of the composites complicates their operation.

For practical considerations, the contact capacitance and inductance are usually very small as reported by Wang et al. (2005). Therefore, the emphasis is placed here only on modeling the CB/polymer composites as resistive elements.

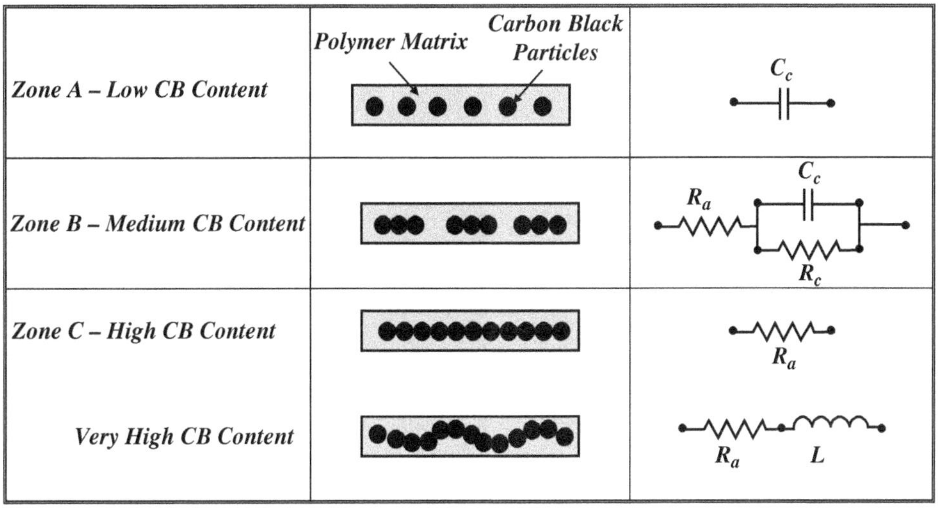

Figure 11.26 Equivalent electrical circuits of a CB/polymer composite.

11.4.2 Modeling of the Piezo-Resistance of CB/Polymer Composites

The electrical resistance R, under no loading conditions, of CB/Polymer composites can be determined from:

$$\rho_s = RA/l_c \tag{11.60}$$

where ρ_s, A, and l_c denote the resistivity, cross sectional area, and thickness of the composite, respectively.

It is important to note that the electrical resistance of the composite is usually called "piezo-resistance." Such a resistance defines changes in the resistance of the CB/polymers due to the application of load.

Generally, the piezo-resistance depends on the properties of the polymer matrix, filler properties and filler concentration, and applied load. An excellent account that describes the interactions between all these parameters and their effect on the piezo-resistivity of the conducting polymer composite is given by Zhang et al. (2000, 2001). In their work, Zhang et al. developed a physics-based mathematical model to predict the piezo-resistivity of polymers impregnated with 11 different fillers.

The total resistance R of one conducting path is given by:

$$R = L(R_c + R_a)/S \tag{11.61}$$

where R_c = resistance between two adjacent filler particles, R_a = resistance across filler particle, L = number of particles forming one path, and S = number of conducting paths.

If the inter-particle separation is very large, no current flows. However, as the separation becomes adequately small, a tunneling current $\boldsymbol{I}$ will flow due to the application of a voltage $\boldsymbol{V}$ such that (Simmons 1963):

$$I = \frac{3\sqrt{2m\phi}}{2s}\left(\frac{e}{h_P}\right)^2 V e^{\left(-\frac{4\pi s}{h_P}\sqrt{2m\phi}\right)} \tag{11.62}$$

where m = electron mass, e = electron charge, h_P = Planck constant, s = separation between two adjacent particles, and ϕ = height of potential barrier between adjacent particles.

Assuming a^2 = the cross sectional area of the conducting particle, then the resistance R_c can be determined from:

$$R_c = \frac{V}{a^2 I} = \frac{8\pi h_P s}{3a^2 \gamma e^2} e^{\gamma s} \tag{11.63}$$

where

$$\gamma = \frac{4\pi}{h_P}\sqrt{2m\phi} \tag{11.64}$$

Because the conductivity of the particles is very large compared with that between two adjacent particles, then $R_a \cong 0$, reducing Eq. (11.61) to:

$$R = \frac{L}{S}\frac{8\pi h_P s}{3a^2 \gamma e^2} e^{\gamma s} \tag{11.65}$$

Equation (11.65) can be used to predict the resistance of the conducting polymer composite and it is clear that it varies exponentially with the separation distance s between the particles, which is a function of the applied load or strain experienced by the composite. Now, let us assume that the inter-particle separation changes from s_0 to s due to the application of stress, then the fractional resistance change $(-\Delta R/R_0)$ can be predicted from:

$$-\Delta R/R_0 = 1 - \frac{s}{s_0} e^{-\gamma(s_0 - s)} \tag{11.66}$$

where R_0 is the original resistance. Note that s and s_0 can be related to the strain ε and the stress σ by the following relationships:

$$s = s_0(1-\varepsilon) = s_0\left(1 - \frac{\sigma}{E}\right) \tag{11.67}$$

where E = modulus of elasticity of the polymer matrix.

The initial separation distance s_0 is estimated from Han and Choi (1998) as follows:

$$s_0 = D\left[\left(\frac{\pi}{6}\right)^{\frac{1}{3}} v_f^{-\frac{1}{3}} - 1\right] \tag{11.68}$$

where D = particle diameter and v_f = volume fraction of filler.

Accordingly, Eq. (11.66) reduces to:

$$-\Delta R/R_0 = 1 - \left(1 - \frac{\sigma}{E}\right) e^{-\gamma D\left[\left(\frac{\pi}{6}\right)^{\frac{1}{3}} v_f^{-\frac{1}{3}} - 1\right]\frac{\sigma}{E}} \tag{11.69}$$

Equation (11.69) predicts the piezo-resistance changes of conducting polymer composite as function of the applied stress σ, modulus of elasticity of the polymer matrix E, filler particle diameter D, filler volume fraction v, and the parameter γ. Equation (11.69) can be rewritten as:or

$$\frac{[1 + \Delta R/R_0]}{\left(1 - \frac{\sigma}{E}\right)} = e^{-\gamma\frac{\sigma}{E}} \tag{11.70}$$

where

$$\bar{\gamma} = \gamma D\left[\left(\frac{\pi}{6}\right)^{\frac{1}{3}} v_f^{-\frac{1}{3}} - 1\right] \tag{11.71}$$

For $\sigma/E < 0.01$, Eq. (11.71) can be simplified to:

$$[1 + \Delta R/R_0] \cong e^{-\bar{\gamma}\frac{\sigma}{E}} \tag{11.72}$$

Expanding the exponential in Eq. (11.72), yields:

$$\Delta R/R_0 \cong -\bar{\gamma}\frac{\sigma}{E} = -\bar{\gamma}\,\varepsilon,$$

or

$$R_\varepsilon = R_0\,(1 - \bar{\gamma}\,\varepsilon). \tag{11.73}$$

where R_ε is the resistance of the CB/polymer composites following the application of a strain ε. Equation (11.73) indicates that the relationship between R_ε/R_0 and the σ/E (or ε) is a straight line with a slope of $\bar{\gamma}$ as shown in Figure 11.27.

The value of $\bar{\gamma}$ can be determined, for CB particle-filled polymer, using the following parameters: mass of the electron m = 9.180938E-31 kg, h_P = Planck constant = 6.626E-34 $m^2\,kg\,s^{-1}$, and ϕ = 0.05 EV.

These parameters yield a slope $\bar{\gamma}$ of the piezo-resistive characteristics of 54.97.

11.4.3 The Piezo-Resistivity of CB/Polymer Composites

The electrical resistivity ρ_s of the CB/polymer composites can be determined by rewriting Eq. (11.65) as follows:

$$\rho_s = \frac{R\left(Sa^2\right)}{(LD)} = \frac{8\,\pi\,h_P\,s}{3D\gamma\,e^2}e^{\gamma s} \tag{11.74}$$

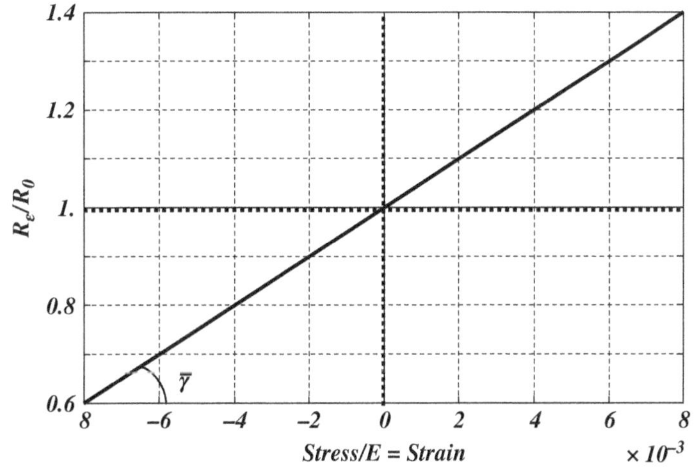

Figure 11.27 The piezo-resistance vs. strain characteristics of CB/polymer composites.

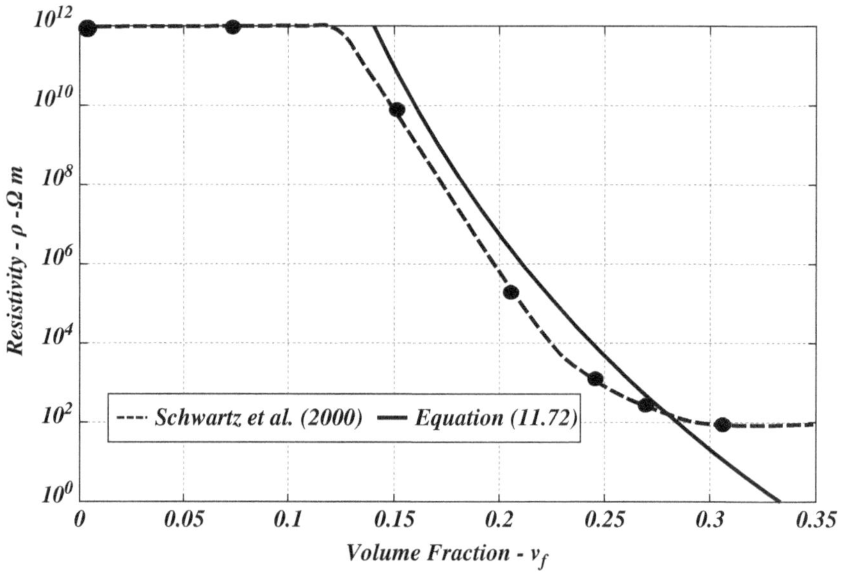

Figure 11.28 The resistivity of the CB/polymer composite.

In Eq. (11.74), it is assumed that Sa^2 and LD denote the area and the length of the conducting path, respectively.

Figure 11.28 displays the resistivity of the CB/polymer composite as predicted by Eq. (11.74), with CB particles of an average size of 30 nm (Kaiser 1993), in comparison with the prediction of Schwartz et al. (2000). In their prediction approach, Schwartz et al. determined the electrical resistivity of a composite of CB in an insulating matrix in terms of the volume fraction of the CB and aggregate size and distribution. It is assumed that the resistance of a random lattice composite between sites different aggregates varies exponentially with the gap.

11.5 CB/Polymer Composite as a Shunting Resistance of Piezoelectric Layers

11.5.1 Finite Element Model

In this section, the CB/polymer is utilized as a means for vibration damping. Consider a rod consisting of an assembly of CB/polymer layers sandwiching, in a periodic manner, piezoelectric layers as shown in Figure 11.29.

The interaction between the dynamics of the piezoelectric layer that is shunted by the resistive CB/polymer layer is described using the constitutive equation of the piezo-layer as given in Chapter 9 as follows:

$$\begin{Bmatrix} T_3 \\ E_3 \end{Bmatrix} = \begin{bmatrix} c_{33}^D & -h_{33} \\ -h_{33} & \dfrac{1}{\varepsilon_{33}^S} \end{bmatrix} \begin{Bmatrix} S_{3p} \\ D_3 \end{Bmatrix} \tag{11.75}$$

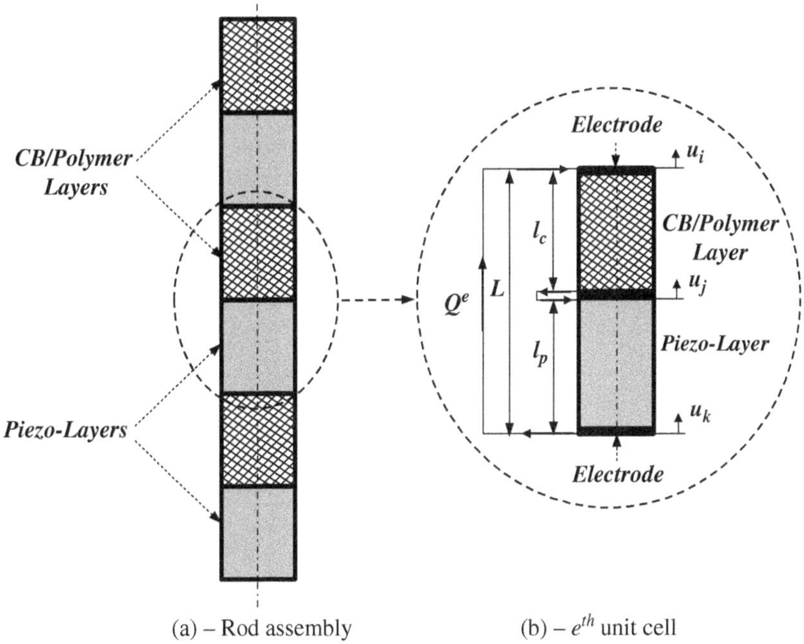

Figure 11.29 Finite element of a rod/shunted piezo-network. (a) Rod assembly and (b) *e*th unit cell.

where T_3 and S_{3p} are the stress and strain along the composite rod. Also, E_3 and D_3 denote the electric field and the electrical displacement, respectively. Furthermore, $k_{33}^2 = d_{33}^2/\left(s_{33}^E \varepsilon_{33}^T\right)$, $h_{33} = d_{33}/\left[s_{33}^E\left(1-k_{33}^2\right)\varepsilon_{33}^T\right]$, $\varepsilon_{33}^S = \varepsilon_{33}^T\left(1-k_{33}^2\right)$, and $c_{33}^D = 1/s_{33}^D = 1/s_{33}^E\left(1-k_{33}^2\right)$.

This form of constitutive relationship is used to generate the potential and kinetic energies of an element of the CB/polymer composite and shunted piezo-network in terms of the *DOF* $\{\Delta^e\} = \{u_i\ u_j\ u_k\}^T$ and the electrical DOF Q^e (the charge of the *e*th cell). Note that the longitudinal displacements of the composite $u_c(x)$ and piezo-layer $u_p(x)$, at any position x, are given by the following shape function

$$u_c(x) = \{N_c\}\{\Delta^e\} \text{ and } u_p(x) = \{N_p\}\{\Delta^e\} \tag{11.76}$$

where $\{N_c\} = \{(1-x/l_c)\ x/l_c\ 0\}$ and $\{N_p\} = \{0\ (1-x/l_p)\ x/l_p\}$ where l_c and l_p are equal the length of the composite layer and the piezo-layer, respectively.

The kinetic energy KE of the *e*th unit cell can now be determined from:

$$KE = \frac{1}{2}A_p\rho_p\int_0^{l_p}\dot{u}_p^2\,dx + \frac{1}{2}A_c\rho_c\int_0^{l_c}\dot{u}_c^2\,dx = \frac{1}{2}\{\Delta^e\}^T\left[[M_p]+[M_c]\right]\{\Delta^e\} = \frac{1}{2}\{\Delta^e\}^T[M^e]\{\Delta^e\} \tag{11.77}$$

where

$$[M_p] = \rho_p A_p \int_0^{l_p} [N_p]^T [N_p]\,dx = \text{mass matrix of the piezo-layer,}$$

$[M_c] = \rho_c A_c \int_0^{l_c} [N_c]^T [N_c]\, dx$ = mass matrix of the CB/polymer layer,

and $[M^e] = [M_p] + [M_c]$ = total mass of the eth unit cell.

Also, ρ_l and A_l denote the density and cross sectional area of the lth layer ($l = p$ for the piezo-layer and $l = c$ for the CB/polymer layer, respectively).

The potential energy PE of the eth unit cell is given by:

$$PE = \frac{1}{2}\int_V (S_{3p} + S_{3c})\, T_3 dv + \frac{1}{2}\int_V D_3 E_3 dv \tag{11.78}$$

where S_{3p} = strain in the piezo-layer = $u_{,xp}$, $T_3 = S_{3p}/s_{33}^D - h_{33}D_3$ = stress on piezo-layer, S_{3c} = strain in the CB/polymer layer = $u_{,xc}$, and $T_3 = E_c^* S_{3c}$ = stress on CB/polymer layer with E_c^* denoting the complex modulus of the composite layer.

Substituting from Eq. (11.75) into Eq. (11.78) gives:

$$PE = \frac{1}{2}A_p \int_0^{l_p} S_{3p}\left[c_{33}^D S_{3p} - h_{33}D_3\right] dx + \frac{1}{2}A_c \int_0^{l_c} S_{3c}E_c^* S_{3c} dx + \frac{1}{2}A_p \int_0^{l_p} D_3\left[-h_{33}S_{3p} + \frac{1}{\varepsilon_{33}^S} D_3\right] dx \tag{11.79}$$

Note that: $S_{3p} = u_{,xp} = \{N_{,xp}\}\{\Delta^e\}$, $S_{3c} = u_{,xc} = \{N_{,xc}\}\{\Delta^e\}$, and $D_3 = Q^e/A_p$ = constant along the piezo-layer, then Eq. (11.79) reduces to

$$PE = \frac{1}{2}\{\Delta^e\}^T [K^e]\{\Delta^e\} - h_{33}Q^e\{0\ 1\ -1\}\{\Delta^e\} + \frac{1}{2}\frac{(Q^e)^2 l_p}{A_p\ \varepsilon_{33}^S} \tag{11.80}$$

where $[K^e] = A_p c_{33}^D \int_0^{l_p} \{N_{,xp}\}^T \{N_{,xp}\} dx + A_c E_c^* \int_0^{l_c} \{N_{,xc}\}^T \{N_{,xc}\} dx$ = stiffness matrix of piezo-layer and CB/polymer assembly.

The virtual work δW associated with the inductive-resistive-capacitive shunting network of the CB/polymer composite and the external load $\{F_i\}$ is given by:

$$\delta W = -\left(L_e \ddot{Q}^e + R_e \dot{Q}^e + \frac{1}{C_e} Q^e\right)\delta Q^e + \{F^e\}\,\delta\{\Delta^e\} \tag{11.81}$$

Now, the equations governing the dynamics of the/piezo-patch system can be extracted using the Lagrangian dynamics to give

$$\frac{d}{dt}\frac{\partial KE}{\partial\{\dot{\Delta}^e\}} + \frac{\partial PE}{\partial\{\Delta^e\}} = \{F^e\},$$

and

$$\frac{d}{dt}\frac{\partial KE}{\partial\{\dot{Q}^e\}} + \frac{\partial PE}{\partial\{Q^e\}} = -\left(L_e \ddot{Q}^e + R_e \dot{Q}^e + \frac{1}{C_e} Q^e\right). \tag{11.82}$$

where L_e, R_e, and C_e are the components of the inductive-resistive-capacitive shunting network of the CB/Polymer composite. Note that $R_e = R_\varepsilon$ where it is assumed here that the effect of the strain on the resistance of the CB/Polymer composite is negligible, that is, $R_e = R_\varepsilon \approx R_0$. This assumption is considered only to simplify the analysis and makes the finite element model linear to enable direct solution. But, when the resistance $R_\varepsilon \neq R_0$, the analysis becomes iterative in nature and requires checking of the convergence at every time or frequency step.

These equations can be written in the following matrix form:

$$\begin{bmatrix} [M^e] & 0 \\ 0 & L_e \end{bmatrix} \begin{Bmatrix} \{\ddot{\Delta}^e\} \\ \ddot{Q}^e \end{Bmatrix} + \begin{bmatrix} 0 & 0 \\ 0 & R_e \end{bmatrix} \begin{Bmatrix} \{\dot{\Delta}^e\} \\ \dot{Q}^e \end{Bmatrix} + \begin{bmatrix} [K^e] & -h_{33}\begin{Bmatrix} 0 \\ 1 \\ -1 \end{Bmatrix} \\ -h_{33}\begin{Bmatrix} 0 \\ 1 \\ -1 \end{Bmatrix}^T & \dfrac{1}{C_e} + \dfrac{1}{C^s} \end{bmatrix} \begin{Bmatrix} \{\Delta^e\} \\ Q^e \end{Bmatrix} = \begin{Bmatrix} \{F^e\} \\ 0 \end{Bmatrix} \tag{11.83}$$

or

$$[M]\{\ddot{X}\} + [C]\{\dot{X}\} + [K]\{X\} = \{F\} \tag{11.84}$$

where $[M] = \begin{bmatrix} [M^e] & 0 \\ 0 & L_e \end{bmatrix}$, $[C] = \begin{bmatrix} 0 & 0 \\ 0 & R_e \end{bmatrix}$, $[K] = \begin{bmatrix} [K^e] & -h_{33}\begin{Bmatrix} 0 \\ 1 \\ -1 \end{Bmatrix} \\ -h_{33}\begin{Bmatrix} 0 \\ 1 \\ -1 \end{Bmatrix}^T & \dfrac{1}{C_e} + \dfrac{1}{C^s} \end{bmatrix}$,

$$\{X\} = \begin{Bmatrix} \{\Delta^e\} \\ Q^e \end{Bmatrix}, \{F\} = \begin{Bmatrix} \{F^e\} \\ 0 \end{Bmatrix} \text{ and } C^s = A_p \varepsilon_{33}^s / l_p. \tag{11.85}$$

The equation of motion of the entire CB composite/shunted piezo-network assembly can then be determined by assembling the mass and stiffness matrices of the individual cells to yield:

$$[M_o]\{\ddot{X}\} + [C_o]\{\dot{X}\} + [K_o]\{X\} = \{F_o\} \tag{11.86}$$

where $[M_o]$, $[C_o]$, and $[K_o]$ are the overall mass, damping, and stiffness matrices of the entire assembly. Also, $\{X\}$ and $\{F_o\}$ denote the structural and electrical DOF as well as load vectors of the entire assembly.

Note that bonding the shunted piezo-layer to the CB composite has resulted in adding damping to the assembly as quantified by the damping matrix $[C_o]$. It has also modified the stiffness and mass matrices.

11.5.2 Condensed Model of a Unit Cell

As the CB/polymer composite can be treated electrically as a resistive element, then L_e and C_e are set equal to zero in Eq. (11.81). Further, using the static condensation method to condense the DOF of the charge Q^e leads to:

$$Q^e = C^s h_{33} \begin{Bmatrix} 0 \\ 1 \\ -1 \end{Bmatrix}^T \{\Delta^e\} = T\{\Delta^e\} \tag{11.87}$$

i.e. then:

$$\begin{Bmatrix} \{\Delta^e\} \\ Q^e \end{Bmatrix} = \begin{Bmatrix} I \\ T \end{Bmatrix} \{\Delta^e\} = \bar{T}\{\Delta^e\} \tag{11.88}$$

Let $e = \{0 \;\; 1 \;\; -1\}$, then $\bar{T}$ is given by:

$$\bar{T} = \begin{bmatrix} I \\ C^s h_{33}\, e \end{bmatrix} \tag{11.89}$$

Substituting Eq. (11.88) into Eq. (11.83) and pre-multiplying by $\bar{T}^T$ gives:

$$[M^e]\left\{\ddot{\Delta}^e\right\} + \left[R_e (C^s h_{33})^2 e^T e\right]\left\{\dot{\Delta}^e\right\} + \left[[K^e] - C^s h_{33}^2 e^T e\right]\{\Delta^e\} = \{F^e\}$$

or

$$[\bar{M}^e]\left\{\ddot{\Delta}^e\right\} + [\bar{C}^e]\left\{\dot{\Delta}^e\right\} + [\bar{K}^e]\{\Delta^e\} = \{F^e\} \tag{11.90}$$

where $[\bar{M}^e] = [M^e]$, $[\bar{C}^e] = R_e (C^s h_{33})^2 e^T e$, and $[\bar{K}^e] = [K^e] - C^s h_{33}^2 e^T e$.

For the unit cell, shown in Figure 11.29, the nodal deflection vector $\{\Delta^e\}$ is defined as:

$$\{\Delta^e\} = \{u_k \;\; u_j \;\; u_i\}^T \tag{11.91}$$

where u_k, u_j, and u_i denote the bottom, internal, and top deflection vectors as shown in Figure 11.29b.

This vector is condensed to support the Bloch wave propagation (Hussein 2009). Hence, the displacements at the boundaries are related as follows:

$$u_i = e^{-ikL} u_k \tag{11.92}$$

where k and L denote the wave number and the length of the unit cell, respectively.

Hence, define an independent nodal deflection vector $\{\bar{\Delta}^e\}$ such that:

$$\{\bar{\Delta}^e\} = \{u_k \; u_j\}^T \tag{11.93}$$

The deflection vectors $\{\Delta^e\}$ and $\{\bar{\Delta}^e\}$ are related as follows:

$$\{\Delta^e\} = \widetilde{T}\{\bar{\Delta}^e\} \tag{11.94}$$

where $\widetilde{T}$ is a transformation matrix such as:

$$\widetilde{T} = \begin{bmatrix} 1 & 0 & e^{-ikL} \\ 0 & 1 & 0 \end{bmatrix}^T \tag{11.95}$$

Substituting Eqs. (11.94) and (11.95) into Eq. (11.90), it reduces to:

$$\left[\widetilde{M}^e\right]\left\{\ddot{\bar{\Delta}}^e\right\} + \left[\widetilde{C}^e\right]\left\{\dot{\bar{\Delta}}^e\right\} + \left[\widetilde{K}^e\right]\left\{\bar{\Delta}^e\right\} = \left[\widetilde{F}^e\right] \tag{11.96}$$

where $\left[\widetilde{M}^e\right] = \widetilde{T}^*\left[M^e\right]\widetilde{T}$, $\left[\widetilde{C}^e\right] = \widetilde{T}^*\left[C^e\right]\widetilde{T}$, $\left[\widetilde{K}^e\right] = \widetilde{T}^*\left[K^e\right]\widetilde{T}$, and $\left[\widetilde{F}^e\right] = \widetilde{T}^*\left[F^e\right]$.

Equation (11.95) is now cast in the following state-space form (Meirovitch 2010):

$$\begin{bmatrix} \left[\widetilde{K}^e\right] & 0 \\ 0 & \left[\widetilde{M}^e\right] \end{bmatrix}\dot{Y} + \begin{bmatrix} 0 & -\left[\widetilde{K}^e\right] \\ \left[\widetilde{K}^e\right] & \left[\widetilde{C}^e\right] \end{bmatrix} Y = \left\{ \begin{matrix} 0 \\ \left[\widetilde{F}^e\right] \end{matrix} \right\} \tag{11.97}$$

where $Y = \left\{\{\bar{\Delta}^e\}\ \left\{\dot{\bar{\Delta}}^e\right\}\right\}^T$.

Let the state-space solution be assumed as

$$\{Y\} = e^{\lambda t}\{Y^c\} \tag{11.98}$$

where $\{Y\}^c = \{\{x\} + i\{z\}\}^c$ and $\lambda = i\omega$.

Equation (11.98) reduces to the following eigenvalue problem:

$$\left\{ i\omega \begin{bmatrix} \left[\widetilde{K}^e\right] & 0 \\ 0 & \left[\widetilde{M}^e\right] \end{bmatrix} + \begin{bmatrix} 0 & -\left[\widetilde{K}^e\right] \\ \left[\widetilde{K}^e\right] & \left[\widetilde{C}^e\right] \end{bmatrix} \right\} \{\{x\} + i\{z\}\}^c = \{0\} \tag{11.99}$$

In a compact form, Eq. (11.99) reduces to

$$[i\omega[M^*]^c + [D^*]^c]\{\{x\} + i\{z\}\}^c = 0 \tag{11.100}$$

where

$$[M^*]^c = \begin{bmatrix} \left[\widetilde{K}^e\right] & 0 \\ 0 & \left[\widetilde{M}^e\right] \end{bmatrix} \quad \text{and} \quad [D^*]^c = \begin{bmatrix} 0 & -\left[\widetilde{K}^e\right] \\ \left[\widetilde{K}^e\right] & \left[\widetilde{C}^e\right] \end{bmatrix} \tag{11.101}$$

Equating the real and imaginary coefficients on both sides of Eq. (11.100) yields:

$$\mathit{Real}: [D^*]^c\{x\}^c = \omega\,[M^*]^c\{z\}^c \tag{11.102}$$

$$\mathit{Imaginary}: -[D^*]^c\{z\}^c = \omega\,[M^*]^c\{x\}^c \tag{11.103}$$

Equations (11.102) and (11.103) can be rewritten in compact and standard eigenvalue problem form such that:

$$[A]^c\{z\}^c = \omega^2\{z\}^c \tag{11.104}$$

Table 11.4 The physical and geometrical properties of the composite.

	Geometry		Piezo-layer				Viscoelastic layer				
PROPERTY	$l_c = l_p$ (m)	$A_c = A_p$ (m^2)	s^E (m^2 N^{-1})	d_{33} (m V^{-1})	k_{33}	ρ_p (kg m^{-3})	E_v (N m^{-2})	α	ω_n (rad s^{-1})	ζ	ρ_v (kg m^{-3})
VALUE	0.1	0.01	16×10^{-12}	296×10^{-12}	0.37	7600	1E4	5	25,000	10	1100

where $[A]^c = [M^*]^{c-1} [\ [D^*]^{Tc}[M^*]^{c-1}[D^*]^c\ ]$.

Note that all the entries of the matrix $[A]^c$ are functions of the dimensionless wave number kL. Therefore, the eigenvalues of the matrix $[A]^c$ can be determined for different values of the wave number kL.

Then, the eigenvalues λ_s are given by

$$\lambda_s(kL) = \omega_s \tag{11.105}$$

where $s = 1$ to n.

The dispersion characteristics of the unit cell are constructed by plotting the resonant frequency ω_s against the wave number kL. The resulted dispersion curves further define the zones of stop and pass bands of the cell as described in Chapter 10.

Example 11.5 Consider the hunted by the piezoelectric-CB/polymer composite described in Section 11.5.1 and shown in Figure 11.29. The composite has the physical and geometrical properties listed in Table 11.4.

In Table 11.4, the parameters E_v, α, ω_n, and ζ are the parameters of a single GHM mini-oscillator, which is used to describe the dynamic behavior of the VEM layer.

Determine the dispersion characteristics of a unit cell of the composite when the VEM layer is:

a) non-conducting (R_e = 0 ohm).
b) conducting using *CB* with volume fraction v = 0.25 and R_e = 10 kohm.
c) conducting using *CB* with volume fraction v = 0.25 and R_e = 1 Mohm.

Determine also the frequency response of a cantilevered composite consisting of 10 cells of the piezo/CB conducting polymer when it is subjected to a unit load at its free end. Compare the response with that of a composite consisting of 10 piezo/non-conducting polymer cells.

Solution

Figure 11.30 parts a–c displays the dispersion characteristics of the non-conducting VEM, conducting VEM with R_e = 10 kohm, and VEM with R_e = 1 Mohm, respectively.

Comparing Figure 11.30a,b indicates that making the VEM layer conduct with the CB filler, with an effective resistance of R_e = 10 kohm, has resulted in enhancing the stop band characteristics, particularly at the high frequency end. Further increase of the effective resistance by adding an additional external resistance to augment the

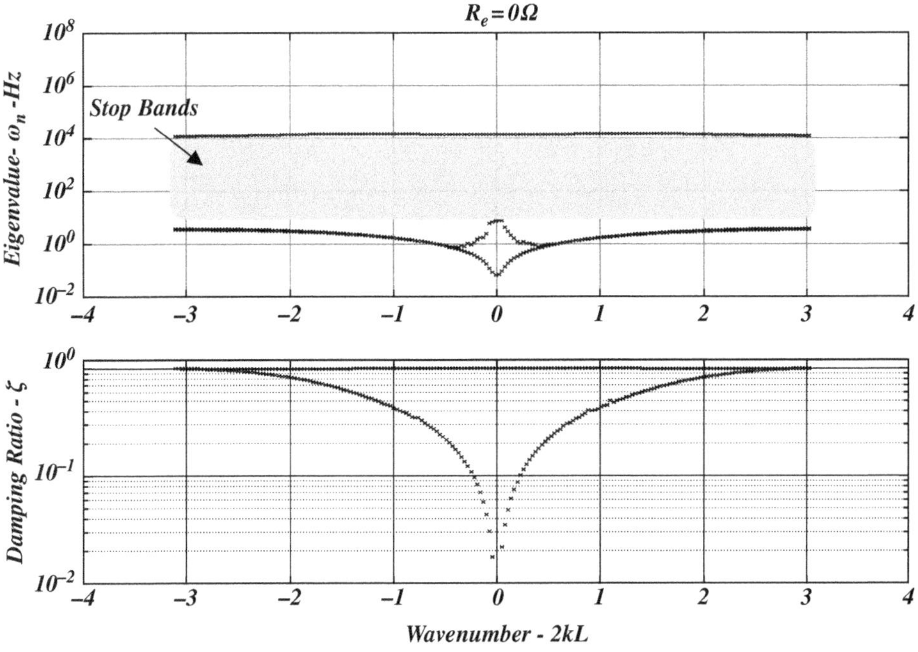

(a)

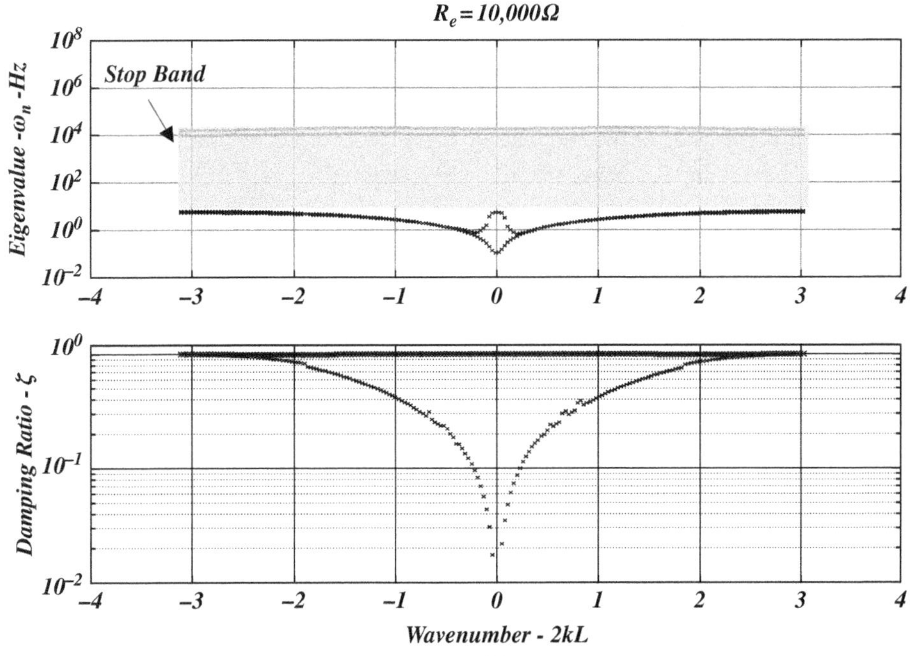

(b)

Figure 11.30 (a–c) Dispersion characteristics of the CB/polymer-piezo-unit cell.

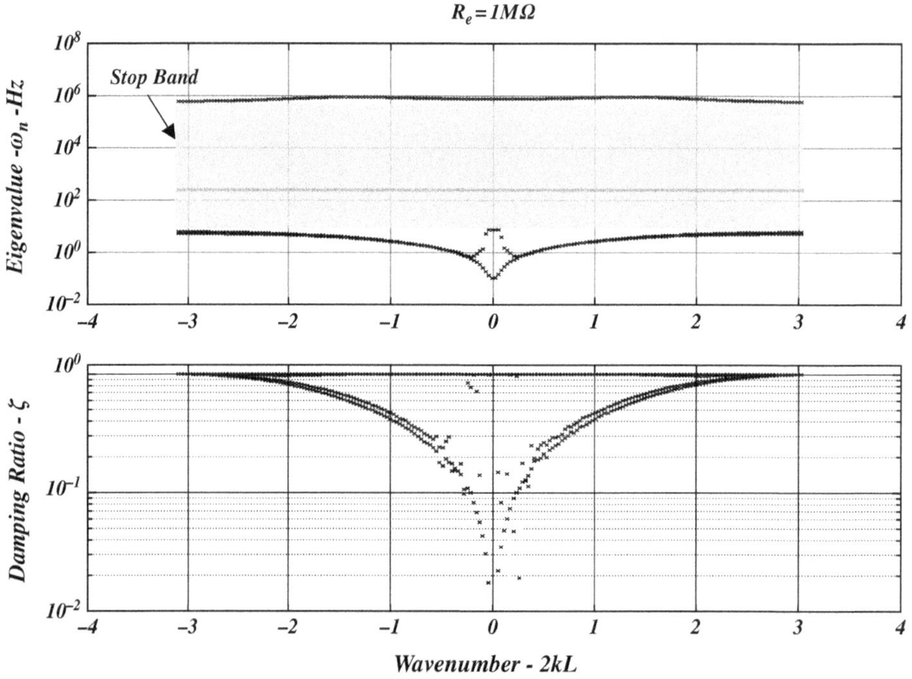

(c)

Figure 11.30 (Continued)

piezo-resistance of the VEM conducting layer to reach 1 Mohm has significantly improved the stop band characteristics and extended it to 1 MHz as indicated in Figure 11.30c.

The effect of such improvements in the stop band characteristics is demonstrated clearly in the frequency response behavior of the composite as shown in Figure 11.31a,b.

Figure 11.31a indicates that the embedding the CB filler in the pristine VEM layer has introduced two important effects when the effective piezo-resistance is R_e = 10 Kohm. The first effect appears in the form of the reinforcement of the polymer, as predicted by the reinforcement theory of Smallwood (1944), which has resulted in stiffening the VEM. Such a stiffening effect is manifested by the shift of the resonant frequencies to higher frequency ranges that, in turn, produced a significant attenuation of the transmitted vibration.

The second effect manifests itself by the vibration attenuation at high frequencies, particularly at 12.93 KHz, which is attributed to the energy dissipation due to shunting of the piezo-layer by the piezo-resistance of the conducting VEM/CB layer.

Increasing further the effective shunting resistance to R_e = 1 Mohm, as indicated in Figure 11.31b resulted in complete blockage of the high frequency vibration around 12 KHz. Such a blockage is due to the enhanced stop band characteristics shown in Figure 11.30c.

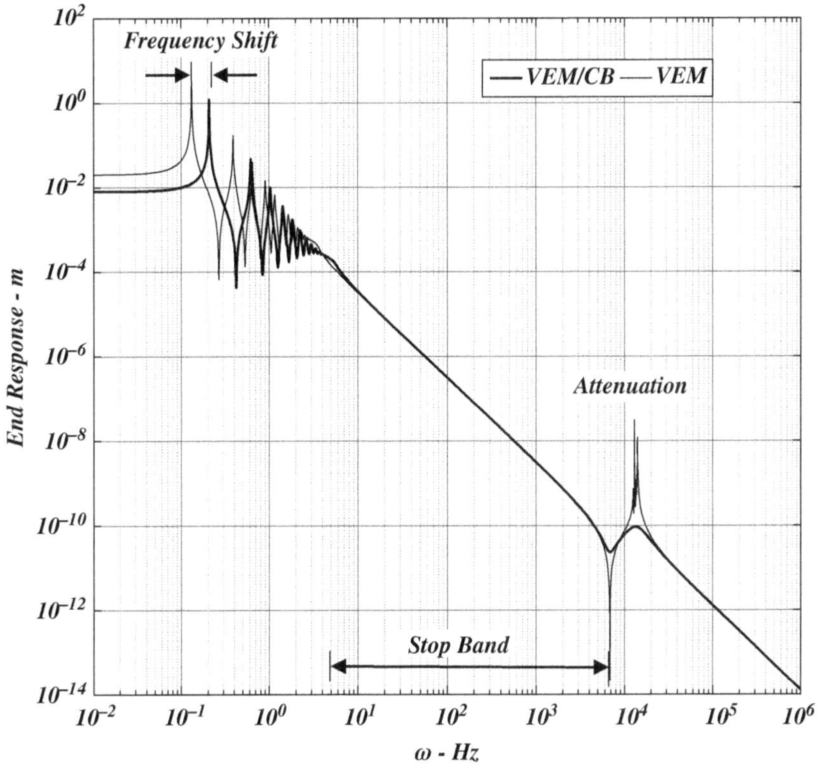

(a) – R_e=10kΩ

Figure 11.31 Effect of shunting resistance on the frequency response characteristics of the CB/polymer-piezo-composite. (a) R_e = 10 k and (b) R_e = 1 M.

11.6 Hybrid Composites with Shunted Piezoelectric Particles

11.6.1 Composite Description and Assumptions

The hybrid composite considered in this section, shown in Figure 11.32, consists of a polymer matrix, electrically conducting particles, and piezoelectric inclusions. All the ingredients of the composite are assumed to be perfectly bonded. Further, the inclusions are assumed to be spheroidal in shapes that are arbitrarily oriented within the matrix. The considered piezoelectric inclusions are assumed to be transversely isotropic with poling directions and electrodes in the three direction. The state of stress and electrode distributions of all the piezoelectric inclusions are assumed identical and uniform. The electrically conducting particles and the matrix are assumed to form resistive electric current paths $\boldsymbol{R}$, which are adequate to load the piezoelectric inclusions (Aldraihem 2011; Aldraihem et al. 2007) as displayed in Figure 11.32b. Finally, the composite is assumed to be homogeneous on the macro-scale and obeys the basic assumptions used in micromechanics analysis.

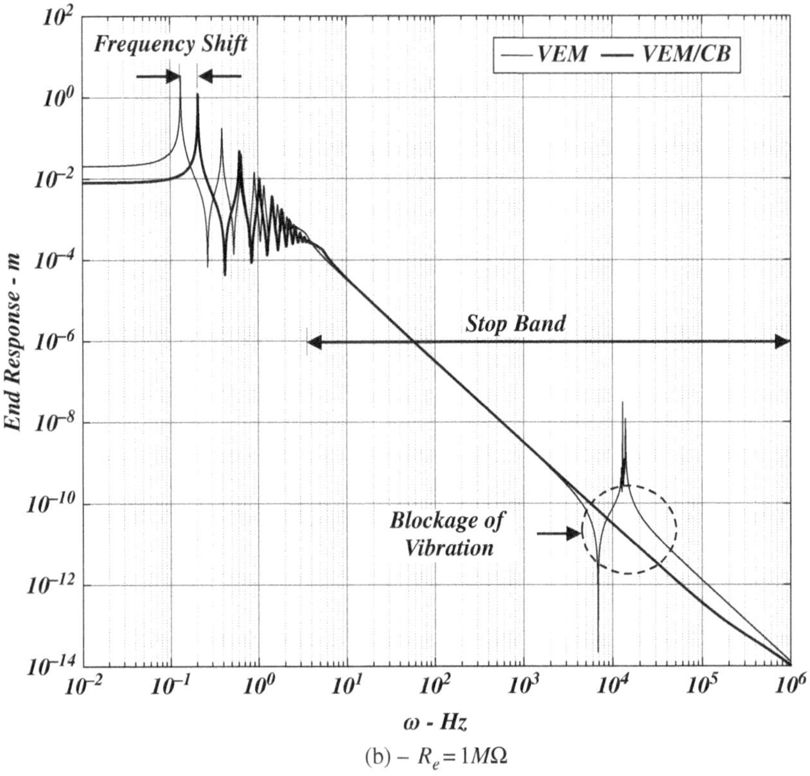

Figure 11.31 (Continued)

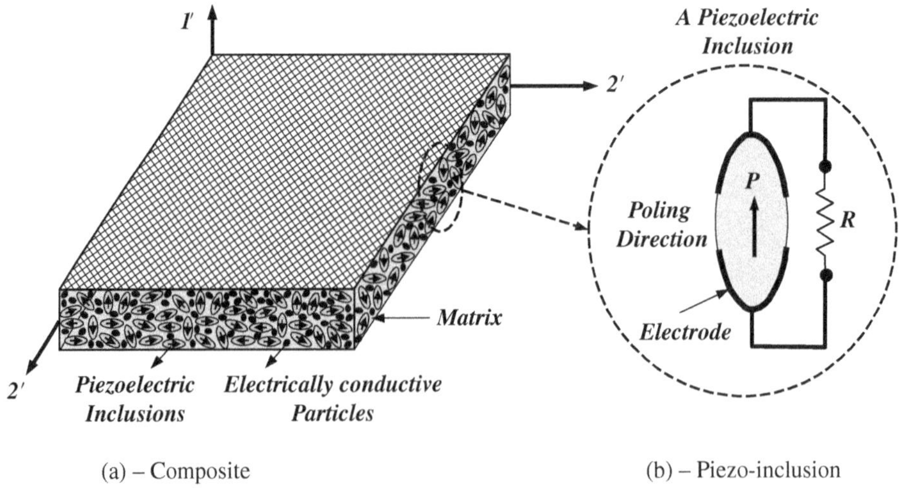

Figure 11.32 The structure and global coordinate system of a hybrid composite. (⬭ piezoelectric inclusions ● conducting particles ▒ matrix). (a) Composite and (b) piezo-inclusion.

11.6.2 Shunted Piezoelectric Inclusions

Consider a hybrid composite that is subjected to a far-field stress in the composite global (*1′-2′-3′*) coordinate system. Then, each of the embedded piezoelectric inclusions generates an electric energy that is dissipated into the resistance of the surrounding conducting polymer matrix. As outlined in Chapter 9, a resistively shunted piezoelectric element behaves like a VEM with the nonzero compliances, in the principle material (*1-2-3*) coordinate system, which are defined as follows:

$$s_{11}^* = s_{11}^E\left(1 - k_{31}^2 Z_3^{EL}\right), s_{22}^* = s_{11}^* \tag{11.106a}$$

$$s_{12}^* = s_{12}^E - s_{11}^E k_{31}^2 Z_3^{EL} \tag{11.106b}$$

$$s_{13}^* = s_{13}^E - \sqrt{s_{11}^E s_{33}^E}\, k_{31} k_{33} Z_3^{EL}, s_{23}^* = s_{13}^* \tag{11.106c}$$

$$s_{33}^* = s_{33}^E\left(1 - k_{33}^2 Z_3^{EL}\right) \tag{11.106d}$$

$$s_{44}^* = s_{44}^E\left(1 - k_{15}^2\right), s_{55}^* = s_{44}^*, \tag{11.106e}$$

$$s_{66}^* = s_{66}^E \tag{11.106f}$$

$$\text{with } Z_3^{EL} = \left(\frac{\Omega}{\Omega^2 + 1}\right)(\Omega + i); i = \sqrt{-1},\ \Omega = R\, C^T \omega, \text{ and } C^T = \frac{A}{2a_3}\varepsilon_{33}^{\sigma} \tag{11.107}$$

In Eqs. (11.106a)–(11.106f) and (11.107), s_{ij}^E denotes the elements of elastic compliance matrix at constant electric field. Also, k_{31}, k_{33}, and k_{15} denote the piezoelectric coupling factors, $\varepsilon_{33}^{\sigma}$ is the dielectric permittivity at constant stress, R is the shunt resistance, and ω is the frequency. Furthermore, a_3 defines the inclusion half-length and A is the electrodes area with normal along the three-axis as shown in Figure 11.7.

For the considered shunting, Eqs. (11.106a)–(11.106f) indicates that the out-of-plane shear compliances exhibit open-circuit conditions, the in-plane shear compliance exhibits short-circuit conditions, whereas the remaining compliances exhibit shunted-circuit conditions. Hence, the shear compliances primarily behave in an elastic manner and have limited damping capacity.

Substituting Eq. (11.107) into Eqs. (11.106a)–(11.106f), yields the complex compliance matrix that can be separated into real and imaginary parts as follows:

$$[s^*] = [s'] + i[s''] \tag{11.108}$$

where $[s']$ and $[s'']$ denote the real and imaginary part of the compliance, respectively.

Equations (11.106a)–(11.106f) or (11.107) represent the compliances in the principle material (*1-2-3*) coordinate system. Moreover, the compliance can be used to extract the stiffness as the stiffness is inverse of the compliance; that is, $[c^*] = [s^*]^{-1}$ and $[c^E] = [s^E]^{-1}$.

11.6.3 Typical Performance Characteristics of Hybrid Composites

In this section, the micromechanics model presented in Section 11.6.2 is utilized to predict the mechanical properties of a hybrid composite consisting of an epoxy matrix (Hercules 3501-6) (Lesieutre et al. 1993), CB particles, and piezoelectric particles

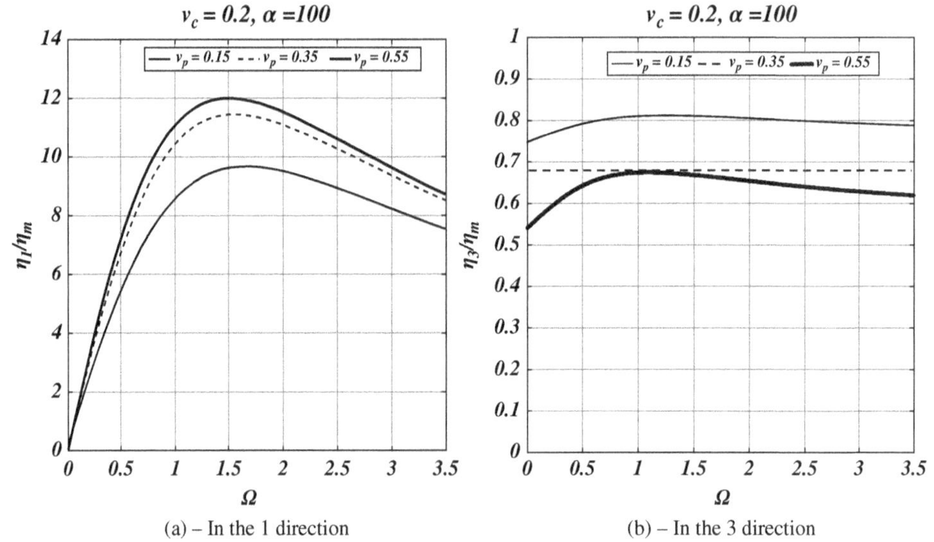

Figure 11.33 Effect of the shunt frequency on the loss factor of the hybrid composite. (a) In the one direction and (b) in the three direction.

(Noliac® – PCM51). The physical properties of the three ingredients of the composite are listed in Table 11.3. The epoxy has a Poisson's ratio, v, of 0.38, and the CB particles are embedded with a volume fraction 0.20. It is assumed that the piezo-particles have aligned orientations.

It should be mentioned that both the inclusions shape and shunting characteristics are unknown. Hence, in the model predictions the inclusions aspect ratio, α, and the shunt frequency, Ω, are investigated.

Figure 11.33a displays the damping ratios of the composite in direction 1 and 2, as function of the shunt frequency, $\Omega = R\ C^T\omega$ for different values of the volume fraction v_p of the piezo-particles and for aspect ratio $\alpha = 100$. The figures indicate that there are optimal values of *ohm*, that is, optimal shunting resistance, for each volume fractions v_p at which the loss factor attain maximum values. It can be seen that these values of *ohm* are 1.68, 1.54, and 1.43 when the volume fractions v_p are equal to 0.15, 0.35, and 0.55. The corresponding maximum loss factors are 9.67, 11.45, and 11.99, respectively. The values of the loss factors are normalized with respect to that of the epoxy matrix. This indicates that using the shunting of the piezo-particles enhances the damping characteristics by almost an order of magnitude in both the one and two directions as the hybrid composite is transversally isotropic.

In direction 3, that is, during compression loading, the corresponding maximum loss factors are 0.81, 0.68, and 0.67 when the volume fractions v_p are equal to 0.15, 0.35, and 0.55, respectively, as shown in Figure 11.33b. The corresponding optimal shunt frequencies ohm are 1.28, 1.17, and 1.035.

Figures 11.34 and 11.35 display the effects of the volume fraction of the piezo-particles v_p and aspect ratio α, respectively. The figures indicate that increasing the volume

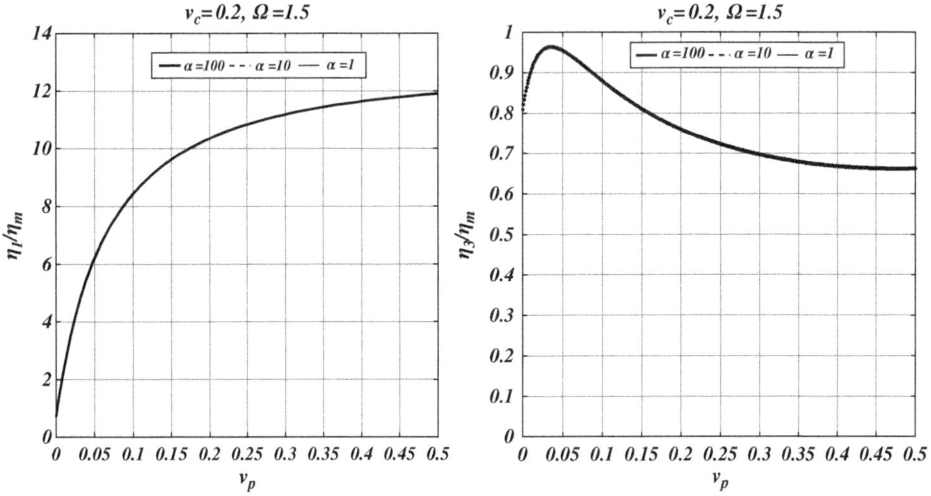

(a) – In the 1 direction (b) – In the 3 direction

Figure 11.34 Effect of the volume fraction of piezo-particles on the loss factor of the hybrid composite. (a) In the one direction and (b) in the three direction.

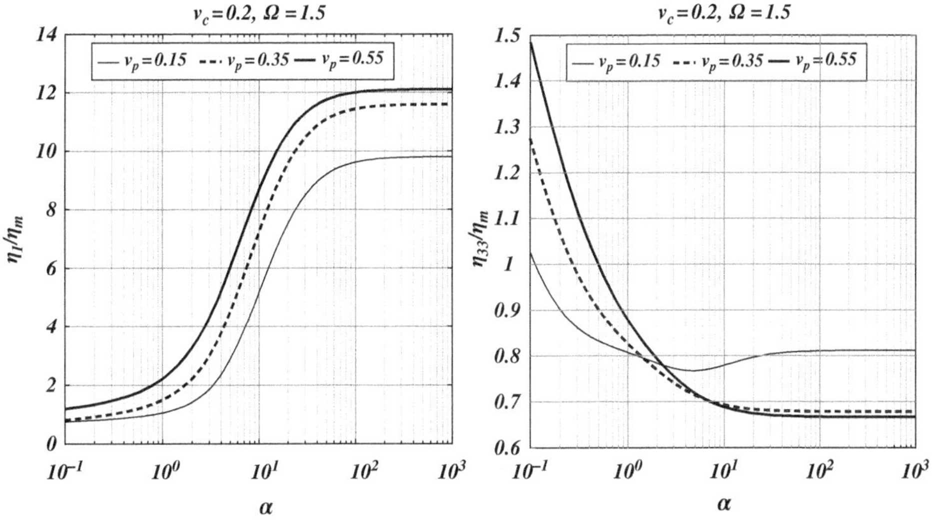

(a) – In the 1 direction (b) – In the 3 direction

Figure 11.35 Effect of the aspect ratio of the piezo-particles on the loss factor of the hybrid composite. (a) In the one direction and (b) in the three direction.

fraction and the aspect ratio result in a significant increase in the transverse loss factor. The opposite is true for the compressive loss factors.

The displayed results in Figures 11.33" through 11.35 indicate that the loss factors of the hybrid composite depend on the volume fraction, the orientation distribution, the aspect ratio (shape), and the shunt frequency-resistance parameter. Table 11.5

Table 11.5 Values of Ω_{max} and η_{max} for various orientations and aspect ratios when $v_c = 0.20$ and $v_p = 0.35$.

Inclusion aspect ratio (α)	Aligned orientation ($\theta_0 \to 0,\ \phi_0 \to 0$)				2D random orientation ($\theta_0 \to \pi,\ \phi_0 \to 0$)						3D random orientation ($\theta_0 \to \pi,\ \phi_0 \to \pi/2$)			
	Ω_{max}	$\eta_{1_{max}}$	Ω_{max}	$\eta_{3_{max}}$	Ω_{max}	$\eta_{1_{max}}$	Ω_{max}	$\eta_{3_{max}}$	Ω_{max}	$\eta_{6_{max}}$	Ω_{max}	$\eta_{1_{max}}$	Ω_{max}	$\eta_{6_{max}}$
0.0	1.16	0.62	1.13	2.36	1.75	3.71	1.16	2.42	2.03	4.41	1.78	3.71	2.02	4.42
0.5	1.07	1.18	1.06	0.92	1.03	1.23	1.08	0.92	1.16	0.71	1.01	1.10	1.10	0.71
1.0	1.07	1.53	1.07	0.84	1.11	1.37	1.07	0.84	1.34	0.75	1.06	1.23	1.28	0.75
100	1.58	11.45	1.13	0.68	2.12	8.45	1.16	0.68	2.43	7.52	2.05	8.47	2.44	7.52
1000	1.59	11.61	1.19	0.68	2.15	8.60	1.16	0.68	2.49	7.68	2.02	8.47	2.51	7.68

summarizes a listing of the shunt frequency parameter ohm at which the loss factor ratio, η/η_m, attains maximum values for different orientation distributions and various aspect ratios. The listed values are calculated for piezo-particles *PCM*51 with volume fraction of 35% and CB volume fraction of 25%.

Figures 11.36–11.38 display the effect of the shunt frequency, piezo-particle volume fraction, and piezo-particles aspect ratio on the compliance coefficients of the hybrid composite, respectively.

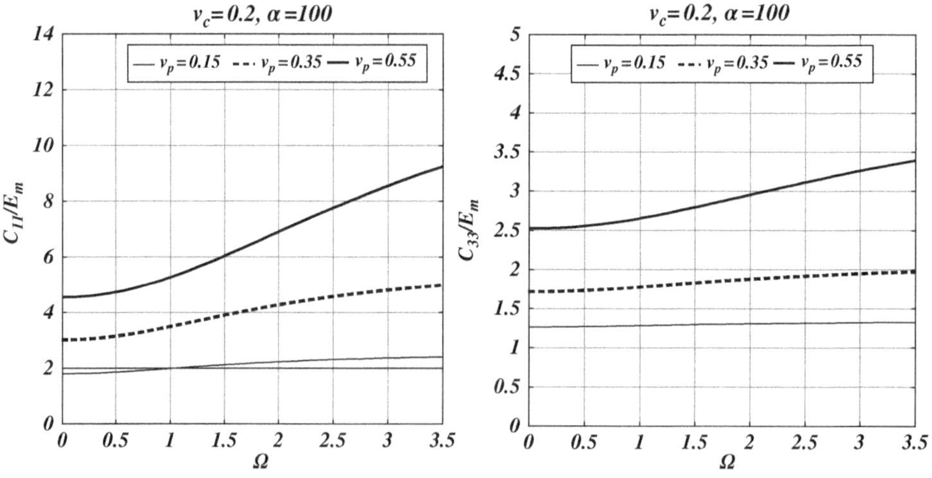

(a) – In the 1 direction (b) – In the 3 direction

Figure 11.36 Effect of the shunt frequency on the compliance coefficients of the hybrid composite. (a) In the one direction and (b) in the three direction.

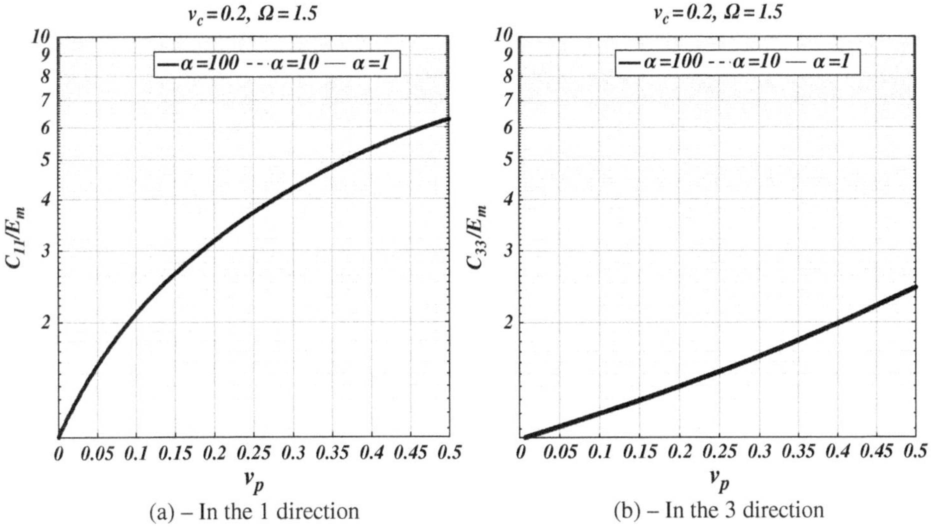

Figure 11.37 Effect of the volume fraction of piezo-particles on the compliance coefficients of the hybrid composite. (a) In the one direction and (b) in the three direction.

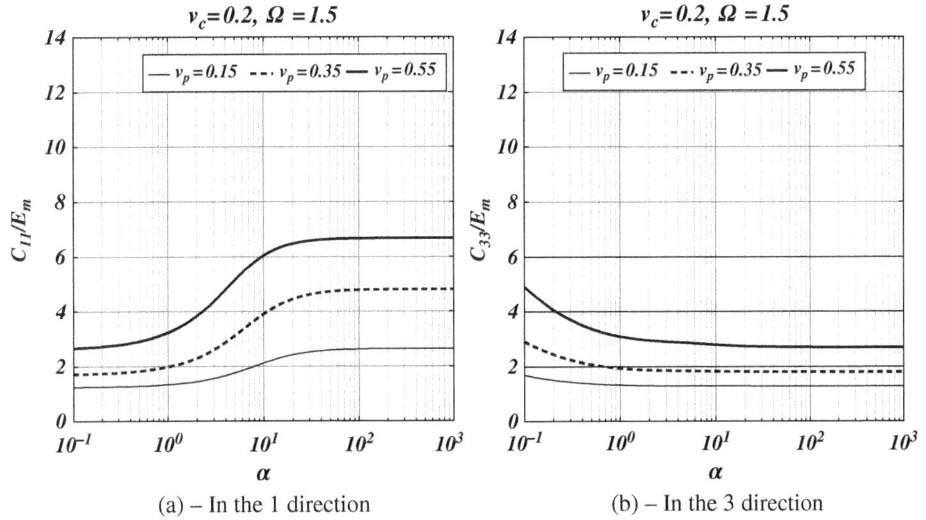

Figure 11.38 Effect of the aspect ratio of piezo-particles on the compliance coefficients of the hybrid composite. (a) In the one direction and (b) in the three direction.

11.7 Summary

This chapter has presented the prediction of the mechanical properties of composites consisting of a polymer matrix filled with nanoparticles as a function of different volumetric fractions and physical properties of the constituents. The presented prediction approaches are based on the MTM, which is rooted on Eshelby's equivalent inclusion technique, the classical filler reinforcement methods that are either analytical in nature, and finite element modeling using the RVE approach.

In this chapter, particular emphasis is placed on electrically conducting composites, such as CB/polymer composites, because of the possibility of integrating these composites with piezoelectric particles or elements in order to enhance their damping properties through appropriate electric shunting.

References

Advani, S.G. and Tucker, C.L. III (1987). The use of tensors to describe and predict fiber orientation in short fiber composites. *Journal of Rheology* 31: 751–784.

Aldraihem, O.J. (2011). Micromechanics modeling of viscoelastic properties of hybrid composites with shunted and arbitrarily oriented piezoelectric inclusions. *Mechanics of Materials* 43: 740–753.

Aldraihem, O.J., Baz, A., and Al-Saud, T.S. (2007). Hybrid composites with shunted piezoelectric particles for vibration damping. *Mechanics of Advanced Materials and Structures* 14: 413–426.

ANSI/IEEE Std 176-1987 (1987)., Standards Committee of the IEEE Ultrasonic, Ferroelectrics, and Frequency Control Society, American National Standard IEEE Standard on Piezoelectricity, Institute of Electrical and Electronics Engineers, Inc., New York, USA, 1987.

Barber, J. R. (2010). *Solid Mechanics and Its Applications*, 3rd ed. Springer.

Barbero, E.J. (2013). *Finite Element Analysis of Composite Materials Using ANSYS®*, 2e. CRC Press.

Brinson, L.C. and Lin, W.S. (1998). Comparison of micromechanics methods for effective properties of multiphase viscoelastic composites. *Composite Structures* 41: 353–367.

Ding, N., Wang, L., Zuo, P. et al. (2013). Study on electrical properties of activated carbon black filled polypropylene composites using impedance analyser. *Advanced Materials Research* 712–715: 175–181.

Dvorak, G.J. and Benveniste, Y. (1992). On transformation strains and uniform fields in multiphase elastic media. *Proceedings of the Royal Society: Mathematical and Physical Sciences* 437: 291–310.

Electro Ceramic Division. (n.d.) "Data for designers," Morgan Matroc Inc., 232 Forbes Road, Bedford, OH 44146.

Eshelby, J.D. (1957). The determination of elastic field of an ellipsoidal inclusion and related problems. *Proceedings of Royal Society of London* 276–396.

Gonella, S. and Ruzzene, M. (2008). Homogenization of vibrating periodic lattice structures. *Applied Mathematical Modelling* 32 (4): 459–482.

Gusev, A., Heggli, M., Lusti, H.R., and Hine, P.J. (2002). Orientation averaging for stiffness and thermal expansion of short fiber composites. *Advanced Engineering Materials* 4: 931–933.

Guth, E.J. (1945). Theory of filler reinforcement. *Journal of Applied of Physics* 16: 20–25.
Han, D.G. and Choi, G.M. (1998). Computer simulation of the electrical conductivity of composites: the effect of geometrical arrangement. *Solid State Ionics* 106: 71–87.
Hashin, Z. and Shtrikman, S. (1962). On some variational principles in anisotropic and nonhomogeneous elasticity. *Journal of the Mechanics and Physics of Solids* 10: 335–342.
Hine, P.J., Lusti, H.R., and Gusev, A. (2004). On the possibility of reduced variable predictions for the thermoelastic properties of short fiber composites. *Composites Science and Technology* 64: 1081–1088.
Hine, P.J., Lusti, H.R., and Gusev, A.A. (2002). Numerical simulation of the effects of volume fraction, aspect ratio and fiber length distribution on the elastic and thermoelastic properties of short fiber composites. *Composites Science and Technology* 62: 1445–1453.
Hussein, M.I. (2009). Theory of damped bloch waves in elastic media. *Physical Reviews B* 80: 212301.
Kaiser, J.H. (1993). Microwave evaluation of the conductive filler particles of carbon black-rubber composites. *Applied Physics A* 56: 299–302.
Leblanc, J.L. (2002). Rubber–filler interactions and rheological properties in filled compounds. *Progress in Polymer Science* 27 (4): 627–687.
Lesieutre, G.A., Yarlagadda, S., Yoshikawa, S. et al. (1993). Passively damped structural composite materials using resistively shunted piezoceramic fibers. *Journal of Materials Engineering and Performance* 2: 887–892.
Marenić E., Brancherie D., and Bonnet M., "Multiscale asymptotic-based modeling of local material inhomogeneities", Proceedings of the 8th International Congress of Croatian Society of Mechanics, 29 September - 2 October 2015, Opatija, Croatia, 2015.
Meirovitch, L. (2010). *Fundamentals of Vibration*. Long Grove, IL: Waveland.
Mori, T. and Tanaka, K. (1973). Average stress in matrix and average elastic energy of materials with misfitting inclusion. *Acta Metallurgica* 21: 571–574.
Nemat-Nasser, S. and Hori, M. (1999). *Micromechanics: Overall Properties of Heterogeneous Materials*, Second Revised Edition. Amsterdam: Elsevier.
Noliac, Piezoelectric Particles: (2018)Available online at http://www.noliac.com (accessed June 2018).
Pascault, J.P., Sauterau, H., Verdu, J., and Williams, R.J.J. (2002). *Thermosetting Polymers*. New York, NY: Marcel Dekker.
Prasad J. and Diaz A. R. "A concept for a material that softens with frequency", Paper No. DETC2007-34299, pp. 761–768, Proceedings of the ASME 2007 International Design Engineering Technical Conferences and Computers and Information in Engineering Conference, Vol. 6, 33rd Design Automation Conference, Las Vegas, NV, USA, September 4–7, 2007.
Reddy, J.N. (2013). *An Introduction to Continuum Mechanics*. Cambridge University Press.
Reuss, A. (1929). Berechnung der Fließgrenze von Mischkristallen auf Grund der Plastizitatsbedingung für Einkristalle. *Zeitschrift für Angewandte Mathematik und Mechanik* 9: 49–58.
Schwartz, G., Cerveny, S., and Marzocca, A.J. (2000). A numerical simulation of the electrical resistivity of carbon black filled rubber. *Polymer* 41: 6589–6595.
Sejnoha, M. and Zeman, J. (2013). *Micromechanics in Practice*. Southampton, UK: WIT Press.
Simmons, J.G. (1963). Generalized formula for the electric tunnel effect between similar electrodes separated by a thin insulating film. *Journal of Applied Physics* 34 (6): 1793–1803.

Smallwood, H.M. (1944). Limiting law of the reinforcement of rubber. *Journal of Applied Physics* 15: 758–766.
Song, Y. and Zheng, Q. (2016). Concepts and conflicts in nanoparticles reinforcement to polymers beyond hydrodynamics. *Progress in Materials Science* 84: 1–58.
Tandon, G.P. and Weng, G.J. (1986). Average stress in the matrix and effective moduli of randomly oriented composites. *Composites Science and Technology* 27: 111–132.
Tian, S. and Wang, X. (2008). Fabrication and performances of epoxy/multi-walled carbon nanotubes/piezoelectric ceramic composites as rigid piezo-damping materials. *Journal of Materials Science* 43: 4979–4987.
Torquato, S. (2001). *Random Heterogeneous Materials: Microstructure and Macroscopic Properties*. New York, NY: Springer.
Tucker, C.L. III and Liang, E. (1999). Stiffness predictions for unidirectional short-fiber composites: review and evaluation. *Composites Science and Technology* 59: 655–671.
Voigt, W. (1887). Theorie des Lichts für bewegte Medien. *Göttinger Nachrichten* 8: 177–238.
Wang, Y.-J., Pan, Y., Zhang, X.-W., and Tan, K. (2005). Impedance spectra of carbon black filled high-density polyethylene composites. *Journal of Applied Polymer Science* 98: 1344–1350.
Wetherhold, R.C. (1988). *Elastic Plates Theory and Application, Riesmann, H.* New York, NY: Wiley, Chapter 10.
Zhang, J.F. and Yi, X.S. (2002). Dynamic rheological behavior of high-density polyethylene filled with carbon black. *Journal of Applied Polymer Science* 86: 3527–3531.
Zhang, X.W., Pan, Y., Zheng, Q., and Yi, X.S. (2000). Time dependence of piezoresistance for conductor-filled polymer composites. *Journal of Applied Polymer Science Part B: Polymer Physics* 38: 2739–2749.
Zhang, X.W., Pan, Y., Zheng, Q., and Yi, X.S. (2001). Piezoresistance of conductor filled insulator composites. *Polymer International* 50: 229–236.

11.A Transformation Matrix

The transformation matrix shown in Eqs. (11.43) and (11.44) is defined by Wetherhold (1988):

$$[R_1] = \begin{bmatrix} m_1^2 & n_1^2 & p_1^2 & 2n_1p_1 & 2m_1p_1 & 2m_1n_1 \\ m_2^2 & n_2^2 & p_2^2 & 2n_2p_2 & 2m_2p_2 & 2m_2n_2 \\ m_3^2 & n_3^2 & p_3^2 & 2n_3p_3 & 2m_3p_3 & 2m_3n_3 \\ m_2m_3 & n_2n_3 & p_2p_3 & n_2p_3+p_2n_3 & m_2p_3+p_2m_3 & m_2n_3+n_2m_3 \\ m_1m_3 & n_1n_3 & p_1p_3 & n_1p_3+p_1n_3 & m_1p_3+p_1m_3 & m_1n_3+n_1m_3 \\ m_1m_2 & n_1n_2 & p_1p_2 & n_1p_2+p_1n_2 & m_1p_2+p_1m_2 & m_1n_2+n_1m_2 \end{bmatrix} \tag{11.A.1}$$

with

$$\begin{bmatrix} m_1 & n_1 & p_1 \\ m_2 & n_2 & p_2 \\ m_3 & n_3 & p_3 \end{bmatrix} = \begin{bmatrix} \cos\theta\cos\phi & -\sin\theta & \cos\theta\sin\phi \\ \sin\theta\cos\phi & \cos\theta & \sin\theta\sin\phi \\ -\sin\phi & 0 & \cos\phi \end{bmatrix} \tag{11.A.2}$$

where θ and φ are rotation Euler angles defined in Figure 11.3.

The transformation matrix $[R_2]$ shown in Eqs. (11.43) and (11.44) can be obtained by the relationship:

$$[R_1]^{-1} = [R_2]^T, [R_2]^{-1} = [R_1]^T \tag{11.A.3}$$

11.B Reinforcement Mechanics of Particle-Filled Polymers

11.B.1 Basics

In this appendix, the reinforcement of a polymer by virtue of embedding solid particles in it is analyzed using the approach developed by Smallwood (1944) in an attempt to determine the effect of particle size and volume fraction on the modulus of the resulting composite.

Consider a rigid, spherical filler particle of radius R embedded in an isotropic polymer of known storage modulus as shown in Figure 11.B.1. The deflections and the resulting stresses near the particle are determined when the polymer is subjected to unidirectional tensile force T along the y direction.

The governing equilibrium equations as obtained from the theory of elasticity are given, for example by Barber (2010), as follows:

$$\frac{\lambda + \mu}{\mu}\frac{\partial \Delta}{\partial x} + \nabla^2 u = 0, \frac{\lambda + \mu}{\mu}\frac{\partial \Delta}{\partial y} + \nabla^2 v = 0, \text{and} \frac{\lambda + \mu}{\mu}\frac{\partial \Delta}{\partial z} + \nabla^2 w = 0 \tag{11.B.1}$$

where u, v, w are the deflections along the x, y, z directions. Also, Δ denotes the cubical dilation ($\Delta = \varepsilon_{xx} + \varepsilon_{yy} + \varepsilon_{zz}$) and ∇ is the Laplacian. Note that the equilibrium Eq. (11.B.1)

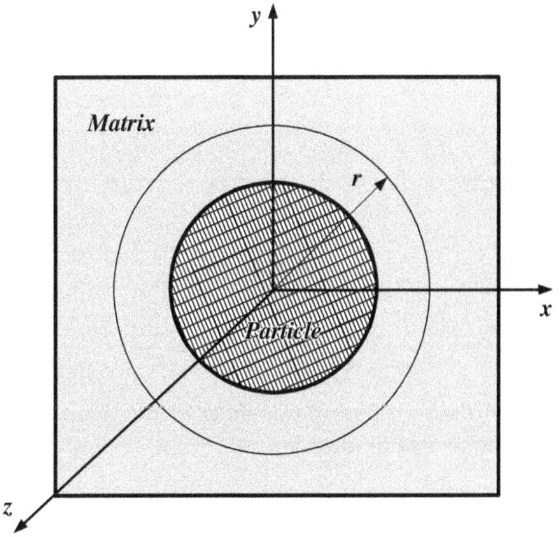

Figure 11.B.1 Polymer with an embedded filler particle.

are written in terms of the Lamé constants λ and μ that are related to Young's modulus E and Poisson's ratio, υ, such that (Reddy 2013):

$$E = \mu(3\lambda + 2\mu)/(\lambda + \mu) \quad \text{and} \quad \upsilon = \lambda/[2(\lambda + \mu)] \tag{11.B.2}$$

Note that Lamé constants λ and μ are also given by:

$$\lambda = \frac{E}{(1+\upsilon)(1-2\upsilon)} \quad \text{and} \quad \mu = G \tag{11.B.3}$$

where G is the shear modulus.

Equation (11.B.1) are subjected to the following boundary conditions:

$$\text{At}\, r = R : u = v = w = 0 \quad \text{(as the particle is assumed rigid)} \tag{11.B.4}$$

$$\text{and} \quad \text{At}\, r = \infty : u = -\frac{\lambda T}{2\mu(3\lambda + 2\mu)}x = -Cx, \tag{11.B.5}$$

$$v = \frac{(\lambda+\mu)T}{\mu(3\lambda + 2\mu)}y \; = Ay, \tag{11.B.6}$$

and

$$w = -\frac{\lambda T}{2\mu(3\lambda + 2\mu)}z = -Cz. \tag{11.B.7}$$

The boundary conditions given by Eqs. (11.B.4) through (11.B.7), at $r = \infty$, indicate that far away from the particle, the deflections (u, v, w) are caused only by the tension T.

Solutions of Eq. (11.B.1) subject to the boundary conditions are as follows:

$$u = -\left[C\left(1-\bar{r}^{-3}\right) - B\bar{r}^{-3}\left(1-\bar{r}^{-2}\right)\left(1-5\bar{y}^2/\bar{r}^2\right)\right]x, \tag{11.B.8}$$

$$v = \left[A\left(1-\bar{r}^{-3}\right) - B\bar{r}^{-3}\left(1-\bar{r}^{-2}\right)\left(3-5\bar{y}^2/\bar{r}^2\right)\right]y, \tag{11.B.9}$$

and

$$w = -\left[C\left(1-\bar{r}^{-3}\right) - B\bar{r}^{-3}\left(1-\bar{r}^{-2}\right)\left(1-5\bar{y}^2/\bar{r}^2\right)\right]z, \tag{11.B.10}$$

where $\bar{r} = r/R$, $\bar{y} = y/R$, $r^2 = x^2 + y^2 + z^2$, and $B = \frac{3}{4}\frac{T}{\mu}\frac{\lambda+\mu}{3\lambda+8\mu}$.

Assuming that Poisson's ratio μ of the composite is close to that of the polymer, which is nearly 0.5. Eq. (11.B.2) can be rewritten as:

$$\upsilon = 1/[2(1+\mu/\lambda)] \rightarrow \upsilon \cong 1/2 \;\; if\, \mu/\lambda \cong 0 \tag{11.B.11}$$

For such a limit ($\mu/\lambda \cong 0$), the parameters A, B, and C reduce to:

$$A = \frac{1}{3}\frac{T}{\mu}, B = \frac{1}{4}\frac{T}{\mu}, \quad \text{and} \quad C = \frac{1}{6}\frac{T}{\mu}. \tag{11.B.12}$$

In order to quantify the effect of the filler particle, the strain energy function W_T is determined as follows:

$$W_T = \int_V W dV = \frac{T^2}{2\mu}\left[\frac{1}{3}\left(V - \frac{4}{3}\pi R^3 N\right) + \frac{14}{23}\pi R^3 N\right] \tag{11.B.13}$$

where N is the number of particles and V is the total volume of the polymer and the particles.

Then, the strain energy per unit volume is given by:

$$\bar{W}_T = \frac{T^2}{2}\frac{1}{3\mu}\left(1 + \frac{1}{2}\nu\right) \tag{11.B.14}$$

where ν is the volume fraction of the particles.

Substituting Eqs. (11.B.3) and (11.B.12) into Eq. (11.B.14) yields:

$$\begin{aligned}\bar{W}_T &= \frac{T^2}{2}\frac{1}{3\mu}\left(1 + \frac{1}{2}\nu\right) = \frac{AT}{2}\left(1 + \frac{1}{2}\nu\right) = \frac{A^2(T/A)}{2}\left(1 + \frac{1}{2}\nu\right) \\ &= \frac{A^2 3\mu}{2}\left(1 + \frac{1}{2}\nu\right) = \frac{A^2 3G}{2}\left(1 + \frac{1}{2}\nu\right) = \frac{A^2 E}{2}\left(1 + \frac{1}{2}\nu\right)\end{aligned} \tag{11.B.15}$$

The deflection w, at any point, due to the effect of all the filler particles is given by modifying Eq. (11.B.7) to take the following form:

$$w = \bar{A}z \tag{11.B.16}$$

where $\bar{A}$ accounts for the effect of the particles and takes the following form (Smallwood 1944):

$$\bar{A} = A(1 - \nu) \tag{11.B.17}$$

Then, the associated strain energy per unit volume is given by:

$$\bar{W}_T^c = \frac{\bar{A}^2 E_c}{2} = \frac{A^2(1-\nu)^2 E_c}{2} \tag{11.B.18}$$

The equivalence of Eqs. (11.B.15) and (11.B.18) yields:

$$\frac{A^2 E}{2}(1 + \nu) = \frac{A^2(1-\nu)^2 E_c}{2}, \text{that is}, E_c = \frac{E(1+\nu)}{(1-\nu)^2} \tag{11.B.19}$$

Expanding Eq. (11.B.19) into a first order Taylor series yields:

$$E_c = E(1 + 2.5\nu) \tag{11.B.20}$$

Song and Zheng (2016) summarized a list of other mathematical models that intend to improve the approximation of Smallwood Eq. (11.B.20) by adding higher order terms. In their summary, the coefficient of second order Taylor series expansion ν^2 is reported by to vary between 2.5 and 15.6 depending on the adopted mathematical model. In here, the approach developed by Smallwood (1944) and then modified by Guth (1945) yields the following commonly used reinforcement relationship:

$$E_c = E\left(1 + 2.5\nu + 14.1\nu^2\right) \tag{11.B.21}$$

Problems

11.1 Consider the electrical circuit shown in Figure P11.1 that simulates the electro-dynamic response of a CB/polymer composite in the percolation zone as shown in Figure 11.28.

Show that the electrical impedance Z of the circuit is given by:

$$Z = Z_1 + iZ_2$$

where Z_1 and Z_2 denote the real and imaginary components of the impedance given by:

$$Z_1 = R_a + \frac{R_c}{1+\omega^2 R_c^2 C_s^2}, \quad Z_2 = -\omega \frac{R_c^2 C_s}{1+\omega^2 R_c^2 C_s^2}$$

By eliminating the frequency ω from Z_1 and Z_2, show that the equation of the Cole–Cole plot, displayed in Figure P11.2, is given by:

$$(\bar{Z}_1 - r_o)^2 + (\bar{Z}_2)^2 = 1,$$

where $\bar{Z}_1 = \frac{Z_1}{R_c/2}, \bar{Z}_2 = \frac{Z_2}{R_c/2}$, $r_o = 2\bar{R} + 1$, and $\bar{R} = \frac{R_a}{R_c}$.

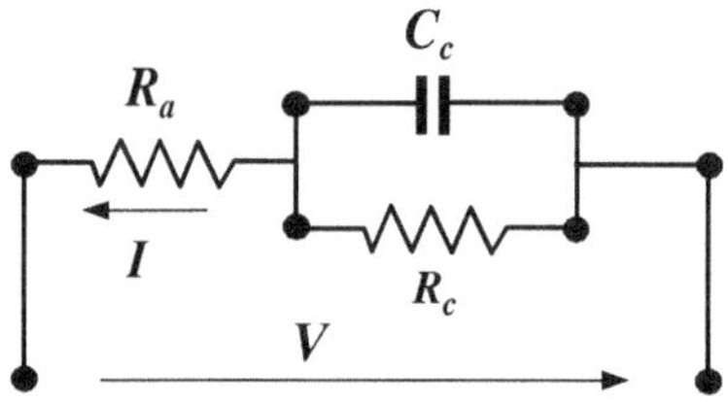

Figure P11.1 Electric circuit simulating response of carbon black/polymer composite in percolation zone

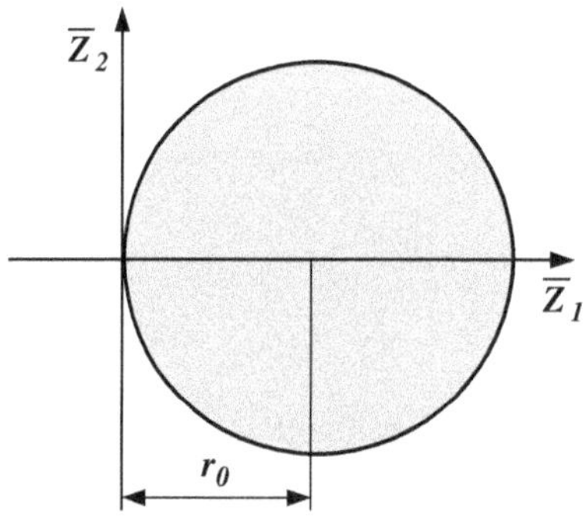

Figure P11.2 Cole–Cole plot of CB/polymer electrical impedance characteristics.

11.2 Consider the composite shown in Figure P11.3a, which consists of two components A and B arranged in a square array. The component A is a polymer matrix in which particles of component B are embedded. The structure of each of these

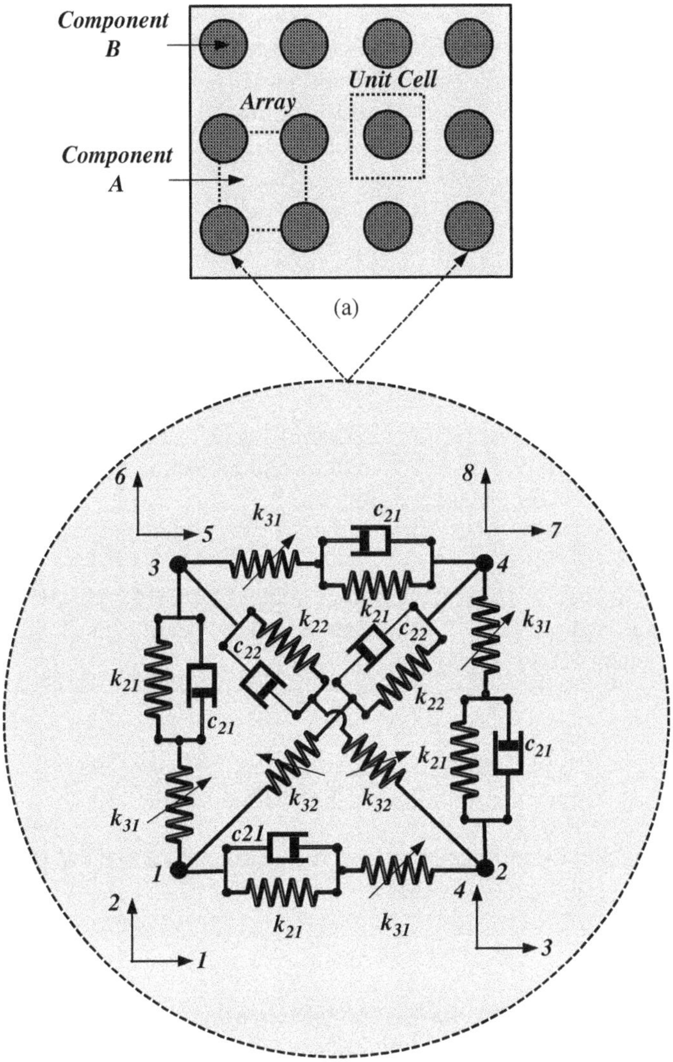

Figure P11.3 A composite with positive and negative stiffness components and damping.

particles, as shown in Figure P11.3b, is an assembly of positive springs (k_{21}, k_{22}), negative springs (k_{31}, k_{32}), and dampers (c_{21} and c_{22}) (Prasad and Diaz 2007b).

The stiffness matrix C_A of the polymer matrix A, in a 2D state of plane-strain, is given by:

$$C_A = \frac{E_A^*}{(1+\upsilon_A)(1-2\upsilon_A)}\begin{bmatrix} 1-\upsilon_A & \upsilon_A & 0 \\ \upsilon_A & 1-\upsilon_A & 0 \\ 0 & 0 & \frac{1}{2}(1-2\upsilon_A) \end{bmatrix}$$

where E_A^* and υ_A are the complex modulus and Poisson's ratio of the polymer, respectively. Assume that $E_A^* = E_A(1+1i)$ where E_A is arbitrary and $\upsilon_A = 0.4995$.

Show that the complex modulus stiffness matrix (C_B) of component B corresponding to the arrangement shown in Figure P11.3 is given by:

$$C_B = \begin{bmatrix} f_{11} & f_{12} & 0 \\ f_{21} & f_{22} & 0 \\ 0 & 0 & f_{33} \end{bmatrix}$$

The component f_{11} denotes the reaction force along the six and eight directions. It is determined by restraining the DOF 1, 2, 3, 4, 5, and 7 while the DOF 6 and 8 are subjected to unit displacements. This yields f_{11} as follows:

$$f_{11} = 2s_1 + s_2 = f_{22}$$

Also, f_{12} denotes the reaction force along the DOF 1 and 5, which is obtained by restraining the DOF 2, 3, 4, 6, 7, and 8 while subjecting the DOF 1 and 5 to unit displacements. This yields f_{12} as follows:

$$f_{12} = s_2 = f_{21}$$

Finally, f_{33} denotes the reaction force along the DOF 5 and 7, which is obtained by restraining the DOF 1, 2, 3, 4, 6, and 8 while subjecting the DOF 5 and 7 to unit displacements. This yields f_{33} as follows:

$$f_{33} = s_2$$

where $s_1 = \dfrac{k_{31}(k_{21} + i\omega c_{21})}{(k_{31} + k_{21} + i\omega c_{21})}$ and $s_2 = \dfrac{k_{32}(k_{22} + i\omega c_{22})}{(k_{32} + k_{22} + i\omega c_{22})}$

Assume that the stiffness matrix of the component B is set to represent an isotropic material, then, show that this implies that:

$$\upsilon_B = {}^1/_4, E_B^* = \frac{5}{2}s^1 = \frac{5}{2}s^2, \text{ and } s_1 = s_2.$$

Let $k_{22} = k_{21}$, $k_{32} = k_{31}$, and $c_{21} = c_{22}$ to ensure that $s_1 = s_2$ and let $c_{21}/k_{21} = 0.2\text{E-}3$, then show that if $k_{21} = 8.45E_A$ and $k_{31} = -0.304k_{21}$, the complex modulus of component B is given by:

$$E_B^*/E_A = -6.42\frac{(1 + 0.0002\,\omega i)}{(0.696 + 0.0002\,\omega i)}$$

11.3 Consider the composite shown in Figure P11.3a and the established properties of its two ingredients A and B as given by the stiffness matrices C_A and C_B, respectively.

Determine the equivalent dimensionless storage modulus E^*/E_A and loss factor η/η_A of the composite as function of the frequency and for volume fraction $\nu_B = 0.15$ of the ingredient B. Compare the obtained results with those corresponding to positive $k_{31} = 0.304\,k_{21}$. Show that the storage moduli and loss factors in the 11, 12, and 33 directions are as displayed in Figure P11.4a–c, respectively. Observe that the storage moduli of the composite with negative k_{31} are lower than those of the composite with positive k_{31} whereas the loss factors of the composite with negative k_{31} are higher.

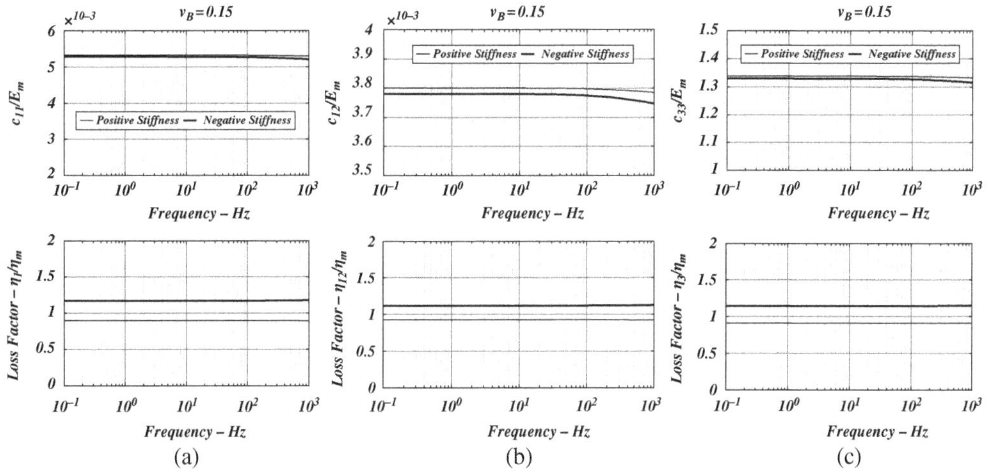

Figure P11.4 (a–c) The moduli and loss factor of a composite as function of frequency.

Note that for such a 2D composite, the associated Eshelby strain tensor is given (Marenić et al. 2015) as follows:

$$S = \begin{bmatrix} S_{1111} & S_{1122} & 0 \\ S_{2211} & S_{2222} & 0 \\ 0 & 0 & 2S_{1212} \end{bmatrix}$$

where

$s_{1111} = A(3+\gamma) = s_{2222}$, $s_{1122} = A(1-\gamma) = s_{2211}$, and $s_{1212} = A(1+\gamma)$ with $A = 1/[8(1-\upsilon)]$ and $\gamma = 2(1-2\upsilon)$.

11.4 Determine the reinforcement effect resulting from embedding CB particles, of different volume fractions, in HDPE. Use the finite element modeling of a RVE of the CB/polymer composite as shown in Figure P11.5a for the displayed unit cell of the

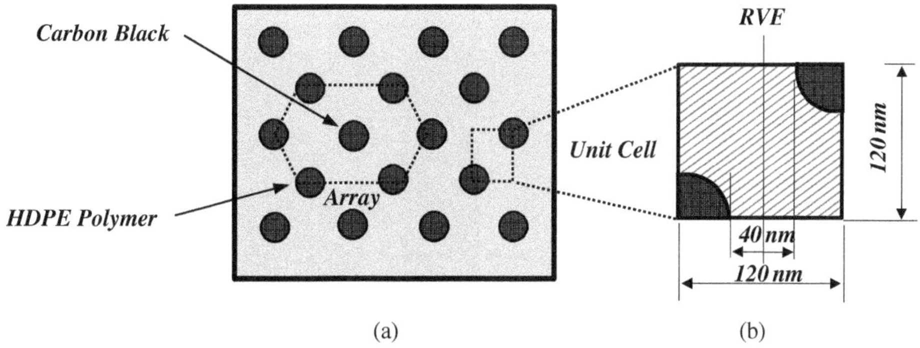

Figure P11.5 Polymer composite in hexagonal array arrangement. (a) Whole arrangement and (b) inset.

hexagonal array arrangement. The properties of the polymer and the CB are as listed in Example 11.4. The geometrical parameters of the RVE are shown in Figure P11.5b.

11.5 Consider the polymer/hexagonal inclusion composite shown in Figure P11.6. The composite is arranged in a square array arrangement as shown in Figure P11.6a. The geometrical parameters of the RVE are shown in Figure P11.6b. According

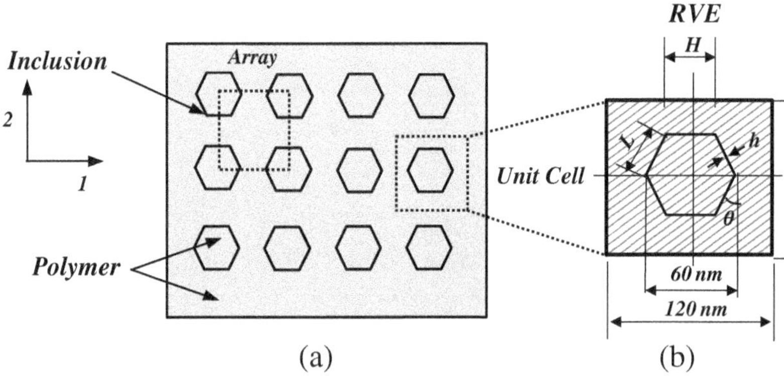

Figure P11.6 Polymer composite in square array arrangement. (a) Whole arrangement and (b) inset.

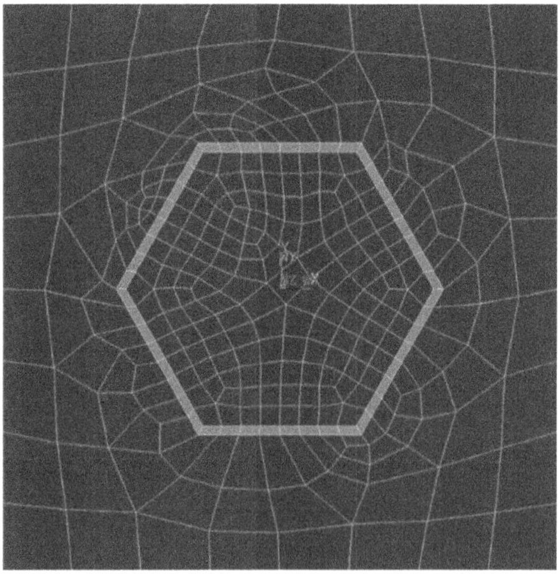

Figure P11.7 Finite element mesh of rectangular array of the composite.

to Gonella and Ruzzene (2008), the mechanical properties of the inclusion are given by:

$$\frac{E_1}{E_B}=\beta^3\frac{\cos\theta}{(\alpha+\sin\theta)\sin^2\theta},\ \frac{E_2}{E_B}=\beta^3\frac{(\alpha+\sin\theta)}{\cos^3\theta},$$

$$\frac{G_{12}}{E_B}=\beta^3\frac{(\alpha+\sin\theta)}{\alpha^2(1+2\alpha)\cos\theta}\quad\text{and}\quad\upsilon_{12}=\frac{\cos^2\theta}{(\alpha+\sin\theta)\sin\theta}.$$

where $\alpha = H/L$ and $\beta = h/L$.

Assuming Young's modulus E_A and Poisson's ratio υ_A of the polymer are such that $E_A = 10$ MPa and $\upsilon_A = 0.49$, whereas Young's modulus E_B and Poisson's ratio υ_B of the inclusion are such that $E_B = 210$ GPa and $\upsilon_B = 0.30$. If $\alpha = 1$, $\beta = 1/15$, and $\theta = 30°$, then determine the equivalent dimensionless moduli c_{ij}/E_A of the composite using *ANSYS* analysis of the *RVE* as shown in Figure P11.7. Show that $c_{11}/E_A = 604.69$, $c_{21}/E_A = 96.35$, $c_{31}/E_A = 81.09$, $c_{12}/E_A = 5.36$, $c_{22}/E_A = 5.66$, $c_{32}/E_A = 5.39$, $c_{13}/E_A = 5.03$, $c_{23}/E_A = 4.98$, and $c_{33}/E_A = 5.40$.

11.6 Consider the polymer/hexagonal mass-in inclusion composite shown in Figure P11.8. The composite is arranged in a square array arrangement as shown in Figure P11.8a. The geometrical parameters of the RVE are shown in Figure P11.8b.

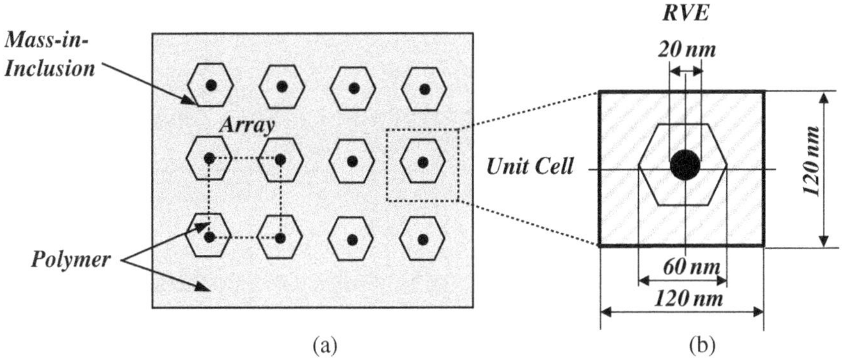

Figure P11.8 Polymer composite of polymer/hexagonal mass and inclusion. (a) Whole arrangement and (b) inset.

Assuming Young's modulus E_A and Poisson's ratio υ_A of the polymer are such that $E_A = 10$ MPa and $\upsilon_A = 0.49$, whereas Young's modulus E_B and Poisson's ratio υ_B of the inclusion are such that $E_B = 210$ GPa and $\upsilon_B = 0.30$. Note that the mass-in-inclusion is made of steel with diameter 20 nm. If $\alpha = 1$, $\beta = 1/15$, and $\theta = 30°$, then determine the equivalent dimensionless moduli c_{ij}/E_A of the composite using ANSYS analysis of the RVE as shown in Figure P11.9. Show that $c_{11}/E_A = 1294.24$, $c_{21}/E_A = 335.84$, $c_{31}/E_A = 190.57$, $c_{12}/E_A = 5.36$, $c_{22}/E_A = 5.66$, $c_{32}/E_A = 5.39$, $c_{13}/E_A = 5.03$, $c_{23}/E_A = 4.98$, and $c_{33}/E_A = 5.40$.

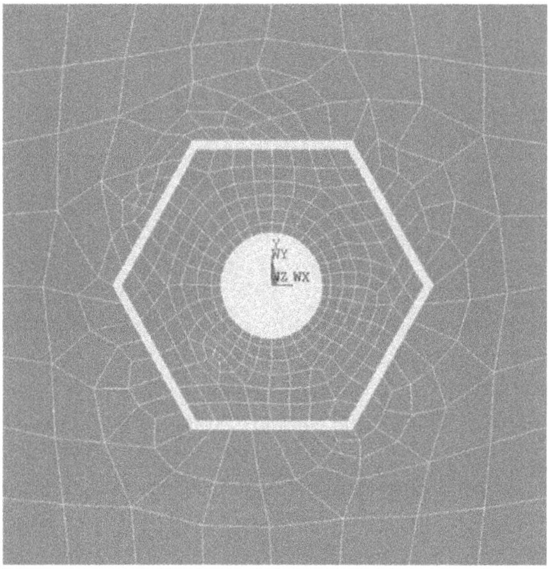

Figure P11.9 Finite element mesh of polymer/ hexagonal mass and inclusion composite.

11.7 Consider the polymer/reentrant hexagonal inclusion composite shown in Figure P11.10. The composite is arranged in a square array arrangement as shown in Figure P11.10a. The geometrical parameters of the RVE are shown in Figure P11.10b.

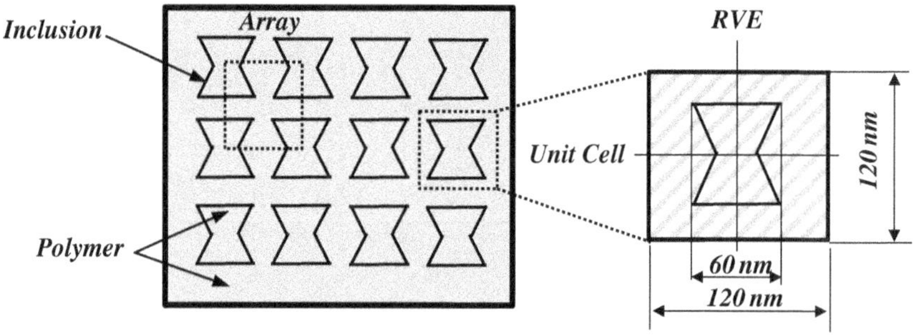

Figure P11.10 Polymer composite of polymer/reentrant hexagonal mass inclusion.

Assuming Young's modulus E_A and Poisson's ratio υ_A of the polymer are such that $E_A = 10$ MPa and $\upsilon_A = 0.49$, whereas Young's modulus E_B and Poisson's ratio υ_B of the inclusion are such that $E_B = 210$ GPa and $\upsilon_B = 0.30$. If $\alpha = 1, \beta = 1/15$, and $\theta = 30°$, then determine the equivalent dimensionless moduli c_{ij}/E_A of the composite using ANSYS analysis of the RVE as shown in Figure P11.11. Show that $c_{11}/E_A = 555.94$, $c_{21}/E_A = 84.01$, $c_{31}/E_A = 93.15$, $c_{12}/E_A = 5.35$, $c_{22}/E_A = 5.66$, $c_{32}/E_A = 5.39$, $c_{13}/E_A = 5.05$, $c_{23}/E_A = 4.97$, and $c_{33}/E_A = 5.44$.

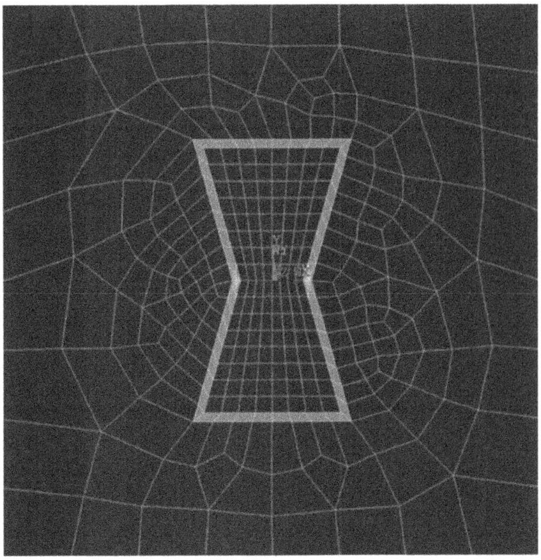

Figure P11.11 Finite element mesh of polymer/re-entrant hexagonal mass with inclusion composite.

11.8 Consider the polymer/reentrant hexagonal inclusion with mass-in-inclusion composite shown in Figure P11.12. The composite is arranged in a square array arrangement as shown in Figure P11.12a. The geometrical parameters of the RVE are shown in Figure P11.12b.

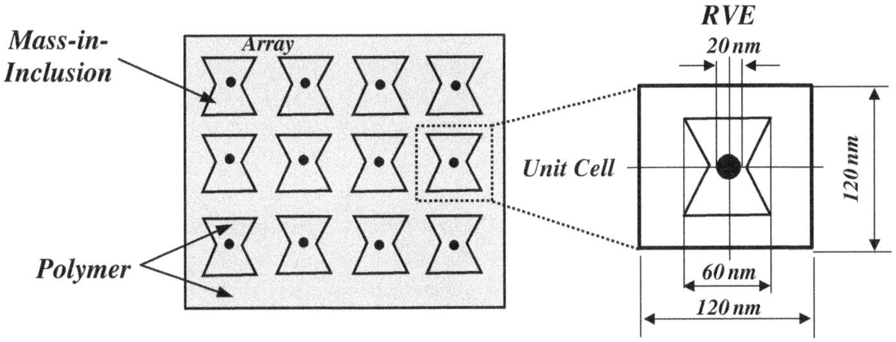

Figure P11.12 Polymer composite of polymer/reentrant hexagonal mass and inclusion.

Assuming Young's modulus E_A and Poisson's ratio υ_A of the polymer are such that E_A = 10 MPa and υ_A = 0.49, whereas Young's modulus E_B and Poisson's ratio υ_B of the inclusion are such that E_B = 210 GPa and υ_B = 0.30. Note that the mass-in-inclusion is made of steel with diameter 20 nm. If $\alpha = 1$, $\beta = 1/15$, and $\theta = 30°$, then determine the equivalent dimensionless moduli c_{ij}/E_A of the composite using ANSYS analysis of the RVE as shown in Figure P11.13. Show that $c_{11}/E_A = 630.04$, $c_{21}/E_A = 102.78$, $c_{31}/E_A = 85.15$, $c_{12}/E_A = 5.33$, $c_{22}/E_A = 5.64$, $c_{32}/E_A = 5.25$, $c_{13}/E_A = 5.01$, $c_{23}/E_A = 4.93$, and $c_{33}/E_A = 5.38$.

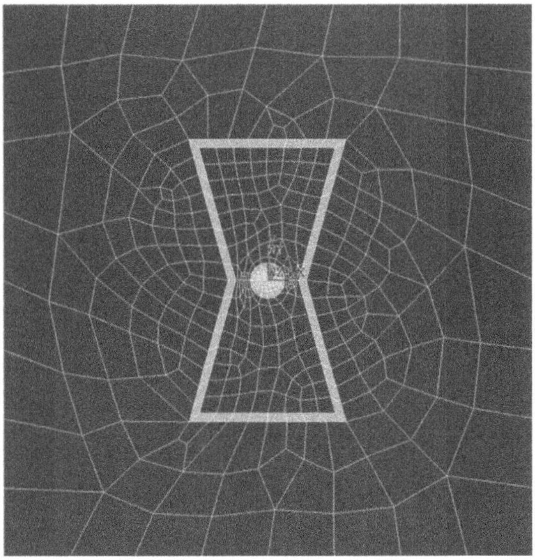

Figure P11.13 Finite element mesh of polymer/ reentrant hexagonal mass with inclusion composite.

11.9 Predict the mechanical properties of a hybrid composite consisting of an epoxy matrix (Hercules 3501-6) (Lesieutre et al. 1993), multi-walled carbon nanotubes (MWCNT) particles, and piezoelectric particles (Noliac-PCM51). The physical properties of the three ingredients of the composite are listed in Table 11.3. The epoxy has Poisson's ratio, ν of 0.38, and the MWCNT particles are embedded with a volume fraction 0.20. It is assumed that the piezo-particles have aligned orientations. Generate the mechanical properties as influenced by the volume fraction of PCM51, aspect ratio, and shunt frequency on the elastic moduli and loss factors of the hybrid composite as outlined in Section 11.6.3. Discuss the basic differences with the case of using CB instead of MWCNT as the conducting particles.

11.10 Predict the mechanical properties of a hybrid composite consisting of an epoxy matrix (Hercules 3501-6) (Lesieutre et al. 1993), MWCNT particles, and piezoelectric particles (PZT5H). The physical properties of the three ingredients of the composite are listed in Table 11.3. The epoxy has Poisson's ratio, ν of 0.38, and the MWCNT particles are embedded with a volume fraction 0.20. It is assumed that the piezo-particles have aligned orientations. Generate the mechanical properties as influenced by the volume fraction of PZT5H, aspect ratio, and shunt frequency on the elastic moduli and loss factors of the hybrid composite as outlined in Section 11.6.3. Discuss the basic differences with the case of using PCM51 instead of PZT5H as the piezoelectric particles.

12

Power Flow in Damped Structures

12.1 Introduction

Extensive efforts have been exerted recently to quantify the power flow and energy transmission paths in vibrating structures. Such a quantification process is essential to the design of appropriate passive/active vibration control systems for these structures. In these control systems, the emphasis is placed on altering/confining the transmission paths to uncritical zones of the vibrating structures or minimizing the power flow across paths encircling the disturbance zones.

Several approaches have been considered for identifying the power flow, which is also known as structural or mechanical intensity, in vibrating structures. Distinct among these methods are: the Statistical Energy Analysis (SEA) method (Lyons 1975), the finite element method (FEM) (Garvic and Pavic 1993; Pavic 1987, 1990, 2005; Alfredsson 1997; Alfredsson et al. 1996), the FEM with heat conduction analogy (Nefske and Sung 1987; Wohlever and Bernhard 1992; Bouthier and Bernhard 1995), and a wide variety of experimental methods (Noiseux 1970; Pavic 1976; Williams et al. 1985; Williams 1991; Linjama and Lahti 1992; Gibbs et al. 1993; Halkyard and Mace 1995). Comprehensive and critical reviews of structural power flow are given by Mandal et al. (2003) and Mandal and Biswas (2005).

It is important here to note that the SEA is particularly suitable for computation of the average spatial power flow in the high frequency domain. The classical FEM is, however, more suitable for low-frequency predictions of the spatial distribution of the power flow. For medium frequencies, the FEM with heat conduction analogy has been shown to accurately predict the power flow in beams, membranes, and plates.

In here, our focus is devoted to the use of the classical FEM as the basic tool for computing the power flow in structures. The prediction accuracy of the FEM is checked against the predictions of classical distributed-parameters methods (Wohlever and Bernhard 1992).

Also, particular attention is given here to the control of the power flow in damped structures.

12.2 Vibrational Power

12.2.1 Basic Definitions

Consider the following vibrating structure:

$$[M]\{\ddot{X}\} + [C]\{\dot{X}\} + [K]\{X\} = \{F\} \tag{12.1}$$

Active and Passive Vibration Damping, First Edition. Amr M. Baz.

where $[M]$, $[C]$, and $[K]$ are the mass, damping, and stiffness matrices. Also, $\{X\}$ and $\{F\}$ are the vectors of nodal deflections and external loads.

For sinusoidal excitations, the mobility Y_F of this structure is given by:

$$Y_F = \{\dot{X}\}/\{F\} = j\omega\left[[K] + i\omega[C] - \omega^2[M]\right]^{-1} = G_F + iB_F \tag{12.2}$$

where G_F is the "conductance" and B_F the "susceptance" matrices.

The complex vibrating power, S_P, supplied by the external loads $\{F\}$ is given by:

$$S_P = \frac{1}{2}\{F\}^*\{\dot{X}\} = \frac{1}{2}\{F\}\{\dot{X}\}^* \tag{12.3}$$

where $\{.\}*$ is the Hermitian transpose of $\{.\}$, that is, the transpose and complex conjugate of $\{.\}$.

From Eq. (12.2), one can write $\{\dot{X}\}$ as follows:

$$\{\dot{X}\} = G_F\{F\} + iB_F\{F\} \tag{12.4}$$

Substituting Eq. (12.4) into Eq. (12.3) yields:

$$S_P = \frac{1}{2}\{F\}^*G_F\{F\} + i\frac{1}{2}\{F\}^*B_F\{F\} = P_F + iQ_F \tag{12.5}$$

$$\text{where } P_F = \text{Active Power} = real\,[S_P] = \frac{1}{2}\{F\}^*G_F\{F\}$$

and

$$Q_F = \text{Reactive Power} = imag\,[S_P] = \frac{1}{2}\{F\}^*B_F\{F\} \tag{12.6}$$

12.2.2 Relationship to System Energies

Equation (12.2) can be rewritten as:

$$\{F\} = \frac{1}{i\omega}\left[[K] + i\omega[C] - \omega^2[M]\right]\{\dot{X}\},$$

Hence,

$$\{F\}^* = \frac{i}{\omega}\{\dot{X}\}^*\left[[K] - i\omega[C] - \omega^2[M]\right]^*, \tag{12.7}$$

But, as the $[K]$, $[C]$, and $[M]$ matrices are symmetric and real, then:

$$\{F\}^* = \frac{i}{\omega}\{\dot{X}\}^*\left[[K] - i\omega[C] - \omega^2[M]\right], \tag{12.8}$$

From Eqs. (12.3) and (12.8), the power S can be determined from:

$$\begin{aligned}
S_P &= \frac{1}{2}\{F\}^*\{\dot{X}\} \\
&= \frac{1}{2}\frac{i}{\omega}\{\dot{X}\}^*\left[[K] - i\omega[C] - \omega^2[M]\right]\{\dot{X}\} \\
&= \frac{1}{2}\{\dot{X}\}^*[C]\{\dot{X}\} + \frac{1}{2}\omega\left[\{X\}^*[K]\{X\} - \{\dot{X}\}^*[M]\{\dot{X}\}\right]i \\
&= P_F + iQ_F
\end{aligned} \tag{12.9}$$

Hence, $P_F = \frac{1}{2}\{\dot{X}\}^*[C]\{\dot{X}\}$,

and

$$Q_F = \frac{1}{2}\omega\left[\{X\}^*[K]\{X\} - \{\dot{X}\}^*[M]\{\dot{X}\}\right] = \omega[PE - KE]. \quad (12.10)$$

where

$$PE = \text{potential energy} = \frac{1}{2}\{X\}^*[K]\{X\} \text{ and } KE = \text{kinetic energy} = \frac{1}{2}\{\dot{X}\}^*[M]\{\dot{X}\}.$$

From Eq. (12.10), the active power, P_F, depends on the damping matrix $[C]$ so it defines the power dissipated in the structures. The reactive power, Q_F, is proportional to the difference between the potential and kinetic energy.

12.2.3 Basic Characteristics of the Power Flow

The reactive power, Q_F, vanishes at natural system frequencies

Proof.
Eq. (12.10) can be rewritten as:

$$\begin{aligned} Q_F &= \frac{1}{2}\omega\left[\{X\}^*[K]\{X\} - \{\dot{X}\}^*[M]\{\dot{X}\}\right] \\ &= \frac{1}{2}\omega\{X\}^*\left[[K]\{X\} - \omega^2[M]\{X\}\right] \end{aligned} \quad (12.11)$$

Hence, $Q_F = 0$ implies that:or,

$$[K]\{X\} = \omega^2[M]\{X\},$$

that is,

$$[M]^{-1}[K]\{X\} = \omega^2\{X\}, \quad (12.12)$$

which means that ω^2 is an eigenvalue of $[M]^{-1}[K]$ and $\{X\}$ is the corresponding eigenvector.

Hence, the total power flow at resonance is reduced to the active power flow component P_F, that is,

$$S_P = P_F = \frac{1}{2}\{\dot{X}\}^*[C]\{\dot{X}\} \quad (12.13)$$

Example 12.1 Consider the following spring-damper-mass system:

$$m\ddot{x} + c\dot{x} + kx = F$$

where m, c, and k are the mass, damping coefficient, and stiffness. If F denotes a sinusoidally applied force, determine the active and reactive powers of the system.

Solution

The system response x is given by:

$$\frac{x}{F}=\frac{1}{m\left[\omega_n^2-\omega^2+i2\zeta\omega\omega_n\right]}$$

where ω, ω_n, and ζ are the excitation frequency, the natural frequency $=\sqrt{k/m}$ and the damping ratio $=c/(2m)\omega_n$.

Then,

$$\frac{\dot{x}}{F}=\frac{i\omega}{m\left[\omega_n^2-\omega^2+i2\zeta\omega\omega_n\right]},$$

or,

$$\dot{x}=\frac{F}{m\omega_n}\frac{i\Omega}{\left[1-\Omega^2+i2\zeta\Omega\right]}$$

with

$$\Omega=\omega/\omega_n.$$

$$\text{Power } S_P=\frac{F^2}{2m\omega_n}\left[\frac{2\zeta\Omega^2}{\left[\left(1-\Omega^2\right)^2+(2\zeta\Omega)^2\right]}+\frac{\Omega\left(1-\Omega^2\right)}{\left[\left(1-\Omega^2\right)^2+(2\zeta\Omega)^2\right]}i\right] \tag{12.14}$$

which gives:

$$P_F=\frac{F^2}{2m\omega_n}\left[\frac{2\zeta\Omega^2}{\left[\left(1-\Omega^2\right)^2+(2\zeta\Omega)^2\right]}\right]$$

and

$$Q_F=\frac{F^2}{2m\omega_n}\frac{\Omega\left(1-\Omega^2\right)}{\left[\left(1-\Omega^2\right)^2+(2\zeta\Omega)^2\right]} \tag{12.15}$$

Figure 12.1 displays the active and reactive powers as function of the frequency ratio Ω for $\zeta = 1$, $\omega_n = 1$, $m = 1$ and $F = 1$.

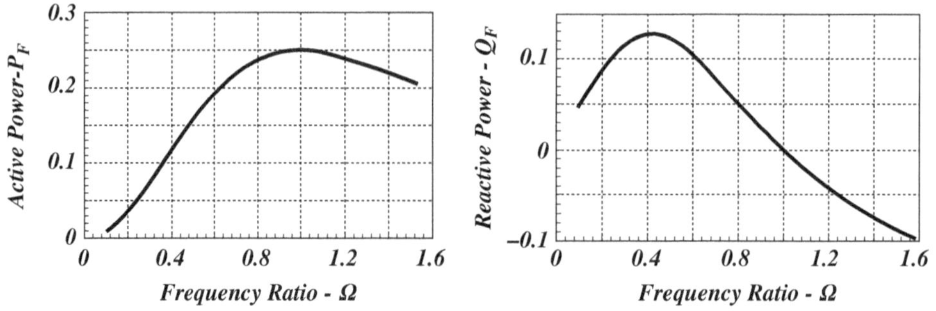

Figure 12.1 Active and reactive power of a spring-damper-mass system.

Note that the reactive power vanishes when $\Omega = 1$, that is, the system is at resonance. Also, when $\Omega = 1$, the active power attains a maximum value of:

$$P_F = \frac{F^2}{4m\omega_n\zeta} \tag{12.16}$$

that is, for a given input excitation F, the active power P_F decreases as the damping ratio ζ increases.

Example 12.2 Considerthe two spring-damper-mass system shown in Figure 12.2. If $k = 1$, $m = 1$ and the damping matrix = 0.1 stiffness matrix, determine the active and reactive powers when the first mass is excited sinusoidally by a unit force.

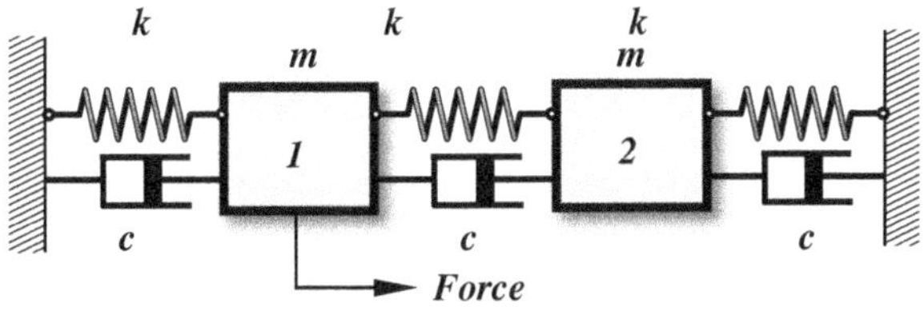

Figure 12.2 Multi-spring-damper-mass system.

Solution

For this system, the mass, stiffness, and damping matrices are:

$$[M] = \begin{bmatrix} 1 & 0 \\ 0 & 1 \end{bmatrix}, [K] = \begin{bmatrix} 2 & -1 \\ -1 & 2 \end{bmatrix}, \text{ and } [C] = 0.1\begin{bmatrix} 2 & -1 \\ -1 & 2 \end{bmatrix}.$$

Also, the system responses $\{X\}$ and $\{\dot{X}\}$ are given by:

$$\{X\} = \{x_1\ x_2\}^T = [[K] + i\omega[C] - \omega^2[M]]^{-1}\{1\ 0\}^T$$

and

$$\{\dot{X}\} = \{\dot{x}_1\ \dot{x}_2\}^T = i\omega[[K] + i\omega[C] - \omega^2[M]]^{-1}\{1\ 0\}^T$$

where x_1 and x_2 are the displacements of masses 1 and 2.

Accordingly, the active and reactive powers can be determined from:

$$P_F = \frac{1}{2}\{\dot{X}\}^*[C]\{\dot{X}\}, \text{and } Q_F = \frac{1}{2}\omega\{X\}^*[[K]\{X\} - \omega^2[M]\{X\}]$$

Figure 12.3 displays the active and reactive powers as function of the excitation frequency ω.

Note that the system has eigenvalues of 1 and 3, that is, natural frequencies of 1 and 1.732 rad s^{-1} and that the reactive power vanishes at 1 and 1.732 rad s^{-1} as shown in Figure 12.3.

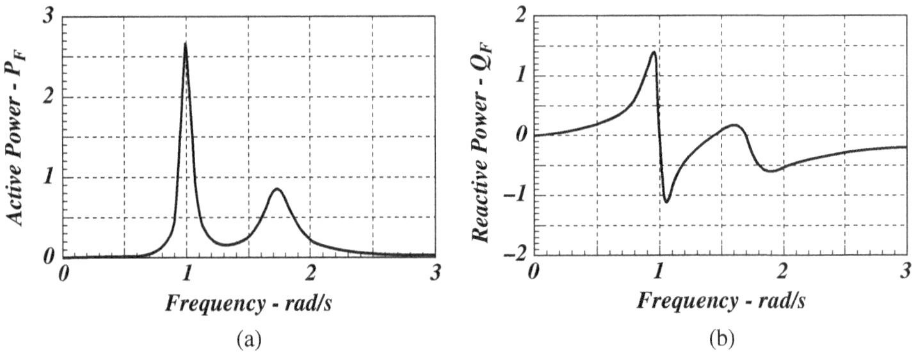

Figure 12.3 Active and reactive power of multi-spring-damper-mass system.

12.3 Vibrational Power Flow in Beams

The power flow in beams can be predicted using a finite element model of the entire beam system, in the global coordinate system, is used to generate the overall stiffness $[K]$, mass $[M]$, and damping $[C]$ matrices of the beam assembly after accounting for the boundary conditions.

The resulting model is then utilized along with the applied load vector $\{F\}$ to compute the nodal deflection vector $\{X\}$ such that:

$$\{X\} = \left[[K] + i\omega[C] - \omega^2[M]\right]^{-1}\{F\} \tag{12.17}$$

where ω is the frequency of a sinusoidal excitation acting on the beam.

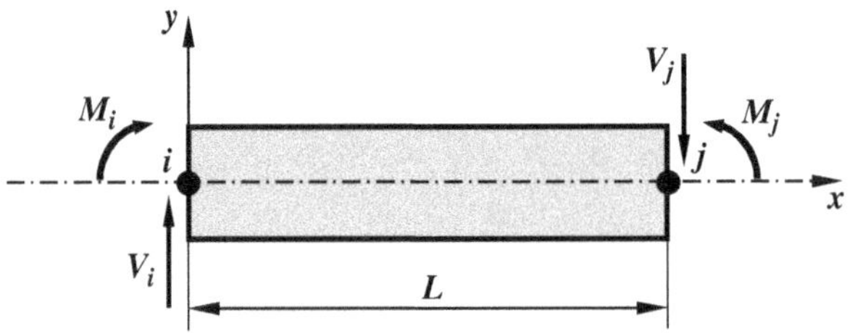

Figure 12.4 Finite element of a beam.

The nodal deflection vector $\{X\}$ is then used to extract the deflection vectors $\{X^e\}$ for the individual elements where $e = 1, \ldots, N$ with N denoting the total number of elements. Note that $\{X^e\} = \{w_i \; w_{i,x} \; w_j \; w_{j,x}\}^T$ with w and $w_{,x}$ denoting the transverse and angular deflection. Also, i and j define the nodes bounding the eth element as shown in Figure 12.4.

Now, the local force vector $\{F^e\}$ acting on the eth beam element can be determined from:

$$\{F^e\} = \left[[K^e] + i\omega[C^e] - \omega^2[M^e]\right]\{X^e\} \tag{12.18}$$

where $[K^e]$, $[C^e]$, and $[M^e]$ are the stiffness, damping, and mass matrices of the eth element. The stiffness and mass matrices are as given in Chapter 4.

Note also that $\{F^e\} = \{V_i\ M_i\ V_j\ M_j\}^T$ with $V_{i,j}$ and $M_{i,j}$ denoting the shear and bending moments acting at nodes i and j as shown in Figure 12.4.

Then, the structural intensity S_P^e of the eth element is given by:

$$S_P^e = \frac{1}{2} i\omega \left[V_i^* w_i + M_i^* w_{i,x} \right] = P_F^e + Q_F^e i \tag{12.19}$$

with P_F^e and Q_F^e denoting the active and reactive powers of the eth element.

Example 12.3 Considerthe steel cantilever beam fully treated with an active constrained layer damping (ACLD) patch, shown in Figure 12.5, such that the physical and geometrical properties of the beam, viscoelastic material (VEM), and the piezoelectric layer are given in Tables 12.1 and 12.2. The shear modulus of the VEM is $G = 20\ (1 + i)$ MPa.

Determine the active power flow distribution along the beam (Alghamdi and Baz 2002).

Solution

The beam/ACLD system is divided into 32 elements. The first three natural frequencies of the uncontrolled system are 48.9, 248.4, and 667.8 Hz, respectively.

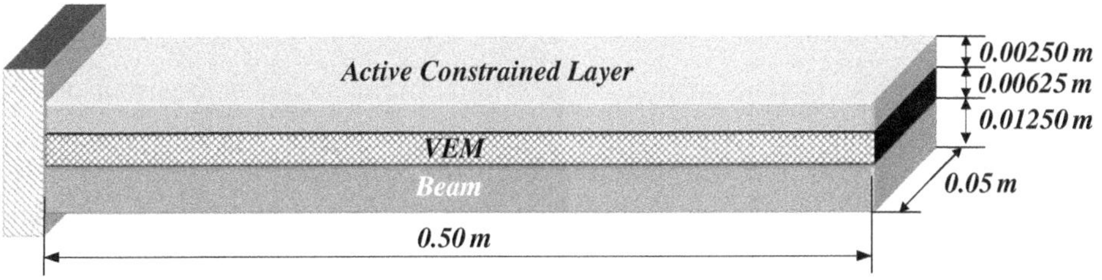

Figure 12.5 Beam with ACLD treatment.

Table 12.1 Physical properties of the base beam and the viscoelastic materials.

Material	Length (m)	Thickness (m)	Width (m)	Density (kg m^{-3})	Young's modulus (MPa)
Beam	0.5	0.012 5	0.05	7800	210,000
VEM	0.5	0.006 25	0.05	1104	60

Table 12.2 Physical properties of the piezoelectric constraining layer.

Length (m)	Thickness (m)	Width (m)	Density (kg m^{-3})	Young's modulus (MPa)	d_{31} (m V^{-1})	k_{31}	g_{31} (Vm N^{-1})	k_{3t}
0.5	0.0025	0.05	7600	63	186E-12	0.34	116E-2	1950

Table 12.3 Physical and geometrical parameters of beam/VEM system.

Layer	Thickness (m)	Width (m)	Young's modulus (GPa)	Density (kg m^{-3})	Poisson's ratio
PZT constraining layer	$h_1 = 0.0025$	0.025	63	7600	0.3
VEM	$h_2 = 0.0025$	0.025	*GHM*	1100	0.5
Beam	$h_3 = 0.0025$	0.025	70	2700	0.3

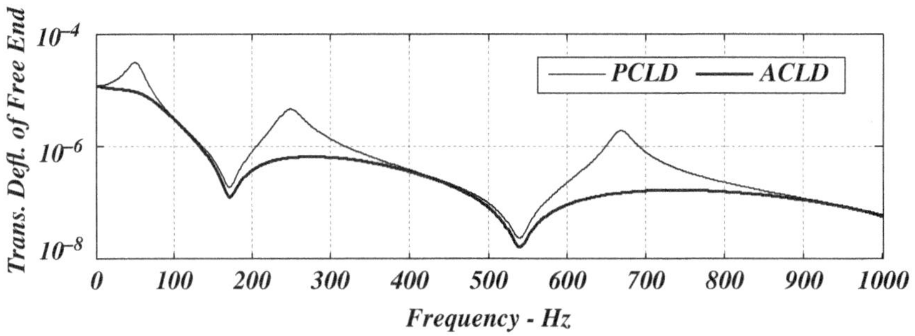

Figure 12.6 Open and closed-loop frequency response functions of a fully treated beam.

Figure 12.6 shows the frequency response of the beam when it is fully treated with the ACLD and subject to unit end transverse load. The figure displays a comparison between the response of the uncontrolled beam and that of the beam when it is controlled using a velocity feedback gain $K_D = 7.4$. It is evident that considerable attenuation of the first three modes is achieved.

Figure 12.7 shows the corresponding active power flow of the fully treated beam as a function of the excitation frequency. The figure indicates that the power flow attains a maximum at the resonant frequencies. Furthermore, it is clear that the power flow is considerably reduced with the velocity feedback. Such a reduction is attributed to the attenuation of the beam vibration.

The power flow, along the beam, are shown in Figures 12.8 through 12.10 for three different excitation frequencies (48.9, 248.4, and 667.8 Hz) that correspond to the first three natural frequencies of the beam/ACLD composite. The figures show also comparisons between the open-loop ($K_D = 0.0$) and closed-loop ($K_D = 7.4$) characteristics. It is important to note that it is evident that the spatial profile of the power flow, at any particular mode of vibration, is to some extent similar to the corresponding mode shape of the beam. Also, note that there is a significant attenuation of the net power flow due to the activation of the ACLD controller when $K_D = 7.4$.

Furthermore, it is evident that the computed power flow distributions are in close agreement with the distributions as predicted by using ANSYS software package.

Computation of the power flow over the beam/passive constrained layer damping (PCLD) assembly is carried out according to the procedure outlined in Appendix 12.A.

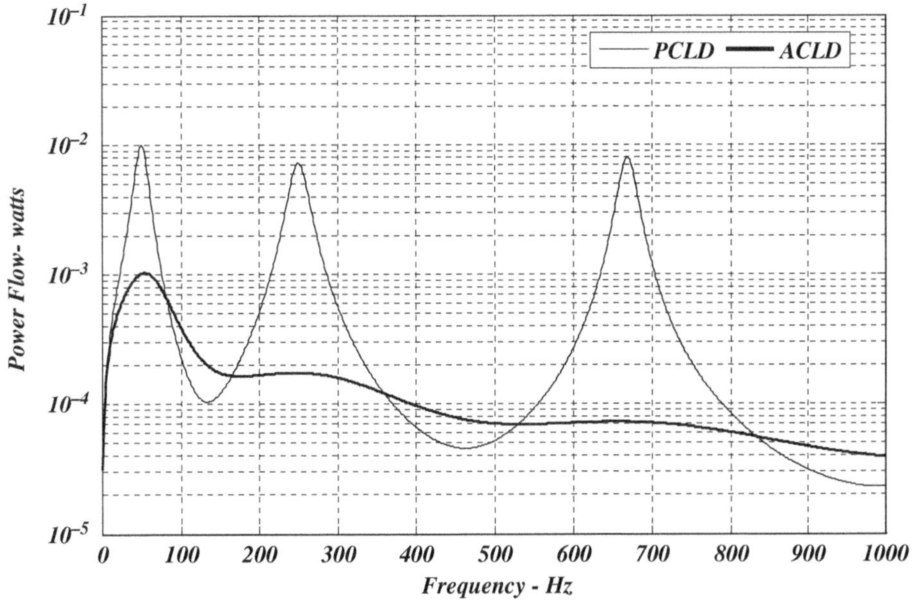

Figure 12.7 Open and closed-loop active power flow of fully treated beam.

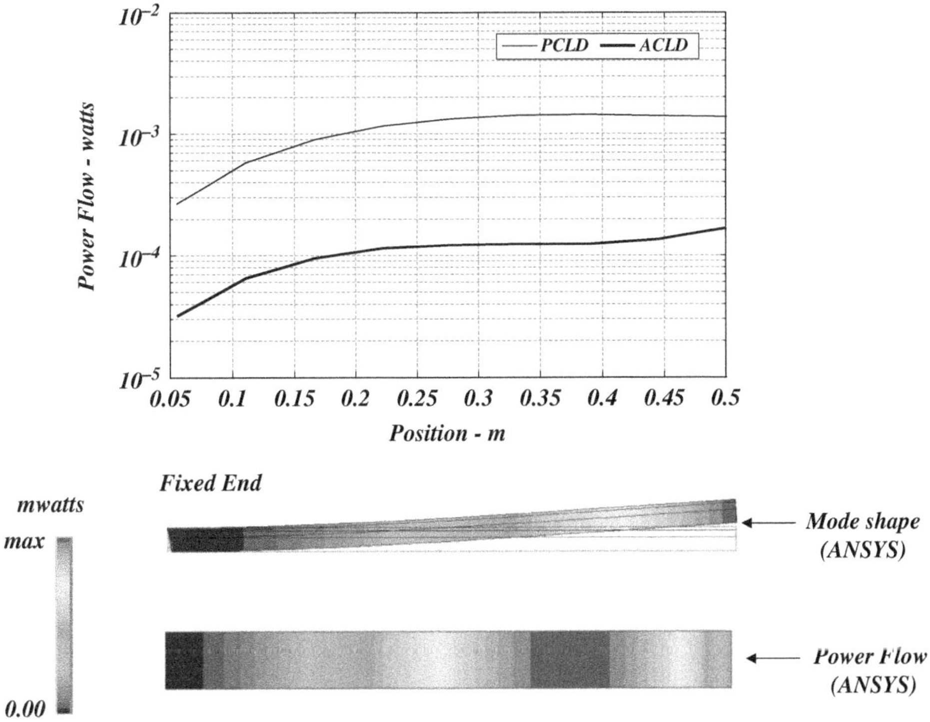

Figure 12.8 Open and closed-loop power flow of fully treated beam at the first mode (48.9 Hz).

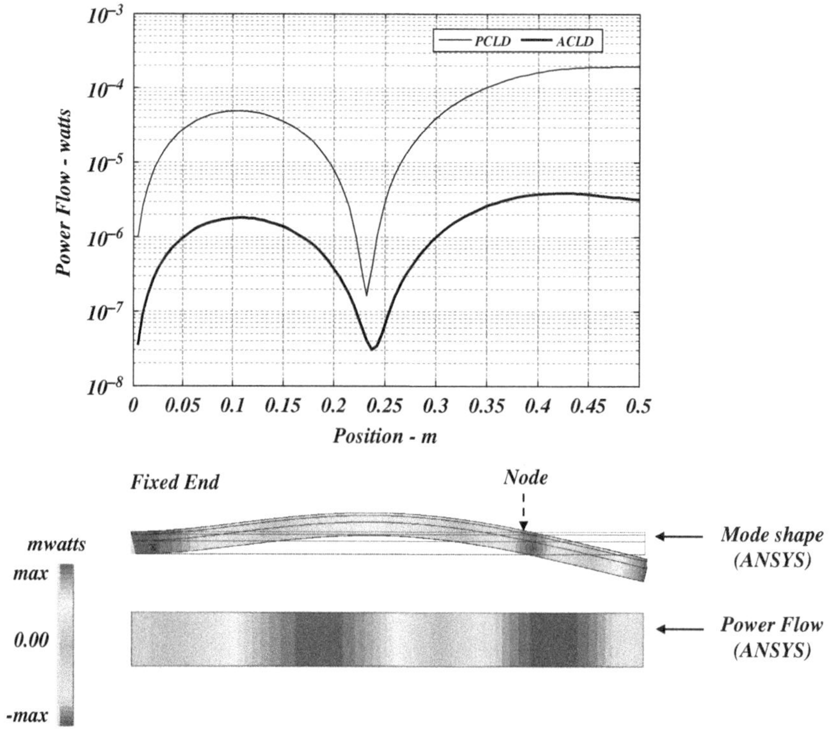

Figure 12.9 Open and closed-loop power flow of fully treated beam at the second mode.

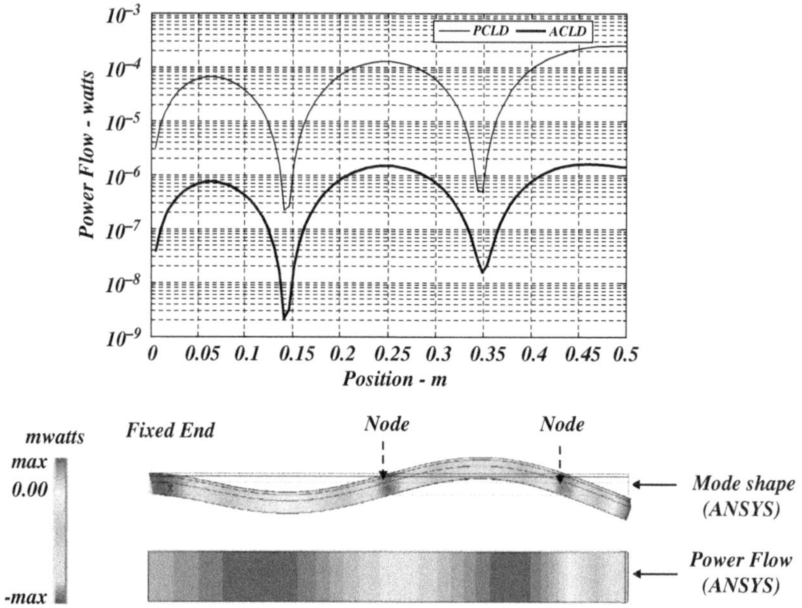

Figure 12.10 Open and closed-loop power flow of fully treated beam at the third mode.

12.4 Vibrational Power of Plates

12.4.1 Basic Equations of Vibrating Plates

In this section, the emphasis is placed on presenting the basic equations that govern the vibration of a thin flat plate based on the classical Kirchhoff assumptions (Rao 2007). Under these assumptions, the plate thickness h is assumed to be small as compared to the plate's length and width (a and b), whereas the transverse deflection w is considered comparable to h. Furthermore, the plate thickness is assumed to be uniform and symmetric about the mid-surface so that three-dimensional stress effects are ignored.

Figure 12.11 displays a schematic drawing of a Kirchhoff plate along with the in-plane normal forces N_i, in-plane shear forces N_{ij}, bending moments M_i, twisting moments M_{ij} acting on it, and transverse shear forces Q_i.

Expressions of these forces and moments are given by (Rao 2007) as follows:

In-Plane Normal Forces (N_i)

$$N_x = -\frac{1}{2}\frac{E}{1-\nu^2} z w_{,x}^2 - \frac{1}{2}\frac{\nu E}{1-\nu^2} z w_{,y}^2,$$
$$N_y = -\frac{1}{2}\frac{E}{1-\nu^2} z w_{,y}^2 - \frac{\nu E}{1-\nu^2} z w_{,x}^2. \tag{12.20}$$

where E is Young's modulus of the plate, ν is its Poisson's ratio, z is the distance from the mid-surface, and $w_{,ii}$ is the second spatial derivative of the deflection with respect to i ($i = x, y$).

In-Plane Shear Forces (Nij)

$$N_{xy} = N_{yx} = -2zGw_{,xy}. \tag{12.21}$$

where G is the shear modulus of the plate.

Bending Moments (M_i)

$$M_x = -D\left(w_{,xx} + \nu w_{,yy}\right), M_y = -D\left(w_{,yy} + \nu w_{,xx}\right). \tag{12.22}$$

where D is the flexural rigidity of the plate, given by: $D = Eh^3/(12[1 - \nu^2])$.

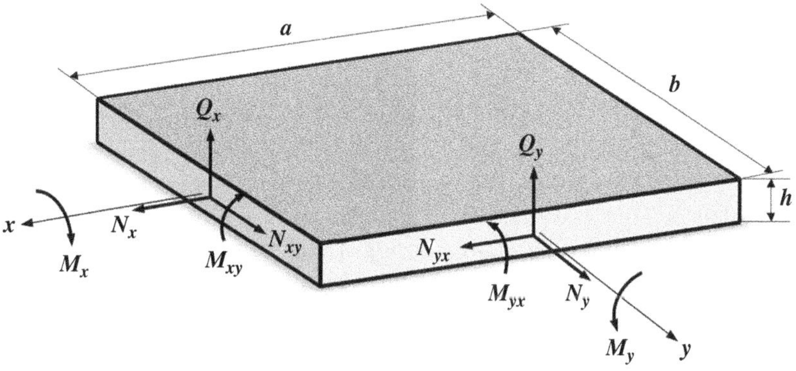

Figure 12.11 Forces and moments acting on a Kirchhoff plate.

Twisting Moments (M_{ij})

$$M_{xy} = M_{xy} = -(1-\nu)Dw_{,xy}. \tag{12.23}$$

Transverse Shear Forces (Q_i)

$$\begin{aligned} Q_x &= -D\frac{\partial}{\partial x}\left(w_{,xx} + w_{,yy}\right), \\ Q_y &= -D\frac{\partial}{\partial y}\left(w_{,yy} + w_{,xx}\right). \end{aligned} \tag{12.24}$$

Note that all these forces and moments are function of the transverse deflection w and its spatial derivatives with respect to x and y. In a finite element formulation, the transverse deflection can be replaced by the classical interpolation representation $w = [N]\{\Delta^e\}$ in terms of the interpolating function $[N]$ and the nodal deflection vector $\{\Delta^e\}$.

12.4.2 Power Flow and Structural Intensity

The structural intensity (I) is the vibrational power flow (S_P) per unit cross-sectional area of a dynamically loaded elastic plate. The net power flow or active intensity in a two-dimensional plate-like structure with a steady-state vibration is given by (Xu et al. 2005):

$$\begin{aligned} I_x(\omega) &= -\frac{1}{2}R_e\left(\sigma_{xk}(\omega)\, V_k^*(\omega)\right), \\ I_y(\omega) &= -\frac{1}{2}R_e\left(\sigma_{yk}(\omega)\, V_k^*(\omega)\right), \; where\, k = x,y,z \end{aligned} \tag{12.25}$$

where $I_i(\omega)$, $\sigma_{ik}(\omega)$, and $V_k^*(\omega)$ denote the structural intensity in the ith direction, the stress in the ikth directions, and the complex conjugate of the velocity in the kth direction, respectively. Also, $R_e(.)$ denotes the real part of (.).

The structural intensity for a quadrilateral plate element can be expressed in the form of power flow per unit width. The x and y components of structural intensities can be expressed as

$$\begin{aligned} I_x(\omega) &= -\frac{\omega}{2}I_m\left(N_x u^* + N_{xy} v^* + Q_x w^* + M_x w^*_{,y} - M_{xy} w^*_{,x}\right), \\ I_y(\omega) &= -\frac{\omega}{2}I_m\left(N_y v^* + N_{yx} u^* + Q_y w^* - M_y w^*_{,x} + M_{yx} w^*_{,y}\right). \end{aligned} \tag{12.26}$$

where I_m, N_i, M_i, M_{ij}, and Q_i denote the imaginary part, in-plane forces, bending moment, twisting moment, and transverse shear forces per unit width, respectively. Also, u^*, v^*, and w^* are the complex conjugate velocities in the x, y, and z directions. Furthermore, $w^*_{,i}$ is the complex conjugate of the angular displacement about the ith direction.

For a particular plate-like structure, the structural intensity components (I_x and I_y) can be computed and maps of the contours of the structural intensity can be plotted.

Equally important are "the streamline" maps that display the flow as lines everywhere parallel to the velocity field. The relative spacing of the lines indicates the speed of the flow. The structural intensity streamline can be mathematically expressed as (Xu et al. 2005):

$$d\mathbf{r} \times \mathbf{I}(\mathbf{r},t) = \begin{vmatrix} \mathbf{i} & \mathbf{j} & \mathbf{k} \\ dx & dy & dz \\ I_x & I_y & I_z \end{vmatrix} = 0 \tag{12.27}$$

where **r** is the energy flow particle position vector.

Thus, for the case of the two-dimensional plate-like structures, the differential equation describing a streamline, in a plane perpendicular to the k direction, is given by:

$$\left(I_y dx - I_x dy\right)\mathbf{k} = 0, \text{ or } \frac{dx}{I_x} = \frac{dy}{I_y} \tag{12.28}$$

Equation (12.28) defines the differential equation $\frac{dy}{dx} = \frac{I_y}{I_x}$ that describes the streamline.

Example 12.4 Considerthe damped aluminum cantilever plate shown in Figure 12.12. The plate is damped using two point viscous dampers with damping coefficients C_1 and C_2. The plate is excited near its fixed end, as shown in Figure 12.12, by a unit transverse sinusoidal load at the first natural frequency of the plate.

Determine the structural intensity and the streamline plots of the system when:

a) C_1 = 2000 Ns m^{-1} and C_2 = 2000 Ns m^{-1}.
b) C_1 = 2000 Ns m^{-1} and C_2 = 20,000 Ns m^{-1}.
c) C_1 = 20,000 Ns m^{-1} and C_2 = 2000 Ns m^{-1}.

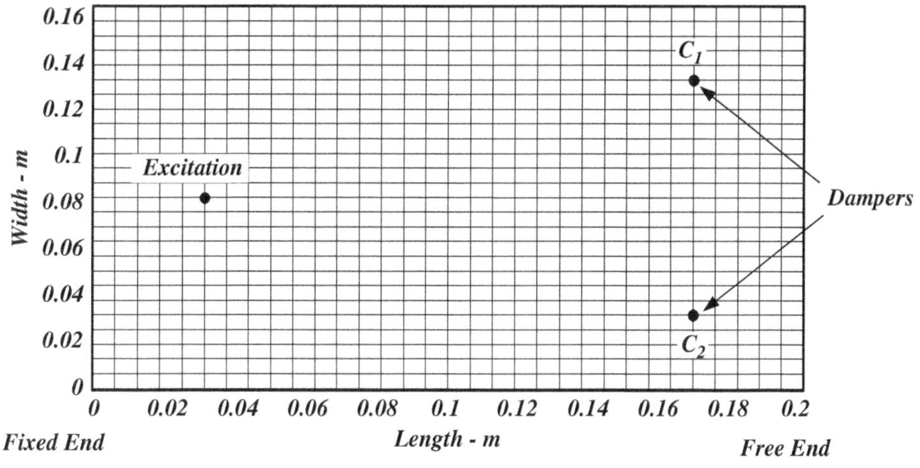

Figure 12.12 Schematic drawing of a damped cantilever plate system.

Solution

For the cantilever plate under consideration, the first natural frequency as computed by the FEM is found to be 28.73 Hz. The corresponding mode shape is displayed in Figure 12.13.

a) C_1 = 2000 Ns m^{-1} and C_2 = 2000 Ns m^{-1}.

For this case, the transverse deflection distribution map of the plate is shown in Figure 12.14. The figure indicates that the maximum deflection is confined to the area close to the excitation. However, the deflection is minimized near and past the locations of the dampers C_1 and C_2.

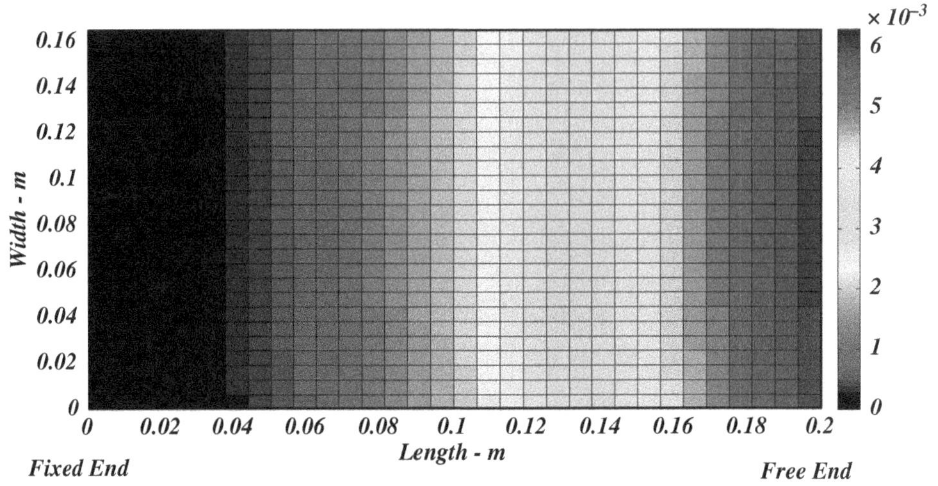

Figure 12.13 The mode shape of the first natural frequency of the cantilever plate system.

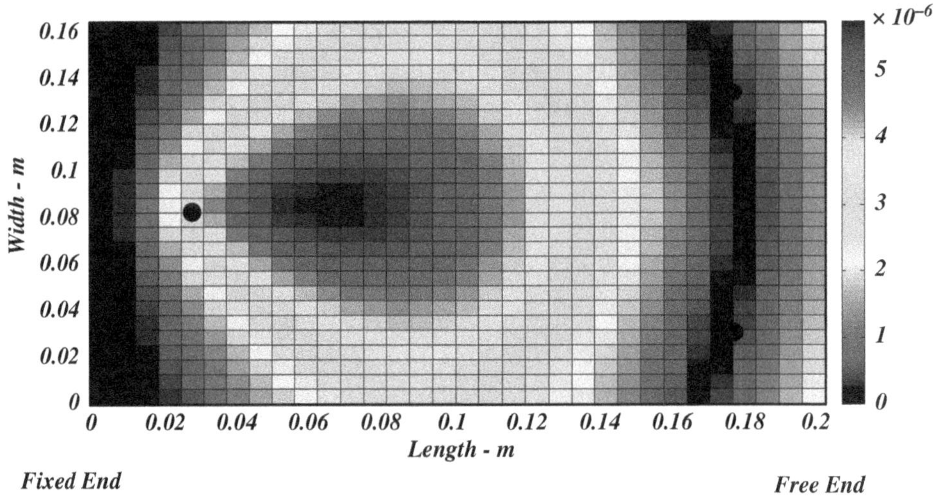

Figure 12.14 The transverse deflection map of the cantilever plate system at the first natural frequency.

Figures 12.15 and 12.16 display the corresponding structural intensity and streamline maps. The maps indicate that the vibration energy flows from the excitation source to the dampers that act as energy sinks. Note that as the values of C_1 and C_2 are equal, then the energy flows are distributed equally from the energy source to the sinks as is displayed clearly in Figures 12.15 and 12.16.

b) $C_1 = 2000$ Ns m^{-1} and $C_2 = 20{,}000$ Ns m^{-1}.

Figures 12.17 and 12.18 display the corresponding power flow and streamline maps for this case. The power flow map of Figure 12.17 indicates that the vibration energy flows from the excitation source primarily to the damper with the low

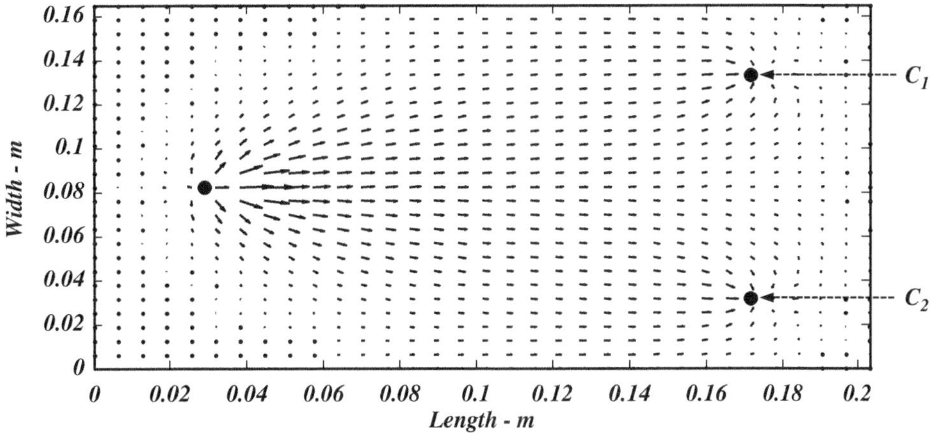

Figure 12.15 The structural intensity distribution map of the cantilever plate system at the first natural frequency (C_1 = 2000 Ns m^{-1} and C_2 = 2000 Ns m^{-1}).

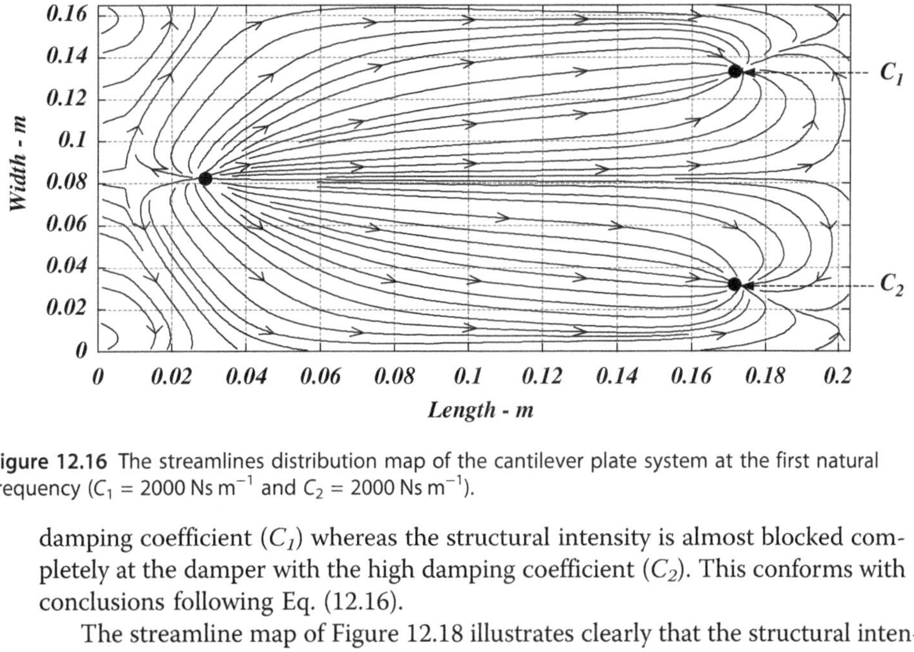

Figure 12.16 The streamlines distribution map of the cantilever plate system at the first natural frequency (C_1 = 2000 Ns m^{-1} and C_2 = 2000 Ns m^{-1}).

damping coefficient (C_1) whereas the structural intensity is almost blocked completely at the damper with the high damping coefficient (C_2). This conforms with conclusions following Eq. (12.16).

The streamline map of Figure 12.18 illustrates clearly that the structural intensity has been steered toward the damper with the low damping coefficient (C_1). Hence, if a critical object, such as a payload or an instrument, is located near the damper C_2, then increasing its damping coefficient tends to redirect the vibration energy away from it to avoid undesirable excitation from reaching it.

c) C_1 = 20,000 Ns m^{-1} and C_2 = 2000 Ns m^{-1}.

Figures 12.19 and 12.20 display the corresponding structural intensity and streamline maps for this case. The maps indicate that increasing the value of C_1 and decreasing that of C_2 results in reversing the wave transmission path as compared to case (b).

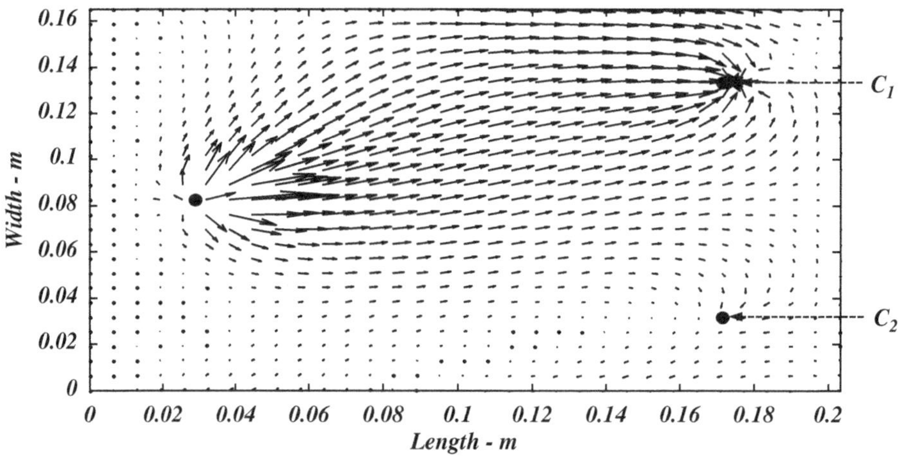

Figure 12.17 The structural intensity distribution map of the cantilever plate system at the first natural frequency (C_1 = 2000 Ns m^{-1} and C_2 = 20,000 Ns m^{-1}).

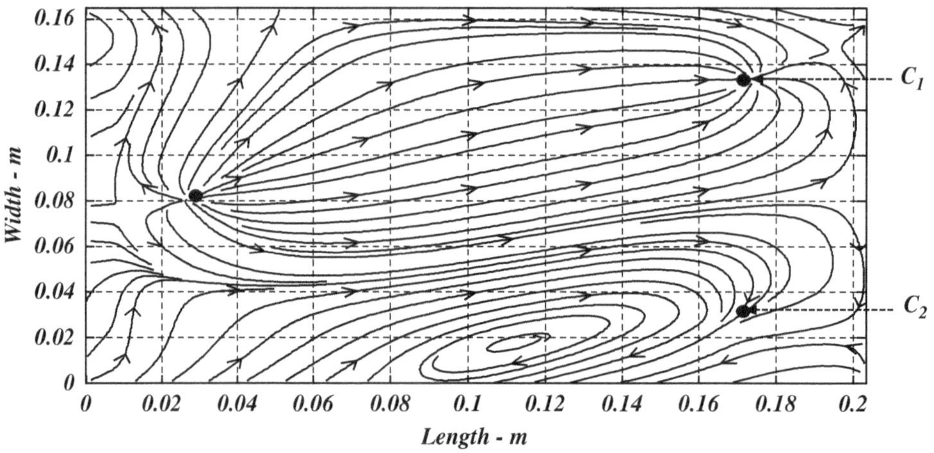

Figure 12.18 The streamlines distribution map of the cantilever plate system at the first natural frequency (C_1 = 2000 Ns m^{-1} and C_2 = 20,000 Ns m^{-1}).

This suggests that, with appropriate control of the damping coefficients, it would be possible to steer the wave propagation direction as needed. More importantly, by monitoring the instantaneous power flow at a particular location, it is possible to adaptively control the damping coefficients of the dampers to drive the power flow at this location to a minimum value using, for example, the *Least Mean Square* (LMS) control algorithm.

Example 12.5 Comparethe power flow distributions over the plate of Example 12.4 as computed by the MATLAB FEM described in Sections 12.4.1 and 12.4.2 as well as those predicted by ANSYS software package.

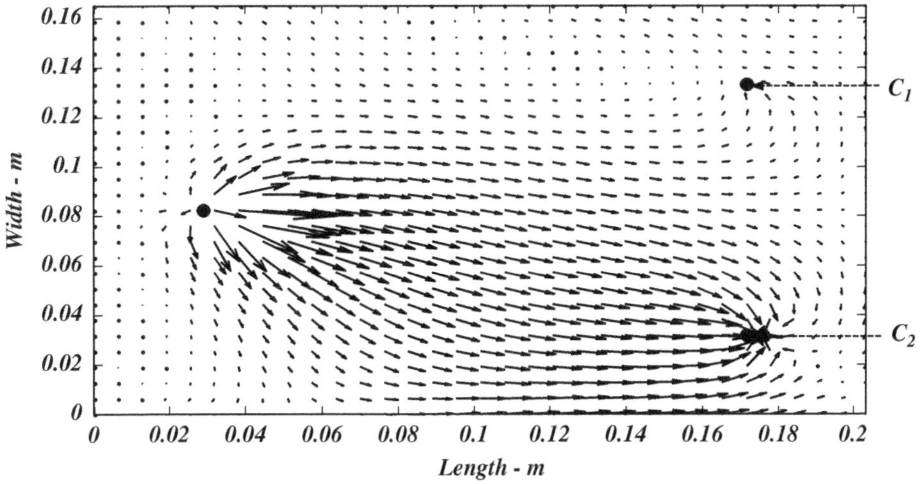

Figure 12.19 The structural intensity distribution map of the cantilever plate system at the first natural frequency (C_1 = 20,000 Ns m^{-1} and C_2 = 2000 Ns m^{-1}).

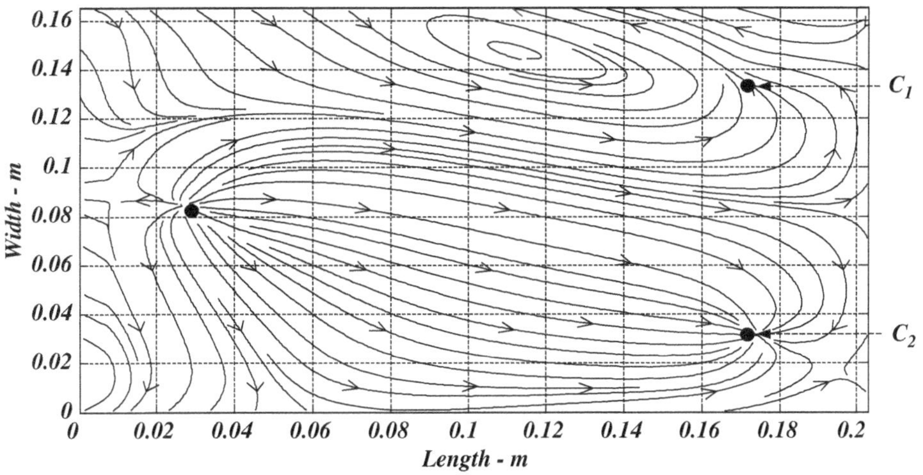

Figure 12.20 The streamlines distribution map of the cantilever plate system at the first natural frequency (C_1 = 20,000 Ns m^{-1} and C_2 = 2000 Ns m^{-1}).

Solution

Figure 12.21a,b show the power flow distributions over the plate as computed by the MATLAB FEM and ANSYS software packages, respectively, when the damping coefficients C_1 = 5 Ns m^{-1} and C_2 = 5 Ns m^{-1}.

It is evident that there is a close agreement between the predictions of the MATLAB FEM and ANSYS software.

When the damping coefficients become C_1 = 1 Ns m^{-1} and C_2 = 5 Ns m^{-1}, Figure 12.22a, b emphasizes the close agreement between the power flow distributions over the plate as predicted by the MATLAB FEM and ANSYS software packages, respectively.

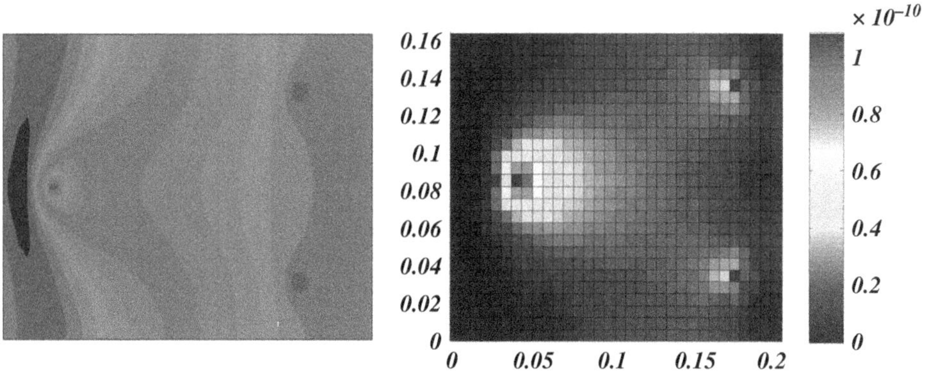

(a) – *ANSYS* (b) – *MATLAB*

Figure 12.21 The power flow distribution maps of the cantilever plate system at the first natural frequency (C_1 = 5 Ns m^{-1} and C_2 = 5 Ns m^{-1}).

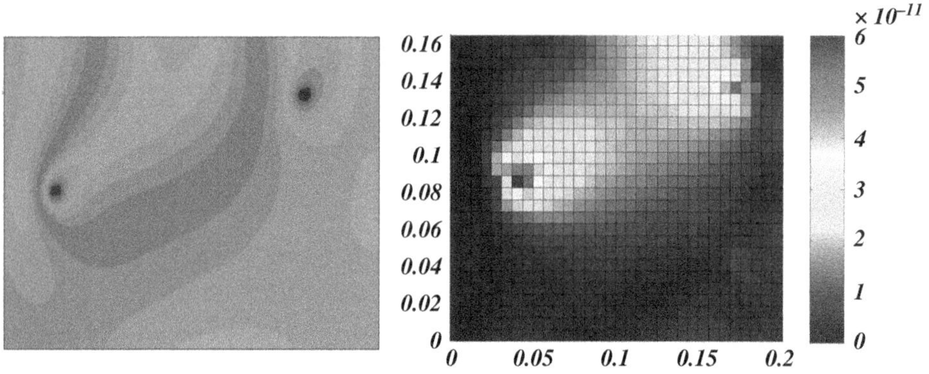

(a) – *ANSYS* (b) – *MATLAB*

Figure 12.22 The power flow distribution maps of the cantilever plate system at the first natural frequency (C_1 = 1 Ns m^{-1} and C_2 = 5 Ns m^{-1}).

12.4.3 Control of the Power Flow and Structural Intensity

The active vibrational power flow (P), at a particular location, can be controlled by adaptively adjusting the damping coefficients of the dampers, as shown schematically in Figure 12.23, to drive the power flow at this location to a minimum value using, for example, the LMS control algorithm.

In order to develop the control algorithm, the equation of motion of a vibrating system excited at one of its natural frequencies, as given by Eq. (12.8), reduces to:

$$\{F\}^* = \{\dot{X}\}^*[C] \text{ or } \{\dot{X}\}^* = \{F\}^*[C]^{-1} \tag{12.29}$$

Substituting Eq. (12.29) into Eq. (12.13), yields the following expression of the active power flow P_{Fi}:

$$P_{F_i} = \frac{1}{2}\{\dot{X}\}^*[C]\{\dot{X}\} = \frac{1}{2}[\{F\}^*[C]^{-1}]\,[C]\,[[C]^{-1}\{F\}] = \frac{1}{2}\{F\}^*[C]^{-1}\{F\} \tag{12.30}$$

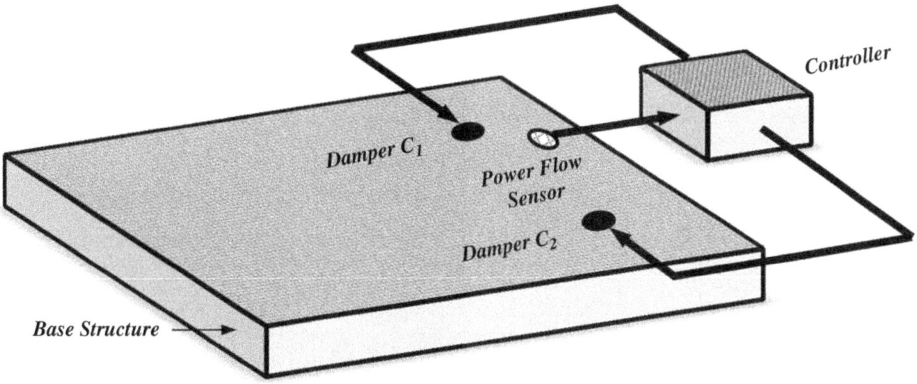

Figure 12.23 Controller of the power flow of a damped plate system.

Hence, for a constant excitation $\{F\}$, the active power flow is inversely proportional to the damping matrix $[C]$. This statement is supported by the conclusions of Example 12.1 as quantified by Eq. (12.16).

For a plate system that is damped by N discrete dampers, the damping matrix [C] is a diagonal matrix, and hence Eq. (12.30) reduces to:

$$P_{Fi} = \sum_{j=1}^{N} \frac{a_j}{C_j} \tag{12.31}$$

where C_j denotes the damping coefficient of the jth discrete damper. Also, the coefficients $a_j = \frac{1}{2} f_j^2 > 0$ where f_j denotes the coefficient of the excitation force vector$\{F\}$.

In order to design the power flow controller, a performance index J is defined such that:

$$J = e^2 = (P_{Fr} - P_{Fi})^2 \tag{12.32}$$

where e denotes the power flow error. Also, P_{Fr} defines a desirable final value of the power flow.

The damping coefficients of the dampers are selected to minimize the performance index resulting in effect in a LMS control action. The LMS algorithm is a gradient-based algorithm that is commonly used in adaptive signal processing (Widrow and Stearns 1985). The LMS controller updates iteratively the damping coefficients of the dampers by moving along the direction of the negative gradient ∇J of J such that the damping coefficients at the $k + 1$th iteration are determined in terms of the damping coefficients at the kth iteration as follows:

$$C_{i_{k+1}} = C_{i_k} - \alpha \nabla J \tag{12.33}$$

where α is the step size such that $\alpha > 0$. From Eqs. (12.31) and (12.32), Eq. (12.33) reduces to:

$$\begin{aligned} C_{i_{k+1}} &= C_{i_k} + 2\alpha (P_{Fr} - P_{Fi}) \frac{\partial P_{Fi}}{\partial C_i} \\ &= C_{i_k} - \frac{A_i}{C_{i_k}^2} (P_{Fr} - P_{Fi}) = C_{i_k} - \frac{A_i}{C_{i_k}^2} e_i \end{aligned} \tag{12.34}$$

where $A_i = 2\alpha\, a_i =$ constant > 0.

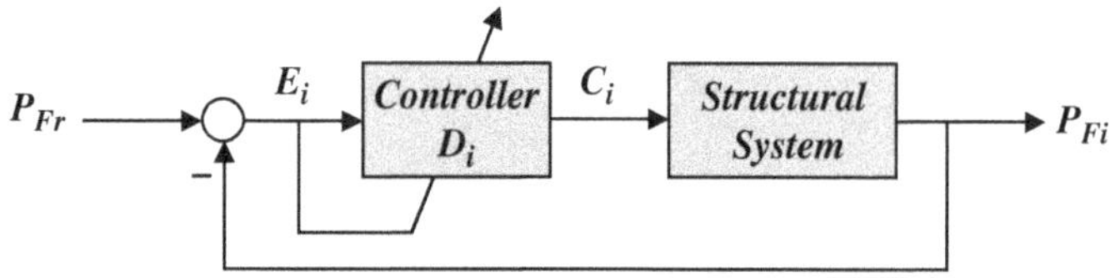

Figure 12.24 Block diagram of the controller of the power flow of a damped plate system.

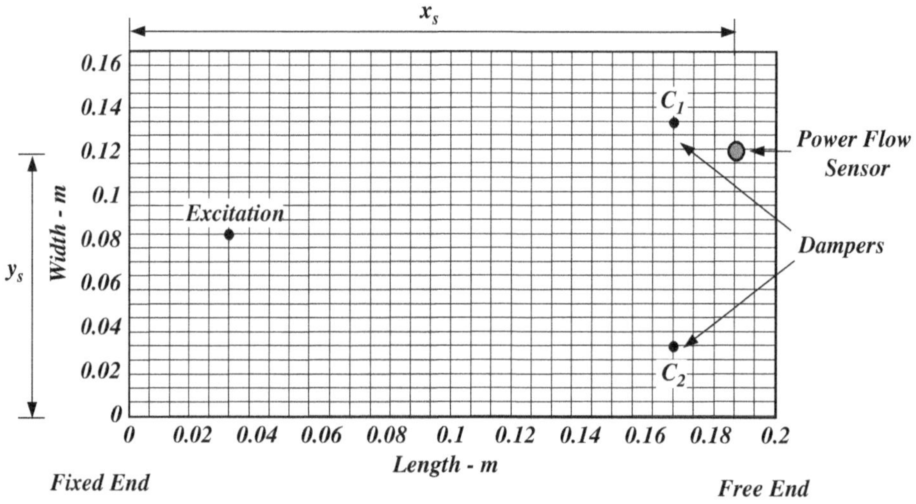

Figure 12.25 Schematic drawing of the location of the dampers and power flow sensor for a damped cantilever plate system.

Accordingly, Eq. (12.34) can be used to generate the block diagram of the power flow control system as shown in Figure 12.24.

Example 12.6 Considerthe damped aluminum cantilever plate described in Example 12.3. Design a power flow controller to minimize the power flow at the location of the sensor (x_s = 0.19 m, y_s = 0.12 m) as shown in Figure 12.25.

Solution

The LMS controller presented in Section 12.4.3 is implemented using A_i = 0.5E20, P_{Fr} = 5E-14 Nm s^{-1}, C_1 = 100 Ns m^{-1}, and C_2 = 1000 Ns m^{-1}.

Figure 12.26 displays the effect of the iteration number k on the adaptive characteristics of the controller as quantified by the value of the coefficient of damping C_1 and the resulting objective function J. It can be seen that, as the iteration number increases, the objective function decreases, and the damping coefficient C_1 increases. More importantly, as k increases, the direction of the power flow can be steered and redirected from flowing toward the damper C_1 (when k = 0 and C_1 = 100 Ns m^{-1}) to flowing toward the damper C_2 (when k = 5000 and C_1 = 6000 Ns m^{-1}). In this manner, the power flow

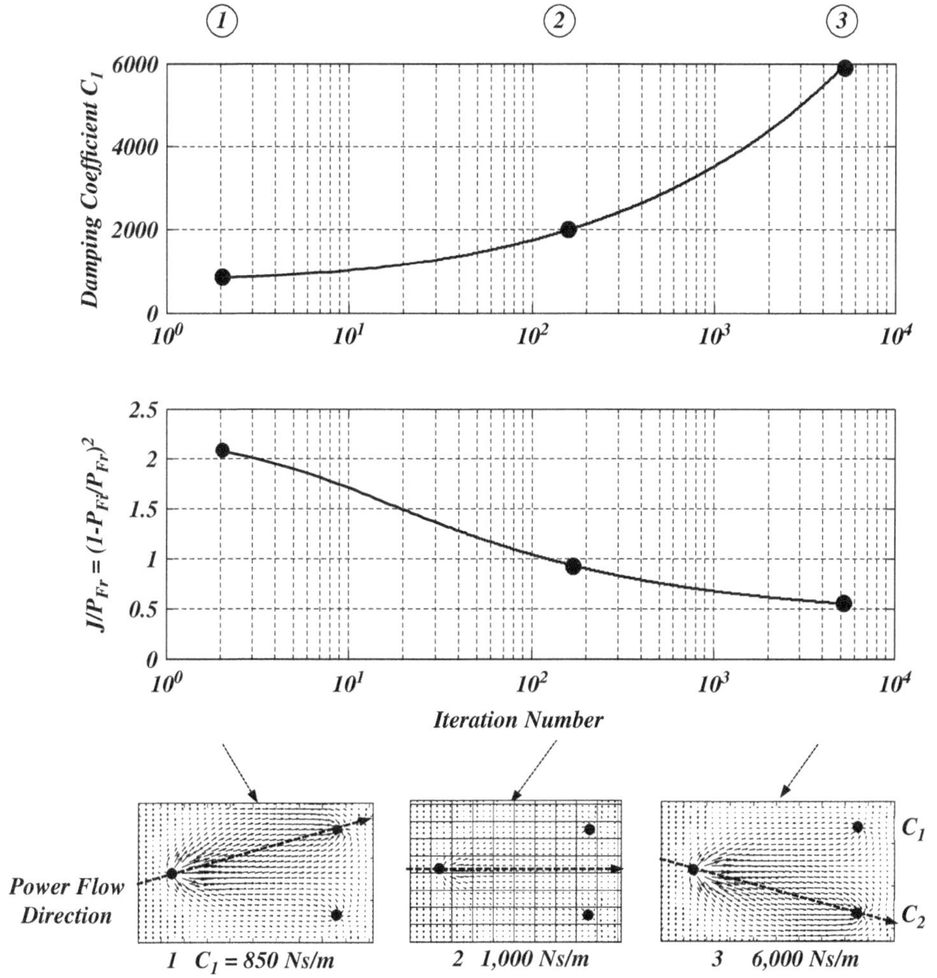

Figure 12.26 Schematic drawing of the location of the dampers and power flow sensor for a damped cantilever plate system.

direction at the sensor location is not only altered, but also minimized to assume a minimum value of $(1 - P_{Fi}/P_{Fr})^2 = 0.5$.

12.4.4 Power Flow and Structural Intensity for Plates with Passive and Active Constrained Layer Damping Treatments

In this section, the power flow analysis presented in Section 12.4.2 is extended and applied to predict the energetics of plates treated with different configurations of passive (PCLD) and ACLD treatments. Such an approach can be employed to the design of optimal layout and control strategies of the treatments in order to steer the vibration energy and the wave propagation as deemed essential to avoid undesirable excitations from reaching critical locations over the plate structure.

The presented approach is guided by the work of Castel et al. (2012) for structures with PCLD treatments and Alghamdi and Baz (2002) for structures with ACLD treatments.

Three examples are presented here to illustrate the applicability and utility of power flow analysis in studying the energetics of structures treated with PCLD and ACLD treatments. The first example deals with a plate with a single PCLD treatment, the second example considers two PCLD treatments whereas the third example presents the performance of the plate when it is treated with two separately controlled ACLD treatments.

Example 12.7 Considerthe aluminum cantilever plate shown in Figure 12.27. The plate is treated with one PCLD patch placed at the fixed end of the plate. The dimensions of the plate and the PCLD are shown in the figure with $h_1 = h_2 = h_3 = 0.005m$. The *VEM* is modeled using a Golla–Hughes–MacTavish (GHM) model with three mini-oscillators such that:

$$G_2(s) = G_0\left[1 + \sum_{i=1}^{3} \alpha_i \frac{s^2 + 2\zeta_i\omega_i s}{s^2 + 2\zeta_i\omega_i s + \omega_i^2}\right]$$

with $G_0 = 0.5$ MPa.

The parameters α_i, ζ_i, and ω_i are listed in Table 4.7. Assume that the constraining layer is made of aluminum. Determine the map of the streamlines of the power flow when the plate is excited at its first natural frequency by a unit force acting at the middle of the free end.

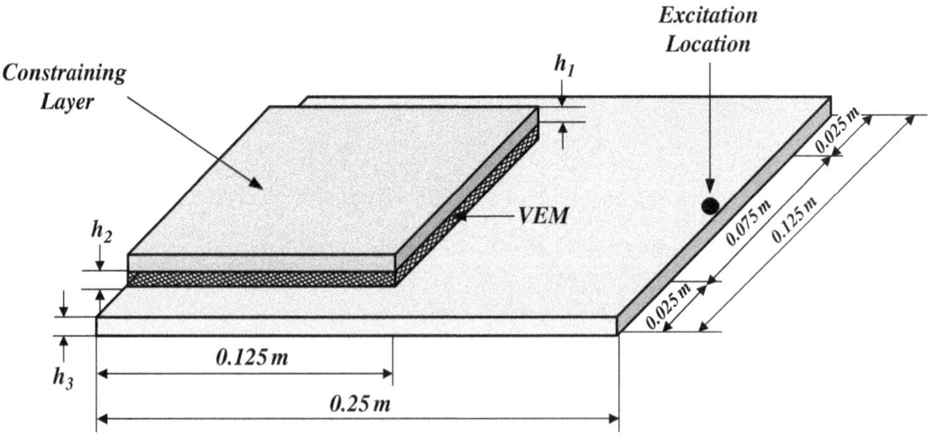

Figure 12.27 Schematic drawing of a cantilever plate treated with a PCLD patch.

Solution

Figure 12.28 displays the first three modes and corresponding mode shapes of vibration of the plate/PCLD treatment. Figure 12.29 shows the map of the streamlines of the power flow when the plate is excited at its first natural frequency of 72 Hz.

Note that the active power flows toward the front of the PCLD patch and then stopped near the front edge of the patch and redirected to flow inward toward the side edges of the patch.

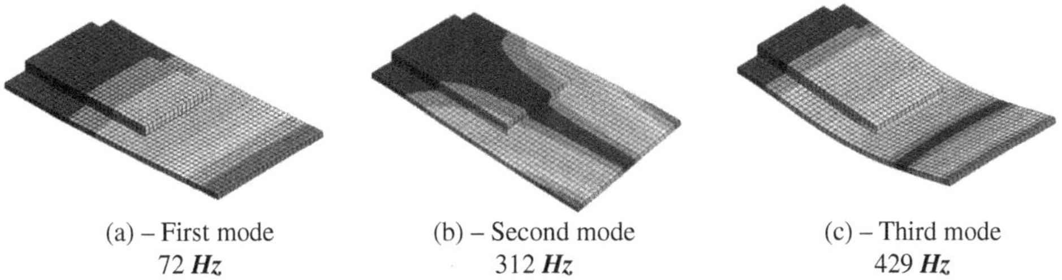

Figure 12.28 Mode shapes of a cantilever plate treated with a PCLD patch at the first three modes.

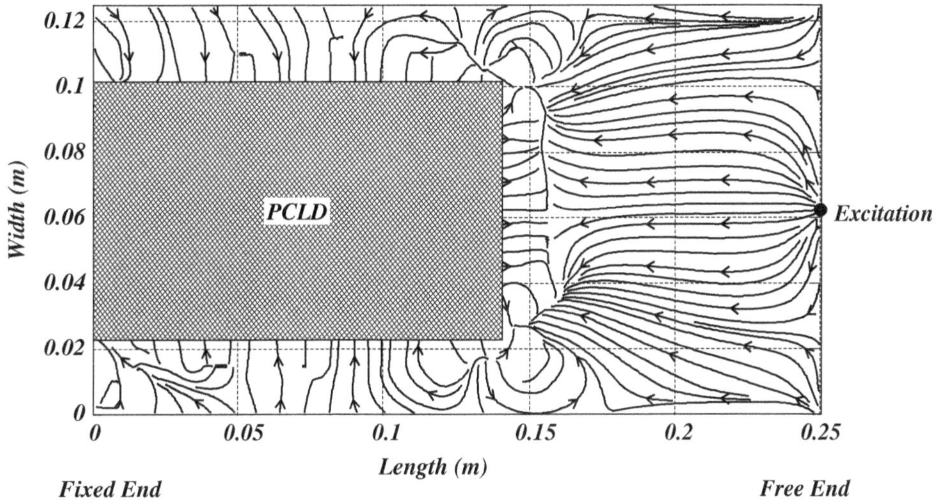

Figure 12.29 The streamlines distribution map of the cantilever plate/PCLD system at the first natural frequency.

Example 12.8 Considerthe aluminum cantilever plate treated with two PCLD patches placed at the fixed end of the plate shown in Figure 12.30. The dimensions of the plate and the PCLD patches are shown in the figure with $h_1 = h_2 = h_3 = 0.005m$.

The VEM is modeled using a GHM model with three mini-oscillators such that:

$$G_2(s) = G_0\left[1 + \sum_{i=1}^{3}\alpha_i \frac{s^2 + 2\zeta_i\omega_i s}{s^2 + 2\zeta_i\omega_i s + \omega_i^2}\right]$$

with G_0 = 0.5 MPa

The parameters α_i, ζ_i, and ω_i are listed in Table 4.7. Assume that the constraining layer is made of aluminum. Determine the map of the streamlines of the power flow when the plate is excited at its first natural frequency by a unit force acting at the middle of the free end.

Solution

Figure 12.31 shows the map of the streamlines of the power flow when the plate is excited at its first natural frequency of 72 Hz. With two PCLD treatments, the streamlines are symmetric and flow from the excitation source to the PCLD treatments.

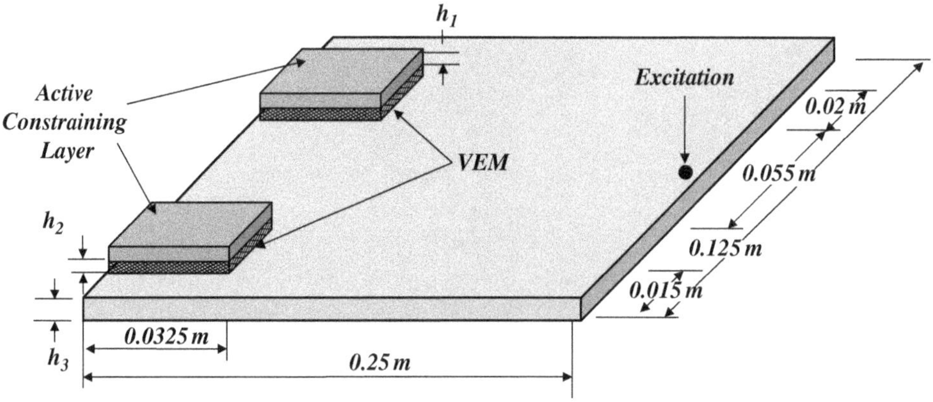

Figure 12.30 Schematic drawing of a cantilever plate treated with two PCLD patches.

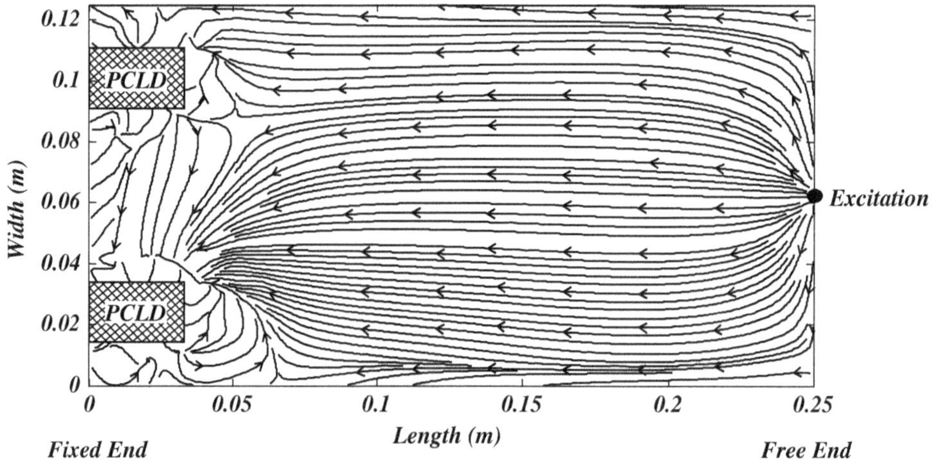

Figure 12.31 The streamlines distribution map of the cantilever plate treated with two PCLD patches at the first natural frequency.

Example 12.9 Consider the aluminum cantilever plate shown in Figure 12.30. Two PCLD treatments are replaced by ACLD patches such that: $h_1 = h_2 = h_3 = 0.005$ m. Also, the constraining layer is made of piezoelectric material lead zirconate titanate (PZT-4) with $d_{31} = d_{32} = -123 \times 10^{-12}$ m V^{-1} and $1/s_{31}^E = 1/s_{32}^E = 78.3GPa$.

The VEM is modeled using a GHM model with three mini-oscillators such that:

$$G_2(s) = G_0\left[1 + \sum_{i=1}^{3} \alpha_i \frac{s^2 + 2\zeta_i\omega_i s}{s^2 + 2\zeta_i\omega_i s + \omega_i^2}\right]$$

with $G_0 = 0.5$ MPa

The parameters α_i, ζ_i, and ω_i are listed in Table 4.7. Determine the map of the streamlines of the power flow, for different piezoelectric control voltages, when the plate is

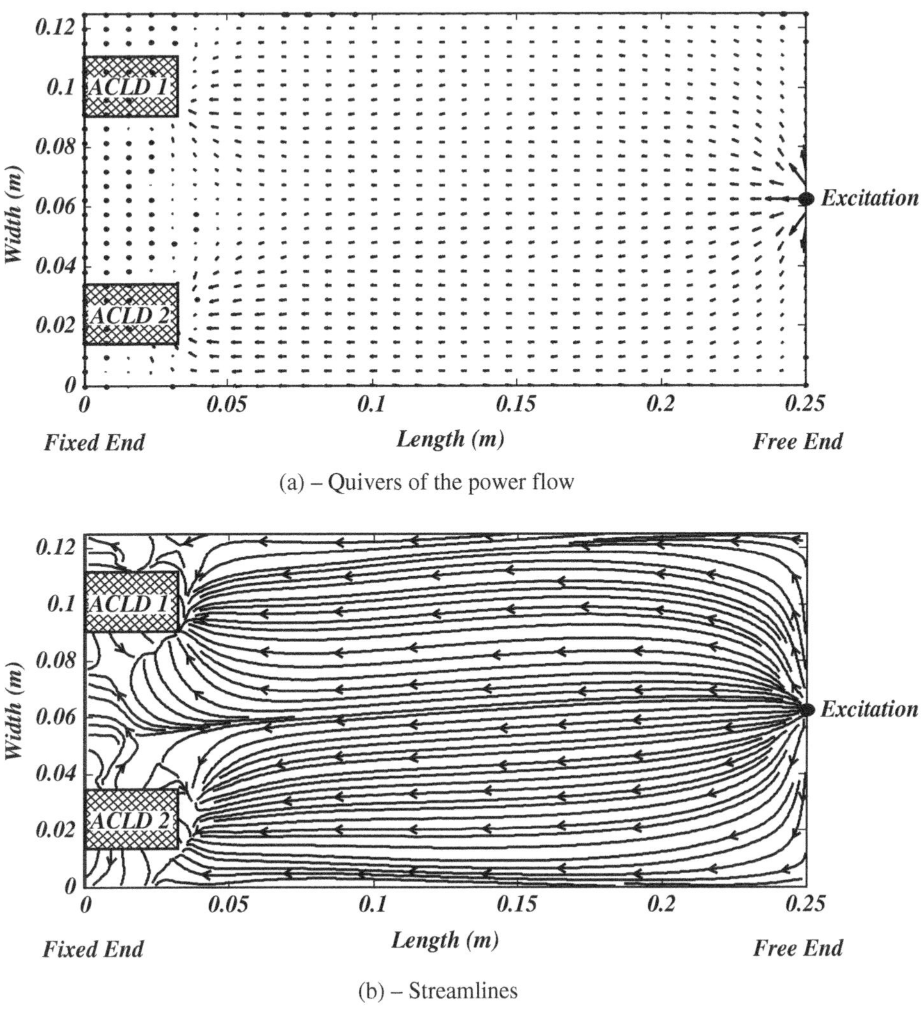

Figure 12.32 The power flow (a) and streamlines (b) distribution maps of the cantilever plate treated with two ACLD patches at the first natural frequency when $K_1 = K_2 = 500\ \mathrm{V\,s\,m^{-1}}$.

excited at its first natural frequency by a unit force acting at the middle of the free end. The control voltage V_c is generated using a velocity feedback control law such that:

$$V_c = -K_g\,\dot{w}_{ACLD}$$

where K_g denotes the control gain and $\dot{w}_{ACLD}$ defines the transverse velocity of the middle point of the ACLD patch.

Solution

Figure 12.32 shows the maps of the quivers and streamlines of the power flow when the plate is excited at its first natural frequency of 72 Hz. The maps are obtained when both patches are controlled with $K_1 = K_2 = 500\ \mathrm{V\,s\,m^{-1}}$. Under such control strategy, the

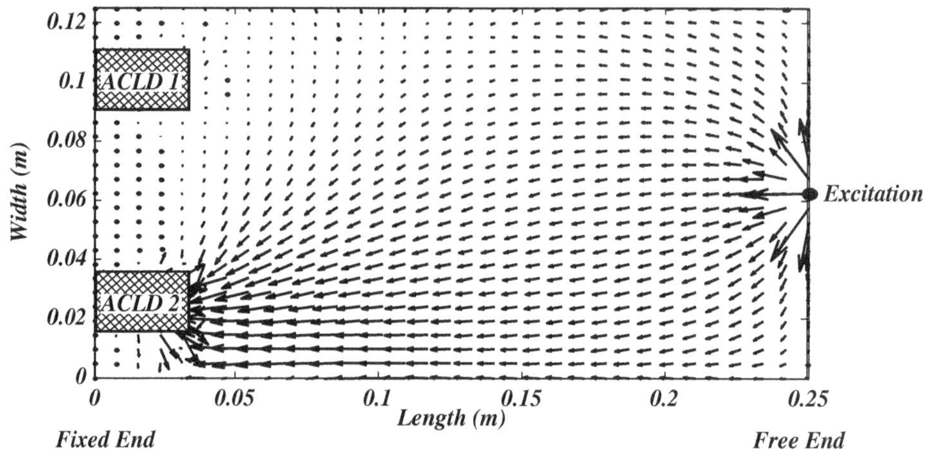

(a) – Quivers of the power flow

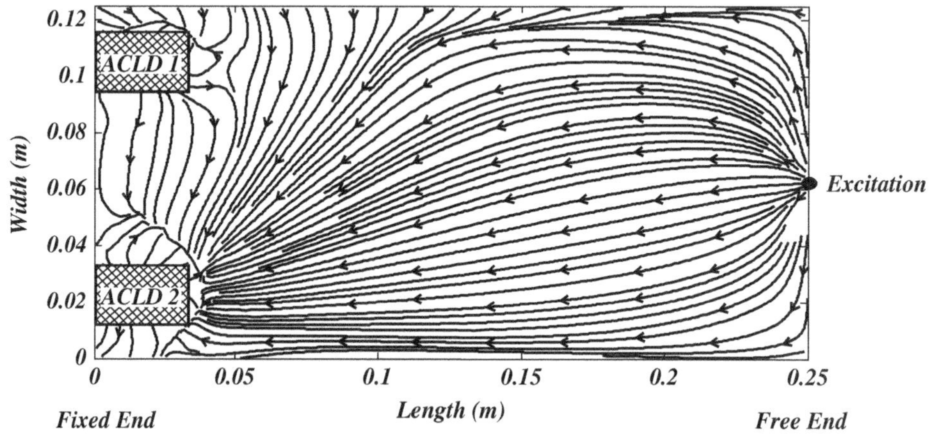

(b) – Streamlines

Figure 12.33 The power flow (a) and streamlines (b) distribution maps of the cantilever plate treated with two ACLD patches at the first natural frequency when $K_1 = 0$ and $K_2 = 500\ \mathrm{V\,s\,m^{-1}}$.

streamlines are also symmetric, as in the case of the PCLD in Example 12.6, and the vibration energy flows from the excitation source directly to the ACLD treatments.

Figure 12.33 shows the corresponding maps of the quivers and streamlines of the power flow when $K_1 = 0$ and $K_2 = 500$ V s m^{-1} with the excitation maintained at the first natural frequency of 72 Hz. Under such a control strategy, the power flow and streamline flows from the excitation source directly to the bottom ACLD treatment.

The reverse occurs when the control gains are switched such that $K_1 = 500$ and $K_2 = 0$ V s m^{-1} as shown in Figure 12.34. In this case, the power flow and streamline flows from the excitation source directly toward the top ACLD treatment.

Hence, by proper control of the ACLD patches, it is demonstrated that the wave propagation, as quantified by the power flow and streamlines, can be steered over the plate surface as desired.

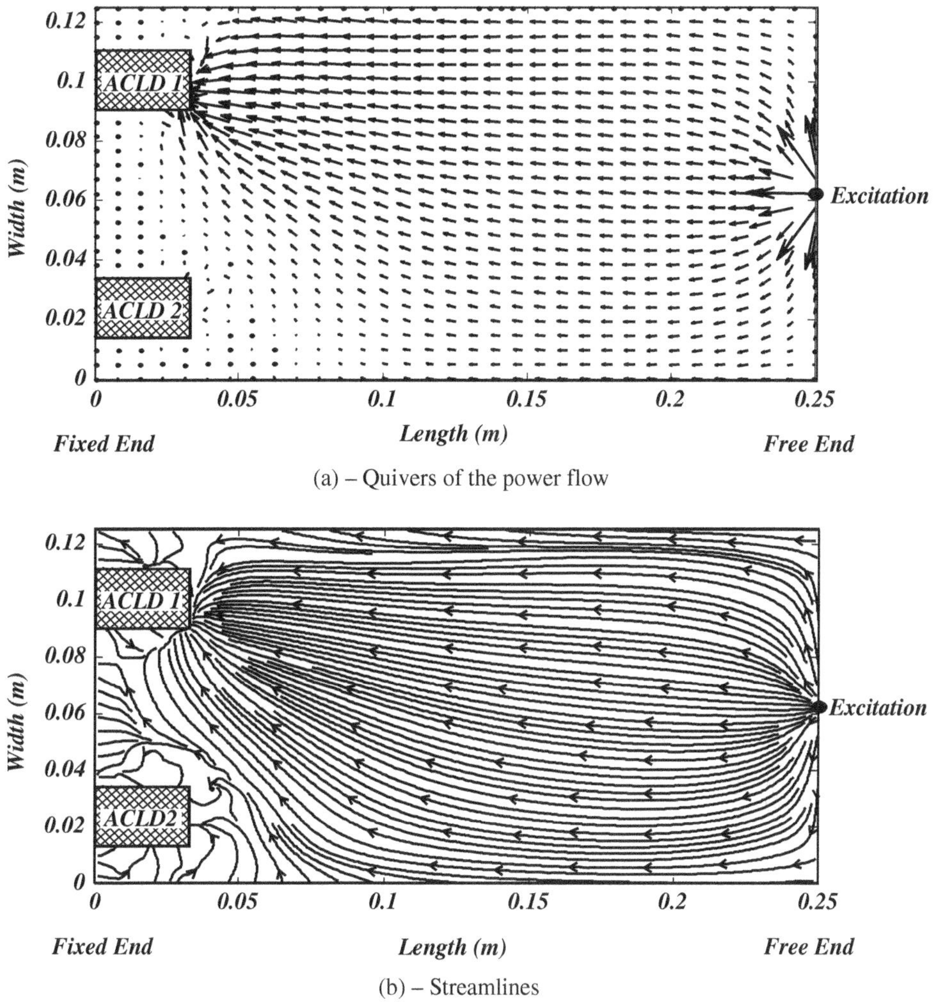

Figure 12.34 The power flow (a) and streamlines (b) distribution maps of the cantilever plate treated with two ACLD patches at the first natural frequency when $K_1 = 500$ and $K_2 = 0\,\mathrm{V\,s\,m^{-1}}$.

Example 12.10 Consider the cantilevered plate shown in Figure 9.24 and described in Example 9.7. The plate is provided with two sets of resistively shunted piezo-patches are placed symmetrically near the fixed end to control the first mode of vibration of the plate.

Determine the power flow distribution over the plate for different resistive shunting strategies when the plate is sinusoidally excited at its first mode of vibration (24.14 Hz) by a unit transverse load placed at its axis of symmetry at a distance of 8″ away from the fixed end as shown in Figure 12.35. Use ANSYS software package.

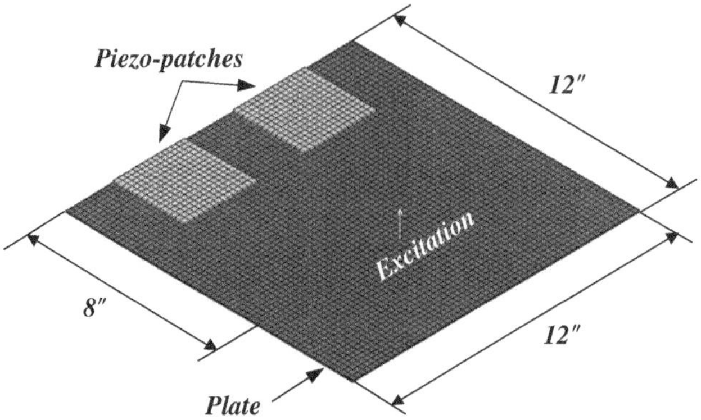

Figure 12.35 The plate with shunted piezo-patches along with the location of the input excitation.

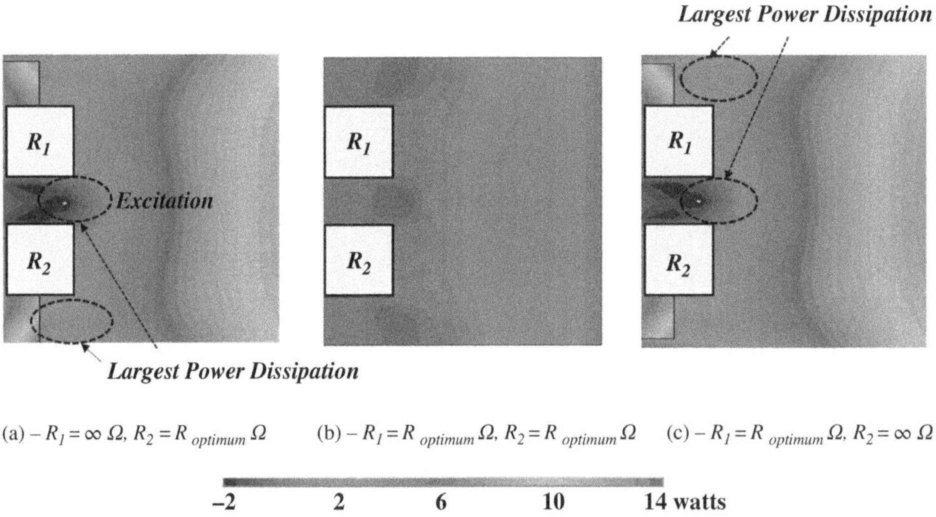

Figure 12.36 The power flow distribution maps of the cantilever plate treated with two piezo-patches with different shunting strategies.

Solution

Figure 12.36 displays the power flow distributions maps for three different combinations of the shunting resistances connected to the two piezo-patches as predicted by using ANSYS.

Figure 12.36a shows the power flow distribution over the plate surface when the two patches are shunted with resistors $R_1 = \infty\ \Omega$ (i.e., open circuit) and R_2 is set equal to the optimum value $\mathrm{R}_{\mathrm{optimum}} = \sqrt{1-k_{31}^2}/[C^S\omega]$ that maximizes the damping ratio corresponding to the first mode of 24.14 Hz. The figure indicates that the power flow distribution near the bottom piezo-patch with R_2 shunting attains its minimal levels. It is evident that the power flows from the excitation source to the optimized energy sink.

The reverse becomes true when the top piezo-patch is shunted with the optimal resistance R_1 that maximizes the damping ratio of the first mode while the resistors $R_2 = \infty\ \Omega$ as shown in Figure 12.36c.

When both patches are shunted with optimal resistors, the power flow distribution is minimized all over the plate surface as indicated in Figures 12.36b and 12.37.

12.5 Power Flow and Structural Intensity for Shells

In this section, the power flow analysis presented in Section 12.4.2 is extended and applied to predict the distribution of power flow over the surface of shell treated with patches of PCLD.

The equations of the structural power flow can be calculated using the approach outlined by Pavic (1990), Gavric and Pavic (1993), Williams (1991), and Ruzzene and Baz (2000). In this approach, the structural intensity for the quadrilateral shell element, shown in Figures 4.32 and 7.37, can be expressed in the form of power flow per unit width. The x and y components of structural intensities can be expressed as

$$
\begin{aligned}
I_x(\omega) &= -\frac{\omega}{2} I_m\left(N_x u^* + N_{xy} v^* + Q_x w^* + M_x w^*_{,y} - M_{xy} w^*_{,x} + M_x u^*/R + M_{xy} v^*/R\right),\\
I_y(\omega) &= -\frac{\omega}{2} I_m\left(N_y v^* + N_{yx} u^* + Q_y w^* - M_y w^*_{,x} + M_{yx} w^*_{,y} + M_y v^*/R + M_{yx} u^*/R\right).
\end{aligned} \tag{12.35}
$$

where N_i, M_i, M_{ij}, and Q_i denote the in-plane forces, bending moment, twisting moment, and transverse shear forces per unit width, respectively. Also, u^*, v^*, and w^* are the complex conjugate velocities in the x, y, *and* z directions. Furthermore, $w^*_{,i}$ is the complex conjugate of the angular displacement about the ith direction.

Alternatively, the power flow S_P can also be calculated directly from the finite element model using Eq. (12.9) such that:

$$
\begin{aligned}
S_P &= \frac{1}{2}\{F\}^*\{\dot{X}\}\\
&= \frac{1}{2}\{\dot{X}\}^*[C]\{\dot{X}\} + \frac{1}{2}\omega\left[\{X\}^*[K]\{X\} - \{\dot{X}\}^*[M]\{\dot{X}\}\right]j,\\
&= P_F + jQ_F
\end{aligned} \tag{12.36}
$$

Note that with the excitation at one of the modes of vibration the reactive power Q_F vanishes and the active power P_F quantifies the total structural power. This approach enables the use of both the *FEM* developed in Sections 4.8 and 7.5 as well as the ANSYS FEM in Appendix 12.A.

Example 12.11 Consider the clamped-free shell/PCLD assembly shown in Figure 12.37. The main physical and geometrical parameters of the shell and PCLD are listed in Table 4.10. The shell has an internal radius R is 0.1016 m. The PCLD treatment consists of two patches as displayed in Figure 12.37. The patches are bonded 180° apart on the outer surface of the cylinder with each of which subtending an angle of 90° angle at the center of the shell as shown in Figure 12.37.

Determine the power flow distribution over the shell surface when it is excited sinusoidally by a unit force, at the first natural frequency of the assembly, at a location L_l = 0.9 m using the FEM of Sections 4.8 and 7.5. Compare the obtained distribution with that predicted by using ANSYS.

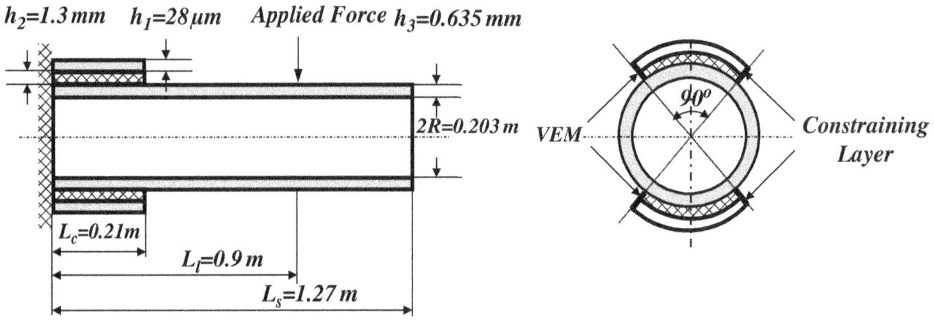

Figure 12.37 Configuration of the shell/PCLD assembly.

Solution

Figure 12.38a displays the power flow distributions maps for the shell/PCLD as predicted by using the FEM described in Sections 4.8 and 7.5.

Figure 12.38b shows the corresponding mode shape of the shell at the first mode of vibration of 60 Hz. It is evident the close correspondence between the power flow distribution and the mode shape of the vibrating shell.

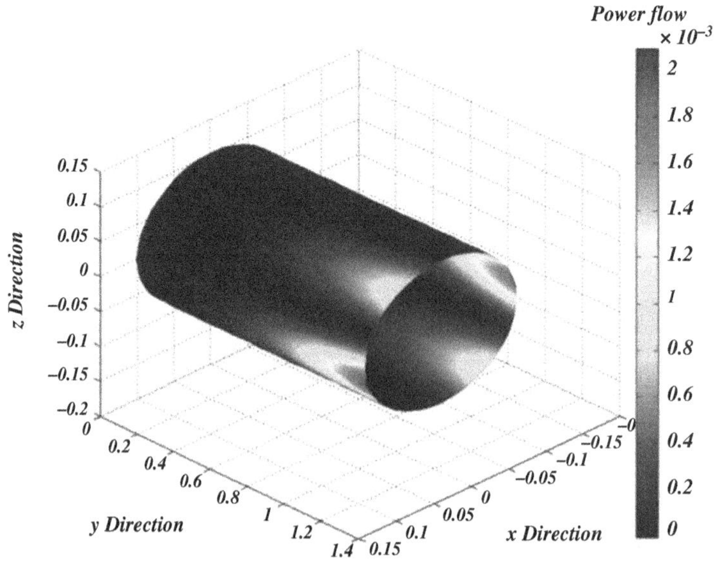

(a) – Power flow distribution

Figure 12.38 The power flow distribution map over the cantilever shell treated with two PCLD patches using MATLAB FEM approach.

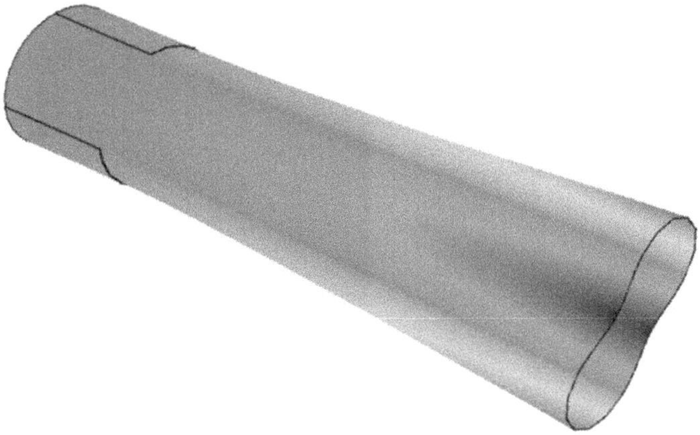

(b) – Mode shape

Figure 12.38 (Continued)

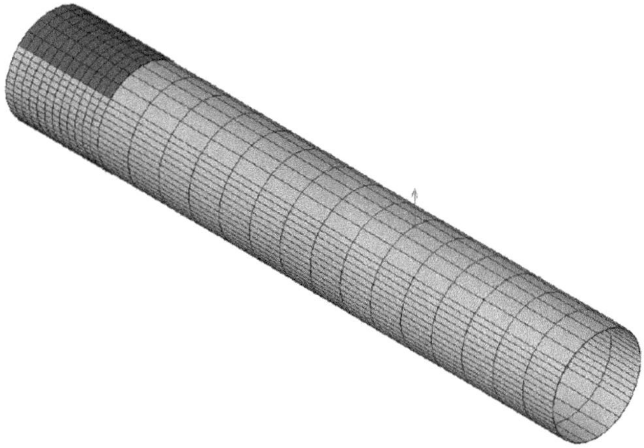

Figure 12.39 The finite element mesh of the shell/PCLD assembly using ANSYS.

Figure 12.39 displays the finite element mesh of ANSYS software package indicating the shell, the PCLD patches, and the location of the external excitation.

Figure 12.40 displays the power flow distributions maps for the shell/PCLD as predicted by using ANSYS. It is evident the close agreement between the MATLAB predictions shown in Figure 12.38a and ANSYS predictions shown in Figure 12.40.

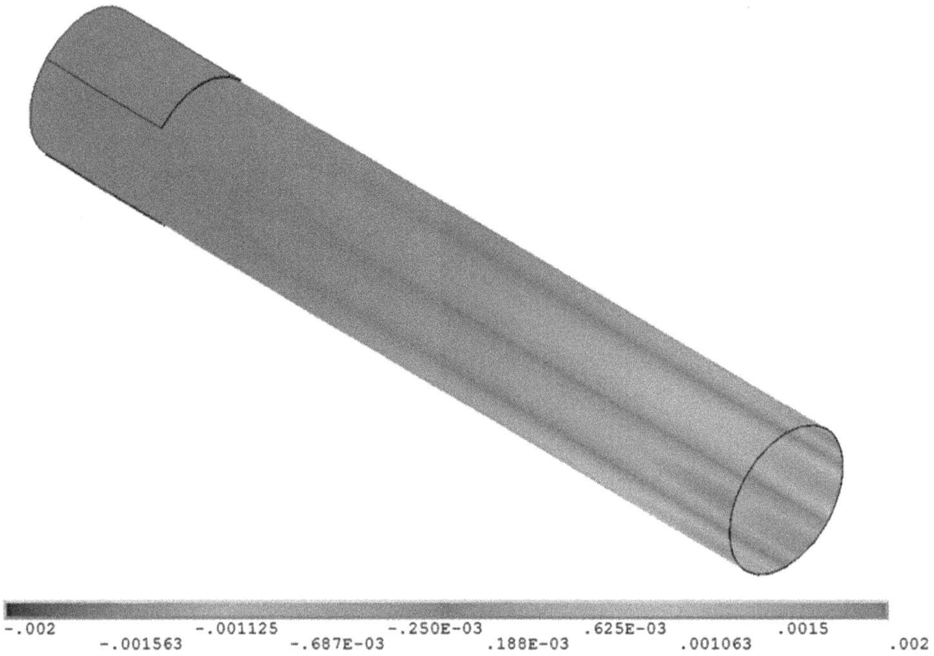

Figure 12.40 The power flow distribution map over the cantilever shell treated with two PCLD patches using ANSYS.

12.6 Summary

This chapter has presented the analysis of the power flow over vibrating structures in the presence of various types of damping including: classical viscous damping, VEM in constrained configurations, passive as well as ACLD, and shunted piezoelectric patches.

The chapter has presented also the concept of using vibration damping element in steering the power flow over the structure passively or actively using adaptive control algorithm.

References

Alfredsson, K. (1997). Active and reactive structural energy flow. *ASME Journal of Vibration and Acoustics* 119: 70–79.

Alfredsson, K., Josefson, B., and Wilson, M. (1996). Use of the energy flow concept in vibration design. *AIAA Journal* 34 (6): 750–755.

Alghamdi A. A. A. and Baz A., "Power flow in beams treated with active constrained layer damping," Proceedings of the 6th Saudi Engineering Conference, KFUPM, Dhahran, Vol. 5, pp. 445–460, 2002.

Bouthier, O. and Bernhard, R. (1995). Simple models of energy flow in vibrating membranes. *Journal of Sound and Vibration* 182 (1): 79–147.

Castel, A., Loredo, A., El Hafidi, A., and Martin, B. (2012). Complex power distribution analysis in plates covered with passive constrained layer damping patches. *Journal of Sound and Vibration* 331 (11): 2485–2498.

Gavric, L. and Pavic, G. (1993). A finite element method for computation of structural intensity by the normal mode approach. *Journal of Sound and Vibration* 164 (1): 29–43.

Gibbs, G., Fuller, C., and Silcox R. (1993). Active Control of Flexural and Extensional Power Flow in Beams Using Real Time Wave Vector Sensors. *Second Conference on Recent Advances in Active Control of Sound and Vibration,* April 28–30.

Halkyard, C. and Mace, B. (1995). Structural intensity in beams – waves, transducer systems and the conditioning problem. *Journal of Sound and Vibration* 185 (2): 279–298.

Linjama, J. and Lahti, T. (1992). Estimation of bending wave intensity in beams using the frequency response technique. *Journal of Sound and Vibration* 153 (1): 21–36.

Lyons, R. (1975). *Statistical Energy Analysis of Dynamical Systems: Theory and Applications.* Cambridge, MA: MIT Press.

Mandal, N.K. and Biswas, S. (2005). Vibration power flow: a critical review. *The Shock and Vibration Digest* 37 (1): 3–11.

Mandal, N.K., Rahman, R.A., and Leong, M.S. (2003). Structure-borne power transmission in thin naturally orthotropic plates: general case. *Journal of Vibration and Control* 9: 1189–1199.

Nefske, D.J. and Sung, S.H. (1987). Power flow finite element analysis of dynamic systems: Basic theory and applications to beams. *Statistical Energy Analysis* 3: 47–54.

Noiseux, D. (1970). Measurement of power flow in uniform beams and plates. *The Journal of The Acoustical Society of America* 47 (1): 238–247.

Nouh, M., Aldraihem, O., and Baz, A. (2015). Wave propagation in metamaterial plates with periodic local resonances. *Journal of Sound and Vibration* 341: 53–73.

Pavic, G. (1976). Measurements of structure borne wave intensity, part I: formulation of the methods. *Journal of Sound and Vibration* 49 (2): 221–230.

Pavic, G. (1987). Structural surface intensity: an alternative approach in vibration analysis and diagnosis. *Journal of Sound and Vibration* 115 (3): 405–422.

Pavic, G. (1990). Vibrational energy flow in elastic circular cylindrical shells. *Journal of Sound and Vibration* 142 (2): 293–310.

Pavic, G. (2005). The role of damping on energy and power in vibrating systems. *Journal of Sound and Vibration* 281 (1–2): 45–71.

Rao, S.S. (2007). *Vibration of Continuous Systems.* New Hoboken, NJ: Wiley.

Ruzzene, M. and Baz, A. (2000). Active control of power flow in ribbed and fluid-loaded shells. *Journal of Thin Walled-Structures* 38 (1): 17–42.

Widrow, B. and Stearns, S. (1985). *Adaptive Signal Processing.* Englewood Cliffs, NJ: Prentice-Hall, Inc.

Williams, E. (1991). Structural intensity in thin cylindrical shells. *The Journal of The Acoustical Society of America* 89 (4): 1615–1622.

Williams, E., Dardy, H., and Fink, R. (1985). A technique for measurements of structure-borne intensity in plates. *The Journal of the Acoustical Society of America* 78 (6): 2061–2068.

Wohlever, J. and Bernhard, R. (1992). Mechanical energy flow models of rods and beams. *Journal of Sound and Vibration* 153 (1): 1–19.

Xua, X.D., Lee, H.P., Lu, C., and Guo, J.Y. (2005). Streamline representation for structural intensity fields. *Journal of Sound and Vibration* 280 (1–2): 449–454.

12.A Calculation of Power Flow in ANSYS

In order to estimate the power flow S_P in structures, the internal forces and deflections at the different finite element nodes, subject to external excitation, are calculated and substituted in:

$$S_P = F^* \cdot \dot{\delta}, \tag{12.A.1}$$

where F is the force and δ is the deflection vector.

In case of harmonic excitation, at a frequency ω, S_P can be rewritten as

$$S_P = i\omega\delta^* \mathbf{K}^* \cdot \delta \tag{12.A.2}$$

where the force F is replaced by $\mathbf{K}\ \delta$ with $\mathbf{K}$ denoting the stiffness matrix.

Accordingly, the following procedure illustrates how to develop the power flow distribution in a finite element model of a composite (ACLD) beam, which is excited with harmonic force tuned at the first natural frequency. The procedure is presented using ANSYS APDL (ANSYS Parametric Design Language).

The procedure applied follows the following sequence:

Step 1: Define all the model parameters
Step 2: Enter the model creation pre-processor (/prep7)
Step 3: Enter the solution processor (/solu)

a) Carry out "Modal Analysis" to calculate the natural frequencies.
b) Carry out "Harmonic Analysis" by defining excitation load at specified location and with specified amplitude and is oscillating at the natural frequencies calculated in the "Modal Analysis" step.

Step 4: Enter the database results post-processor

a) Select the load step to post-process.
b) Calculate the elements results for the entire model (stresses and deflections).
c) Use the "Element Table" feature to carry out mathematical calculations on the extracted stress/deflection results.
d) Plot the power flow distribution over the entire structure.

Step 1: Prepare the pre-processor/solver/post-processor environment:

```
/title, Power Flow in PCLD beam
/UNITS,SI                  !Specifies the unit system to be used in
                           the model
/SHOW,WIN32C               !Specifies the device and other
                           parameters for graphics displays
/CONT,1,32,AUTO            !Specifies the uniform contour values
                           on stress displays.
/page,15000,,15000,,0,     !Defines number of lines per page
```

Step 2: Start the pre-processor where the model is being developed; materials are defined and element types selected

```
/PREP7
!*** Define variables to be used in the FE model later
!*********************************************************
! Geometrical parameters
!***********************
inch=25.4e-3

Lp=0.5              ! length of base beam
Wp=0.05             ! width of base beam
Tp=0.0125           ! thickness of base beam

Lpz=Lp              ! length of PZT constraining layer
Wpz=Wp              ! width of PZT constraining layer
Tpz=0.0025          ! thickness of PZT constraining layer

Lv=Lp               ! length of Viscoelastic (VEM) layer
Wv=Wp               ! width of Viscoelastic (VEM) layer
Tv=0.00625          ! thickness of Viscoelastic (VEM) layer

! Material properties parameters
!*******************************
rho_p=7800          ! Density of base beam
rho_pz=7600         ! Density of PZT layer
rho_v=1104          ! Density of VEM layer

E_p=210e9         ! Young's Modulus of base beam
E_v=60e6          ! Young's Modulus of VEM layer

damp_p=0          ! Damping ratio for base beam material
damp_pz=0         ! Damping ratio for PZT material
damp_v=0.5        ! Damping ratio for VEM material characterized
                  with loss factor = 1
```

```
!*** Element Type and Material Definition
!****************************************
! Element Types
!*************
et,1,solid226,1001      ! Piezoelectric element
et,2,solid186           ! Solid Element for beam and VEM
!********************************************************

! Material Definitions
!*********************
! Due to the anisotropic nature and coupling effect of
  piezoelectric materials, they have specific way in defining
  the structural/electrical material properties matrices. This
  is done using tabular data for the different matrix entries
!********************************************************
/COM, MATERIAL PROPERTIES OF LEAD ZIRCONATE TITANATE (PZT-5A)
/COM,
/COM, - MATERIAL MATRICES (POLAR AXIS ALONG Y-AXIS): IEEE INPUT
/COM,
/COM,  [s11 s13 s12  0  0  0 ]     [ 0 d31  0 ]    [ep11  0   0 ]
/COM,  [s13 s33 s13  0  0  0 ]     [ 0 d33  0 ]    [ 0  ep33  0 ]
/COM,  [s12 s13 s11  0  0  0 ]     [ 0 d31  0 ]    [ 0   0  ep11]
/COM,  [ 0   0   0  s44 0  0 ]     [ 0  0  d15]
/COM,  [ 0   0   0   0 s66 0 ]     [ 0  0    0 ]
/COM,  [ 0   0   0   0  0 s44]     [d15 0    0 ]
/COM,
!********************************************************
S11=15.874E-12
S12=-4.25E-12
S13=-9.49E-12
S33=15.87E-11
S44=48.9E-12
S66=47.3E-12

D15=6.67E-10
D31=-1.86E-10
D33=7.5E-10

EP11=4140
EP33=4500
!********************************************************
! Piezoelectric layers material properties
!****************************************
TB,ANEL,1,,,1           ! ANISOTROPIC ELASTIC COMPLIANCE MATRIX
TBDA,1,S11,S13,S12
TBDA,7,S33,S13
TBDA,12,S11
```

```
TBDA,16,S44
TBDA,19,S44
TBDA,21,S66

TB,PIEZ,1,,,1          ! PIEZOELECTRIC STRAIN MATRIX
TBDA,2,D31
TBDA,5,D33
TBDA,8,D31
TBDA,10,D15

TB,DPER,1,,,1      !  DIELECTRIC  PERMITTIVITY  AT  CONSTANT
                      STRESS
TBDA,1,EP11,EP33,EP11

MP,DENS,1,rho_pz       ! DENSITY of PZT layer
MP,DMPRAT,1,damp_pz
!*******************************************************

! Base Beam Material Definition
!******************************
MP,DENS,2,rho_p
MP,EX,2,E_p
MP,PRXY,2,0.3
MP,DMPRAT,2,damp_p


! VEM layer Material Definition
!******************************
MP,DENS,3,rho_v
MP,EX,3,E_v
MP,PRXY,3,0.499
MP,DMPRAT,3,damp_v
!**********************************************
!**********************************************

!*** MODELING
!************
! Base Beam
!***********
BLOCK,0,Lp,0,Tp,-Wp/2, Wp/2

! VEM Layer
!**********
BLOCK,0,Lp,Tp,Tp+Tv,-Wv/2, Wv/2

! PZT Layer
!**********
BLOCK,0,Lp,Tp+Tv,Tp+Tv+Tpz,-Wv/2, Wv/2
```

```
allsel,all
vglue,all
nummrg,KP
numcmp,all
!**********************************************

!*** MESHING
!***********
! Define number of FE divisions for the model lines
!************************************************
lsel,s,,,all
lsel,u,loc,x,0
lsel,u,loc,x,Lp
lesize,all,,,48

lsel,s,loc,x,0
lsel,a,loc,x,Lp
lsel,u,loc,z,Wp/2
lsel,u,loc,z,-Wp/2
lesize,all,,,1

lsel,s,loc,x,0
lsel,a,loc,x,Lp
lsel,r,loc,z,Wp/2
lesize,all,,,1

lsel,s,loc,x,0
lsel,a,loc,x,Lp
lsel,r,loc,z,-Wp/2
lesize,all,,,1
allsel,all
!************************************************
! Meshing Beam
!*************
allsel,all
vsel,s,,,1
VATT, 2,,2,,
vsweep, all                ! Mesh Volume # 1 (Beam)
allsel,all

! Meshing VEM
!************
allsel,all
vsel,s,,,2
VATT, 3,,2,,
vsweep, all                ! Mesh Volume # 2 (VEM)
```

```
allsel,all

! Meshing PZT
!************
allsel,all
vsel,s,,,3
VATT, 1,,1,11,
vsweep, all              ! Mesh Volume # 3 (PZT)
allsel,all
!****************************************************

!*** Apply Boundary Conditions
!*****************************
! Structural Boundary Conditions
!*******************************
nsel,s,loc,x,0
D,all,ux,0
D,all,uy,0
D,all,uz,0
allsel,all

! Electrical Boundary Conditions for the PZT layer
!**************************************************
vsel,s,,,3
aslv
asel,r,loc,y,Tp+Tv
nsla,s,1
cp,1,volt,all
allsel,all

vsel,s,,,3
aslv
asel,r,loc,y,Tp+Tv+Tpz
nsla,s,1
cp,2,volt,all
allsel,all
finish
!**************************************************
```

Step 3: Define the solver parameter, type of analysis, and carry out the solution

```
/SOLU
! Modal Analysis
!***************
ANTYPE,2             ! Modal analysis
MODOPT,LANB,10
```

```
EQSLV,SPAR
MXPAND,10, , ,1
OUTPR,ALL,ALL,
OUTRES,ALL,ALL
SOLVE

*GET,Mode1_Freq,MODE,1,FREQ,REAL
*GET,Mode2_Freq,MODE,2,FREQ,REAL
*GET,Mode3_Freq,MODE,3,FREQ,REAL
*GET,Mode4_Freq,MODE,4,FREQ,REAL
*GET,Mode5_Freq,MODE,5,FREQ,REAL
*GET,Mode6_Freq,MODE,6,FREQ,REAL
*GET,Mode7_Freq,MODE,7,FREQ,REAL
*GET,Mode8_Freq,MODE,8,FREQ,REAL
*GET,Mode9_Freq,MODE,9,FREQ,REAL
FINISH
!*****************************************************
! Harmonic Analysis
!*****************
/SOLU
antype,3            ! Harmonic Analysis
HARFREQ,Mode1_Freq,Mode1_Freq,
NSUBSET,1
KBC,1
outres,all,all

! Apply Force Excitation
!**********************
esel,s,mat,,2
nsle,s,1
nsel,r,loc,x,Lp
F,all,Fy,1
allsel,all

solve
FINISH
!*****************************************************
```

Step 4: Post-process the calculated results

```
/post1
SET,FIRST

! Calculate the deflection in the x-direction for the structure
  elements
!********************************
```

```
AVPRIN,0, ,
ETABLE,disp_x,U,X

! Calculate the deflection in the y-direction for the structure
  elements
!*********************************
AVPRIN,0, ,
ETABLE,disp_y,U,Y

! Calculate the deflection in the z-direction for the structure
  elements
!*********************************
AVPRIN,0, ,
ETABLE,disp_z,U,Z

! Calculate Fx
!*************
AVPRIN,0, ,
ETABLE,f_xx,F,X

! Calculate Fy
!*************
AVPRIN,0, ,
ETABLE,f_yy,F,Y

! Calculate Fz
!*************
AVPRIN,0, ,
ETABLE,f_zz,F,Z

! Carry out arithmetic operations
!********************************
SMULT,en_xx,F_XX,DISP_X,1,1,     ! Fxux
SMULT,en_yy,F_YY,DISP_Y,1,1,     ! Fyuy
SMULT,en_zz,F_ZZ,DISP_Z,1,1,     ! Fzuz

! Calculate Ix=SQRT[(Fxux)^2 + (Fyuy)^2 + (Fzuz)^2]
!******************************
SEXP,Ixd2,EN_XX,,2,,
SEXP,Iyd2,EN_YY,,2,,
SEXP,Izd2,EN_ZZ,,2,,
SADD,II1,IXD2,IYD2,1,1, ,
SADD,II,II1,IZD2,1,1, ,
SEXP,IISQR,II,,0.5,,
!***********************************************

! Plot the power flow distribution
```

```
!********************************
PLETAB,IISQR,AVG
!***************
! Bottom View
!***********
/VIEW,1,,-1
/ANG,1
/REP,FAST
```

Problems

12.1 Consider the four spring-damper-mass system shown in Figure P12.1.

If $k = 1$, $m = 1$, and the damping matrix = 0.1 stiffness matrix, determine the active and reactive power distributions over the four masses when the first mass is excited sinusoidally by a unit force at the first system natural frequency.

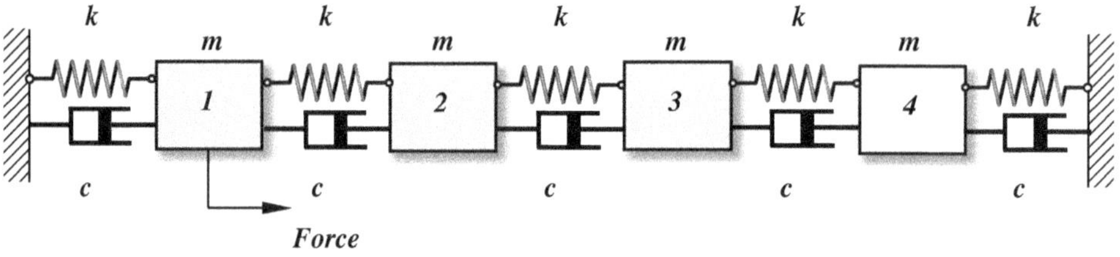

Figure P12.1 Multi-spring-damper-mass system

12.2 Consider the fixed-free rod/VEM system shown in Figure P12.2. The rod is made of aluminum with width of 0.025 m, thickness of 0.025 m, and length of 1 m. The VEM has width of 0.025 m, thickness = 0.025 m, and density of 1100 kg m^{-3}. The storage modulus and loss factor of the VEM are predicted by GHM model with one mini-oscillator ($E_0 = 15.3MPa, \alpha_1 = 39, \zeta_1 = 1, \omega_1 = 19,058rad/s$). Determine the power flow over the rod when it is subjected to a unit sinusoidal force acting at node 3 and the excitation occurs at the first natural frequency of the assembly.

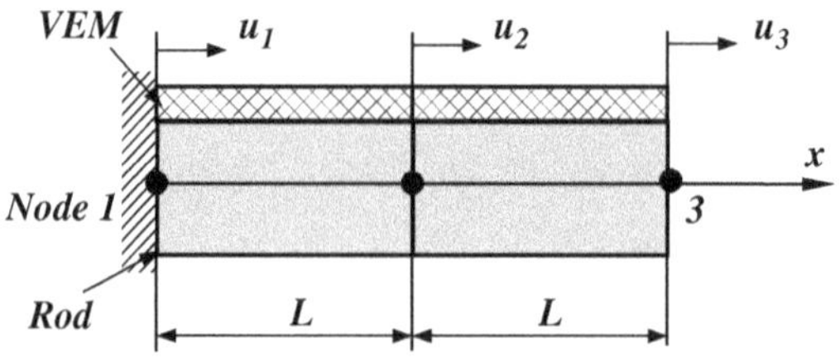

Figure P12.2 Two-element model a rod/unconstrained VEM.

12.3 Consider the fixed-free rod/VEM system shown in Figure 4.11. The rod is made of aluminum with width of 0.025 m, thickness of 0.025 m, and length of 1 m. The VEM has width of 0.025 m, thickness = 0.025 m, and density of 1100 kg m^{-3}.

The storage modulus and loss factor of the VEM are as given in Example 4.2. The VEM is constrained by an aluminum constraining layer that is 0.025 m wide and 0.0025 m thick. Using the GHM modeling approach of the *VEM* with one mini-oscillator ($E_0 = 15.3MPa, \alpha_1 = 39, \zeta_1 = 1,\ \omega_1 = 19,058rad/s$). Determine the power flow over the rod when it is subjected to a unit sinusoidal force acting at node 3 and the excitation occurs at the first natural frequency of the assembly (Figure P12.3).

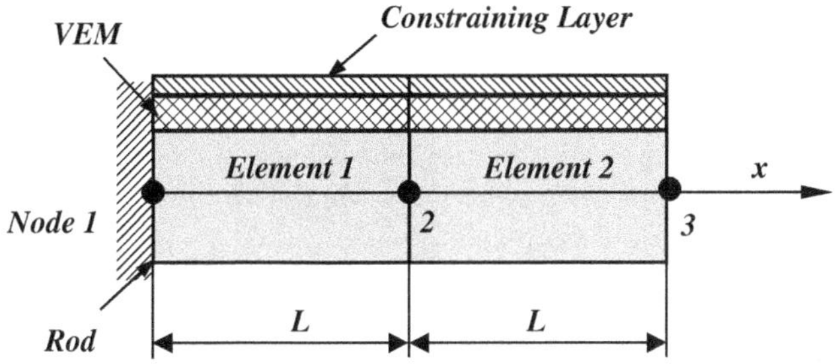

Figure P12.3 Two-element model a rod/constrained VEM.

12.4 Consider the fixed-free beam/VEM system shown in Figure P12.4. The physical and geometrical parameters of the beam/VEM system are listed in Table 4.4. The beam is divided into 10 finite elements and is subjected to a force *F* acting along the transverse direction. Determine the power flow over the beam when it is subjected to a unit sinusoidal force acting at node 11 and the excitation occurs at the first natural frequency of the assembly.

Assume that the complex shear modulus G_2 of the VEM is represented by a GHM model with four mini-oscillators that have the properties listed in Table 4.5 such that:

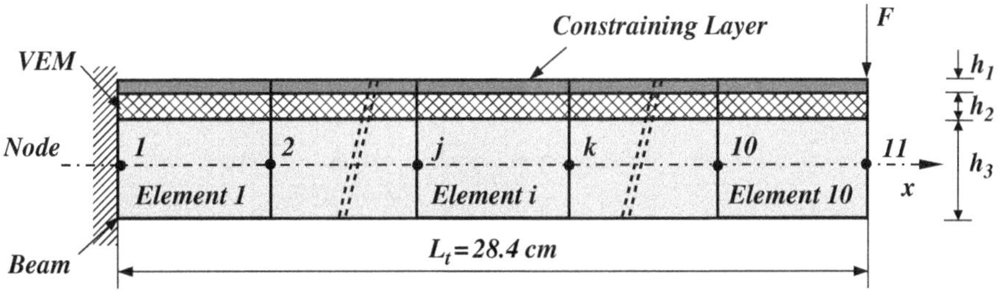

Figure P12.4 Ten-element cantilever beam with a constrained VEM damping treatment.

Table P12.1 Piezo-constraining layer – PZT.

Parameter	d_{31} – (m V^{-1})	k_{31}	g_{31} – (Vm N^{-1})	k_{3t}
Value	186E-12	0.34	1.16	1950

$$G_2(s) = G_0\left[1 + \sum_{i=1}^{4} \alpha_i \frac{s^2 + 2\omega_i s}{s^2 + 2\omega_i s + \omega_i^2}\right] \text{ with } G_0 = 2.72\text{ MPa}$$

Compare the power flow distributions when only two, four, six, and eight elements of the beam are treated with the constrained VEM treatment.

12.5 Consider the fixed-free beam/VEM system of Problem 12.4 and shown in Figure P12.4. The properties of the beam, VEM, and an active piezoelectric constraining layer are given in Tables 12.3 and P12.1. Assume that the beam is treated with ACLD over the first five elements. Determine the power flow over the beam, with open-loop ACLD, when it is subjected to a unit sinusoidal force acting at node 11 and the excitation occurs at the first natural frequency of the assembly.

Determine the power flow distribution also when the ACLD is controlled with a derivative feedback of the position of node 6. Use different values of the control gains and comment on the results.

12.6 Consider the cantilevered flat plate shown in Figure P12.5 that is damped using potentially eight point viscous dampers with damping coefficients C_1 through C_8. The plate is excited near its fixed end, as shown in the figure, by a unit transverse sinusoidal load at the first natural frequency of the plate.

Determine the structural intensity and the streamline plots of the system for the three arrangements of the point dampers which are listed in Table P12.2.

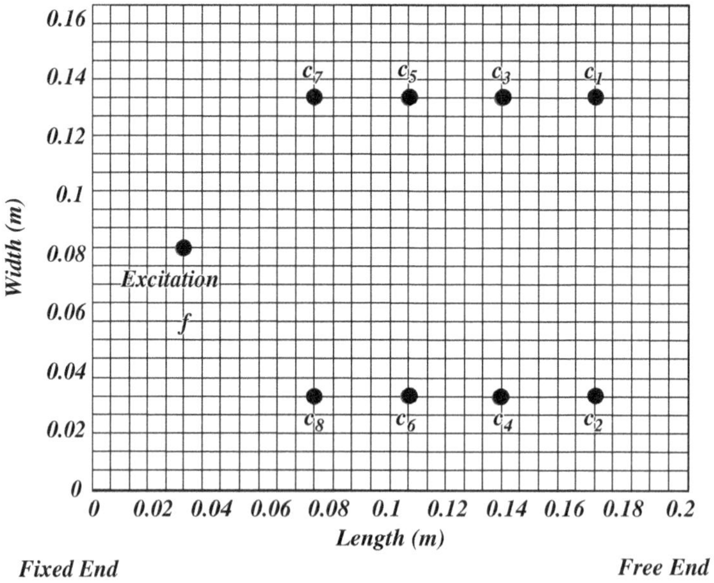

Figure P12.5 Cantilevered plate with eight point dampers.

Table P12.2 Arrangement of point dampers.

Arrangement	C_1	C_2	C_3	C_4	C_5	C_6	C_7	C_8
1	2000	2000	2000	2000	0	0	0	0
2	2000	2000	2000	2000	2000	2000	0	0
3	2000	2000	2000	2000	2000	2000	2000	2000

12.7 Consider the cantilevered beam shown in Figure P12.6 with six periodic sources of internal resonance. The beam is manufactured of assemblies of periodic cells consisting of cavities filled by viscoelastic membranes that support small masses to form sources of local resonance. The geometrical and physical parameters of the beam and the sources of internal resonance are listed in Tables P12.3 and P12.4, respectively. The beam is excited near its free end, as shown in the figure, by a unit transverse sinusoidal load at the first natural frequency of the assembly.

Determine the power flow quiver and the streamline plots of the system.

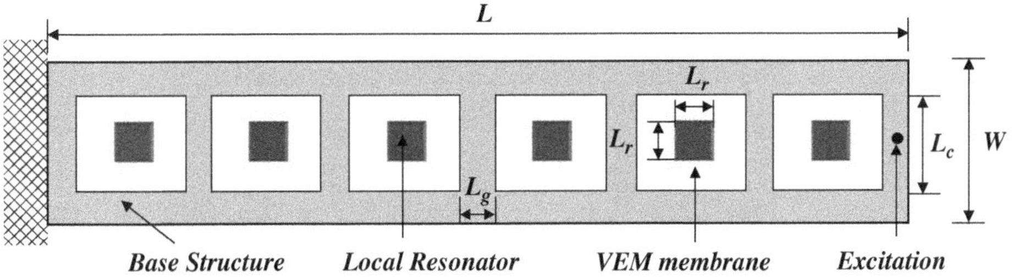

Figure P12.6 Cantilevered beam with periodic sources of internal resonance.

Table P12.3 Geometrical parameters of the beam with periodic sources of internal resonance.

Length	L	w	h	L_c	L_r	L_g
Value (m)	0.3556	0.0635	0.0015	0.0381	0.0127	0.0181

Table P12.4 Physical properties of the beam and damping membranes.

Material	Young's modulus (GPa)	Poisson's ratio (ν)	Density (kg m^{-3})
Aluminum	70	0.30	2700
VEM-Polyurea	$0.02(1 + 0.4i)$	0.49	1018

12.8 Consider the 2D metamaterial plate-like configuration displayed in Figure P12.7. The plate is manufactured of aluminum and consists of assemblies of periodic cells with built-in local resonances. Each cell is made of a base plate-like structure that is provided with cavities filled by a viscoelastic membrane that supports a small mass to form a source of local resonance (Nouh et al. 2015). Table 10.6 lists the main geometrical properties of the aluminum plate and the local resonant masses.

The plate is excited near its fixed end, as shown in the figure, by a unit transverse sinusoidal load at the first natural frequency of the assembly.

Determine the power flow quiver and the streamline plots of the system.

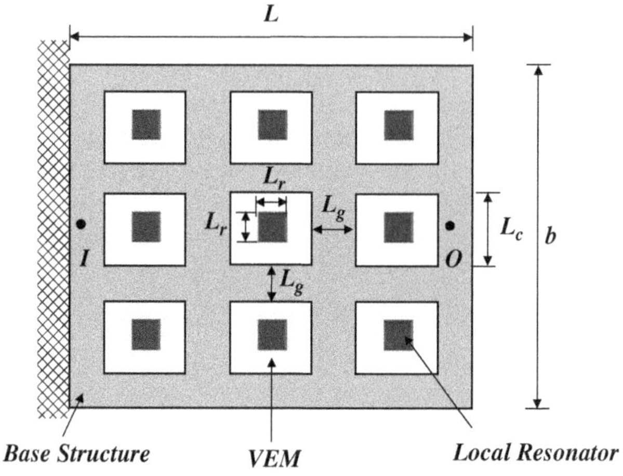

Figure P12.7 Cantilevered plate with periodic sources of internal resonance.

12.9 The 2D metamaterial plate-like configuration displayed in Figure P12.8 is a special case of the plate in Figure 12.7. The plate has only two sources of internal resonance. The main geometrical properties of the aluminum plate and the local resonant masses are as listed in Table 10.6.

The plate is excited near its fixed end, as shown in the figure, by a unit transverse sinusoidal load at the first natural frequency of the assembly.

Determine the power flow quiver and the streamline plots of the system.

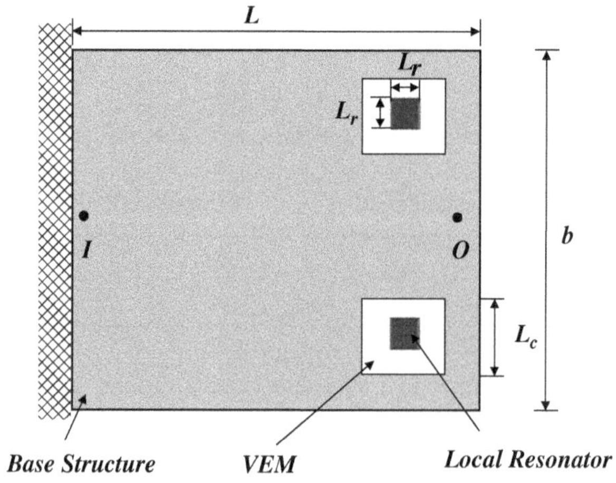

Figure P12.8 Cantilevered plate with two sources of internal resonance.

12.10 Consider the clamped-free shell/PCLD assembly shown in Figure P12.9. The main physical and geometrical parameters of the shell and PCLD are listed in Table P12.5. The shell has an internal radius R is 0.1016 m. The PCLD treatment consists of two patches as displayed in Figure P12.9. The patches are bonded

180° apart on the outer surface of the cylinder with each of which subtending an angle of 90° angle at the center of the shell as shown in Figure P12.9.

Determine the power flow distribution over the shell surface when it is excited sinusoidally by a unit force, at the first natural frequency, at location $L_1 = 0.9$ m using the FEM of Sections 4.8 and 7.5.

Table P12.5 Parameters of the shell/ACLD assembly.

Parameter	Length (m)	Thickness (mm)	Density (kg m^{-3})	Young's modulus (GPa)
Shell	1.270	0.635	7800	210
VEM	0.600	1.300	1140	[a]
PZT constraining layer	0.600	0.028	7600	66

[a] $G_\infty = 292.01$ MPa and $\beta_\infty = 0.007$ for a five-term GMM of the VEM as described in Example 4.6 and Table 4.6.

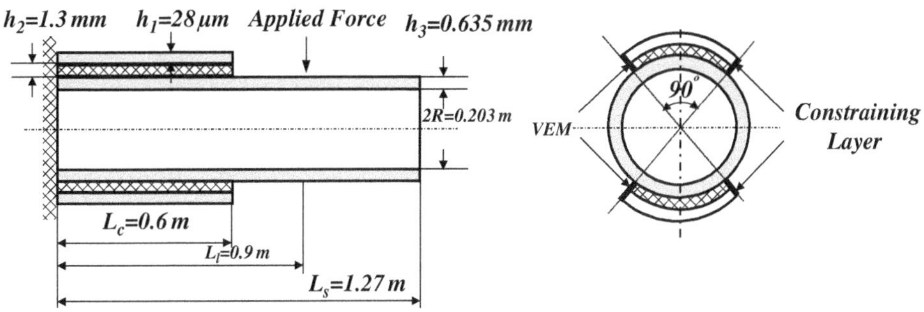

Figure P12.9 Configuration of the shell/PCLD assembly.

Glossary

Active constrained layer damping (ACLD) ACLD is a damping treatment consisting of a viscoelastic layer sandwiched between two piezoelectric layers. The bottom layer acts as a sensor bonded to the structure and the top layer acts as an actuator to enhance the shear strain in response to structural vibration.

Active piezoelectric damping composites (APDC) APDC is a damping treatment consisting of a viscoelastic (VEM) layer with embedded matrix of piezoelectric fibers. The fibers are embedded either perpendicularly or obliquely to the surface of the VEM to enhance compression damping or compression and shear damping, respectively when the piezo-fibers are activated.

Attenuation parameter (α) It is the real part of the propagation parameter (μ) which represents the logarithmic decay of the state vector as the wave propagates between adjacent cells.

Augmented temperature field (ATF) The ATF method is a physically-based approach that models the VEM by introducing a temperature field to interact with the structural field of a vibrating structure. The augmented field is described by a first order differential equation in an internal degree of freedom of the VEM.

Burgers model It consists of a Maxwell VEM model in series with a Kelvin–Voigt VEM model.

Constrained layer damping (CLD) It is a damping treatment consisting of a VEM sandwiched between a base structure and a constraining layer in order to introduce shear deformation in the VEM. This shear deformation enhances significantly the energy dissipation characteristics of the damping treatment.

Cole–Cole plot It is a plot of the storage modulus (E' *or* G') versus the loss modulus (E'' *or* G'') or versus the loss factor (η or tan δ). This plot is also called a "Wicket Plot."

Complex modulus (E^*, G^*) It is a description of the VEM elastic and viscoelastic properties in a complex number format such that: $E^* = E'(1 + \eta i)$ with E' and η denoting the storage modulus and loss factor, respectively. Also, i denote the unit imaginary number which is equal to $\sqrt{-1}$. This is the classical description of VEM when subject to different sinusoidal excitations and temperatures.

Creep It is the physical phenomenon associated with monitoring the continuous drop of the strain of a VEM when subjected to a constant applied stress. Measuring the time history of the strain is used to determine the time-dependent "Creep compliance."

Creep compliance The creep modulus $J(t)$ is determined by measuring the time history of the strain $\varepsilon(t)$ when a VEM is subjected to a step-stress of σ_0, such that: $J(t) = \varepsilon(t)/\sigma_0$.

Damping It is a mechanism for dissipating the energy of a vibrating structure whether in viscous, viscoelastic, or structural manner.

Damping composite It is a damping treatment which is formed by an assembly of viscoelastic material, arranged or mixed with multiple ingredients, layers, particle fillings, and/or fiber mats in order to achieve desirable damping characteristics.

Damping factor It is a quantitative measure of the energy dissipation characteristics of a VEM.

Dynamic mechanical thermal analysis (DMTA) It is a measuring instrument for measuring the storage and loss factor of a VEM at different temperatures and frequencies using sinusoidal excitations in bending, shear, or tension.

Elastomers Elastomers are damping materials that are present in the rubbery phase at room temperature.

Fading memory phenomenon A material exhibits a fading memory phenomenon if the effect of an action on the material response deteriorates and fades away as the time goes by.

Fractional derivatives Fractional calculus, introduced in 1695 by Leibniz, to generalize the meaning of derivatives from integer order to non-integer order derivatives.

Free volume (v_f) The free volume defines the space between the molecules of a viscoelastic material.

Golla–Hughes–McTavish model (GHM) The model describes the shear modulus of viscoelastic materials with a second order differential equation unlike the first order differential equations used to describe the Maxwell, Kelvin–Voigt, Poynting–Thomas, and Zener models. Such a distinction makes it easy to incorporate the dynamics of the viscoelastic materials into finite element models of vibrating structures.

Glass transition temperature (T_g) The glass transition temperature is an important physical property of the VEM at which significant changes in mechanical properties of the VEM takes place. Below T_g the VEM exists in a glassy phase and above T_g it exists in a rubbery state.

Jeffery model It consists of a Kelvin–Voigt VEM model in series with a damping element.

Kelvin–Voigt model A classical model of VEM consisting of a spring and viscous damper in a parallel arrangement.

Loss modulus (E'' *or* G'') This modulus quantifies the dissipative component of the "Complex Modulus" of a VEM. It is commonly denoted also as the "Out of-Phase" and/or "Imaginary Modulus."

Loss factor (or tan δ) It is the ratio of loss modulus to the storage modulus of VEM. It quantifies the energy dissipation characteristics of the *VEM*.

Master curve The master curve is a universal plot of the storage modulus E' (or G') and the loss factor η of a polymer as a function of the reduced frequency $\alpha_T\omega$ and temperature based on the application of the principle of "temperature-frequency" superposition.

Maxwell model A classical model of VEM consisting of a spring and viscous damper in a series arrangement.

Morlet wavelet Morlet wavelet is a sinusoidal function, oscillating at the frequency ω_w, modulated by a Gaussian envelope of unit variance.

Pass band It is a specific frequency band within which waves can propagate along the periodic structures.

Percolation zone It is the zone, where clustering of conductive particles embedded in an insulating matrix are enough to initiate the formation of continuous conductive networks throughout the matrix. In this zone, the conductivity of the composite experiences a significant increase of several orders of magnitude.

Periodic structure It is a structure, whether passive or active, that consists of identical substructures, or cells, which are repeated and connected in an identical manner in 1, 2, and 3D.

Phase parameter (β) It is the imaginary part of the propagation parameter (μ) that defines the phase difference between the adjacent cells.

Poisson's ratio (ν) It is the ratio of the transverse strain to the axial strain of a VEM. This ratio is typically equal to 0.5.

Poynting–Thomson model It consists of a Kelvin–Voigt VEM model in series with an elastic element (spring).

Propagation parameter (μ) It is a complex number whose real part (α) represents the logarithmic decay of the state vector and its imaginary part (β) defines the phase difference between the adjacent cells.

Relaxation It is the physical phenomenon associated with monitoring the continuous drop of the stress of a VEM when subjected to a constant applied strain. Measuring the time history of the stress is used to determine the time-dependent "Relaxation Modulus."

Relaxation modulus The relaxation modulus $E(t)$ is determined by measuring the time history of the stress $\sigma(t)$ when a VEM is subjected to a step-strain of ε_0, such that: $E(t) = \sigma(t)/\varepsilon_0$.

Rayleigh damping model It is a damping model that describes the equivalent viscous damping coefficient of a damping material as linear combination of its mass and stiffness through the use of appropriate mass and stiffness damping parameters.

Static condensation method (Guyan reduction) It is a method that reduces (or condenses) the *DOF* of the system to include only the primary set of *DOF* only without eliminating them.

Stop band It is a frequency band within which the wave propagation is completely blocked.

Strain energy It is the energy stored in a structure as it deforms under the influence of an external load.

Storage modulus (Young's (E'), shear (G')) This modulus quantifies the elastic storage component of the "Complex Modulus" of a VEM that is subjected to sinusoidal in tension or shear loading, respectively. It is also denoted as the "In-Phase" and "Real Modulus."

Viscoelastic material (VEM) Viscoelastic materials have a time-dependent response, even if the loading is constant. Many polymers and biological tissues exhibit such a behavior. Linear viscoelasticity is a commonly used approximation where the stress depends linearly on the strain and its time derivatives.

Viscous damping It is the class of damping in which the damping force is linearly proportional to the velocity of deformation.

Zener model It consists of a Maxwell VEM model in parallel with a damping element.

Appendix A

Complex Modulus of Typical Damping Treatments

This appendix presents a brief summary of the effect of operating temperature and frequency on the complex modulus of three of the most commonly used viscoelastic materials (VEM). These VEMs are manufactured by 3M (Bonding Systems Division, 3M Center, Building 220-7E-01, St. Paul, MN 55144-1000), E.A.R. (Aearo E.A.R. Specialty Composites, 7911 Zionsville Road Indianapolis, IN 46268), and Soundcoat (Soundcoat, 1 Burt Drive, Deer Park, NY 11729).

A.1 3M™ Viscoelastic Damping Polymers

The main characteristics of the ISD series (110,112, 113) of 3M VEM are summarized in Table A.1 and Figures A.1–A.3.

A.2 E.A.R. Viscoelastic Damping Polymers

The main characteristics of the C-1002 and C-2003 of EAR VEM are summarized in Table A.2 and Figures A.4 and A.5.

A.3 Soundcoat Viscoelastic Damping Polymers

The main characteristics of the DYAD series (601, 606, 609) of SOUNDCOAT VEM are summarized in Table A.3 and Figures A.6–A.8.

Note that all the three types of VEM have comparable loss factors of about 1. But, the 3M ISD series is softer that the E.A.R C-1002 which is also softer than the SOUNDCOAT DYAD series. However, E.A.R. C-2003 is the hardest among all the considered VEM treatments. Therefore, for applications that require high damping and medium stiffness, the SOUNDCOAT DYAD series is the most appropriate.

Active and Passive Vibration Damping, First Edition. Amr M. Baz.

Table A.1 Operating temperature range, maximum loss factor, and corresponding storage modulus of 3M VEM (ISD 110, 112, 113).

Polymer[a,b,c]	Operating temperature range (°C)	Maximum loss factor η_{max} at temperature (°C)	Storage shear modulus G' at η_{max} (MPa)
ISD 110	40 to 105 (high)	1.2 at 55	0.12
ISD 112	−20 to 65 (normal)	1.1 at 30	0.15
ISD 113	−40 to 20 (low)	1.2 at −20	0.40

[a] 3M, Bonding Systems Division, 3M Center, Building 220-7E-01, St. Paul, MN 55144-1000.
[b] Poisson's ratio for 110, 112, and 113 viscoelastic damping polymers: approximately 0.49.
[c] Density for 110, 112, and 113 viscoelastic damping polymers: approximately 0.9–1.0 g cm^{-3}.

Major Viscoelastic Material Manufacturers Websites

Soundcoat

Burt Drive, Deer Park, NY 11729
1-800-394-8913 or 631-242-2200
Fax: 631-242-2246
www.soundcoat.com/products.htm

E.A.R. Specialty Composites

650 Dawson Drive Newark, DE 19713
Phone (302) 738-6800 Fax (302) 738-6811
www.earsc.com

3M Industrial Business

Bonding Systems Division
3M Center Bldg. 220-7E-01
St. Paul, MN 55144-1000
1-800-362-3550
www.3m.com/bonding

Roush Anatrol Industries

12447 Levan
Livonia, Michigan 48150
Main Line: 734-779-7006
Toll-Free: 1-800-215-9658
www.roush.com

Damping Technologies, Inc.

12970 McKinley Hwy,
Unit IX
Mishawaka, IN 46545-7518
Tel: 574.258.7916
www.damping.com

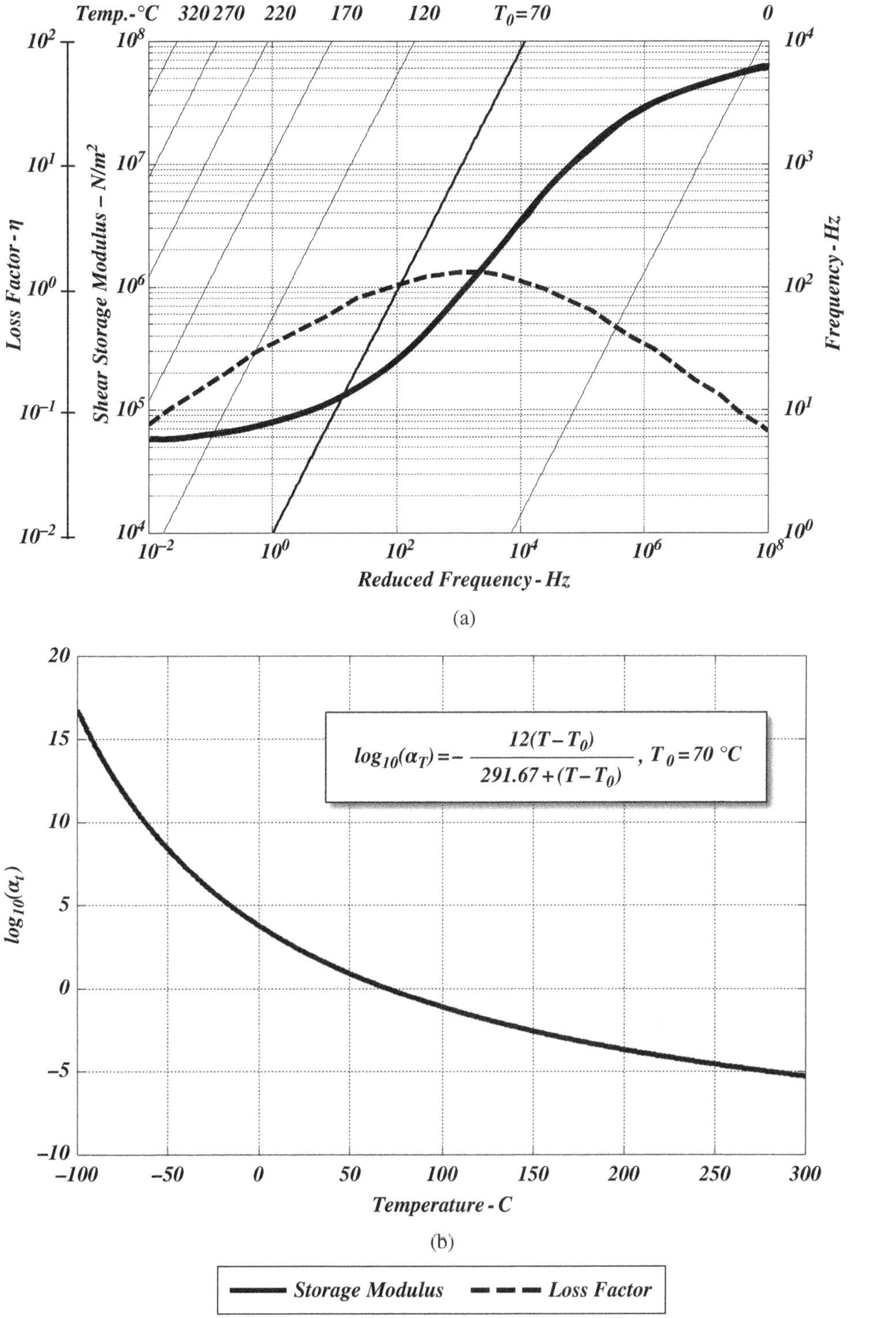

Figure A.1 Complex modulus of ISD-110 (T_0 = 70 °C).

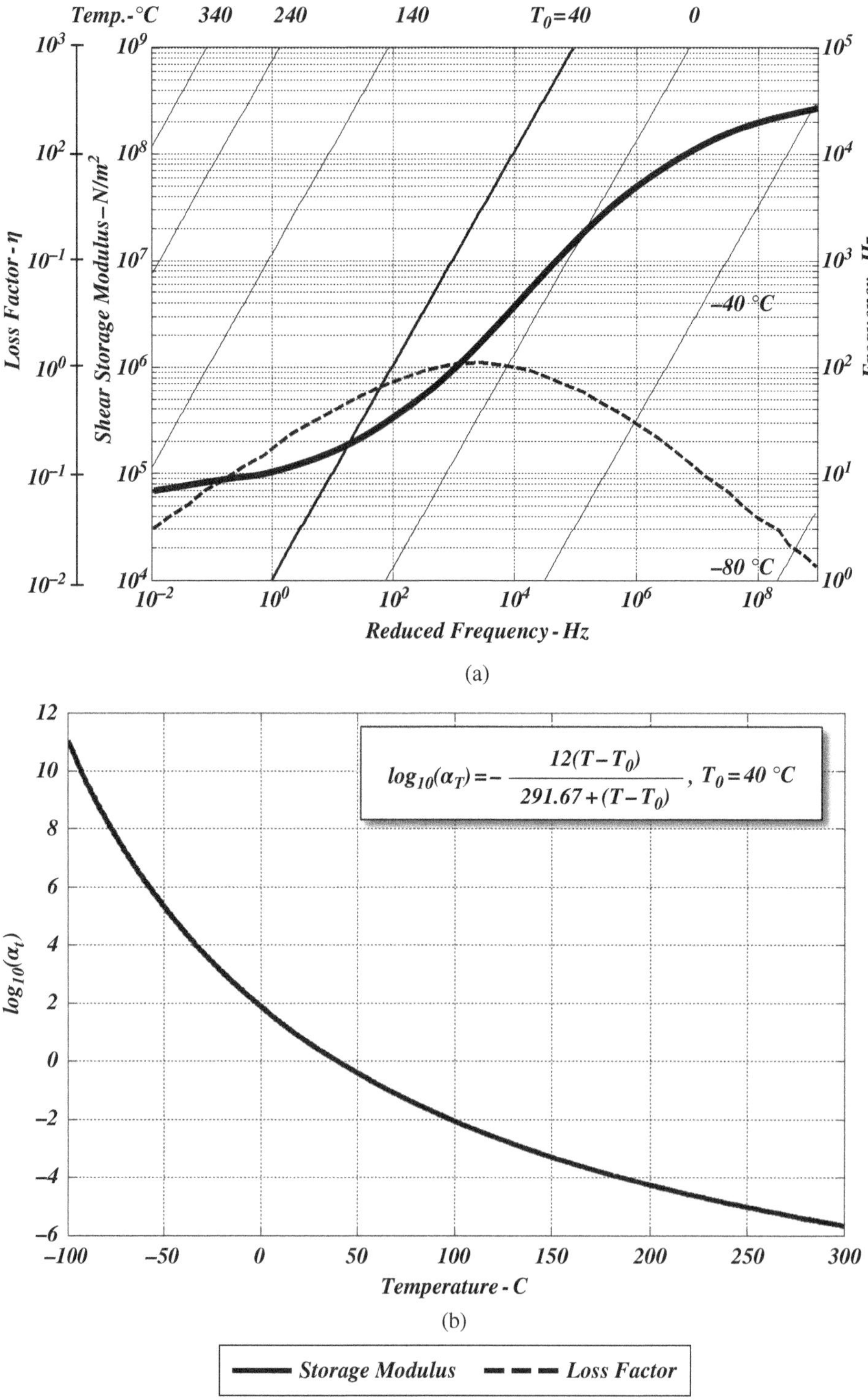

Figure A.2 Complex modulus of ISD-112 (T_0 = 40 °C).

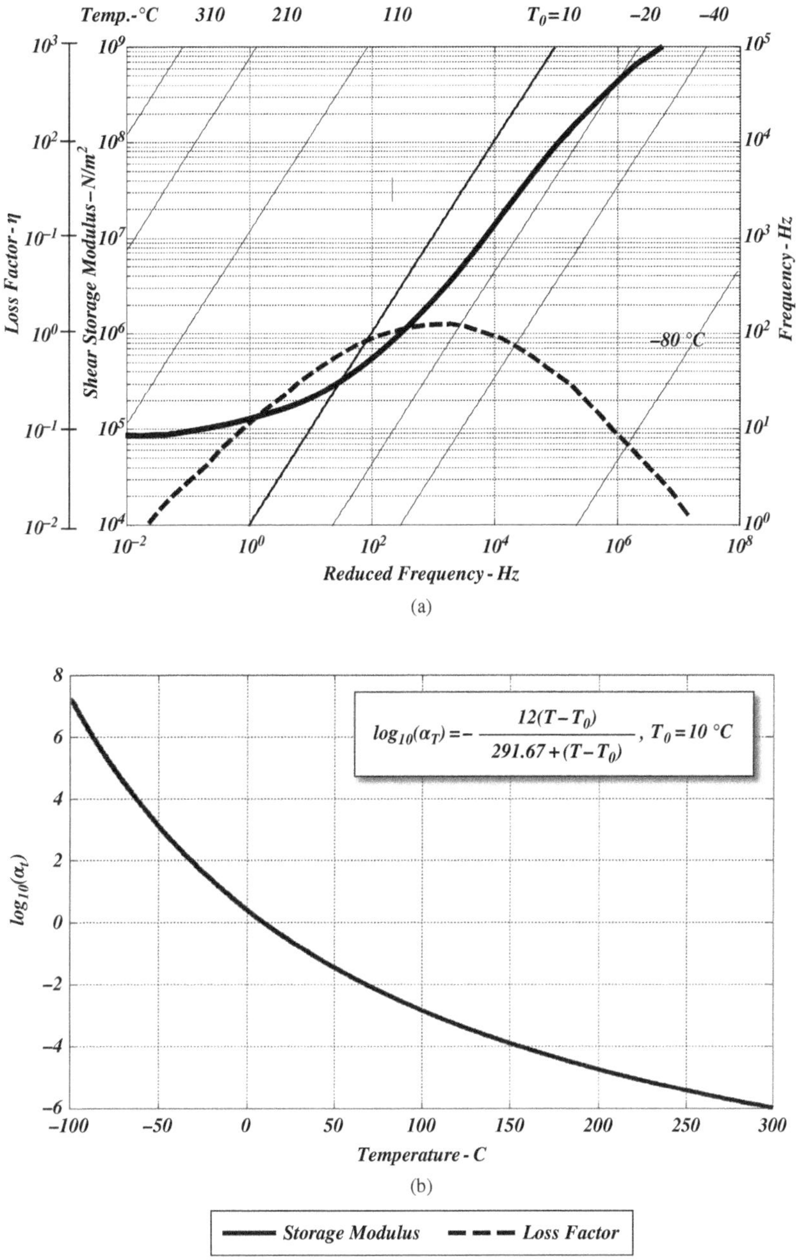

Figure A.3 Complex modulus of ISD-113 (T_0 = 10 °C).

Table A.2 Operating temperature range, maximum loss factor, and corresponding storage modulus of E.A.R. VEM (C-1002, C-2003).

Polymer[a]	Operating temperature range (°C)	Maximum loss factor η_{max} at temperature (°C)	Storage shear modulus G' at η_{max} (MPa)
C-1002	13–41 (low)	1.02 at 15	20.00
C-2003	27–54 (normal)	1.00 at 45	150.00

[a] Aearo E.A.R. Specialty Composites, 7911 Zionsville Road Indianapolis, IN 46268.
[b] Poisson's ratio for EAR viscoelastic damping polymers: approximately 0.49.
[c] Density for EAR C-1002, 1105, and 1100 viscoelastic damping polymers: approximately 1.289–1.282 g cm^{-3}.

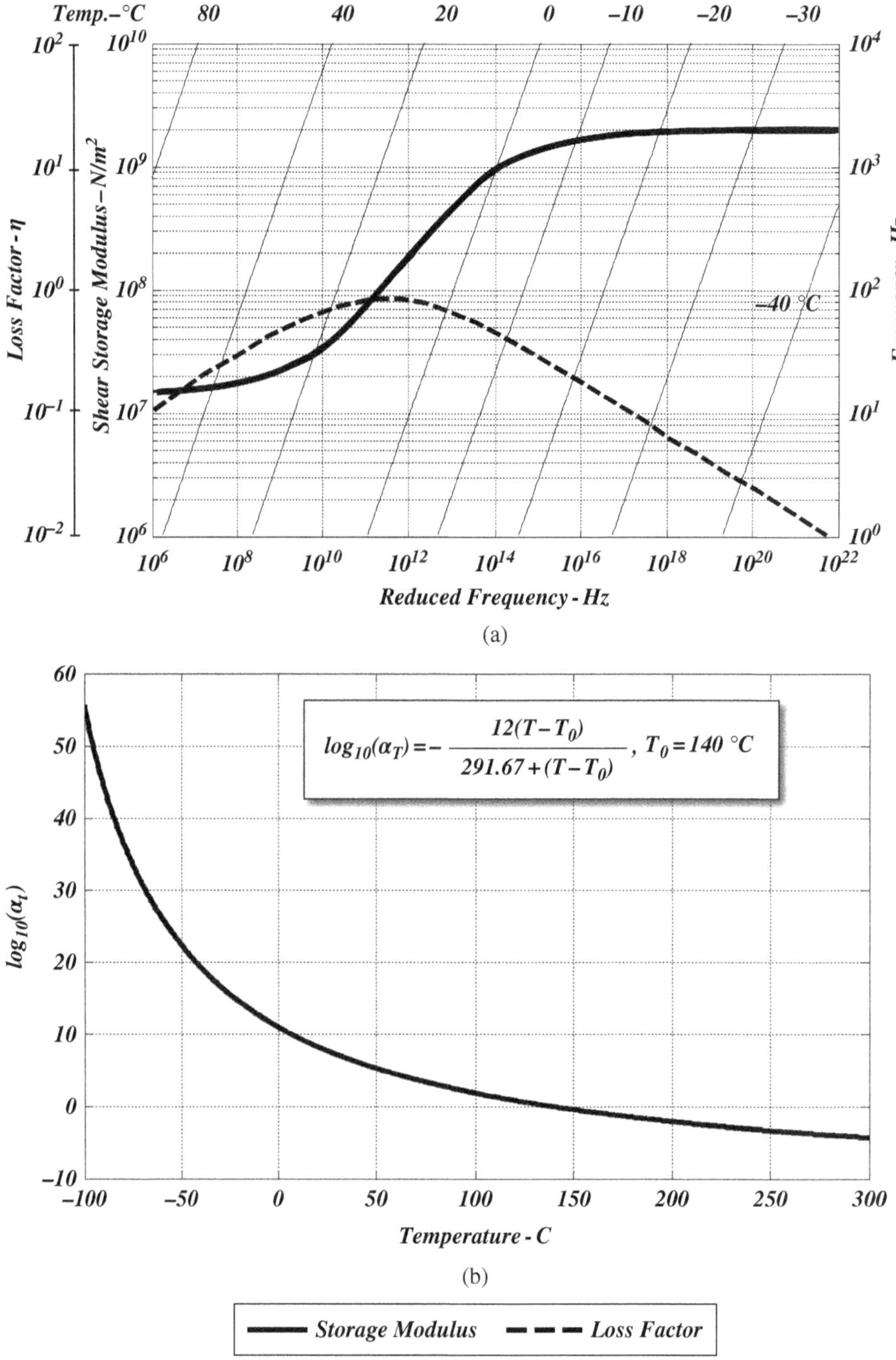

Figure A.4 Complex modulus of EAR-C-1002 (T_0 = 140 °C).

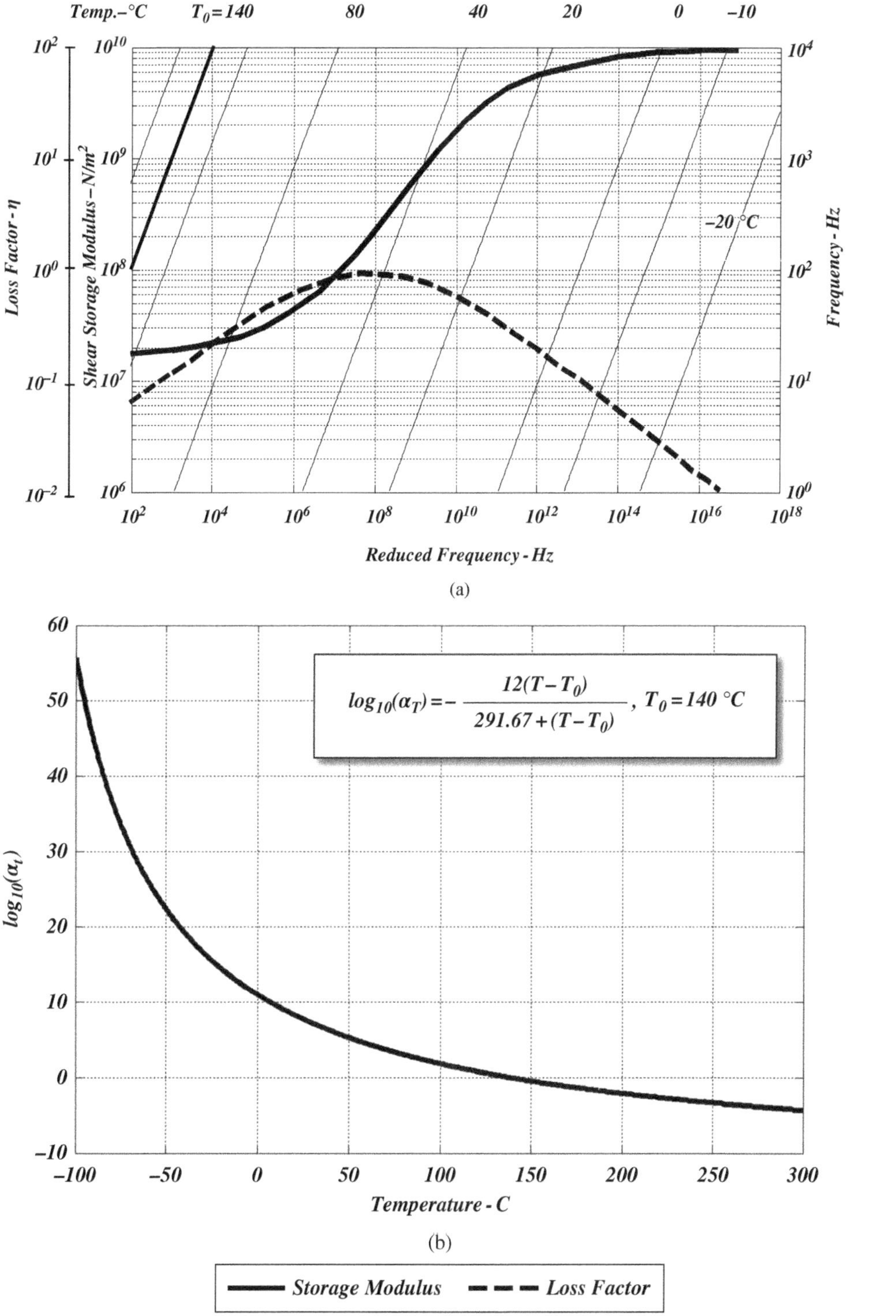

Figure A.5 Complex modulus of EAR-C-2003 (T_0 = 140 °C).

Table A.3 Operating temperature range, maximum loss factor, and corresponding storage modulus of SOUNDCOAT VEM (DYAD – 601, 606, 609).

Polymer[a]	Operating temperature range (°C)	Maximum loss factor η_{max} at Temperature (°C)	Storage shear modulus G' at η_{max} (MPa)
DYAD-601	−10 to 40 (low)	1.00 at 20	5.00
DYAD-606	10 to 80 (normal)	1.05 at 40	10.00
DYAD-609	0 to 50 (high)	0.60 at 20	20.00

[a] Soundcoat, 1 Burt Drive, Deer Park, NY 11729.
[b] Poisson's ratio for SOUNDCOAT viscoelastic damping polymers: approximately 0.49.
[c] Density for DYAD 601, 606, and 609 viscoelastic damping polymers: approximately 1.12–1.3 g cm^{-1}.

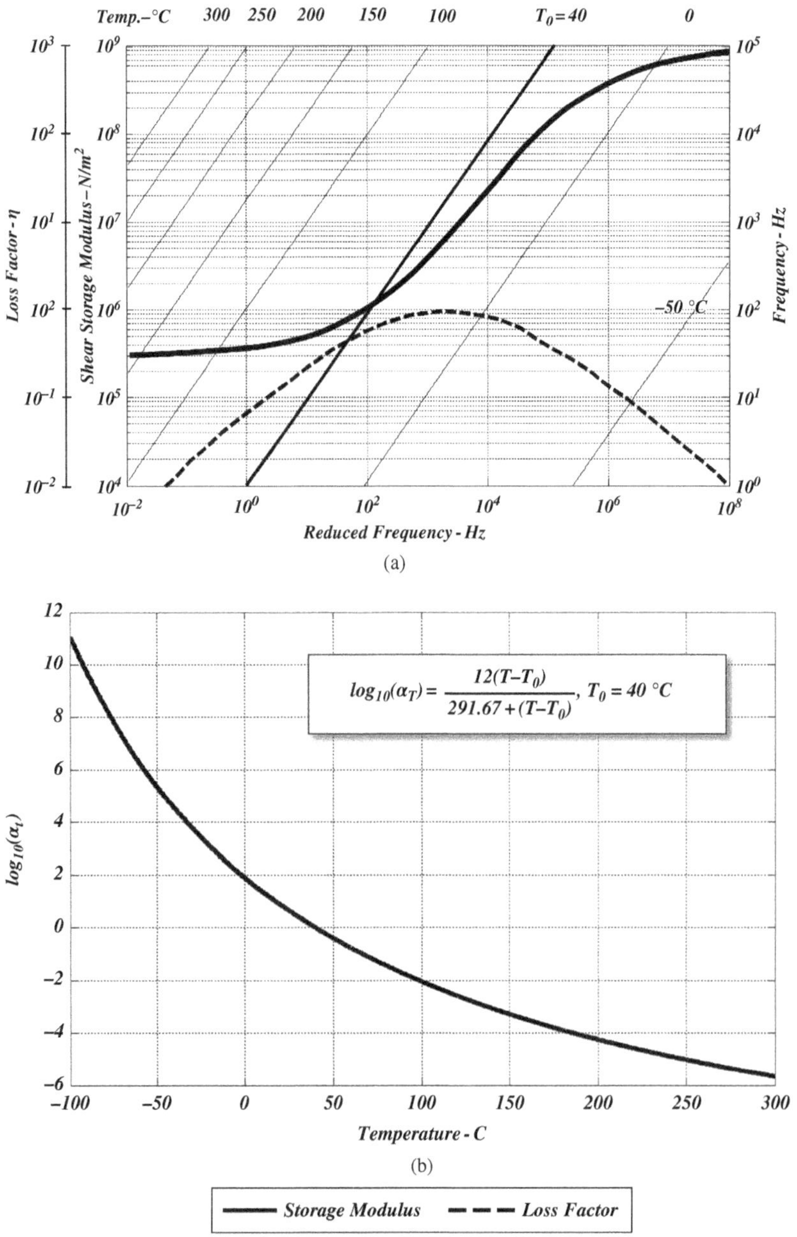

Figure A.6 Complex modulus of Dyad 601 (T_0 = 40 °C).

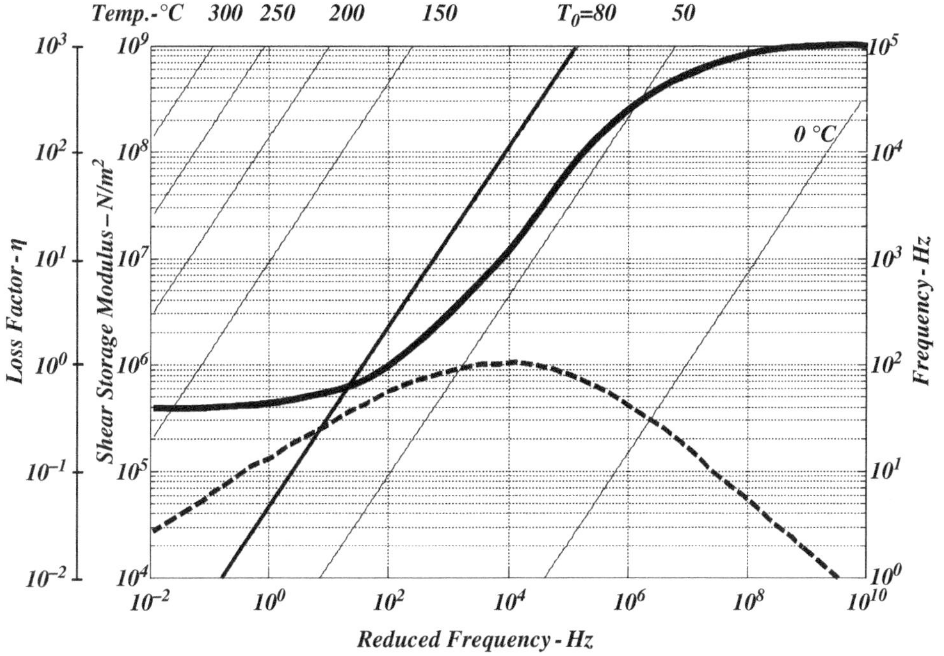

(a)

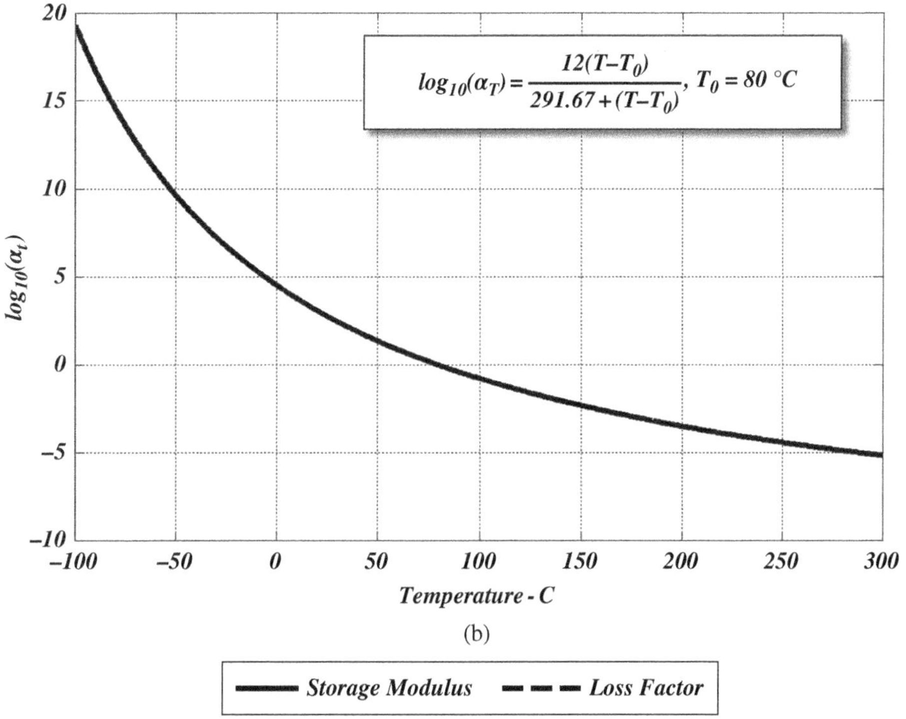

(b)

Figure A.7 Complex modulus of Dyad 606 (T_0 = 80 °C).

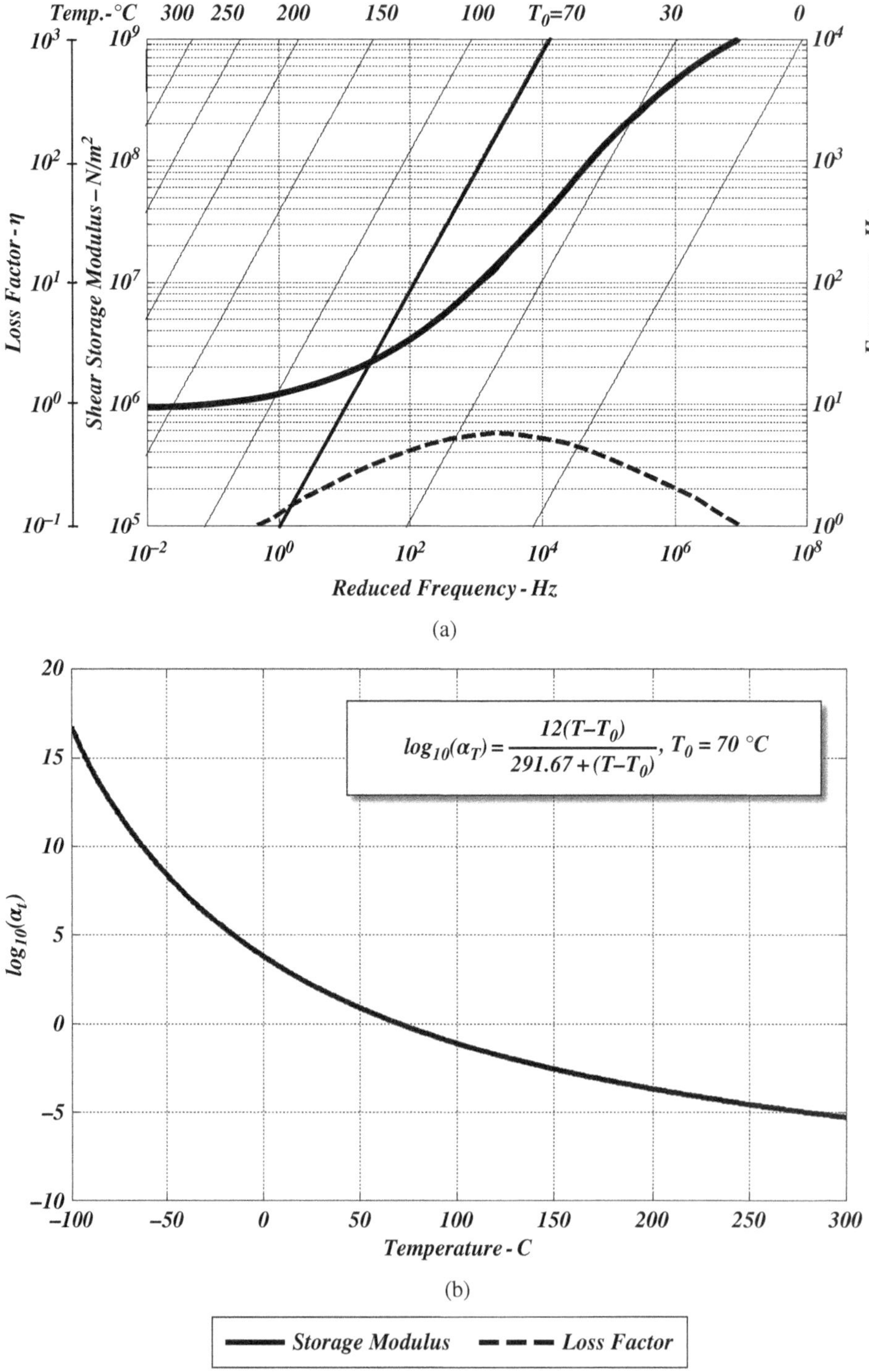

Figure A.8 Complex modulus of Dyad 609 (T_0 = 70 °C).

Further Reading

Adhikari S., *Structural Dynamic Analysis with Generalized Damping Models: Analysis*, ISTE, Ltd/Wiley: London/Hoboken, NJ, 2014.
Beards C., *Structural Vibration, Analysis and Damping*, Butterworth-Heinemann; 1996.
Braun S. G., Ewins D. J., and Rao S. S., *Encyclopedia of Vibration, Volumes I-III*, Academic Press, 2001.
Brown, R., and B. Read, *Measurement Techniques for Polymers Solids*, Elsevier Applied Science Publishers, New York, 1984.
Chen G. and Zhou J., *Vibration and Damping in Distributed Systems, Volume I and II*, CRC Press; 1993.
Christensen R. M., *Theory of Viscoelasticity: An Introduction*, 2, Academic Press Inc., New York, 1982.
Drake M. L. and Terborg G. E., Polymeric Material Testing Procedures to Determine Damping Properties and the Results of Selected Commercial Material, *Technical Report* AFWAL-TR-80-4093, July 1980.
Drozdov A. D., *Mechanics of Viscoelastic Solids*, Wiley, 1998.
Ferry J. D., *Viscoelastic Properties of Polymers* (3), Wiley, 1980.
Findley W. N., Lai J. S., and Onaran K., *Creep and Relaxation of Nonlinear Viscoelastic Materials*, Dover Publications, 1989.
Flugge W., *Viscoelasticity*, Blaisdell Publishing Company, Waltham, MA, 1967.
Garibaldi L. and Onah H. N., *Viscoelastic Material Damping Technology*, Becchis Osiride, Turin, 1996.
Haddad Y. M., *Viscoelasticity of Engineering Materials*, Chapman & Hall, New York, 1995.
Jones D., *Handbook of Viscoelastic Vibration Damping*, Wiley; 2001.
Lakes R., *Viscoelastic Solids*, CRC Press, Boca Raton, FL, 1999.
Lakes R., *Viscoelastic Materials*, Cambridge Press, 2009.
Mead D., *Passive Vibration Control*, Wiley; 1999.
Menard K. P., *Dynamic Mechanical Analysis*, CRC Press, Boca Raton, FL, 1999.
Nashif A., Jones D. and Henderson J., *Vibration Damping*. Wiley, New York, 1985.
Osinski Z., *Damping of Vibrations*, Taylor & Francis; 1998.
Phan-Thien N., *Understanding Viscoelasticity: Basics of Rheology*, Springer Verlag, 2002.
Rivin E. I., *Stiffness and Damping in Mechanical Design*, Marcel Dekker; 1999.
Rivin E. I., *Passive Vibration Isolation*, American Society of Mechanical Engineers; 2003.
Sun C. and Lu Y. P., *Vibration Damping of Structural Elements*, Prentice Hall, Englewood Cliffs, NJ, 1995.
Tschoegl N. W., *The Phenomenological Theory of Linear Viscoelastic Behavior: An Introduction*, Springer Verlag, 1989.
Wineman A. S. and Rajagopal K. R., *Mechanical Response of Polymers: An Introduction*, Cambridge University Press, 2000.
Zener C. M., *Elasticity and Anelasticity of Metals*, University of Chicago Press, Chicago, 1948.

Index

Active and Passive Vibration Damping, First Edition. Amr M. Baz.